SOLIDWORKS 2024
For Beginners

Tutorial Books

SOLIDWORKS 2024 For Beginners

Contents

Introduction

Welcome to the *SOLIDWORKS 2024 For Beginners* book. This book is written to assist students, designers, and engineering professionals in learning SOLIDWORKS. It covers the essential features and functionalities of SOLIDWORKS using relevant examples and exercises.

The author wrote this book for new users, who can use it as a self-study resource to learn SOLIDWORKS. Also, experienced users can use it as a reference. The focus of this book is part modeling, assemblies, and drawings.

Topics covered in this Book

- Chapter 1, "Getting Started with SOLIDWORKS," gives an introduction to SOLIDWORKS. This chapter discusses the user interface and terminology.
- Chapter 2, "Sketch Techniques," explores the sketching commands in SOLIDWORKS. You learn to create parametric sketches.
- Chapter 3, "Extrude and Revolve features," teaches you to create basic 3D geometry using the Extrude and Revolve commands.
- Chapter 4, "Placed Features," covers the features which are created without using sketches.
- Chapter 5, "Patterned Geometry," explores the commands to create patterned and mirrored geometry.
- Chapter 6, "Sweep Features," covers the commands to create swept and helical features.
- Chapter 7, "Loft Features," covers the Loft command and its core features.
- Chapter 8, "Additional Features and Multibody Parts," covers additional commands to create complex geometry. Besides, it covers the multi-body parts.
- Chapter 9, "Modifying Parts," explores the commands and techniques to modify the part geometry.
- Chapter 10, "Assemblies," explains you to create assemblies using the bottom-up and top-down design approaches.
- Chapter 11, "Drawings," covers how to create 2D drawings from 3D parts and assemblies.

Chapter 1: Getting Started with SOLIDWORKS

Introduction to SOLIDWORKS

SOLIDWORKS, a robust computer-aided design (CAD) tool, has significantly revolutionized the engineering and design industry's product development process. Its unique combination of parametric and feature-based technology enables engineers and designers to create precise 3D models efficiently.

One of SOLIDWORKS' standout features is its automation capability, which streamlines repetitive tasks, thereby minimizing errors and enhancing productivity.

Furthermore, SOLIDWORKS is known for its intuitive and customizable interface, providing users with a seamless design experience. Additionally, it supports various file formats, such as STL, OBJ, and IGS, enabling easy data exchange with other CAD applications.

Additionally, SOLIDWORKS has advanced simulation capabilities, allowing users to evaluate product performance before manufacturing, thereby reducing costly errors and improving product quality. Its compatibility with other leading CAD platforms also makes it a popular choice for businesses seeking to integrate their design processes across departments.

SOLIDWORKS is a parametric and feature-based software that allows you to create 3D parts, assemblies, and 2D drawings. The following figure shows the design process in SOLIDWORKS.

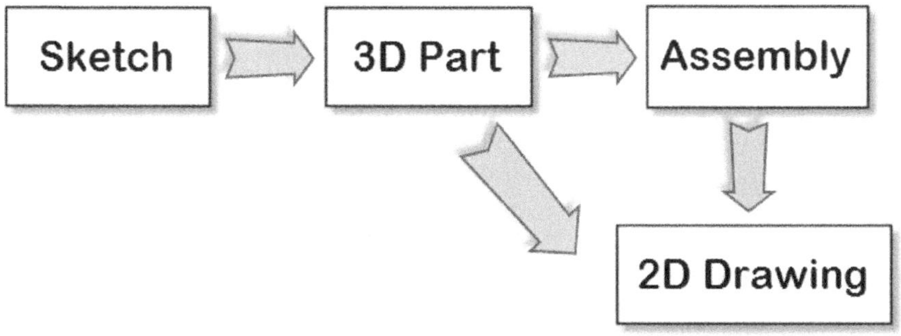

In SOLIDWORKS, you can control the design by using parameters, dimensions, or relationships. For example, if you want to change the slot's position shown in the figure, you need to change the dimension or relation that controls its position.

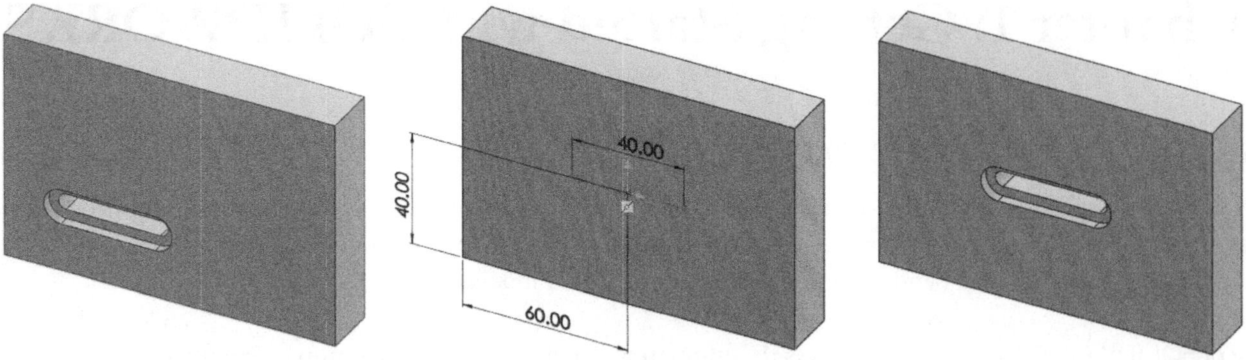

The parameters and relationships you set up allow you to control the design intent. The design intent refers to how the 3D model will behave when you apply dimensions and relationships to it. For example, if you want to position the slot at the center of the block, one way is to add dimensions between the hole and the adjacent edges. However, if you change the block's size, the slot may no longer be at the center unless you have established appropriate design intent through the use of parameters and relationships.

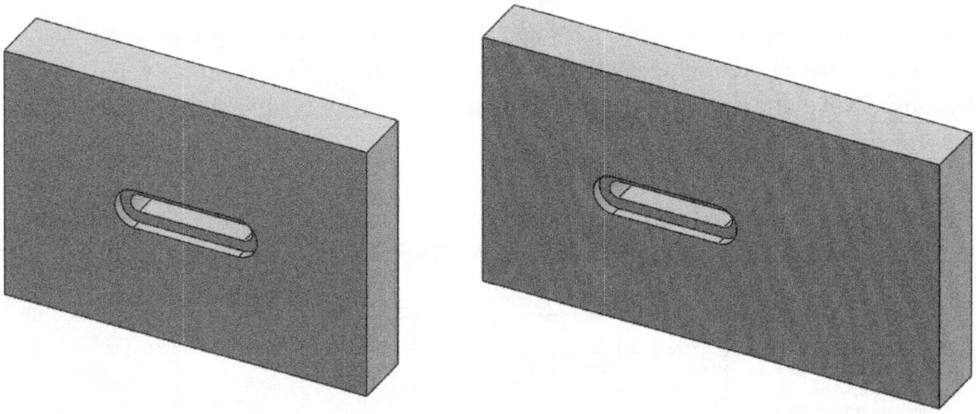

You can make the slot to be at the center, even if the size of the block changes. You need to apply the **Horizontal/Vertical** relationships between the hole and midpoints of the adjacent edges. Even if you change the block's size, the hole will always remain at the center.

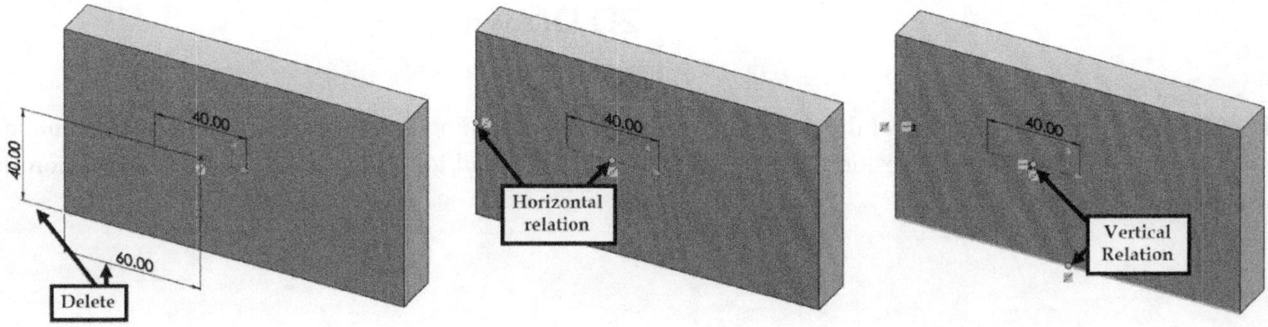

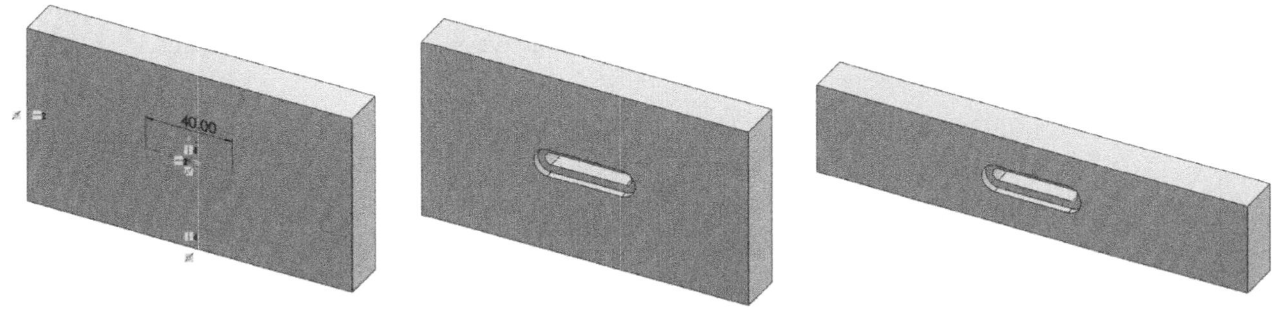

Another significant advantage of SOLIDWORKS is its robust associativity feature, which establishes a dynamic link between parts, assemblies, and drawings. This means that when you make changes to a part's design, these modifications are instantly reflected in any assemblies that include that part. This real-time update ensures that the entire assembly remains consistent and accurate, saving time and reducing the risk of errors during the design process.

Furthermore, SOLIDWORKS' associativity extends to 2D drawings as well. Any changes made to the 3D model automatically propagate to the associated 2D drawings, ensuring that all documentation is consistently updated. This seamless integration between parts, assemblies, and drawings helps designers and engineers maintain design intent, improve collaboration, and streamline the overall product development workflow.

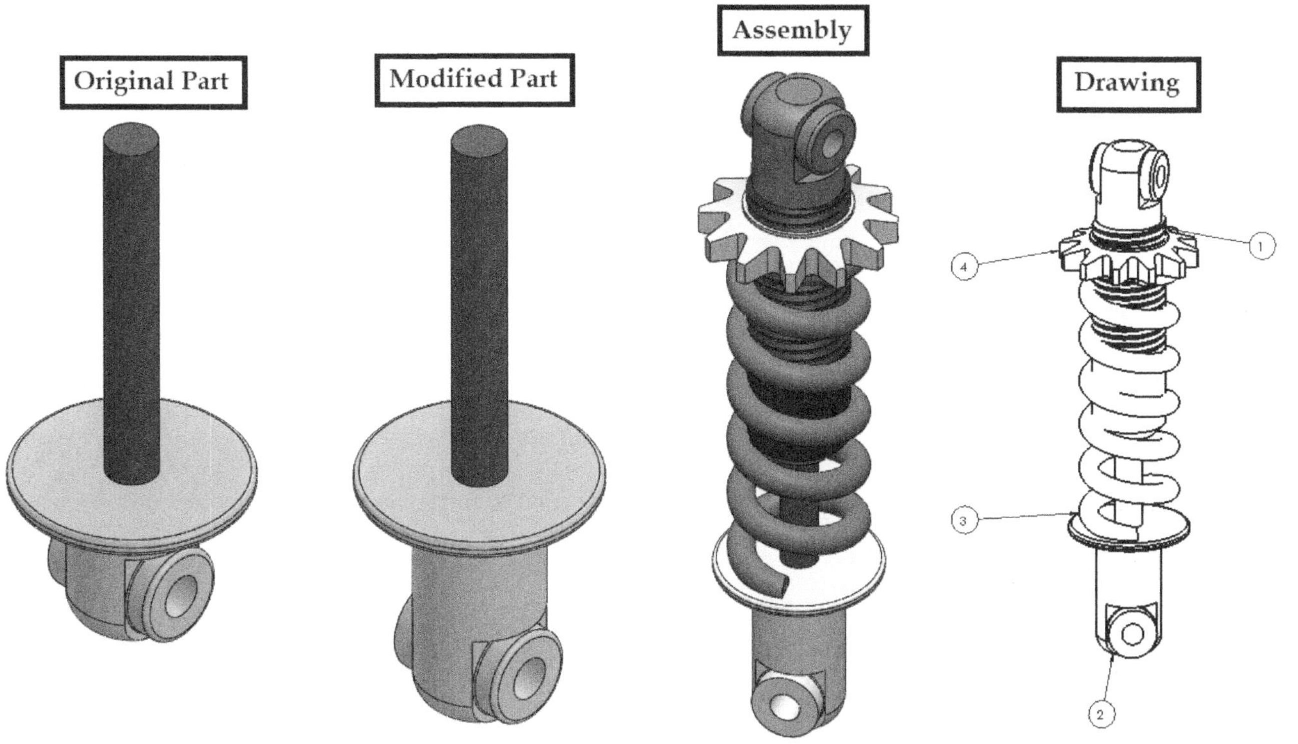

Starting SOLIDWORKS

To start **SOLIDWORKS**, click the **SOLIDWORKS** icon on your computer screen; the **SOLIDWORKS** application window appears. On this window, click the **Welcome to SOLIDWORKS** icon located on the

Quick Access Toolbar; the Welcome window appears. You can use this menu to start a new document, open an existing one, learn SOLIDWORKS, browse recent documents, and access SOLIDWORKS resources.

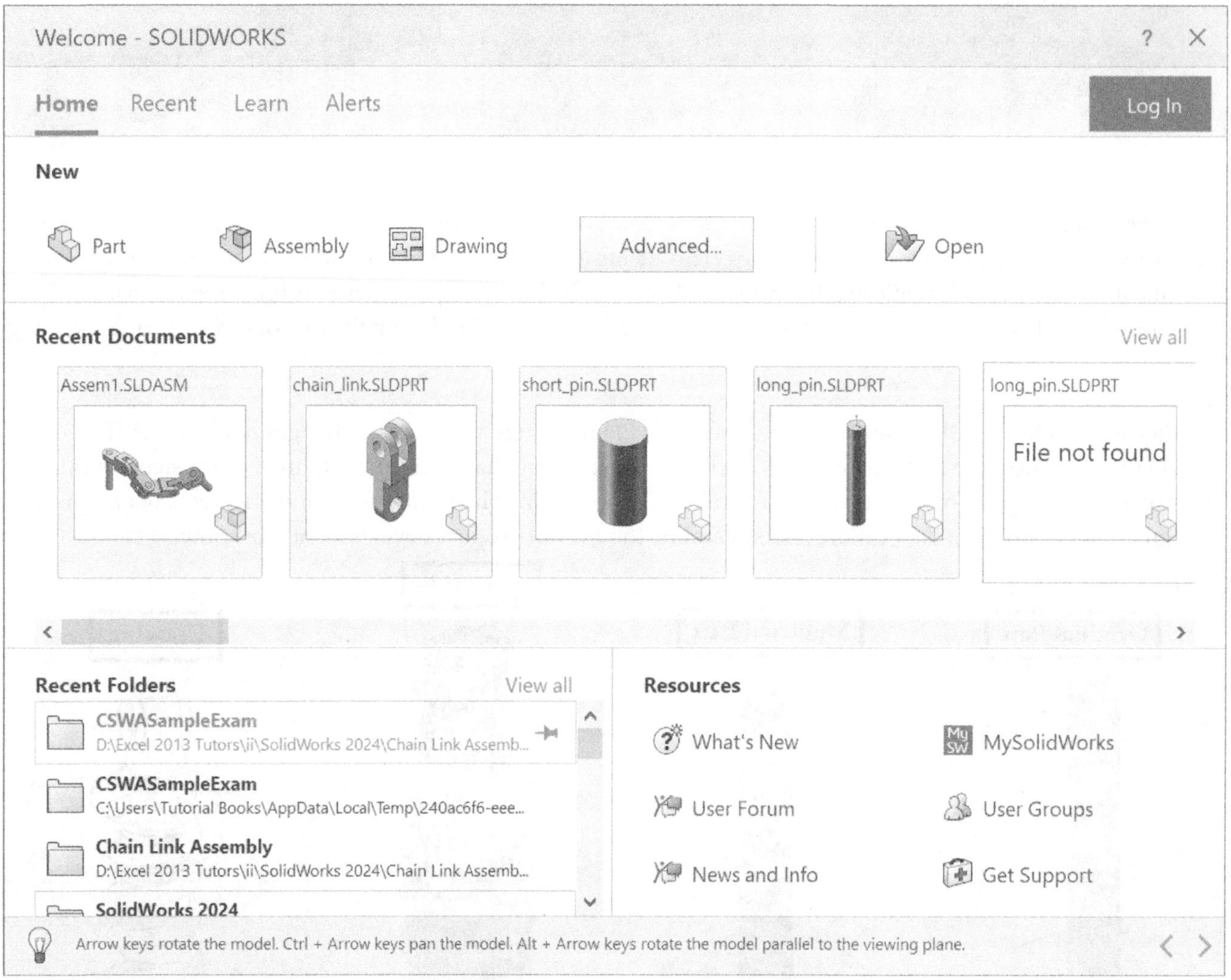

On the **Welcome** window, click the **Part** icon under the **New** section to start a new part document.

File Types

The following list shows various file types that you can create in SOLIDWORKS.

- **Part (.sldprt)**
- **Assembly (.sldasm)**
- **Draft (.slddrw)**

User Interface

The following image shows the **SOLIDWORKS** application window.

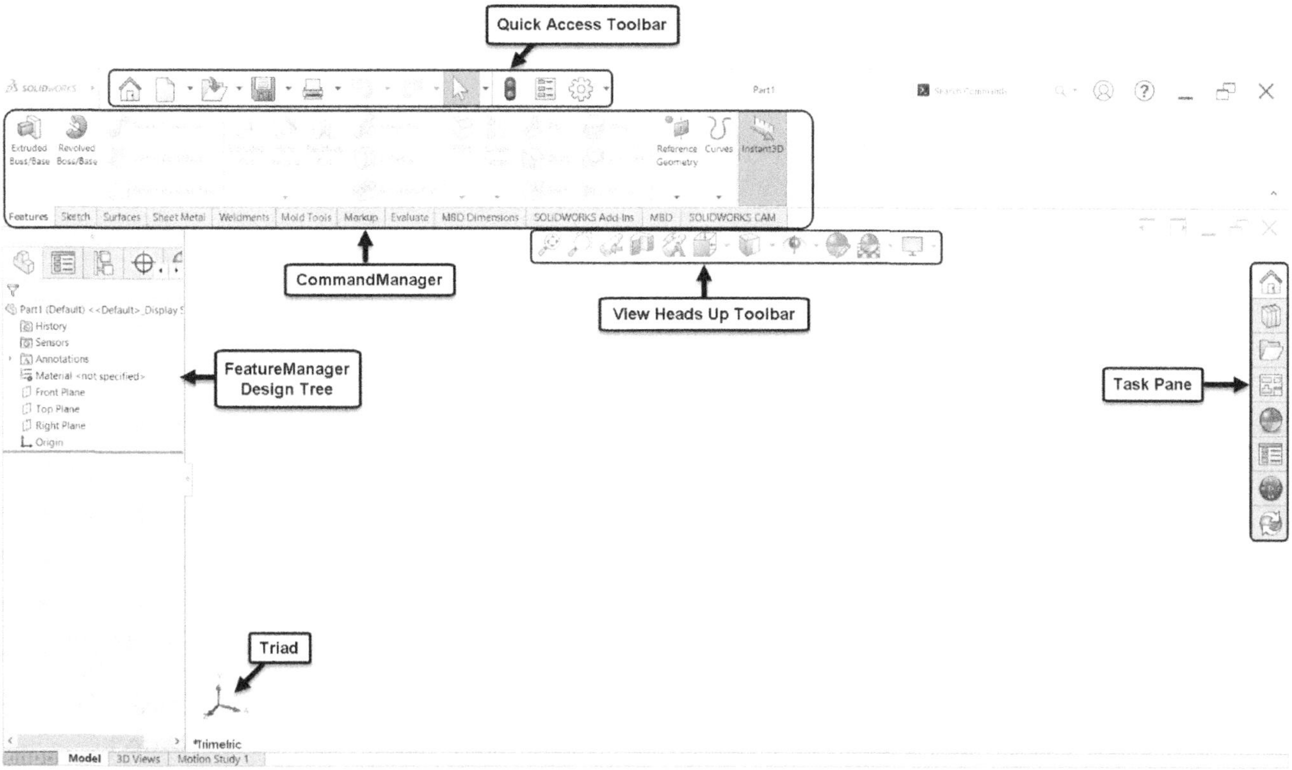

Environments in SOLIDWORKS

There are three main environments available in SOLIDWORKS: **Part**, **Assembly**, and **Draft**.

Part environment

The Part environment provides you with all the necessary tools and commands to create 3D part models of your design. Within this environment, the FeatureManager Design Tree is an essential tool that stores every feature or sketch you create, allowing you to easily edit and modify them at any time. The tree organizes the features and sketches in a hierarchical manner, making it simple to navigate and manage your design's structure.

Located at the top of the screen is the CommandManager, which provides quick and easy access to various commands grouped by their functionality. The CommandManager includes tabs such as **Features**, **Sketch**, and **Surfaces**, which house related commands for creating and manipulating features, sketches, and surface geometry. This organization of commands helps to optimize the workflow and improve productivity, making it simple to find and use the desired tool quickly and efficiently.

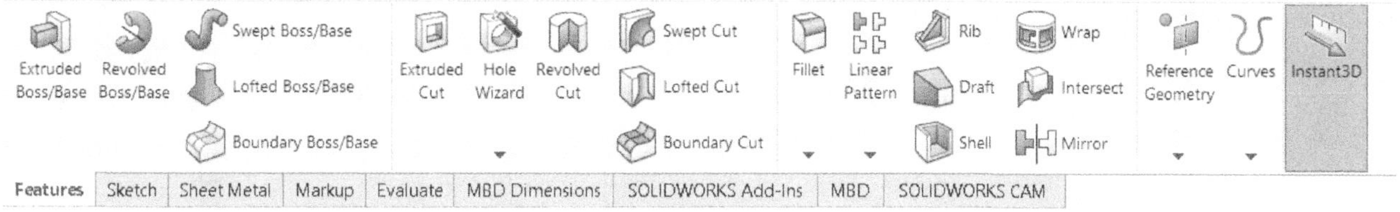

The **Sheet Metal** tab features commands for creating and editing sheet metal parts, accounting for manufacturing considerations like bend radius, material thickness, and edge conditions.

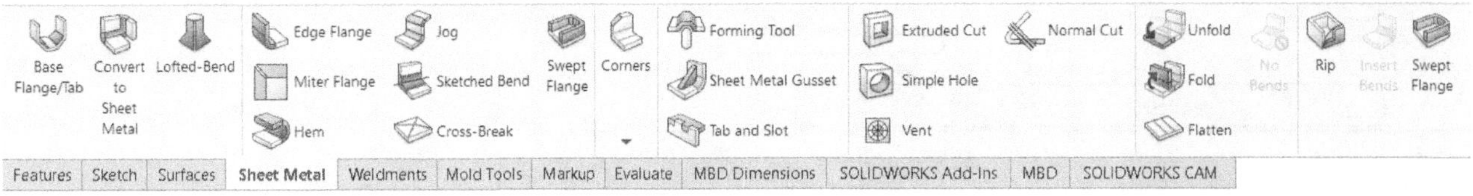

Assembly environment

You can create assemblies using the SOLIDWORKS assembly environment. The **Assembly** tab of the CommandManager provides various commands for assembling and modifying components. This allows you to create complex product structures, analyze motion and interactions, and simulate real-world performance.

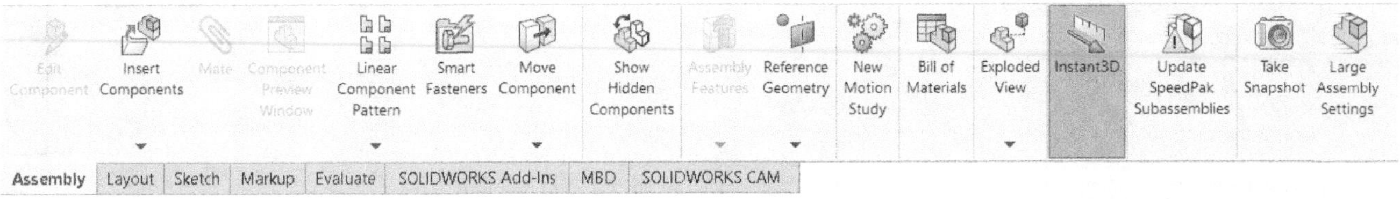

The **Evaluate** tab is a valuable tool for both part and assembly design, providing various commands and tools to inspect and analyze geometry. During part design, the Evaluate tab helps ensure accuracy and integrity by checking for potential interferences, measuring distances and angles, and analyzing part performance. This is particularly useful when performing part-level simulations, such as stress analysis, to ensure the correct functioning of individual components.

During assembly design, the **Evaluate** tab allows you to inspect and analyze assembly geometry, ensuring the accuracy and integrity of the product structure. This tab provides commands and tools to check for potential interferences, measure distances and angles, and analyze assembly performance, making it particularly useful when performing motion studies, assemblies checks, and other forms of real-world simulation.

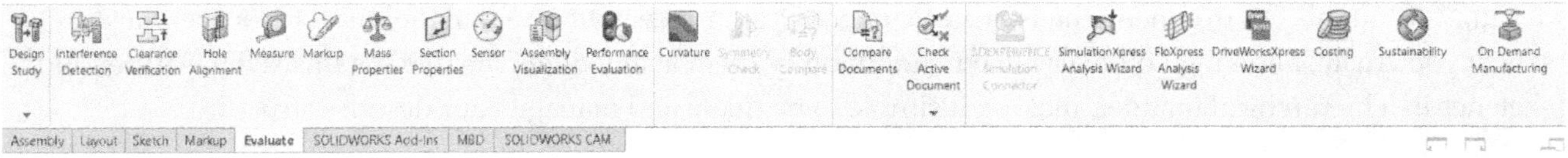

Draft environment

The drafting environment is a useful tool for creating 2D drawings of your part and assembly designs. This environment has all the necessary commands for generating 2D drawings, allowing you to create detailed documentation of your designs. With this tool, you can create various types of drawings, such as detailed drawings, assembly drawings, and orthographic views, while automatically accounting for dimensions, annotations, and other manufacturing requirements.

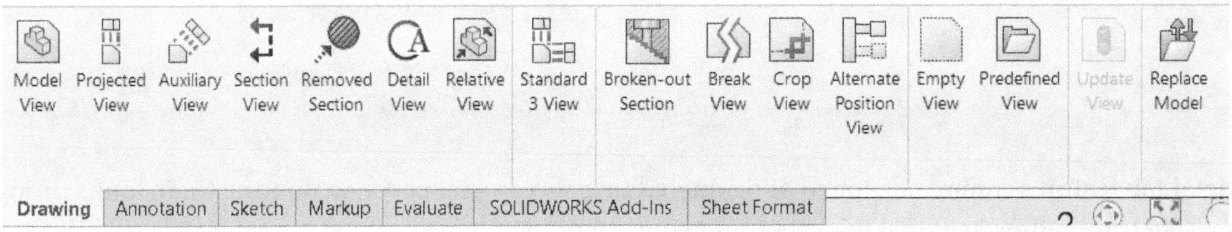

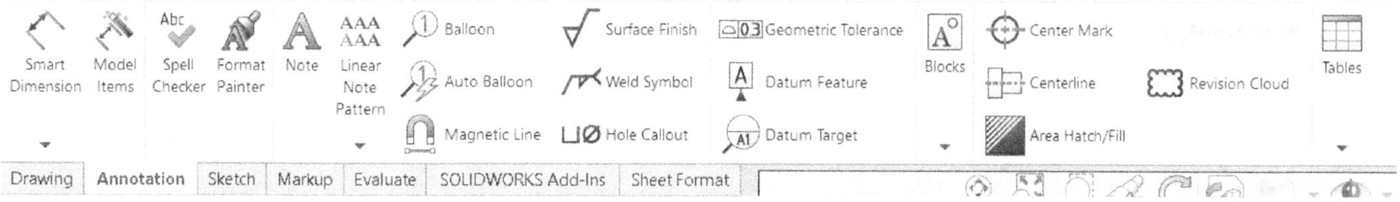

The following sections discuss the remaining components of the user interface.

File Menu

The File menu is a part of the SOLIDWORKS interface that appears when you click on the File option located on the Menu bar. This menu consists of a list of specific options, including the "**Open Recent**" sub-menu. When you place the pointer on the "**Open Recent**" sub-menu, you will see a list of recently opened documents, making it easy to quickly access and resume work on a previous design.

The File menu also provides options for creating new documents, saving and closing existing documents, and accessing other settings. By understanding the options available in the File menu, you can efficiently navigate the SOLIDWORKS interface and access the necessary tools to create and manage your designs.

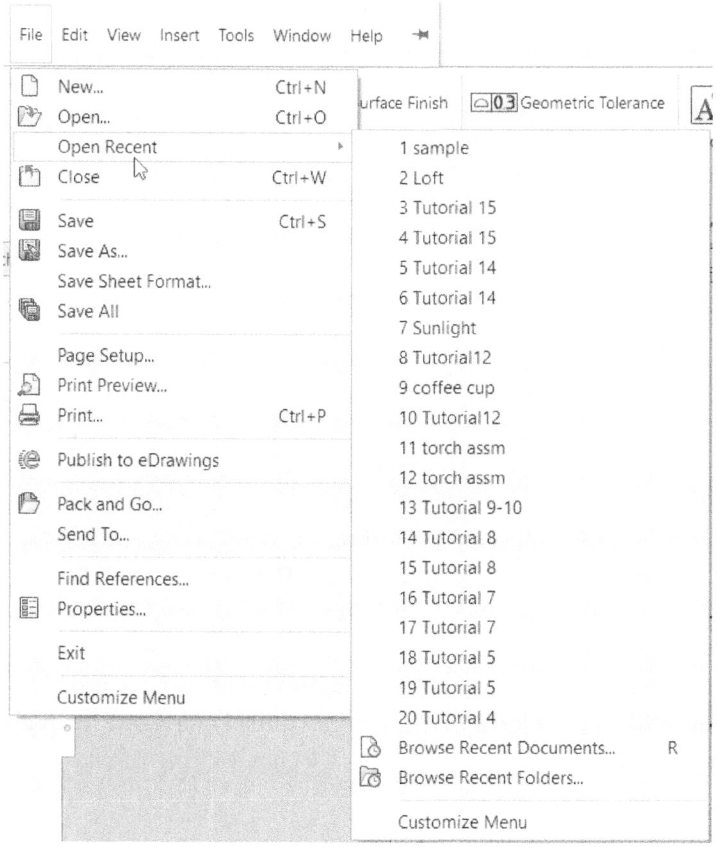

Quick Access Toolbar

The Quick Access Toolbar is a convenient feature that provides quick access to commonly used commands. The Quick Acess Toolbar is located at the top of the window, and consists of commands such as Home, New, Save, Open, Save As, and Print.

Search Commands Bar

The Search Commands bar is a powerful feature that allows you to quickly find any command available in the software. By typing a keyword into the Search Commands bar, you can generate a list of commands related to that keyword, making it easy to locate and access the tools you need without having to navigate through multiple menus or sub-menus.

The Search Commands bar can help reduce the learning curve and improve productivity by providing quick access to necessary tools. Additionally, it can improve efficiency by reducing the time spent searching for commands, allowing you to focus on the design process.

To access the Search Commands bar, simply click on the magnifying glass icon located at the right end of the Menu bar. By taking advantage of the Search Commands bar, you can save time and increase productivity.

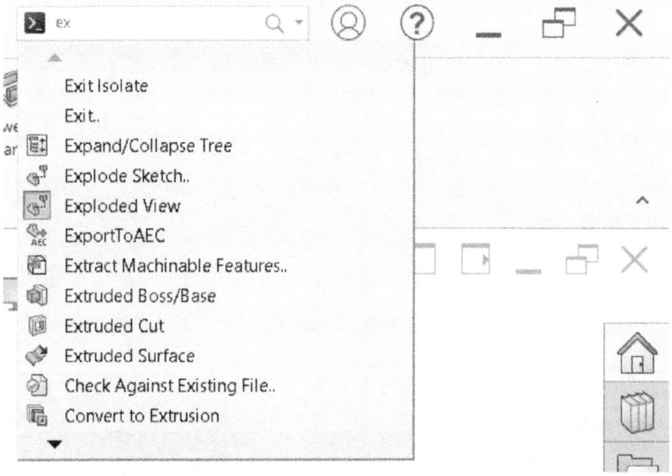

Graphics Window

The Graphics window is a key component of the SOLIDWORKS interface, providing the main space where you can create and modify 3D models and sketches. Located below the CommandManager, the Graphics window is where you will perform most of your design work, from creating sketches to building 3D solids and assemblies.

The FeatureManager Design Tree, located on the left side of the graphics window, allows you to access and manage the features of your 3D model. With the FeatureManager Design Tree, you can quickly navigate through the features of your model, modify existing features, and add new ones.

To get the most out of SOLIDWORKS, it's important to become comfortable with the Graphics window and FeatureManager Design Tree, as they are essential tools for creating and managing 3D models and sketches.

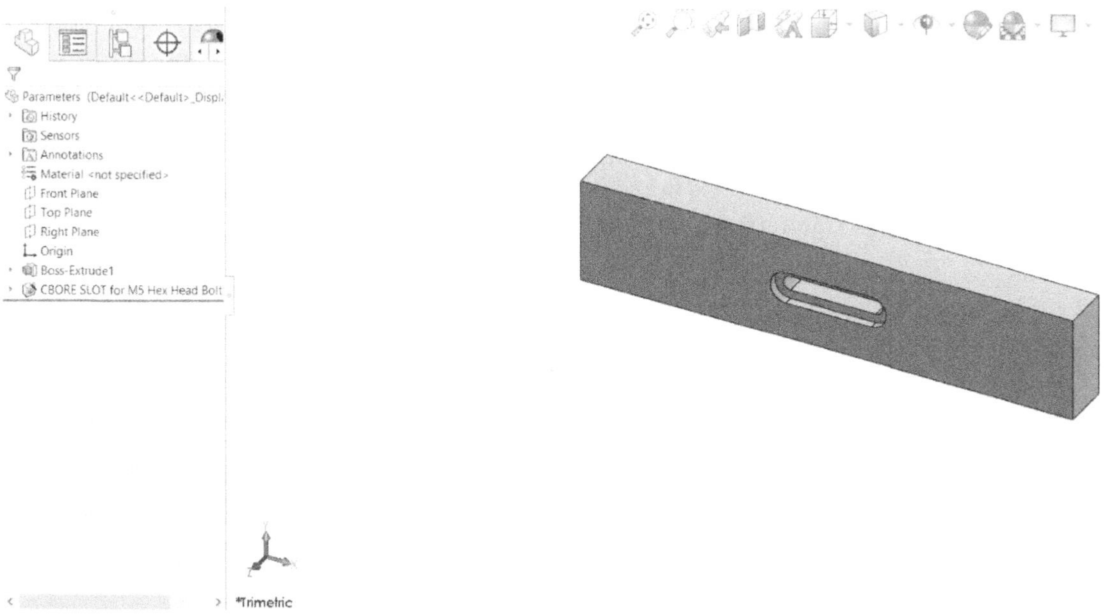

View (Heads Up) Toolbar

The View (Heads Up) Toolbar in SOLIDWORKS contains a variety of tools that help you to visualize your 3D model. This toolbar is typically located at the top of the graphics window and provides quick access to commands for changing the view orientation, zooming, and rotating the model.

Some of the tools available in the View (Heads Up) Toolbar include:

Zoom: Allows you to zoom in or out on the model.

Pan: Allows you to move the model horizontally or vertically in the graphics window.

Rotate: Allows you to rotate the model around a specific axis.

Shaded View: Allows you to change the display mode of the model, such as shaded, wireframe, or hidden lines.

By using the tools in the View (Heads Up) Toolbar, you can quickly and easily adjust the view of your model to better understand its geometry and design intent. This toolbar is useful, as it provides quick access to the necessary tools for visualizing 3D models.

View Selector

The **View Selector** is a tool that allows you to set the view orientation of your 3D model. The **View Selector** can be accessed by clicking on the **View Orientation** drop-down located on the View (Heads Up) toolbar.

The **View Selector** provides a variety of predefined view orientations, including isometric, axonometric, and orthographic views. By selecting a view orientation from the View Selector, you can quickly and easily adjust the view of your model to better understand its geometry and design intent.

In addition to the predefined view orientations, the **View Selector** also allows you to customize the view orientation by using the **New View** option. This allows for more precise control over the view of the model and can be particularly useful when working with complex or detailed models.

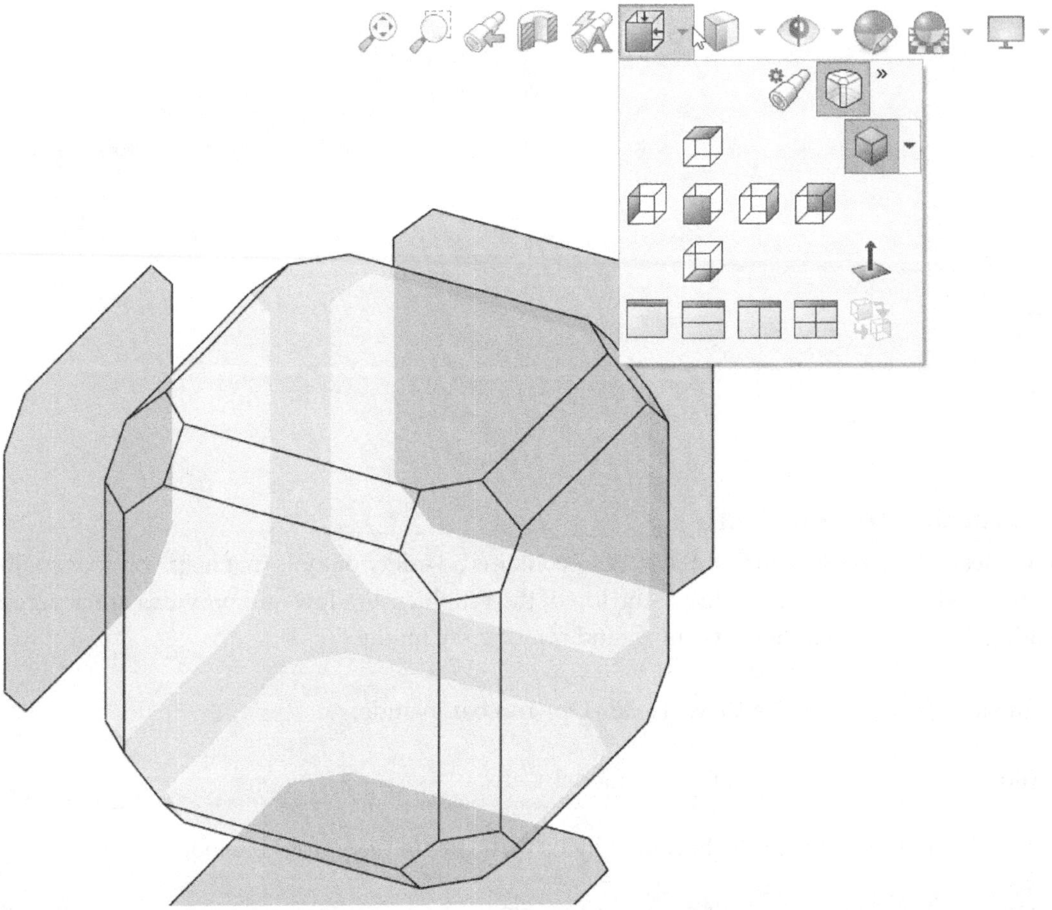

Status Bar

The Status Bar provides important information about the current sketch or model. The Status Bar is located at the bottom of the window and displays various messages and notifications related to the current design.

For example, the Status Bar may display messages related to the selection status, dimension values, or errors in the design. It may also display the, units, and other important information.

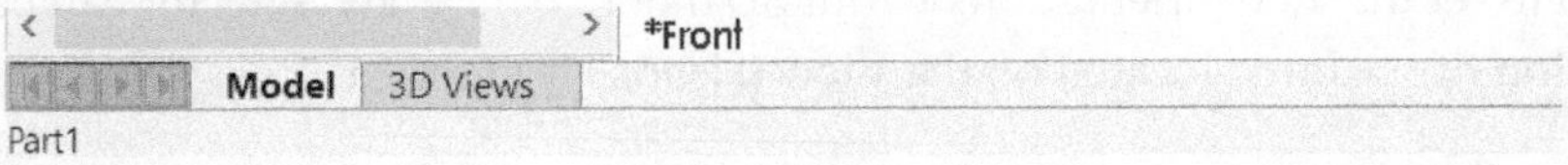

PropertyManager

The PropertyManager is a powerful feature that provides detailed options and settings for various commands. When you activate a command, the related PropertyManager appears on the left side of the screen, displaying the options and steps to complete the execution of the command.

The PropertyManager typically includes various settings and options related to the active command. For example, when creating a new extrusion, the PropertyManager would display options for the direction, distance, and taper of the extrusion, allowing you to customize the design according to your needs.

The PropertyManager is designed to be user-friendly and intuitive, with clear instructions and default settings that make it easy to complete common tasks. However, it also offers advanced options and settings for more complex tasks, allowing you to customize the software according to your needs.

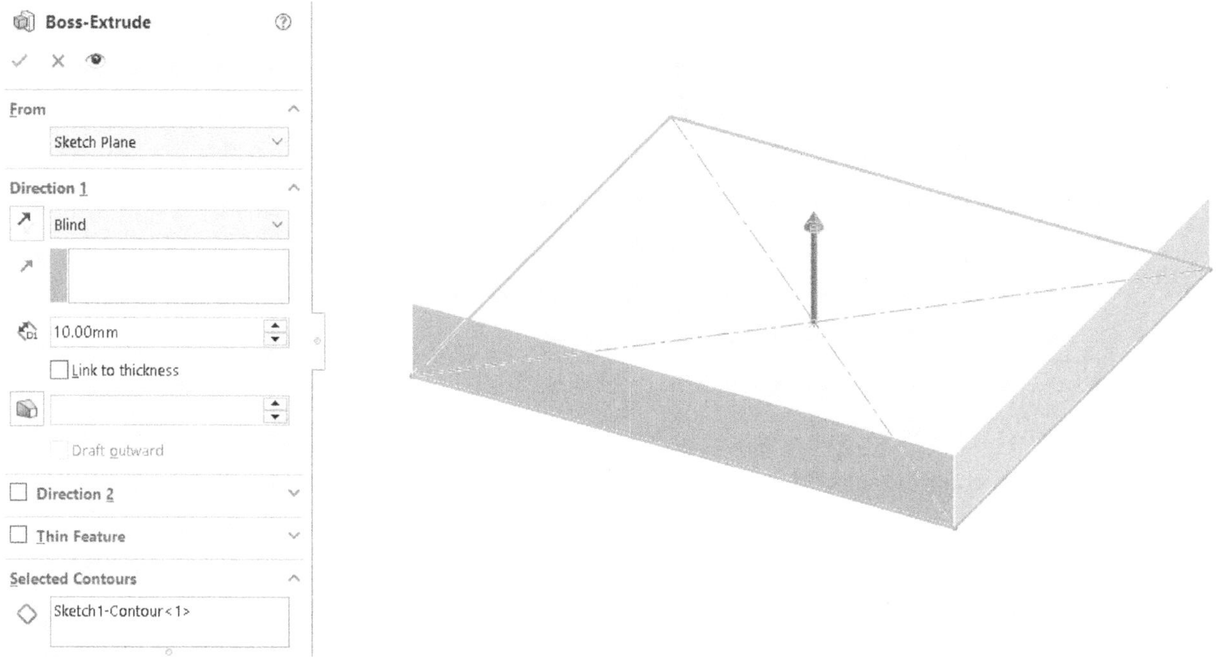

Changing the display of the CommandManager

The CommandManager in SOLIDWORKS can be customized to suit your specific workflow and needs. To customize the CommandManager, right-click on it and select "Customize" from the context menu. This will open the Customize dialog, which provides options for adding, removing, and organizing commands in the CommandManager.

On the **Customize** dialog, click on the **Commands** tab to access the list of available commands. The **Categories** list on the left side displays the various categories of commands, such as **File**, **Edit**, **View**, and so on. Click on a category to display the commands related to it in the **Buttons** area.

To add a command to the CommandManager, click and drag its icon from the **Buttons** area onto the CommandManager. You can also remove commands from the CommandManager by dragging them off or right-clicking and selecting **Delete**.

Customizing the CommandManager can help improve your productivity and efficiency in SOLIDWORKS by making the tools you use most frequently easily accessible. It's a good idea to spend some time customizing the CommandManager to suit your specific needs and workflow.

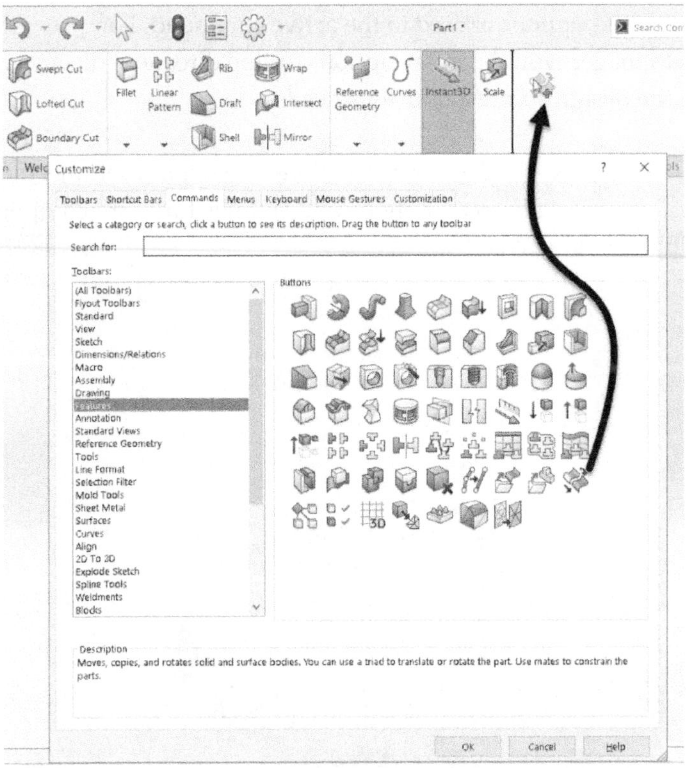

After making the required changes, click **OK** to close the dialog.

Dialogs

Dialogs are an integral part of the SOLIDWORKS user interface, providing a way to easily specify settings and options for various commands and tasks. A dialog is a graphical user interface element that appears on the screen, usually in response to a user action. It typically contains fields, buttons, and other controls that allow the user to enter or select information.

The figure below shows an example of a SOLIDWORKS dialog.

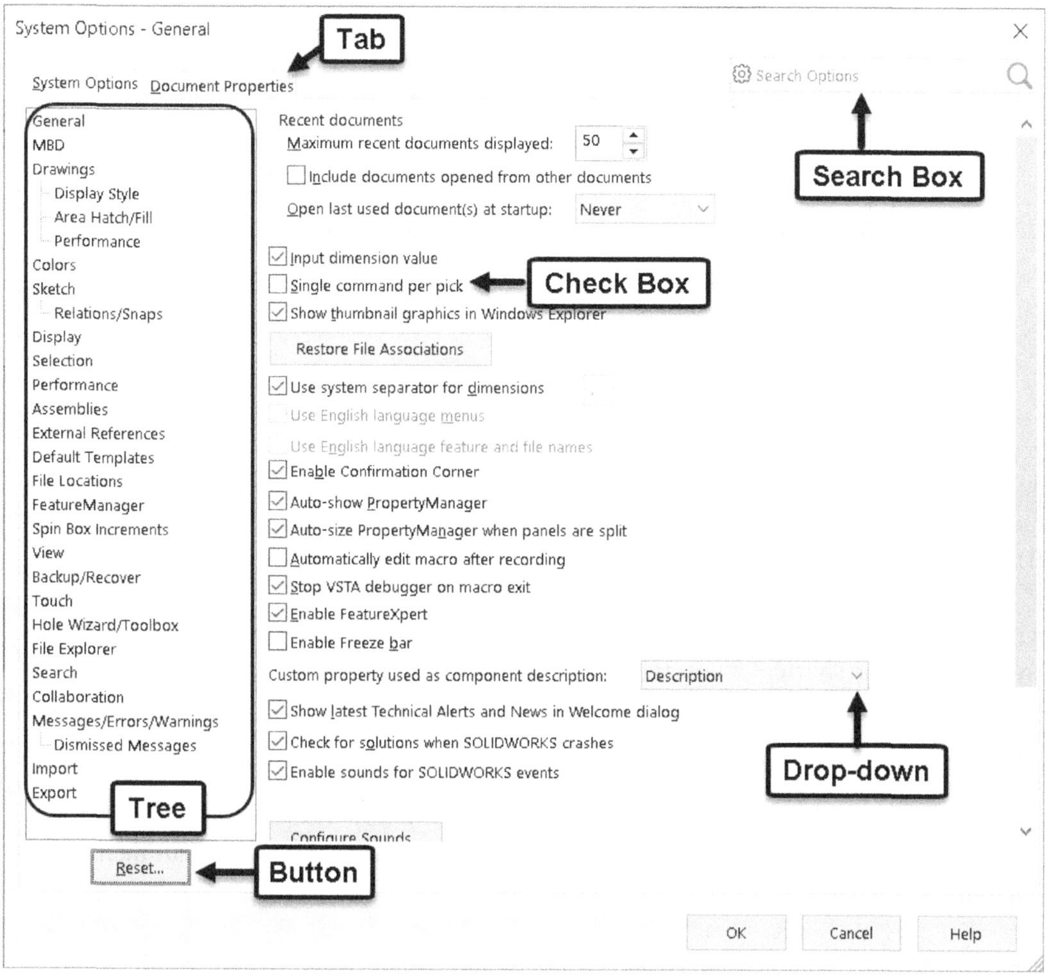

Mouse Gestures

Mouse gestures allows you to activate commands by performing a specific mouse movement. To use a mouse gesture, right-click the mouse button and drag the pointer in a specific direction. This will display a radial menu of commands related to the gesture.

You can customize the mouse gestures by using the **Customize** dialog. To access the **Customize** dialog, right-click the mouse button on the CommandManager and select **Customize** from the context menu. In the **Customize** dialog, click on the **Mouse Gestures** tab to access the options for customizing the mouse gestures.

On the **Mouse Gestures** tab, you can add or remove commands from the radial menu by dragging and dropping the command icons. You can also change the number of mouse gestures to be displayed on the radial menu by adjusting the settings in the dialog.

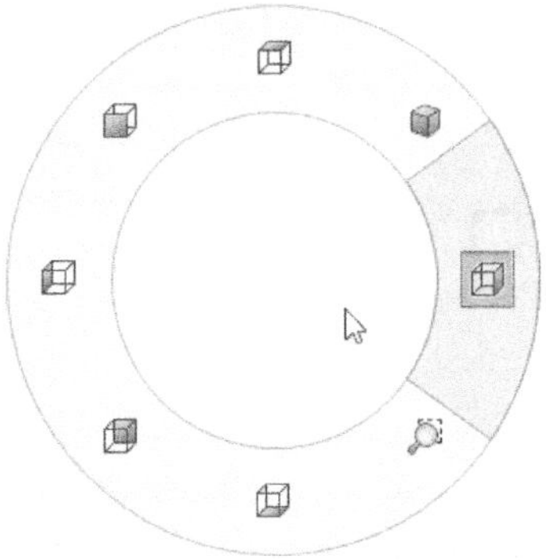

Shortcut Menus

Shortcut menus provide quick access to a set of related commands and options based on the current context or selection. Shortcut menus are displayed when you right-click the mouse button in the graphics window, or in other parts of the SOLIDWORKS interface.

The options in shortcut menus vary depending on the current environment and selection. For example, when you right-click in a sketch environment, the shortcut menu will display options related to sketching, such as adding lines, arcs, or circles. When you right-click in a part environment, the shortcut menu will display options related to creating or modifying 3D geometry.

Shortcut menus can be a helpful way to access frequently used commands and options quickly and easily, without having to navigate through menus or toolbars. By learning the options available in shortcut menus, you can improve your productivity and efficiency.

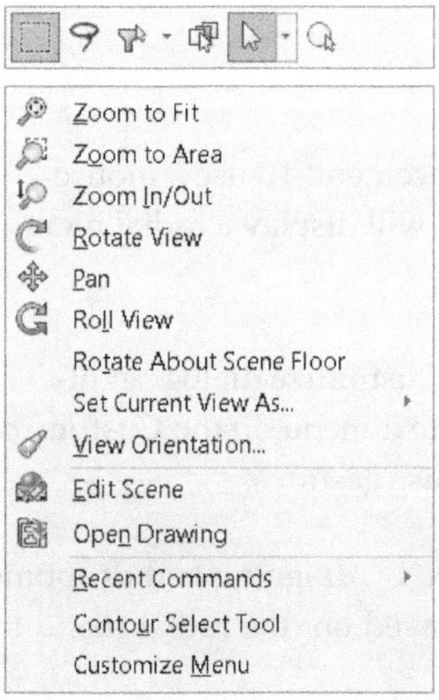

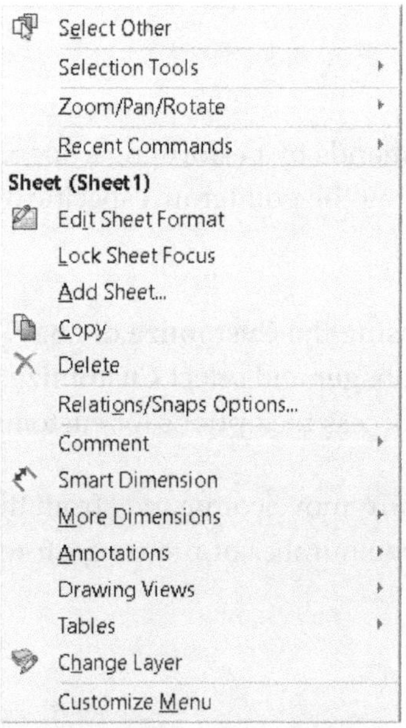

You can customize the options in shortcut menus by using the **Customize menu** option. To do this, right-click and select **Customize Menu**. Next, uncheck the options to be excluded from the shortcut menu, and then click in the graphics window.

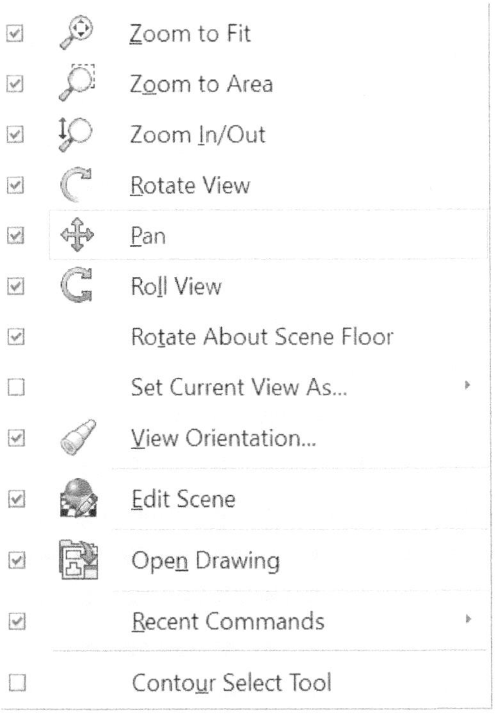

Selection Tools

A selection tool is used to select multiple elements of a model or sketch. You can select multiple elements by using two types of selection tools. The first type is the Box selection tool. You can create this type of selection tool by defining its two diagonal corners. When you define the first corner of the selection box on the left and the second corner on the right side, the elements which fall entirely under the selection box are selected.

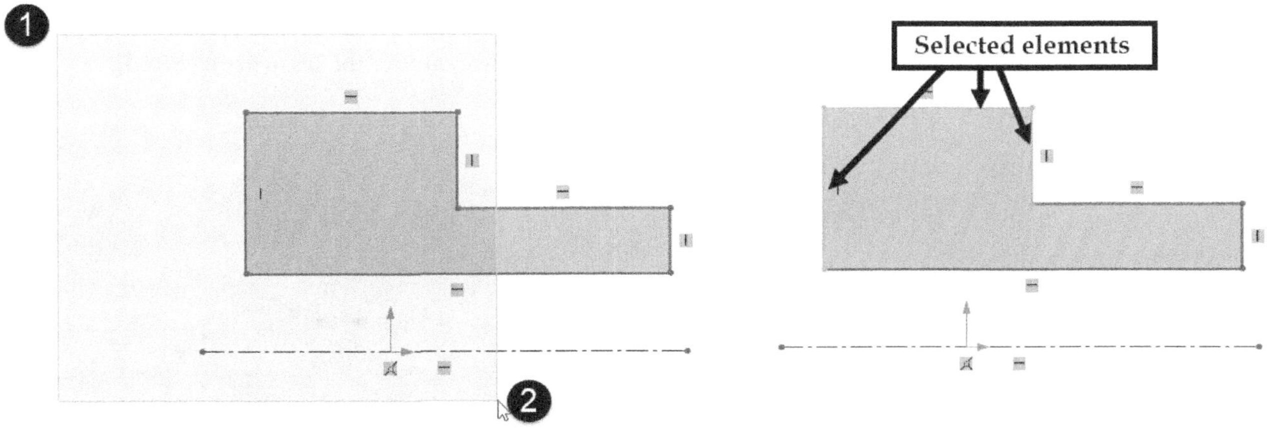

However, if you define the first corner on the right side and the second corner on the left side, the elements, which fall entirely or partially under the selection box, are selected.

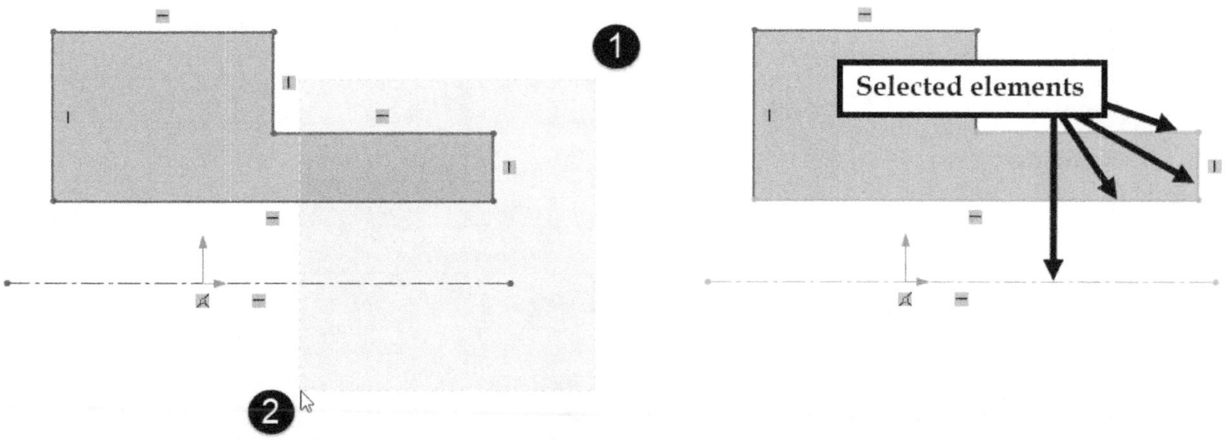

The second type of selection tool is **Lasso** (right-click and select **Selection Tools > Lasso**). Lasso is an irregular shape created by clicking and dragging the pointer across the elements to select. If you drag the pointer from the bottom to the top, the elements cutting the lasso are selected.

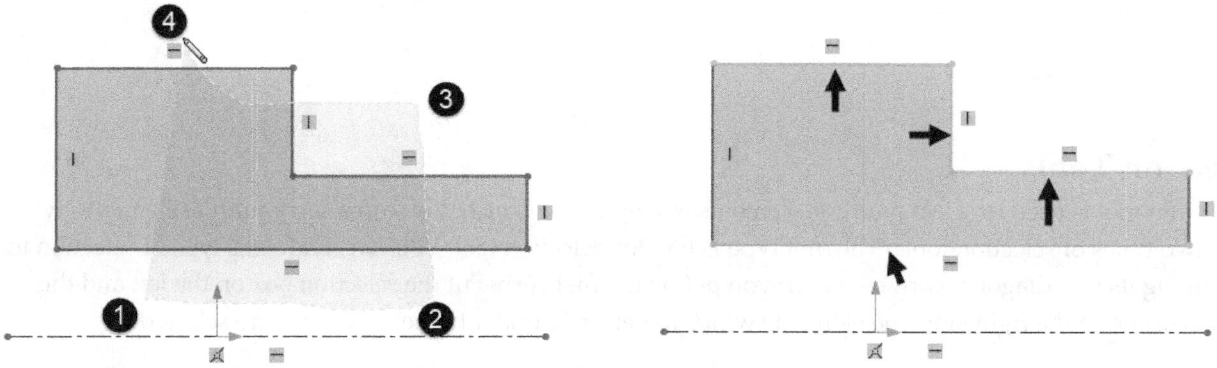

If you drag the pointer from top to bottom, the elements that fall entirely under the lasso are selected.

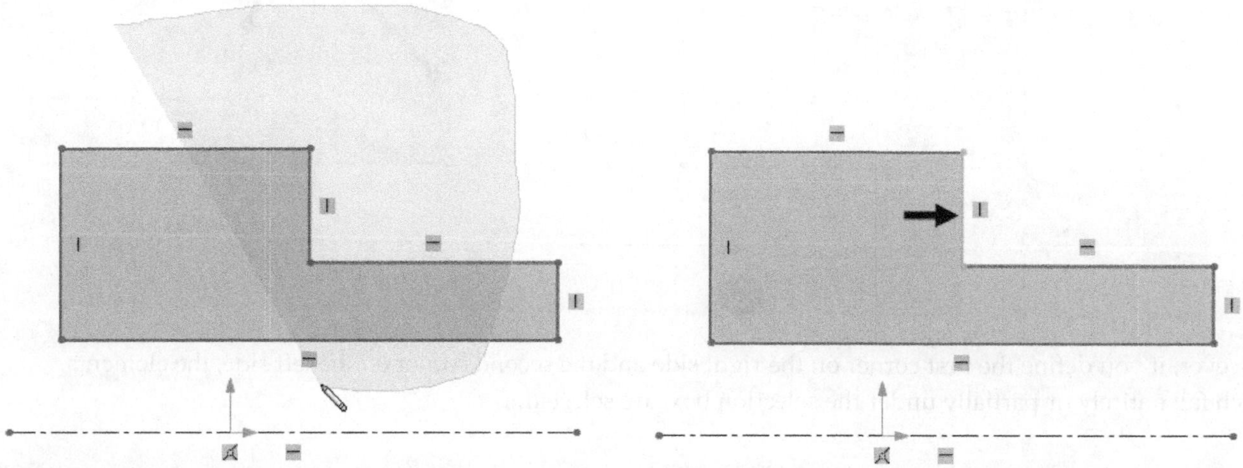

SOLIDWORKS Options

The SOLIDWORKS Options dialog is a powerful tool that allows you to customize various system and document settings. To access the SOLIDWORKS Options dialog, click on the Options icon in the Quick Access Toolbar or select "Options" from the Application Menu.

The SOLIDWORKS Options dialog contains two main sections: System Options and Document Properties. The **System Options** section contains settings that apply to the entire SOLIDWORKS system, such as file locations, colors, and units. The **Document Properties** section contains settings that apply to the current document, such as drafting standards, custom properties, and security.

To set the options, select the desired category from the tree on the left side of the dialog. The options for the selected category will appear on the right side of the dialog. Make the desired changes, and then click "OK" to save the changes.

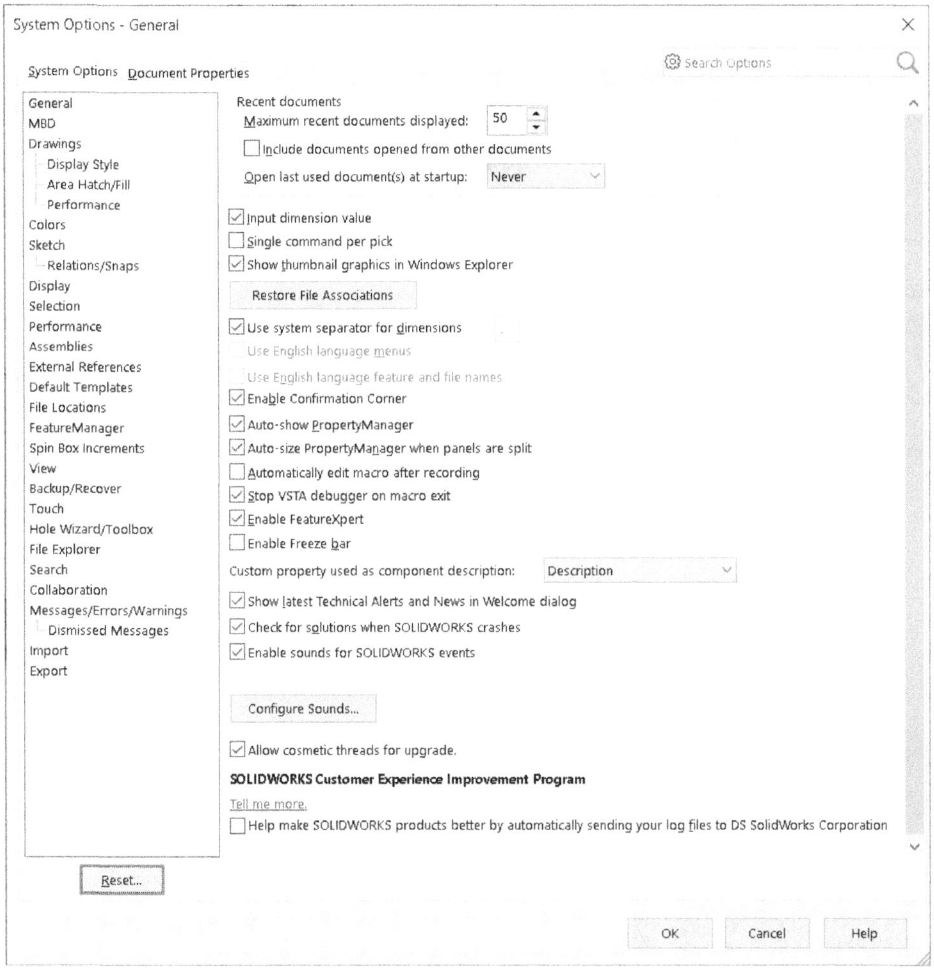

Changing the Background appearance

The **SOLIDWORKS Options** dialogue helps you to change the background color, rendering, and light settings. On this dialogue, click the **System Options** tab and set the **Colors** option from the tree. Next, select the required option from the **Background appearance** section. For example, select **Plain** (Viewport background color above). Next, click the **Edit** button below the color swatch; the **Color** dialog appears. Select the required color (for example, White), and then click **OK** twice.

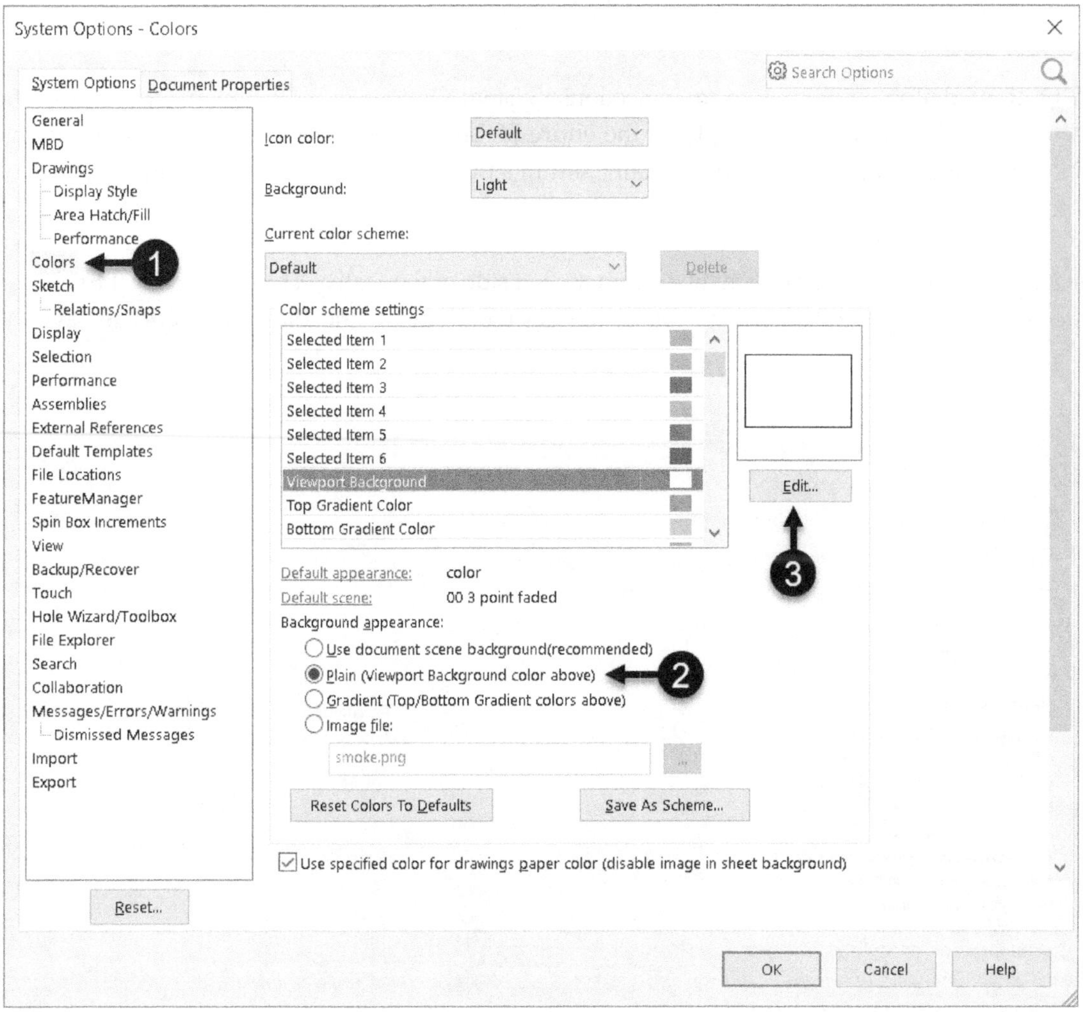

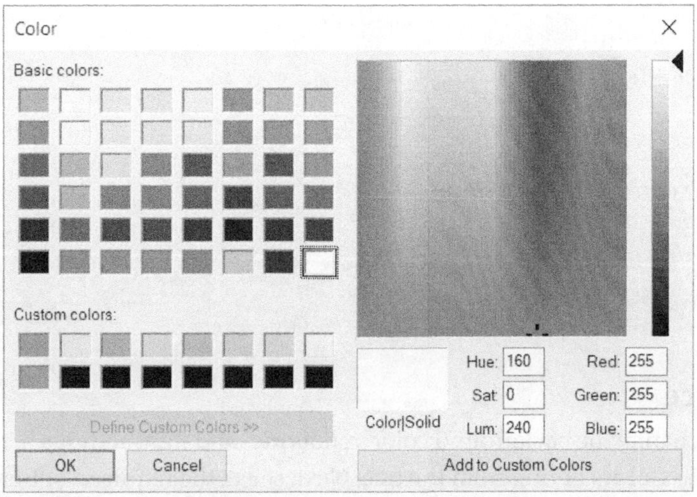

SOLIDWORKS Help

SOLIDWORKS offers a comprehensive help system that goes beyond basic command definitions. Here are some ways to access SOLIDWORKS help.

1. Press F1: This is the default keyboard shortcut to open the SOLIDWORKS Help system.
2. Click the Help button: This button is located in the upper-right corner of the SOLIDWORKS window.
3. Use the Search bar: The Search bar is located at the top of the Help window. You can type in a keyword or phrase related to the topic you need help with.

Questions

1. Explain how to customize the CommandManager.
2. Give one example of where you would establish a relationship between a part's features.
3. Explain the term 'associativity' in SOLIDWORKS.
4. List any two procedures to access SOLIDWORKS Help.
5. How can you change the background color of the graphics window?
6. How can you activate the Mouse Gestures?
7. How is SOLIDWORKS a parametric modeling application?

Chapter 2: Sketch Techniques

This chapter covers the methods and commands to create sketches. In SOLIDWORKS, you create a rough sketch and then apply dimensions and relations that define its shape and size. The dimensions define the length, size, and angle of a sketch element, whereas relations define the relations between the sketch elements.

The topics covered in this chapter are:

- *Learning Sketching commands*
- *Using relations and dimensions to control the shape and size of a sketch*
- *Learning commands and options that help you to create a sketch easily*
- *Learn advanced sketching commands*

Starting a New Part Document

To start a new Part document, click **File** > **New** on the Quick Access Toolbar. Next, click the **Part** icon on the **New SOLIDWORKS Document** dialog. Click **OK** to start the Part document.

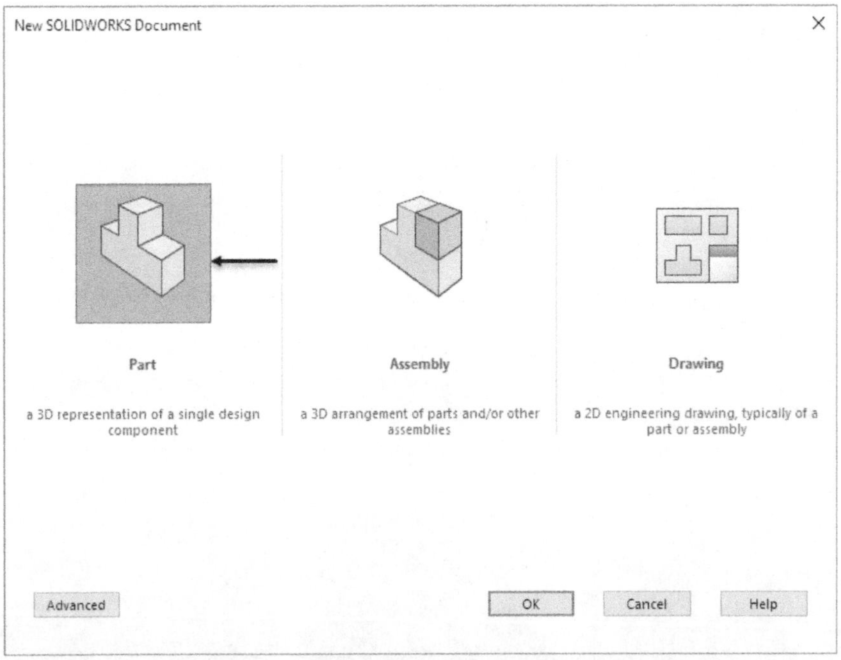

Sketching in the Sketch Environment

Creating sketches in the Sketch environment is very easy. You have to activate the **Sketch** command and select a plane on which you want to create the sketch. To do this, click **Sketch** tab > **Sketch** on the CommandManager. Next, click on any of the planes located at the center of the graphics window; the Sketch environment is active. You can now start drawing sketches on the selected plane. For example, activate the **Line** command and start sketching lines. After completing the sketch, click **Sketch** tab > **Exit Sketch** on the CommandManager (or) click **Exit Sketch** at the top right corner of the graphics window.

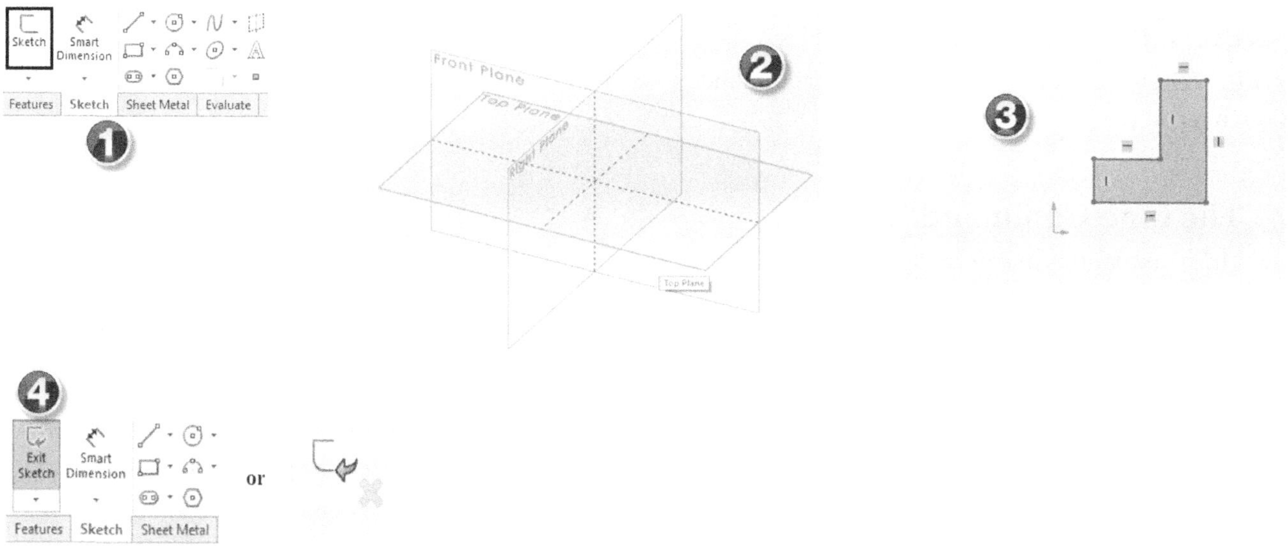

Rapid Sketching

The **Rapid Sketch** command helps you create a sketch directly by activating a draw command and selecting the sketch plane. Activate this command (On the **Sketch** CommandManager, click **Rapid Sketch**) and select a draw command from the **Sketch** CommandManager. Click on a plane or planar face, and then start sketching. Click **Exit Sketch** at the top right corner of the graphics window.

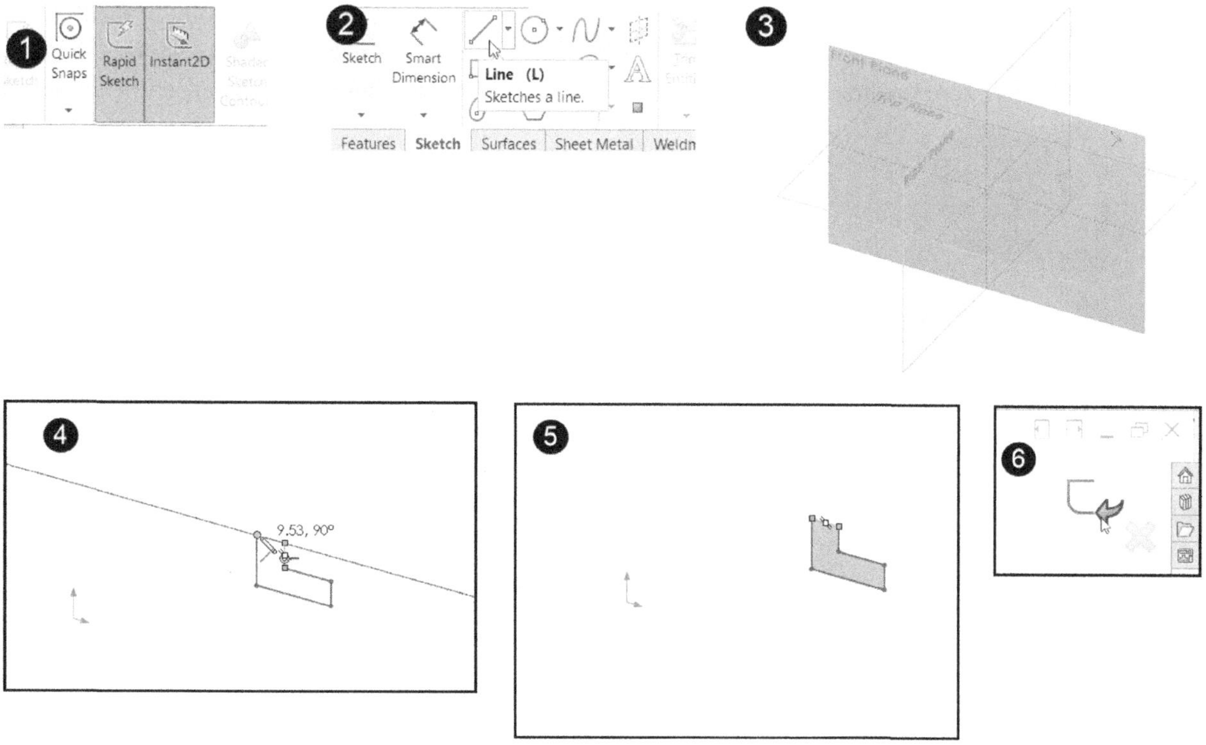

Sketching Commands

SOLIDWORKS provides you with a set of commands to create sketches. You can find these commands on the **Sketch** tab of the CommandManager.

 The Line Command

The **Line** command is one of the most commonly used commands while creating a sketch. To activate this command, you need to click **Sketch** tab > **Line** on the CommandManager. Notice that the mouse pointer changes to a pencil glyph with a line at the bottom  indicating that the **Line** command is active. Also, the **Insert Line** PropertyManager appears. On this PropertyManager, the **Orientation** is set to **As sketched**. As a result, the lines are created based on the points you specify on the screen. If you set the **Orientation** to **Horizontal** or **Vertical**, SOLIDWORKS creates horizontal or vertical lines, respectively. If you set the **Orientation** to **Angle**, SOLIDWORKS creates a line and adds a length and angle dimension.

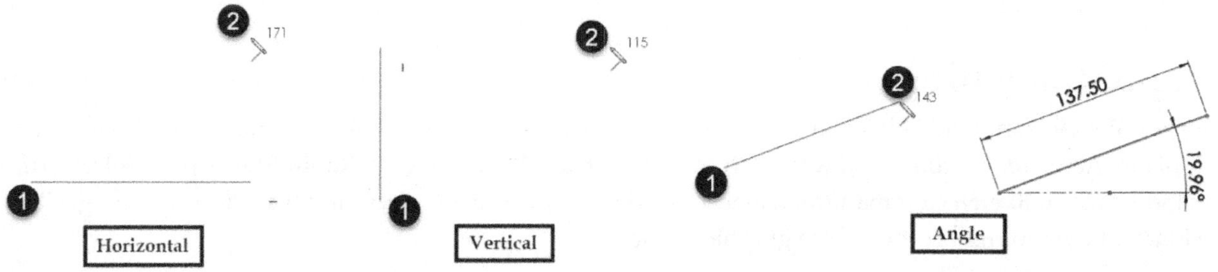

Set the **Orientation** to **As sketched**, and then click in the graphics window to specify the first point of the line. Move the pointer and click to define the endpoint of the line segment. Now, notice the two dotted lines originating from the endpoint of the line segment. These lines help you to create a line that is perpendicular or collinear to the first line.

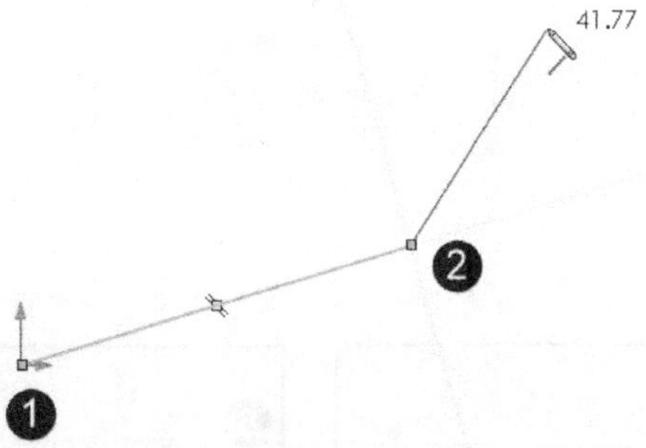

Move the pointer along the dotted line perpendicular to the first line. Notice the **Perpendicular** relation on the line. The relation locks the behavior of the sketch element. You learn about relations later in this chapter. Click to create a line perpendicular to the first line. Besides, the **Perpendicular** relation is created between the first line and the second line.

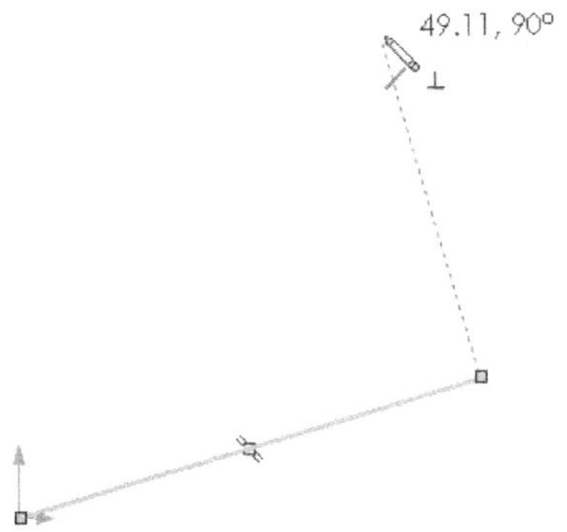

Move the pointer along the dotted line, which is collinear to the second line. The **Collinear** relation appears next to the line. Click to create a line collinear to the second line.

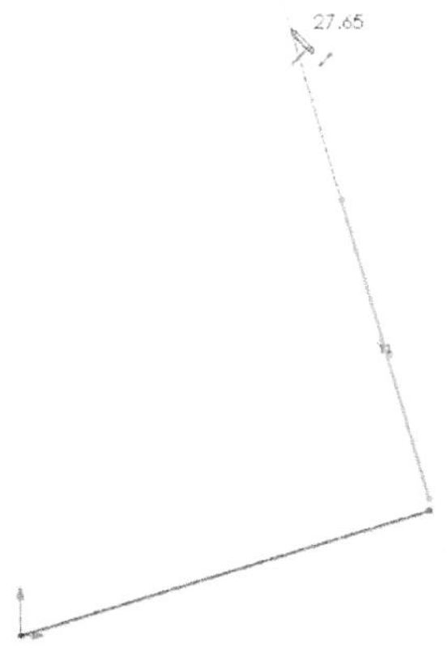

Move the pointer horizontally toward the left and notice the **Horizontal** relation next to the line. Click to create the horizontal line.

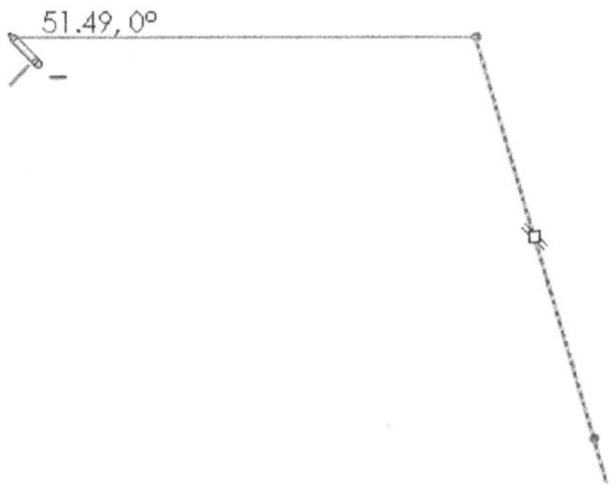

Move the pointer vertically downward and notice the **Vertical** relation. Click to create the vertical line.

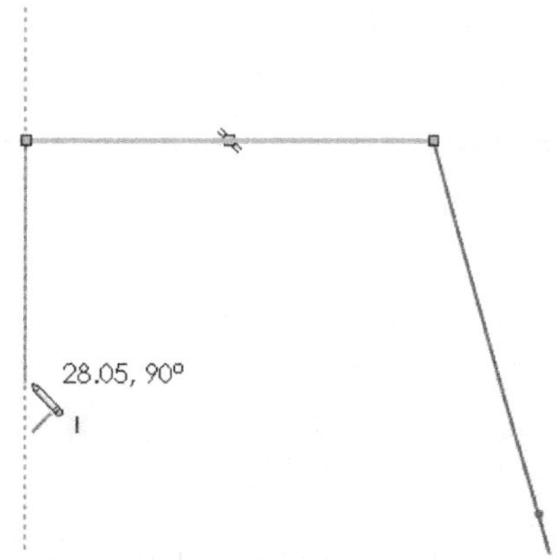

Move the pointer and notice that the **Parallel** relation appears. Also, a line is highlighted. Click to create a line parallel to the highlighted line.

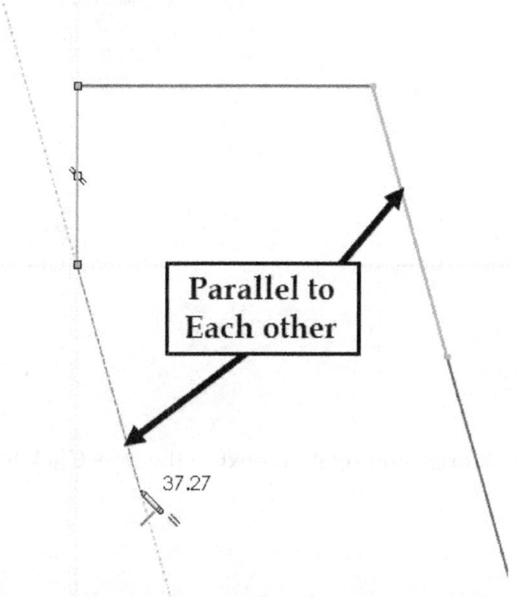

Move the pointer to the first line's midpoint; a dotted line appears between the midpoint and the pointer. Click to specify the endpoint of the line. Notice that no relation is created between the midpoint and endpoint. However, this helps you to specify the points quickly.

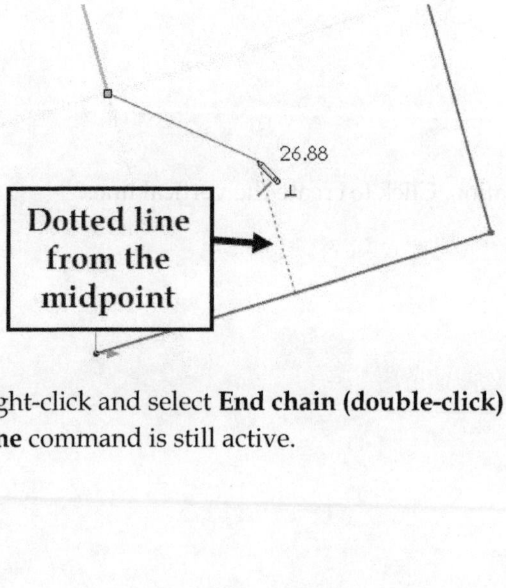

Right-click and select **End chain (double-click)** from the shortcut menu; the line chain is stopped. Notice that the **Line** command is still active.

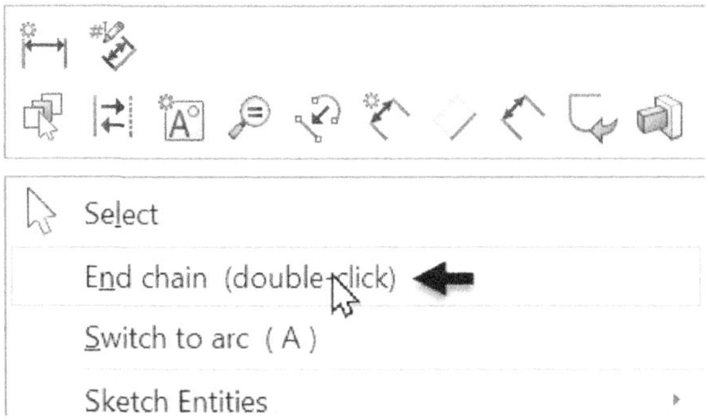

The **Line** command can also be used to draw arcs continuous with lines. Click in the graphics window, move the pointer horizontally toward the right and click. Right-click and select **Switch to arc (A)** from the shortcut menu.

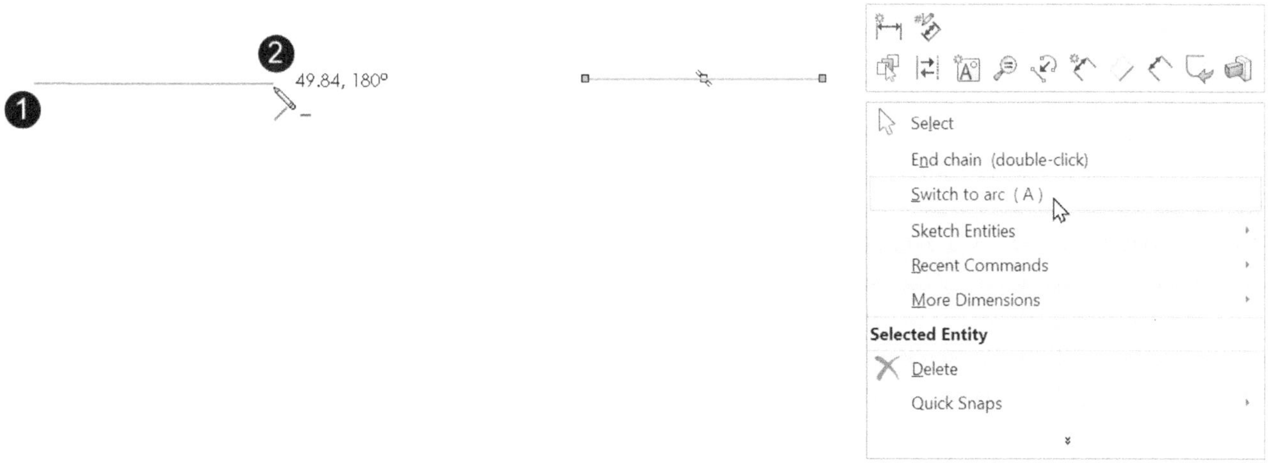

Move the pointer toward the right and then downwards; the arc appears tangent to the horizontal line.

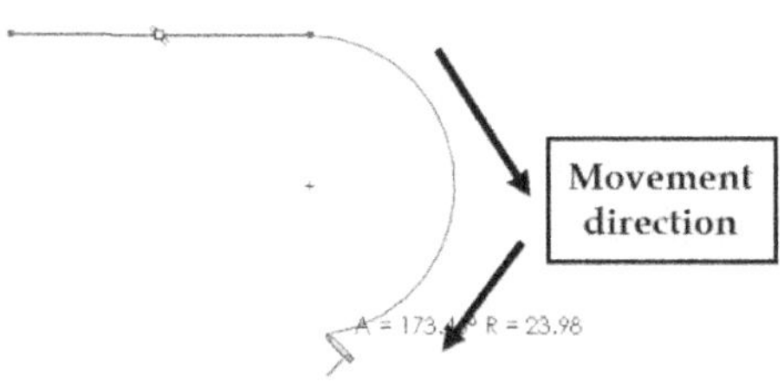

Move the pointer to the endpoint of the horizontal line. Next, move the pointer downwards and then toward the left; the arc appears normal to the horizontal line. Click to create the arc normal to the line. Next, press Esc to deactivate the **Line** command.

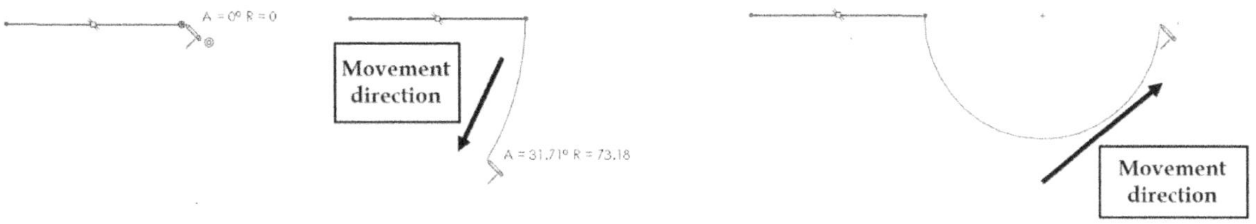

To delete a line, select it and press the **Delete** key. To select more than one line, press and hold the **Ctrl** key and then click on the line segments; the lines will be highlighted. You can also select multiple lines by dragging a box from right to left. Press and hold the left mouse button and drag a box from right to left; the lines inside or crossing the box boundary will be selected. Dragging a box from left to right will only select the lines that are inside the box.

Creating a Midpoint Line

You can create a midpoint line using the **Midpoint Line** command (on the CommandManager, click **Sketch** tab > **Line** drop-down > **Midpoint Line**). Next, specify the midpoint of the line, and then move the pointer. Click to specify the endpoint of the line.

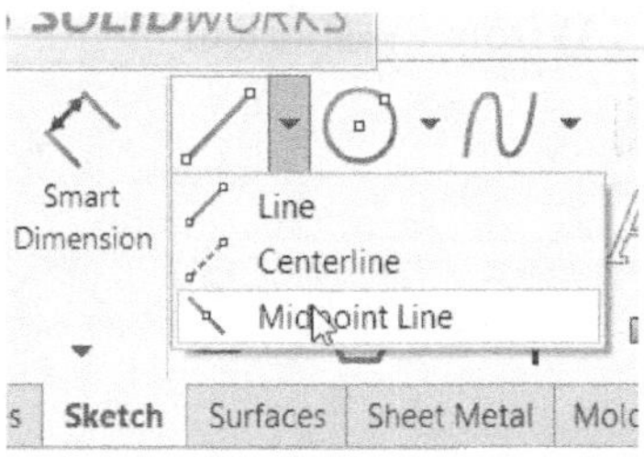

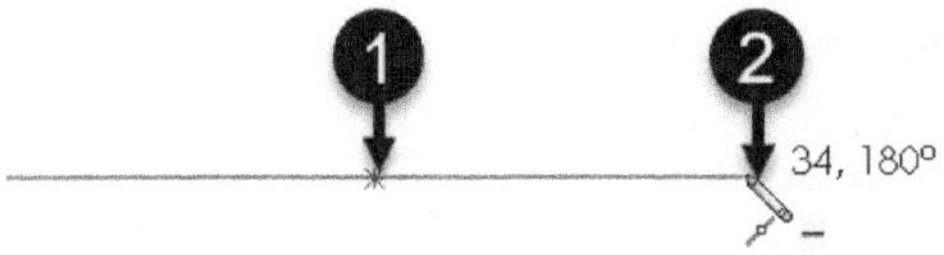

The Tangent Arc Command

This command creates an arc tangent to another entity. The working of this command is the same as the Arc in the **Line** command. You have to select the endpoint of a line and create a tangent or normal arc. On the CommandManager click **Sketch** tab > **Arc** drop-down > **Tangent Arc**. Next, select the endpoint of a line, move the pointer along the direction of the line, and then click to create a tangent arc.

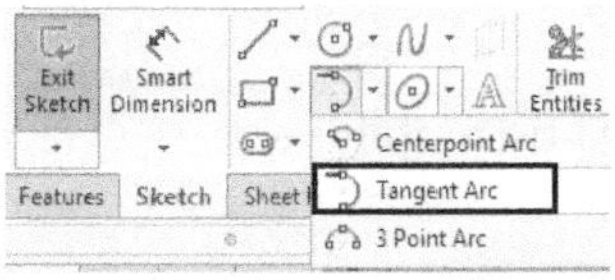

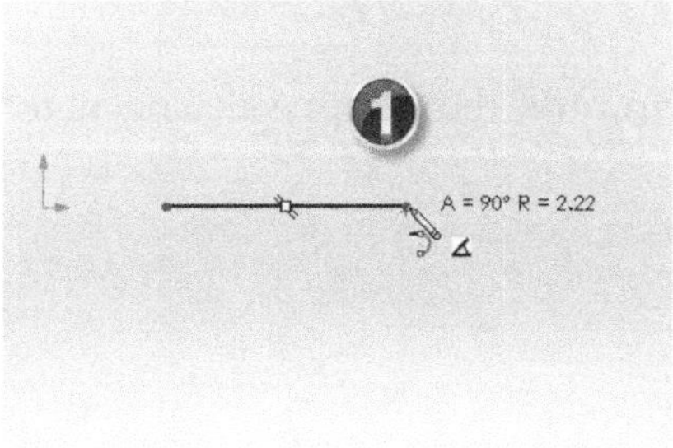

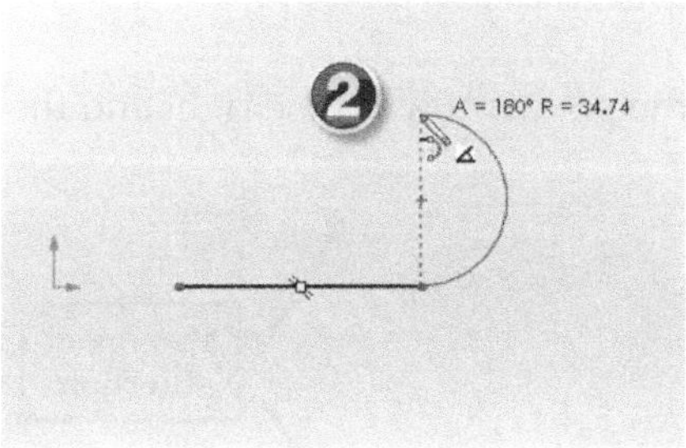

Click **Close Dialog** on the **Arc** PropertyManager.

The 3 Point Arc command

This command creates an arc by defining its start, end, and radius. Activate this command (On the CommandManager, click **Sketch** tab > **Arc** drop-down > **3 Point Arc**) and click to define the start point of the arc. Next, move the pointer and click again to define the endpoint of the arc. After defining the start and endpoints, you have to define the arc's size and position. To do this, move the pointer and click to define the radius and position of the arc. Next, click the **Close Dialog** icon on the **Arc** PropertyManager.

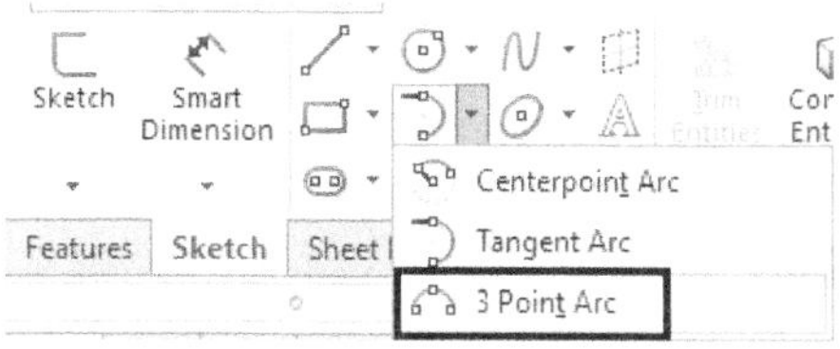

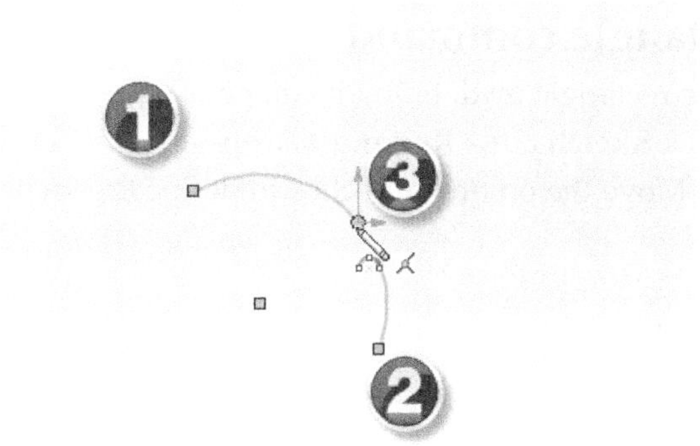

The CenterPoint Arc command

This command creates an arc by defining its center, start and endpoints. Activate this command (On the CommandManager, click **Sketch** tab > **Arc** drop-down > **Centerpoint Arc**) and click to define the center point. Now, move the pointer and notice that a line appears between the center and the mouse pointer. This line is the radius of the arc. Now, click to define the start point of the arc and move the pointer. You will notice that an arc is drawn from the start point. Also, the angle of the arc is displayed. Now, click to define the arc's endpoint and then click the **Close Dialog** icon on the **Arc** PropertyManager.

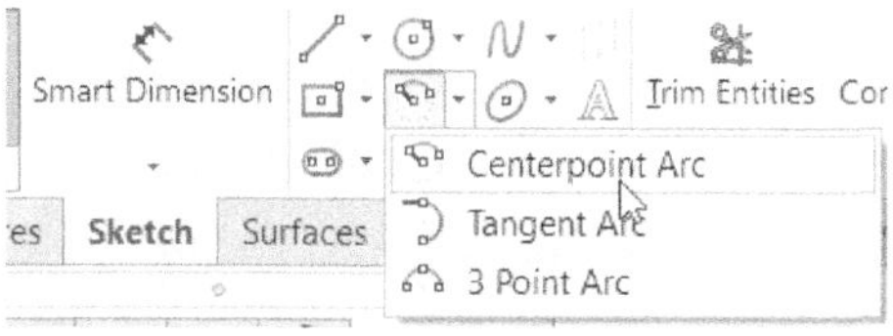

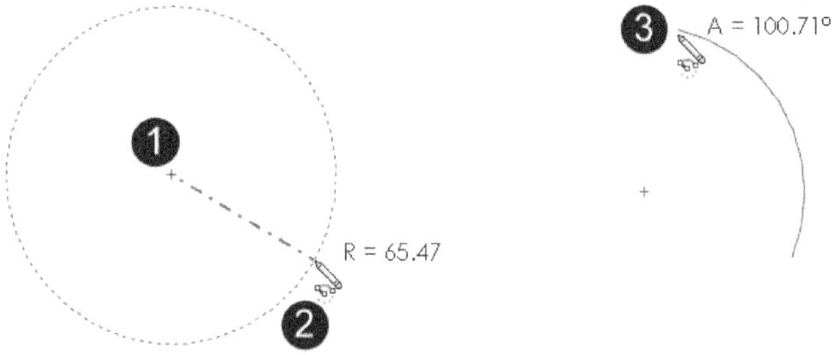

The Center Rectangle command

This command creates a rectangle by defining its center and a corner point. Activate this command (on the CommandManager, click **Sketch** tab > **Rectangle** drop-down > **Center Rectangle**) and click to define the center point of the rectangle. Move the pointer outward and click to define a corner point of the rectangle.

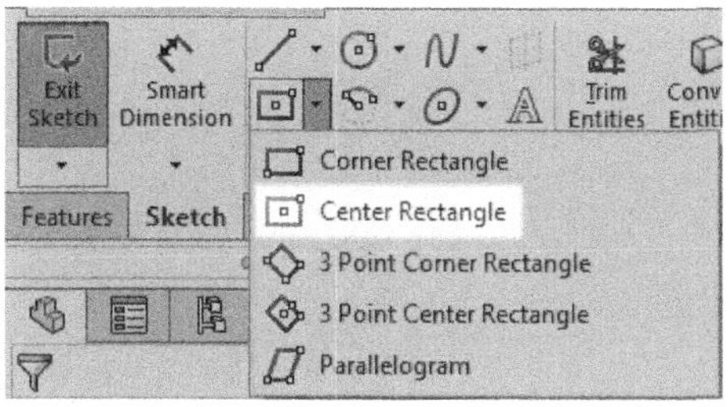

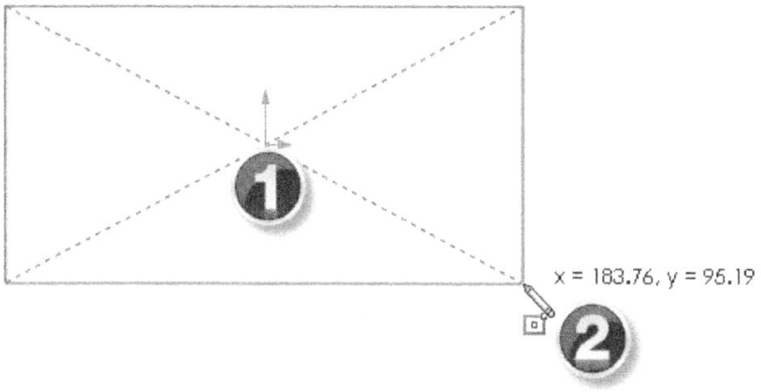

The Corner Rectangle command

This command creates a rectangle by defining its diagonal corners. Activate this command (on the CommandManager, click **Sketch** tab > **Rectangle** drop-down > **Corner Rectangle**) and click to define the first corner of the rectangle. Move the pointer and click to define the second corner.

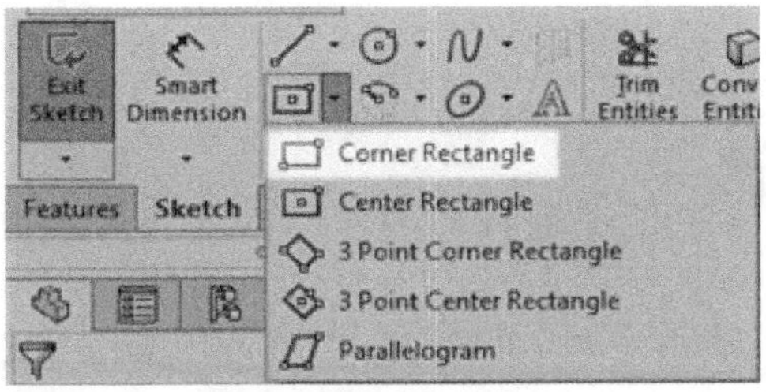

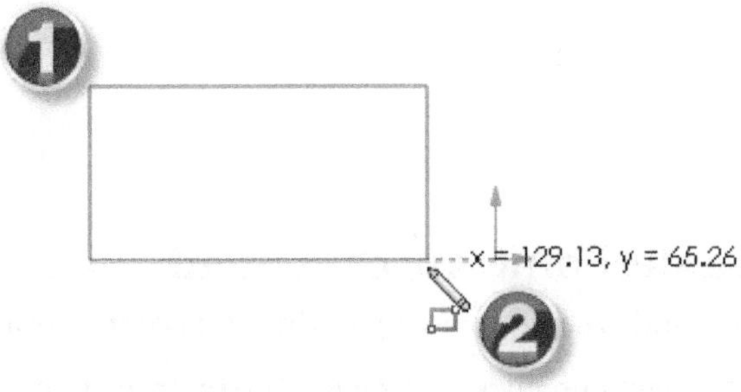

The 3 Point Corner Rectangle command

This command creates an inclined rectangle. The first two points define the length and inclination angle of the rectangle. The third point defines its width. Activate this command (on the CommandManager, click **Sketch** tab > **Rectangle** drop-down > **3 Point Corner Rectangle**) and click to define the first corner of the rectangle. Move the pointer and click to define the second corner. Next, move the pointer in the direction perpendicular to the length and click to define the width.

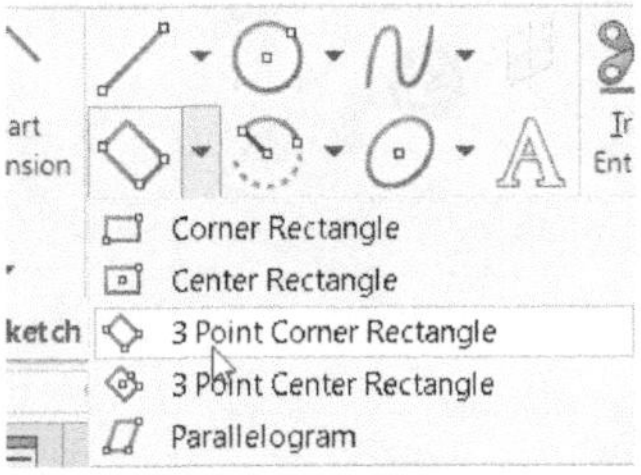

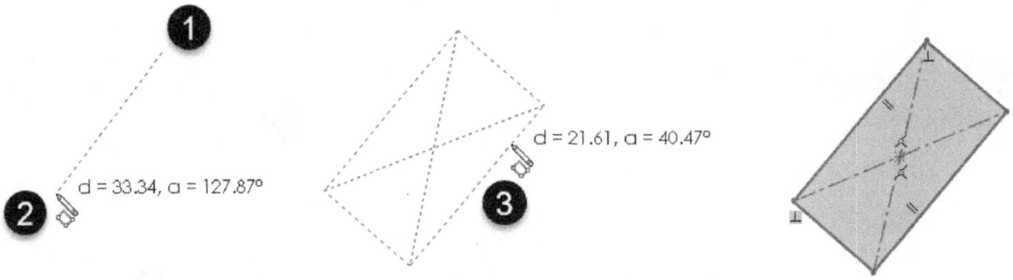

The 3 Point Center Rectangle command

This command creates a rectangle by defining its center, midpoint of one side (length or width), and corner point. The first point defines the center of the rectangle. Whereas the second point defines the midpoint of a side. Also, the angle of inclination of the rectangle is defined by the second point. The third point defines the corner of the rectangle.

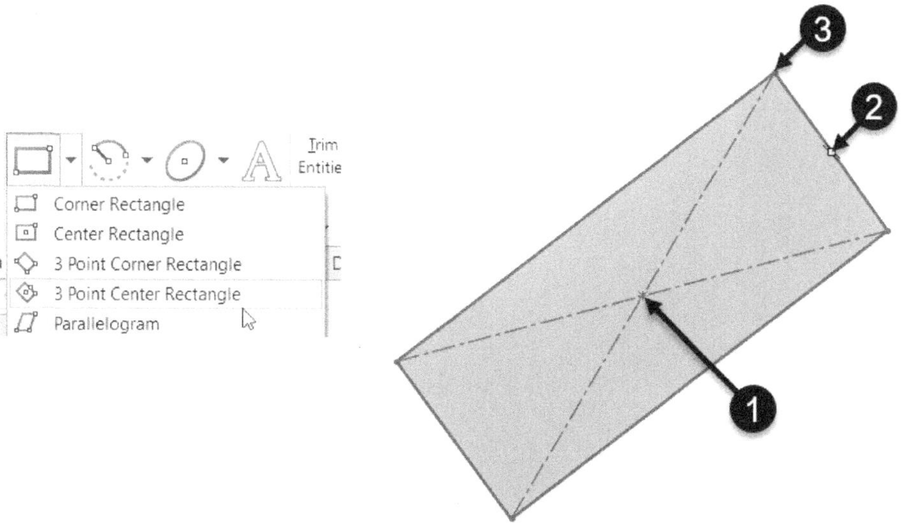

You can notice that the construction lines are displayed in Center Rectangle, 3 Point Corner Rectangle, and 3 Point Center Rectangle. By default, the construction lines are created connecting the corners of the rectangle. However, you can create the construction lines between the midpoints of the sides of the rectangle. To do this, select the **From Midpoints** option from the **Rectangle Type** section of the **Rectangle** PropertyManager,

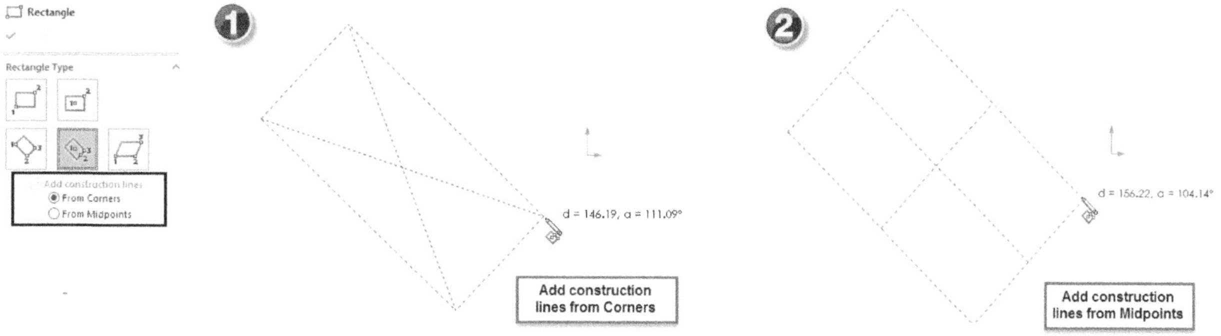

The Parallelogram command

This command provides a simple way to create a parallelogram. A parallelogram is a four-sided object. It has opposite sides parallel to each other. However, the adjacent sides are not perpendicular to each other. Activate this command (On the CommandManager, click **Sketch** tab > **Rectangle** drop-down > **Parallelogram**). Click in the graphics window to specify the first corner of the parallelogram. Move the pointer, and then click to specify the second corner. Also, the length and inclination of one side of the parallelogram are defined. Next, move the pointer and click to define the length and inclination of the other side.

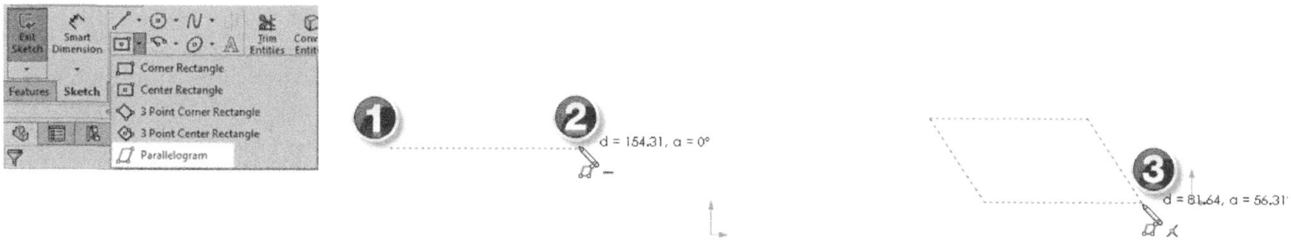

The Circle command

This command provides an easy way to draw a circle. Activate this command (click **Sketch** tab > **Circle** on the CommandManager) and click to locate the circle's center. Next, move the pointer, and then click again to define the radius of the circle.

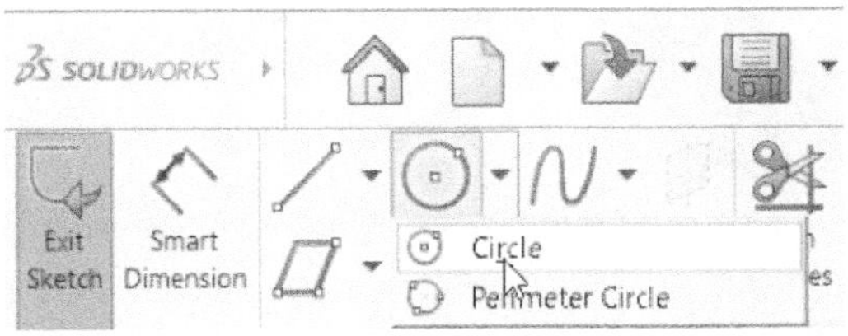

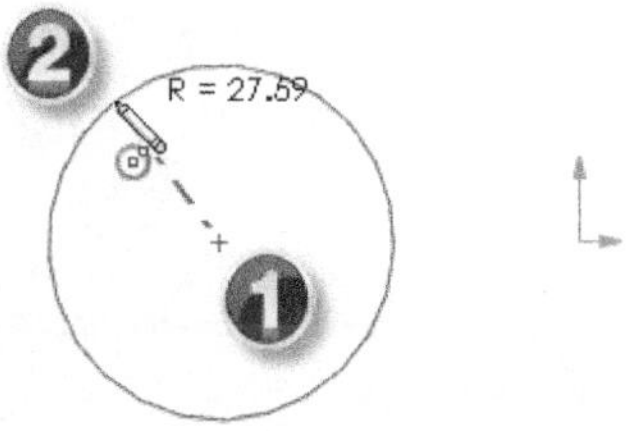

The Perimeter Circle command

This command creates a circle by using three points. Activate this command (On the CommandManager, click **Sketch** tab > **Circle** drop-down > **Perimeter Circle**) and select three points from the graphics window. You can also select existing points from the sketch geometry. The first two points define the location of the circle, and the third point defines its radius.

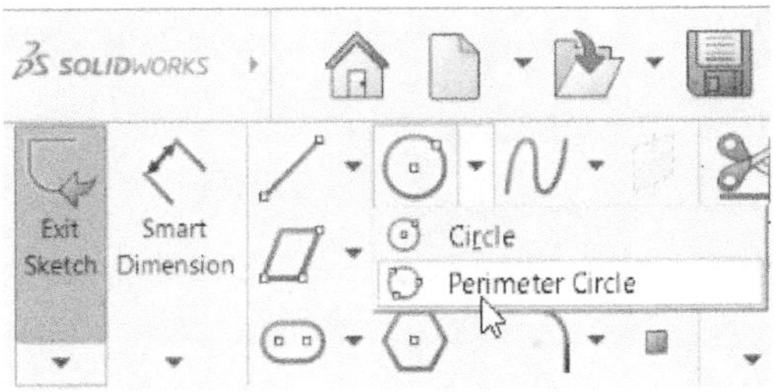

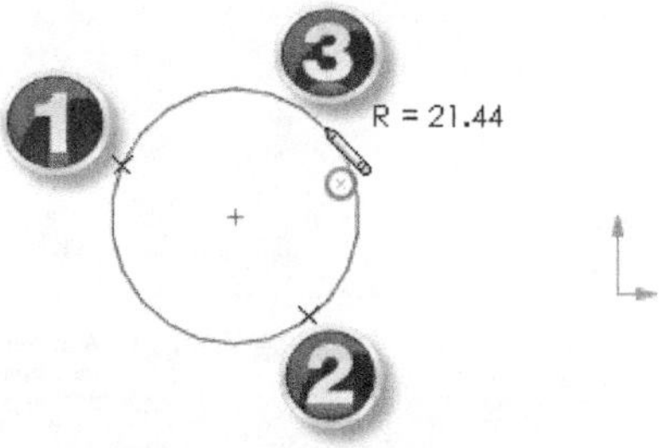

Shaded Sketch Contours

The **Shaded Sketch Contours** button on the CommandManager displays the grey shade inside a closed sketch. You can turn ON or OFF the grey shade inside a closed sketch by clicking on this button. The grey shade helps you to recognize a closed sketch very easily. Also, you can select a closed sketch by merely clicking on the grey shade.

Sketch Shaded Contours ON

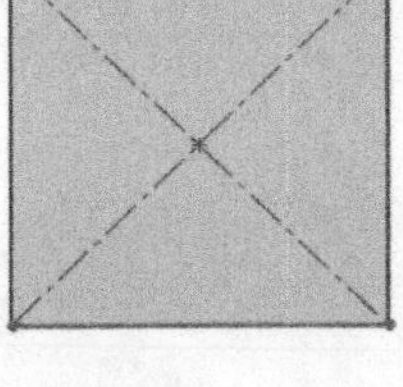

Sketch Shaded Contours OFF

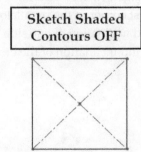

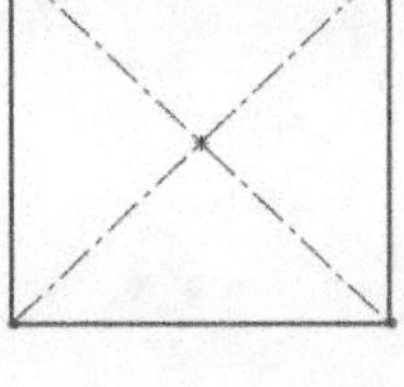

The Straight Slot command

As the name implies, the **Straight Slot** command helps you to sketch a straight slot. Activate this command (On the **Sketch** CommandManager, click **Slot** drop-down > **Straight Slot**) and check the **Add dimensions** option if you want to add dimensions to the slot. Next, you can select the **Center to Center** or **Overall Length** option from the PropertyManager. The **Center to Center** option creates a dimension between the two centers of the slot. Whereas the **Overall Length** option creates an overall length dimension of the slot.

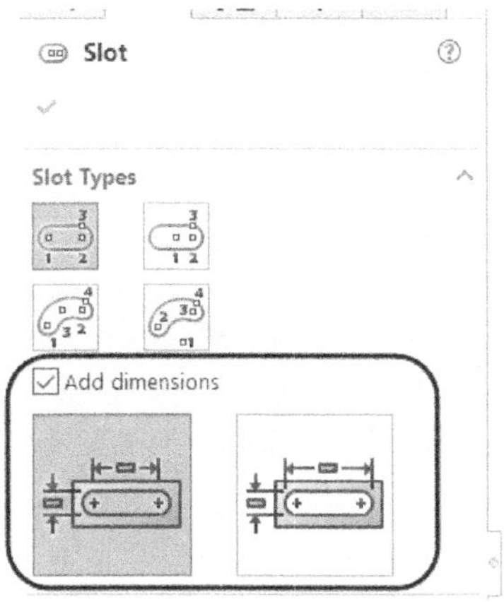

Click to define the first point of the centerline of the slot. Move the pointer and click to define the endpoint of the slot centerline. Move the pointer outward and click to define the slot width.

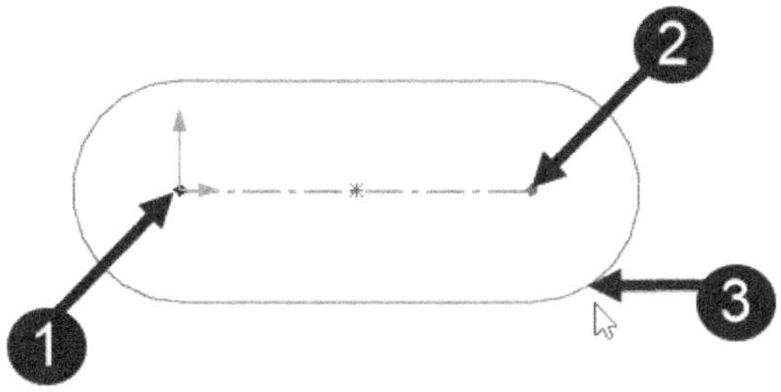

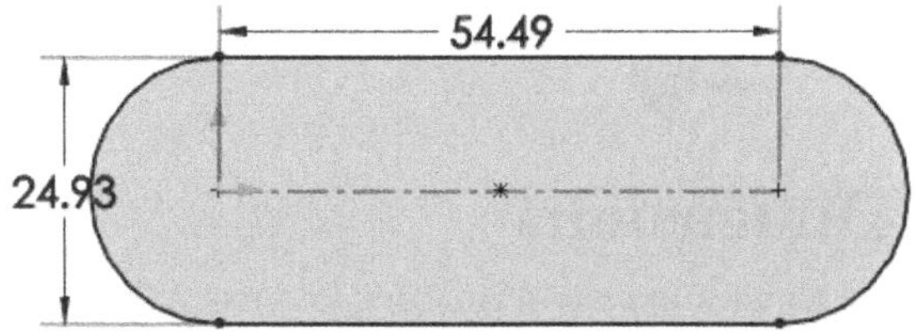

The Centerpoint Straight Slot command

Use this command to create a straight slot by specifying the center, end, and slot width. Activate this command (On the **Sketch** CommandManager, click **Slot** drop-down > **Centerpoint Straight Slot**) and specify the slot center, end, and width.

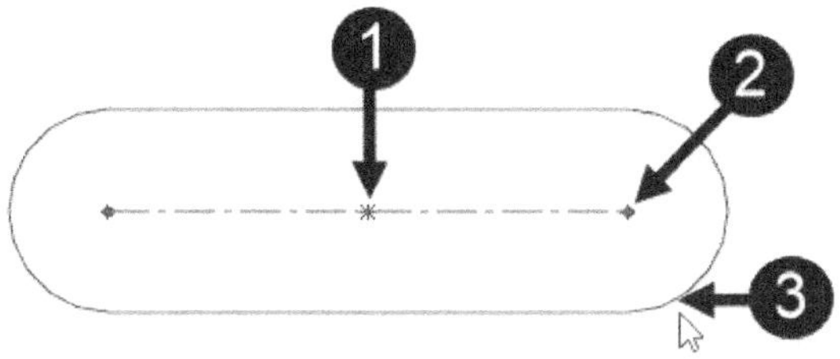

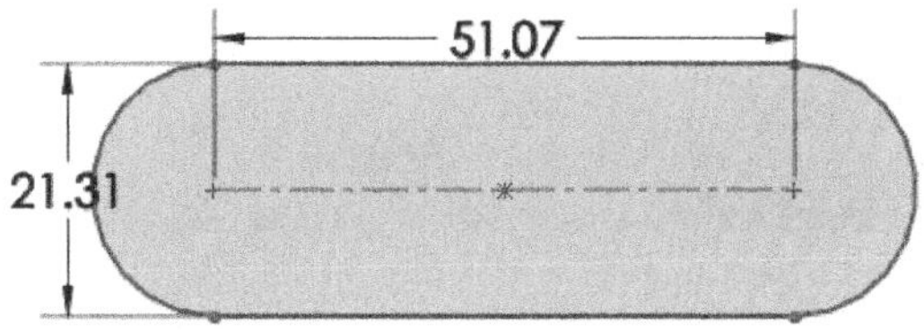

The 3 Point Arc Slot command

Use this command to create a curved slot. Activate this command (On the **Sketch** CommandManager, click **Slot** drop-down > **3 Point Arc Slot**) and click to specify the start point of the center arc. Next, move the pointer and click to specify the endpoint of the center arc. Next, click to define the radius of the center-arc. After defining the center arc, move the pointer outward and click to define the slot width.

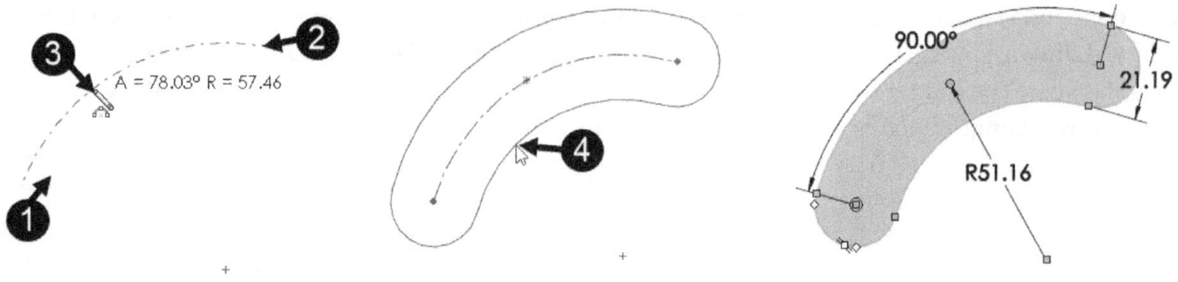

The Centerpoint Arc Slot command

Use this command to create a curved slot using the center point and two endpoints that you specify. Activate this command (On the **Sketch** CommandManager, click **Slot** drop-down > **Centerpoint Arc Slot**) and specify the center point of the center arc. Next, move the pointer outward and click to specify the start point of the arc. Next, move the pointer and click to specify the endpoint of the arc. After defining the center arc, move the pointer outward and click to define the slot width.

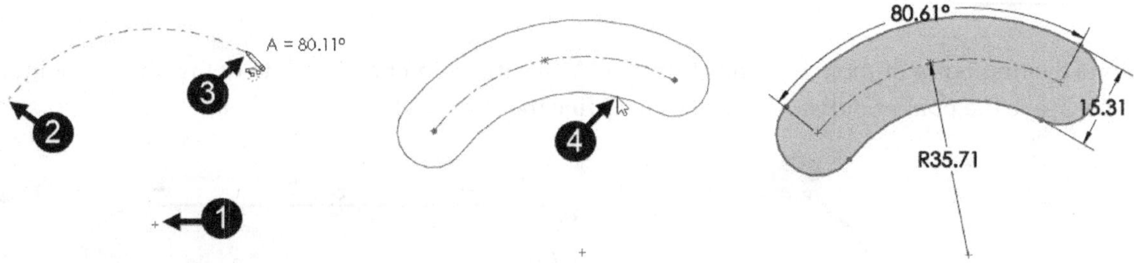

Smart Dimensions

It is generally considered good practice to ensure that every sketch you create is fully-constrained before moving on to creating features. The term 'fully-constrained' means that the sketch has a definite shape and size. You can fully-constrain a sketch by using dimensions and relations. You can add dimensions to a sketch by using the **Smart Dimension** command. You can use this command to add all dimensions such as length, angle, and diameter. This command creates a dimension based on the geometry you select. For instance, to dimension, a circle, activate the **Smart Dimension** command and then click on the circle. Next, move the pointer and click again to position the dimension; notice that the **Modify** box pops up. You can type in a value in this box and then press Enter to update the dimension.

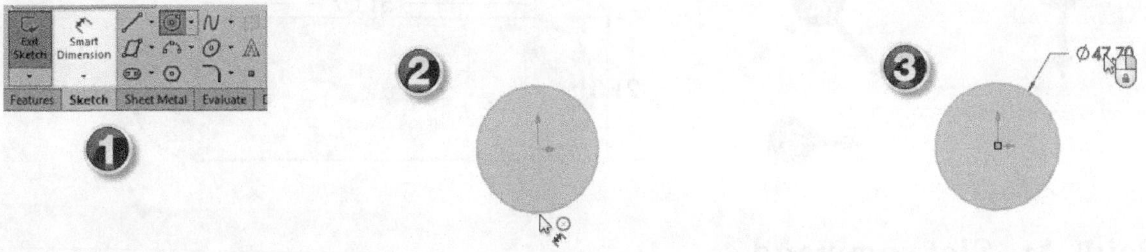

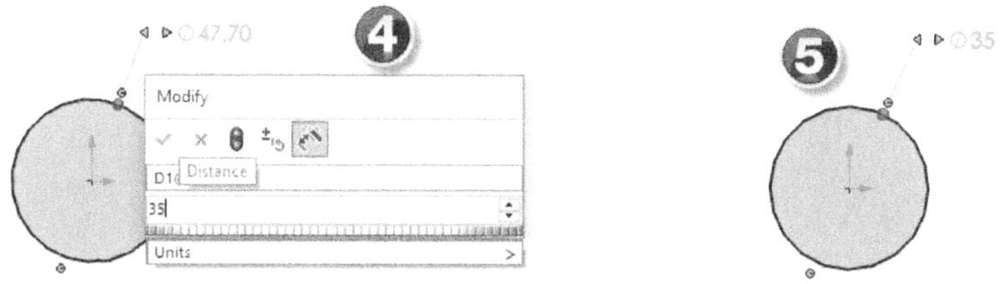

If you select a line, this command automatically creates a linear dimension. Move the pointer and click to position the dimension. Type in a value and press Enter; the dimension will be updated.

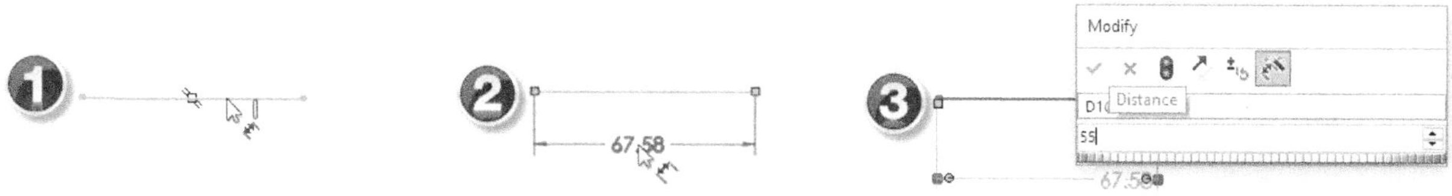

The **Smart Dimension** command can be used to create three different types of linear dimensions between two points. Activate this command and select the center points of the two circles, as shown. Move the pointer vertically downward and notice the horizontal dimension between the two selected points. Move the pointer horizontally and notice the vertical dimension between the selected points.

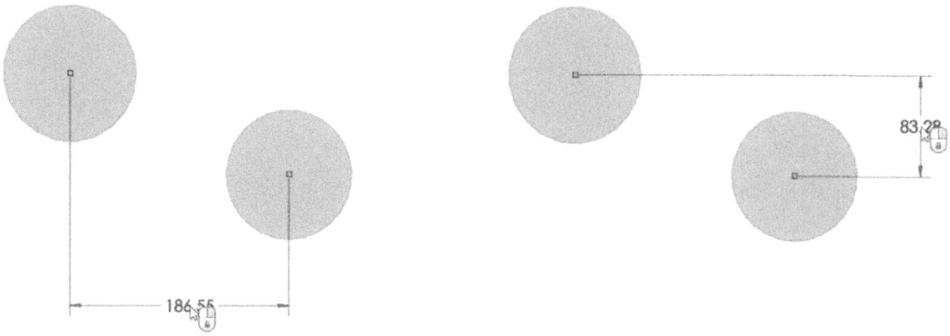

Horizontal Dimension

Vertical Dimension

Move the pointer between the two selected points and notice that the dimension is turned into an aligned dimension. Right-click to lock the dimension; the dimension's orientation remains the same even when you move it in different directions. You can unlock the orientation of the dimension by right-clicking again.

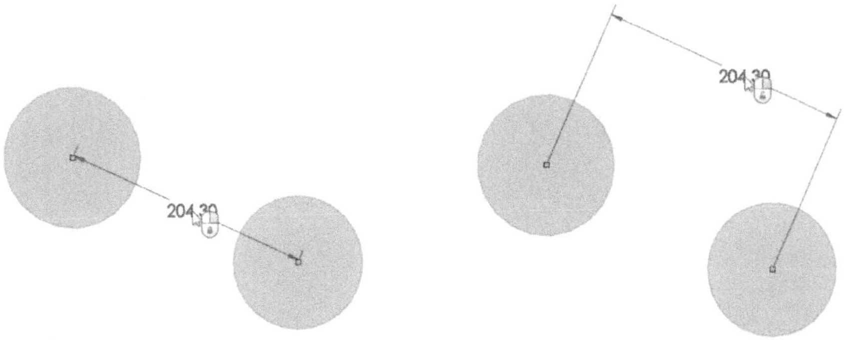

Aligned Dimension

Locked Aligned Dimension

You can also use the **Smart Dimension** command to create an angle dimension. Activate this command and select two lines that are positioned at an angle to each other. Move the pointer between the two lines and click to position the dimension. Type-in a value, and then click the green check on the **Modify** box.

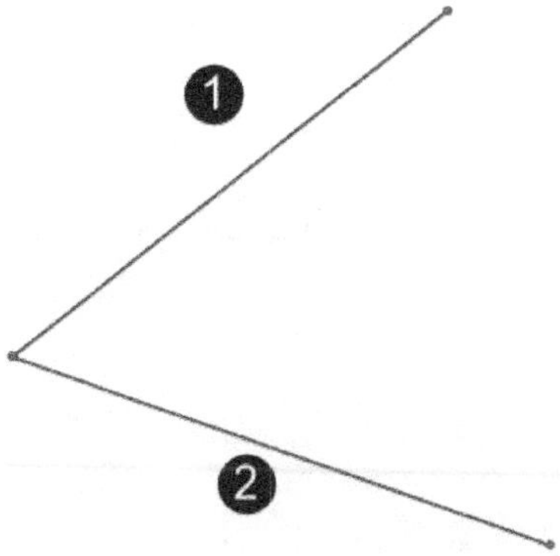

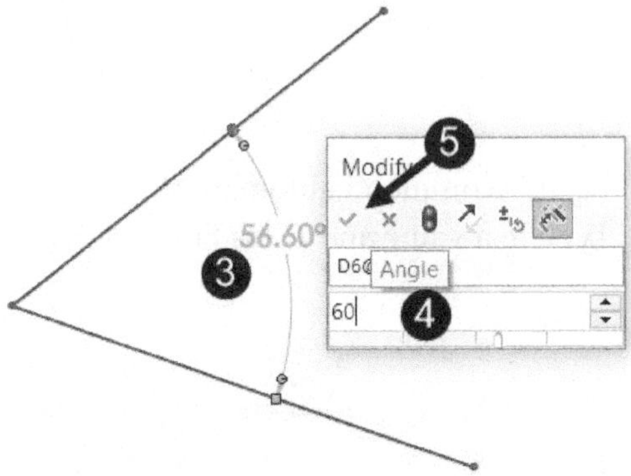

In addition to creating the angled dimension between two lines, you can define an arc angle using the **Smart Dimension** command. Activate this command and select the endpoints of the arc. Next, click on the center point of the arc, move the pointer, and position the angle dimension.

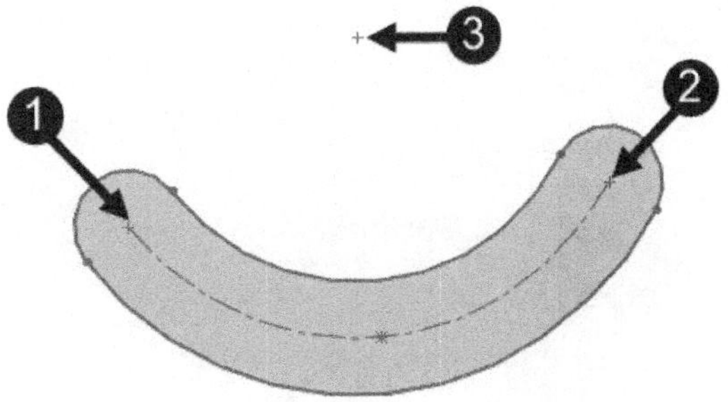

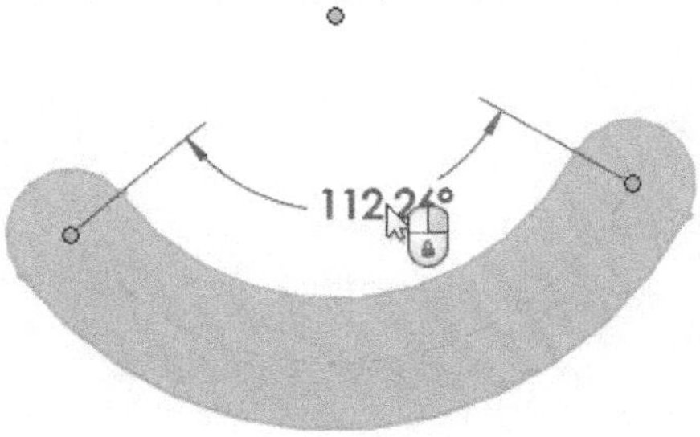

The Horizontal Dimension command

This command creates a horizontal dimension between two points. Activate this command (on the CommandManager, click **Sketch** tab > **Smart Dimension** > **Horizontal Dimension)**. Select two points from the sketch. Move the pointer and click to establish the horizontal dimension. Next, type in a value in the **Modify** box, and then press **Enter**.

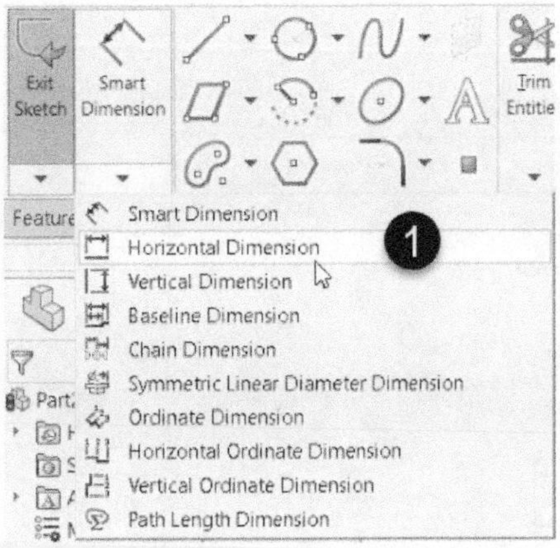

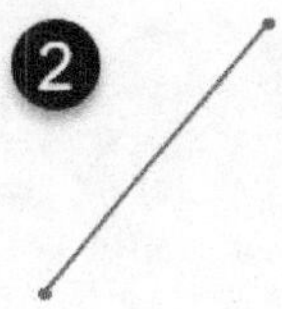

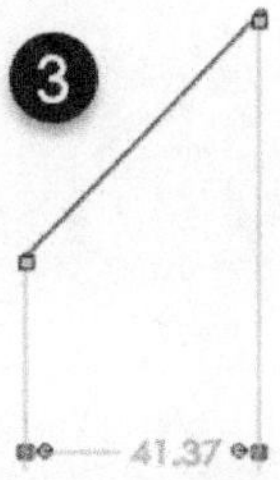

If you want to edit the value, then double click the dimension value and type in a new value in the **Modify** box. Next, press Enter to update the dimension.

The Vertical Dimension command

This command creates a vertical dimension between two points. Activate this command (on the CommandManager, click **Sketch** tab > **Smart Dimension** > **Vertical Dimension**). Select two points or a line from the sketch, move the pointer and click to create the vertical dimension. Next, type in a value in the **Modify** box and press **Enter** to update the dimension.

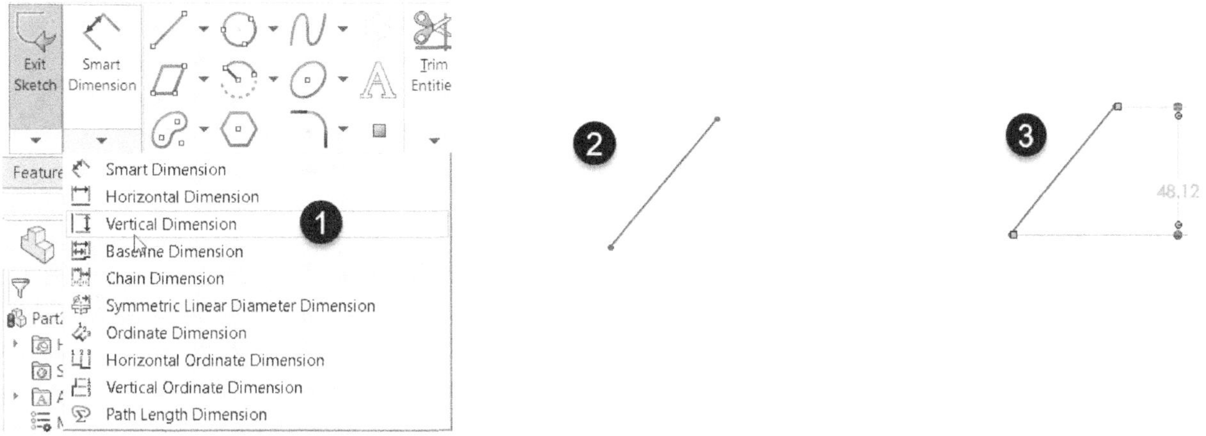

Driving vs. Driven dimensions

When creating sketches for a part, SOLIDWORKS will not allow you to over-define the geometry. The term 'over-define' means adding more dimensions than required. The following figure shows a fully defined sketch. If you add another dimension to this sketch (e.g., diagonal dimension), the **Make Dimension Driven** dialog appears. It shows that "Adding this dimension will make the sketch over define or unable to solve. Do you want to add it as a driven dimension instead?". Select **Make this dimension Driven** and click **OK**; the over-defining dimension is greyed out.

Full-defined sketch

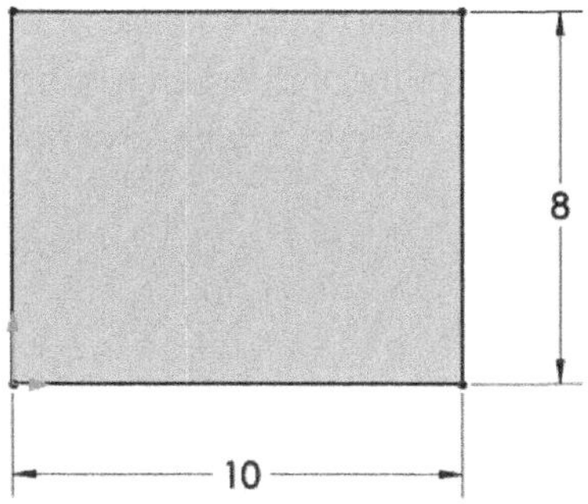

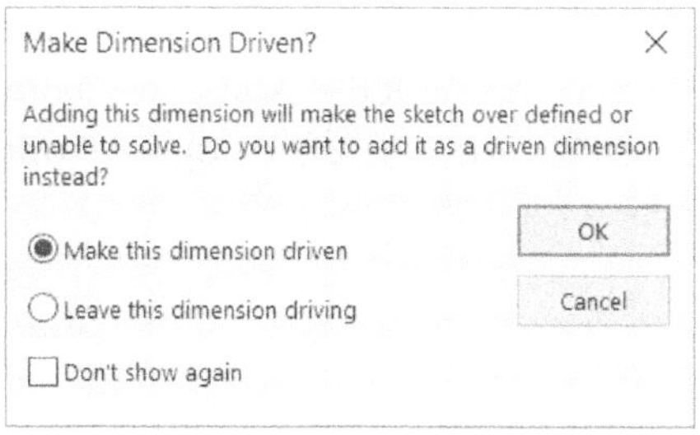

Existing driving dimensions (dimensions in black color) already define the sketch geometry. Driving dimensions are so named because they drive the geometry of the sketch. Double-clicking one of the driving dimensions and changing the value will change the geometry of the sketch. For example, if you change the width's value, the driven dimension along the diagonal updates automatically. Also, note that the dimensions, which are initially created, will be driving dimensions, whereas the dimensions created after fully defining the sketch are driven dimensions.

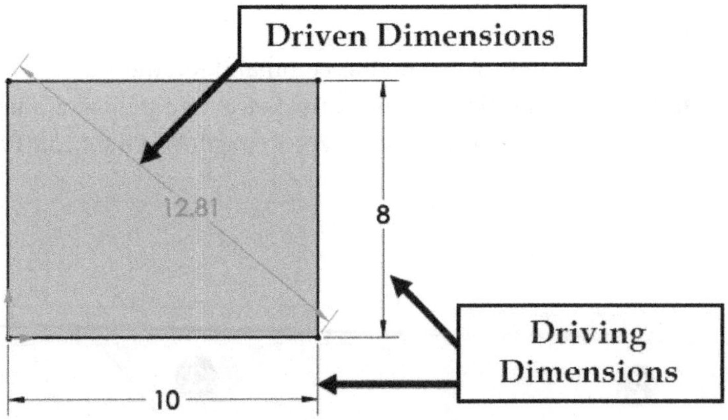

Reverse the direction of the Dimension

The **Reverse Direction** option gives an alternate solution for the dimension applied between sketch entities. For example, if you want to change the side of the linear dimension shown in the figure, select the dimension from the sketch. Next, right click and select **Reverse Direction**. The alternate solution of the sketch appears.

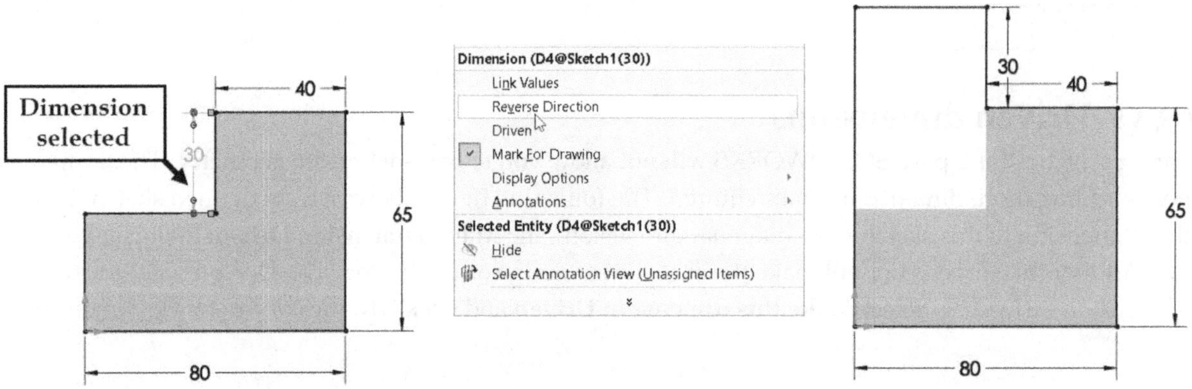

Adding Dimensions by Numeric Input

You can add dimensions to the sketch while creating the entities. To do this, right click and then select **Sketch Numeric Input**. Next, activate a sketching command and notice the **Add dimensions** option on the PropertyManager. Check the **Add Dimension** option, and then specify the start point of the sketch entity. Next, move the pointer and notice the numeric input box attached to the sketch entity. Type in a value and notice that the dimension is created.

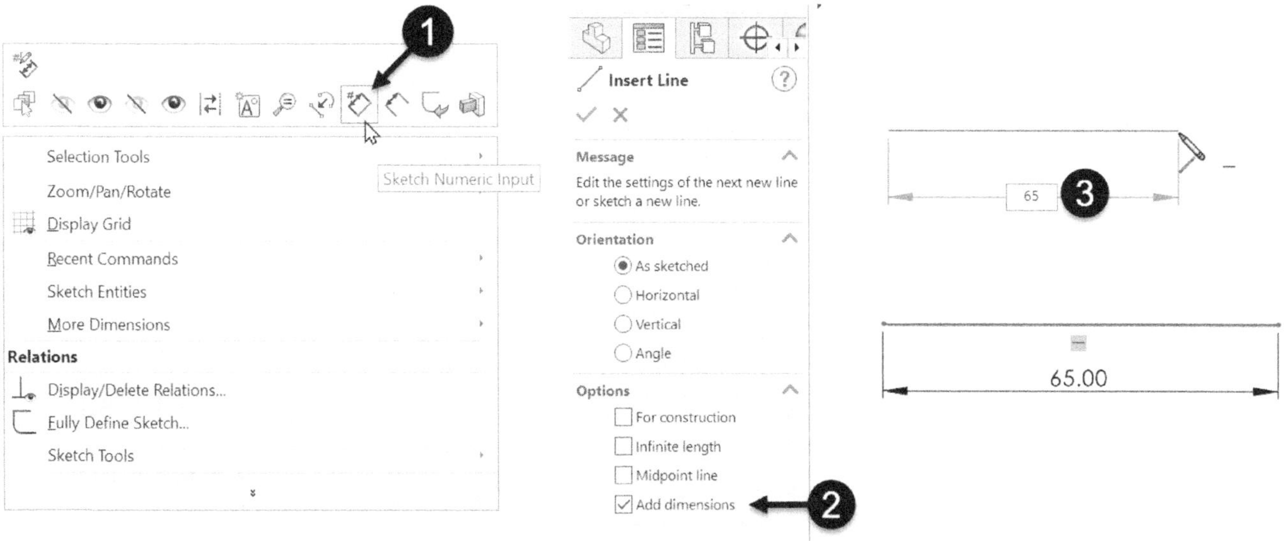

Preview Sketch Dimensions

In SOLIDWORKS 2024, the **Preview Sketch Dimension** option allows you to see a preview of the dimensions when you select a sketch entity. To activate this option, right-click in the graphics window and select **Preview Sketch Dimension**. Now select any sketch entity to preview it dimension.

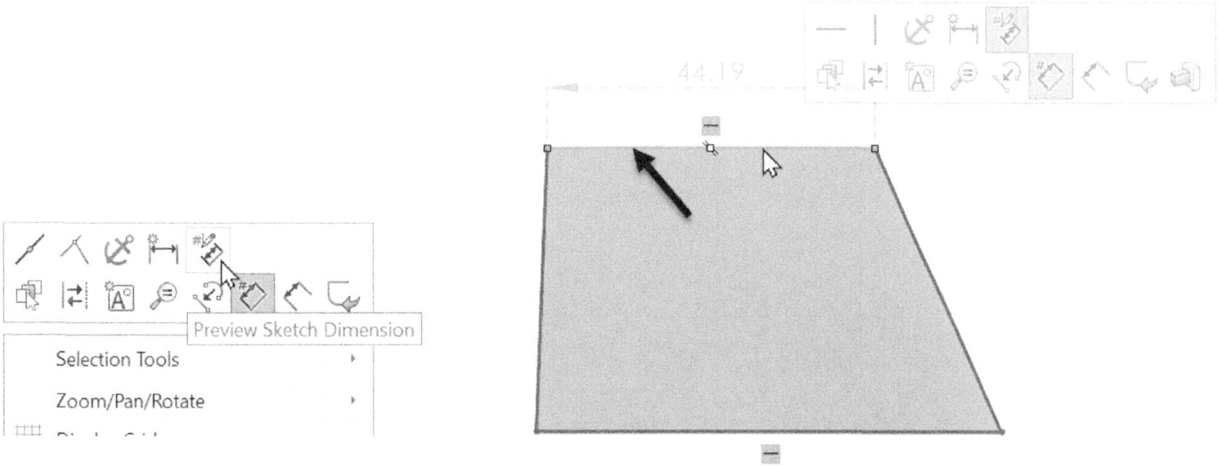

Geometric Relations

Relations are used to control the shape of a sketch by establishing relationships between the sketch elements. You can add relations using the **Add Relation** tool or context toolbar.

Merge

This relation connects a point to another point. On the **Sketch** tab of the CommandManager, click **Delete/Display Relations > Add Relation** (or) click **Tools > Relations > Add** on the menu bar. Select two points. Click **Merge** on the PropertyManager. The selected points will be merged.

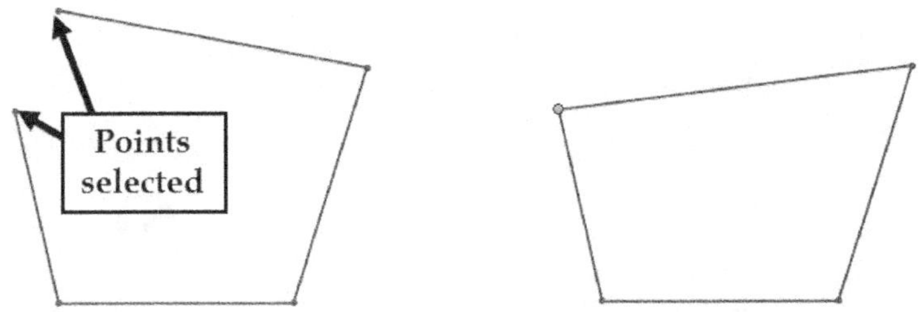

Coincident

This relation makes a point coincident with a line, circle, arc, spline, or ellipse. On the **Sketch** tab of the CommandManager, click **Delete/Display Relations > Add Relation** (or) click **Tools > Relations > Add** on the menu bar. Select a point and any one of the entities mentioned above. Click **Coincident** on the PropertyManager. The selected point and line are made coincident. Click and drag the point and notice that it is movable along the selected line.

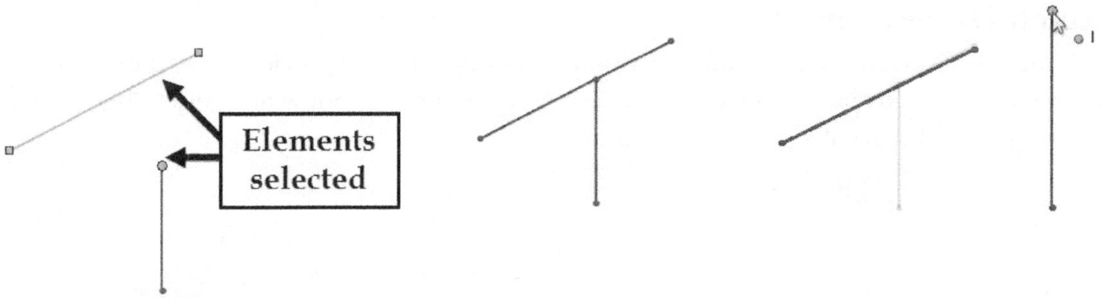

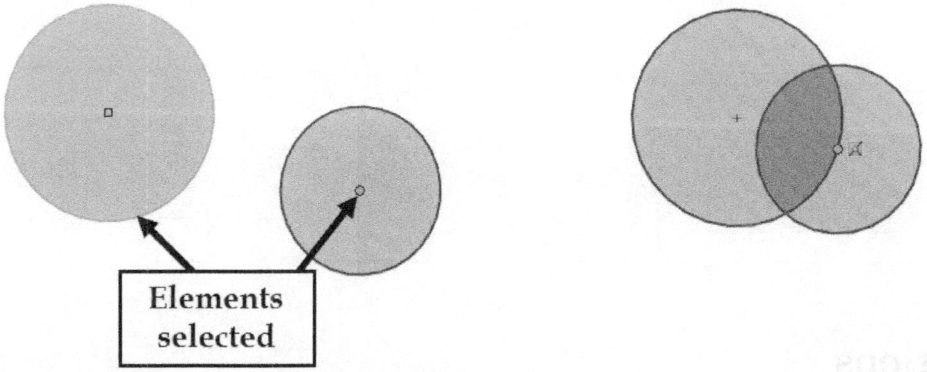

Horizontal

To apply the **Horizontal** relation, click on a line and click the **Make Horizontal** icon on the context toolbar.

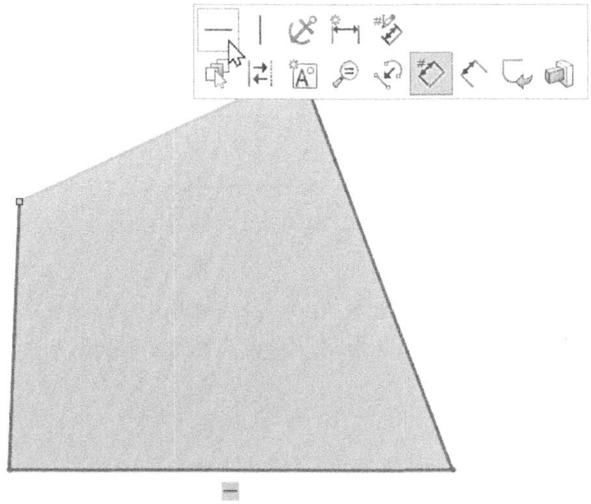

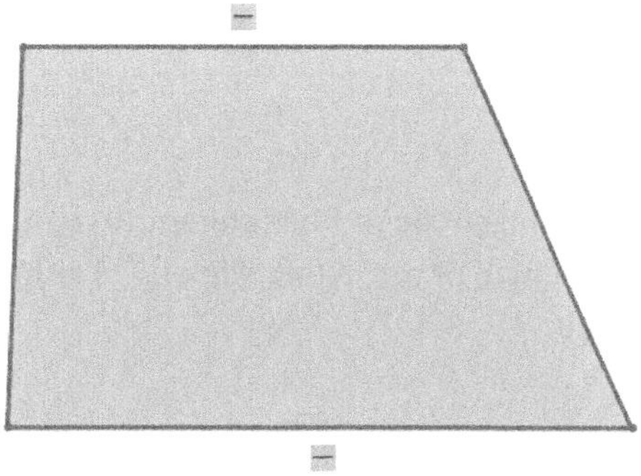

You can also align two points horizontally. Press the Ctrl key and select the two points. Click the **Horizonta**l icon on the PropertyManager.

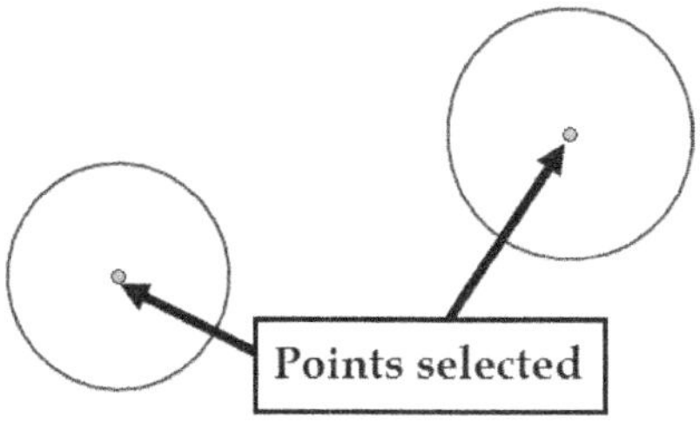

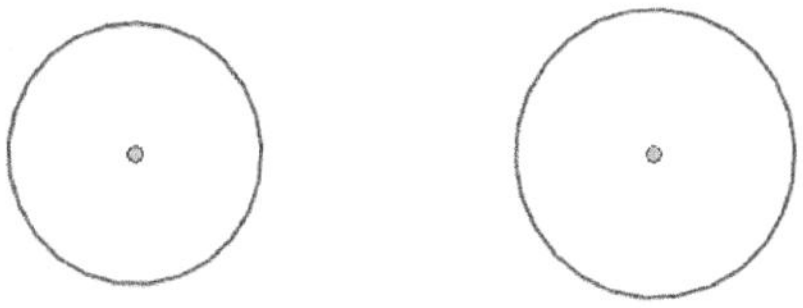

Vertical

Use the **Vertical** relation to make a line vertical. You can also align two endpoints or points vertically by using this relation.

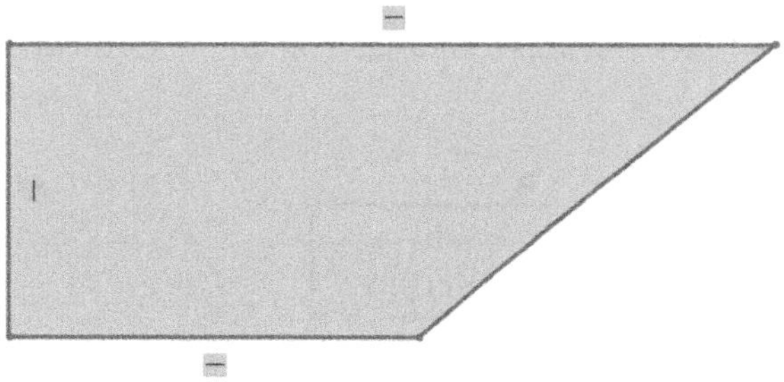

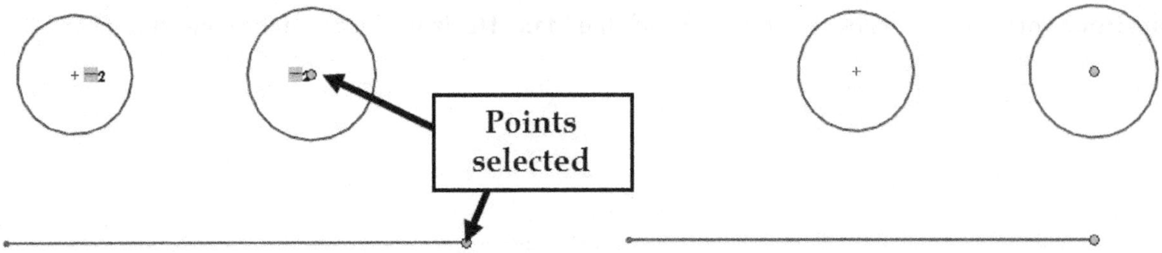

Tangent

This relation makes an arc, circle, or line tangent to another arc or circle. Select a circle, arc, or line. Press the Ctrl key and select another circle or arc. Next, click the **Tangent** icon on the PropertyManager. The two elements will be tangent to each other.

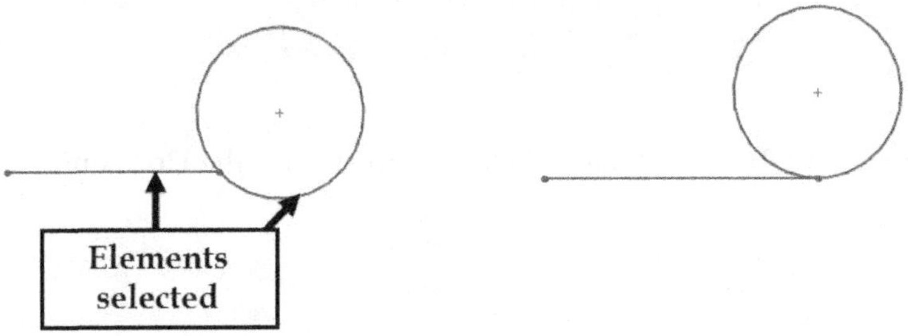

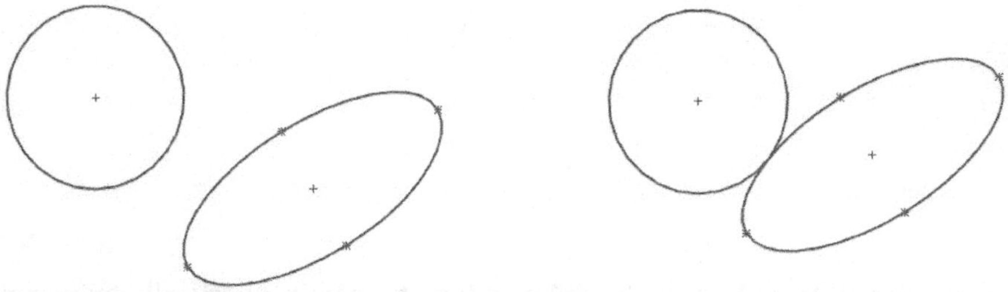

Parallel

Use the **Parallel** relation to make two lines parallel to each other. Press and hold the Ctrl key and select two lines. Next, click the **Make Parallel** icon on the **Context** toolbar.

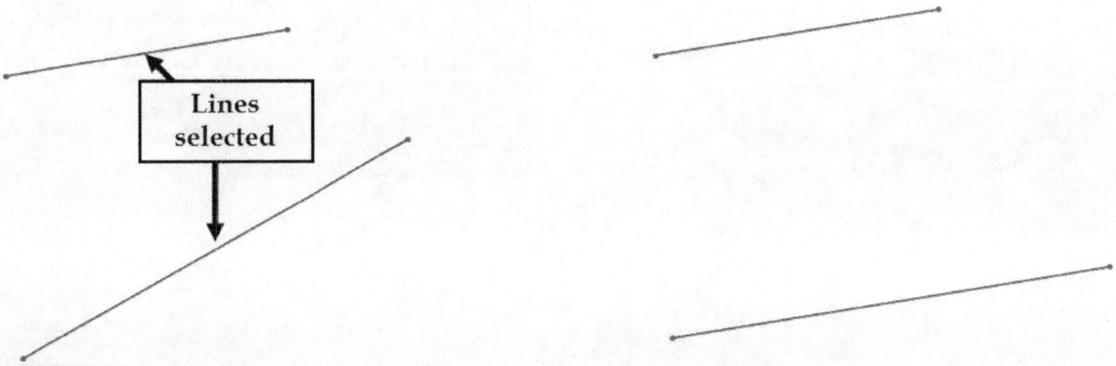

Perpendicular

Use the **Perpendicular** relation to make two entities perpendicular to each other.

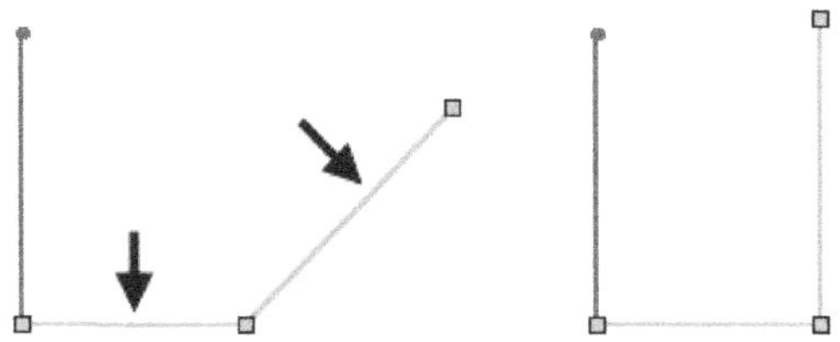

Midpoint

Use the **Midpoint** relation to make a point coincident with the midpoint of a line or arc. To do this, press the Ctrl key, and then select a line/arc and a point. Click the **Midpoint** icon on the PropertyManager.

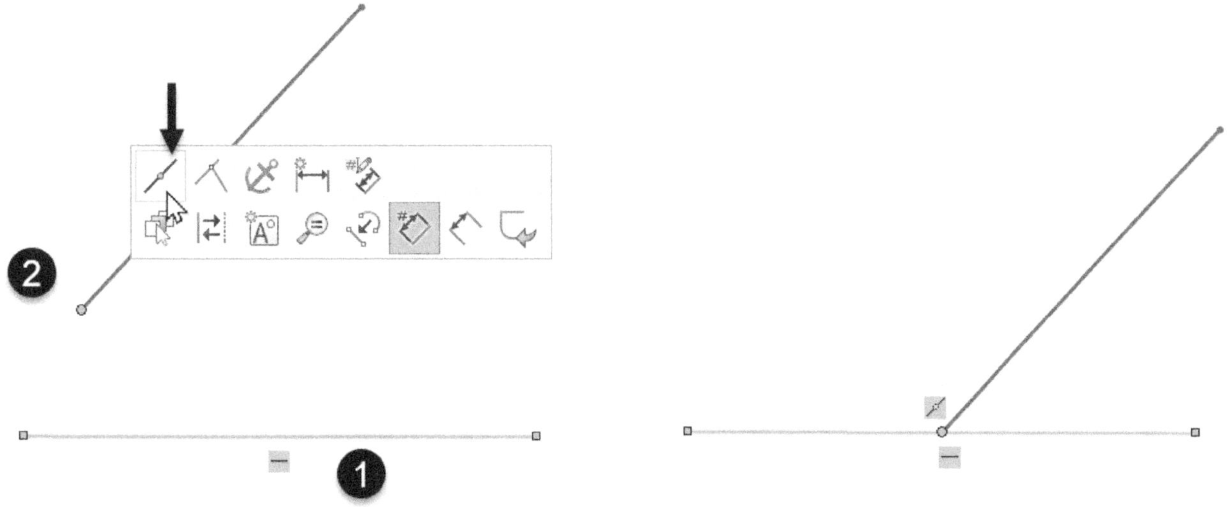

= Equal

This relation makes the selected objects equal in magnitude. For example, if you select two circles, the diameter of the selected circles will become equal. If you select two lines, the length of the two lines will be equal.

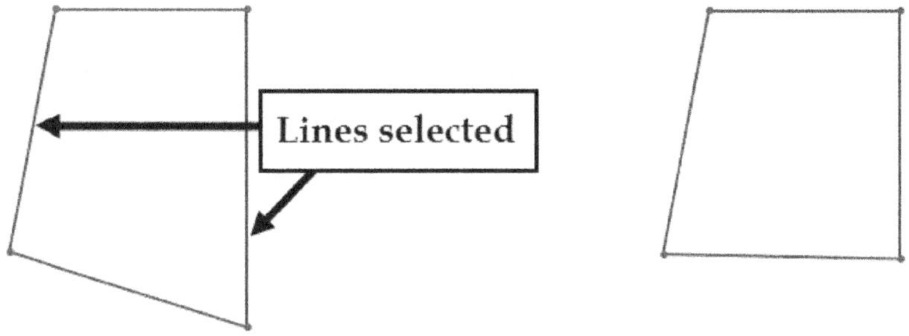

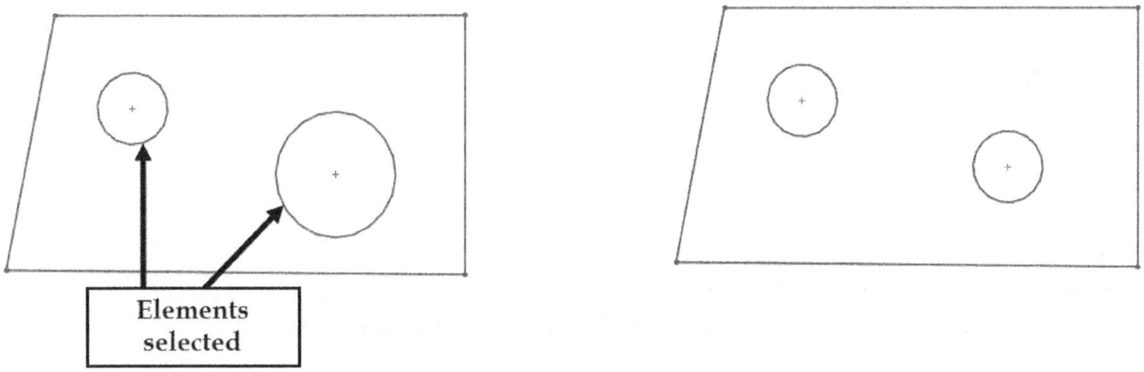

Concentric

This relation makes the center points of arcs, circles or ellipses coincident. Activate the **Add Relation** command and select a circle or arc from the sketch. Next, select another circle or arc, and click the **Concentric** icon on the PropertyManager. The first circle/arc will be concentric with the second circle/arc.

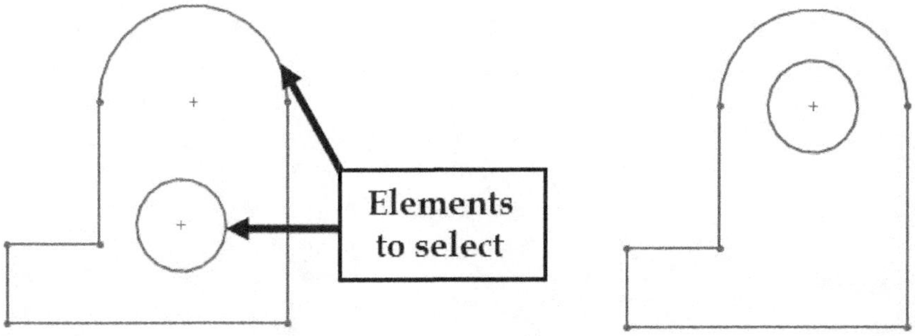

Fix

This relation fixes a sketch element so that it cannot be moved. Select a sketch element and click the **Fix** icon on the PropertyManager; it will be fixed at its current position.

Coradial

Use the **Coradial** relation to make two arcs to share the same center point. Also, the radius of the arcs become equal. Activate the **Add Relation** command and select two arcs. On the PropertyManager, click the **Coradial** icon. Next, click **OK** on the PropertyManager.

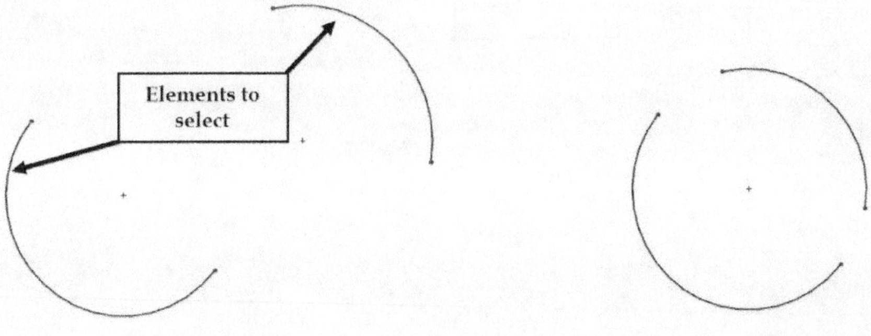

Collinear

This relation forces a line to be collinear to another line. The lines are not required to touch each other. Activate the **Add Relation** command and select the lines to be aligned. Click the **Collinear** icon on the PropertyManager.

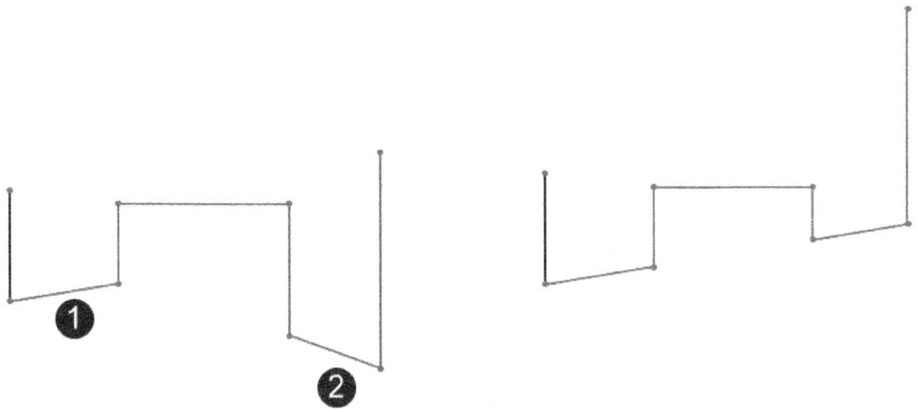

Symmetric

This relation makes two objects symmetric about a line. The objects will have the same size, position, and orientation about a line. For example, if you select two circles about a line, they will become equal in size, aligned horizontally, and positioned at an equal distance from the line. Activate the **Add Relation** command and select a construction line to define the symmetry line. Select two objects from the sketch. Click **Symmetric** on the PropertyManager; the selected objects will be made symmetric about the selected line.

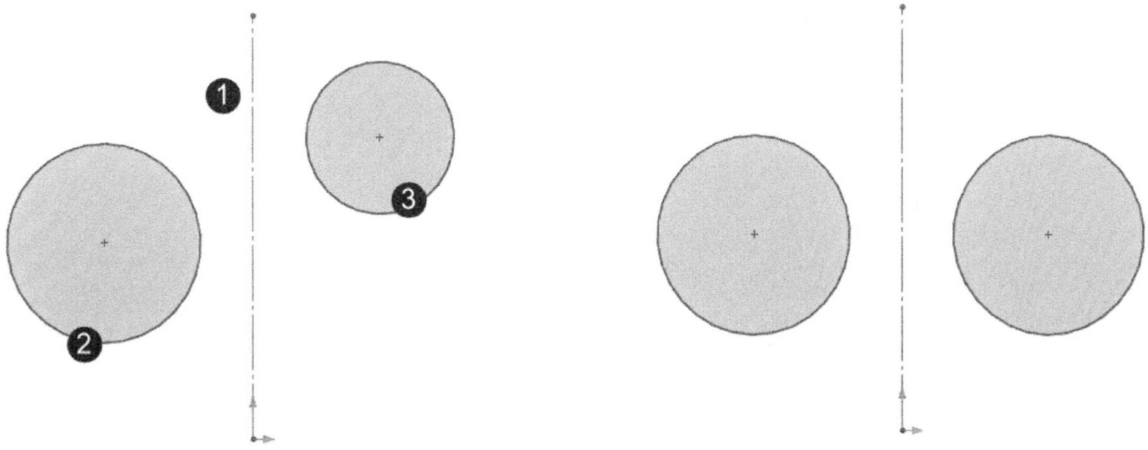

Showing/Hiding Relations

To show/hide sketch relations, click the **Hide/Show Items** drop-down on the **View (Heads-Up)** Toolbar and activate/deactivate the **View Sketch Relations** icon. When dealing with complicated sketches involving many relations, you can deactivate this button to turn off all relations.

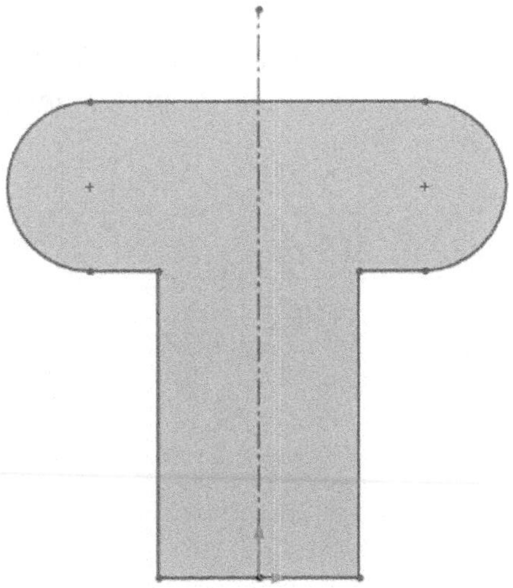

Sketch Relations OFF

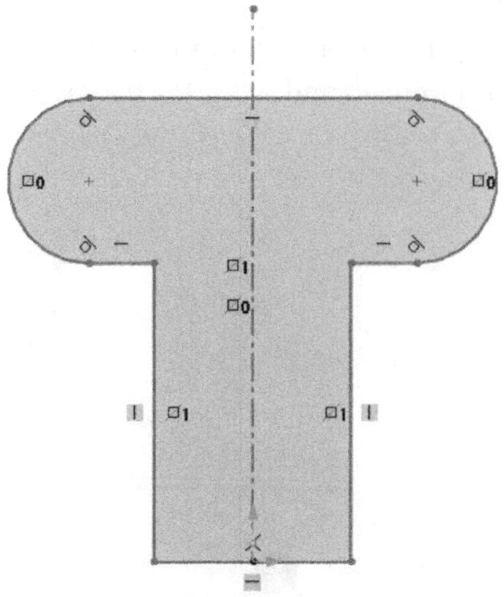

Sketch Relations ON

Fully Define Sketch

In addition to the **Smart Dimension** command and other geometric relations, SOLIDWORKS provides you with the **Fully Define Sketch** command. This command automatically applies relations and dimensions to fully-constrain a sketch. To activate this command, click **Sketch** tab > **Display/Delete Relations** drop-down > **Fully Define Sketch** on the CommandManager. On this PropertyManager, check the **Relations** section, and then expand it. In this section, select the relations to be applied.

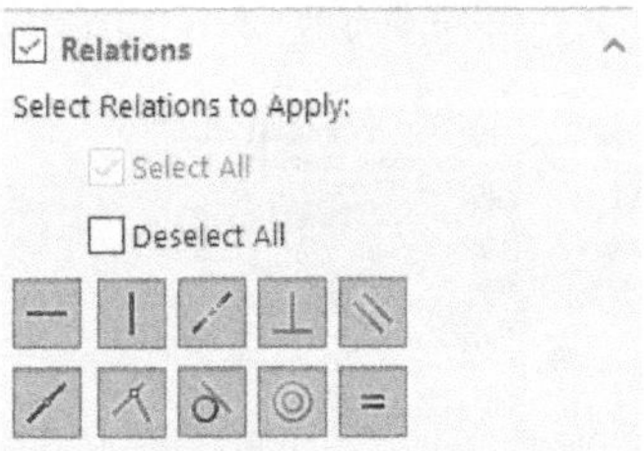

Likewise, check the **Dimension** section and then expand it. Next, select the dimension schemes (**Chain**, **Baseline**, or **Ordinate**) from the **Horizontal Dimensions Scheme** and **Vertical Dimensions Scheme** drop-downs. After specifying the dimension schemes, you need to define the datum of the horizontal and vertical dimensions. By default, the sketch origin is selected as the origin for horizontal and vertical dimensions. You can click in the **Horizontal Datum** box and select a vertical model edge, vertex, vertical line, or point. You can also click in the **Vertical Datum** box and select a horizontal model edge, vertex, horizontal line, or point. However, it is recommended to leave the sketch origin as the datum for both horizontal and vertical dimensions. Select an option from the **Dimension placement** section.

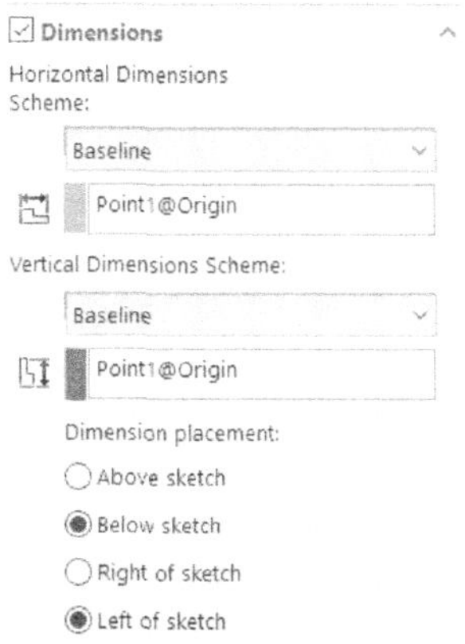

The next step is to select the entities of the sketch. By default, the **All entities in sketch** option is selected. You can also select the **Selected entities** option and then select the entities to apply dimension. Next, click **Calculate** on the PropertyManager. Click **OK** on the PropertyManager to create the relations and dimensions automatically.

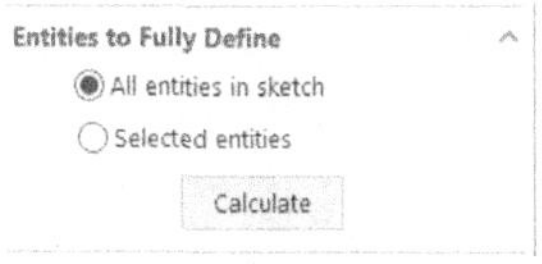

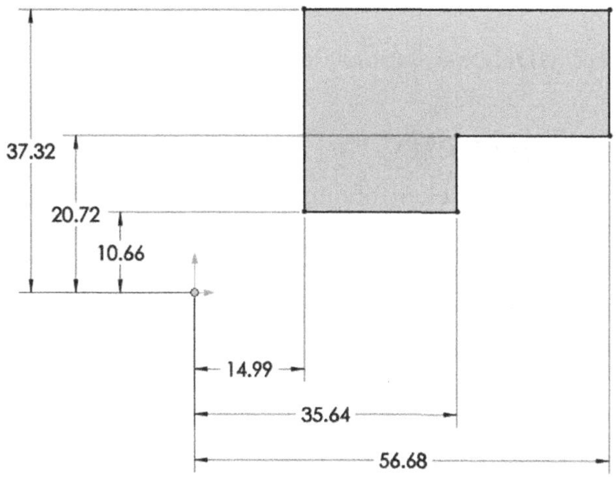

Display/Delete Relations

The **Display/Delete Relations** command helps you view all the sketch's constraints, status, and elements associated with them. Activate the **Display/Delete Relations** command (On the CommandManager, click **Sketch** tab > **Display/Delete Relations**). On the **Display/Delete Relations** PropertyManager, select **Filter > All in this sketch** to view all the relations and dimensions in the active sketch. Next, select an entity from the sketch to view all the relations associated with it. The **Information** ⓘ section displays the status of each entity: **Satisfied, Driven** or **Over Defining**. You can select an over defining sketch entity and check the Suppressed option to suppress it (or) click **Delete** to delete the entity completely.

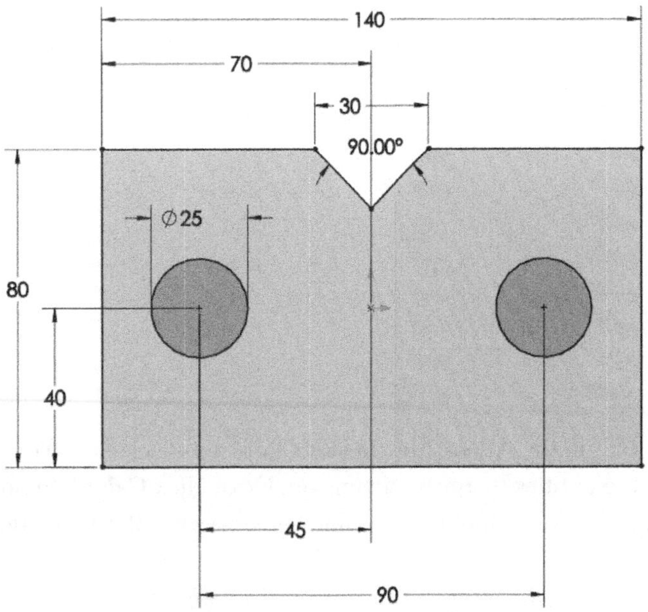

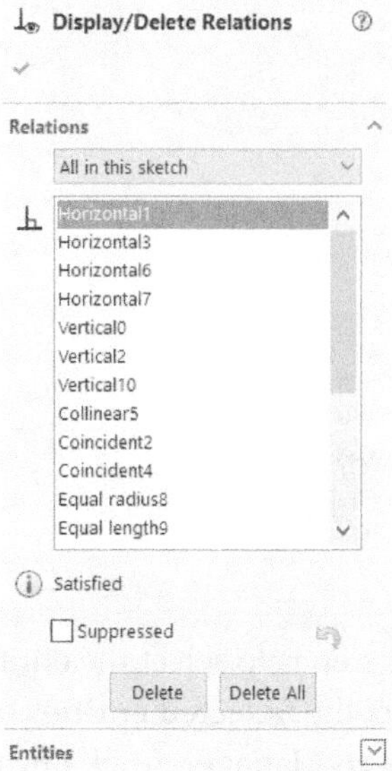

Click **OK** on the **Display/Delete Relations** PropertyManager.

Construction Geometry

This command converts a sketch element into a construction element. Construction elements are reference elements, which support you to create a sketch of desired shape and size. Click on a sketch element and select the **Construction Geometry** command. The selected element will be converted to a construction element. You can also convert the construction element to a sketch element by clicking on the element and selecting the **Construction Geometry** command.

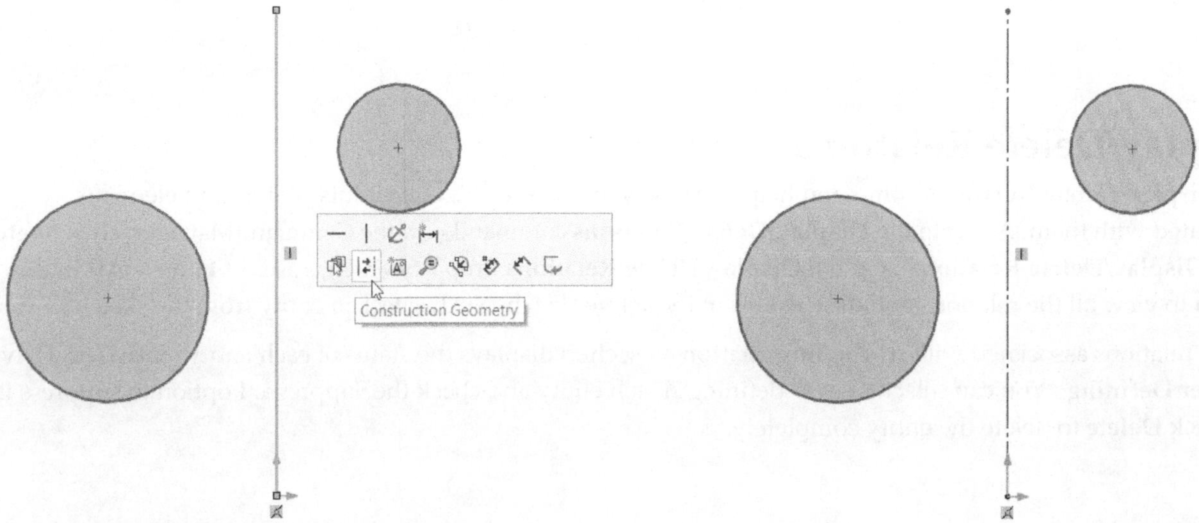

Sketch Editing Commands

Sketch editing commands help you perform various operations such as extending or trimming entities, offsetting, rounding corners, and mirroring. These commands are explained next.

The Sketch Fillet command

This command rounds a sharp corner created by the intersection of two lines, arcs, circles, and rectangle or polygon vertices. Activate this command (On the CommandManager, click **Sketch** tab > **Sketch Fillet**) and select the elements' ends to be filleted. Type in a value in the **Fillet Radius** box and press Enter. The elements to be filleted are not required to touch each other.

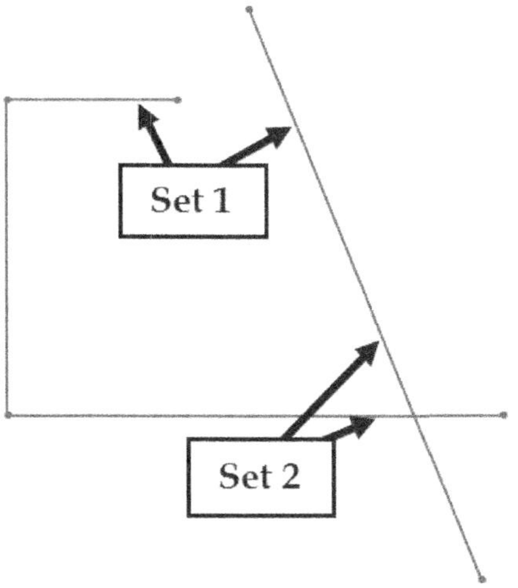

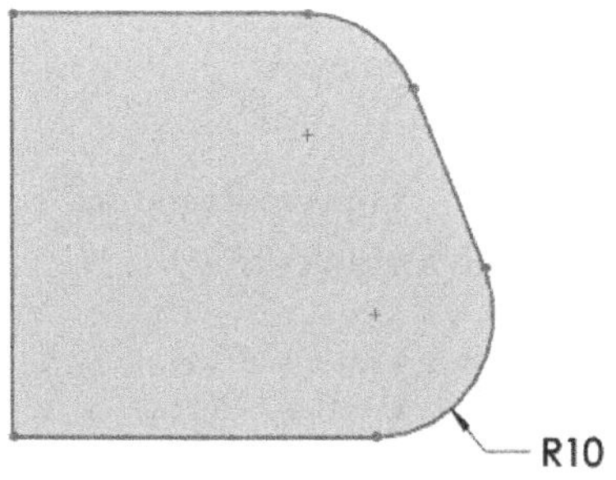

The Sketch Chamfer command

This command replaces a sharp corner with an angled line. Activate this command (On the CommandManager, click **Sketch** tab > **Fillet** drop-down > **Sketch Chamfer**) and type-in the chamfer distance in the **Distance1** box. Select the elements' ends to be chamfered and press **OK**.

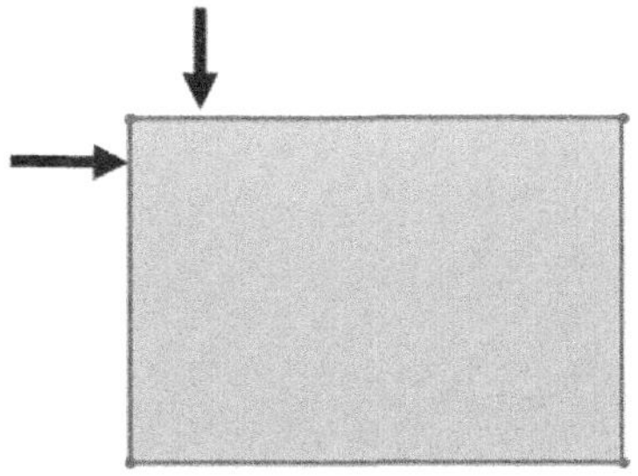

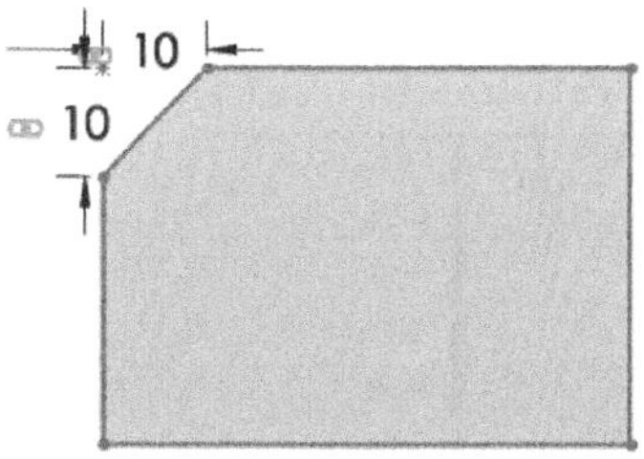

The Trim Entities command

This command trims the end of an element back to the intersection of another element. Activate this command (On the CommandManager, click **Sketch** tab > **Trim Entities**) and notice that there are options on the PropertyManager to trim the sketch entities. Select the **Power trim** option from the PropertyManager and drag the pointer across the elements to trim, as shown.

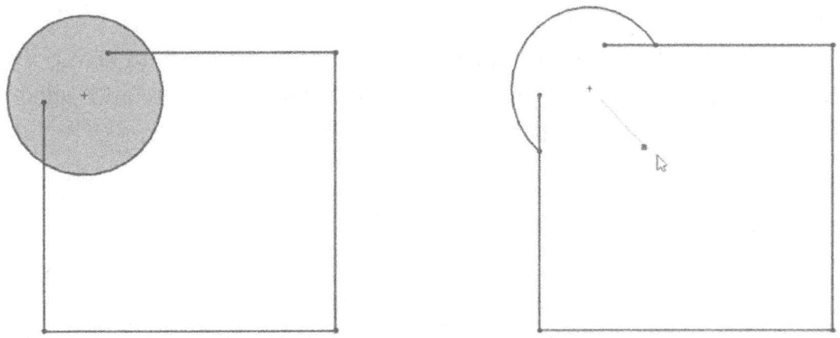

Select the **Corner** option from the PropertyManager, and then select two intersecting elements. The elements will be trimmed to form a corner.

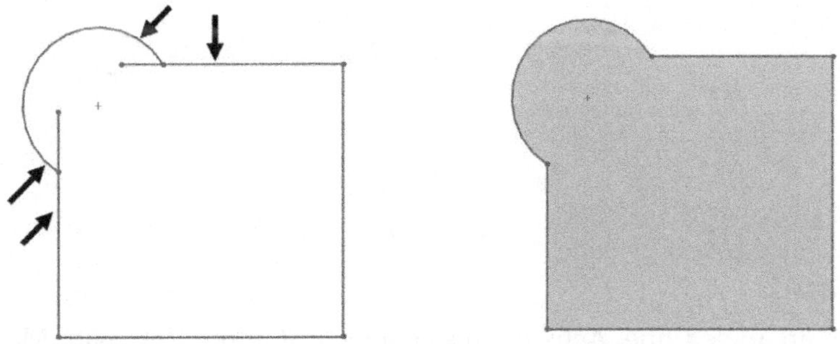

Create four intersecting entities, as shown. Activate the **Trim Entities** command and select the **Trim away inside** option from the PropertyManager. Select the two vertical lines to define the bounding entities. Click on the horizontal lines and notice that the inside portions are trimmed.

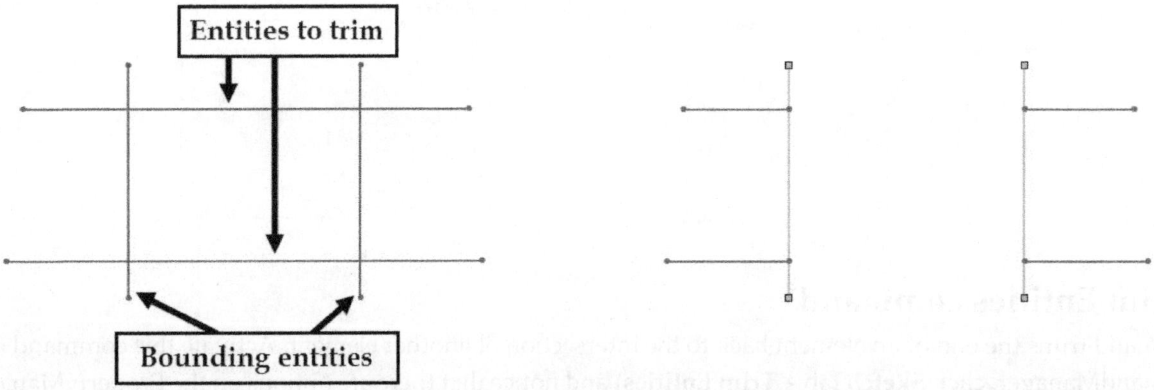

On the PropertyManager, click **Trim away outside** icon and select the bounding entities, as shown. Next, select the horizontal lines, as shown. The outer portions are trimmed, and the inner portions are retained.

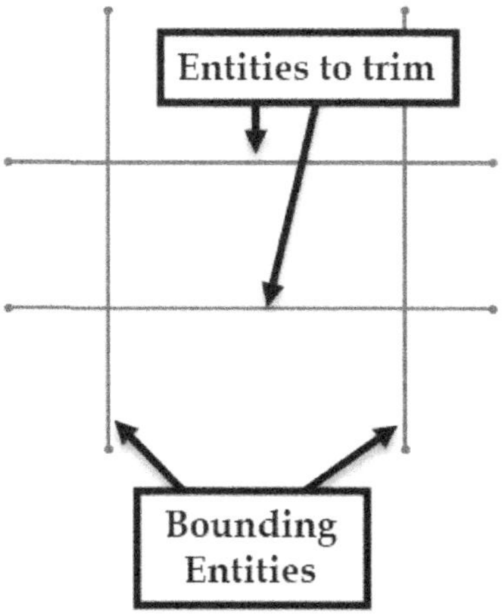

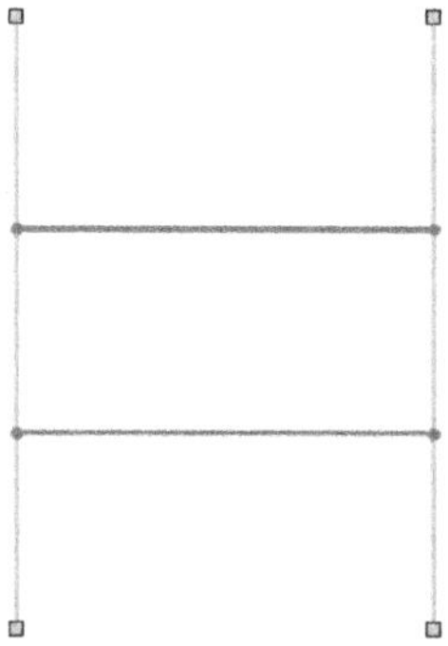

Create the intersecting lines, as shown. Activate the **Trim Entities** command and select the **Trim to closest** option from the PropertyManager. Click on the top portions of the vertical lines; the lines are trimmed up to the nearest edge.

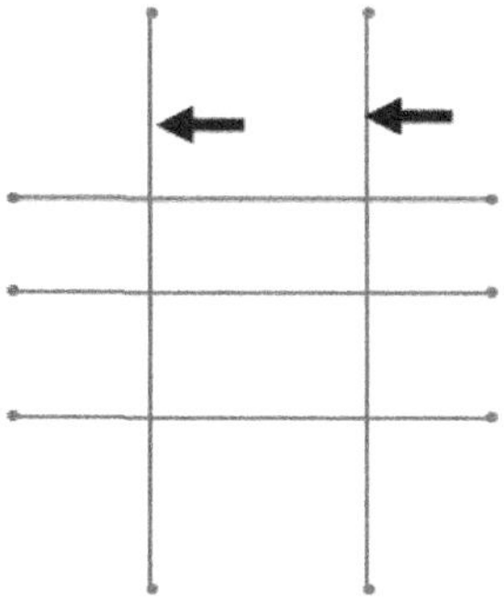

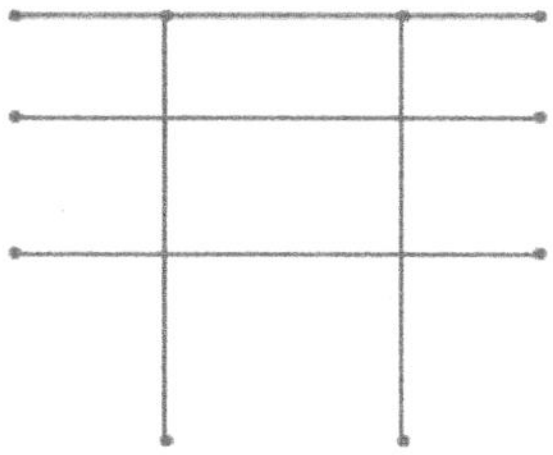

The Extend Entities command

This command extends elements such as lines, arcs, and curves until they intersect another element called the boundary edge. Activate this command (On the CommandManager, click **Sketch** tab > **Trim Entities** > **Extend Entities**) and click on the element to extend. It will extend up to the next element.

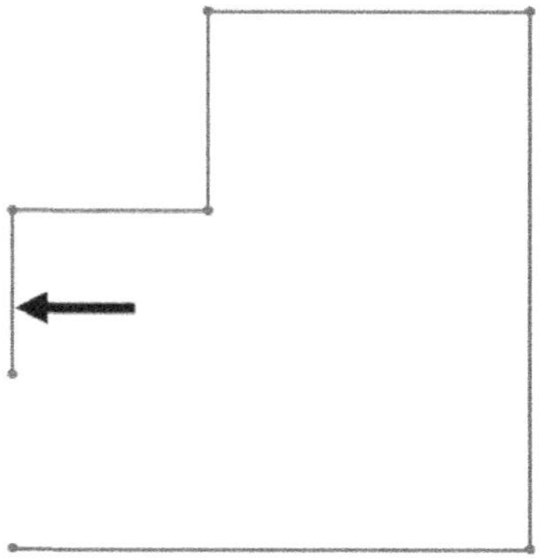

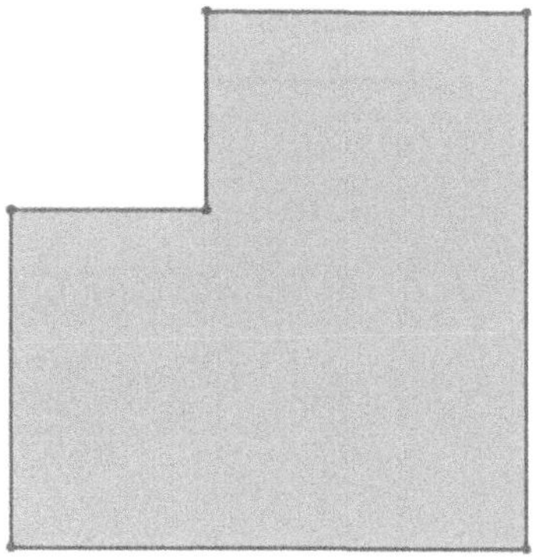

The Offset Entities command

This command creates a parallel copy of a selected element or chain of elements. Activate this command (On the CommandManager, click **Sketch** tab > **Offset Entities**) and select an element or chain of elements to offset. After selecting the element(s), type in a value in the **Offset Distance** box available on the PropertyManager. On the **Offset Entities** PropertyManager, check the **Reverse** option to reverse the side of the offset. Check the **Bi-directional** option to create a parallel copy on both sides. Use the Construction geometry section's options to convert the base or offset entities into construction geometry. On the dialog, click **OK**. A parallel copy of the elements will be created.

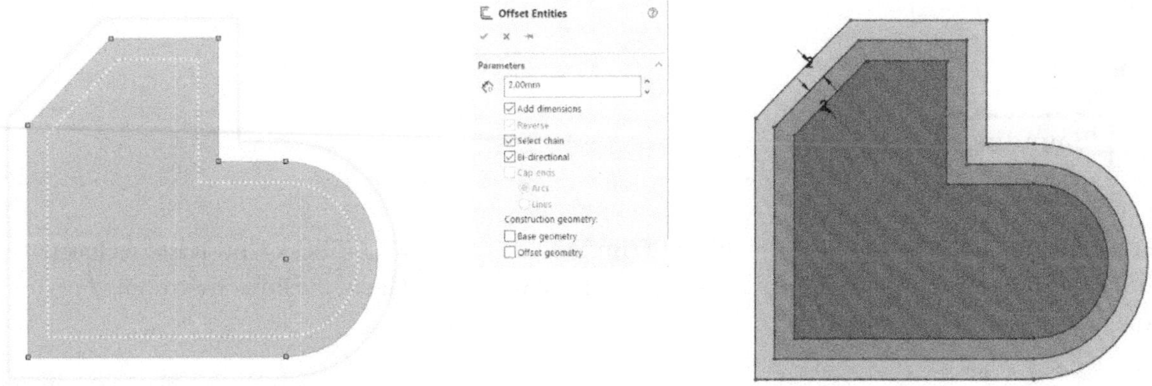

The Mirror Entities command

The **Mirror Entities** command creates a mirrored copy of the selected entities about a line. Create the sketch entities similar to that shown next and activate the **Mirror Entities** command (On the **Sketch** CommandManager, click **Mirror Entities**). Select the entities to mirror, and then click in the **Mirror about** box. Select the centerline from the graphics window. Make sure that the **Copy** option is checked to create the mirrored copy. Otherwise, the source entities will be deleted after mirroring. Click **OK** on the PropertyManager to complete the mirror operation.

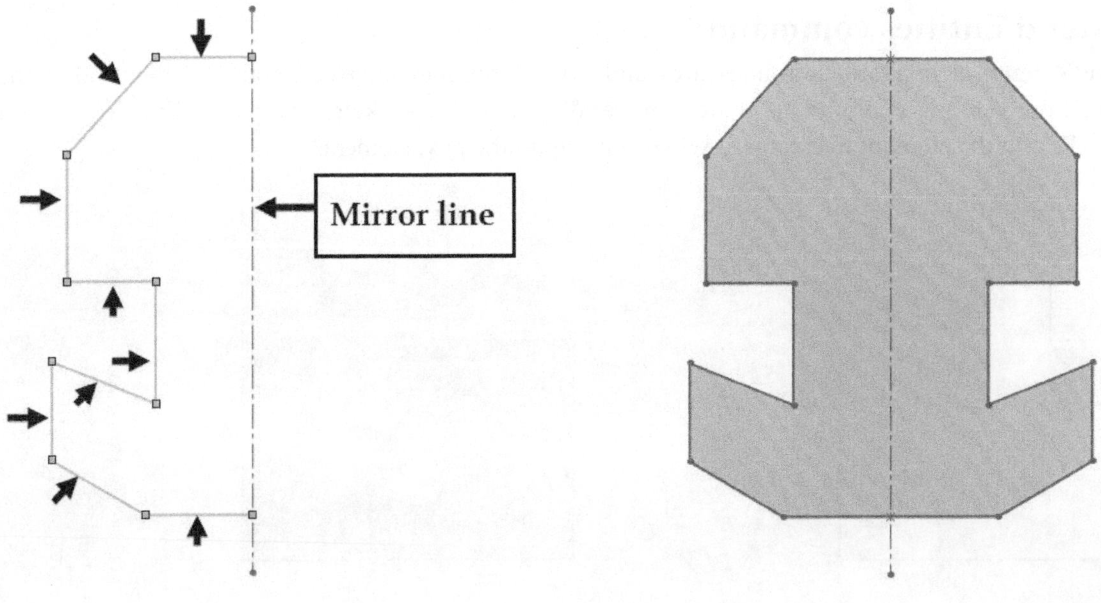

The Dynamic Mirror Entities command

The **Dynamic Mirror Entities** command defines a centerline about which the sketch entities are mirrored immediately after creating them. To activate this command, click **Tools > Sketch Tools > Dynamic Mirror** on the menu bar. You can add it to the CommandManager from the **Customize** dialog. To do this, click the **Options** drop-down > **Customize** on the Quick Access Toolbar. On the **Customize** dialog, click the **Commands** tab and select **Sketch** from the **Categories** list. Next, drag the **Dynamic Mirror Entities** button from the Button area and then release it on the CommandManager. Close the **Customize** dialog.

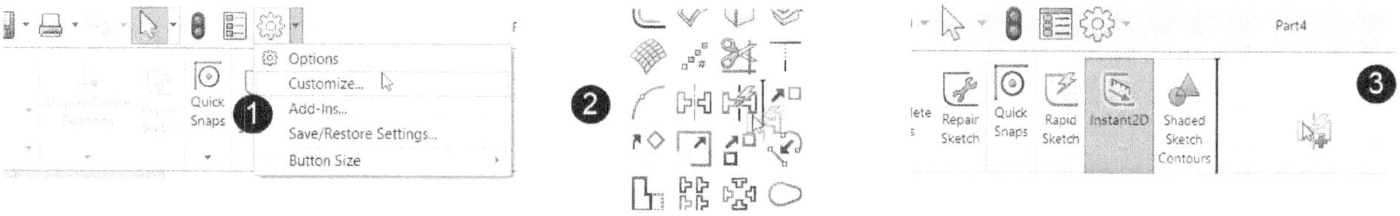

First, create a centerline, and then activate the **Dynamic Mirror Entities** command. Select the centerline and notice the mirror marks at both ends of the centerline. Next, activate a sketch command and draw the entities on one side of the centerline; the entity is automatically mirrored on the centerline's other side.

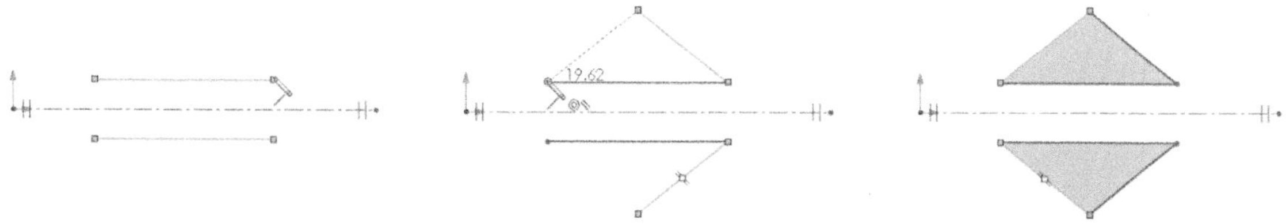

Add dimensions and relations to the sketch entities and notice that the other side is modified automatically. However, you have to trim or delete the entities separately on both sides.

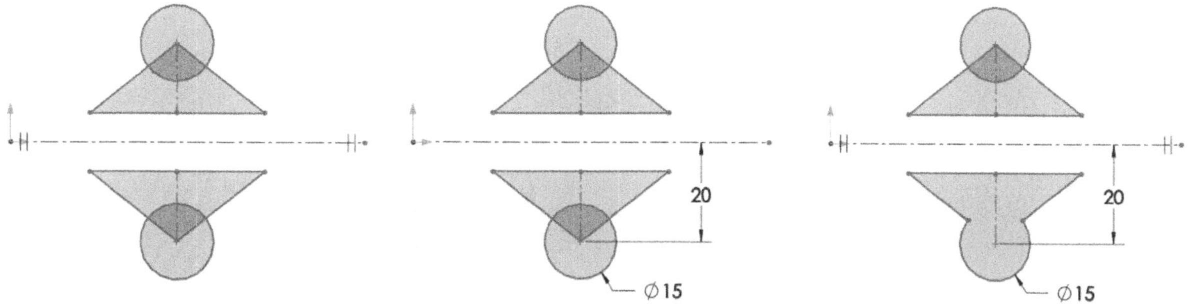

Deactivate the **Dynamic Mirror Entities** command on the CommandManager if you wish to stop mirroring the sketch entities. You can again activate this command and select a centerline if required.

The Linear Sketch Pattern command

This command creates a linear pattern of the sketch elements. Click **Linear Sketch Pattern** icon on the **Sketch** CommandManager. Click in the **Entities to Pattern** box and select the entities to pattern. Next, you need to specify the direction along which the selected entities are to be patterned. By default, the X-axis and Y-axis are selected as **Direction 1** and **Direction 2**. Specify the number of pattern instances and distance between the instances in the **Number of Instances** and **Spacing** boxes, respectively. Click the **Reverse Direction** button next to the **Direction 1** and **Directions 2** boxes, if you want to reverse the direction in which the pattern is created.

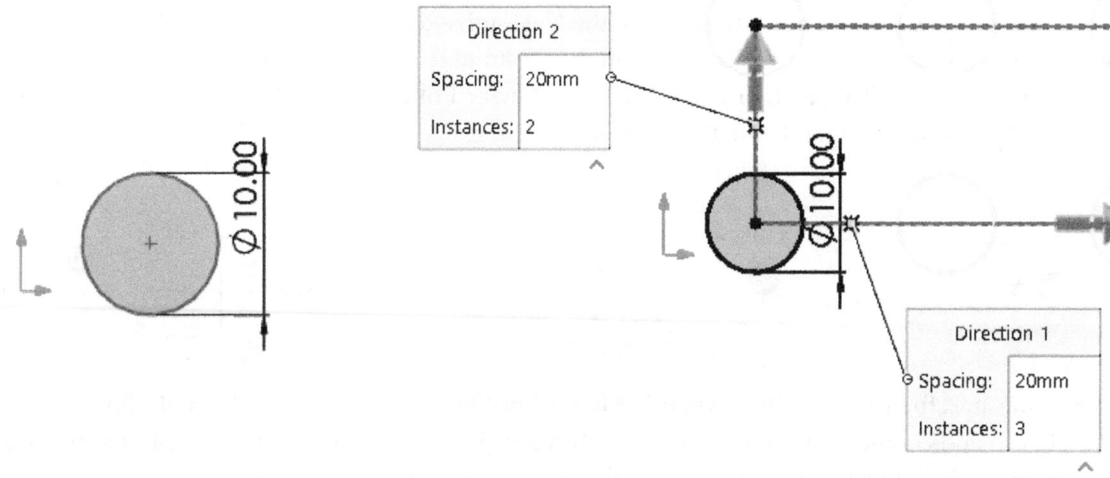

Use the **Angle** box to create a linear pattern at angle to the X-axis.

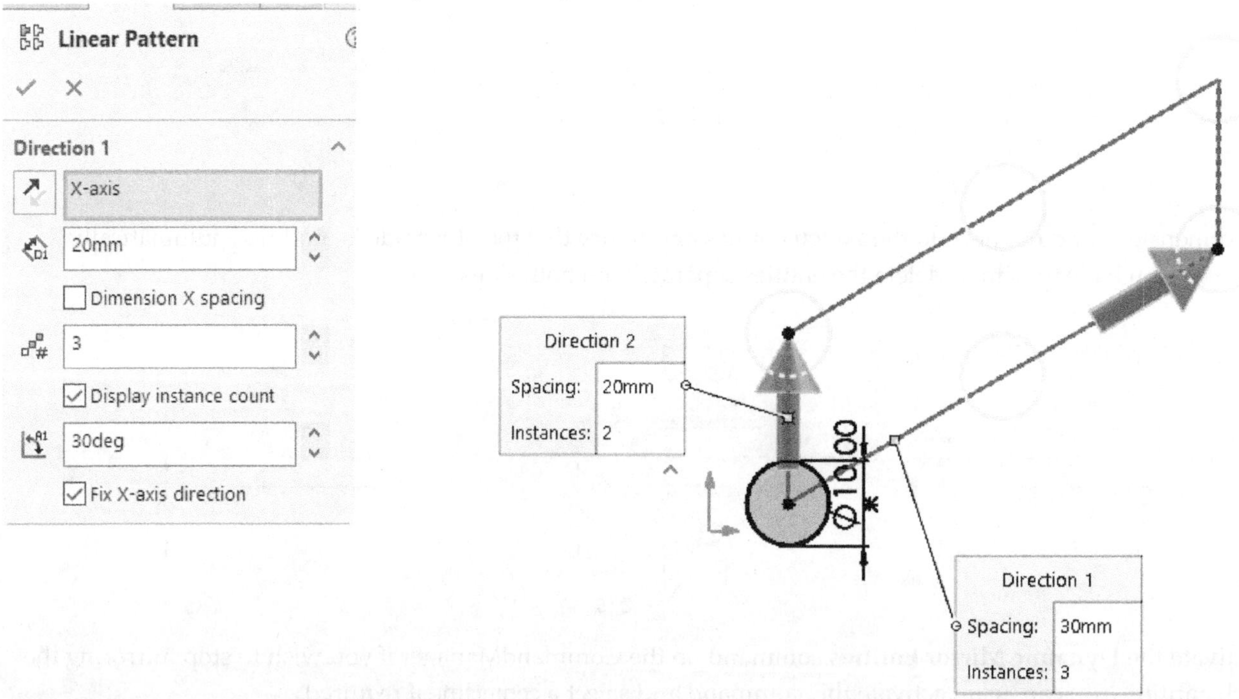

Alternatively, you can select a linear entity to define the direction of the linear pattern. To do this, click in the **Direction 1** box and select the linear sketch entity.

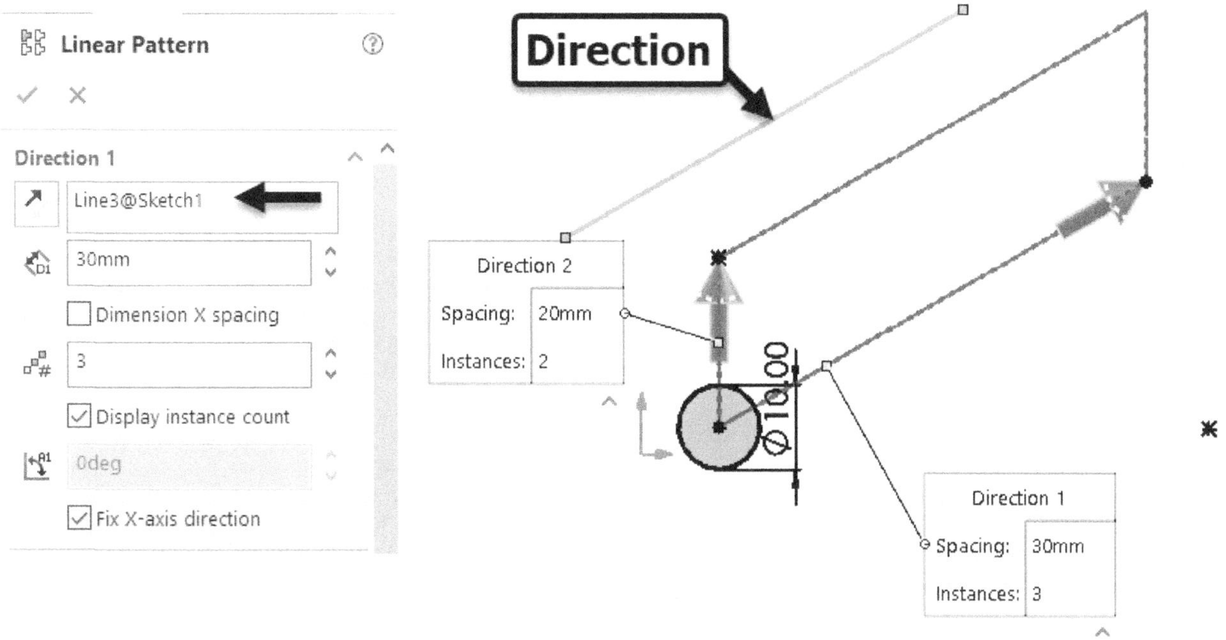

Expand the **Instances to skip** section and click in the Instances to skip selection box. Next, select the dots displayed at the center of the instances to be skipped. The selected instances will be skipped from the linear pattern. Click **OK** on the PropertyManager to complete the linear pattern.

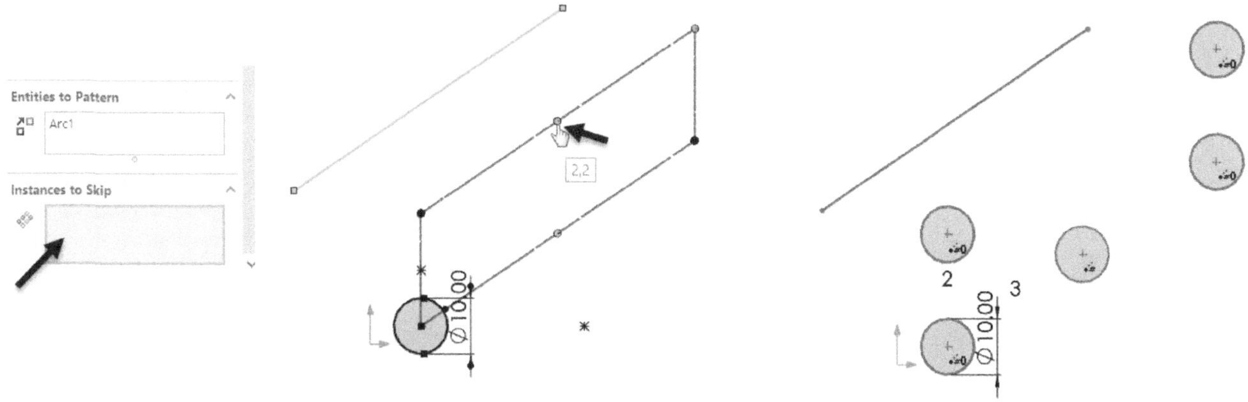

Advanced Sketching commands

SOLIDWORKS provides some advanced drawing commands to create complex sketches. These commands are explained in the following sections.

The Polygon command

This command provides a simple way to create a polygon with any number of sides. As soon as you activate this command, the **Polygon** PropertyManager appears. Now, specify the number of sides in the **Number of Sides** box under the **Parameter** section of the **Polygon** PropertyManager. Now, click in the graphics window to define the center of the polygon. Move the pointer away from the center and click to create the polygon. On the **Polygon** PropertyManager, notice the two options: **Inscribed circle** and **Circumscribed Circle**. If you select the **Inscribed circle** option, a dashed circle appears inside the polygon touching the flat sides. The **Circumscribed circle** option creates a dashed circle outside the polygon touching its vertices. Select the required option and click **OK** on the PropertyManager.

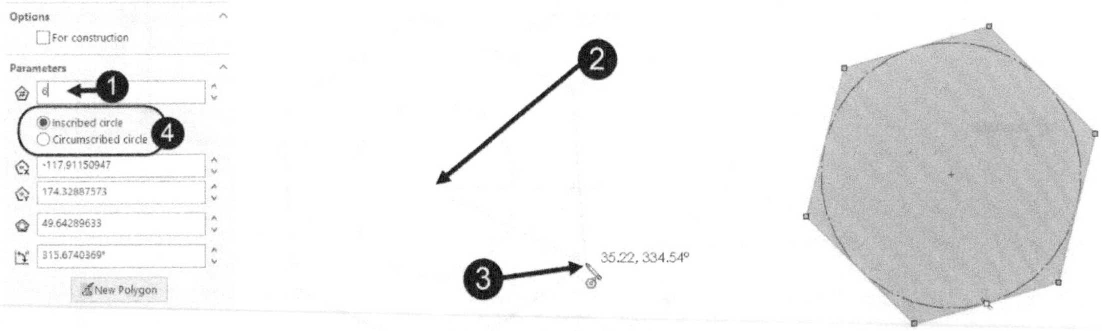

The Ellipse command

This command creates an ellipse using a center point and major and minor axes. Activate this command (click **Sketch** tab > **Ellipse** on the **CommandManager**) and check the **Add construction lines** option on the PropertyManager. Next, click to define the ellipse center. As you move the pointer away from the center, you will notice that a dotted line is displayed. It can be the radius of either the major or the minor axis of the ellipse. Move the pointer and click to define the first axis's radius and orientation; a preview of the ellipse appears. Next, move the pointer and click to define the second axis's radius; the ellipse will be drawn.

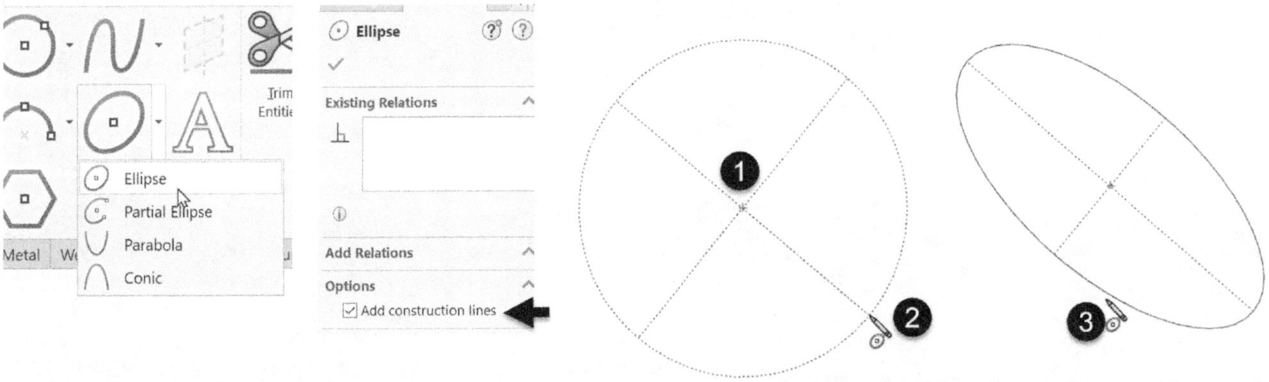

Next, you need to fully-define the ellipse using the relations and dimensions. First, you need to make the ellipse center point coincident with the sketch origin or another sketch entity. You can also make it fixed using the **Fix** relation (or) add dimensions to it. Next, activate the **Centerline** command (on the CommandManager, click **Sketch** tab > **Line** drop-down > **Centerline**) and select the center point of the ellipse. Next, move the pointer horizontally toward the right and click. Make sure that the **Horizontal** relation is applied to the line.

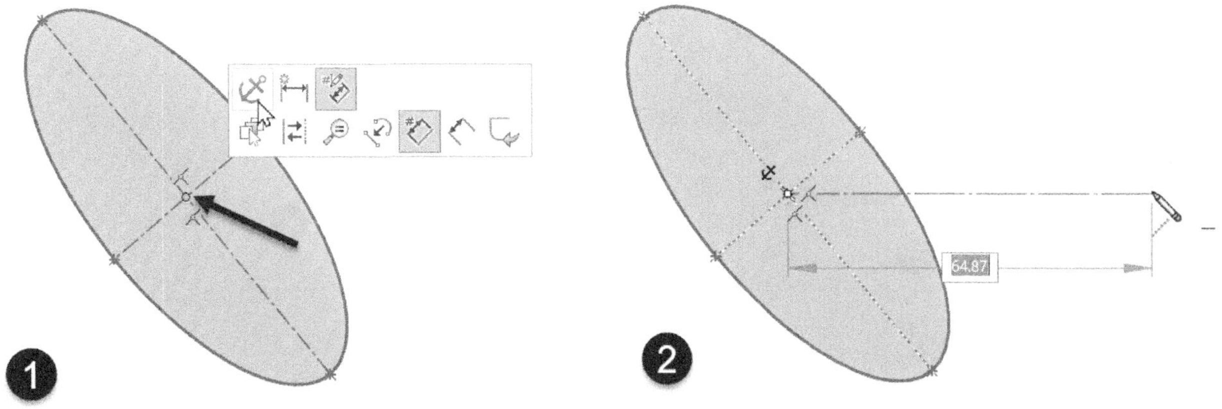

Activate the **Smart Dimension** command and select the horizontal centerline and the major axis of the ellipse. Move the point between the selected lines and click. Type-in the angle value in the **Modify** box and click the green check.

Select the ellipse's major axis and move the pointer along the direction perpendicular to it. Click to position the dimension and type in a value.

Select the ellipse's minor axis, move the pointer along the direction perpendicular to it, and click to position the dimension and type in a value.

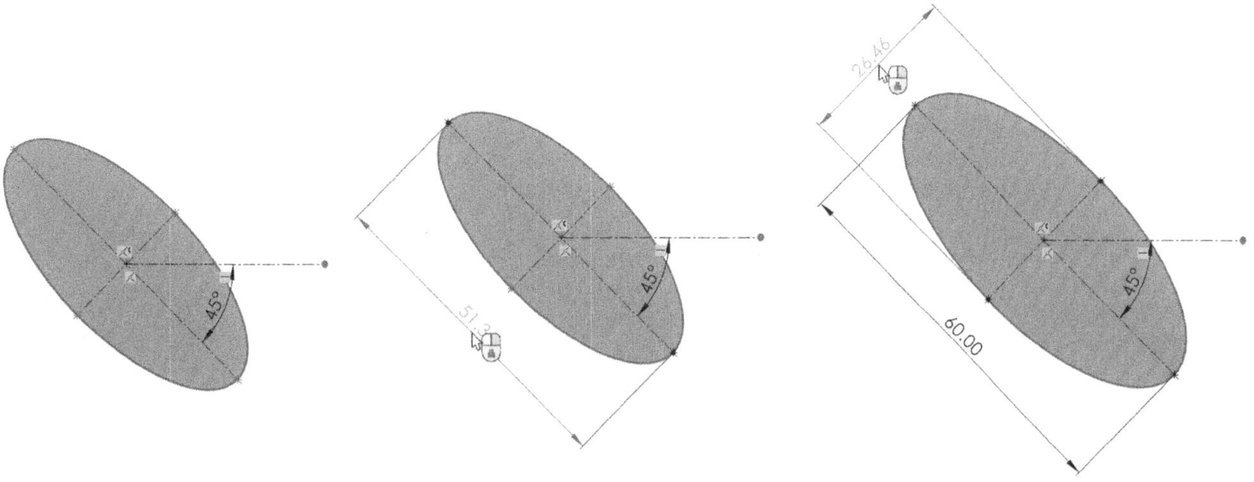

The Partial Ellipse command

This command creates a partial ellipse using a center point, radius and axes. Activate this command (on the CommandManager, click **Sketch** tab **> Ellipse** drop-down **> Partial Ellipse**). Next, click to define the center of the partial ellipse. Move the pointer and click to define the radius and orientation of the first axis. Again, move the pointer and click to define the radius of the second axis. Also, the start point of the ellipse is defined. Move the pointer and click to define the endpoint. Click **Close Dialog** on the PropertyManager.

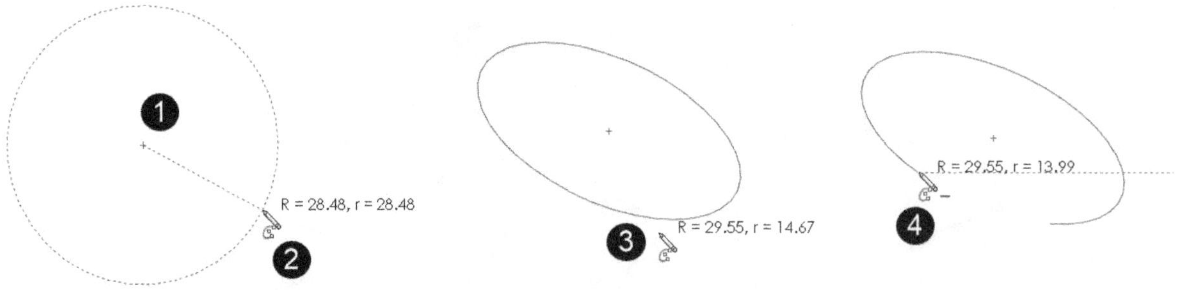

The Conic command

This command creates a conic curve, which is used to connect two open sketch entities. First, you have to create a line and an arc, as shown. Activate the **Conic** command (on the **Sketch** CommandManager, click **Ellipse** drop-down > **Conic**). Select the endpoints of the line and arc.

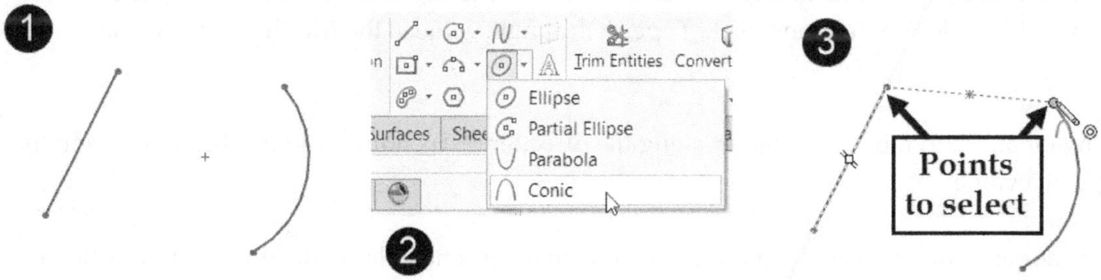

Move the pointer and notice the dotted lines. These lines are tangent to the line and arc. Now, select the intersection point of the dotted tangent lines. Again, move the pointer and notice the rho value displayed. The rho value defines the curvature of conic created. Click to create the conic curve. Now, expand the **Parameters** section of the PropertyManager and specify the rho value (i.e., ρ box) of the conic curve. You can type in a value between 0 and 1. Click **Close Dialog** on the PropertyManager.

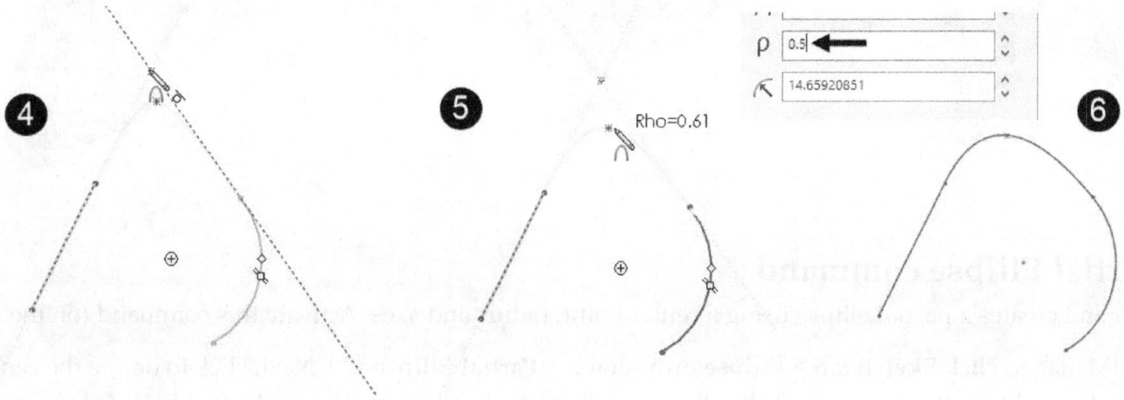

The Parabola command

This command helps you to create a parabola without much effort. Activate this command (On the **Sketch** CommandManager, click **Ellipse** drop-down > **Parabola**) and specify the focal point of the parabola. Move the pointer in the necessary direction and click to define the vertex point. Click at the required location on the dotted curve to define the start point. Move the pointer up to the required curve distance, and then click to define the endpoint.

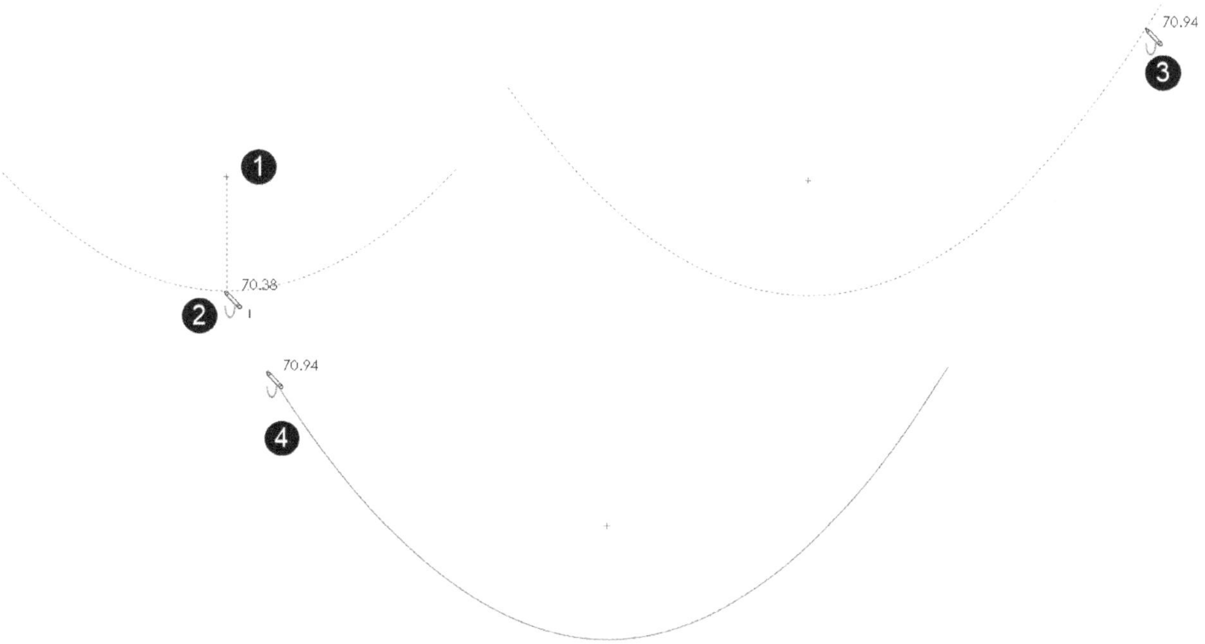

Next, you need to fully-define the parabola to control its shape. First, you need to make the parabola's focal point coincident with the sketch origin or another sketch entity. You can also make it fixed using the **Fix** relation. Next, you need to constrain the parabola's vertex by connecting to the focal point using a construction line.

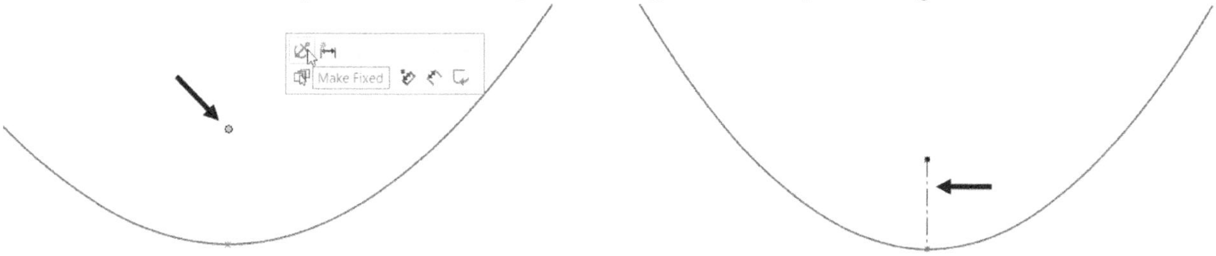

Dimension the construction line and notice that the parabola is turned into black except the endpoints. Click and drag the endpoints and notice that they are movable.

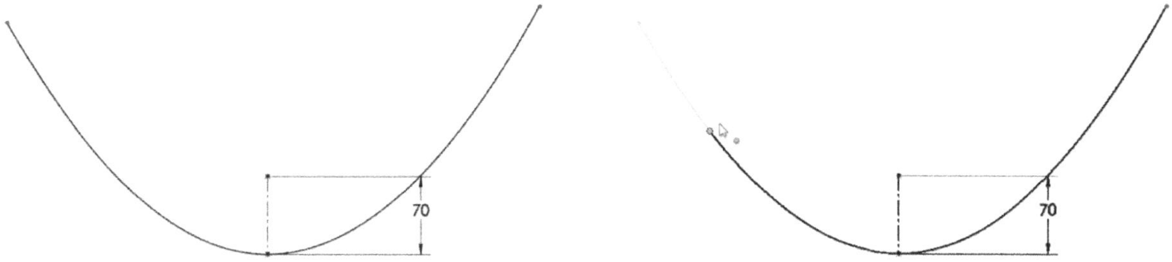

Create the construction lines between the focal point and endpoints of the parabola. Next, create angular dimensions between the construction lines, as shown.

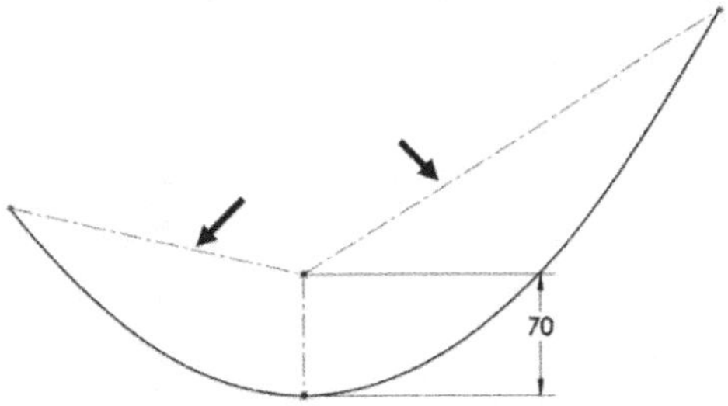

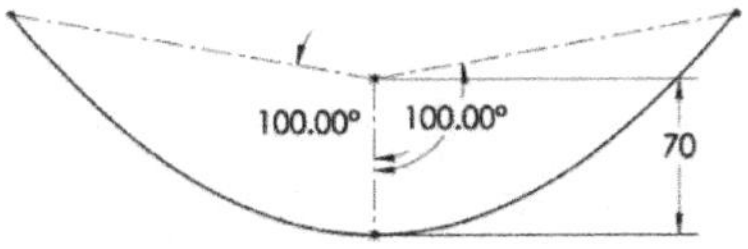

The Spline command

This command creates a smooth B-spline curve along the selected points. B-Splines are non-uniform curves, which are used to create irregular shapes. Activate this command (on the **Sketch** CommandManager, click **Spline**) and click to specify points in the graphics window. The spline is created passing through the points. You can move the pointer and select the spline's first point; a closed spline is created. Press Esc to deactivate the **Spline** command.

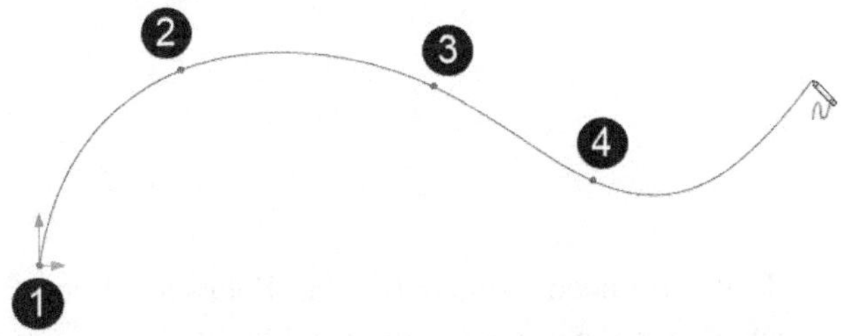

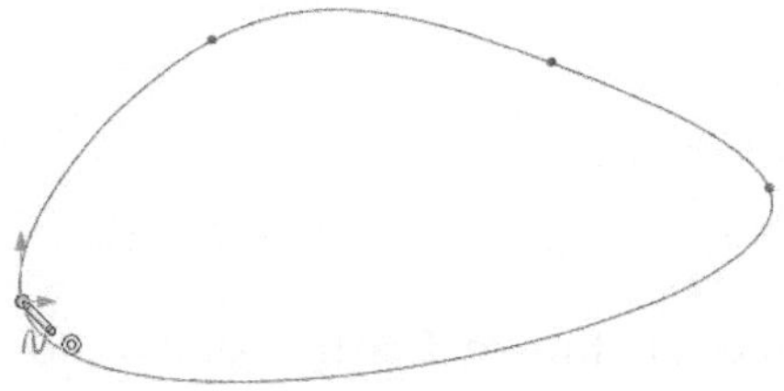

Fully defining a spline is a bit difficult than other sketch entities. For example, create the spline, as shown. Next, activate the **Smart Dimension** command and select the start point of the spline. Select the second point, move the pointer and place the dimension. Type in a value in the **Modify** box and click the green check. Likewise, create the other two dimensions, as shown.

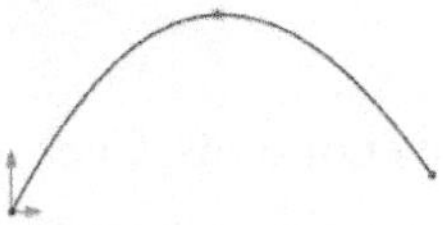

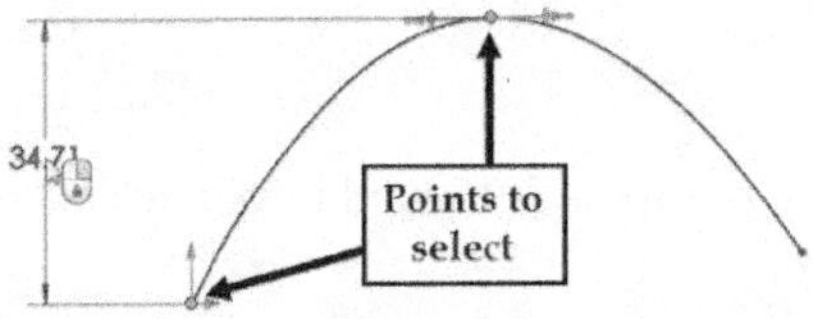

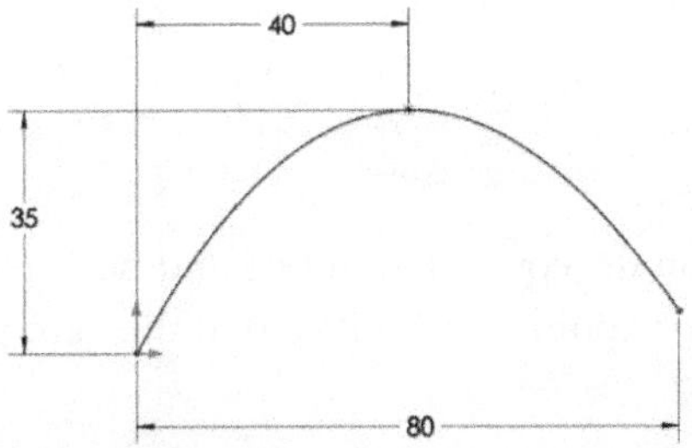

Press the Ctrl key and select the start and endpoints of the spline. Select the **Horizontal** relation from the PropertyManager. Notice that the sketch is turned black, indicating that it is fully defined. However, the spline is modified when you select it and drag the spline handles.

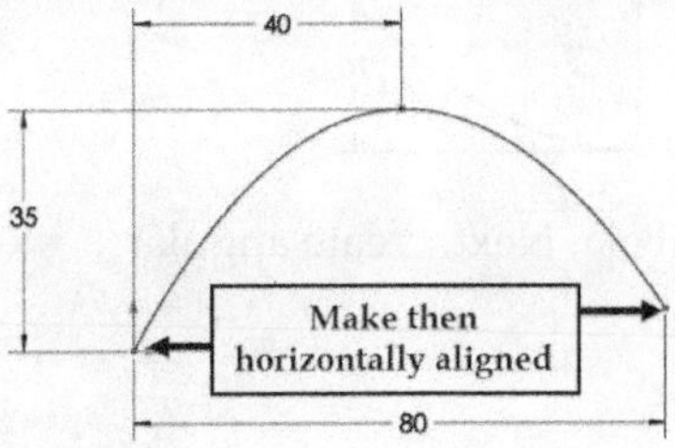

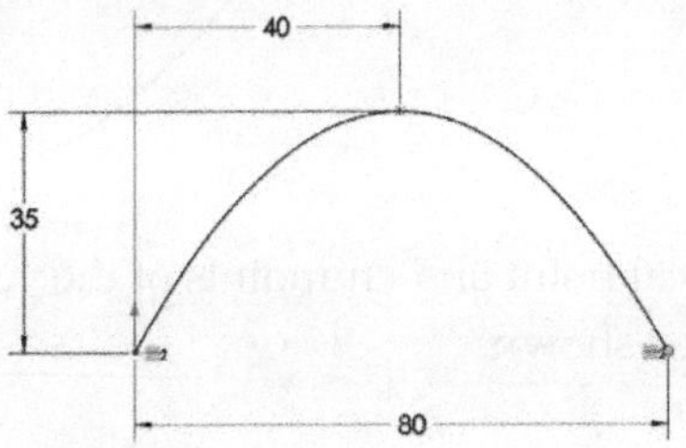

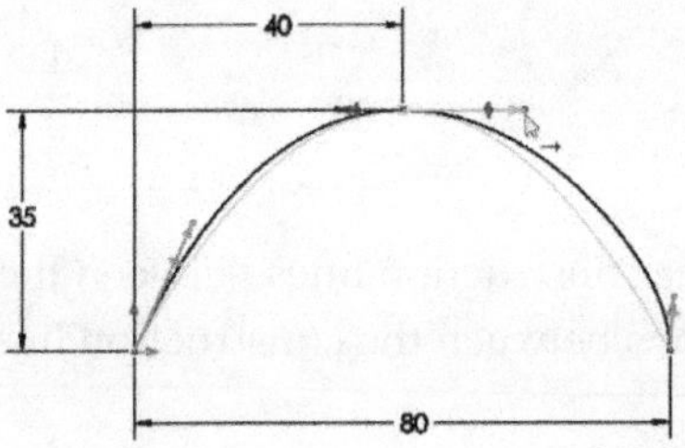

The spline handles can be used to modify the tangency of the spline. The components of a spline handle are shown next.

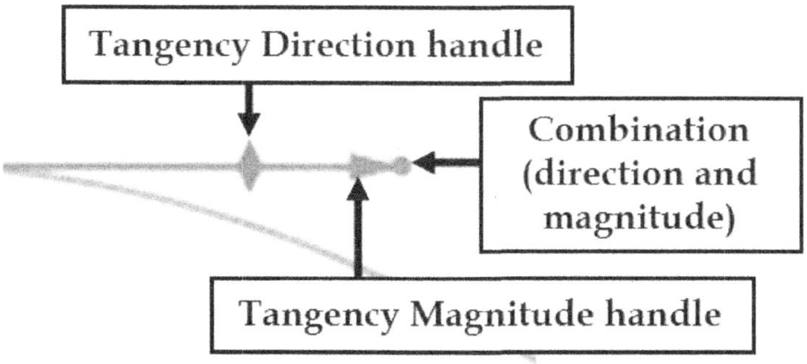

Click and drag the direction handle and notice that the tangency angle is modified. Select the magnitude handle and drag it to change the tangency length. Use the combination handle to change the tangency length and direction simultaneously.

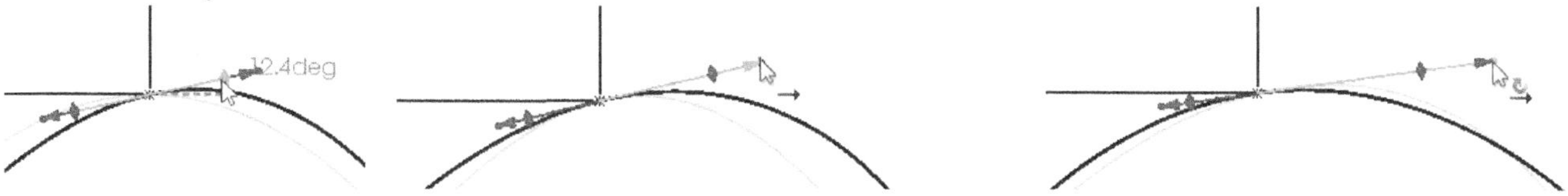

You can control the spline handle by adding dimensions to its direction and magnitude handles. Select the spline to display the spline handles. Click on the magnitude handle of the second point of the spline, and then drag it. Activate the **Smart Dimension** command and select the magnitude. Move the pointer and place the dimension. Next, type in a value in the Modify box and click the green check. Likewise, add dimensions to the other magnitude handles, as shown.

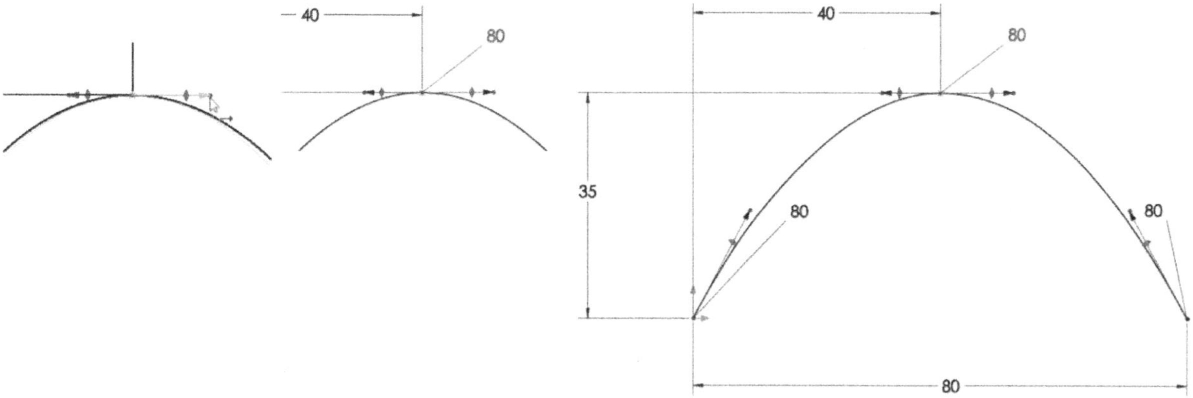

Next, you need to constrain the direction handle either by using dimension or by adding relations. Activate the Centerline command (On the **Sketch** CommandManager, click **Line** drop-down > **Centerline**) and create the line at the origin, as shown.

Activate the **Smart Dimension** command and select the direction handle of the first point. Select the horizontal centerline and place the angular dimension between them. Type-in a value in the **Modify** box and click the green check. Likewise, create the angular dimension between the centerline and the direction handle.

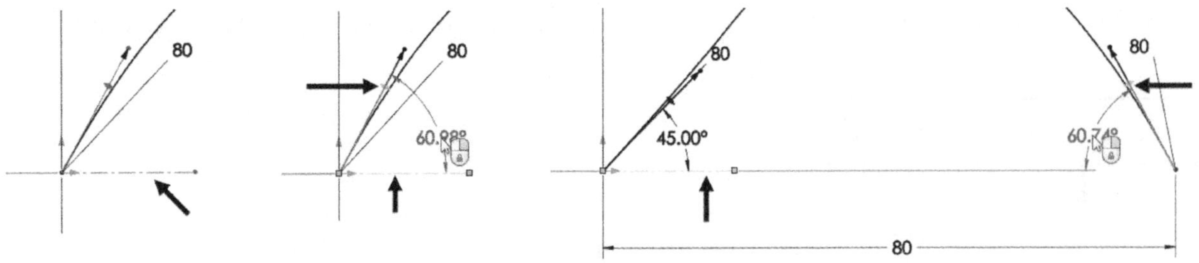

Create a vertical centerline from the second point of the spline. Next, add an angular dimension between the direction handle and vertical centerline. The spline is fully-defined.

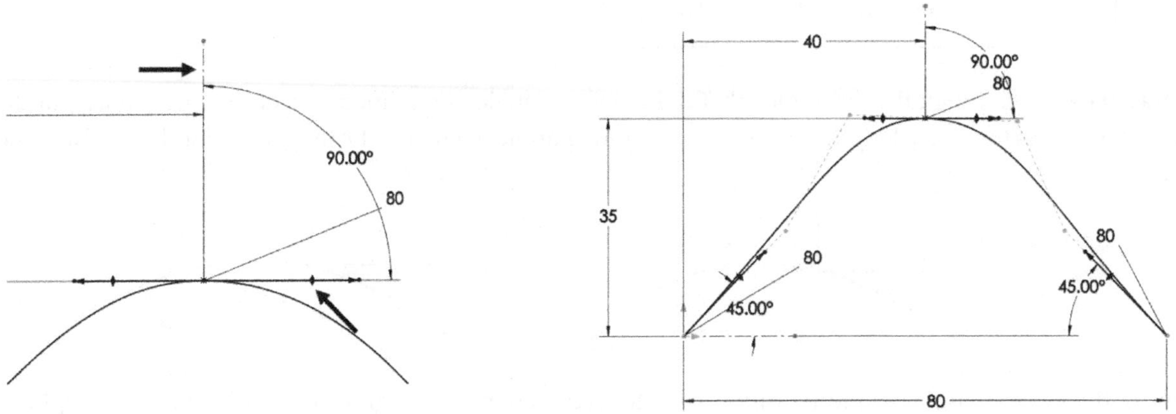

The Style Spline command

The **Style Spline** command creates a spline with control polygons. Controlling this type of spline is easy. Activate this command (on the **Sketch** CommandManager, click **Spline** drop-down > **Style Spline**) and click to specify the control points. The spline is created, as shown. Press Esc to deactivate the command. You can add dimensions and relations to the control polygon (dotted lines).

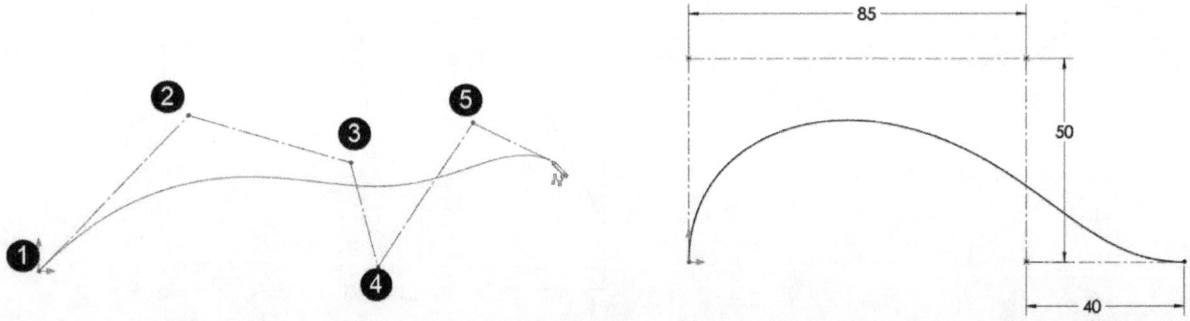

The **Style Spline** command allows you to create four different spline types: Bezier, B-spline: Degree 3, B-spline: Degree 5, B-spline: Degree 7. For example, create the line chain, as shown next. Activate the **Style Spline** command

and select **Bezier** from the **Spline Type** section on the PropertyManager. Select the vertices of the line chain. Press Esc to deactivate the command.

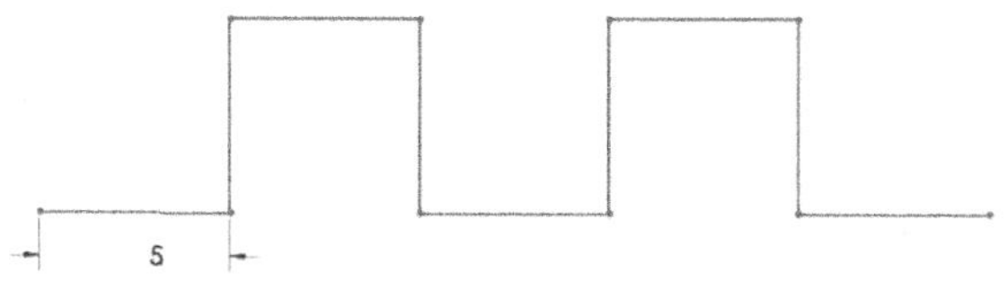

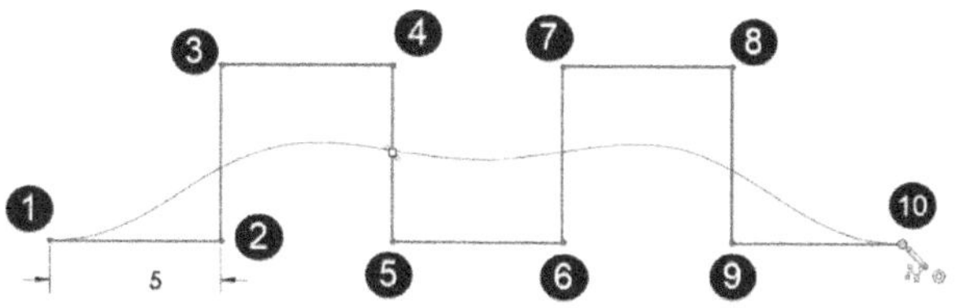

Select the spline to display the **Studio Spline** PropertyManager. On the PropertyManager, select **Curve Type > B-spline: Degree 3**. The spline is modified. Likewise, change the **Curve Type** to **B-spline: Degree 5** and **B-spline: Degree 7**, and then notice the changes.

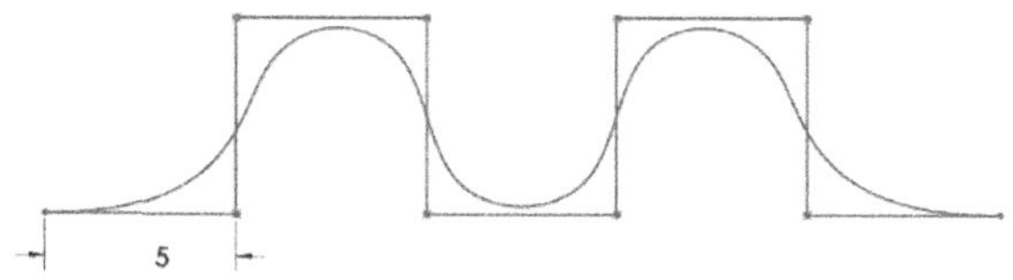

B-spline: Degree 3

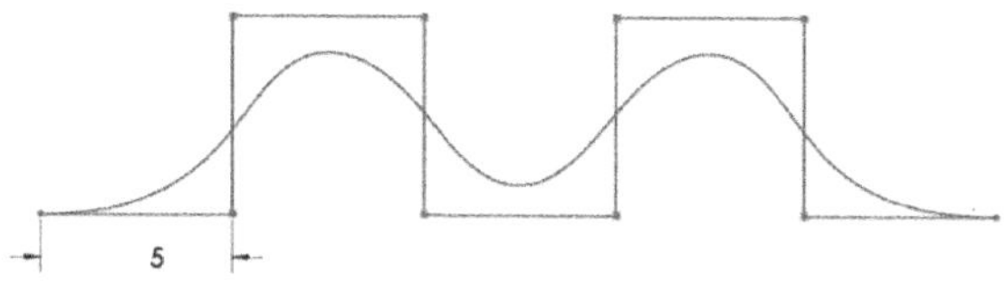

B-spline: Degree 5

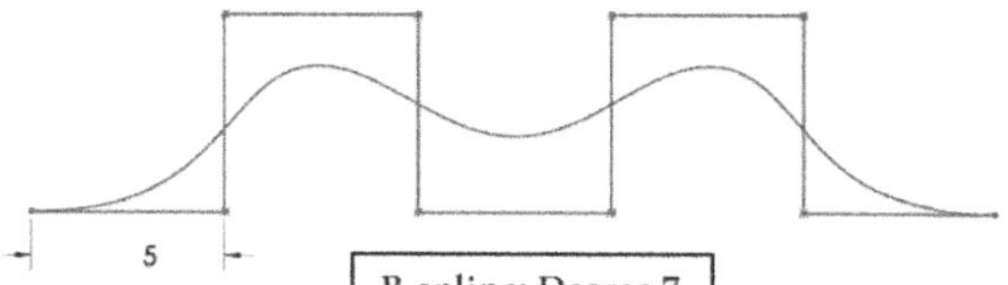

B-spline: Degree 7

Best Practices for Sketching in SOLIDWORKS

1. Proper Units: Set correct units aligning with industry standards.
2. Start with Planes: Define and name planes accurately.
3. Fully Define Your Sketch: Ensure no degrees of freedom are left.
4. Avoid Over-Defining: Maintain flexibility during design iterations.
5. Keep It Simple: Use fewer entities and relations without compromising design intent.
6. Use Automatic Constraints: Leverage constraints for efficient sketching.
7. Use Grid Lines: Utilize grid lines for alignment.
8. Constrain Early, Dimension Later: Apply constraints before specific dimensions.
9. Avoid Excessive Fillets and Chamfers: Minimize use directly in the sketch.

Cautions for Sketching in SOLIDWORKS:

1. Overlapping and Self-Intersecting Geometry: Avoid modeling errors.
2. Incorrect Units: Ensure correct units to prevent errors.
3. Ignoring Warnings and Errors: Address messages promptly.
4. Inadequate Dimensioning: Include relevant dimensions for design intent.
5. Using Dimensions for Symmetric Shapes: Use constraints for symmetry.

Tips for Sketching in SOLIDWORKS:

1. Shortcut Keys: Learn and use shortcuts for efficiency.
2. Constraints: Control sketch elements' behavior effectively.
3. Save Incrementally: Save work regularly.
4. Avoid Underdefined Sketches: Fully define sketches for stability.
5. Regularly Check for Interference: Prevent issues by checking regularly.

By following these streamlined guidelines tailored to SOLIDWORKS, you can enhance your sketching efficiency, adhere to industry standards, and maintain design intent effectively throughout your projects.

Examples

Example 1 (Millimetres)

In this example, you draw the sketch shown below.

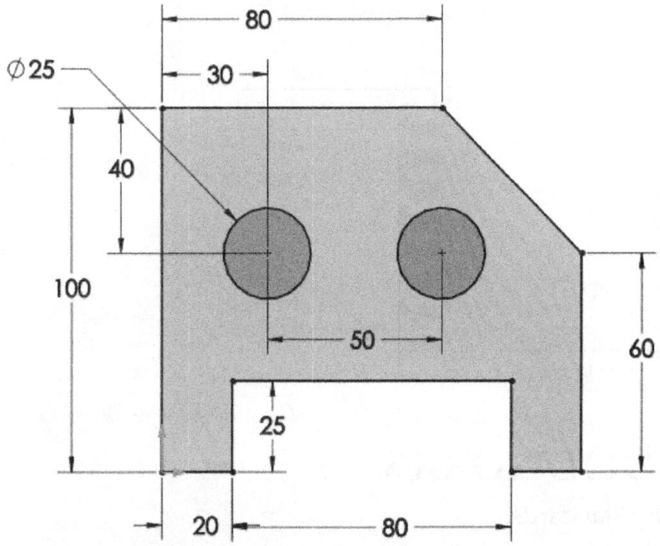

1. Start **SOLIDWORKS 2024** by clicking the **SOLIDWORKS 2024** icon on your desktop.
2. Click the **New** button on the **Quick Access Toolbar**; the **New SOLIDWORKS Document** dialog is opened.

3. On the **New SOLIDWORKS Document** dialog, click the **Part** icon, and then click **OK**; a new part file is opened.
4. At the bottom right corner of the window, click the **Unit System** drop-down and select **MMGS (millimeter, gram, second)**.

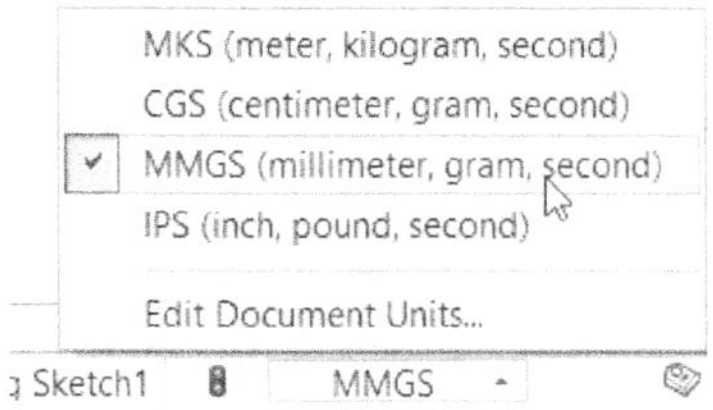

5. To start a new sketch, click **Sketch** tab > **Sketch** on the CommandManager.
6. Click on the Front plane; the selected plane orients normal to the screen.
7. Activate the **Line** command (On the CommandManager, click **Sketch** tab > **Line**)
8. Click on the origin point to define the first point of the line.
9. Move the pointer horizontally toward the right.
10. Click to define the endpoint of the line.
11. Move the pointer vertically upwards. Click to draw the second line.

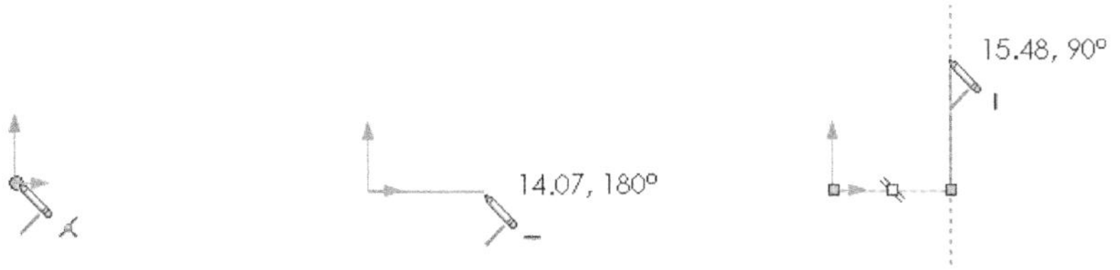

12. Move the pointer horizontally toward the right and click.
13. Move the pointer vertically downward and click when the horizontal trace-lines appear from the sketch origin.
14. Move the pointer horizontally toward the right up to a short distance, and then click.

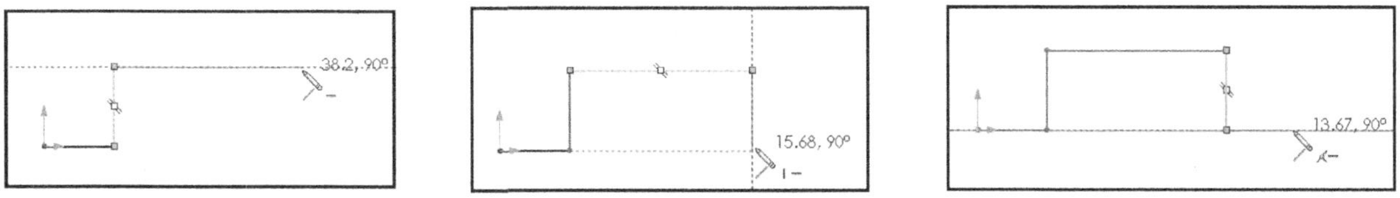

15. Move the pointer vertically upward and click.
16. Move the pointer in the top-left direction and click.
17. Move the pointer horizontally towards the left and click when vertical trace-lines appear from the origin.

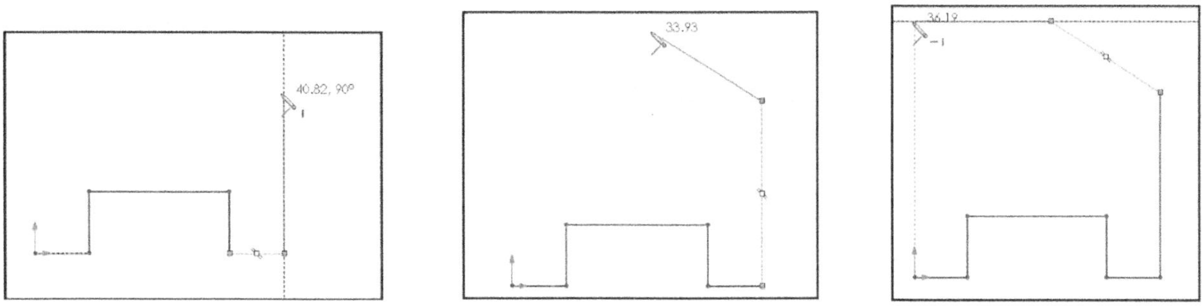

18. Select the start point of the sketch to create a closed contour. Notice the shade inside the contour.

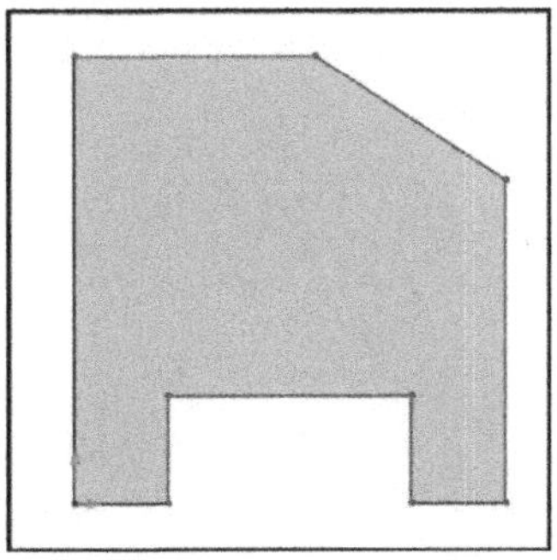

19. Right-click and select **Select** (or) press Esc to deactivate the **Line** command.
20. Click **Sketch** tab > **Display/Delete Relations > Add Relation** on the CommandManager. Next, click on the two horizontal lines at the bottom.

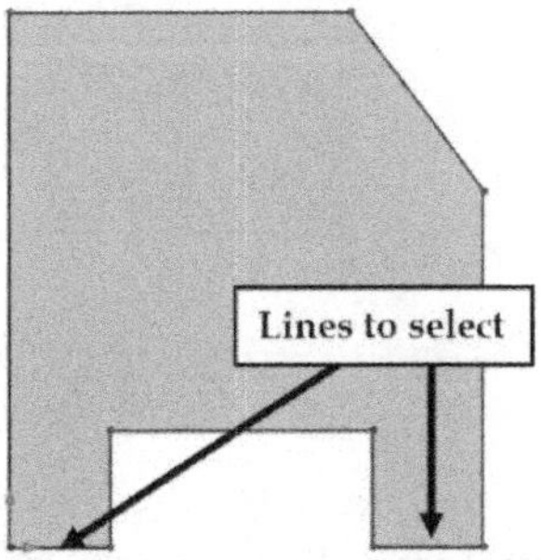

21. On the **Add Relations** PropertyManager, under the **Add Relations** section, select **Equal** = ; the selected lines become equal in length.
22. Press the Ctrl key and select the small vertical lines. Next, click **Equal** = on the **Add Relations** PropertyManager; the selected lines are made equal in length. Click **OK** on the PropertyManager.

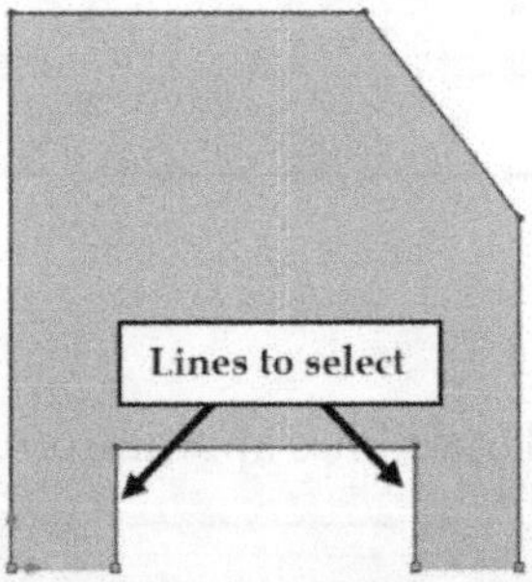

23. Click **Sketch** tab > **Smart Dimension** on the CommandManager and click on the lower left horizontal line. Move the mouse pointer downward and click to locate the dimension.
24. Type-in **20** and click the green check on the **Modify** box.

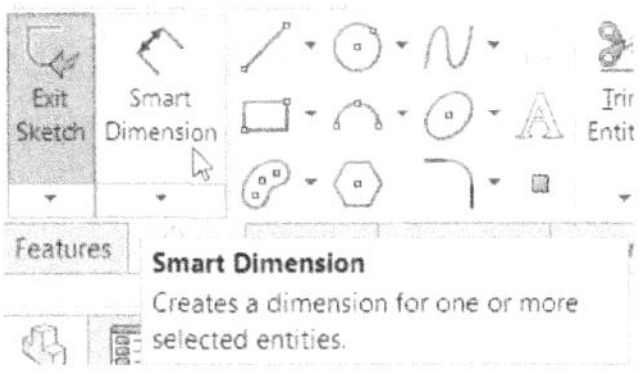

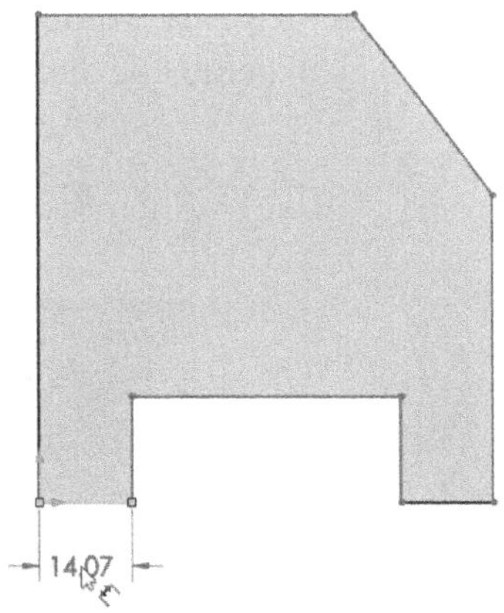

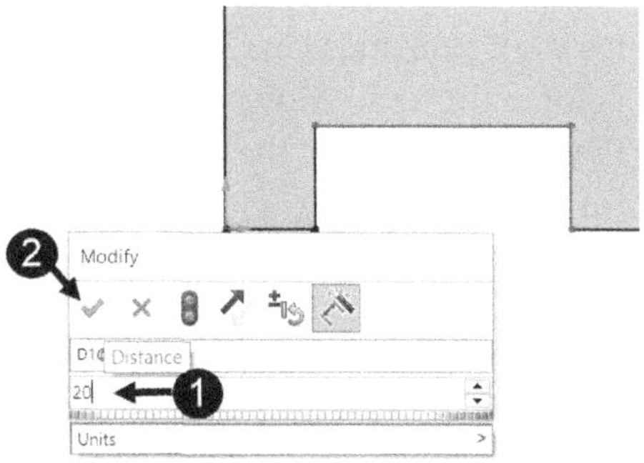

25. Click on the small vertical line located on the left side. Move the mouse pointer towards the right and click to position the dimension.
26. Type-in **25** and click the green check on the **Modify** box.

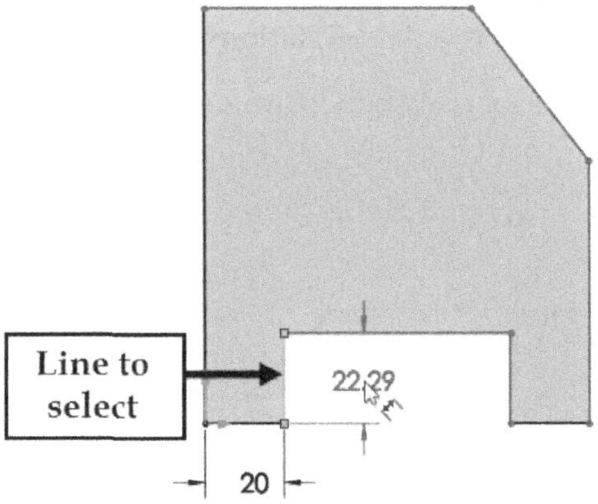

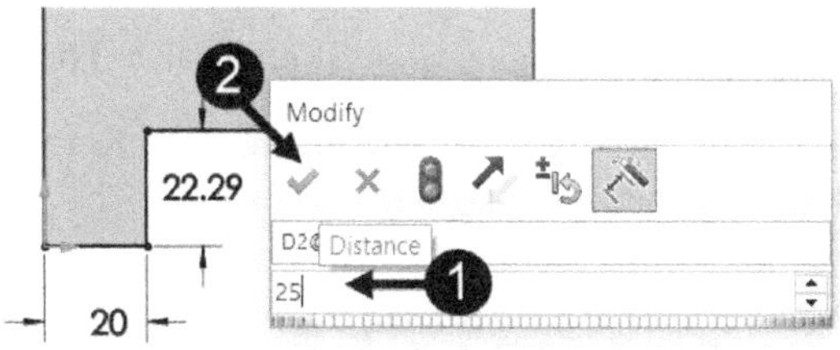

27. Create other dimensions in the sequence, shown below. Press Esc to deactivate the **Smart Dimension** command.

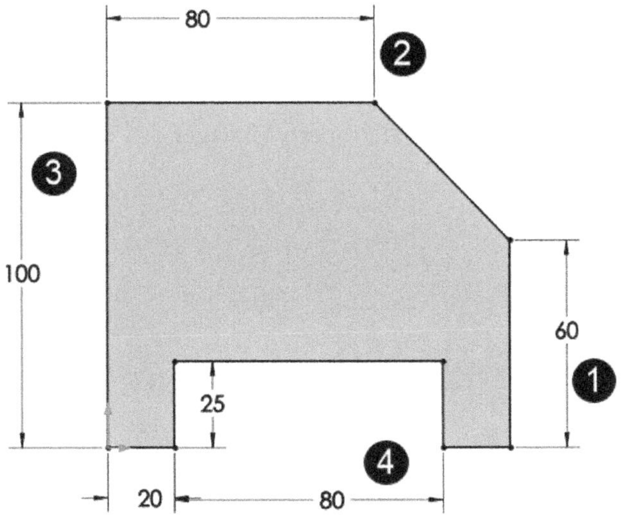

28. On the CommandManager, click **Sketch** tab > **Circle**. Click inside the sketch region to define the center point of the circle. Move the mouse pointer and click to define the diameter. Likewise, create another circle.
29. Press Esc to deselect the circle.

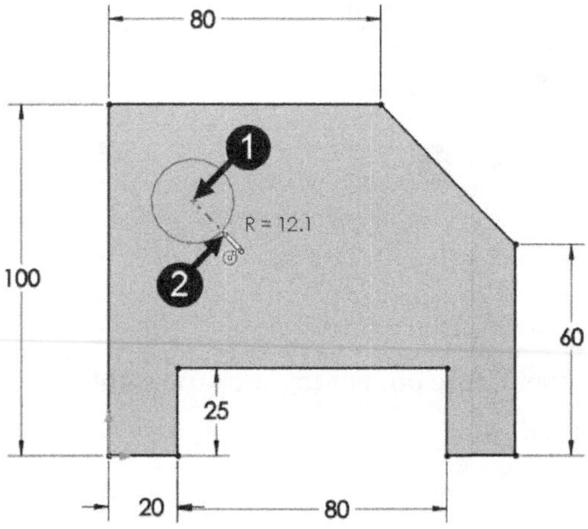

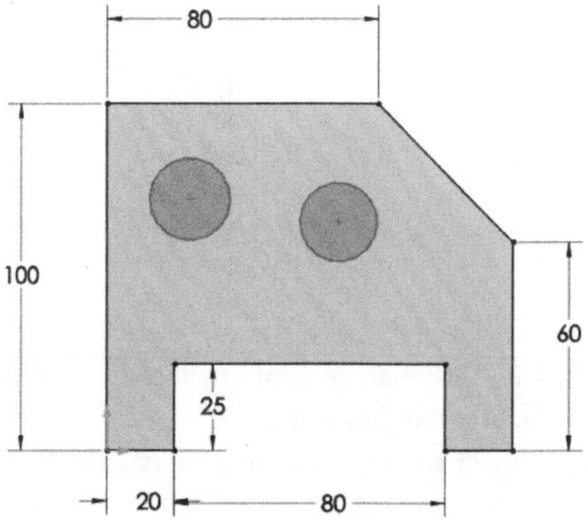

30. On the CommandManager, click **Sketch** tab **> Display/Delete Relations > Add Relation**. Click on the center points of the two circles. On the **Add Relation** PropertyManager, click the **Horizontal** icon to make them horizontally aligned.

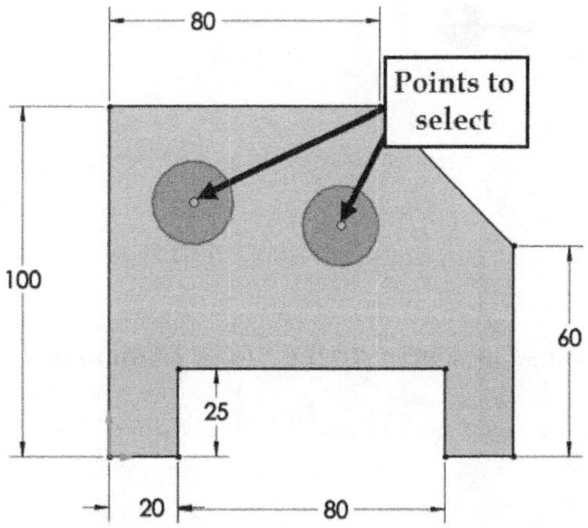

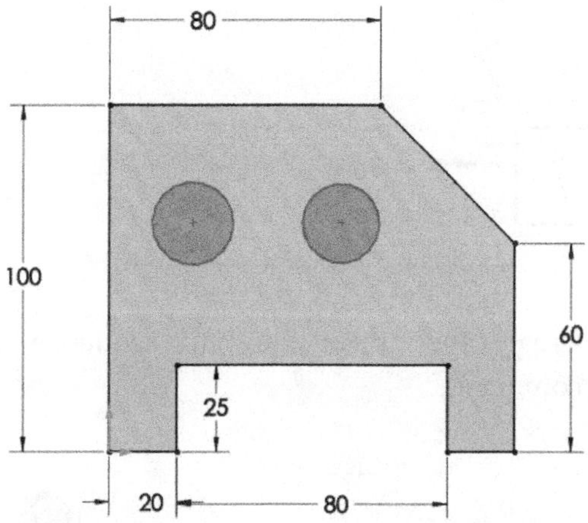

31. Press the Ctrl key, and then click on the two circles. Click the **Equal** icon on the PropertyManager; the diameters of the circles will become equal. Click **OK** on the PropertyManager.

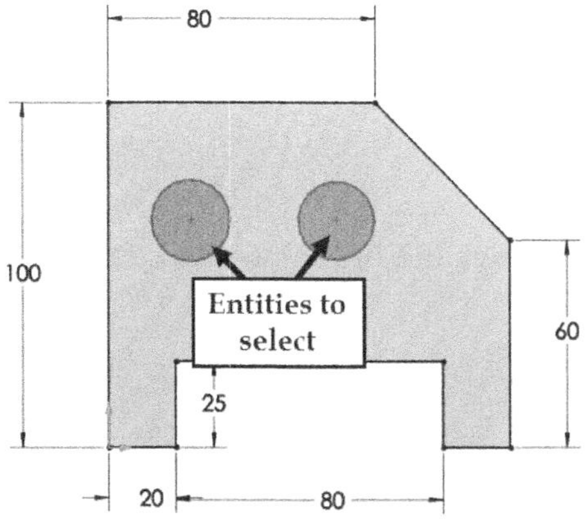

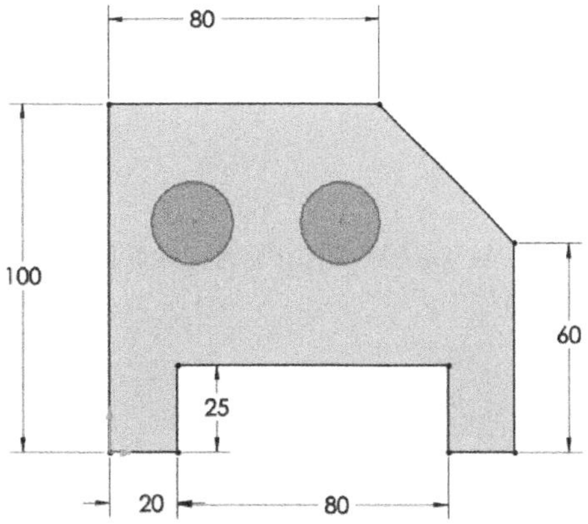

32. Activate the **Smart Dimension** command and click on any one of the circles. Move the mouse pointer and click to position the dimension. Type 25 in the **Modify** box and click the green check.

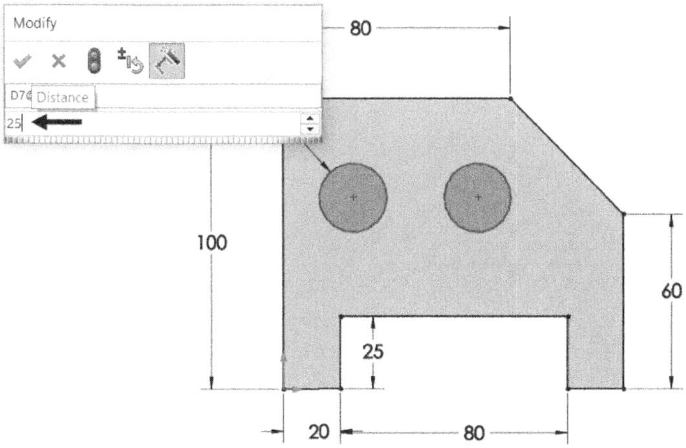

33. Create other dimensions between the circles and the adjacent lines, as shown below.

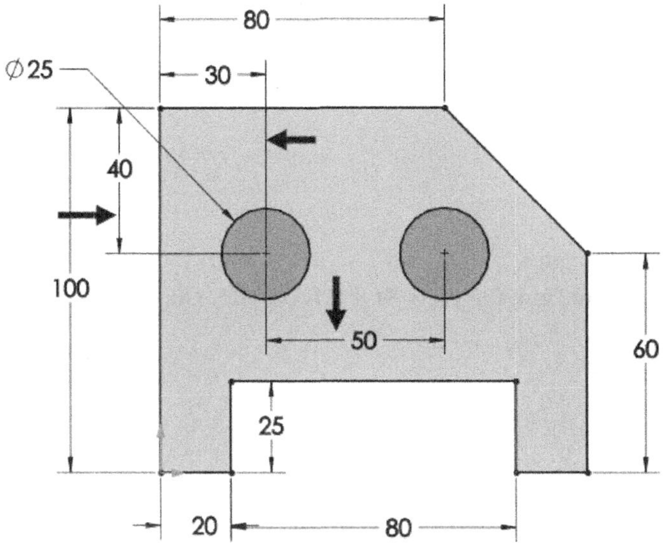

34. Click the **Exit Sketch** icon on the right side of the graphics window to deactivate the Sketch environment.

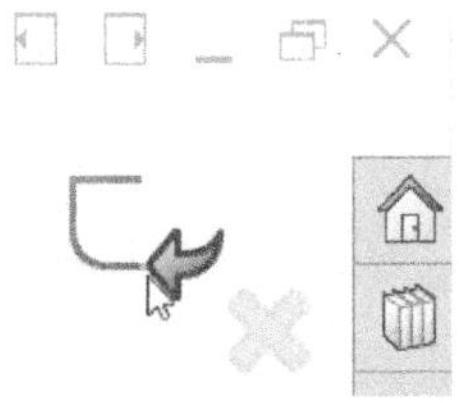

35. Click the **Save** icon on the **Quick Access Toolbar**. Define the location and file name. Next, click **Save** to save the part file.

36. Click **Close** on the top right corner of the graphics window.

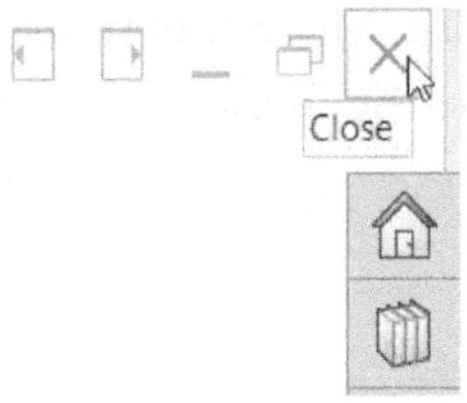

Example 2 (Inches)

In this example, you draw the sketch shown below.

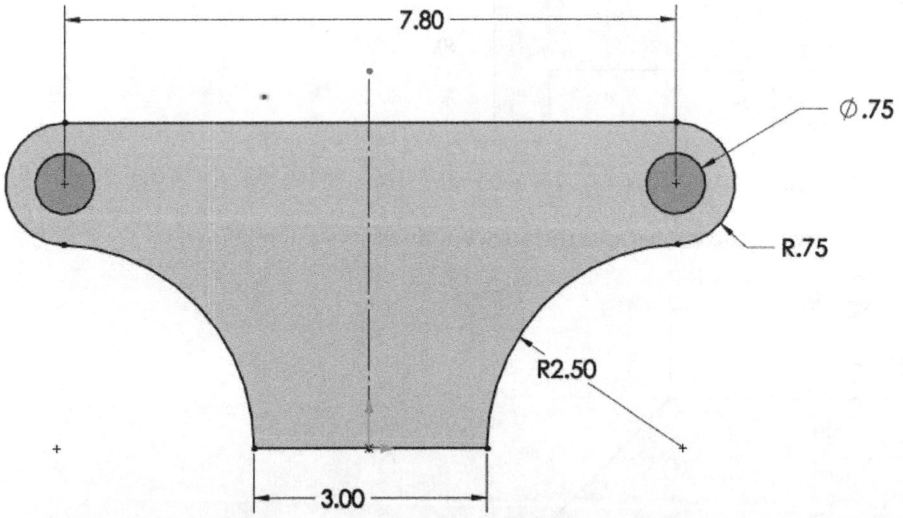

1. Start **SOLIDWORKS 2024**, if not already opened.
2. On the **Quick Access Toolbar**, click the **New** icon; the **New SOLIDWORKS Document** dialog is opened.

3. On the **New SOLIDWORKS Document** dialog, click the **Part** icon. Click **OK** to start a new part file.
4. At the bottom right corner of the window, click the **Unit System** drop-down and select **IPS (inch, pound, second)**.

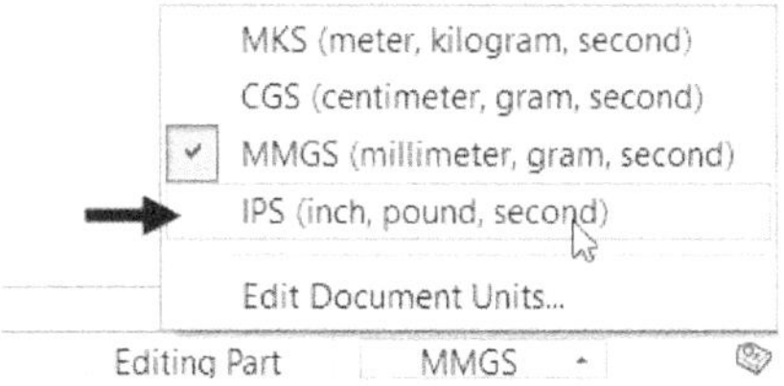

5. To start a new sketch, click **Sketch** tab > **Sketch** on the CommandManager.
6. Click on the Front Plane.
7. On the **Sketch** CommandManager, click **Line** drop-down > **Midpoint Line**.
8. Click on the origin point to define the midpoint of the line. Move the mouse pointer horizontally and click to draw a line.

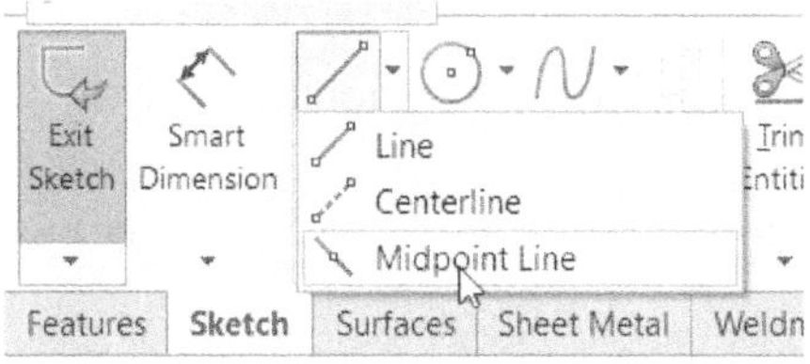

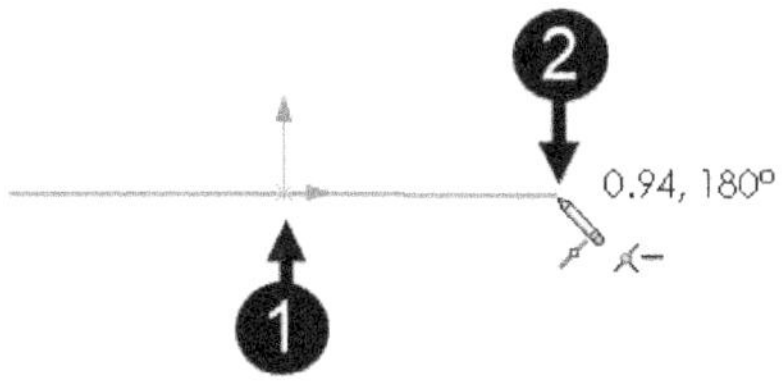

9. Move the pointer away from the line's endpoint and then move it back to the endpoint; the pointer is switched to the arc mode. Also, the start point of the arc is defined.
10. Move the pointer vertically upward, and then move it in the top-right direction.
11. Click to define the endpoint of the arc.

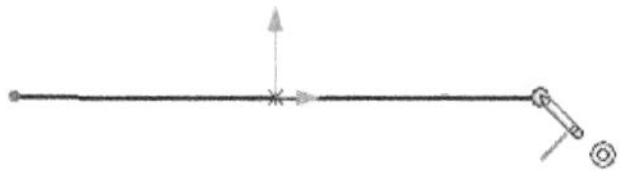

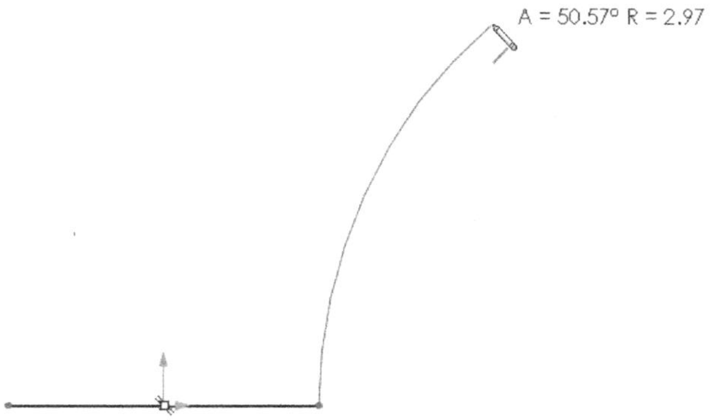

12. Move the pointer away from the endpoint of the arc, and then move it back; the pointer is switched to the arc mode.
13. Move it upwards right.
14. Move the pointer toward the left and click when a vertical dotted line appears, as shown below.

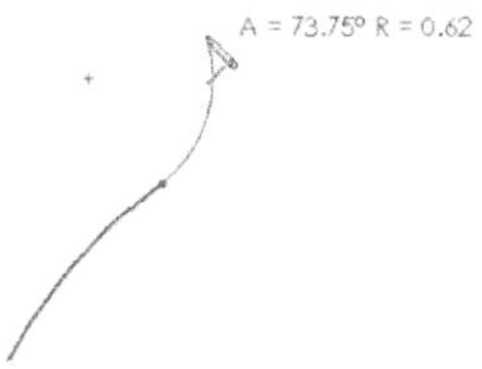

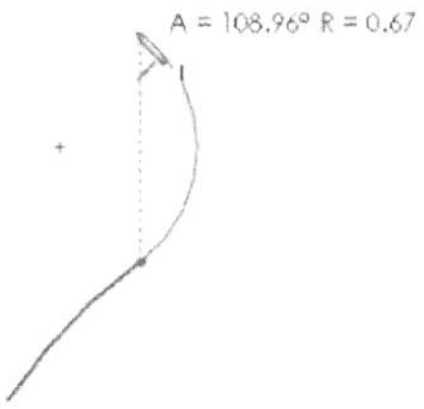

15. Move the mouse pointer toward the left and click to create a horizontal line. Note that the length of the new line should be greater than that of the lower horizontal line.

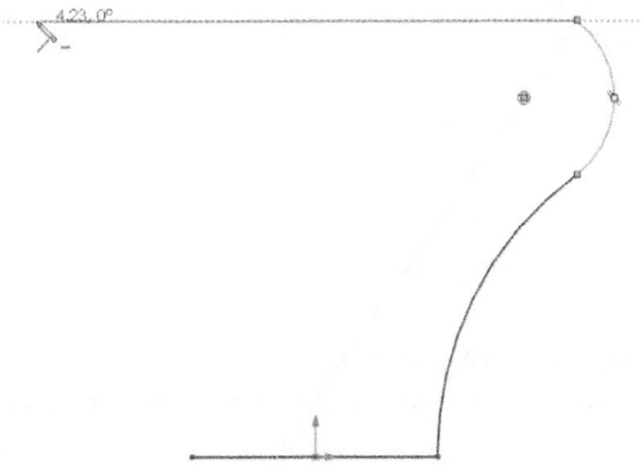

16. Move the pointer away from the line's endpoint, and then move it back; the pointer is switched to the arc mode.
17. Move the pointer toward the right and click when a vertical dotted line appears, as shown below.

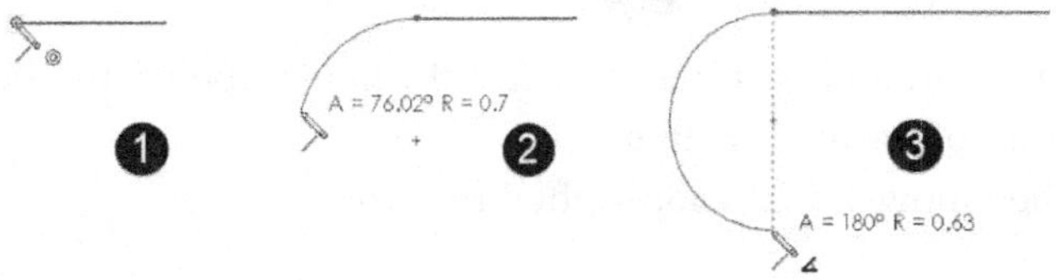

18. Move the pointer away from the endpoint of the arc, and then move it back; the pointer is switched to the arc mode.
19. Move the mouse pointer downward right and click on the origin to close the sketch.

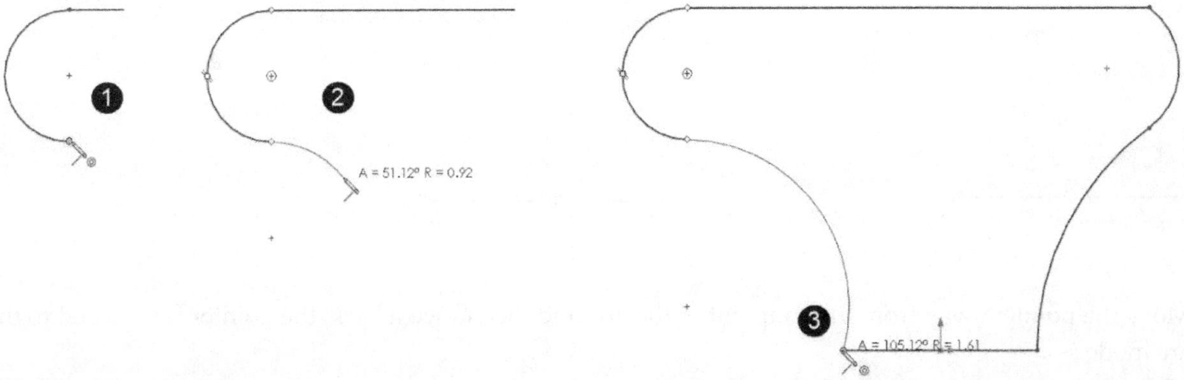

20. Click on the midpoint of the lower horizontal line. Move the mouse pointer vertically up and click to create a vertical line.
21. Right-click and select **Select**.

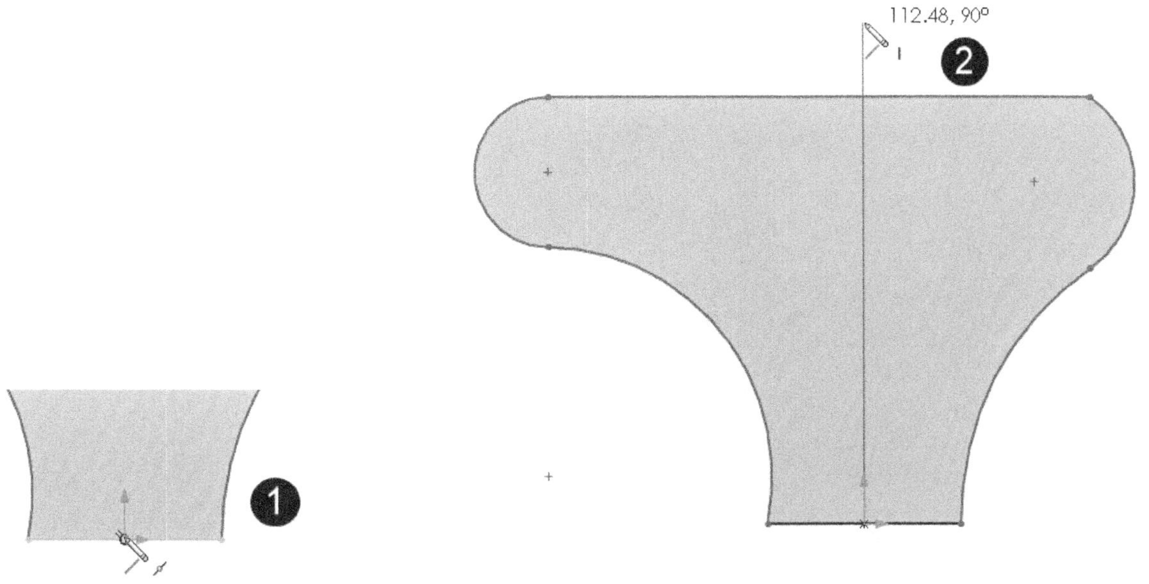

22. Click on the vertical line located at the center, and then select **Construction Geometry** from the context toolbar. The line is converted into a construction element.

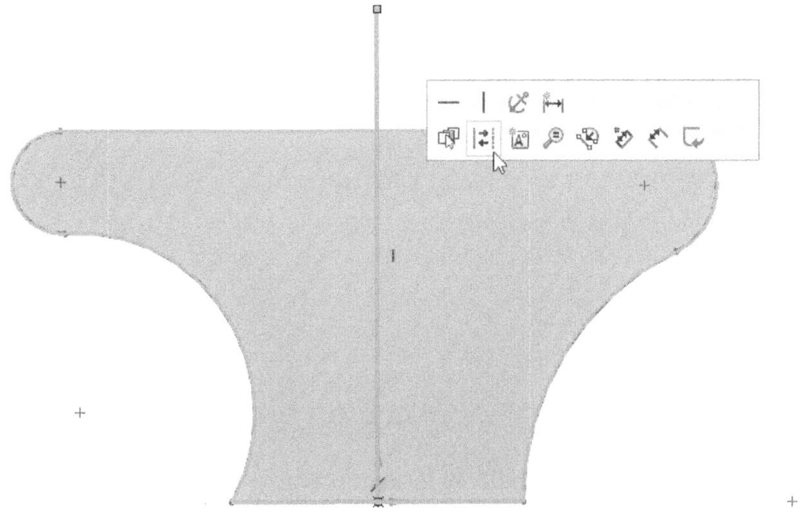

23. Activate the **Circle** command. Click on the right side of the construction line to specify the center point. Move the pointer outward and click to create the circle.
24. Likewise, create another circle on the left side of the construction line.
25. Press **Esc** to deselect the circles.

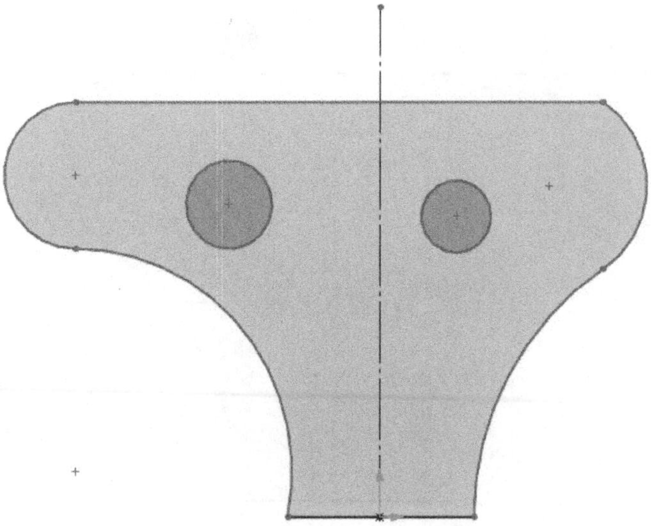

26. On the CommandManager, click **Sketch** tab > **Display/Delete Relations** > **Add Relation**. Click on the circle and the small arc on the right side.
27. Click the **Concentric** icon on the PropertyManager. The circle and arc are made concentric.

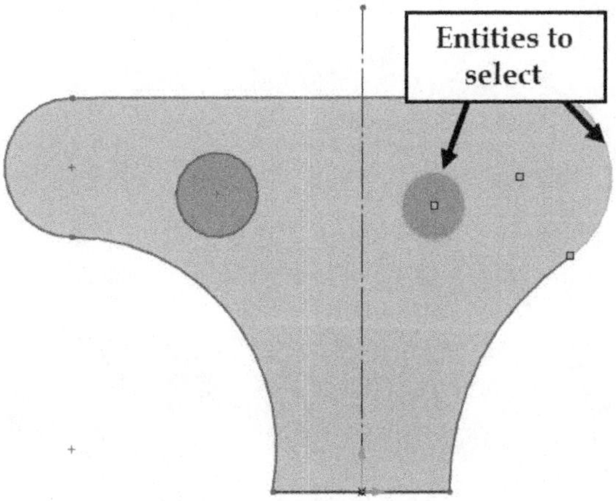

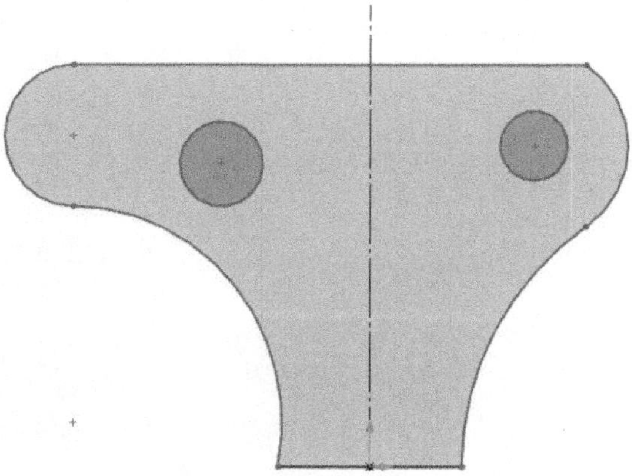

28. Likewise, make the other circle concentric to the small arc located on the construction line's left side.
29. On the CommandManager, click **Sketch** tab > **Display/Delete Relations** > **Add Relation**. Click on the construction line located at the center.
30. Click on the small arcs on both sides of the construction line.
31. On the PropertyManager, click the **Symmetric** icon. The arcs are made symmetric about the construction line. Click **OK** on the PropertyManager.

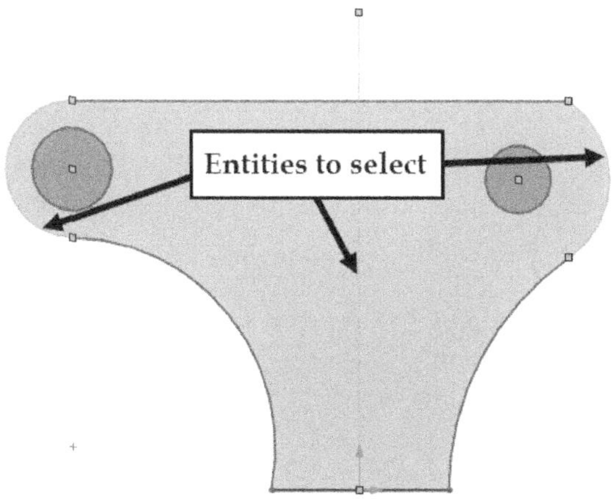

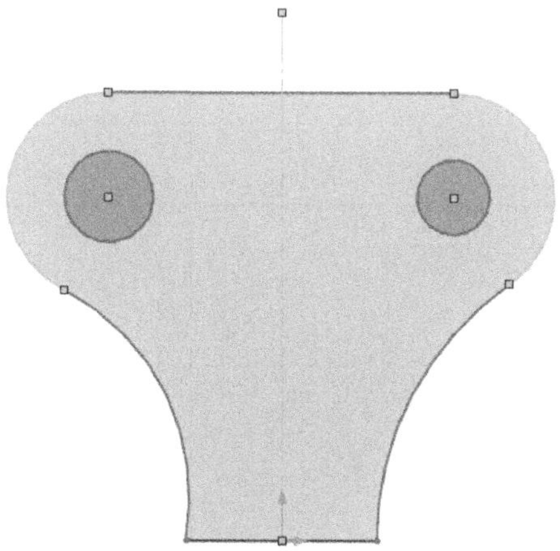

32. Likewise, make the large arcs and circles symmetric about the construction line.

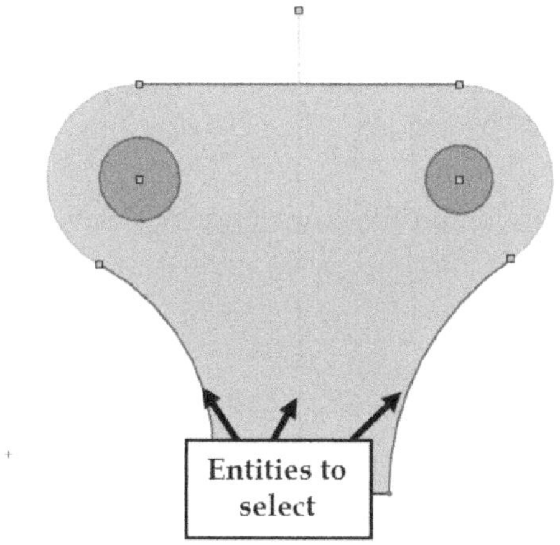

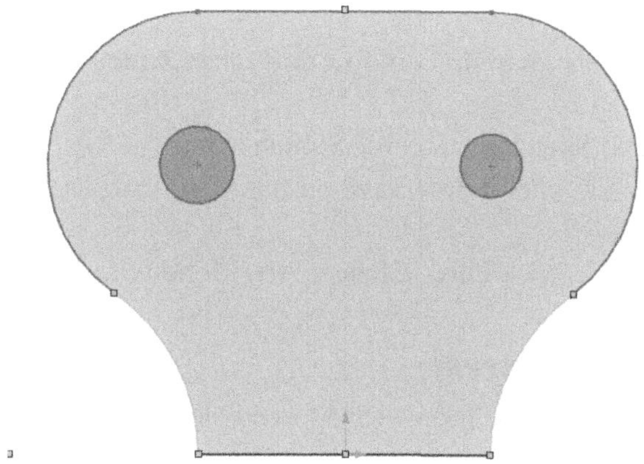

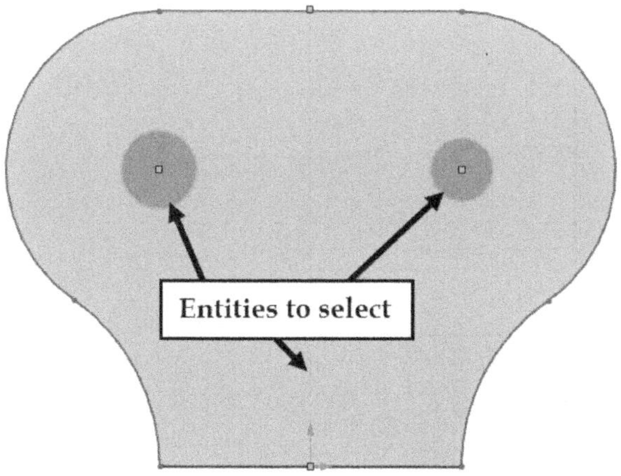

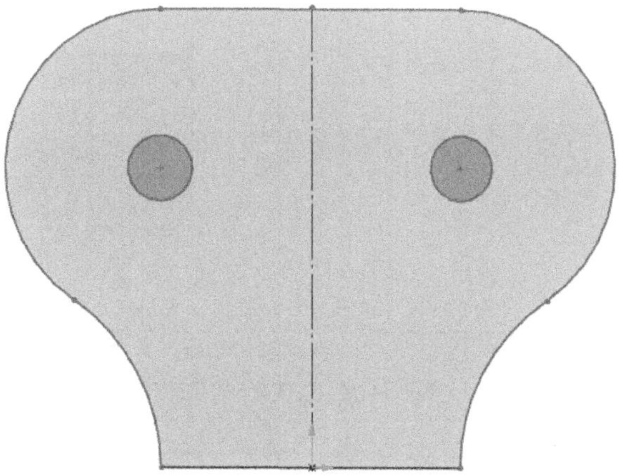

33. Activate the **Smart Dimension** command and apply dimensions to the sketch, as shown below.
34. Press Esc.

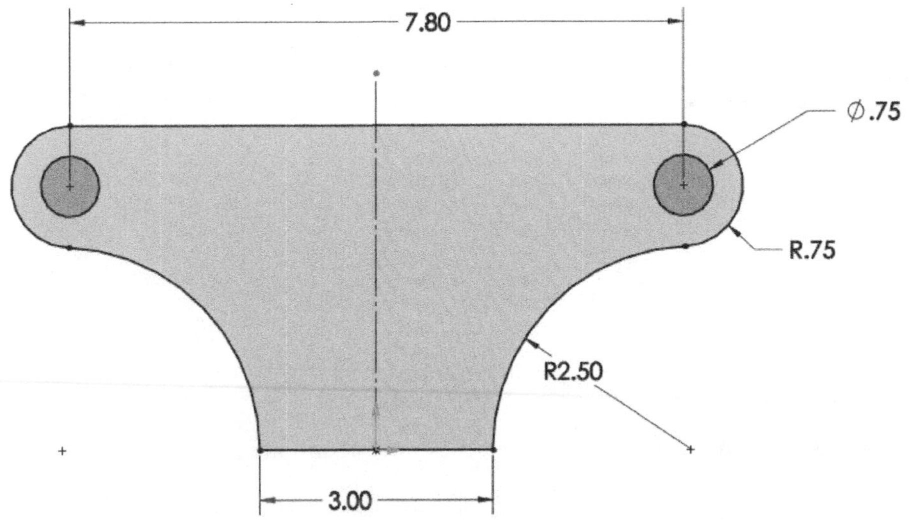

35. Right-click in the graphics window and select **Zoom/Pan/Rotate > Zoom to Fit**; the sketch fits the graphics window.
36. Click the **Exit Sketch** icon on the **Sketch** CommandManager.
37. To save the file, click **File > Save** on the Menu bar. Define the location and file name and click **Save**; the part file is saved.
38. To close the file, click **File > Close** on the Menu bar.

Example 3 (Millimetres)

In this example, you will draw the sketch shown below.

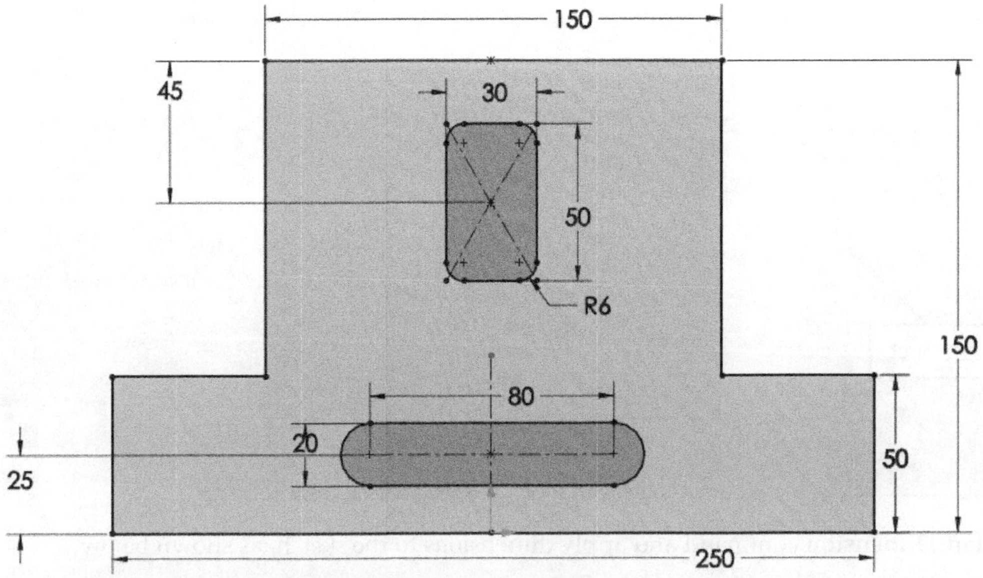

1. Click the **SOLIDWORKS 2024** icon on the desktop.
2. To start a new part file, click **File > New** on the Menu bar.
3. On the **New SOLIDWORKS Document** dialog, click the **Part** icon, and then click **OK**.

4. At the bottom right corner of the window, click the **Unit System** drop-down and select **MMGS (millimeter, gram, second)**.

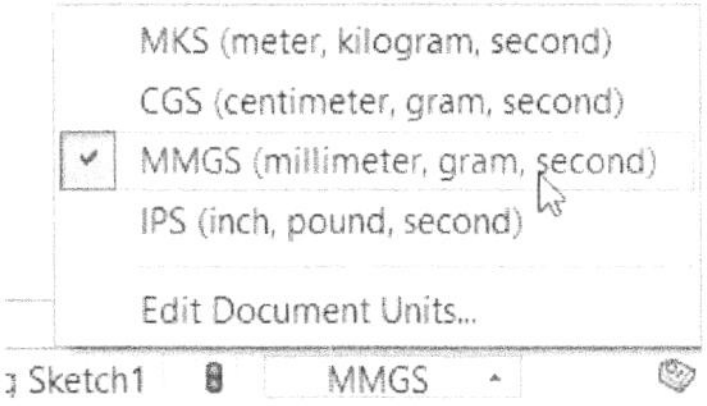

5. On the CommandManager, click **Sketch** tab > **Sketch**. Next, select the Front plane from the graphics window.
6. On the **Sketch** tab of the CommandManager, click **Line** drop-down > **Centerline**.
7. Select the origin and move the pointer vertically up. Click to create a vertical centerline.

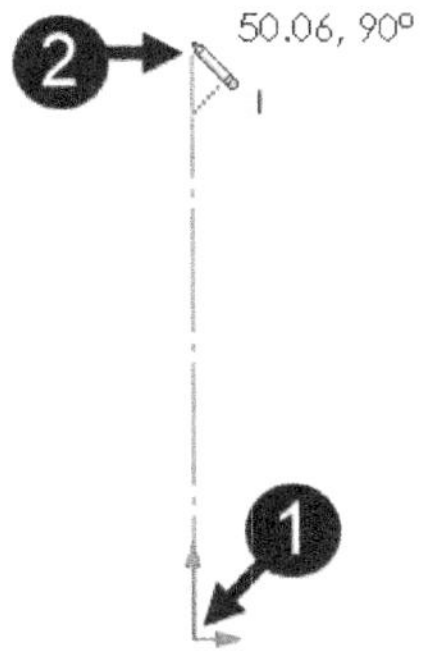

8. On the CommandManager, click **Sketch** tab > **Dynamic Mirror Entities**.

*Tip: If the **Dynamic Mirror Entities** icon is not displayed, use the **Customize** dialog to display it on the CommandManager. The procedure is already discussed in the **Dynamic Mirror Entities** section of this chapter.*

9. Select the vertical centerline to define the mirror line.
10. On the CommandManager, click **Sketch** tab > **Line**. Click on the origin point to define the first point.
11. Move the mouse pointer horizontally toward the right and click to define the second point.

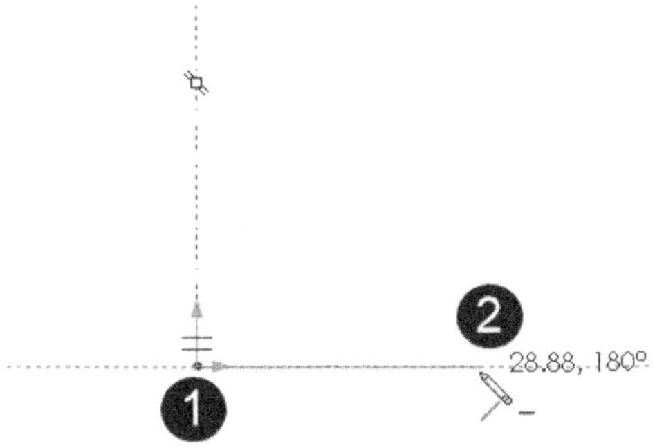

Notice that the line is mirrored on the other side of the mirror line, automatically.

12. Create a closed-loop by clicking points in the sequence shown below. Press Esc to deactivate the **Line** command.

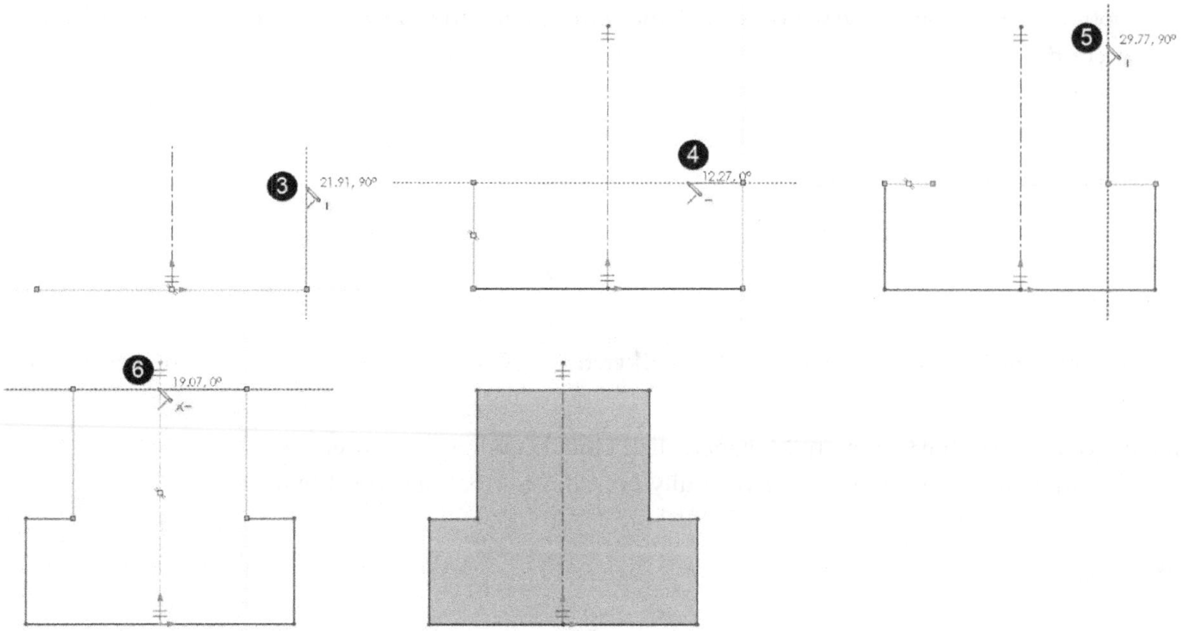

13. Click inside the sketch region to select all the entities. Notice the symmetric relations created on both sides of the mirror line.
14. Click and drag any one of the sketch entities on the right side and notice that its counterpart is also moved.

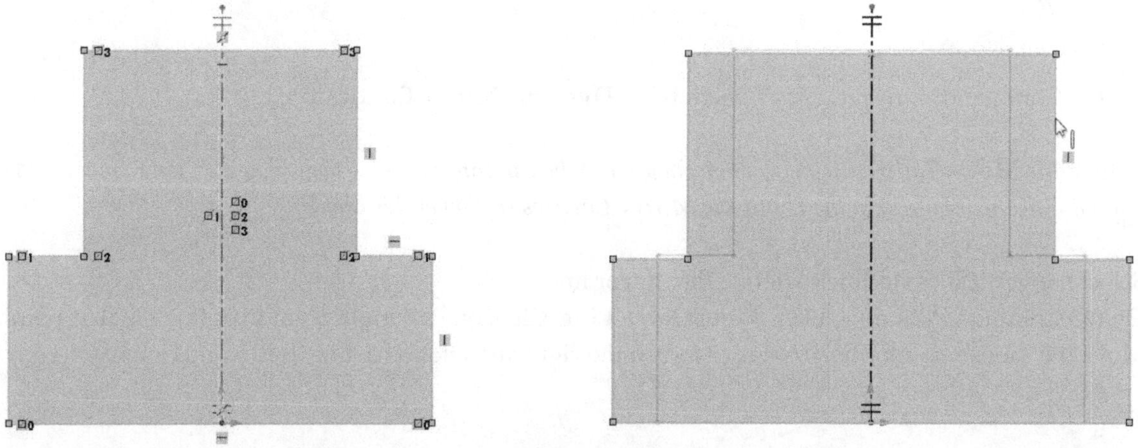

15. Deactivate the **Dynamic Mirror Entities** command on the CommandManager. The **Dynamic mirror** mode is turned OFF, and entities are not mirrored while you create them.
16. Click **Sketch** tab > **Rectangle** drop-down > **Center Rectangle** on the CommandManager. Click on the centerline to define the center of the rectangle.
17. Move the mouse pointer toward the top right and click to define the corner of the rectangle.

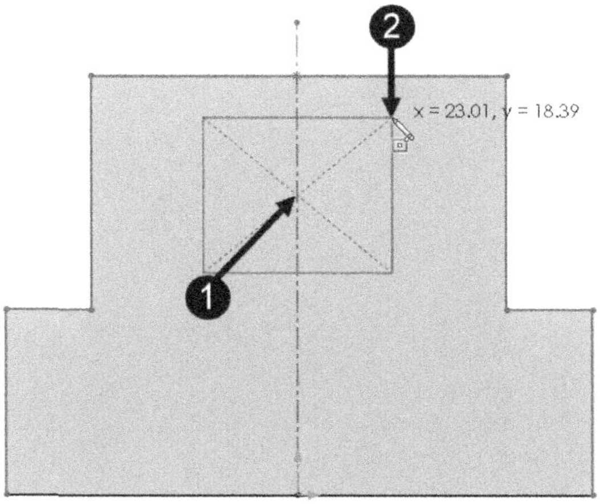

18. On the CommandManager, click **Sketch** tab > **Slot** drop-down > **Centerpoint Straight Slot**.
19. On the PropertyManager, check the **Add dimensions** option, and then select the **Center to Center** option.

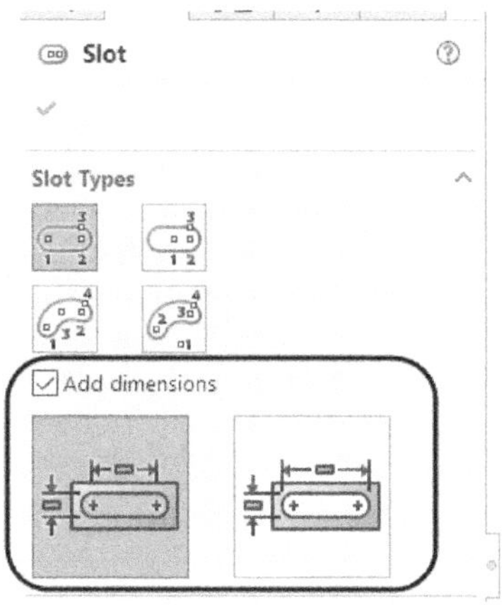

20. Click on the lower portion of the centerline to define the center point of the slot.
21. Move the pointer toward the right and click to define the endpoint of the slot.
22. Move the pointer outward and click to define the width of the slot. Press Esc.

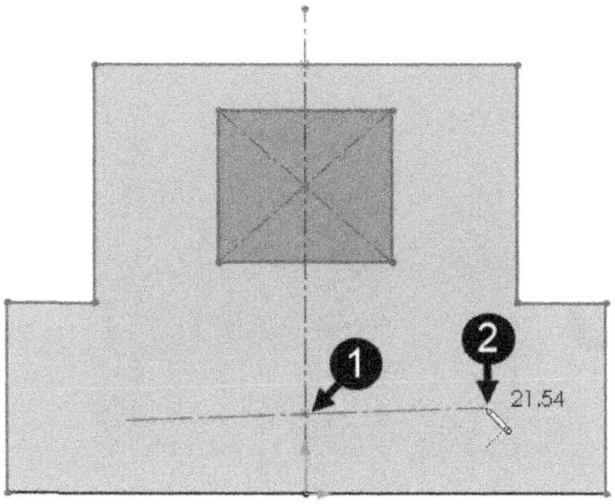

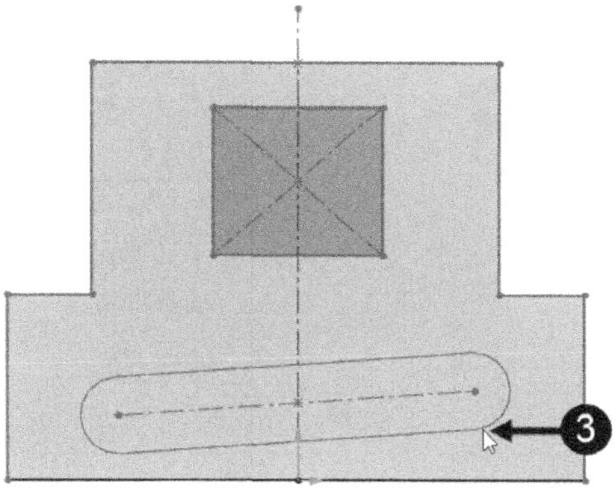

23. Select the centerline of the slot and click the **Make Horizontal** icon on the Context Toolbar.

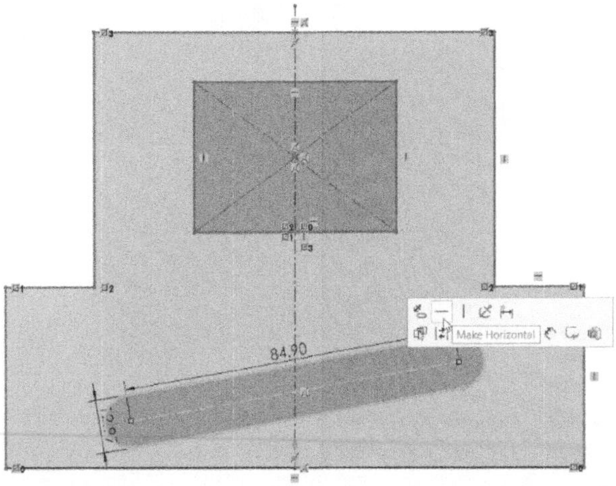

24. Click **Sketch tab > Sketch Fillet** on the CommandManager. On the PropertyManager, type-in **6** in the **Fillet Radius** box and press Enter.
25. Check the **Keep constrained corners** option on the PropertyManager to create constraints after filleting the corners.
26. Create fillets by clicking on the corners of the rectangle. Click **OK** on the PropertyManager.

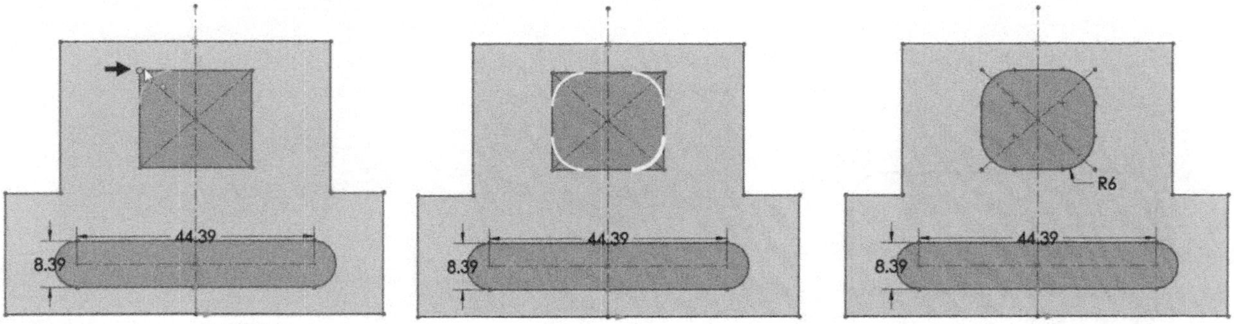

27. Activate the **Smart Dimension** command and apply dimensions in the sequence shown below.

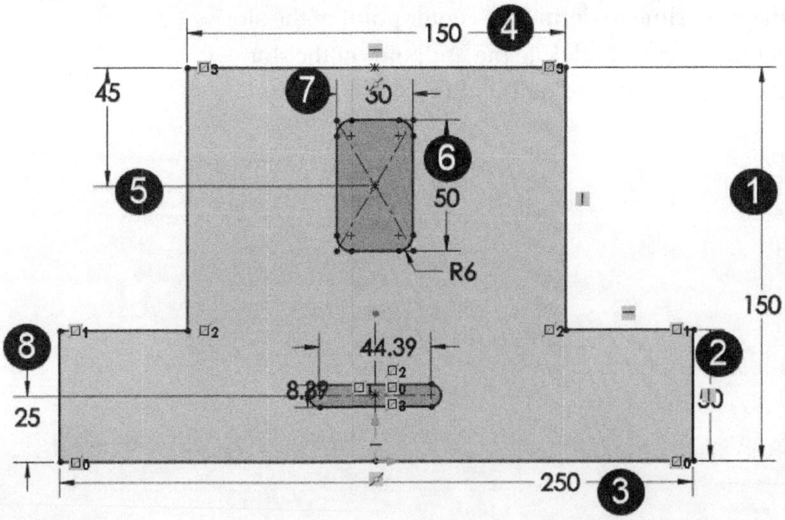

28. Double-click on the slot length dimension, and then type 80 in the **Modify** box. Click the green check to update the dimension.
29. Likewise, change the slot width to 20 mm.

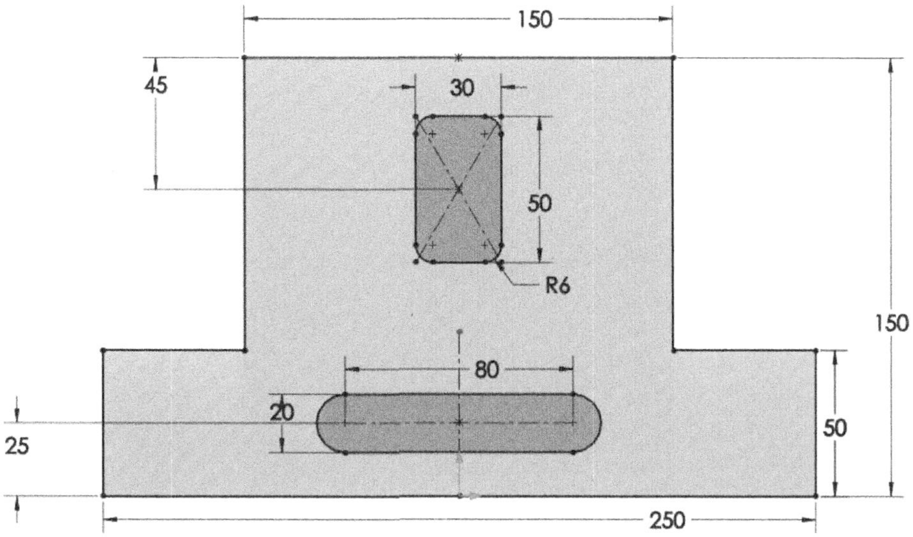

30. Click the **Exit Sketch** icon on the **Sketch** CommandManager.
31. Save and close the file.

Questions

1. What is the procedure to create sketches in SOLIDWORKS?
2. List any two sketch *Relations* in SOLIDWORKS.
3. Which option orients the sketch normal to the screen?
4. What is the procedure to create sketches rapidly?
5. Which command allows you to apply dimensions and relations to a sketch automatically?
6. Describe the method to create an ellipse.
7. How do you define the shape and size of a sketch?
8. How do you create a tangent arc using the **Line** command?
9. Which command is used to apply multiple types of dimensions to a sketch?
10. List any two commands to create circles?

Exercises

Exercise 1

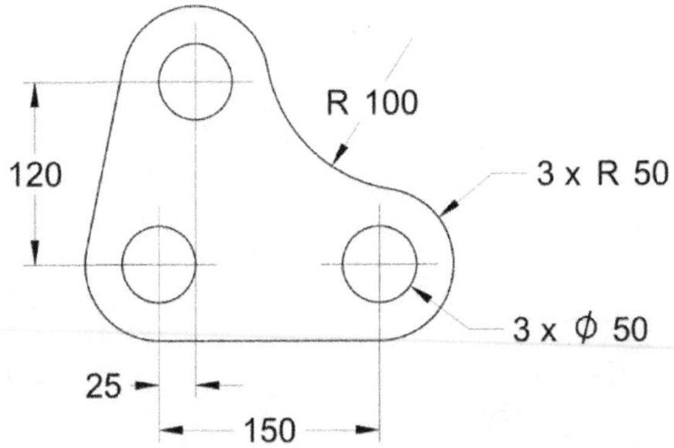

Exercise 2

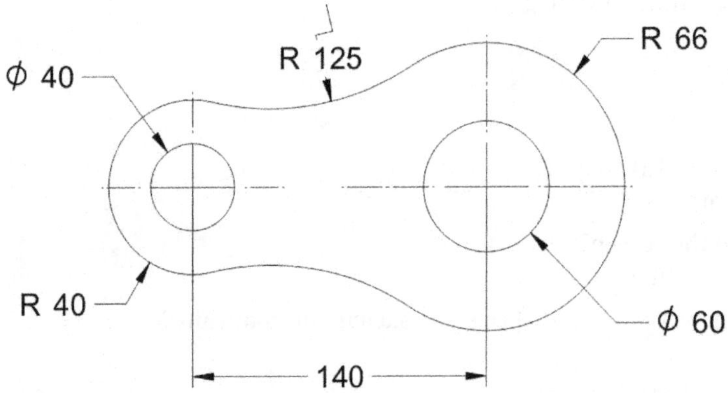

Exercise 3

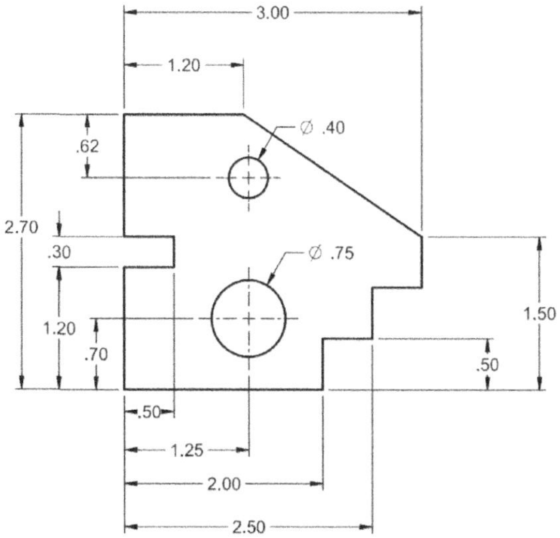

Chapter 3: Extrude and Revolve Features

This chapter covers the methods and commands to create extruded and revolved features.

The topics covered in this chapter are:

- *Constructing Extruded features*
- *Constructing Revolved features*
- *Creating Reference Planes*
- *Creating reference points*
- *Creating reference axes*
- *Additional option while creating Extruded and Revolved features*

Extruded Boss/Base

Extrusion is the process of taking a two-dimensional profile and converting it into 3D by giving it some thickness. A simple example of this would be taking a circle and converting it into a cylinder. To create an Extrude feature, first, draw a sketch, and then activate the **Extruded Boss/Base** command (on the CommandManager, click **Features** tab > **Extruded Boss/Base).** Click on the sketch profile; notice the preview of the extruded boss with the default depth. Type-in a new value in the **Depth** box under the **Direction 1** section on the **Boss-Extrude** PropertyManager. You can also click and drag the arrow that appears on the extrude preview. Next, release the arrow at the required value on the scale that appears on the extrude preview.

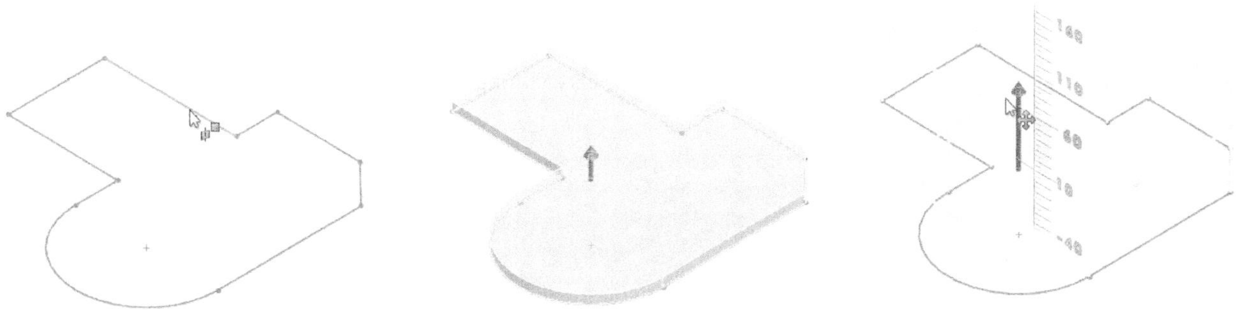

On the PropertyManager, check the **Direction 2** section to extrude the sketch in two opposite directions. Next, you need to define the depth along the second direction. You can type in a value in the **Depth** box under the **Direction 2** section or drag the arrow that appears in the second direction.

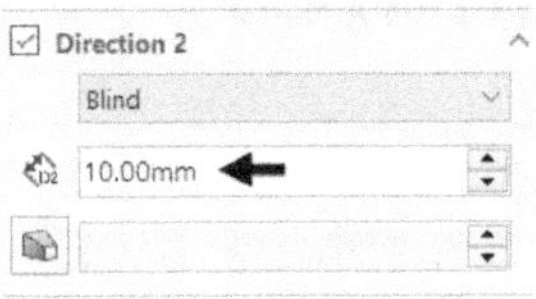

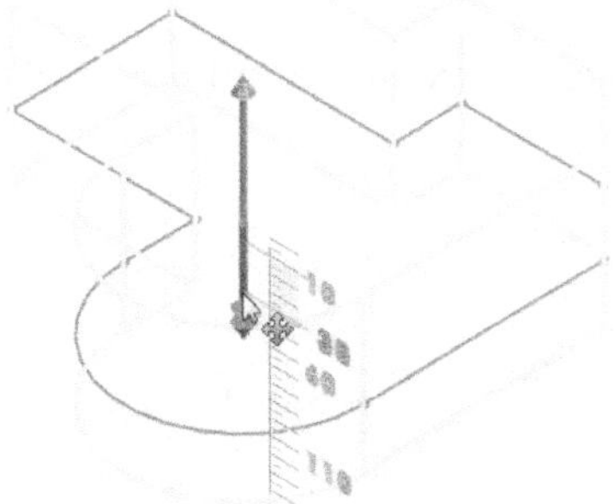

If you want to reverse the sides, click the **Reverse Direction** icon in the **Direction 1** section of the PropertyManager.

Original Direction

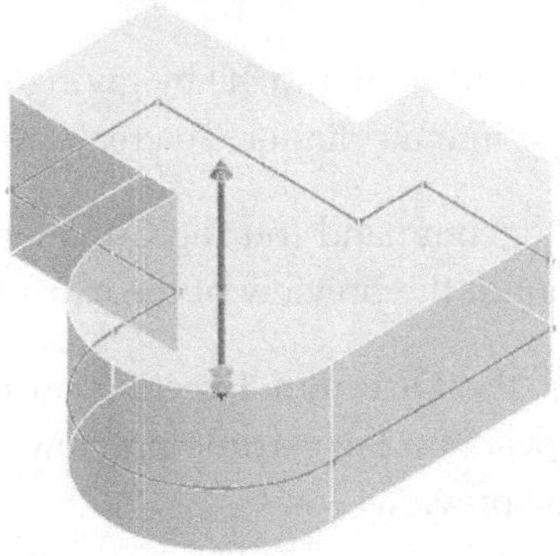

Reversed Direction

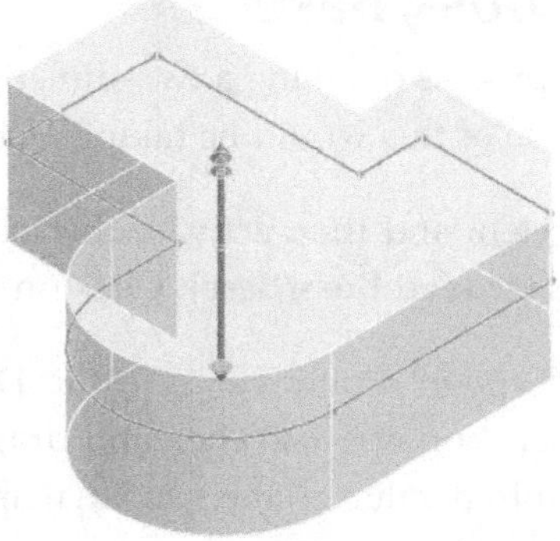

If you want to extrude the sketch with equal depth on both sides, select **Mid Plane** from the **End Condition** drop-down in the **Direction 1** section. Click **OK** on the PropertyManager to complete the **Extrude Boss/Base** feature.

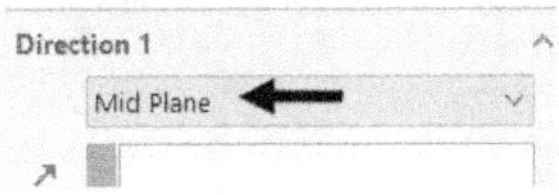

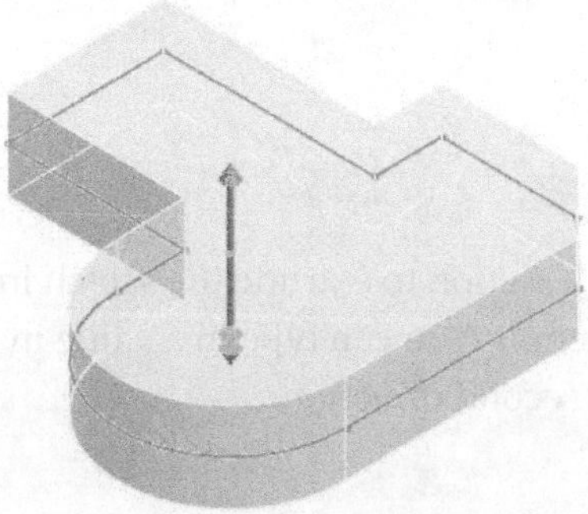

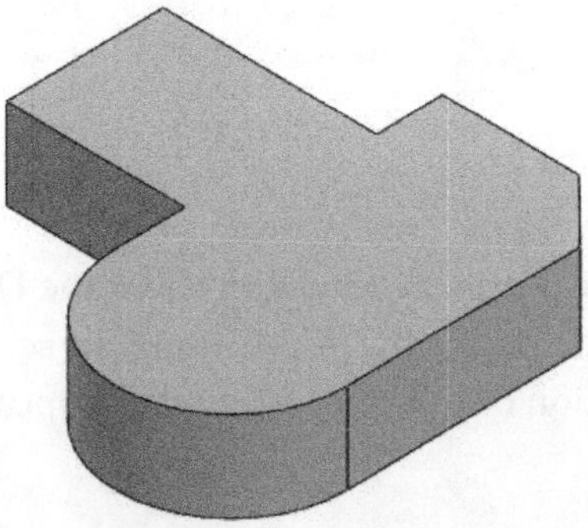

While creating an *Extrude Boss/Base* feature, SOLIDWORKS adds material in the direction normal to the sketch. If you want to manually define the direction in which the material will be added, then click in the **Direction of Extrusion** box on the **Extrude-Boss** PropertyManager, and then select a line.

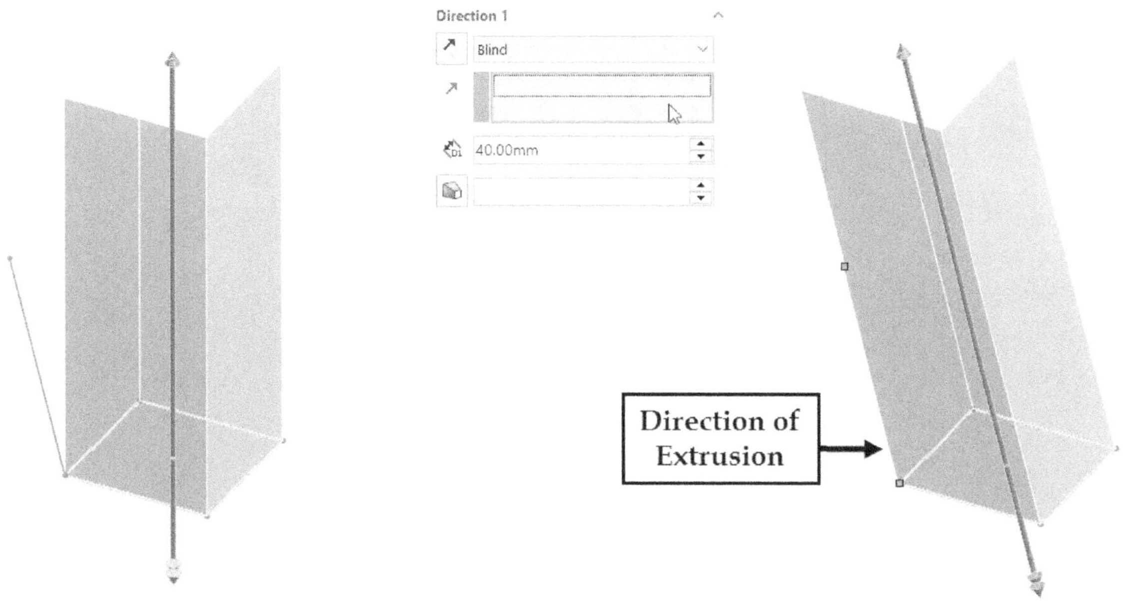

Revolved Boss/Base

Revolving is the process of taking a two-dimensional profile and revolving it about a centerline to create a 3D geometry (axially symmetric shapes). While creating a sketch for the *Revolve* feature, it is important to think about the cross-sectional shape that will define the 3D geometry once it is revolved about an axis. For instance, the following geometry has a hole in the center. You can create it with a separate *Cut* or *Hole* feature. However, to make that hole a part of the *Revolve* feature, you need to sketch the revolution axis to leave a space between the profile and the axis.

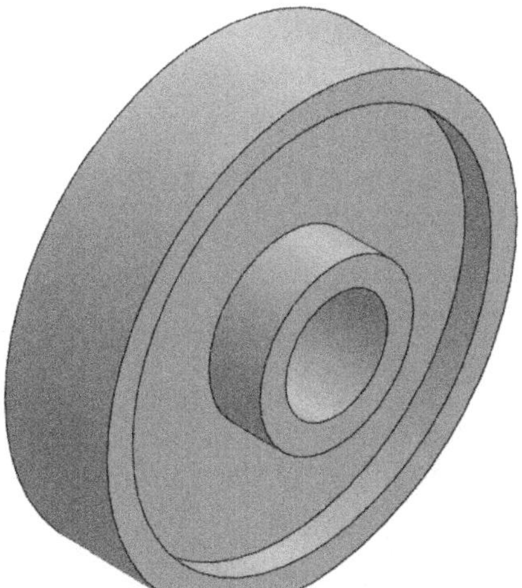

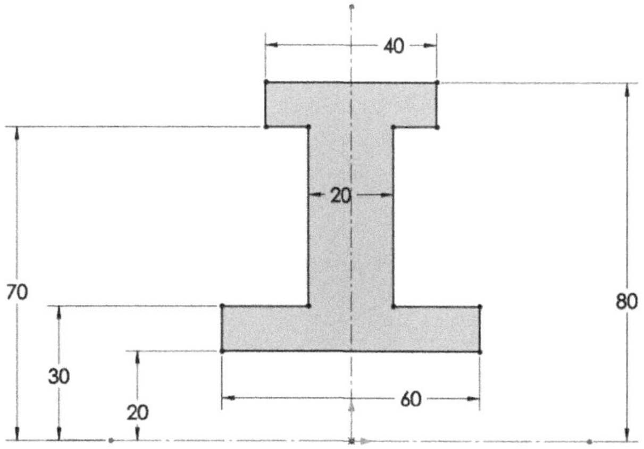

To create a *Revolve* feature, first, sketch a revolve profile along with the centerline. Next, exit the sketch and click **Features** tab > **Revolved Boss/Base** on the CommandManager. The sketch will be revolved by a full 360 degrees.

Suppose you want to enter a different angle of revolution, type in a value in the **Direction 1 Angle** box on the PropertyManager. Click **OK** to complete the *Revolve Boss/Base* feature.

Revolved at 360 degrees

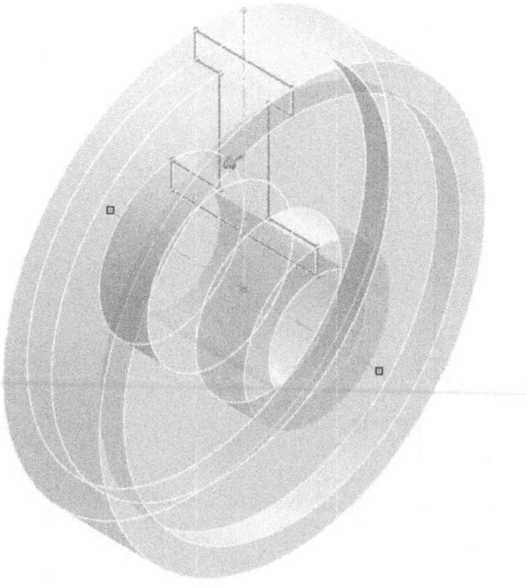

Revolved at 270 degrees

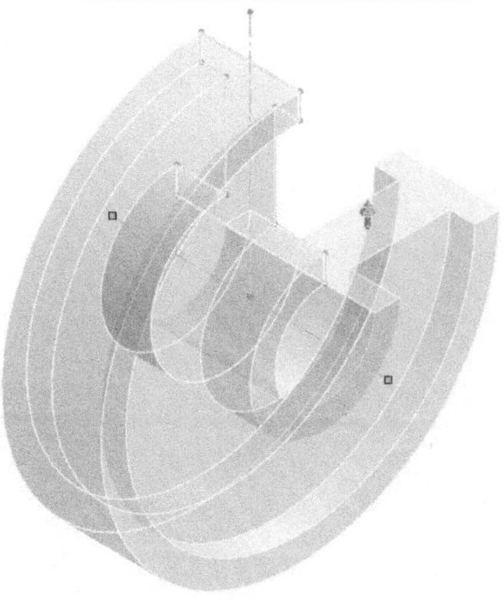

Convert Entities

This command projects the edges of a 3D geometry onto a sketch plane. Activate the Sketch Environment by selecting a plane or a model face. On the CommandManager, click **Sketch** tab > **Convert Entities** (or) click **Tools > Sketch Tools > Convert Entities** on the Menu. Click on the edges of the model geometry to project them onto the sketch plane. Click **OK** on the **Convert Entities** PropertyManager. The projected element will be black and fully constrained. Complete and exit the sketch.

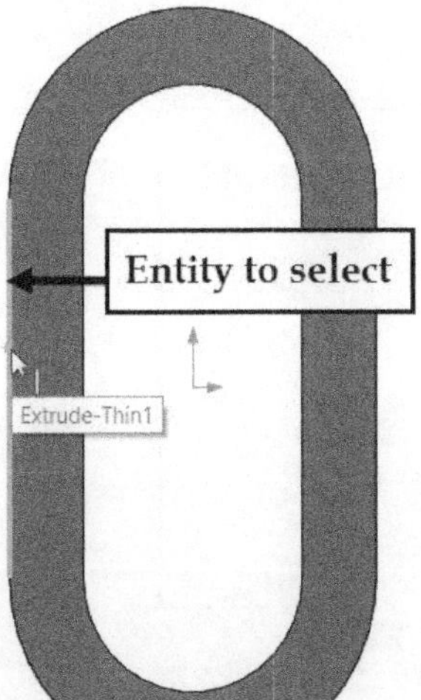

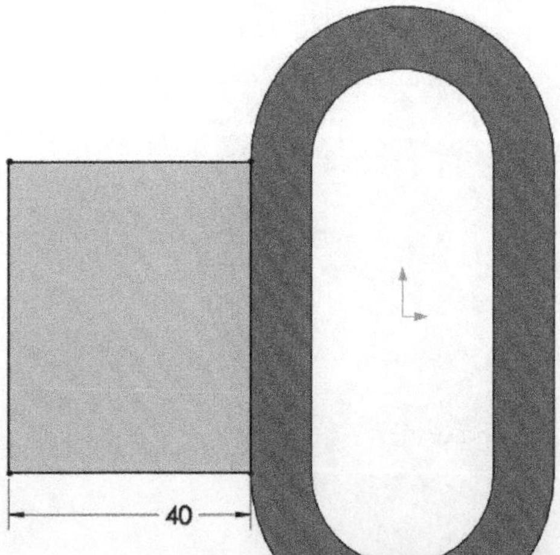

The Extruded Cut command

This command removes material from the geometry by extruding a sketch. It functions on the same lines as the **Extrude Boss/Base** command. Draw a sketch on a plane or a model face. On the **Features** CommandManager, click the **Extruded Cut** icon (or) click **Insert > Cut > Extrude** on the Menu. Select the sketch. On the **Cut-Extrude** PropertyManager, type in a value in the **Depth** box; the extruded cut's preview appears in the direction normal to the sketch.

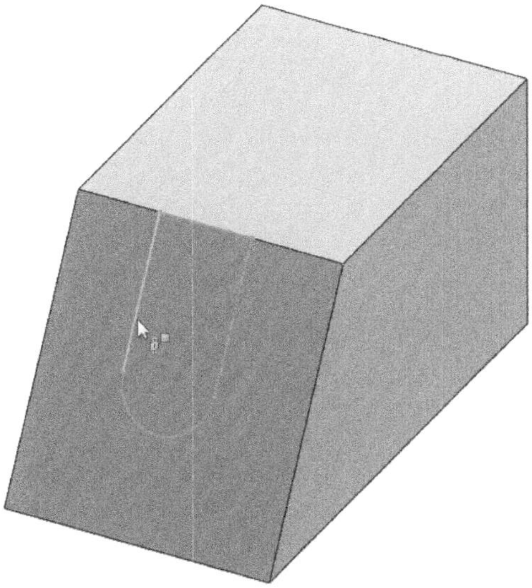

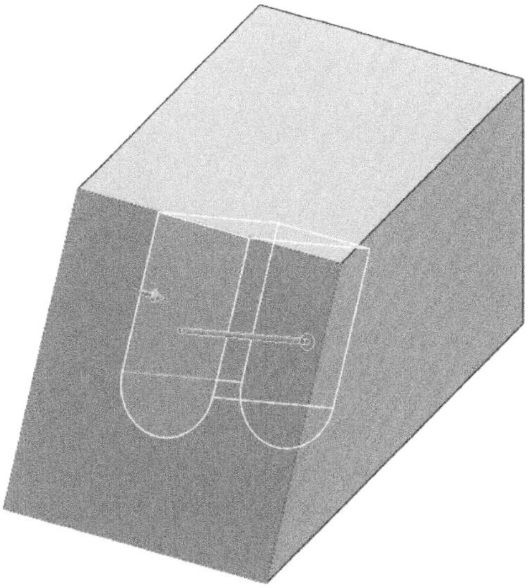

If you want to define the direction of material removal, click in the **Direction of Extrusion** selection box and select an edge or line to define the direction. Click **OK** to complete the **Extruded Cut** feature.

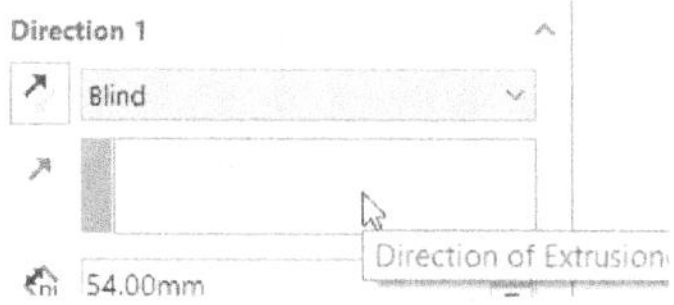

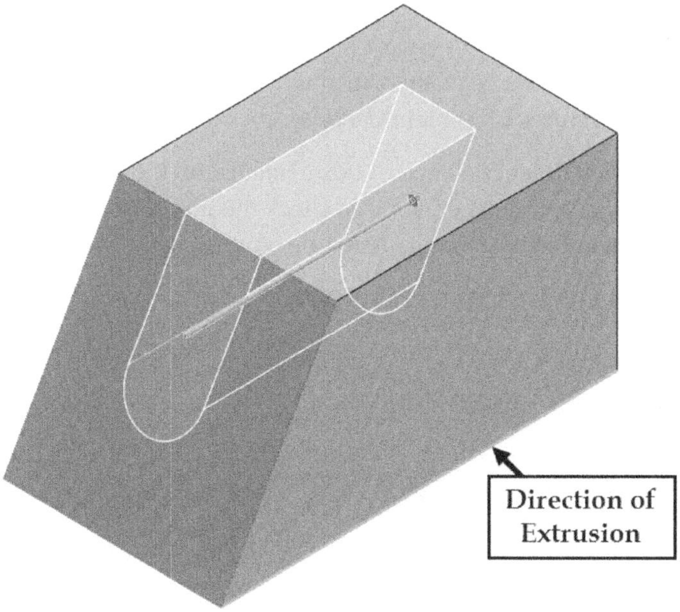

The Revolved Cut command

This command removes material from the geometry by revolving a sketch about an axis. It functions in a way similar to the **Revolve Boss/Base** command. Draw a sketch on a plane or a model face. Also, draw a centerline using the Centerline command. On the Features CommandManager, click the **Revolved Cut** icon (or) **click Insert > Cut > Revolve** on the Menu. Select the sketch. If you have created the centerline, then revolved cut will be created automatically. You can also choose to reverse the cutting side for the revolved cut feature, similar to the cut-extrude operation. This keeps the inside of the sketch and removes the region outside it. In the Cut-Revolve PropertyManager, under the **Direction 1** section, check the **Flip side to cut** option. Click **OK** to complete the revolved cut feature.

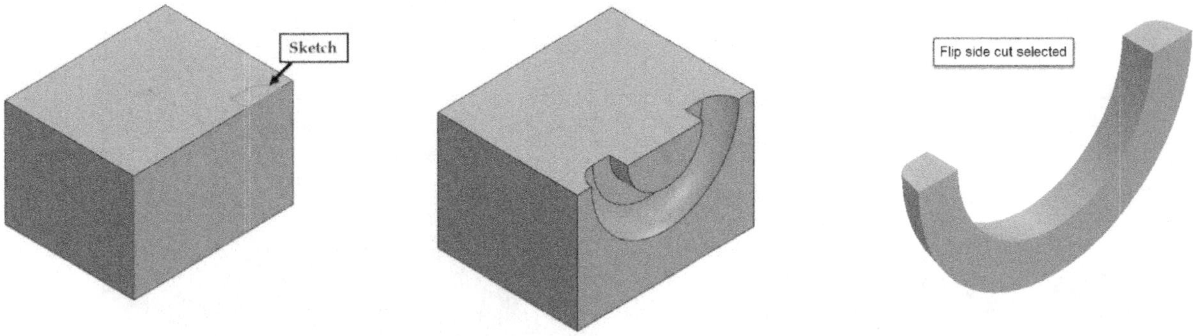

The Plane command

Each time you start a new part file, SOLIDWORKS automatically creates default Reference Planes. Planes are a specific type of element in SOLIDWORKS, known as Reference Elements. These features act as supports to your 3D geometry. In addition to the default Reference features, you can create your additional planes. Until now, you have learned to create sketches on any of the default reference planes (Front, Right, and Top planes). If you want to create sketches and geometry at locations other than default reference planes, you can manually create new reference planes. You can do this by using the **Plane** command.

Offset plane

This method creates a reference plane, which will be parallel to a face or another plane. Activate the **Plane** command (On the **Features** CommandManager, click **Reference Geometry > Plane**). Select a flat face, and then type in a value in the **Offset Distance** box available on the **Plane** PropertyManager to define the offset distance. On the PropertyManager, you can check the **Flip Offset** option to flip the plane to another side of the model face. If you want to create more than one offset plane, then type in a value in the **Number of planes to create** box. Click **OK** to create the offset plane.

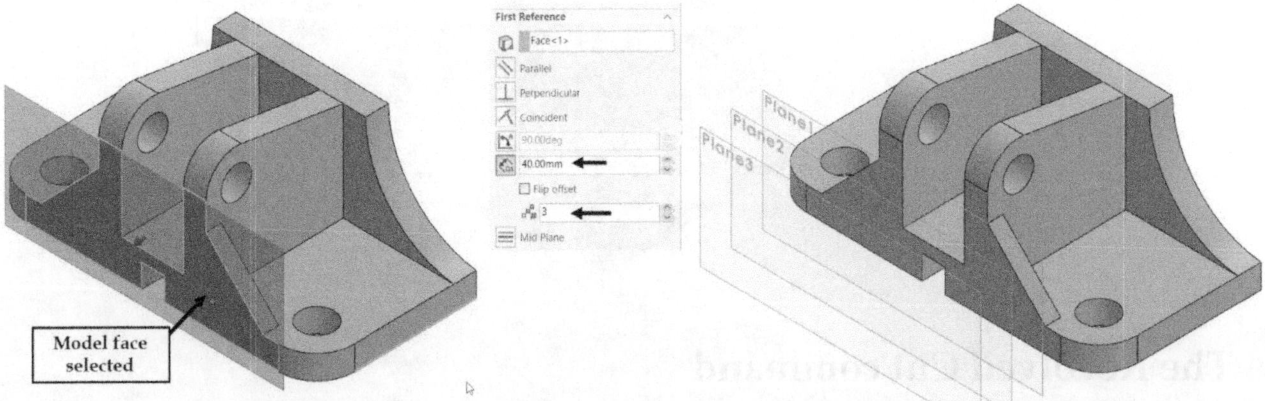

Parallel Plane

This method creates a plane parallel to a flat face at a selected point or edge. Activate the **Plane** command and then select a flat face. On the **Plane** PropertyManager, select the **Parallel** icon. Select a point to define the parallel plane location.

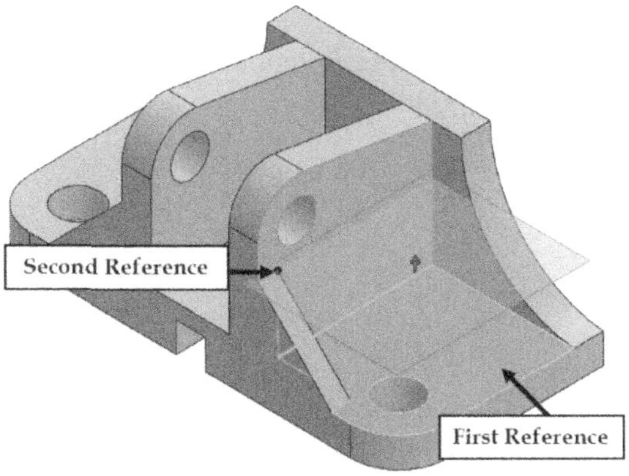

Angle plane

This method creates a plane, which will be positioned at an angle to a face or plane. Activate the **Plane** command.

Select a flat face or plane to define the reference. Select **At angle** from the **Plane** PropertyManager. Click on the edge of the part geometry to define the rotation axis. Type-in a value in the **Angle** box and press the Enter key.

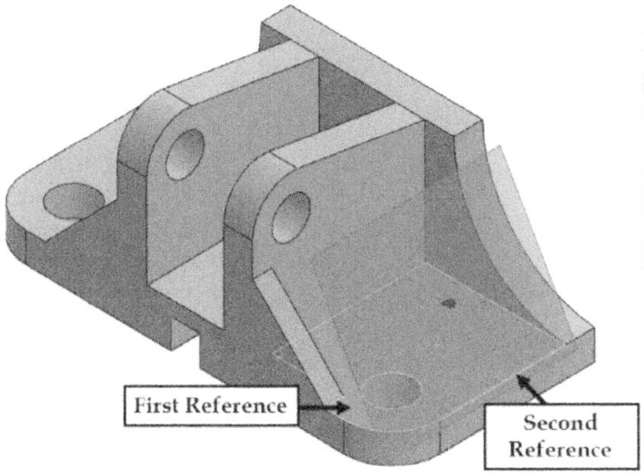

Perpendicular plane

This method allows you to create a plane, which will be positioned perpendicular to a face or plane. Activate the **Plane** command. Select a flat face or plane to define the first reference. On the **Plane** PropertyManager, click the **Perpendicular** icon, and then select an edge; the plane is created perpendicular to the selected face passing through the edge. Click **OK**.

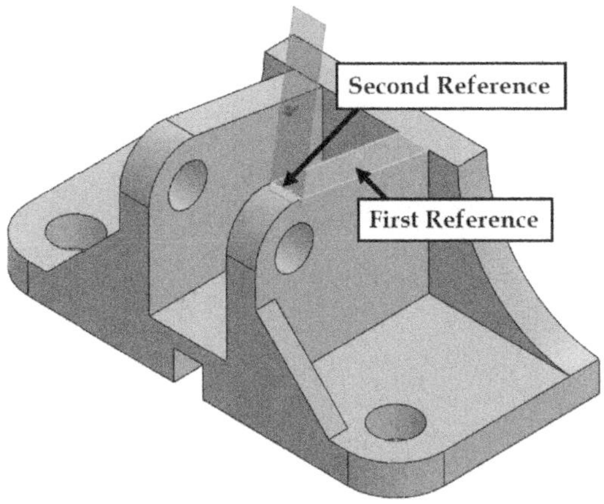

Through three points

This method allows you to create a plane by selecting three points. Activate the **Plane** command. Select three points from the model geometry. Click **OK** to create a plane passing through the points.

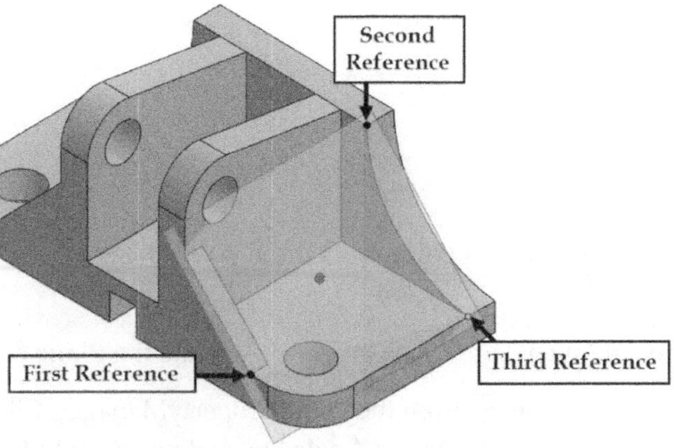

Through two lines

This method allows you to create a plane by selecting two lines or edges parallel to each other. Activate the **Plane** command. Select two lines or edges from the model geometry. Click **OK**.

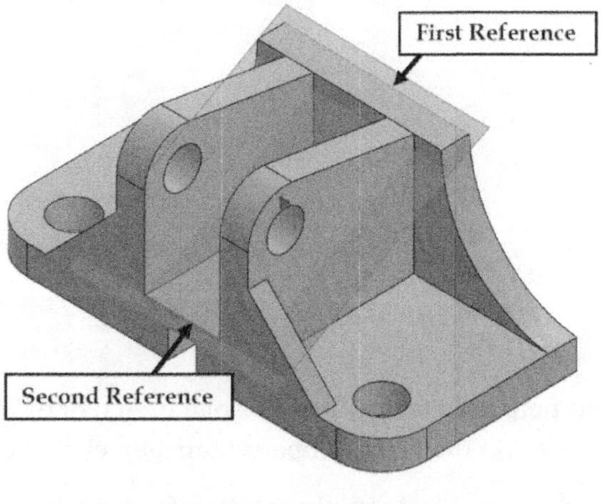

Through point and line

This method allows you to create a plane by selecting a point and line. Activate the **Plane** command. Select a point and line. Click **OK**.

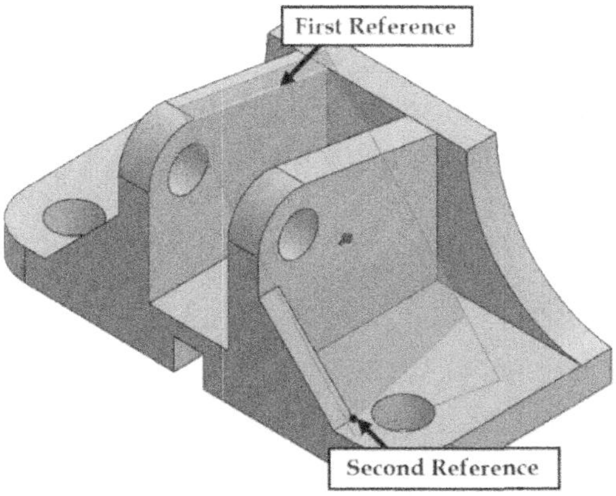

Normal to curve

This method creates a reference plane, which will be normal (perpendicular) to a line, curve, or edge. Activate the **Plane** command and select an edge, line, curve, arc, or circle. Click on a point to define the location of the plane. Click **OK** on the PropertyManager.

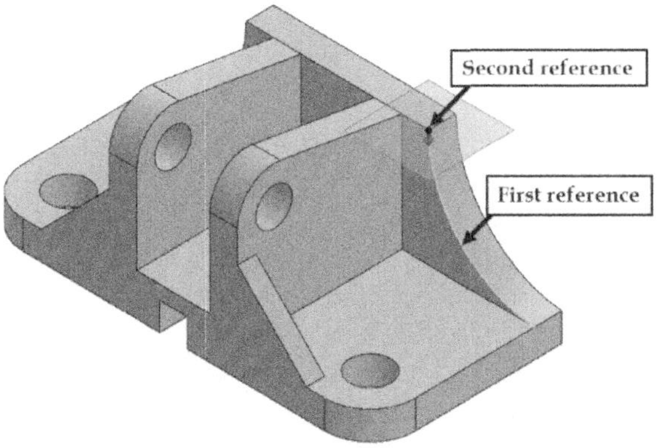

Tangent to surface

This method creates a plane tangent to a curved face. Activate the **Plane** command and select a curved face. Click on a point. A plane tangent to the selected face appears. Click **OK** on the PropertyManager.

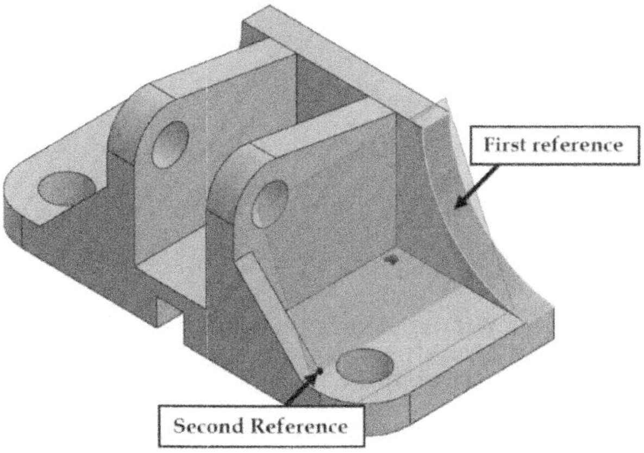

Mid Plane

This method creates a plane, which lies at the midpoint between two selected faces. You can also create a plane passing through the intersection point of the two selected planes or faces. Activate the **Plane** command and select two faces of the model geometry which are parallel to each other. Click **OK** to create the midplane.

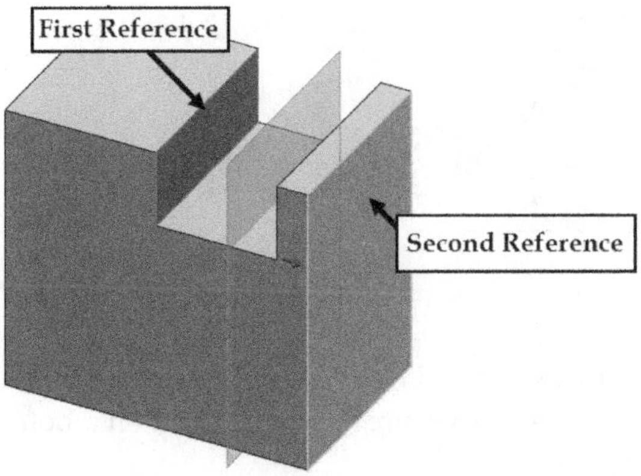

Activate the **Plane** command and select two intersecting faces or planes from the graphics window; the midplane preview appears. On the PropertyManager, check the **Flip Offset** option; the plane orientation changes. Click **OK** to create the midplane.

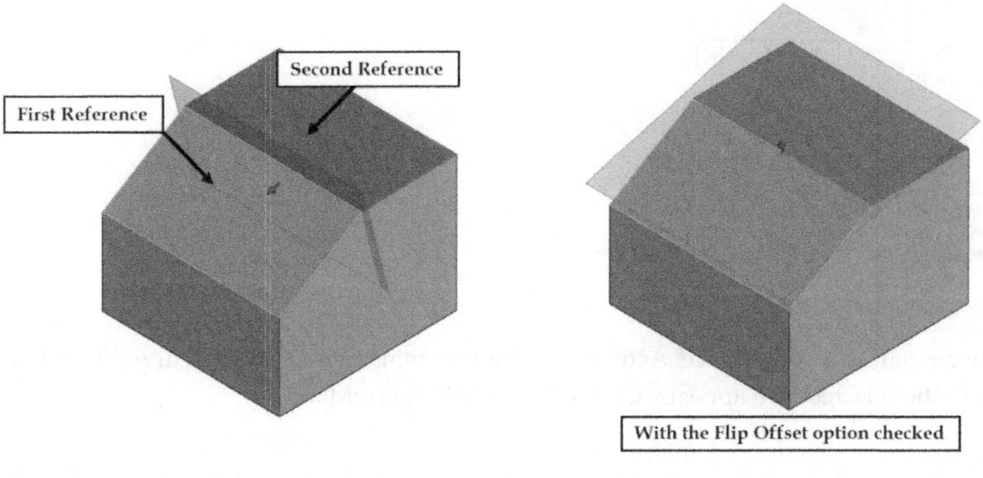

Point

The **Point** command creates points in the 3D space using six different methods. The following sections explain you to create points using these methods.

Along curve

This method creates a point on a curve or edge. Activate the **Point** command and click on a curve or edge. On the **Point** PropertyManager, click the **Along curve distance or multiple reference point** icon. Enter a value in the Number of reference points box. Next, select the **Distance** option and enter the distance of the first point from the start point of the curve or edge. The same distance value also defines the spacing between the points.

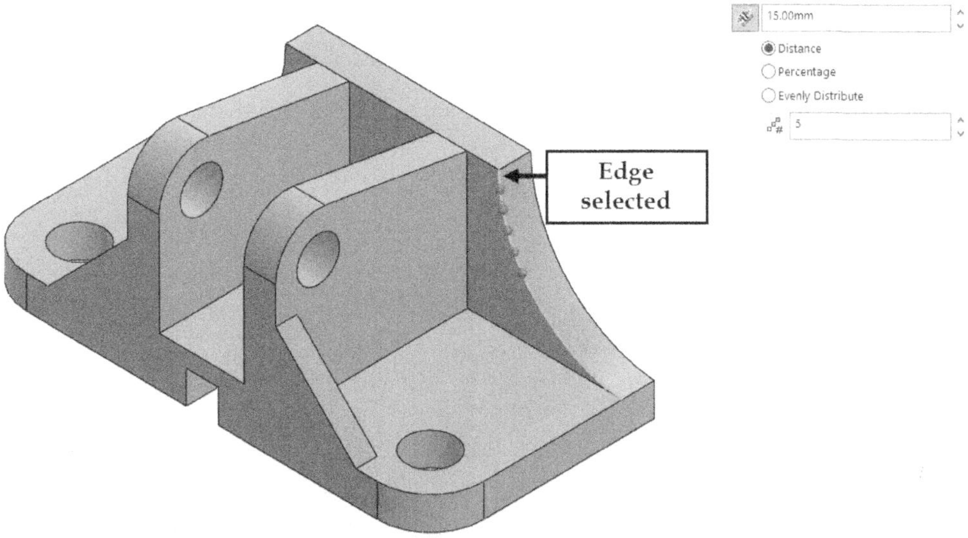

Select the **Percentage** option and enter the percentage value to define the distance between the points. For example, if you enter 15 as the percentage value, the distance between the points will be 15 percent of the edge's total length.

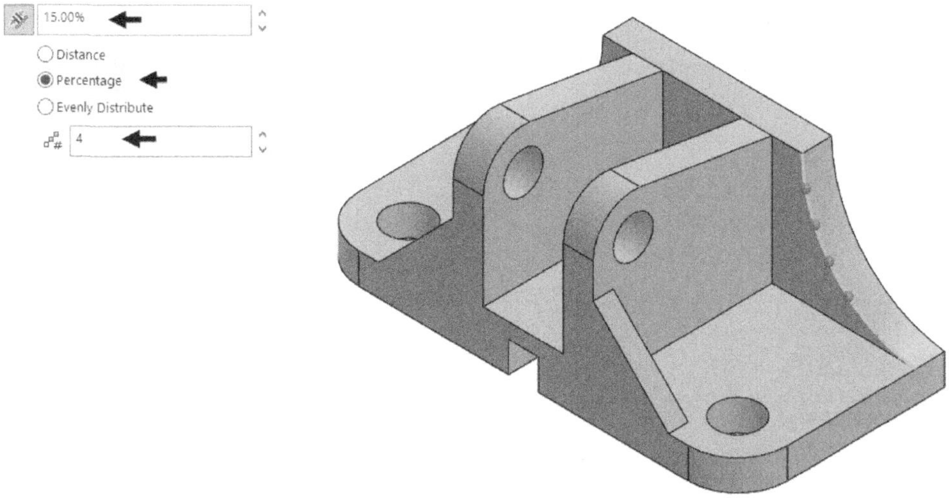

Select the **Evenly Distribute** option and enter the number of points. The points are evenly distributed on the selected curve or edge.

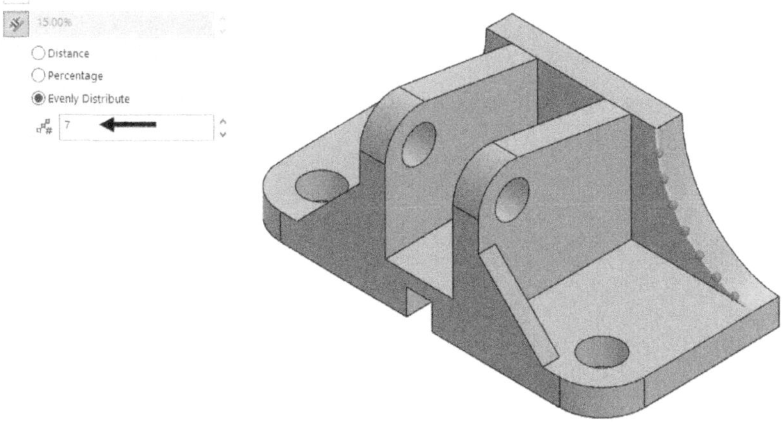

Center of face

This method creates a point at the center of the selected planar or curved face. Activate the **Point** command and click on a face or plane. On the PropertyManager, click the **Center of Face** icon, and then click **OK**.

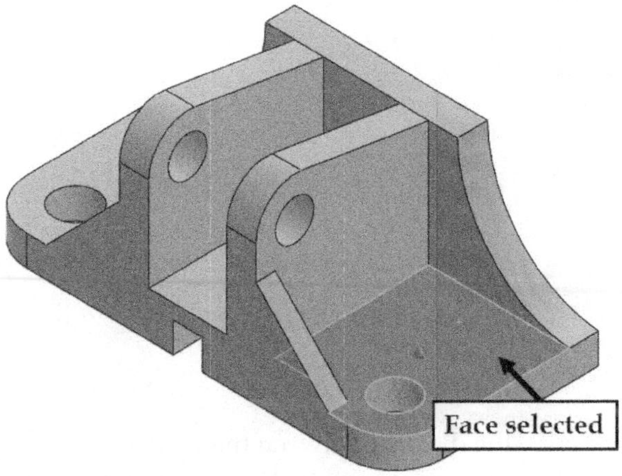

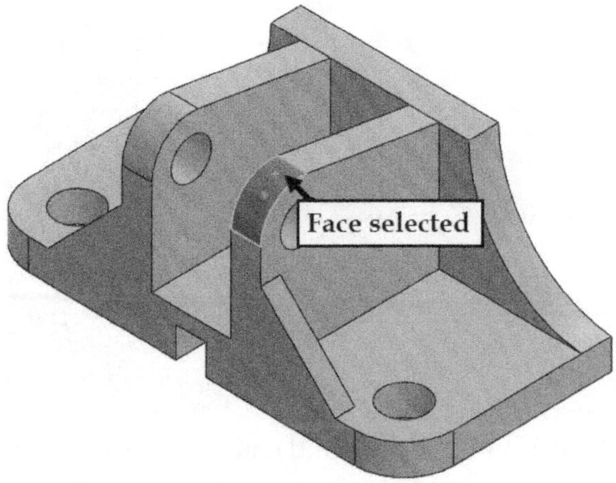

Arc Center

This method creates a point at the center of an arc or circle. Activate the **Point** command (on the CommandManager, click **Reference Geometry** drop-down > **Point**). Click on an arc or a circular entity, and then click **OK**.

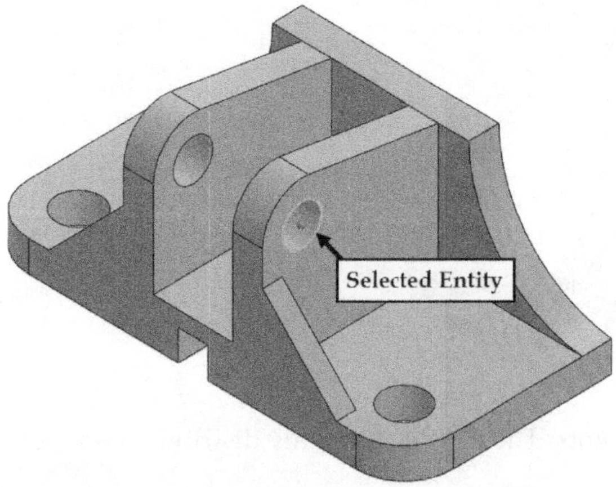

Projection

This method creates a point by projecting a point or vertex onto a surface or plane. Activate the **Point** command and select the **Projection** icon on the PropertyManager. Select the face on which the point is to be created. Click on the point or vertex to be projected, and then click **OK**.

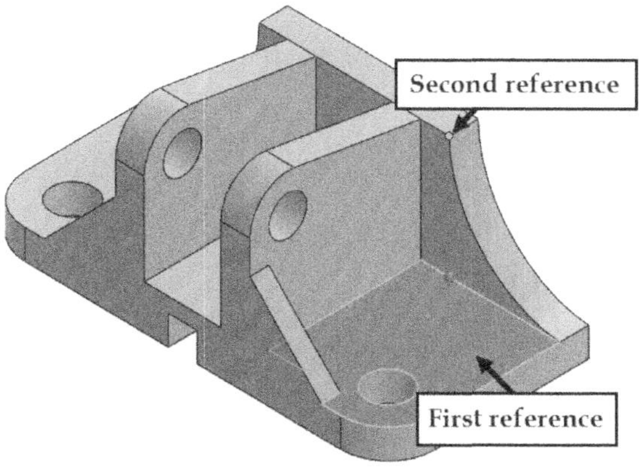

Intersection

This method creates a point at the intersection of two lines, curves, or edges. Activate the **Point** command and select two intersecting lines, curves, or edges. Click **OK** on the PropertyManager; the reference point is created at the intersection.

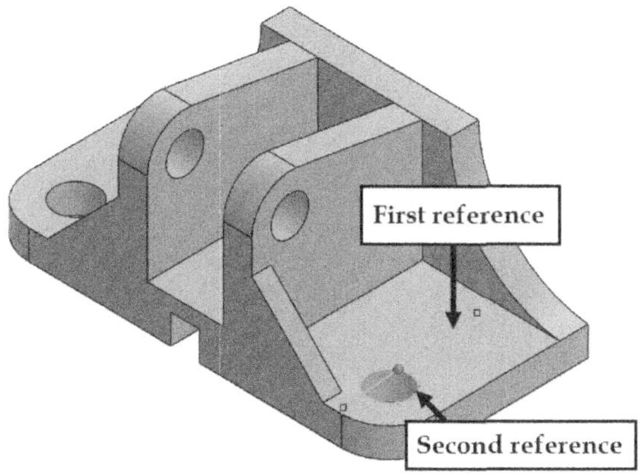

On point

This method creates a point on the selected sketch points such as center points, endpoints, or points. Activate the Point command and select the **On point** icon on the PropertyManager. Next, select a sketch point and click **OK**.

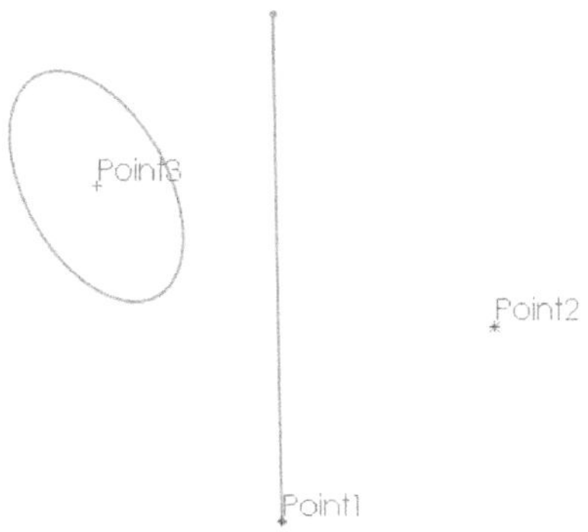

Axis

The **Axis** command (on the **Features** tab, click **Reference Geometry** drop-down > **Axis**) creates an axis in the 3D space. The methods to create axes using this command are shown next.

Two Points/Vertices

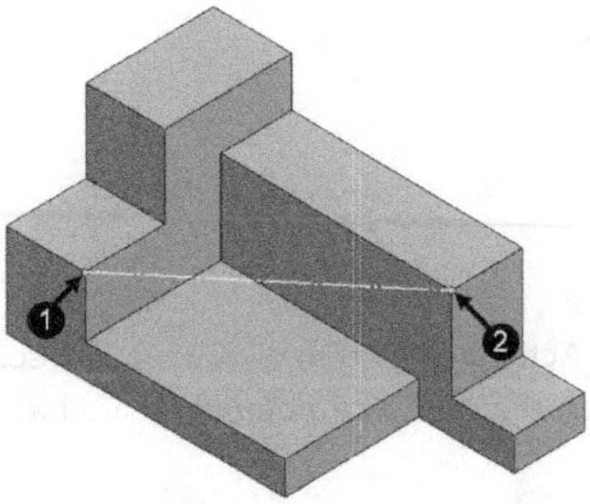

Two Planes

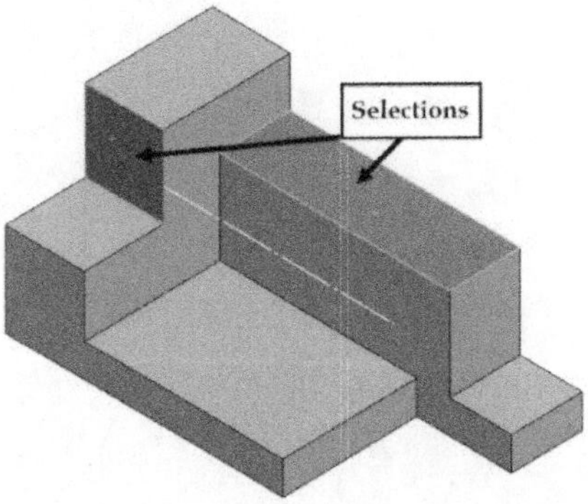

One Line/Edge/Axis

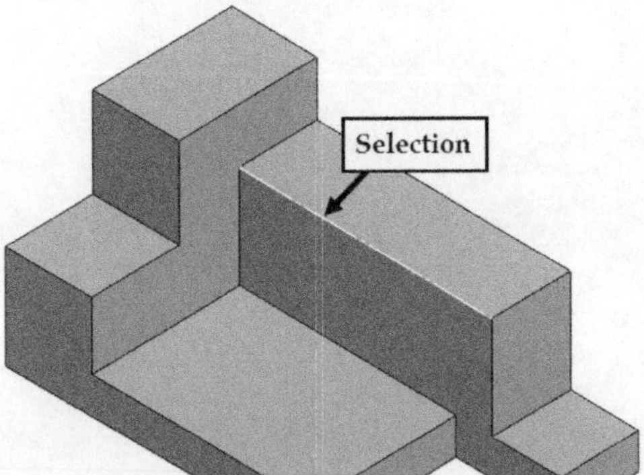

Point on Face/Plane

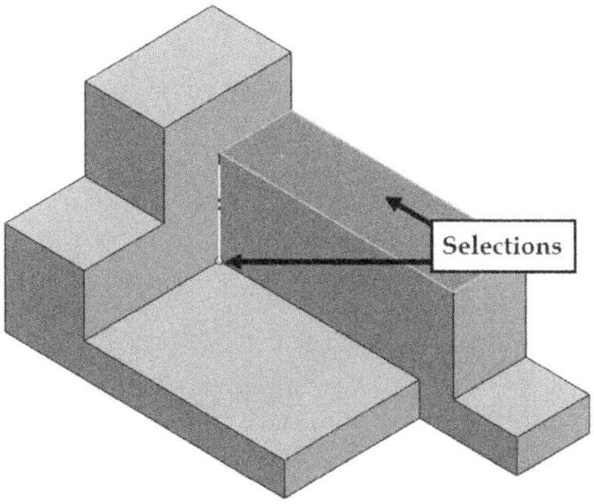

Cylindrical/Conical Face

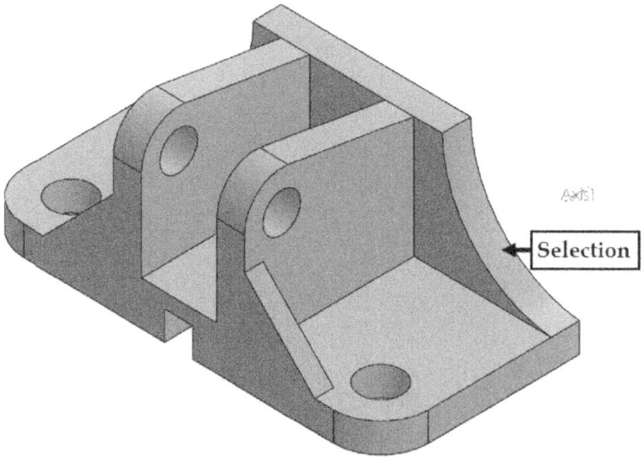

Creating a Coordinate System

SOLIDWORKS provides you with the default coordinate system. It is used as a reference to create sketches, mates and so on. However, you can create your own coordinate system by specifying the X, Y, and Z axis values. To create a new coordinate system, click the **Features** CommandManager **> Reference Geometry** drop-down **> Coordinate System.** On the **Coordinate System** PropertyManager, click in the **Origin** selection box under the **Selections** section, and then select a point as the origin. Click in the **X-Axis Direction reference** selection box and select a line or edge. Likewise, define the Y and Z axes of the coordinate system. Click the **OK** button to create a new coordinate system.

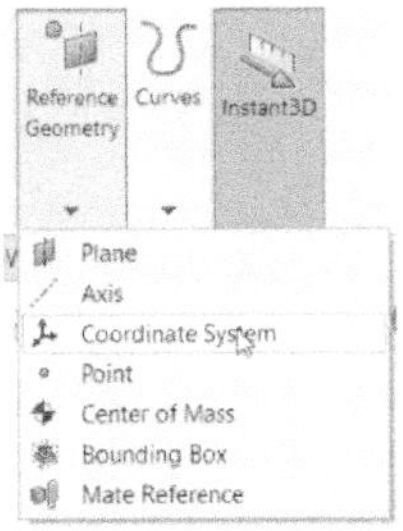

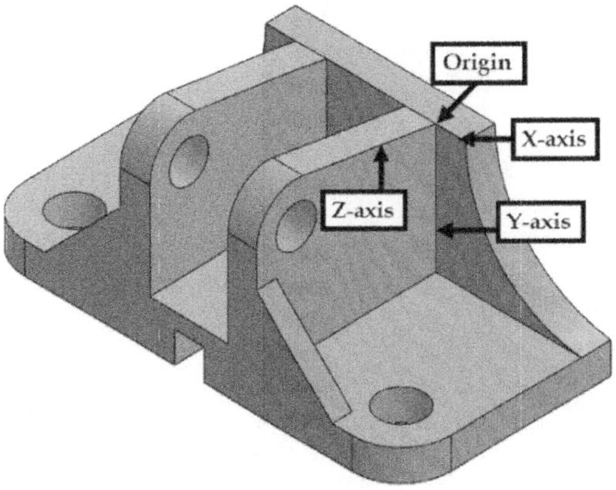

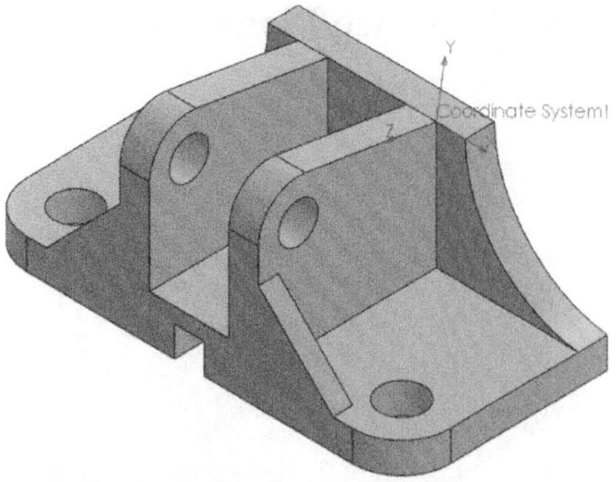

Creating a Coordinate System by entering Numeric Values

SOLIDWORKS allows you to define the position of the new coordinate system by specifying the X, Y, and Z coordinate values. To do this, check the **Define position with numeric values** option and enter values in the **X**, **Y**, and **Z** boxes, respectively.

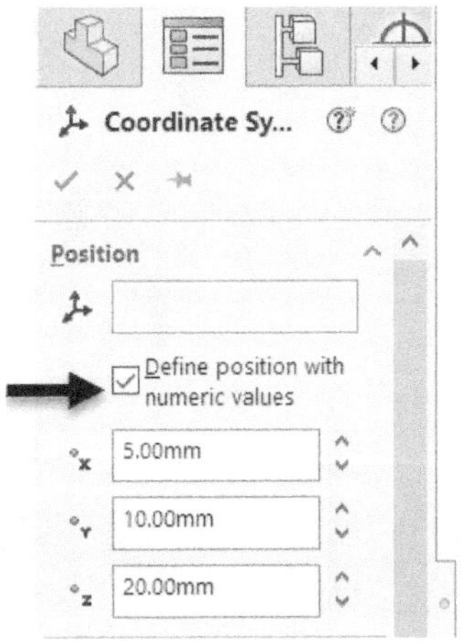

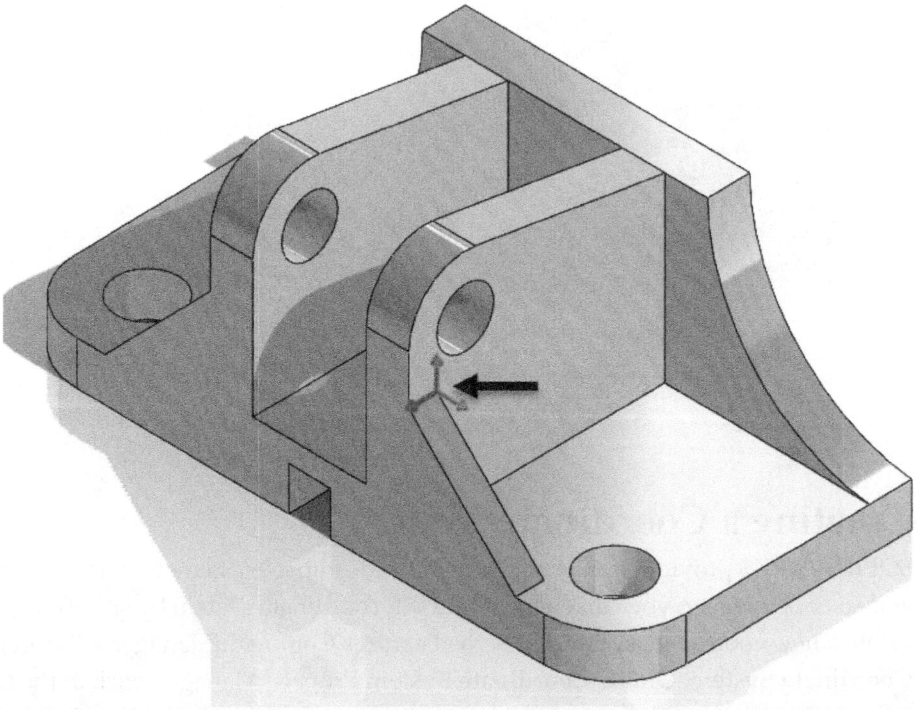

In addition to that, you can rotate the newly created coordinate system about the X, Y, and Z axes. To do this, check the **Define rotation with numeric values** option in the **Orientation** section of the PropertyManager. Next, enter values in the **X**, **Y**, or **Z** boxes, respectively. Next, click **OK** to create the coordinate system.

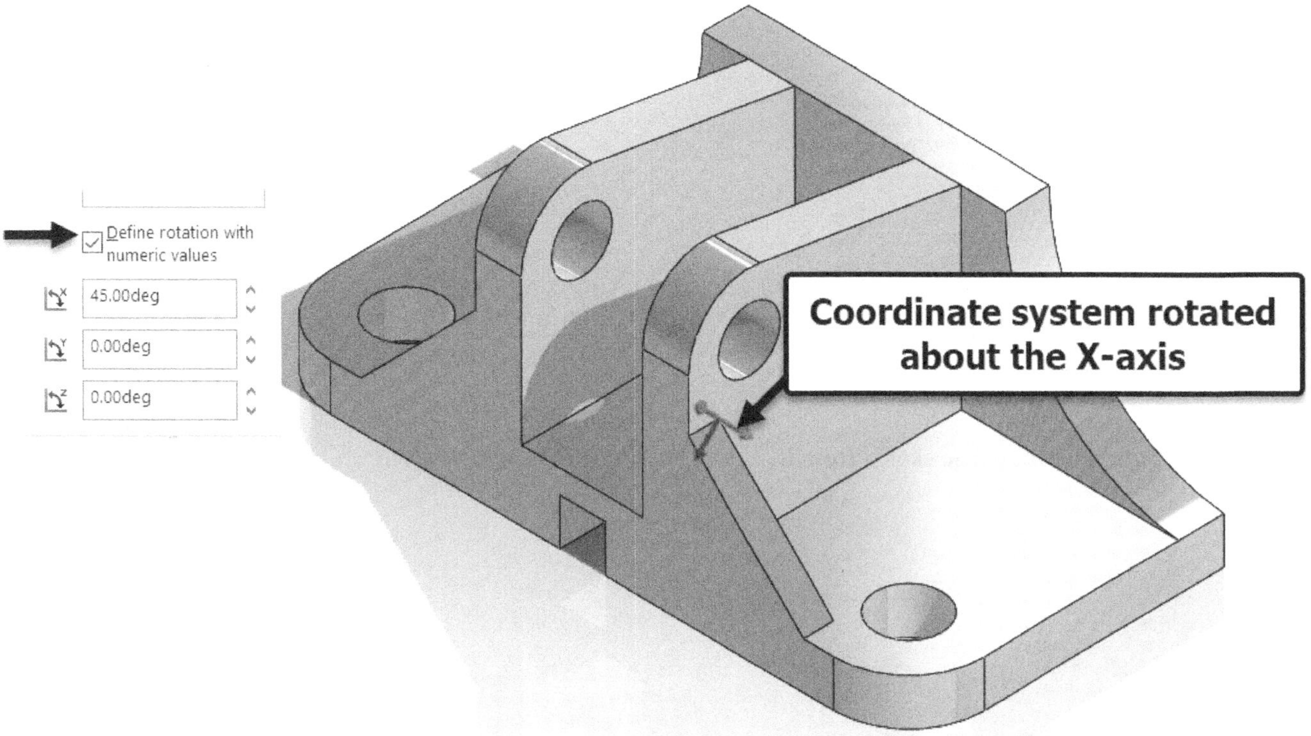

Additional options of the Extruded Boss/Base and Extruded Cut commands

The **Extruded Boss/Base** and **Extruded Cut** commands have some additional options to create 3D geometry and complex features.

Start and End Conditions

On the **Extruded Boss/Base** or **Extruded Cut** PropertyManager, the **Start Condition** and **End Condition** drop-downs have various options to define the start and end limits of the *Extruded Boss* or *Extruded Cut* feature.

Start Conditions

The **Start Condition** drop-down is available in the **From** section of the PropertyManager. The options in this drop-down are explained next.

The **Sketch Plane** option extrudes the sketch from the plane or face on which the sketch was created.

The **Surface/Face/Plane** option extrudes the sketch from a selected surface, face, or plane other than the sketch plane. Note that the selected surface, face, or plane should not be perpendicular to the sketch plane.

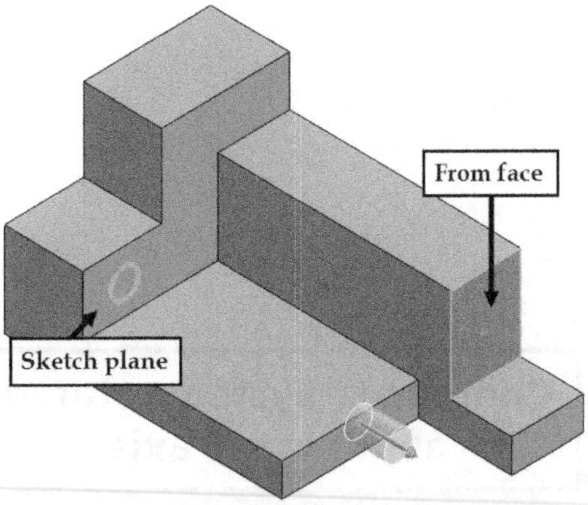

The **Vertex** option extrudes the sketch from the location of a selected vertex point.

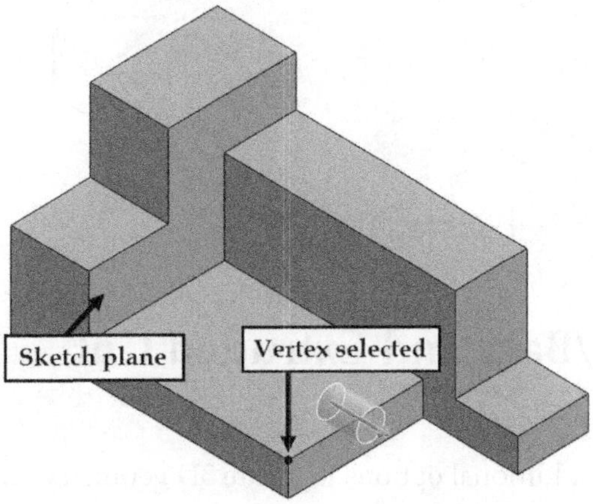

The **Offset** option extrudes the sketch from the offset distance that you specify.

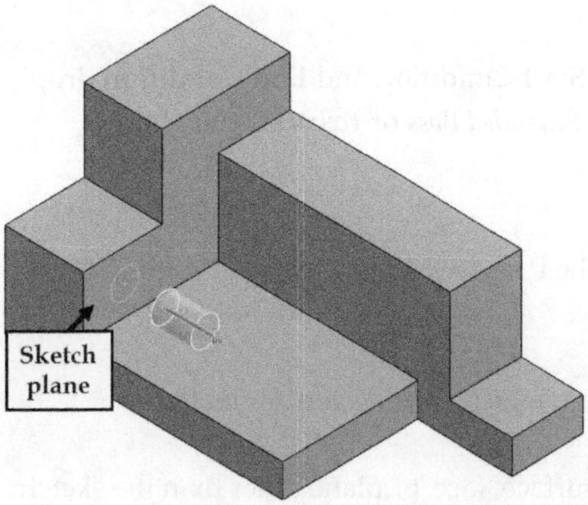

End Conditions

The **End Condition** drop-down is available in the **Direction 1** section of the PropertyManager. The options in this drop-down are explained next.

The **Up to next** option extrudes the sketch through the face next to the sketch plane. This option is especially helpful while creating extruded cuts.

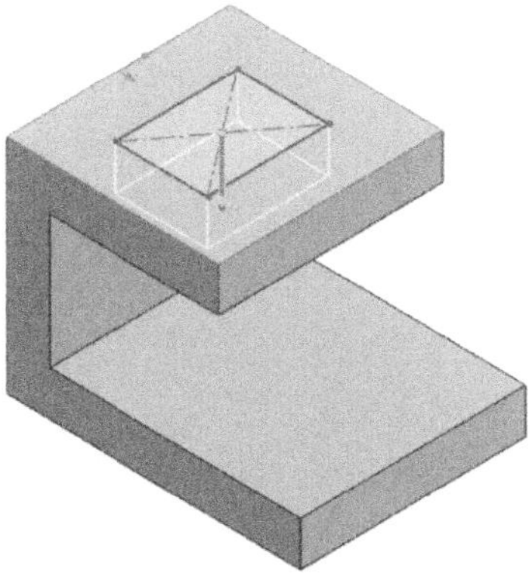

The **Up to surface** option extrudes the sketch up to a selected surface.

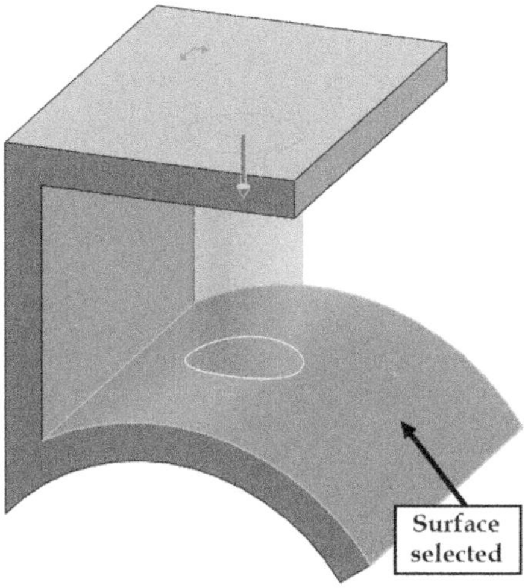

The **Offset From Surface** option extrudes the sketch up to the selected surface but offsets the extrusion by the offset distance you enter. Check the **Reverse offset** option if you want to offset the extrusion beyond the selected surface.

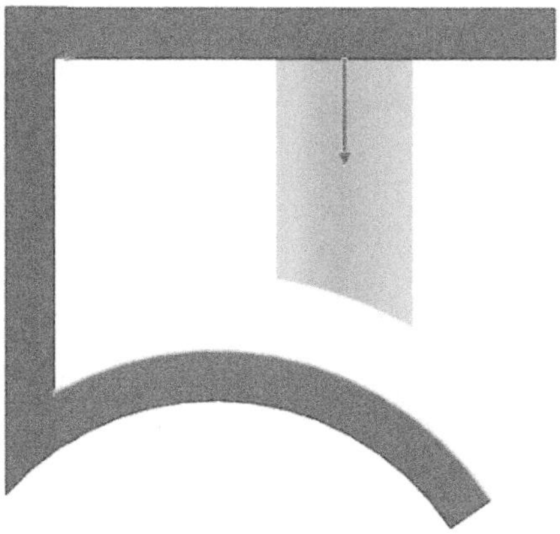

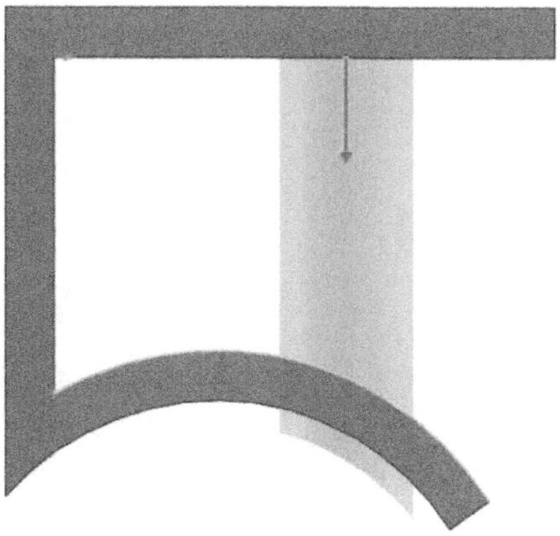

With the Reverse Offset option selected

Check the **Translate Surface** option if you want the end face of the extrusion the same as the selected reference surface. Note that this option is available only when the **Offset From Surface** option is selected from the **End Condition** drop-down.

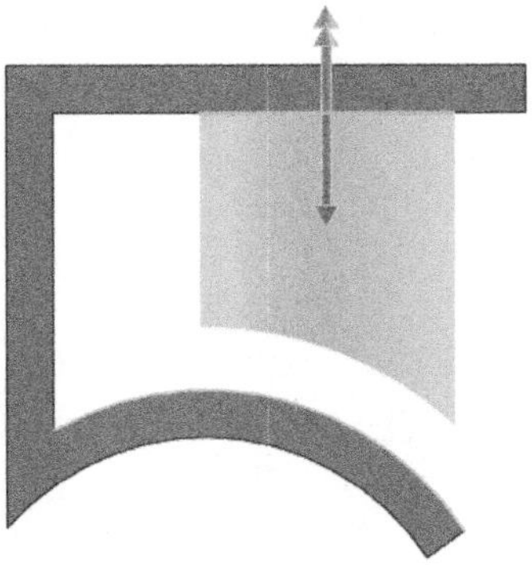

The **Translate Surface** option unchecked

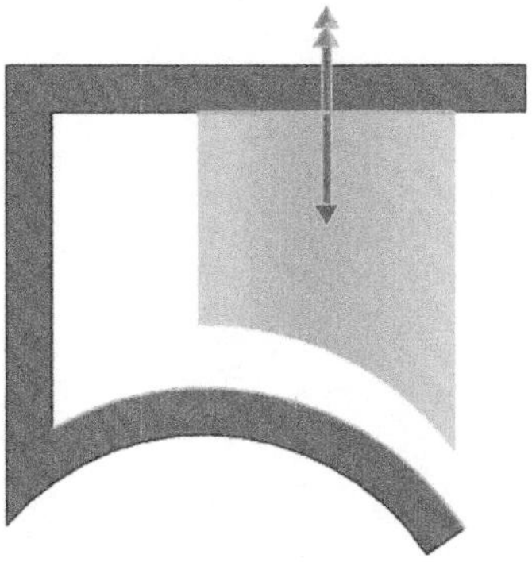

The **Translate Surface** option checked

The **Up to vertex** option extrudes the sketch from the sketch plane up to a selected vertex.

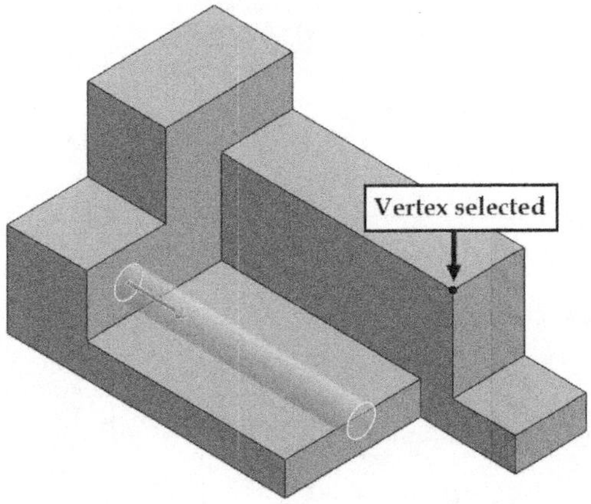

The **Through All** option extrudes the sketch throughout the 3D geometry. The **Through All – Both** option extrudes the sketch throughout the geometry on both sides.

Through All

Through All - Both

Thin Feature

The **Thin Feature** option will help you to add thickness to the selected sketch. Check this option on the **Boss-Extrude** or **Revolve** PropertyManager to add thickness to the sketch. In the **Thin Feature** section, select **One-Direction** from the **Type** drop-down to add thickness to one side of the sketch. Next, type in a value in the **Thickness** box. Use the **Reverse Direction** icon next to the **Type** drop-down if you want to change the direction.

Select the **Mid-plane** option from the **Type** drop-down in the **Thin feature** section if you want to add thickness symmetrically on both sides of the sketch.

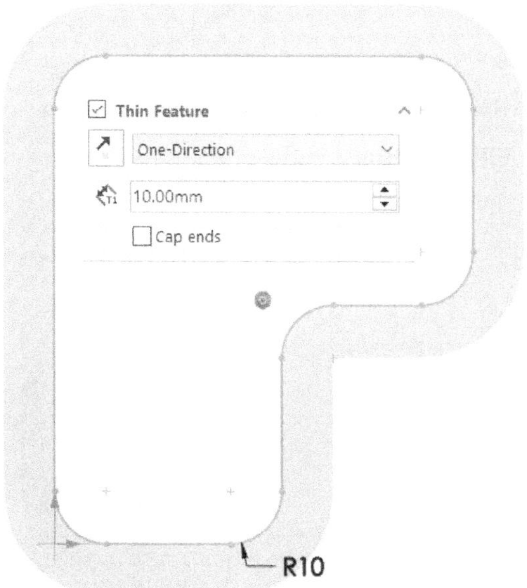

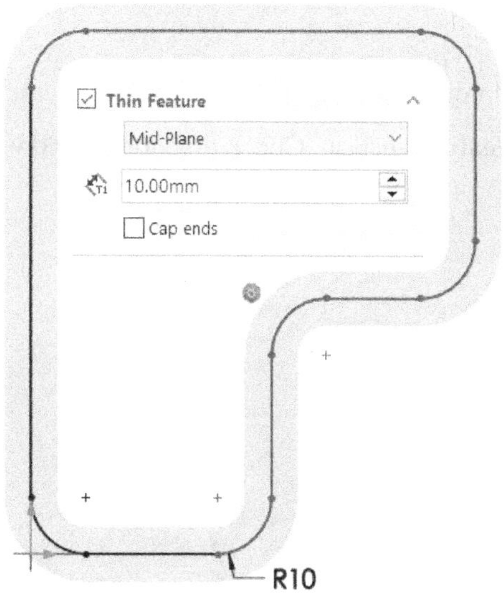

Select the **Two-Direction** option from the **Type** drop-down to add separate thickness on both sides of the sketch. Type-in thickness values in **Direction 1 Thickness** and **Direction 2 Thickness** boxes.

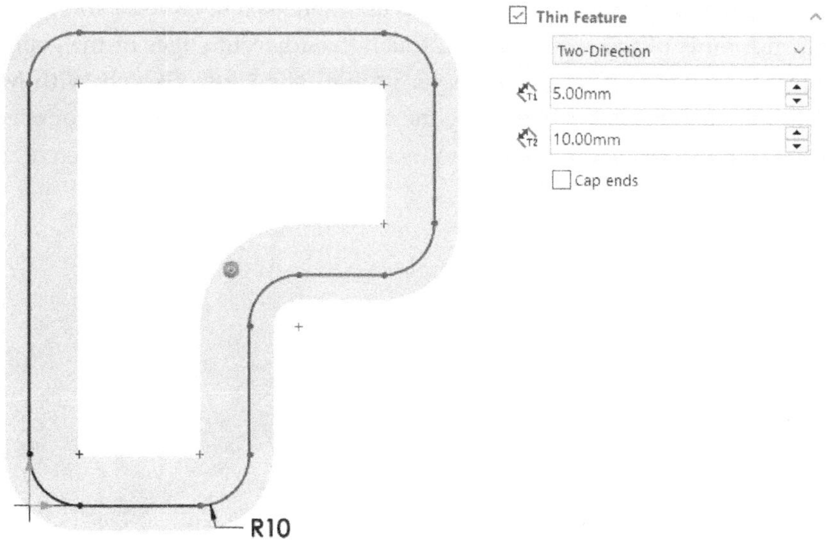

Check the **Cap ends** option if you want to close the ends of the thin feature. Next, you need to specify the **Cap Thickness** value. Note that this option is available only for the closed sketch.

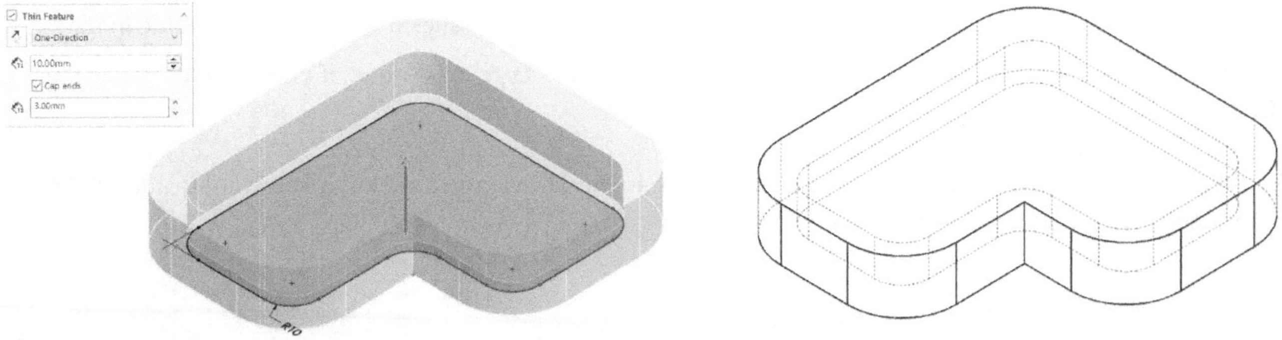

If you are using an open sketch to create the thin feature, the **Auto-fillet corners** option appears in the **Thin Feature** section. Check this option and enter the radius value in the **Fillet Radius** box.

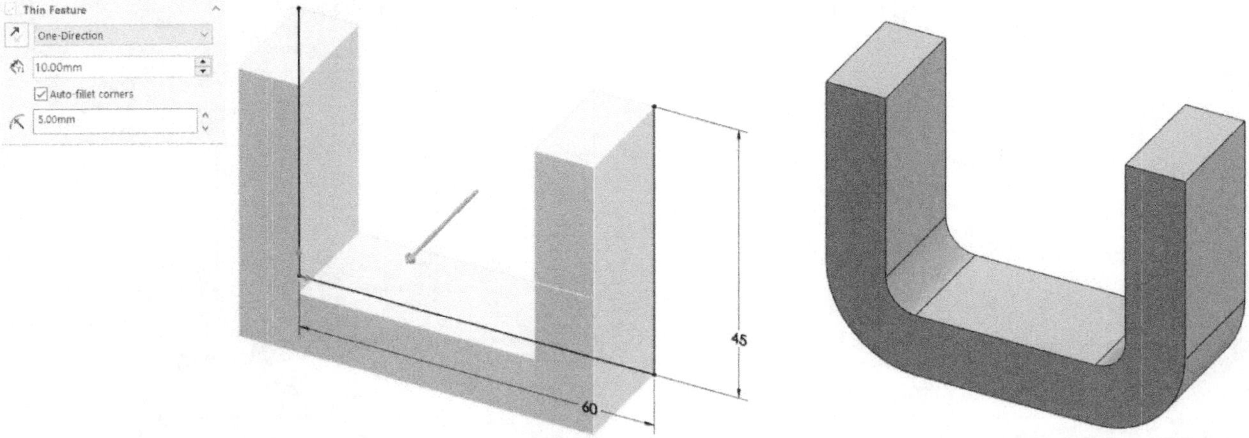

Open Sketch Extrusion

SOLIDWORKS helps you to add an *Extrude* feature to the geometry using an open sketch. It closes the profile by using the adjacent edges. Note that the endpoints of the sketch should touch the adjacent edges of the geometry. Activate the **Extruded Boss/Base** command, and then click on the open profile. The **Close Sketch With Model Edges?** dialog pops up on the screen. On this dialog, click **Yes** to close the sketch using the adjacent model edges. You can also check the **Reverse direction to close the sketch** option to use the model edges on the sketch's other side.

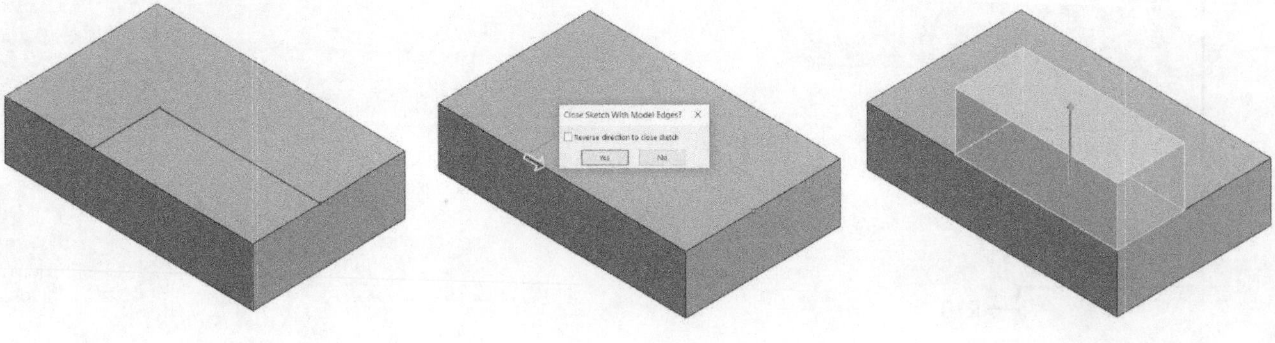

When you are creating an **Extruded Cut** feature, SOLIDWORKS creates the cut without closing the sketch. A preview of the *Extruded cut* feature appears automatically on selecting an open profile. Note that the endpoint of the sketch should meet the adjacent edges. On the PropertyManager, check the **Flip side cut** option to change the material side. Next, type in a value in the **Depth** box and click **OK**.

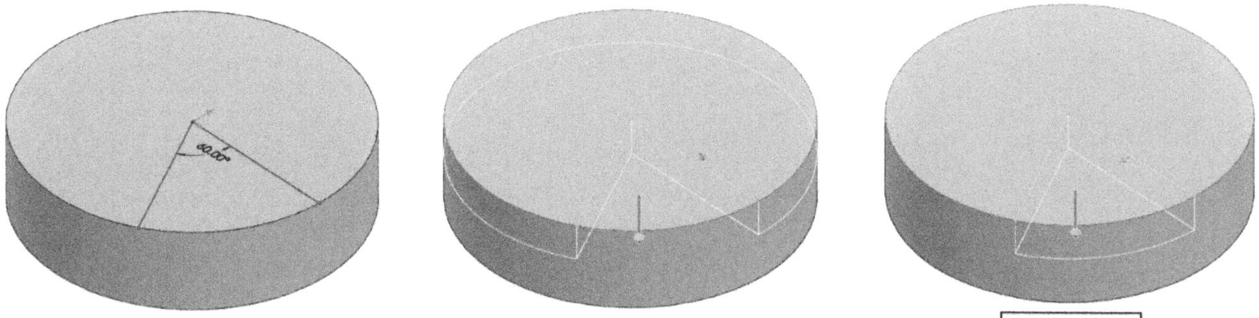

Flip side cut

Draft

The **Draft** option will help you to apply a draft to the extrusion. Click the **Draft** icon on the PropertyManager and type in a value in the **Draft Angle** box. Check the **Draft outward** option if you want to change the draft direction.

Inward Draft

Outward Draft

Contour Select Tool

This tool helps you to create multiple *Extrude* features using a single sketch with internal loops in it. To do this, draw a sketch containing internal loops, and then click **Exit Sketch** on the CommandManager.

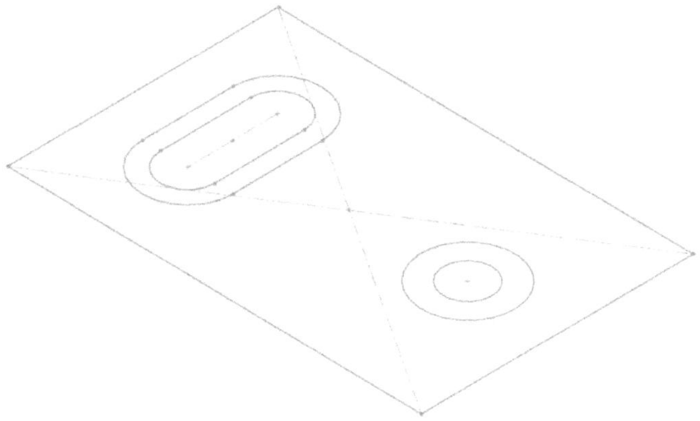

Click on an entity of the sketch, right-click, and then click the arrows pointing downwards; the shortcut menu is expanded. On the expanded shortcut menu, click **Contour Select Tool**.

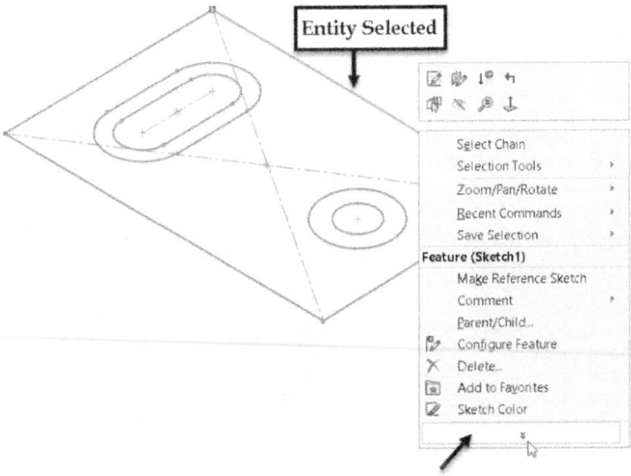

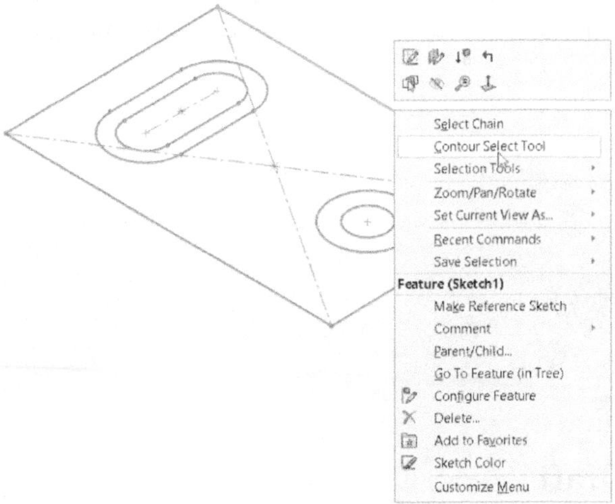

Click on the region of the sketch, as shown. Click and drag the arrow that appears on the selected region; the sketch extrudes. Release the arrow at the required value on the scale.

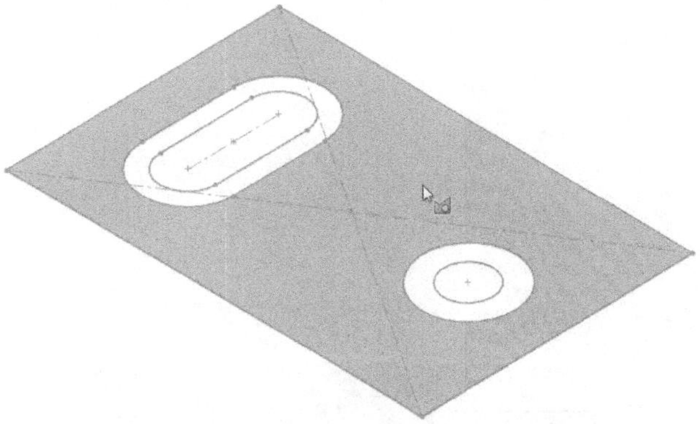

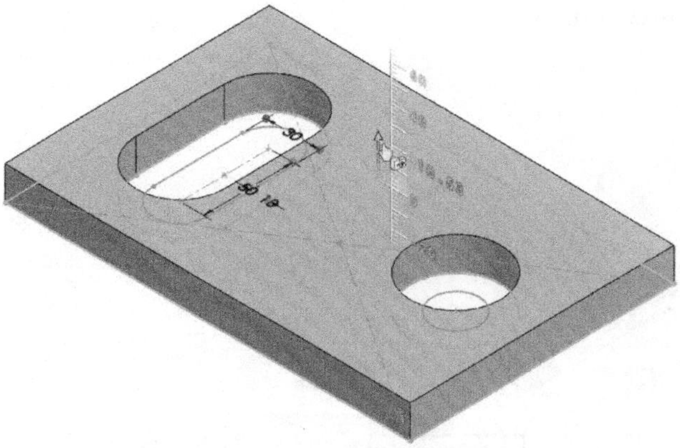

Click on the sketch region, as shown. Next, drag the arrow that appears in the region. Define the extrusion depth using the scale.

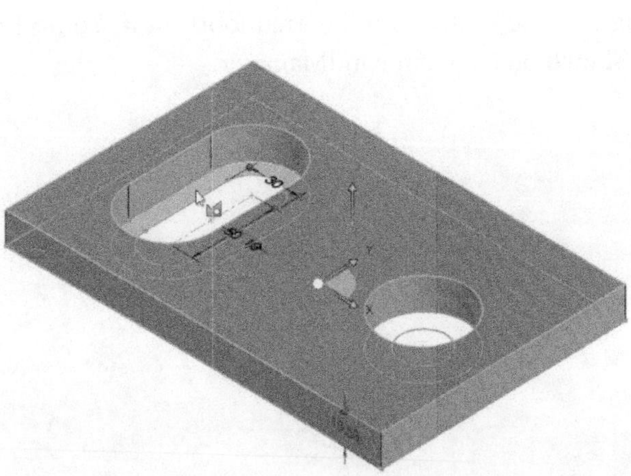

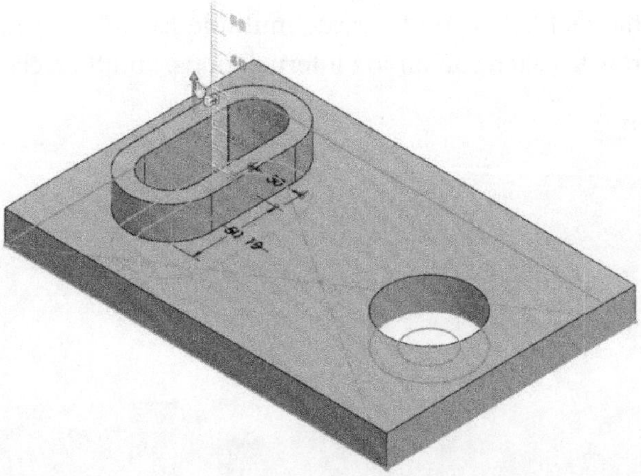

Likewise, extrude the remaining sketch region. Next, click the **Exit Contour Select tool** icon on the top right corner of the graphics window.

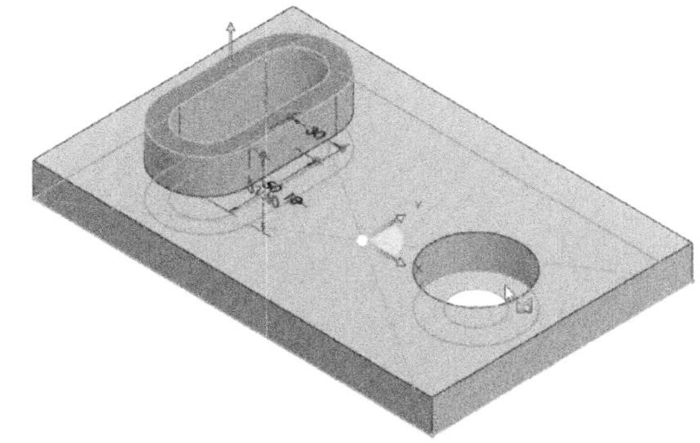 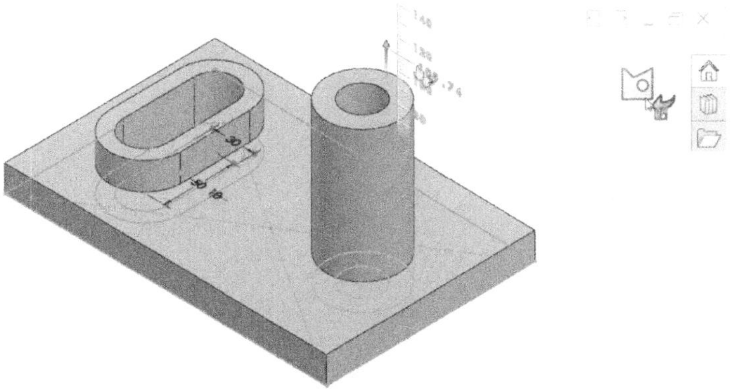

Notice that three separate **Boss-Extrude** features are created in the FeatureManager Design tree. You can edit them individually. Expand the three features and notice the same sketch under them.

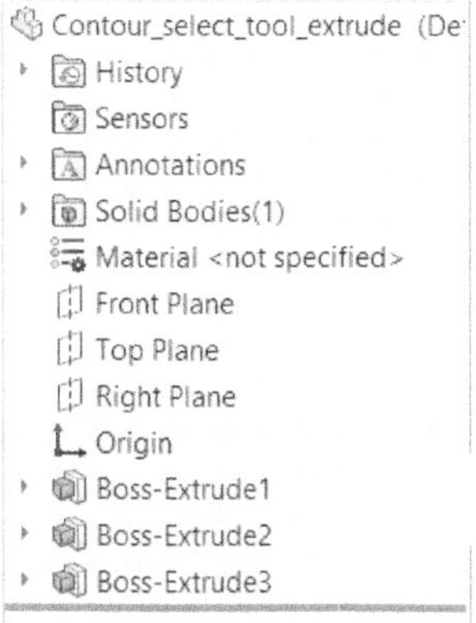

Applying Material to the Geometry

SOLIDWORKS makes it easy to apply a material to the solid geometry. In the FeatureManager Design tree, click the right mouse button on the part and select **Material**. The favorite material list appears. Select a material from the favorite list; the selected material is applied to the geometry. You can also click **Edit Material**; the **Material** dialog appears. On this dialog, expand the **SOLIDWORKS Materials** tree node and notice various material categories under it. Expand a material category and select a material. Click **Apply** to apply the selected material to the geometry. Click **Close** to close the dialog.

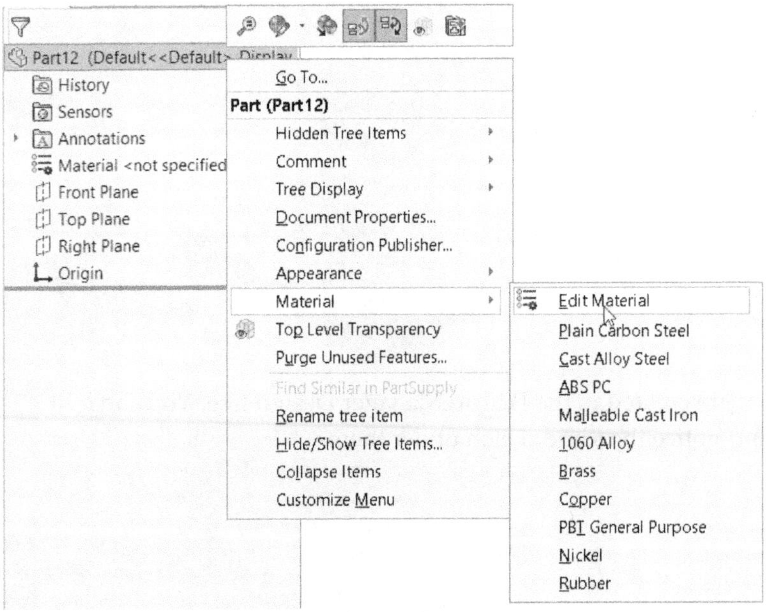

View commands

The model display in the graphics window can be changed using various view commands. Most of these commands are located on the **View Heads-Up** toolbar, or the **View > Modify** menu. You can also access these commands from the shortcut menu.

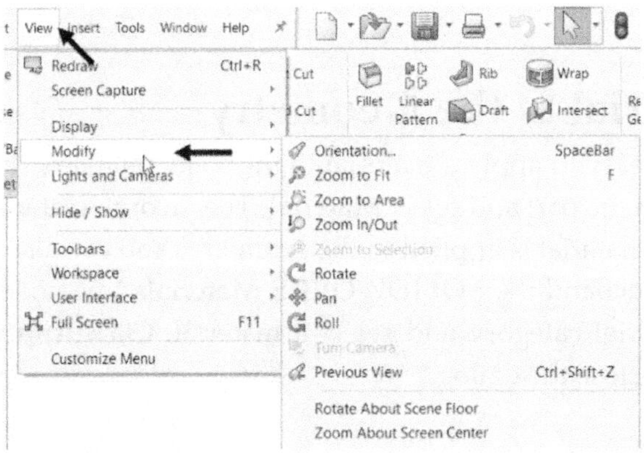

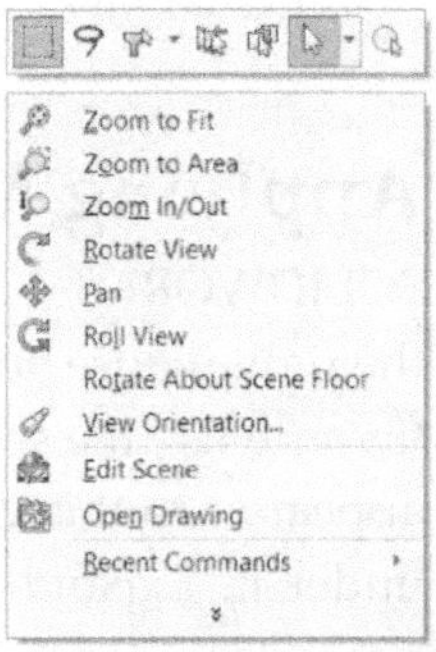

The following are some of the main view commands:

	Zoom to Fit	The model will be fitted in the graphics window's current size to be visible completely.
	Zoom to Area	Activate this command and drag a rectangle. The contents inside the rectangle will be zoomed.
	Zoom In/Out	Activate this command and press the left mouse button. Drag the mouse to vary the size of the objects accordingly.

	Zoom to Selection	This command fits the selected objects in the graphics window.
	Pan	Activate this command and press the left mouse button. Drag the pointer to move the model view on a plane parallel to the screen.
	Rotate	Activate this command and press the left mouse button. Drag the pointer to rotate the model view.
	Roll View	Activate this command and press the left mouse button. Drag the pointer to rotate the model in the view parallel to the screen.

Display Styles

The **Display Styles** drop-down on the **View Heads-Up** toolbar allows you to change the model geometry's representation. These display styles are explained next.

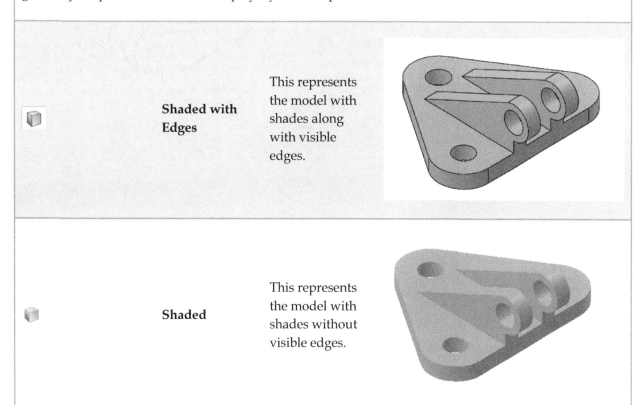

	Shaded with Edges	This represents the model with shades along with visible edges.	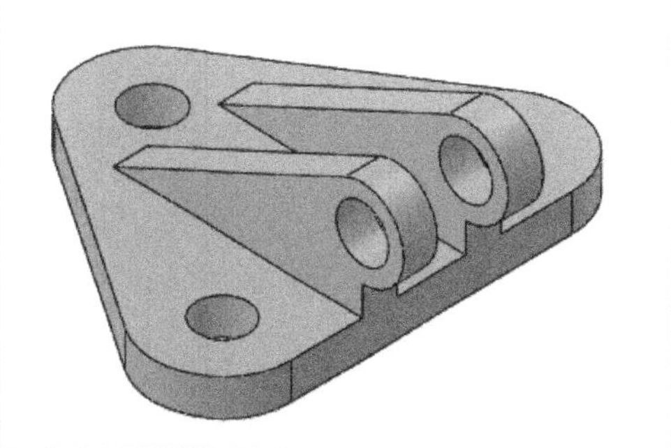
	Shaded	This represents the model with shades without visible edges.	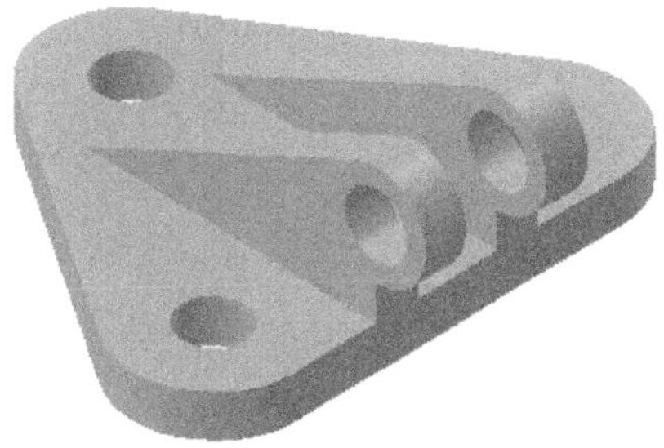

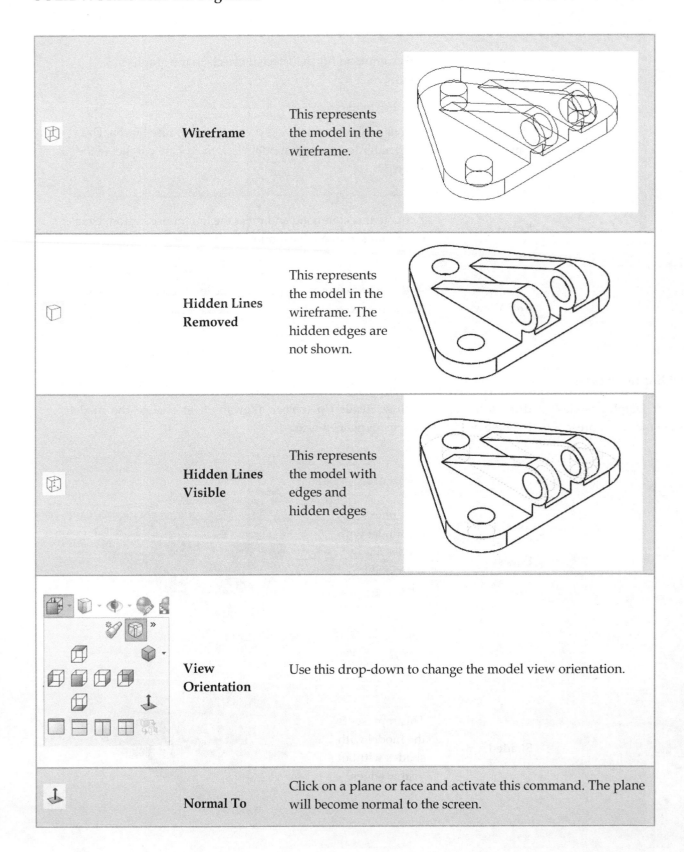

	**Wireframe**	This represents the model in the wireframe.	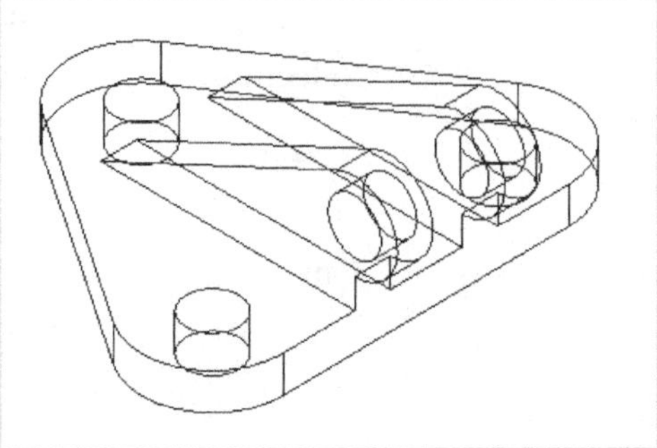
	**Hidden Lines Removed**	This represents the model in the wireframe. The hidden edges are not shown.	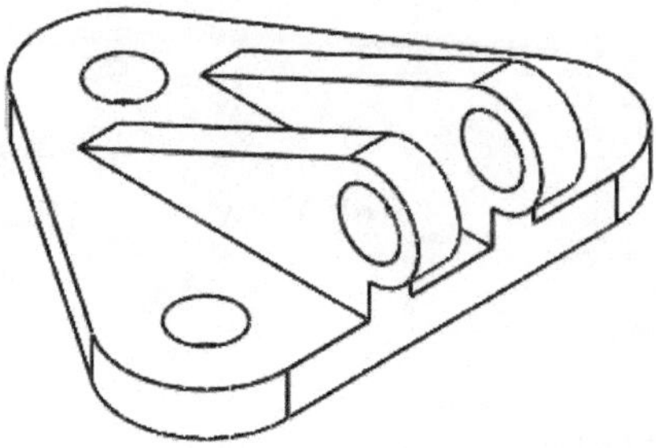
	**Hidden Lines Visible**	This represents the model with edges and hidden edges	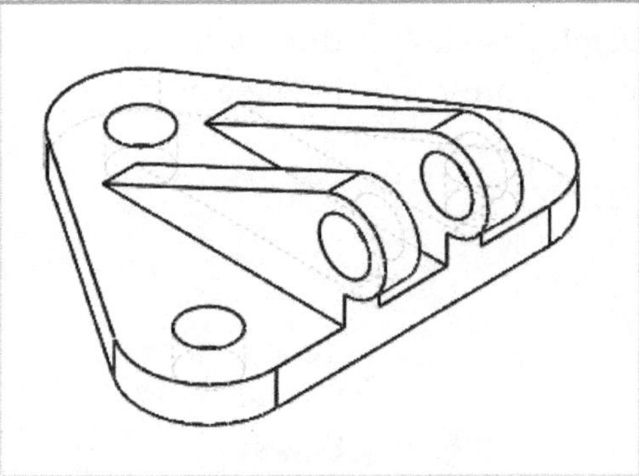
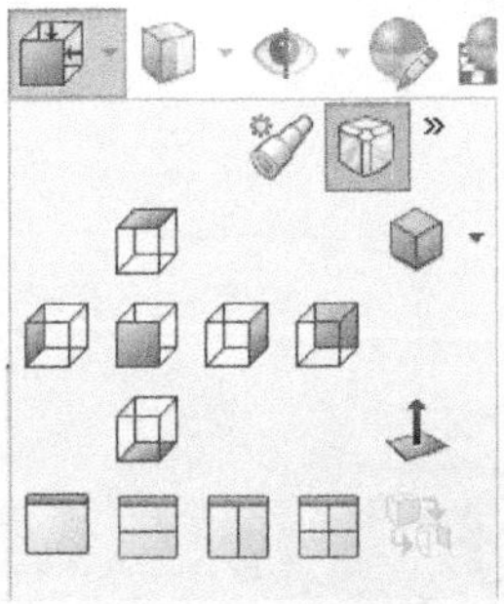	**View Orientation**	Use this drop-down to change the model view orientation.	
	**Normal To**	Click on a plane or face and activate this command. The plane will become normal to the screen.	

Best Practices

1. Start with a Clear Sketch: Begin by developing accurate and well-defined 2D sketches using dimensions and constraints.

2. Sketch on Appropriate Planes: Choose appropriate working planes for your sketches, which affect the orientation of your extrusions. Typically, start with the XY plane and proceed from there.

3. Combine Features: Take advantage of SolidWorks' ability to combine various features, including extrusions, cuts, and fillets. Plan your design accordingly to create complex shapes.

4. Revaluate as You Go: Periodically assess your design to ensure it meets your requirements throughout its development. This proactive approach saves time later on.

Cautions to Keep in Mind

1. Avoid Over-Complexity: While SolidWorks allows intricate designs, excessive complexity may result in poor performance and difficulty managing your models.

2. Check for Intersecting Geometry: Be cautious about intersecting or overlapping geometry during extrusions, as this can cause issues in your model, particularly in intricate designs.

3. Watch for Thin Features: Thin extrusions may present challenges, especially in 3D printing and manufacturing. Ensure your parts have sufficient thickness for their intended use and production methods.

Useful Tips

1. Utilize Symmetry: Apply symmetry whenever possible to reduce workload and ensure balanced designs.

2. Experiment with Type Options: Explore various extrusion types, such as "Up to face" or "To first," to achieve your desired outcomes.

3. Use Reference Geometry: Implement reference geometry, such as planes, axes, and points, to accurately position and orient your sketches and extrusions. Simplify the modeling process.

Examples

Example 1

In this example, you will create the part shown below.

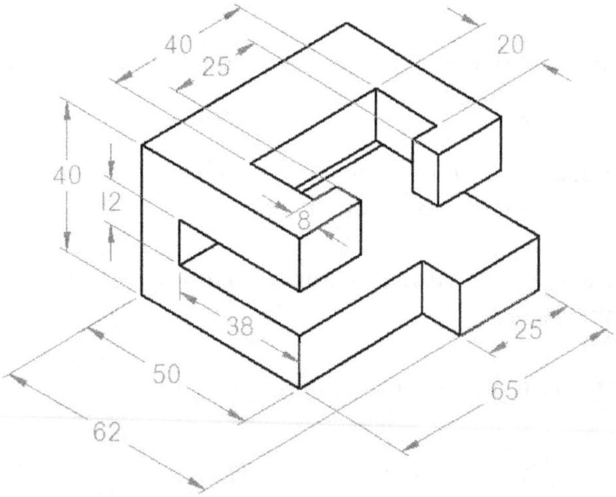

1. Start **SOLIDWORKS 2024**.
2. On the Menu, click **File > New**.
3. On the **New SOLIDWORKS Document** dialog, click the **Part** icon, and then click **OK**.
4. At the bottom right corner of the window, click the **Unit System** drop-down and select **MMGS (millimeter, gram, second)**.

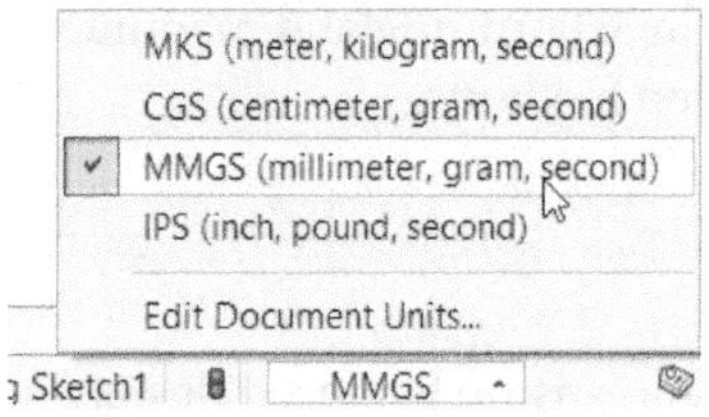

5. On the **Sketch** CommandManager, click the **Sketch** icon. Next, click the Front plane to start the sketch.
6. On the **Sketch** CommandManager, click **Rectangle drop-down > Corner Rectangle**.
7. Click the origin point to define the first corner of the rectangle.
8. Move the pointer toward the top right and click to define the second corner.
9. Activate the **Smart Dimension** command (on the CommandManager, click **Sketch** tab > **Smart Dimension)**.
10. Select the horizontal line, move the pointer upward, and click.
11. Type-in 50 in the **Modify** box and click the green check.
12. Likewise, apply the dimension to the vertical line, as shown.
13. On the **Sketch** CommandManager, click **Exit Sketch**.
14. On the **Features** CommandManager, click **Extruded Boss/Base**.
15. On the **Boss-Extrude** PropertyManager, under the **Direction 1** section, click **End Condition > Mid Plane**.
16. On the PropertyManager, type-in **65** in the **Depth** box.
17. Click **OK** to complete the *Extrude* feature.

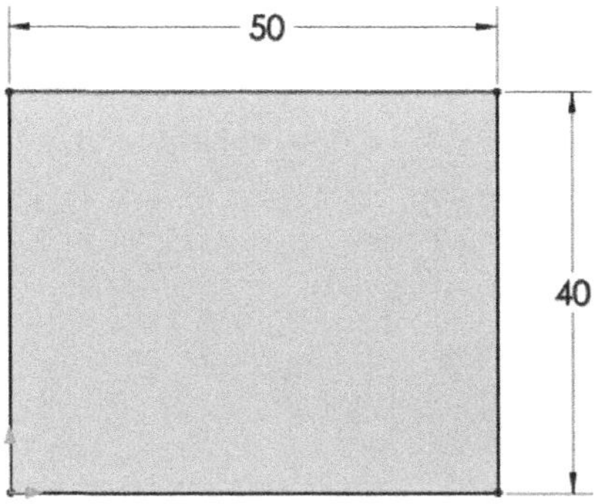

18. On the **Features** CommandManager, click **Extruded Cut**. Click on the front face of the part geometry.
19. On the **View Heads-Up** toolbar, click **View Orientation** drop-down > **Normal To**.
20. Activate the **Corner Rectangle** command (on the CommandManager, click **Sketch** tab > **Rectangle** drop-down > **Corner Rectangle**).
21. Click on the right edge of the model, move the pointer toward the left, and then click.
22. Activate the **Smart Dimension** command. Select the rectangle's horizontal line, move the pointer upward, and then click to position the dimension. Type **38** in the **Modify** box, and then click the green check.
23. Select the bottom edge of the model and the lower horizontal line of the rectangle. Move the pointer toward the right and click to position the dimension. Type **14** in the **Modify** box, and then click the green check.
24. Likewise, apply the remaining dimension to the sketch.

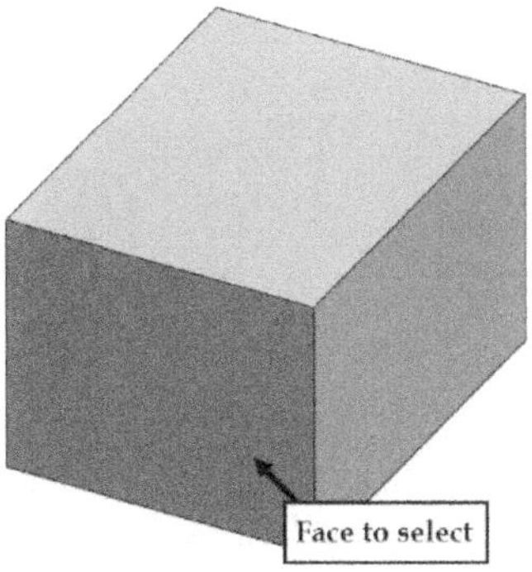

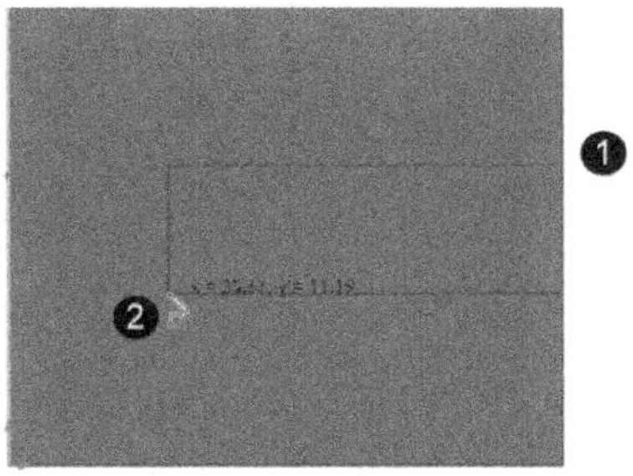

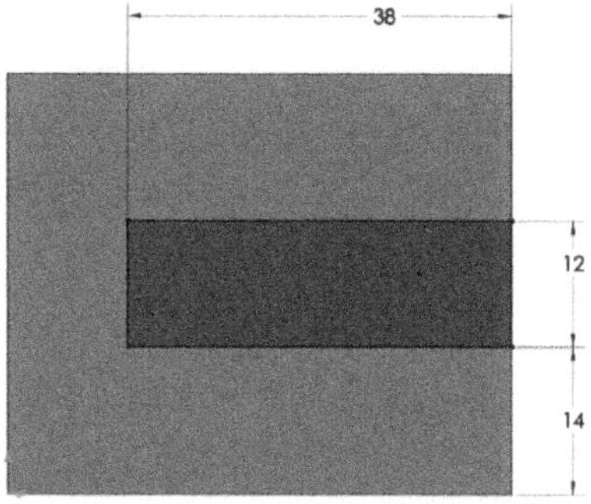

25. On the **Sketch** CommandManager, click **Exit Sketch**.
26. On the **Cut-Extrude PropertyManager**, under the **Direction 1** section, select **End Condition > Through All**.
27. Click **OK** to create the cut throughout the part geometry.
28. Press the Spacebar on your keyboard to display the **Orientation** dialog. On this dialog, click the **Isometric** icon to change the orientation to Isometric view.

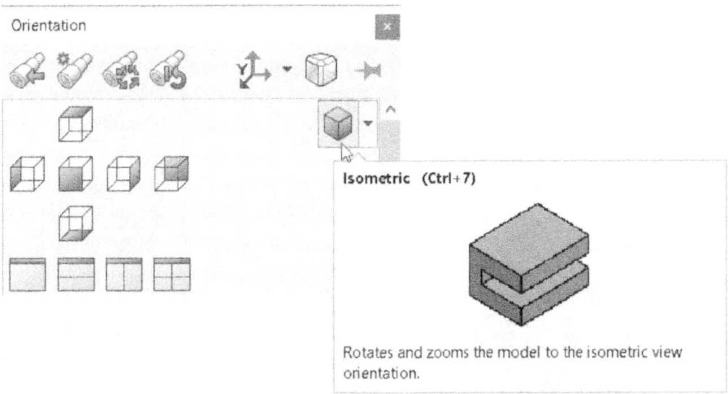

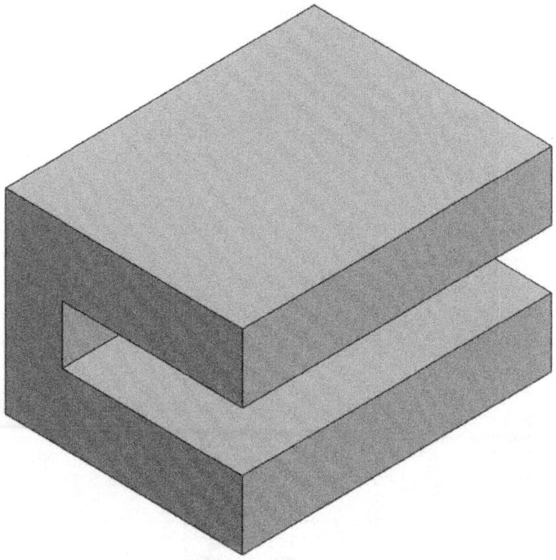

29. Activate the **Extruded Cut** command and click on the top face of the part geometry.
30. On the **View Heads-Up** toolbar, click the **View Orientation** drop-down and select **Normal To** .
31. On the **Sketch** CommandManager, click **Line** drop-down **> Centerline** .
32. Select the origin point of the sketch, move the pointer horizontally toward the right and click. Next, press Esc to deactivate the **Centerline** command.
33. On the menu bar, click **Tools > Sketch Tools > Dynamic Mirror Entities** icon, and then select the centerline.
34. Activate the **Line** command (On the **Sketch** CommandManager, click the **Line** icon) and click on the intersection point between the centerline and the right model edge.
35. Create the closed sketch, as shown.

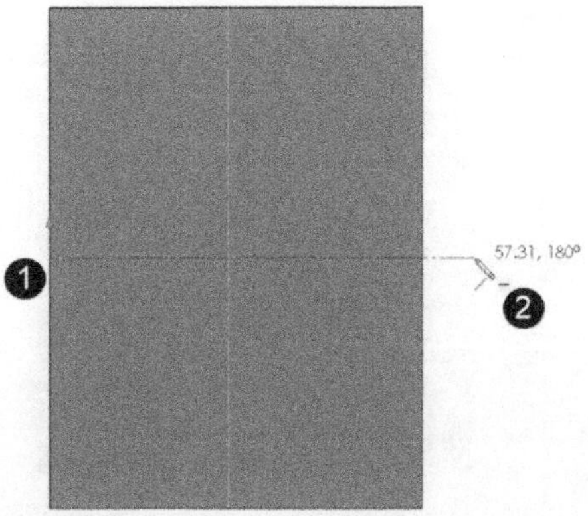

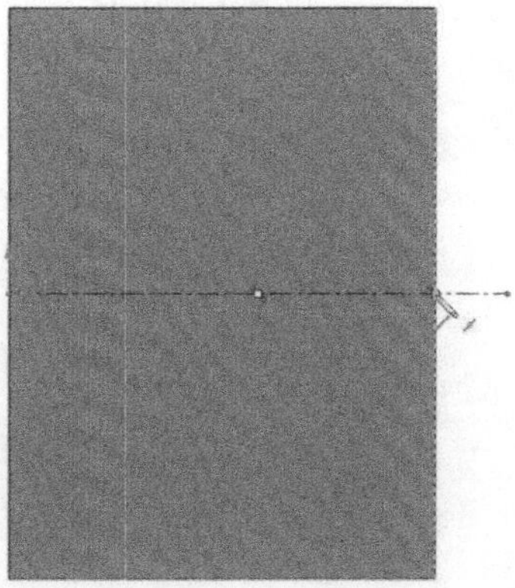

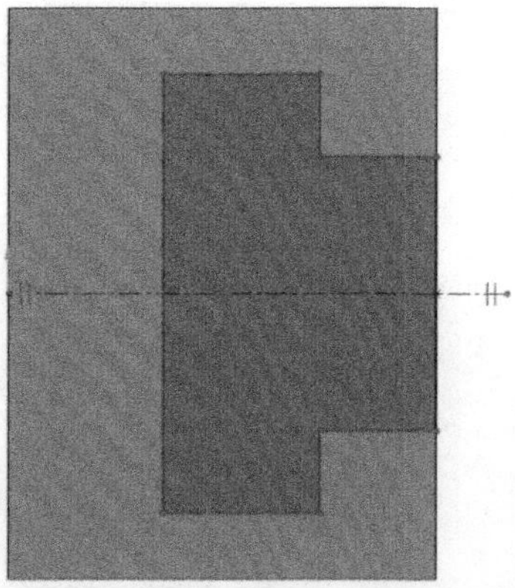

36. Add dimensions to the sketch, as shown.
37. On the **Sketch** CommandManager, click **Exit Sketch**.
38. On the **Cut-Extrude** PropertyManager, click **End Condition > Up to Next**.
39. Click **OK** to create the *Extruded Cut* feature until the surface next to the sketch plane.

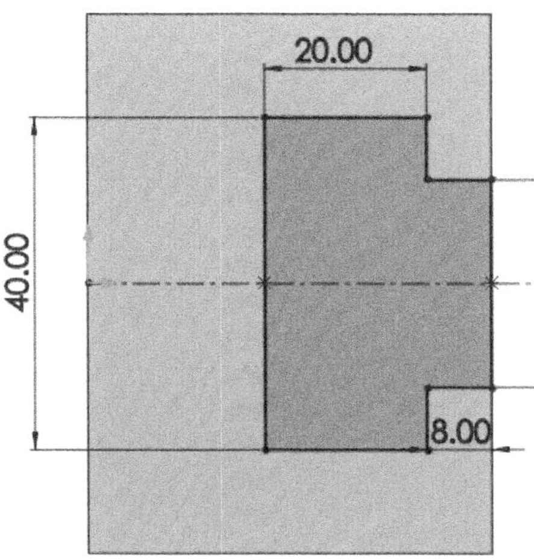

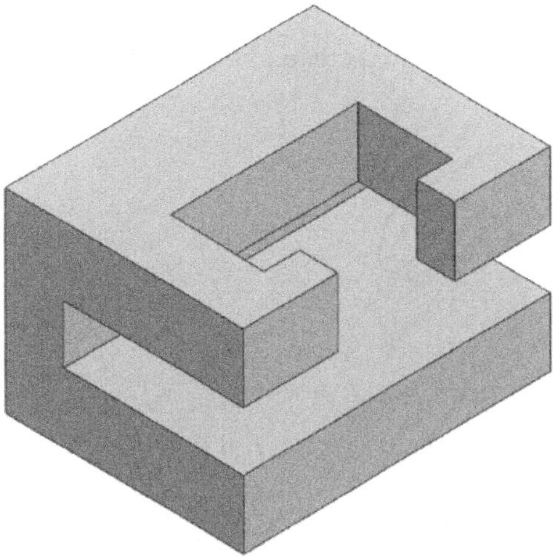

40. Activate the **Extruded Boss/Base** command. Click on the Top plane.
41. Draw a closed sketch. Apply dimensions and exit the sketch.

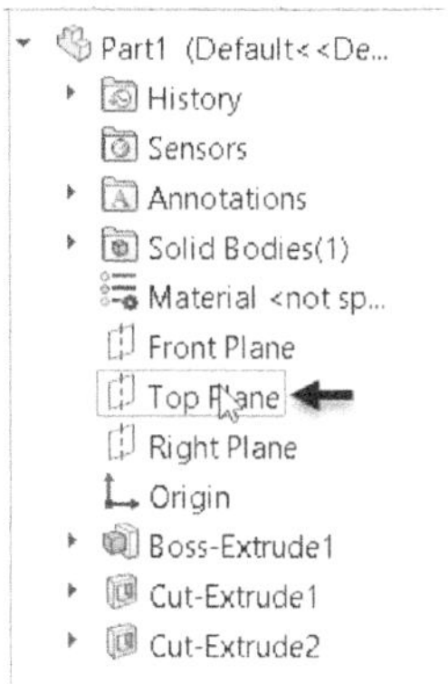

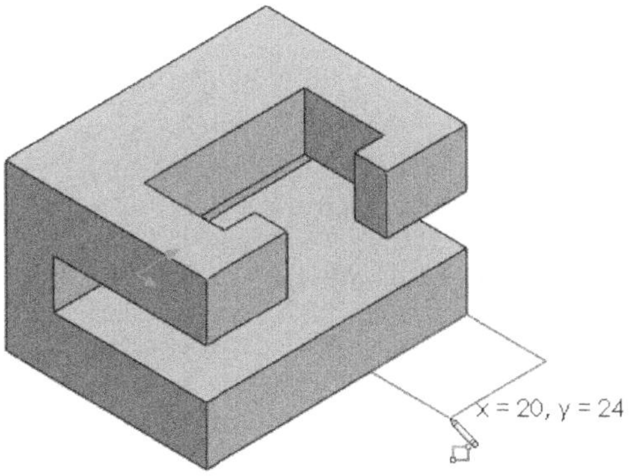

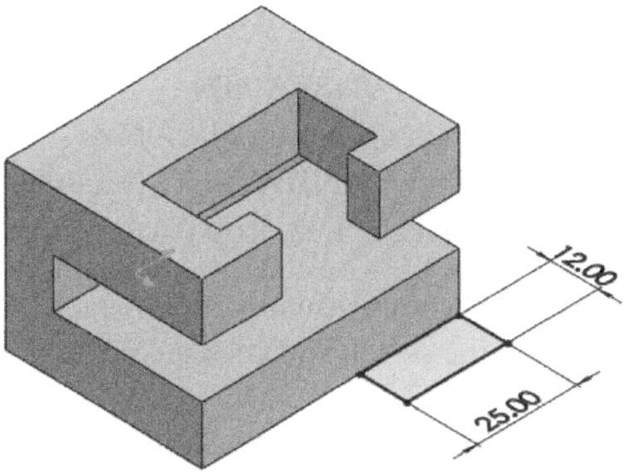

42. On the **Boss-Extrude** PropertyManager, click **Type > Up to Surface** and select the flat face of the part geometry, as shown in the figure. Click **OK** to complete the part.

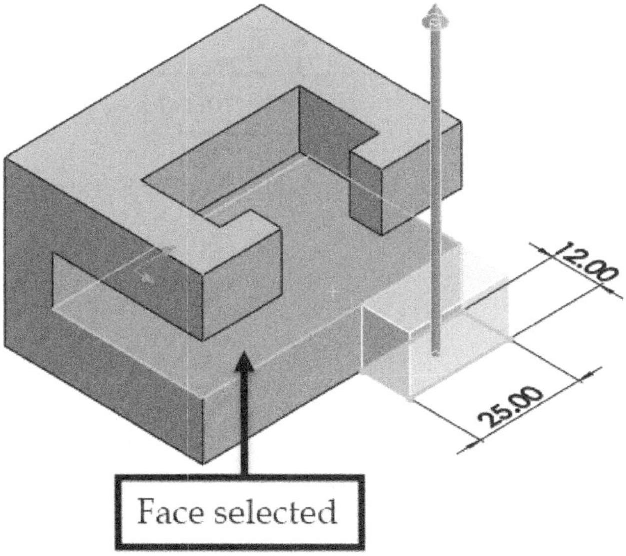

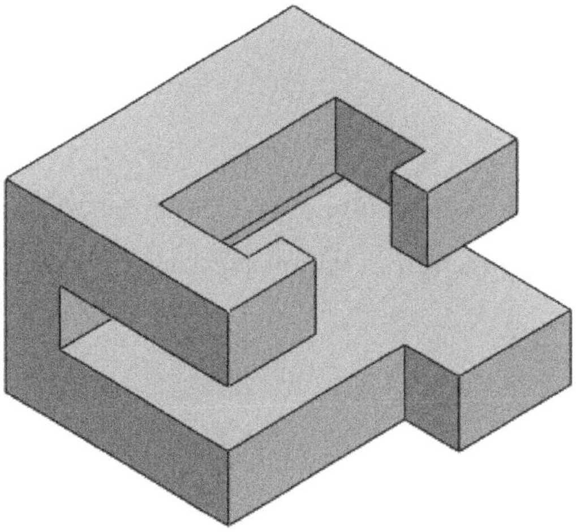

43. Save and close the file.

Example 2

In this example, you will create the part shown below.

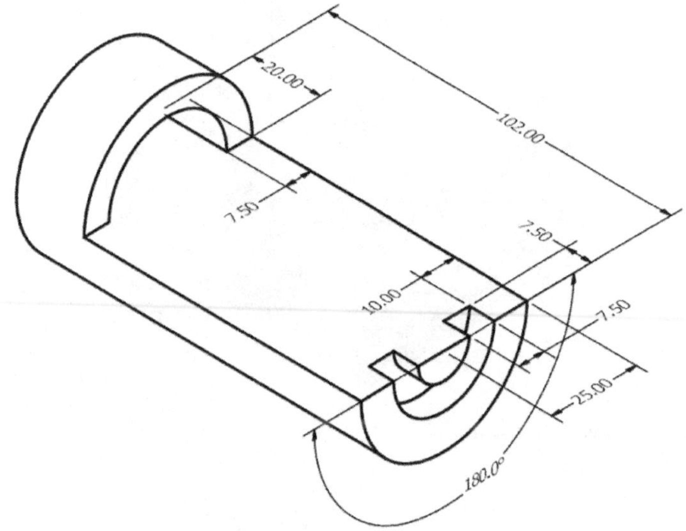

1. Start **SOLIDWORKS 2024**.
2. On the Menu bar, click **File > New**.
3. On the **New SOLIDWORKS Document** dialog, select **Part**, and then click **OK**.
4. Draw a sketch on the Top plane, as shown. Exit the **Sketch** environment.
5. On the **Features** CommandManager, click the **Revolved Boss/Base** icon.
6. Select the sketch and click on the line, as shown.

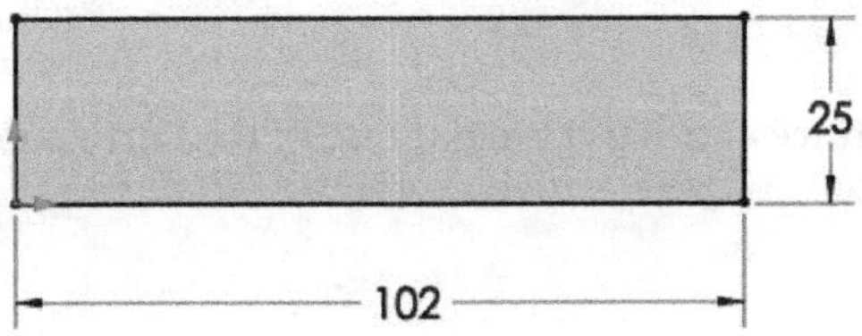

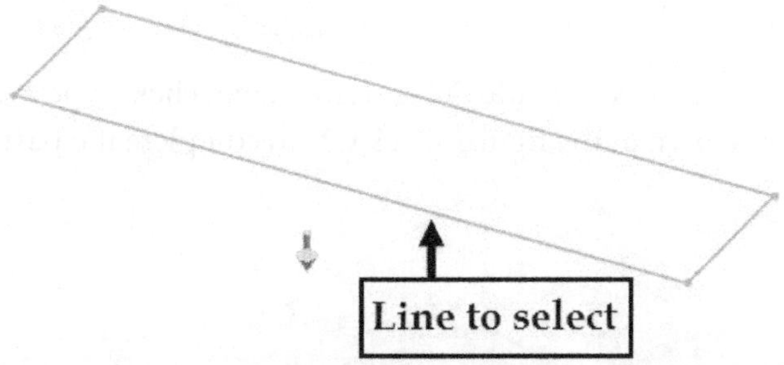

7. On the **Revolve** PropertyManager, type-in 180 in the **Direction 1 Angle** box, and then click the **Reverse Direction** button. Next, click **OK** to create the *Revolved Boss/Base* feature.

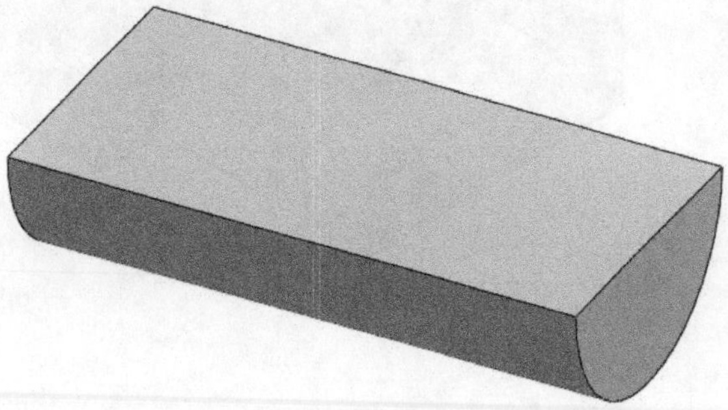

8. On the **Features** CommandManager, click the **Revolved Cut** icon (or) click **Insert > Cut >Revolve** on the Menu. Next, select the top face of the part geometry.
9. Draw the sketch on the top face and apply dimensions. Exit the sketch.
10. Type-in 180 in the **Direction 1 Angle** box. Next, click the **Reverse Direction** button, and then click **OK** to create the revolved cut.

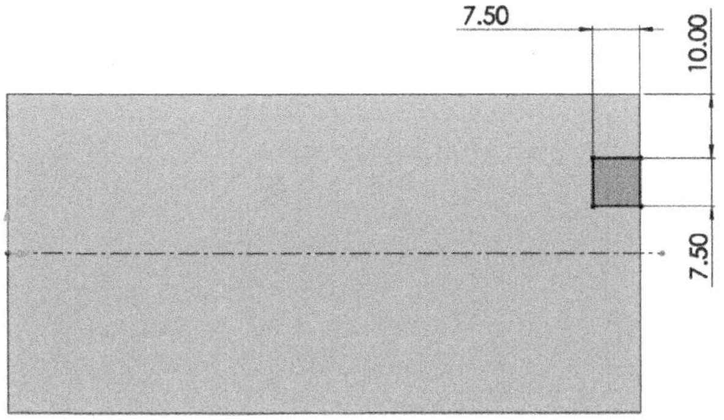

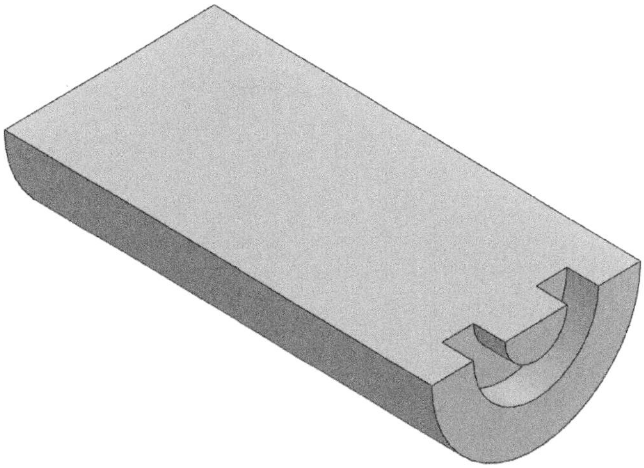

11. Activate the **Revolved Boss/Base** command, and then click on the part geometry's top face. Next, draw a sketch and click **Exit Sketch**.

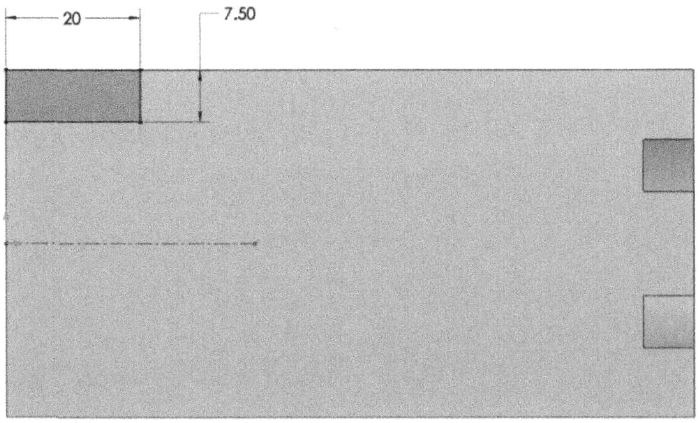

12. Type-in **180** in the **Direction 1 Angle** box. Click **OK** to add the *Revolved* feature to the geometry.

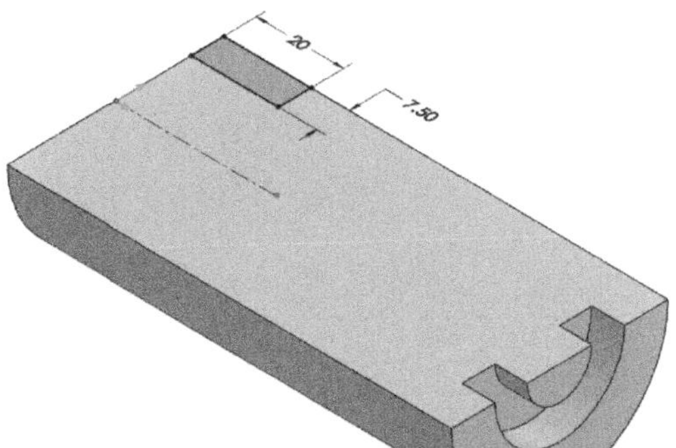

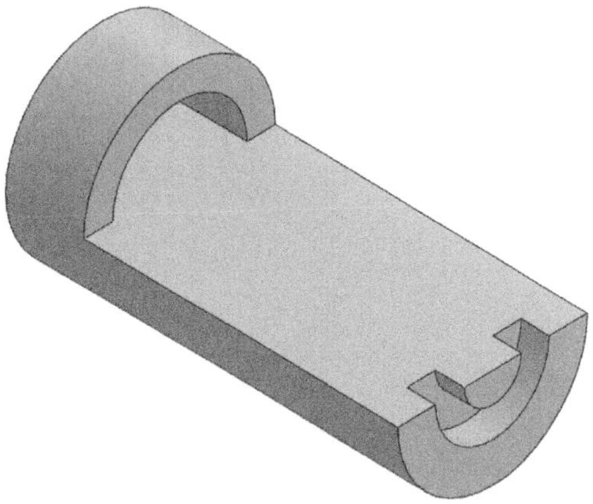

13. Save and close the file.

Questions

1. How do you create parallel planes in SOLIDWORKS 2024?
2. List anyone of the **End Condition** types available on the **Boss-Extrude** PropertyManager.
3. How do you create angled planes in SOLIDWORKS 2024?
4. List the methods to create an axis.
5. How to extrude an open sketch?

Exercises

Exercise 1

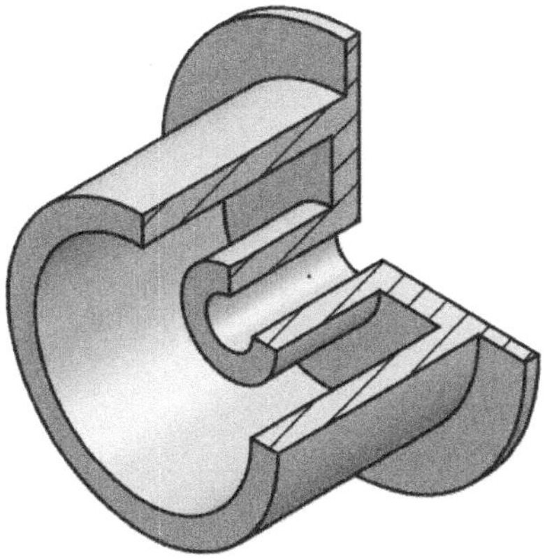

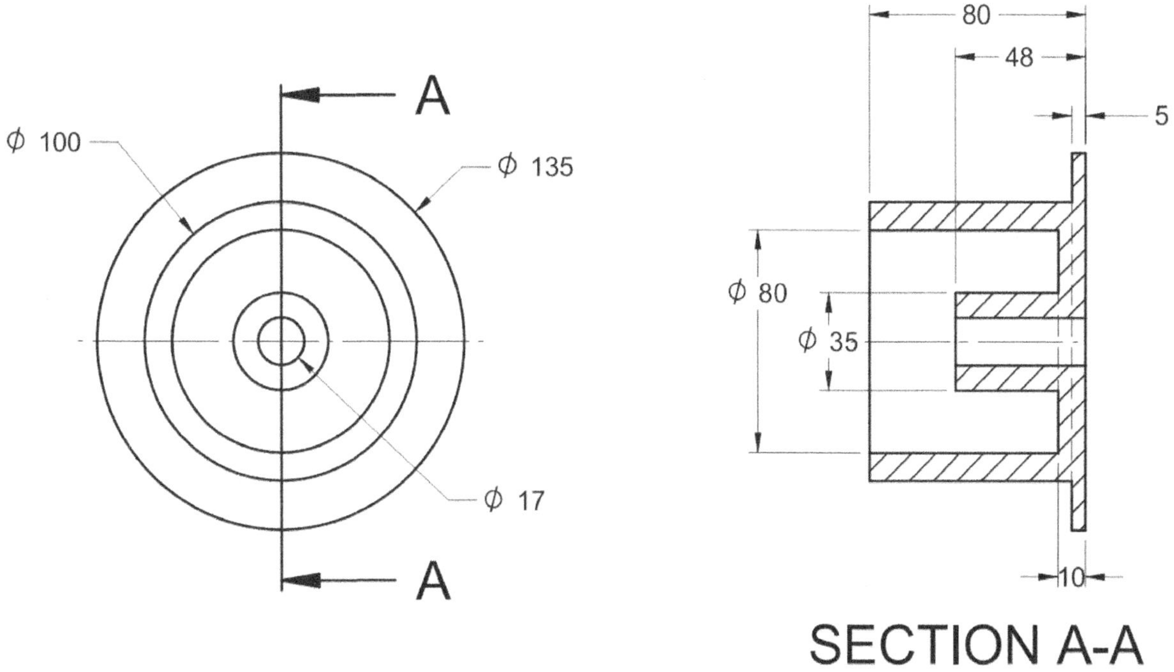

Exercise 2

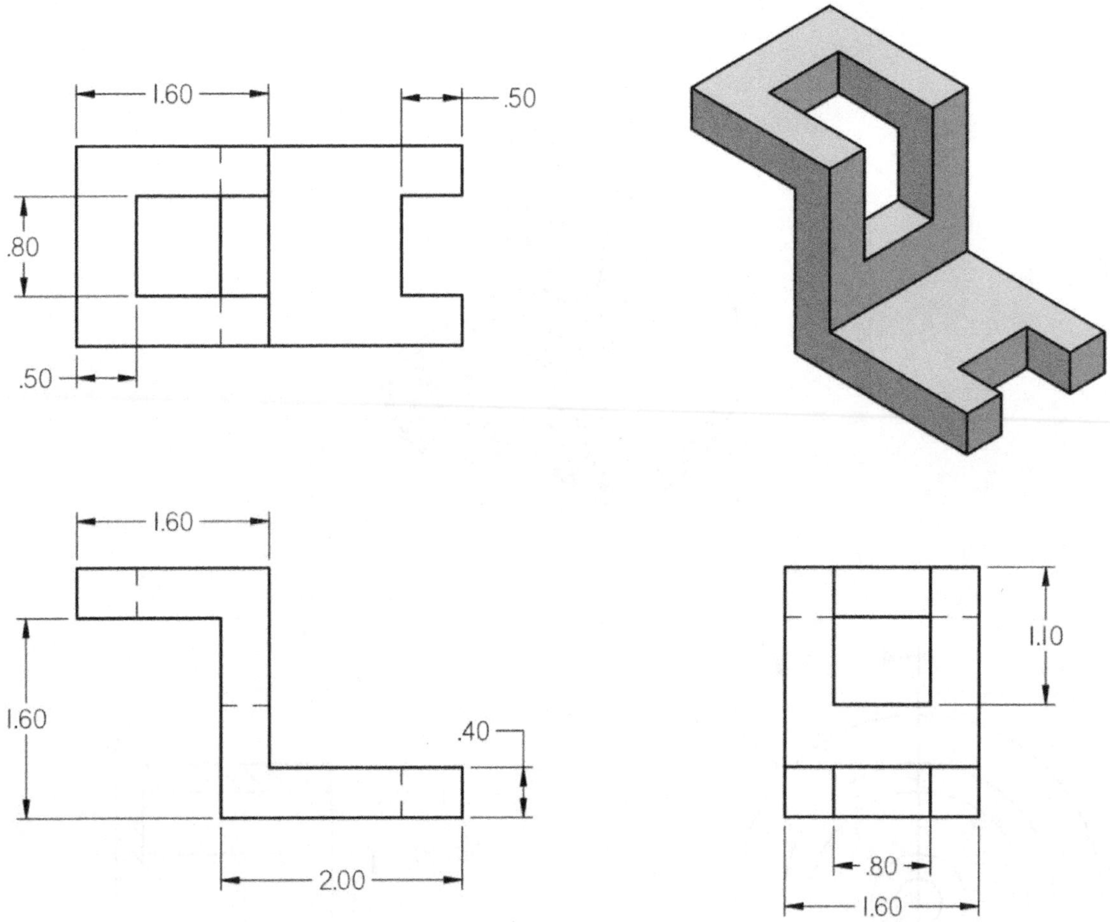
1.60
.50
.80
.50
1.60
1.60
.40
2.00
1.10
.80
1.60

Exercise 3

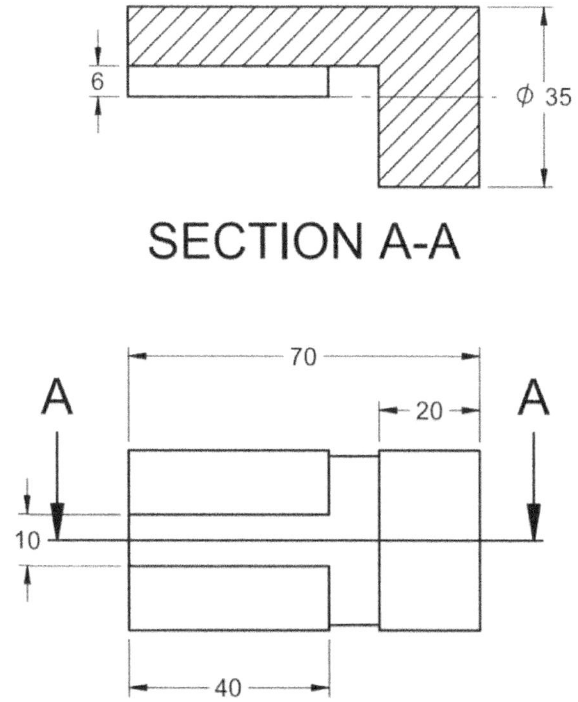

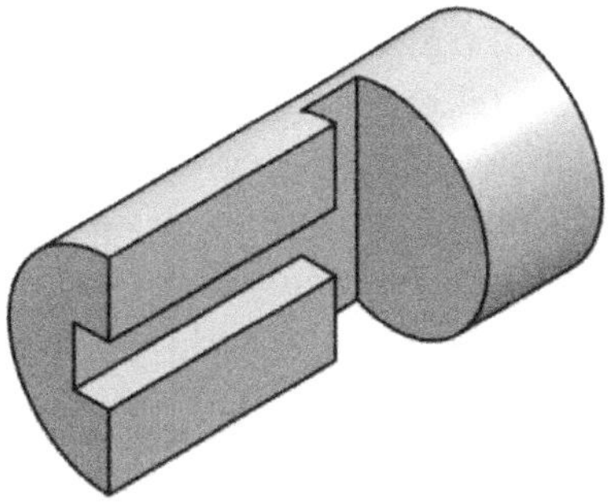

Chapter 4: Placed Features

Until now, all of the features that were covered in the previous chapter were based on two-dimensional sketches. However, there are certain features in SOLIDWORKS that do not require a sketch at all. Features that do not require a sketch are called Placed features. You can just place them on your models. However, you must have some existing geometry to add these features. Unlike a sketch-based feature, you cannot use a Placed feature for a model's first feature. For example, to create a *Fillet* feature, you must have an already existing edge. In this chapter, you will learn how to add Holes and Placed features to your design.

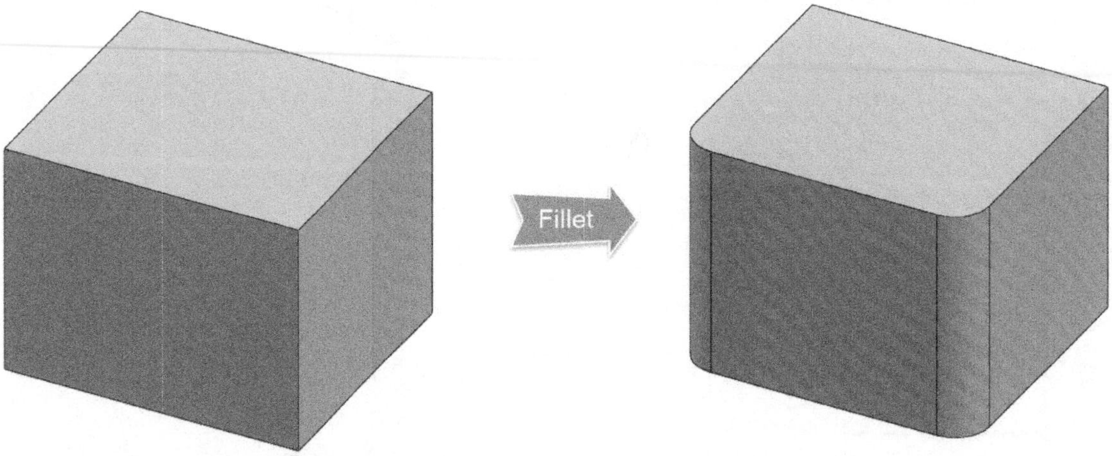

The topics covered in this chapter are:

- *Holes*
- *Threads*
- *Slots*
- *Fillets*
- *Chamfers*
- *Drafts*
- *Shells*

Hole Wizard

You can create holes by merely creating a circular sketch and removing the material using the **Extruded Cut** command. However, if you want to create holes of standard sizes, the **Hole Wizard** command is a better way to do this. The reason for this is it has many hole types already predefined for you. All you have to do is choose the correct hole type and size. The other benefit is when you create a 2D drawing, SOLIDWORKS can automatically place the correct hole annotation. Activate this command (On the **Features** CommandManager, click the **Hole Wizard** icon) and then select a hole-type from the **Hole Specification** PropertyManager. The options in this PropertyManager make it easy to create different types of holes.

Simple Hole

To create a simple hole feature, select **Hole Type > Hole** from the **Hole Specification** PropertyManager. Select the hole standard from the **Standard** drop-down. There are different standards such as ANSI Inch, ANSI Metric, ISO, and DIN. Next, select the **Type** (**Dowel Holes**, **Drill sizes**, **Helical Tap Drills**, **Screw Clearances**, or **Tap Drills**). Under the **Hole Specifications** section, select the hole size from the **Size** drop-down. You can also check the **Show custom sizing** option and then type in a value in the **Through Hole Diameter** box. This creates a hole with a custom diameter.

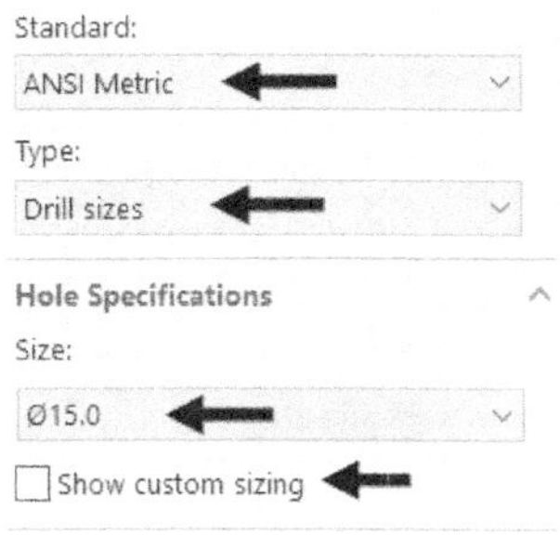

If you want to create a through-hole, click **End Condition** > **Through All**. If you want a blind hole, then select **Blind** from the **End Condition** drop-down. Next, type in a value in the **Blind Hole Depth** box.

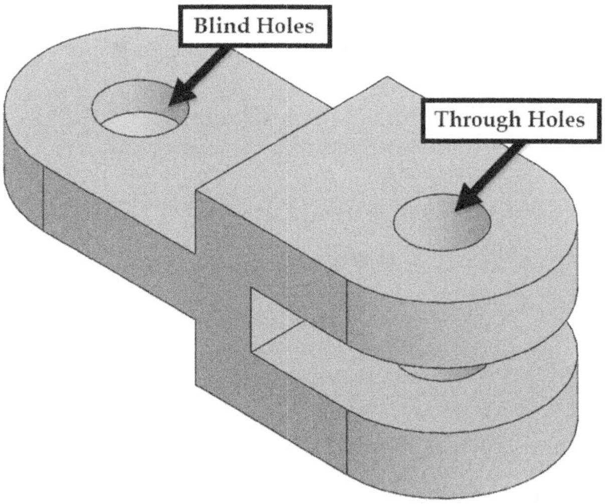

The **Angle at Bottom** box defines the angle of the cone tip at the bottom. Note that this box is available when you select the **Blind** option from the **End Condition** drop-down, and then check the **Show custom sizing** option.

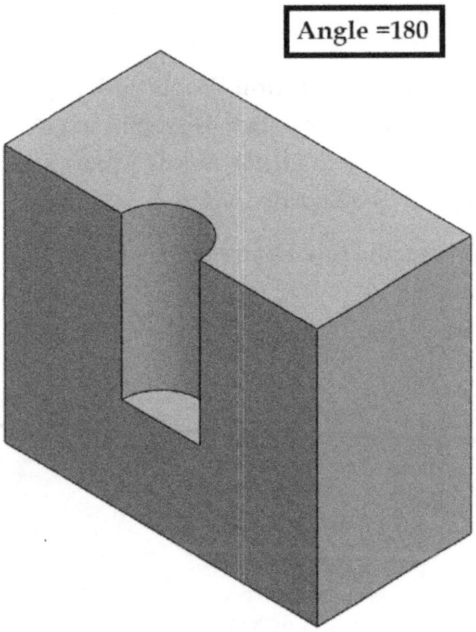

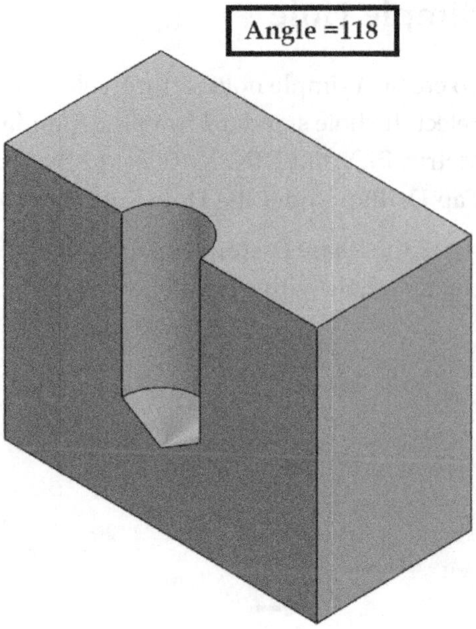

Saving the Hole Settings

While creating a hole, SOLIDWORKS allows you to save the settings specified on the **PropertyManager** for future use. You can use these settings to create a hole with the same specifications' multiple times. On the **PropertyManager**, click the **Add or Update Favorite** icon in the **Favorite** section. Next, type a name for the favorite and click the **OK** button; the settings are saved. You can access the saved settings from the **Favorite** drop-down.

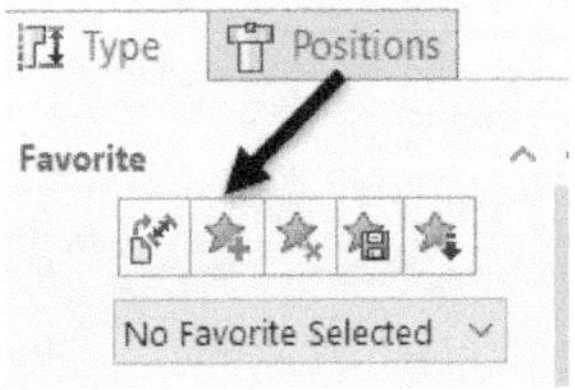

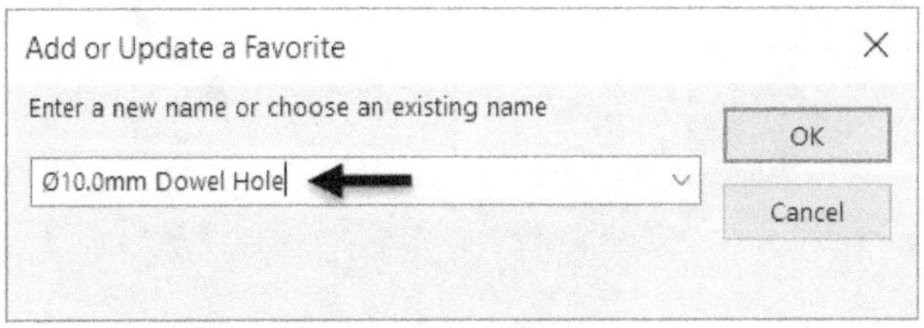

The **Favorite** section has other options such as **Apply Defaults/No Favorites**, **Delete Favorite**, **Save Favorite**, and **Load Favorite**. These options are self-explanatory.

In the PropertyManager, go to the **Positions** tab and choose the hole placement face. You have two options for positioning the hole: you can either specify the position directly and fully define it using relations and dimensions, or you can click on **Existing 2D Sketch** under Hole Positions and select an existing 2D sketch. This automatically creates holes at all endpoints, vertices, and points of the selected sketch geometry, such as lines, rectangles, slots, and splines.

Under Sketch Options, you'll find two choices:

- Choose **Create instances on sketch geometry** to position holes at all endpoints, vertices, and points of the sketch geometry.

- Opt for **Create instances on construction geometry** to position holes at all endpoints, vertices, and points of the construction geometry.

Additionally, you can skip hole instances by selecting them under 'Instances to Skip' directly from the graphics area. Once you've made your selections, click the green checkmark in the top right corner of the graphics window to confirm. Your hole will then be created normal to the selected face.

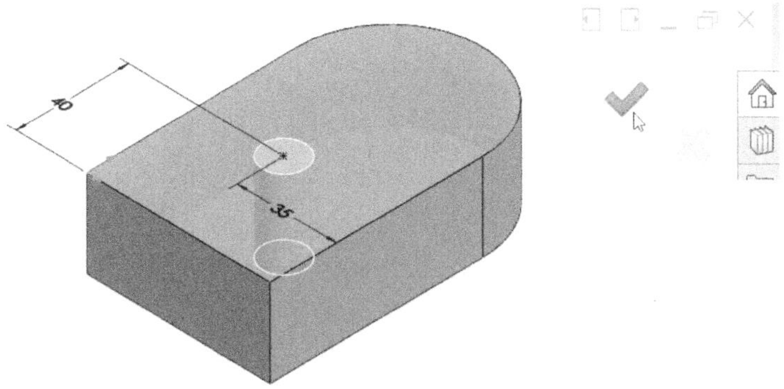

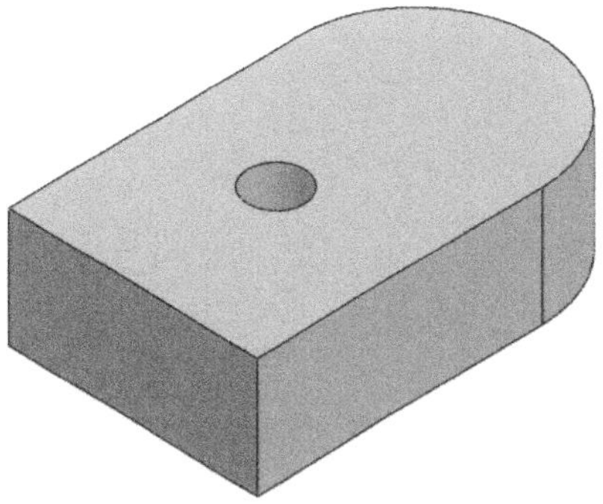

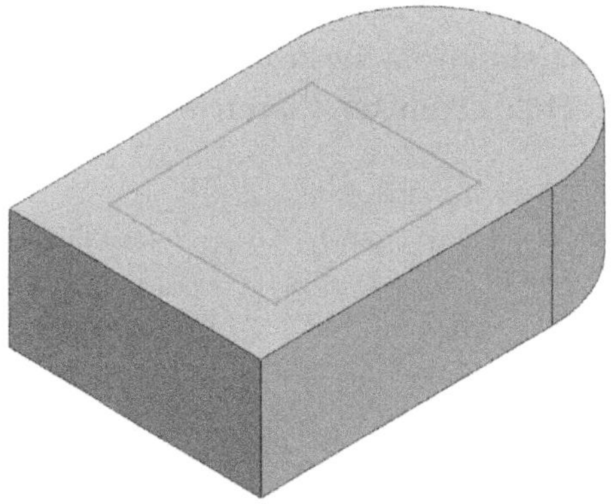

Counterbored Hole

A counterbored hole is a large diameter hole added at the opening of another hole. It is used to accommodate a fastener below the level of the workpiece surface. To create a counterbore hole, Activate the **Hole Wizard** command (click **Features > Hole Wizard > Hole Wizard** on the CommandManager). On the PropertyManager, select **Counterbore** from the **Hole Type** section. Select the required standard from the **Standard** drop-down, and then select the drill size from the **Type** drop-down. In the **Hole Specifications** section, select the hole diameter and fit from the **Size** and **Fit** drop-downs, respectively. You can also check the **Show custom sizing** option and type in the **Through Hole Diameter**, **Counterbore Diameter**, **Counterbore Depth** values. Next, use the **End Condition** section to define the hole depth, and then position it using the **Positions** tab.

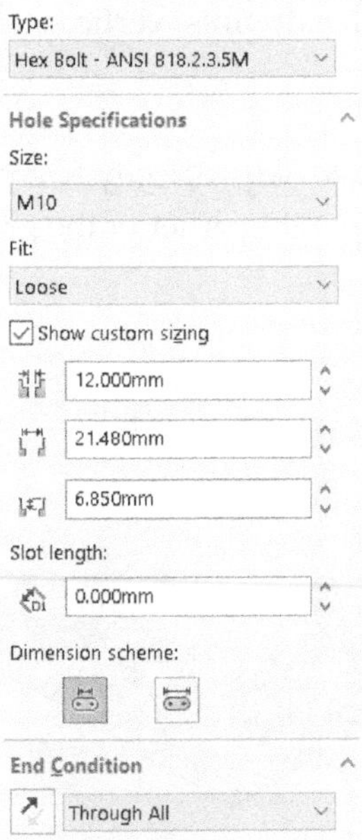

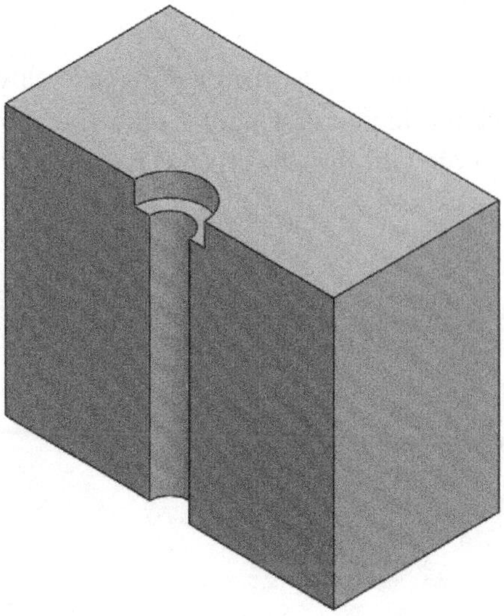

Countersink Hole

A countersunk hole has an enlarged V-shaped opening to accommodate the fastener below the level of the workpiece surface. To create a countersink hole, select **Hole Type > Countersink** from the **PropertyManager**. Check the **Show custom sizing** option and type-in values in the **Through Hole Diameter**, **Countersink Diameter** and **Countersink Angle** boxes. Set the hole depth and end condition.

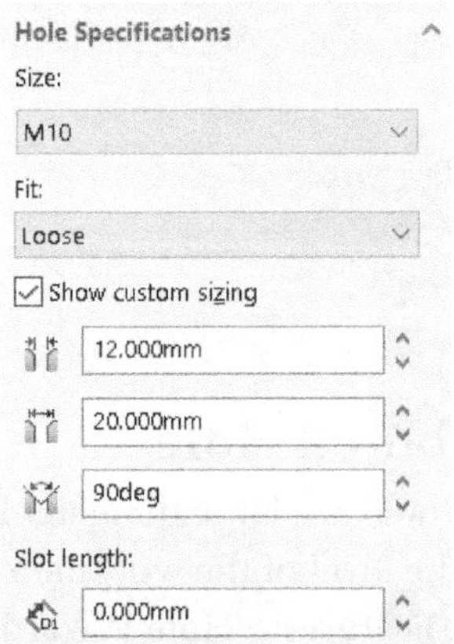

Straight Tap Hole

To create a straight tap hole, select **Hole Type > Straight Tap**. Next, select the thread standard from the **Standard** drop-down. Select the tap (**Straight Pipe Tapped Hole, Bottoming Tapped Hole**, and **Tapped Hole**) and size from the **Type** and **Size** drop-downs, respectively. In the **End Condition** section, select **Thread > Through All** to create the thread for the full depth of the hole. If you select the **Blind (2*DIA)** option, the thread length is calculated automatically using the diameter value. If you want to manually specify the thread length, then deselect the **Automatically calculate thread depth** icon. Next, specify the thread depth in the **Tap Thread Depth** box.

In the **Options** section, select the **Tap Drill diameter** option to create a hole with the tap's diameter. Select the **Cosmetic Thread** option to create a hole with cosmetic thread. You can also display the thread callout by checking the **With Thread Callout** option. Select the **Remove thread** option to create a hole with the diameter of the thread.

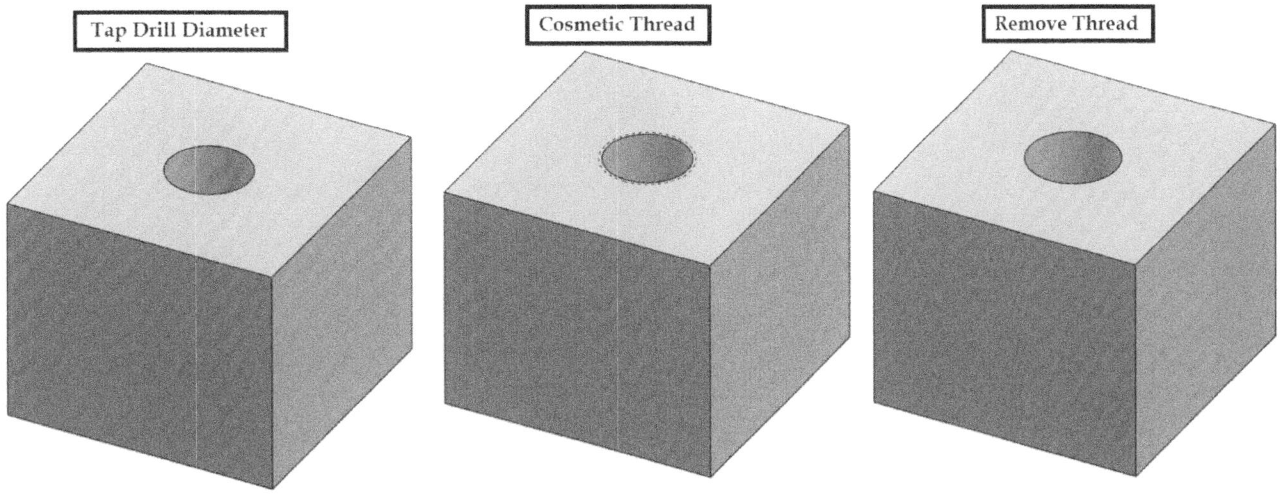

Check the **Thread Class** option and select the class (**1B**, **2B**, or **3B**). Next, check the **Near side countersink** and **Far side countersink** options (optional), and then specify the parameters. Next, specify the hole's position using the **Positions** tab, and then click **OK** to create the straight tapped hole.

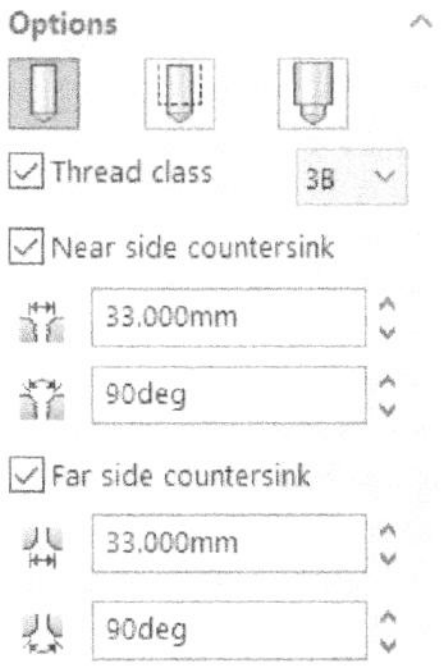

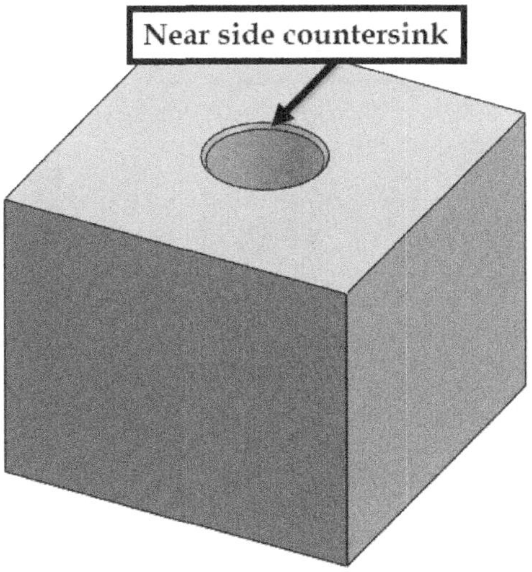

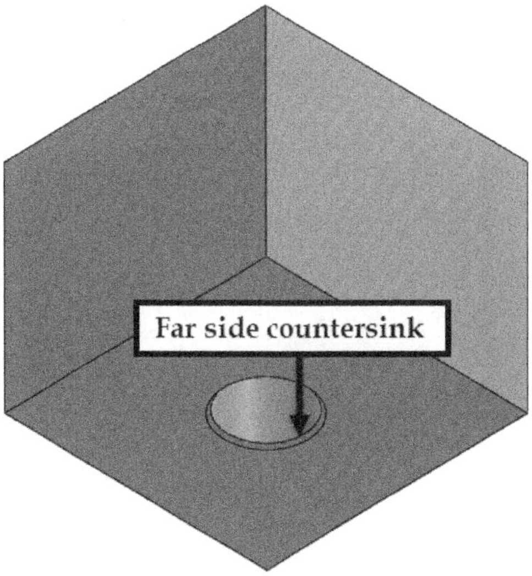

Tapered Tap Hole

Tapering is the process of decreasing the hole diameter toward one end. A tapered tapped hole has a thread, and the diameter gradually becomes smaller towards the bottom. To create a tapered tapped hole, select **Hole Type > Tapered Tapped Hole**. Next, you need to specify the thread **Standard**, **Type,** and **Size**. The tapered thread types and sizes are different from the regular threads.

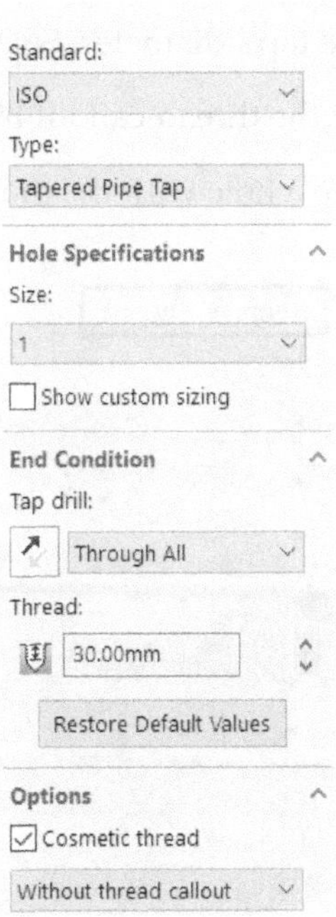

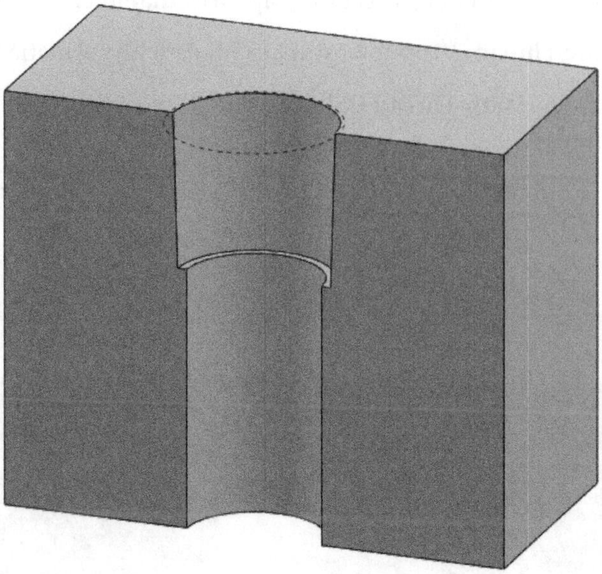

Creating a Legacy Hole

The **Legacy Hole** is the **Hole** feature that was available in the older versions of SOLIDWORKS. It can create different holes like C-Drilled Drilled, C-Sunk Drilled, C-Bored Drilled, Tapered Drilled, Simple Drilled, Counter drilled, Countersunk, Counterbore, Tapered and Simple holes.

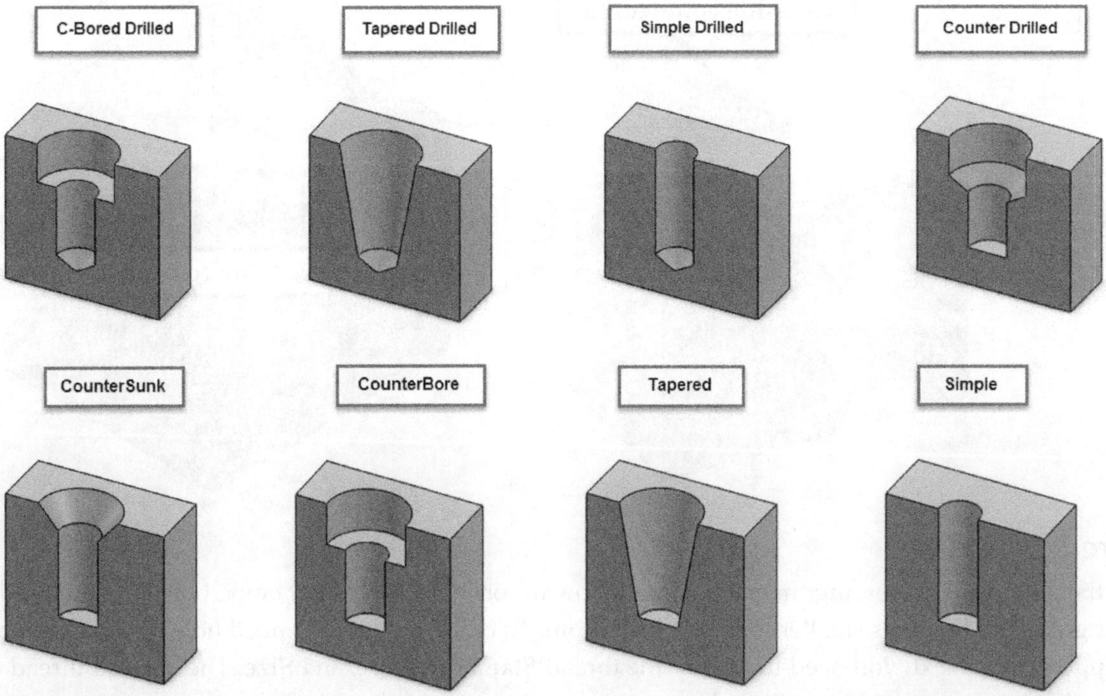

Creating Slots

In SOLIDWORKS, you can create different types of slots, such as Counterbore slot, Countersink slot, and straight slot. To do this, click **Features > Hole Wizard > Hole Wizard** on the CommandManager. Click the **Counterbore slot**, **Countersink slot** or **Slot** button from the **Hole Type** section on the **Hole Specification** PropertyManager.

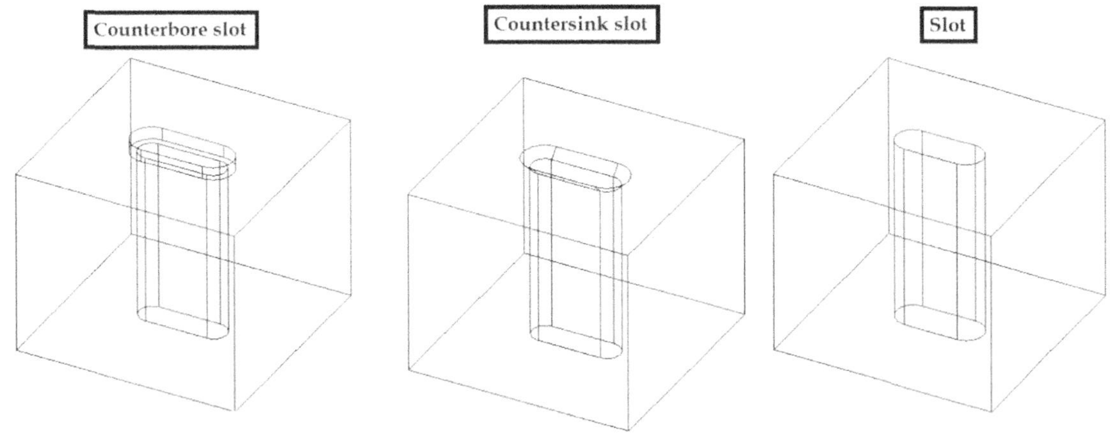

Select the standard from the **Standard** drop-down and select the type from the **Type** drop-down. Next, define the **Size** and **Fit** from the **Hole Specifications** section. Specify the length of the slot in the **Slot Length** box.

Click the **Positions** tab and select a face from the model geometry. Next, press the TAB key, if you want to change the orientation of the slot by 90 degrees. Place the slot on the selected face and fully define it using the dimensions and relations.

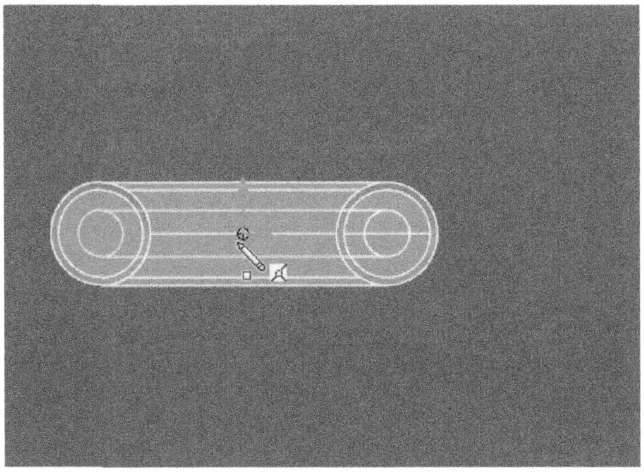

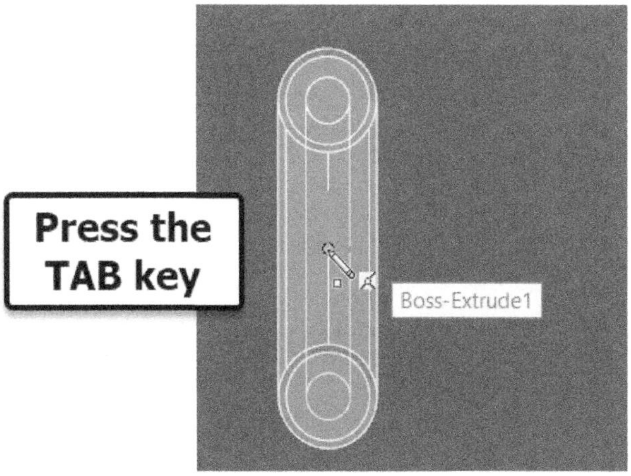

Advanced Hole

The **Advanced Hole** command was added in SOLIDWORKS 2017. This command lets you create holes that are not available in the **Hole Wizard** command. For example, you can create a hole that combines the counterbore, straight tap, and countersink. Activate this command (on the CommandManager, click **Features > Hole Wizard** drop-down > **Advanced Hole**) and notice that the **Advanced Hole** PropertyManager appears along with the **Near side** flyout. This flyout helps you to select elements that form the advanced hole. Select a flat face from the model geometry to specify the near-side face; the preview of the hole appears at the location where you select the face.

The **Near side** flyout has four options: **Insert Element Below Active Element**, **Insert Element Above Active Element**, **Delete Active Element**, and **Reverse Stack Direction**. Notice that the counterbore hole is selected as the first hole type in the drop-down available in the **Near side** flyout. You can click the down-arrow and select another hole type.

Click the **Insert Element Below Active Element** icon on the **Near side** flyout to add a near side element. Click the second drop-down and select the new hole type (for example, select **Hole**).

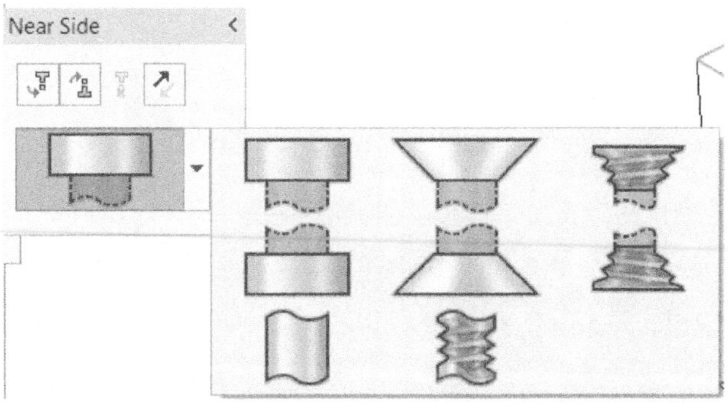

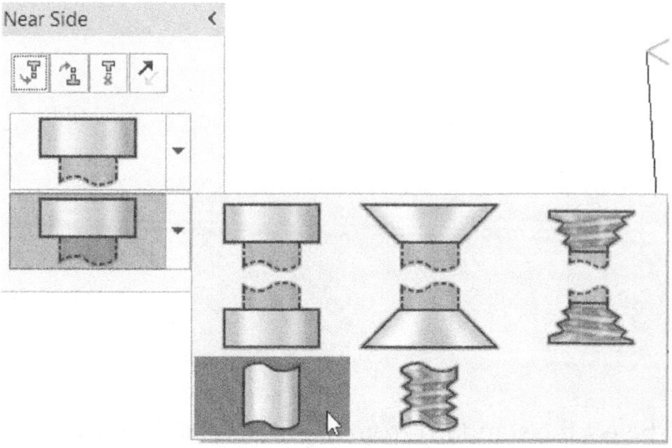

The **Advanced Hole** command allows you to specify a different hole type on the far side. To do this, check the **Far side** option in the **Near And Far sides** section of the **Advanced Hole** PropertyManager. Notice that the **Far Side** flyout is displayed below the **Near side** flyout. Now, select the face on the opposite side of the near side face. Click the drop-down displayed in the **Far side** flyout and select the required hole type (For example, select **Straight Tap**).

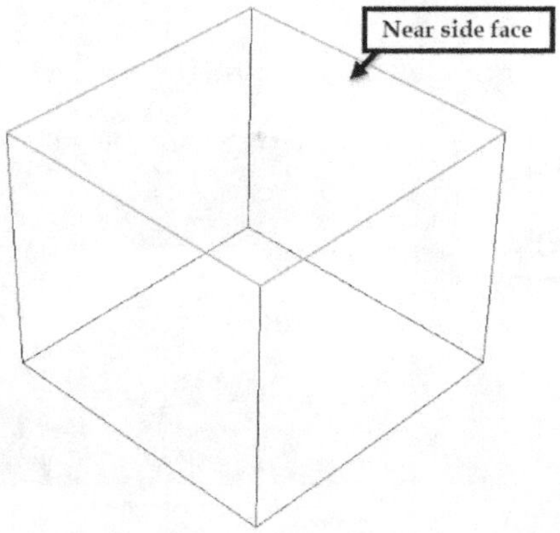

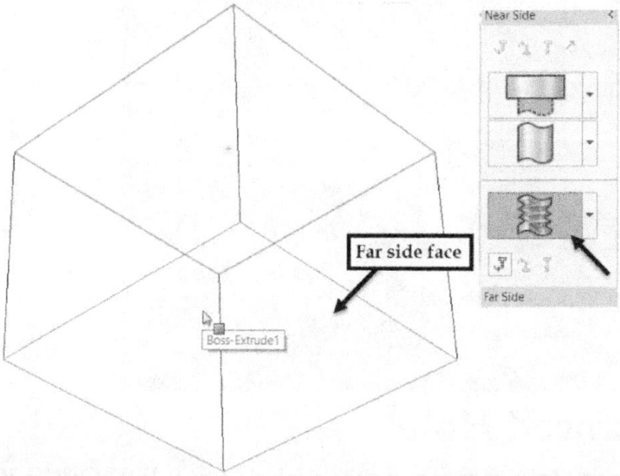

After specifying the hole types, you need to specify the standards and sizes for each of them. Select the first hole type from the **Near side** flyout. In the **Element Specification** section of the PropertyManager, specify the **Standard, Type**, and **Size** (For example, select **Standard > ANSI Inch, Type > Binding Head Screw**, and **Size >#12**). You can also specify a custom size by clicking the **Diameter Override** icon.
Select the second hole type from the **Near side** flyout. Next, specify the settings in the **Element Specification** section (For example, select **Standard > ANSI Inch, Type > Screw Clearances, Size > #12, Fit > Normal, End Condition > Blind,** and **Depth > 1**).

Select the hole type from the **Far side** flyout. Next, specify the settings in the **Element Specification** section (For example, select **Standard > ANSI Inch**, **Type > Tapped Hole**, **Size > #12-24,** and **End Condition > Up To Next Element**).

Click the **Positions** tab and position the hole using dimensions and relations. Next, click the green check to create an advanced hole.

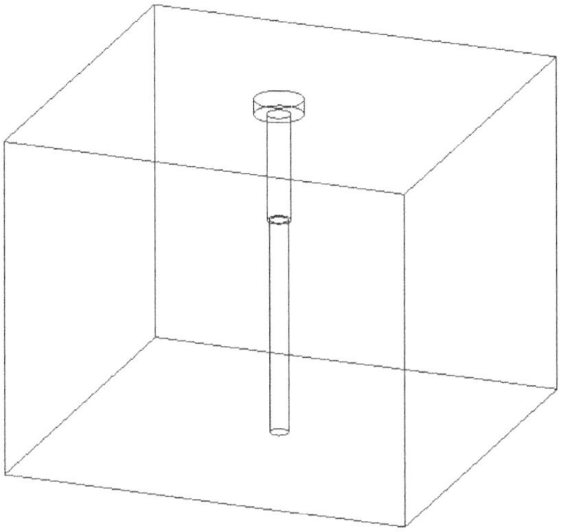

Stud Wizard

The **Stud Wizard** tool is used to place externally threaded studs on the model. Activate this tool (On the **Features** CommandManager, click the **Hole Wizard drop-down > Stud Wizard**) and click the **Creates stud on Cylindrical Body** button on the PropertyManager. Next, select the circular edge of an existing cylindrical feature.

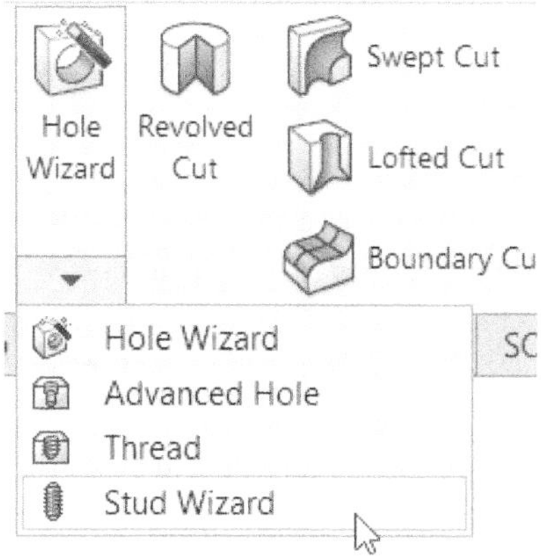

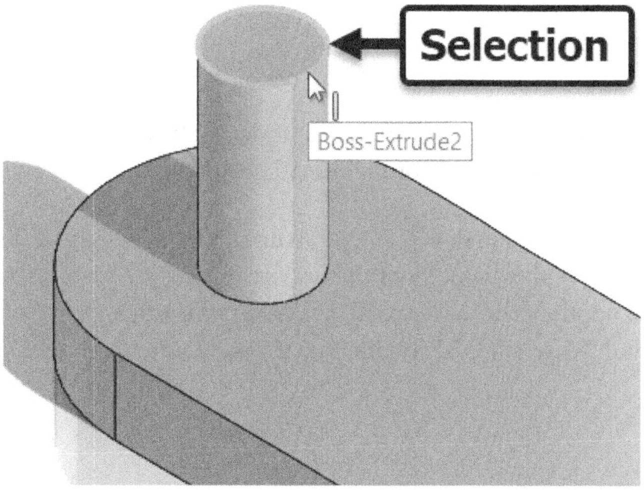

Select the required standard from the **Standard** drop-down, and then select the thread type (**Machined Thread** or **Straight Pipe Tapped Thread**) from the **Type** drop-down. Next, select the thread diameter and pitch combination from the **Size** drop-down. Select an option from the **Thread** drop-down: **Blind, Up to Next,** or **Through**. The **Blind** option defines the thread's length by the value you enter in the **Thread Depth** box. The **Thread Class** option allows you to specify the thread tolerance classification. The selected tolerance classification will appear in the thread callout associated with the stud when you create a 2D drawing. But it will not be displayed in the model geometry.

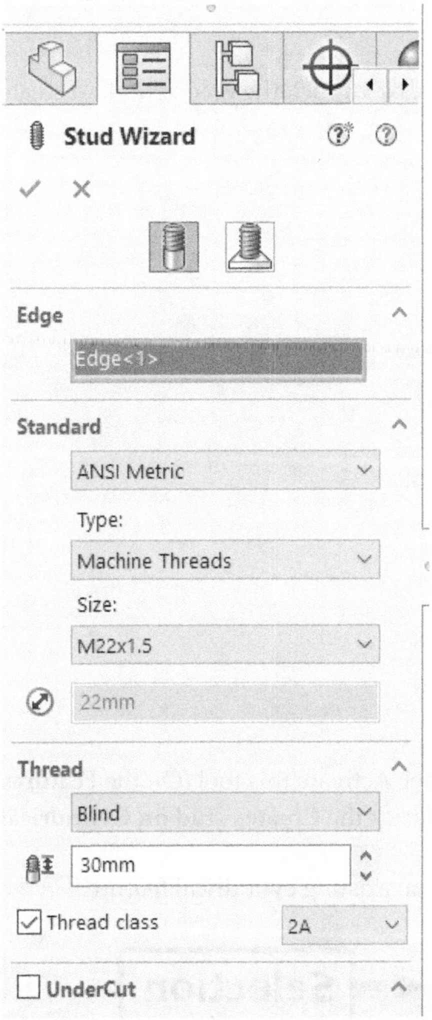

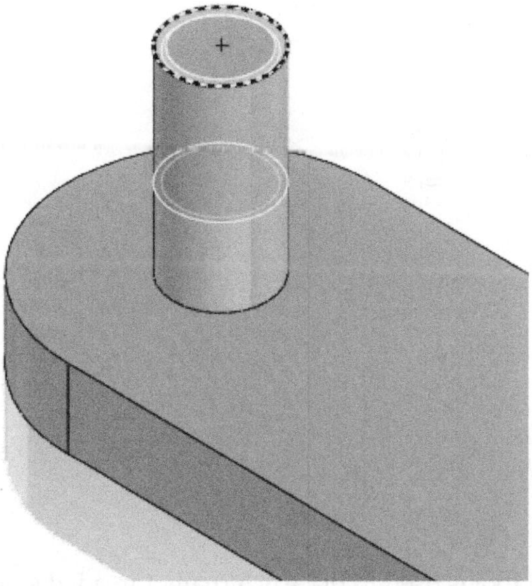

Check the **Undercut** option to create an undercut below the thread. The undercut is basically a revolved cut that is used to provide clearance for the thread cutting tool. Notice that there are three boxes under the **Undercut** section: **Undercut Diameter** , **Undercut Depth** , and **Undercut Radius** . The **Undercut Diameter** box is used to specify the diameter of the undercut after the material is removed. The **Undercut Depth** box is used to specify the depth of the undercut. The **Undercut Radius** box is used to specify the radius of the fillets applies to the undercut. Next, click **OK** on the PropertyManager.

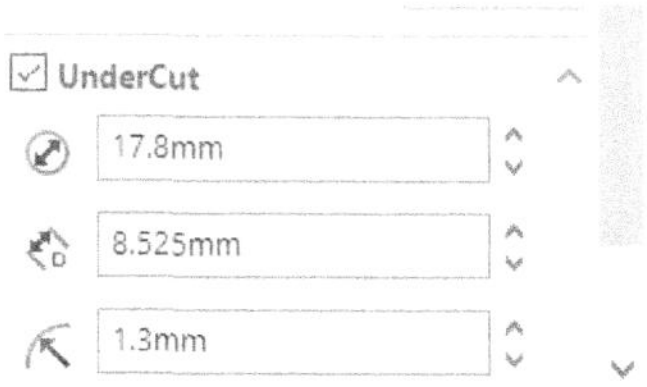

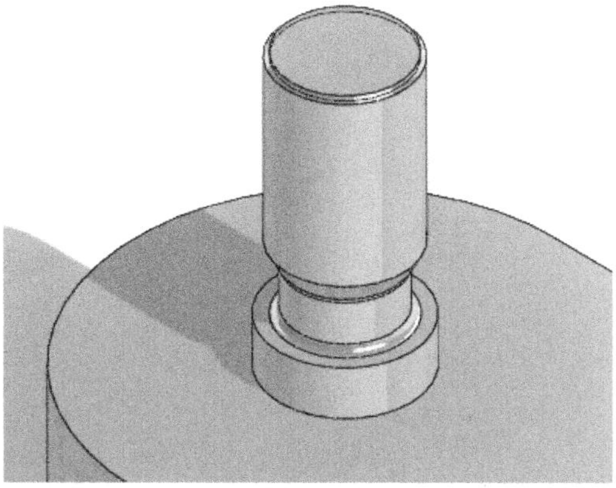

Creating a Stud on Surface

The **Create Stud on a Surface** button helps you to place a stud directly on a planar surface. Activate the **Stud Wizard** tool and click this button on the PropertyManager. Next, click the **Position** tab on the PropertyManager and select the surface on which you want to place the stud. Next, specify the stud position, and then fully define it using the relations and dimensions.

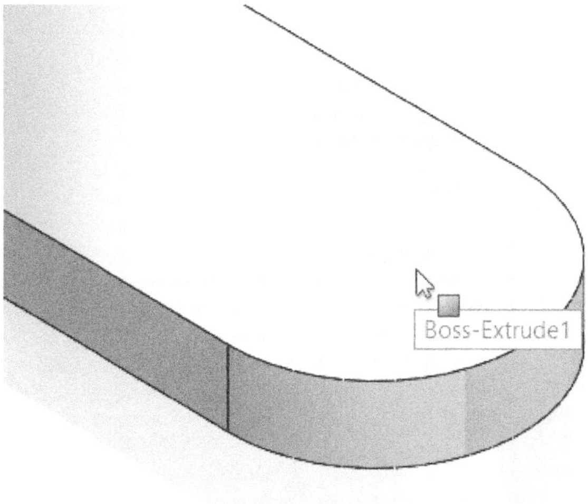

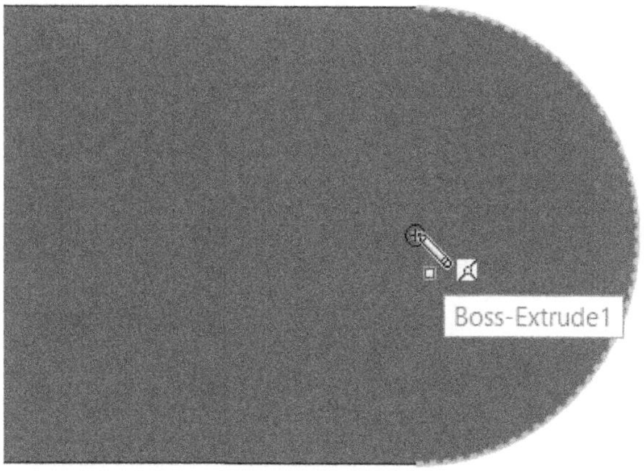

Click the **Stud** tab on the PropertyManager and enter values in the **Shaft length** and **Shaft Diameter** boxes in the **Shaft Details** section. Next, specify the options and parameters in the **Standard**, **Thread**, and **Undercut** sections. The options in these sections are already discussed earlier. Click the green check on the PropertyManager to complete the stud feature.

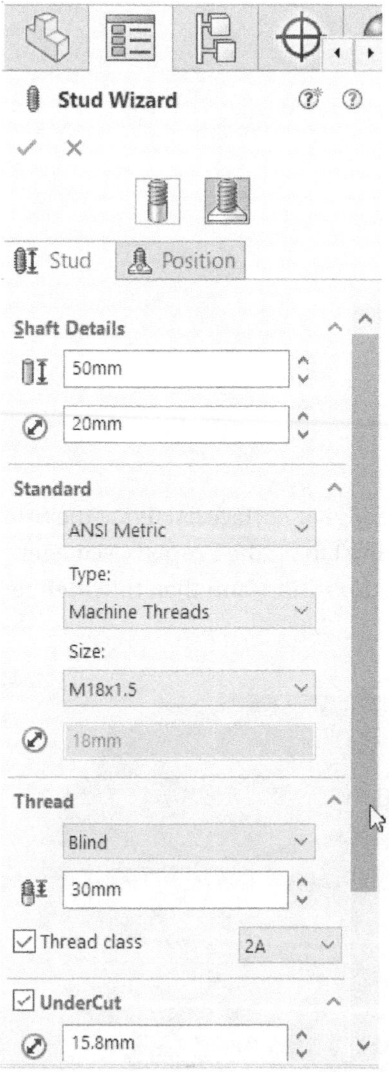

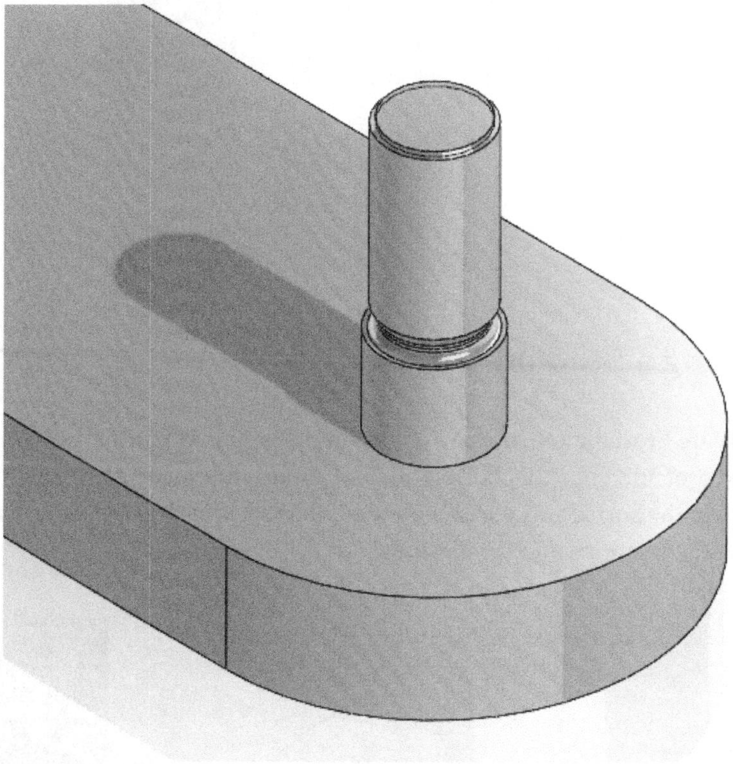

Threads

This command adds a thread/tap feature to a cylindrical face. A thread is added to the outer cylindrical face, whereas a tap is added to the inner cylindrical face (holes). You add a thread feature to a 3D geometry so that when you create a 2D drawing, SOLIDWORKS can automatically place the correct thread annotation. On the **Features** CommandManager, click the **Hole Wizard drop-down > Thread** (or) click **Insert > Features > Thread** on the Menu bar. The **SOLIDWORKS** dialog pops up on the screen, as shown. Click **OK** on the dialog. Click on the round edge of the part geometry to define the diameter of the thread. By default, the selected edge is used as the start point of the thread. However, you can define a different start point by clicking in the **Optional Start Location** box and selecting a face or point. Next, specify the **Start Angle** value or leave it as default. The start angle is the angle at which the thread starts.

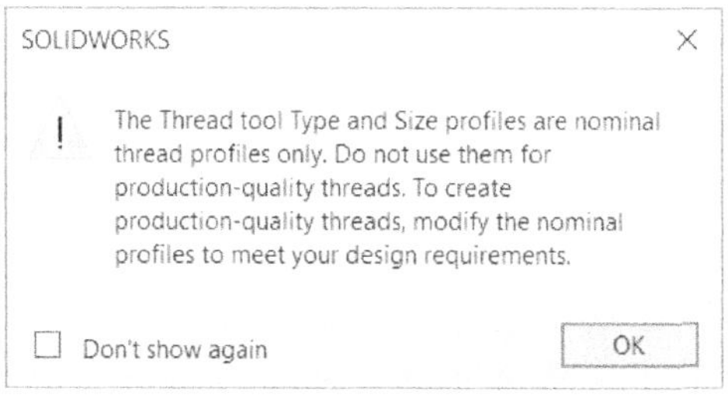

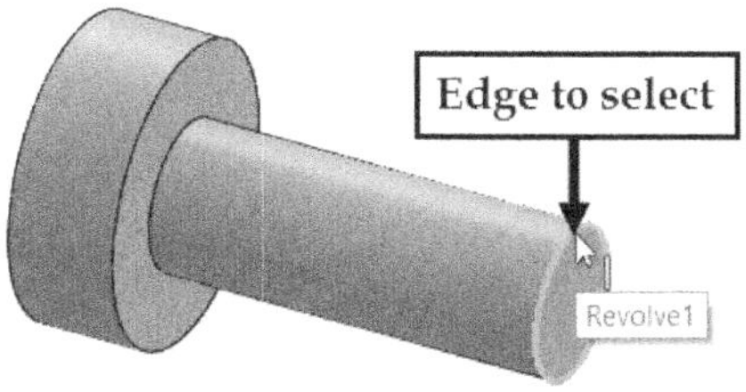

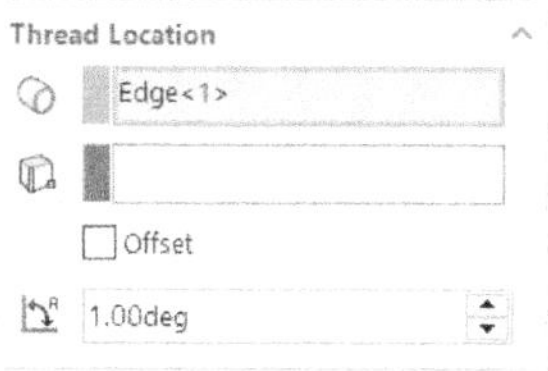

Select an option from the **End Condition** drop-down: **Blind, Revolutions, Up to Selection**. The **Blind** option defines the thread's length by the value you enter in the **Depth** box. The **Revolutions** option defines the thread length by the value that you enter in the **Revolutions** box. The **Up to Selection** option creates the thread up to the face or point that you select.

Under the **Specification** section, select the die from the **Type** drop-down. Next, select the thread size from the **Size** drop-down. You can also override the thread's diameter and pitch values by clicking the Override Diameter and Override Pitch buttons and then enter values in the corresponding boxes. However, a large difference between the cylinder diameter and **Override Diameter** value results in an error.

In the **Thread Options** section, select the **Right-hand thread** or **Left-hand thread** option.

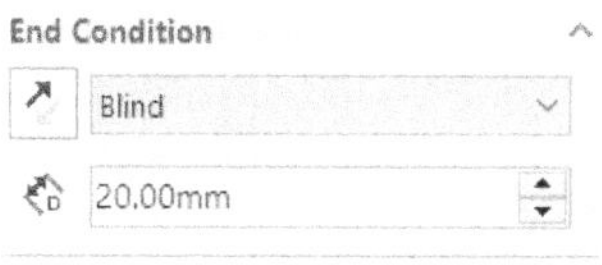

Click **OK** to create the thread. Notice that the start location of the thread is inside the model geometry. Also, when the thread is created up to the flat face, it results in an undercut.

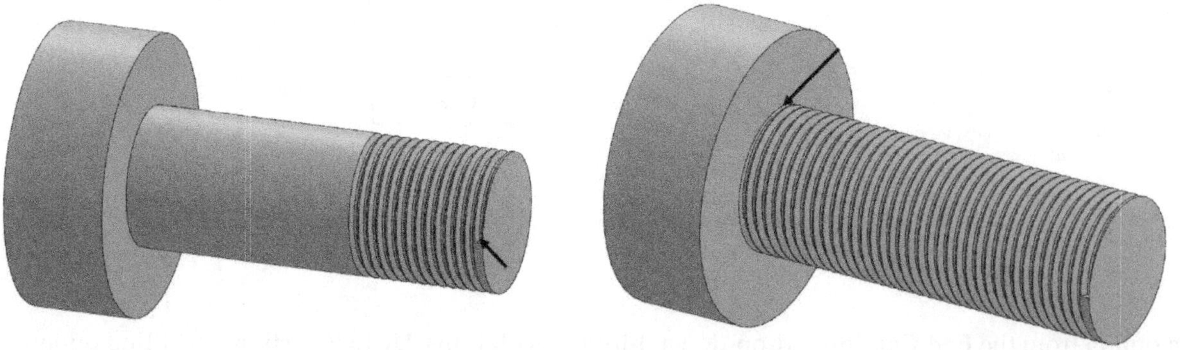

SOLIDWORKS provides you with the Offset, Trim with start face, and Trim with end face options to avoid the above problems. In the FeatureManager Design Tree, click the right mouse button on the **Thread** feature and select **Edit Feature**. On the **Thread** PropertyManager, check the **Offset** option from the **Thread Location** section. Enter the pitch value of the selected thread size in the **Offset Distance** box, and then click the **Reverse Direction** icon next to it. In the **Thread Options** section, check the **Trim with start face** option.

To avoid the undercut at the end face of the thread, select the **Up to Selection** option from the **End Condition** drop-down and select the flat face, as shown. Next, check the **Offset** option and enter the pitch value in the **Offset Distance** box. Click the **Reverse Direction** icon to reverse the offset. Click **OK** on the PropertyManager.

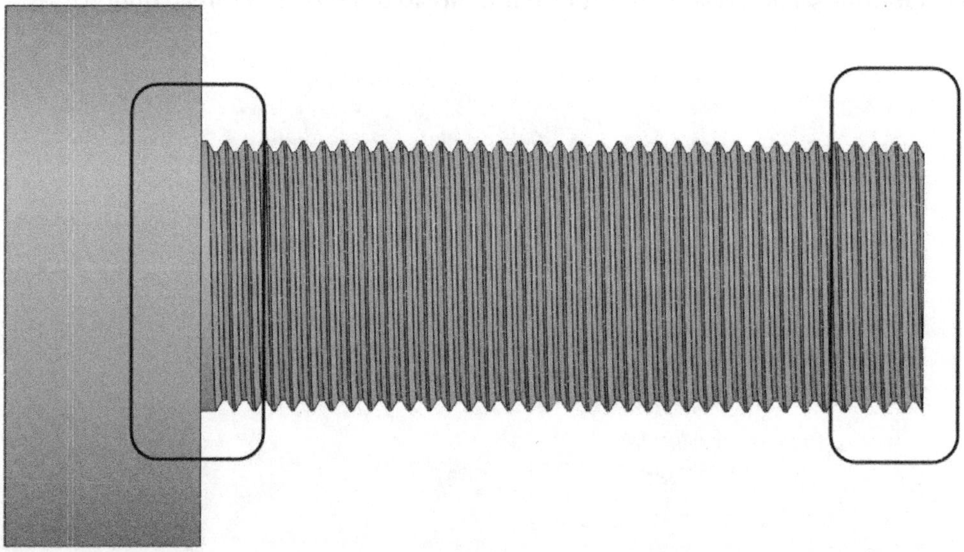

Fillets

This command breaks the sharp edges of a model and blends them. You do not need a sketch to create a fillet. All you need to have is model edges. Click **Features > Fillet > Fillet** on the CommandManager. On the PropertyManager, click the **Manual** tab and select **Constant Size Fillet** from the **Fillet Type** section. Click on the **Items To Fillet** box and select the edges. You can also select all the edges of a face by simply clicking on the face. You can select the edges located at the back of the model without rotating it. To do this, check the **Select Through Faces** option in the **Fillet Options** section. By mistake, if you have selected a wrong edge, you can deselect it by selecting the edge again. You can change the radius by typing a value in the **Radius** box available in the **Fillet Parameters** section (or) clicking in the **Radius** callout attached to the selected edges and typing a new value. As you change the radius, all the selected edges will be updated. This is because they are all part of one instance. If you want the edges to have different radii, check the **Multi Radius Fillet** option in the **Fillet Parameters** section; separate **Radius** callouts are attached to each selected edge. Type in values in the individual **Radius** callouts and click **OK** to finish this feature. The *Fillet* feature will be listed in the FeatureManager Design Tree.

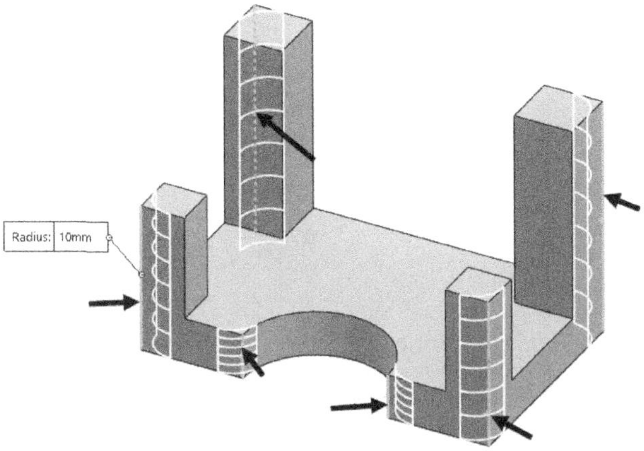

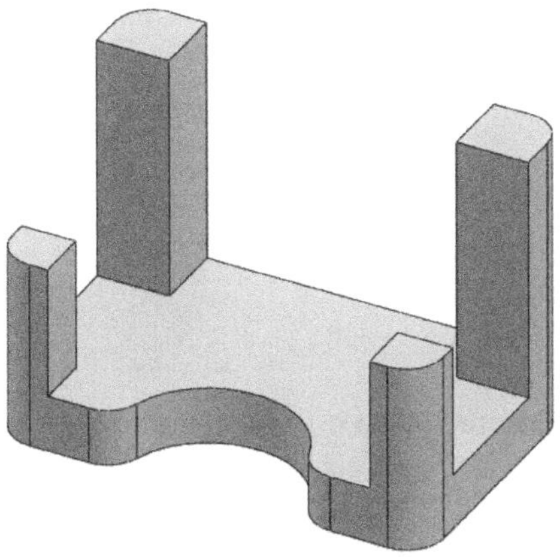

If you want to select all the edges tangentially connected, check the **Tangent Propagation** option on the PropertyManager. If you uncheck this option, the selected edge will be filleted, ignoring the connected ones.

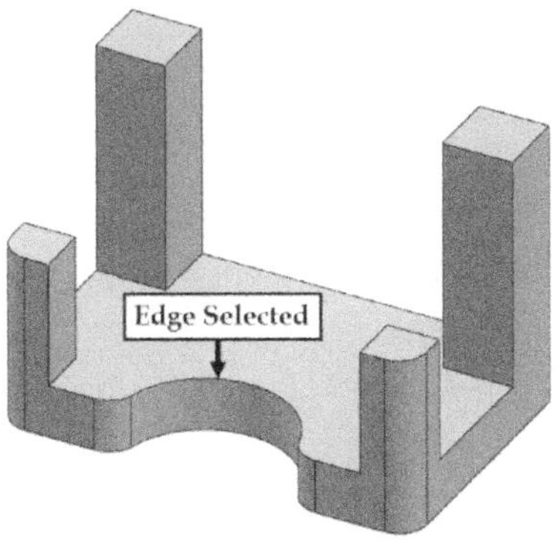

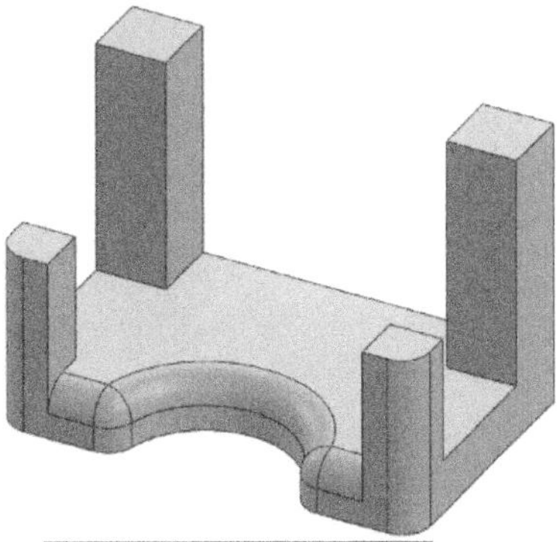

Tangent Propagation ON

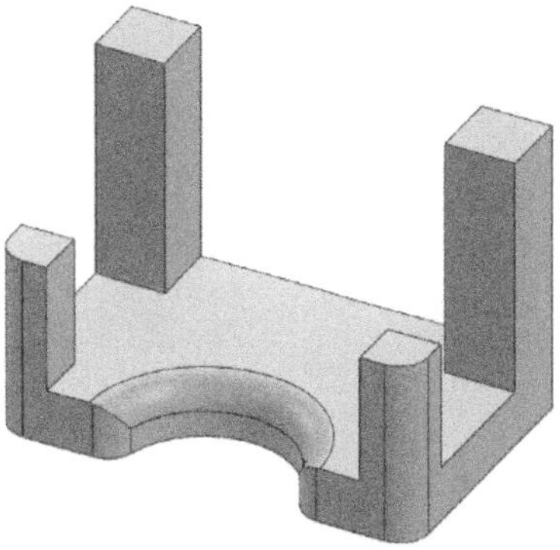

Tangent Propagation OFF

Selecting Multiple Edges using the Selection Toolbar

SOLIDWORKS provides you the Selection toolbar to select multiple edges at a time. Activate the **Fillet** command and check the **Show selection toolbar** option on the PropertyManager. Next, select the vertical edge of the model, as shown. The fillet preview is displayed along with the selection toolbar. Select the **Connected to start face** icon on the selection toolbar; all the edges connected to the start face of the initially selected edge are selected.

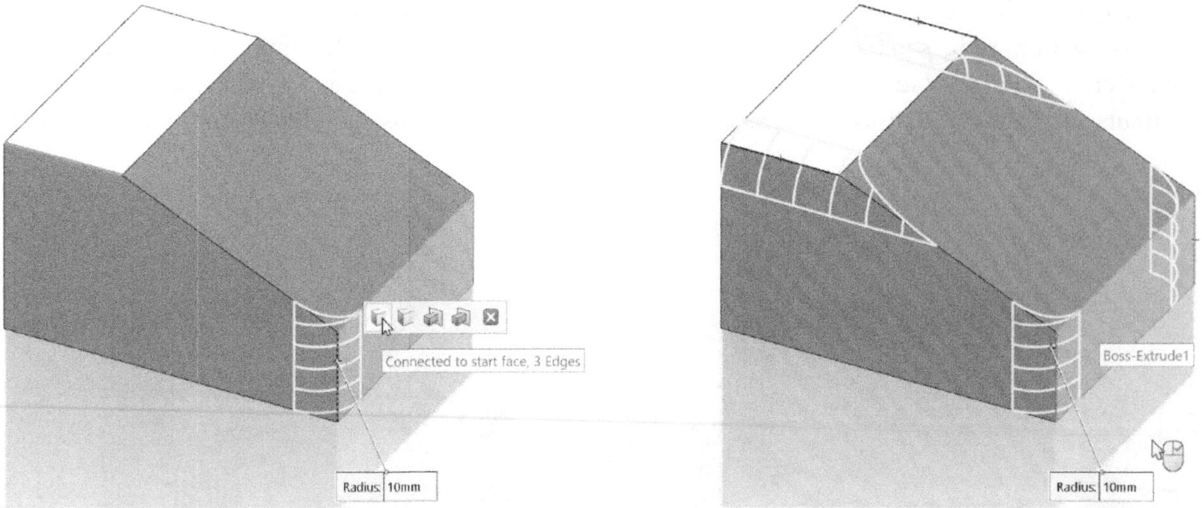

Click the **Connected to end face** icon on the selection toolbar; the edges connected to the end face of the initially selected edge will be selected.

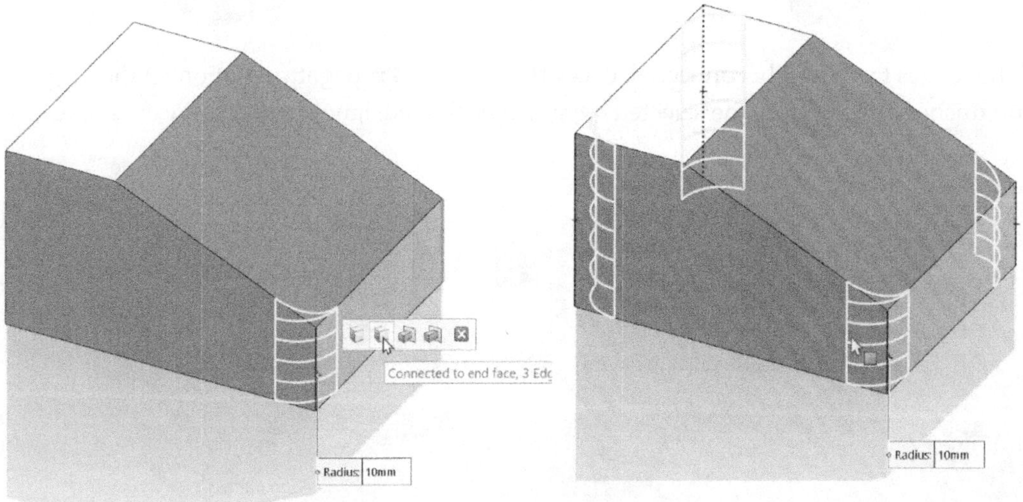

Click the **Connected** icon on the selection toolbar to select all the edges connected to the initially selected edge.

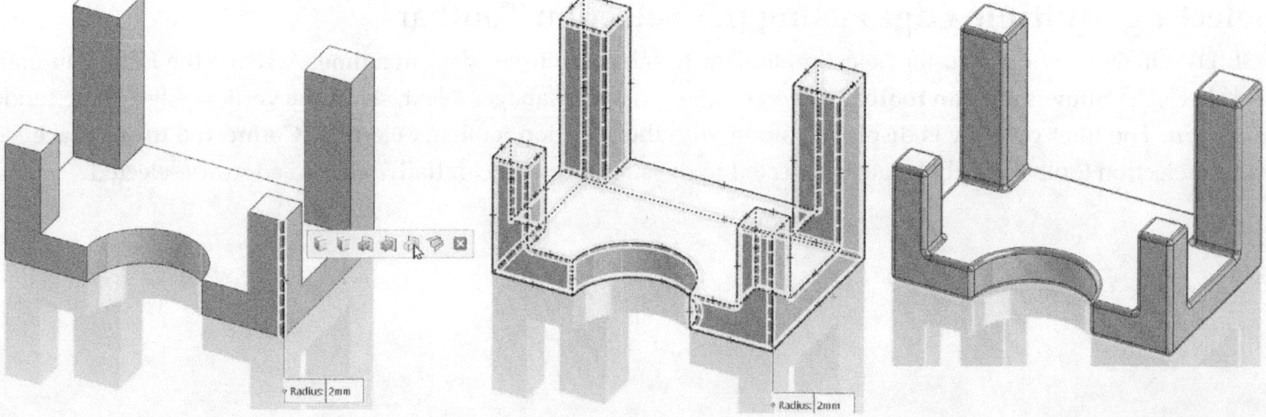

Select an inner edge of a Cut feature and click on the Internal to feature icon on the selection toolbar; all the internal edges of the Cut feature are selected.

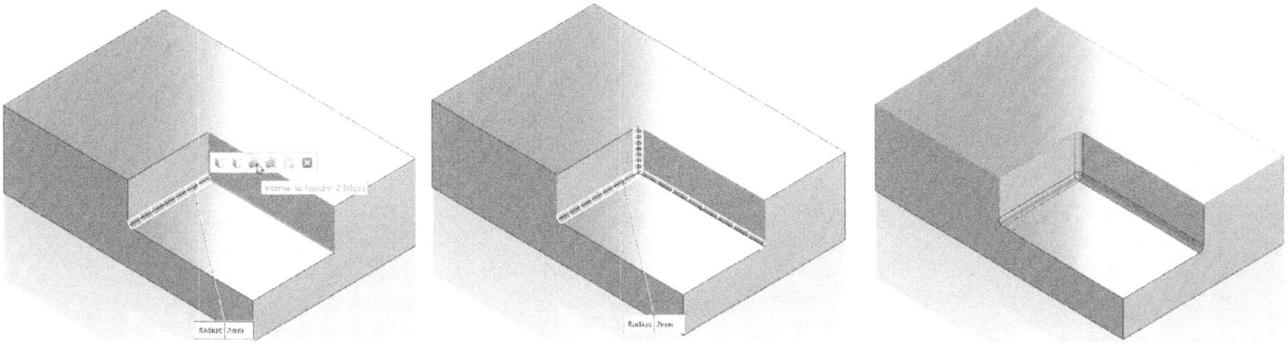

Select an inner edge of the Cut feature and select the Right feature icon on the selection toolbar; all the edges of the Cut feature and its adjacent edges are selected.

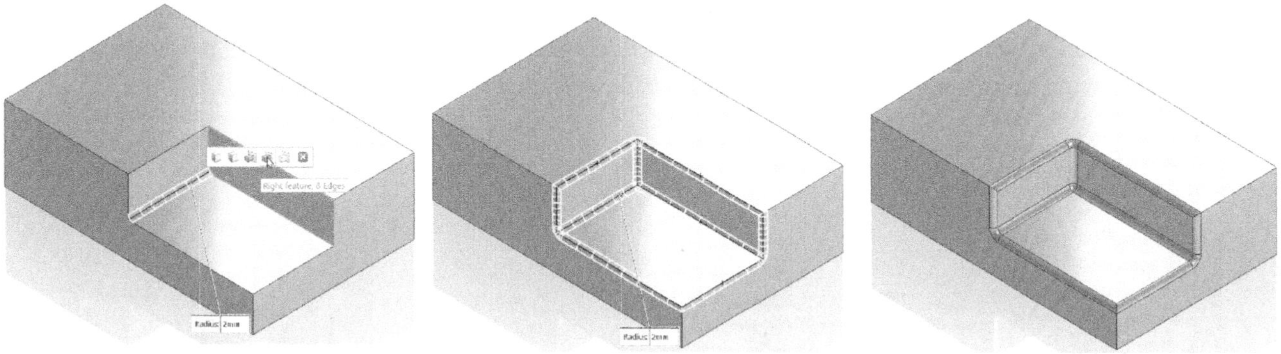

Multiple Radius Fillets

The **Multiple Radius Fillet** option allows you to create multiple radius fillet under a single Fillet feature. Activate the **Fillet** command and click the **Manual** tab on the PropertyManager. Select the edges from the model geometry, as shown. Next, check the **Multiple Radius Fillet** option in the **Fillet Parameters** section; callouts are displayed on the selected edges. Change the radius values of each callout.

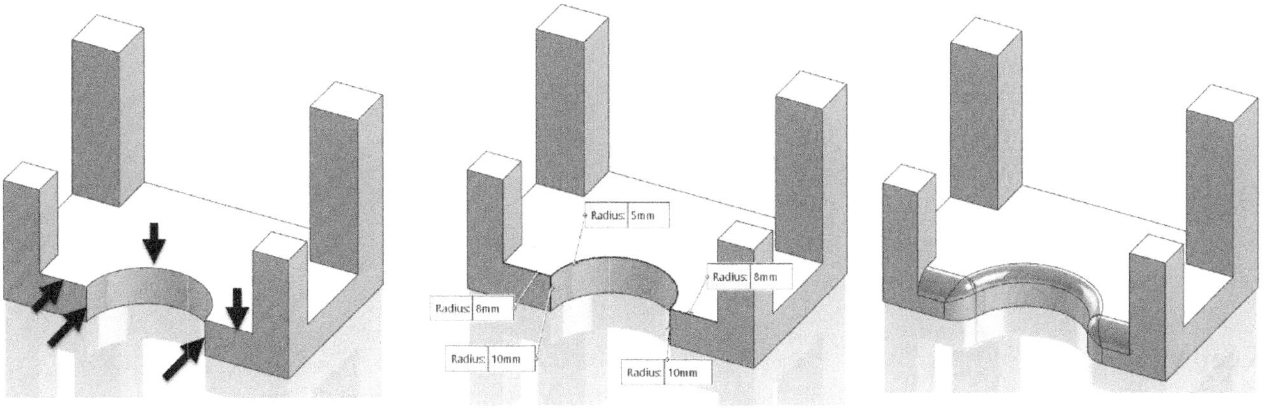

Conic fillets

By default, the fillets have a circular arc profile. However, if you want to create a fillet with a conical arc profile, select **Conic Rho** from the **Profile** drop-down on the PropertyManager. Next, type in a value in the **Rho** box. The fillets with different rho values are shown below. The size of the fillet decreases as you increase the Rho value.

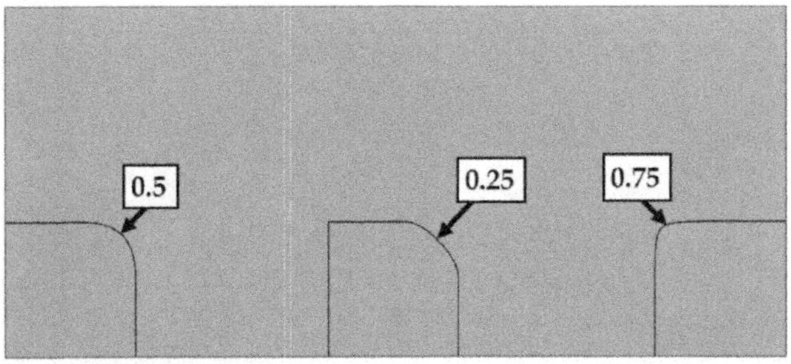

You can also select the Conic Radius option from the Profile drop-down to create a conic fillet. Next, specify the Conic Radius value to define the curvature of the conic fillet. Solidworks creates a circular fillet if the conic radius value is equal to the fillet radius. It creates a flat chamfer like fillet if the Conic Radius value in proportionally larger than the Radius value. The curvature of the conic fillet becomes sharp as you decrease the Conic Radius value. The following figure shows the conic fillet with three different Conic Radius values.

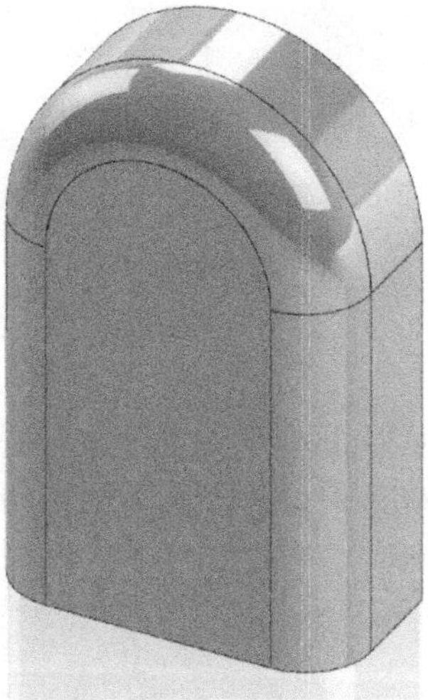

Conic Radius = Radius =10

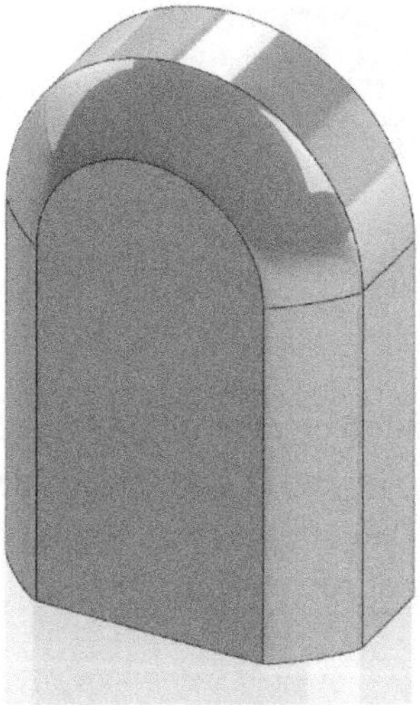

Conic Radius = 50
Radius = 10

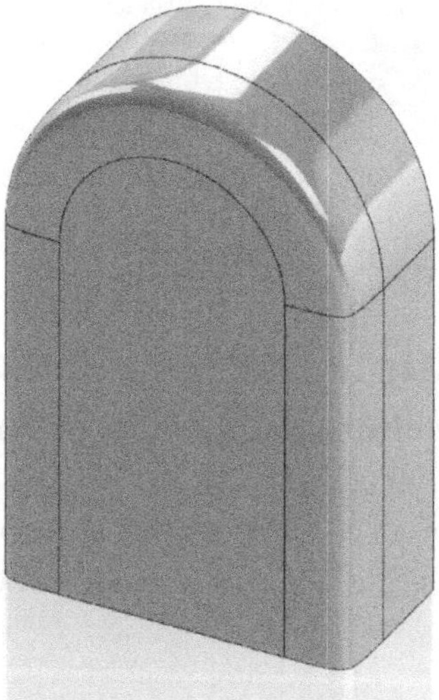

Conic Radius = 2
Radius =10

Overflow type

If you create fillets, which intersect with the adjacent edges, the **Overflow type** options in the **Fillet Options** section will help you control the geometry.

Keep edge

This option helps you to preserve the adjacent edges by trimming the overflowing fillets.

Keep surface

This option helps you to preserve the fillet by distorting the intersecting edges.

Default

This option applies the overflow type depending upon the geometry.

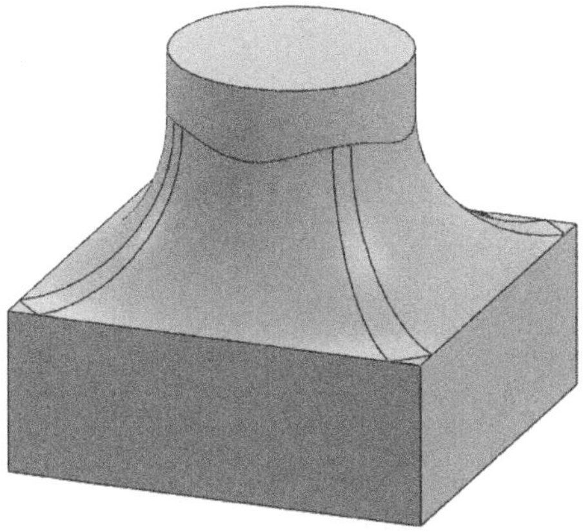

Keep edges

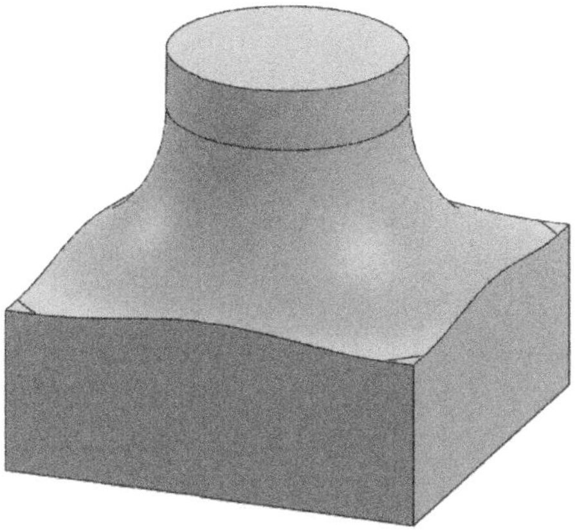

Keep surface

Round Corners

When you fillet two adjacent edges, a sharp juncture appears at the corner. You can avoid this by checking the **Round Corners** option in the **Fillet Options** section. It creates a smooth transition between the selected edges.

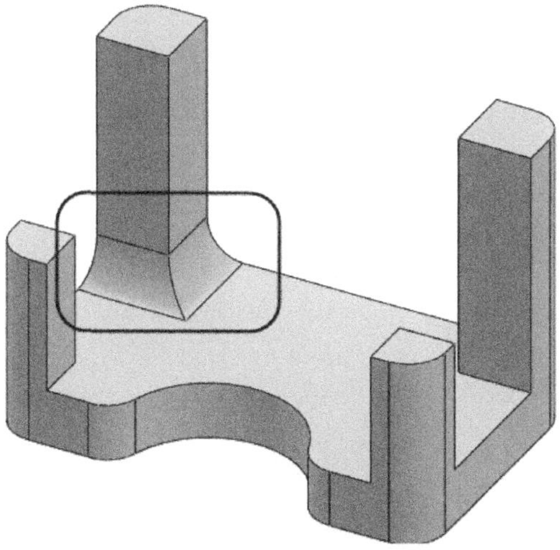

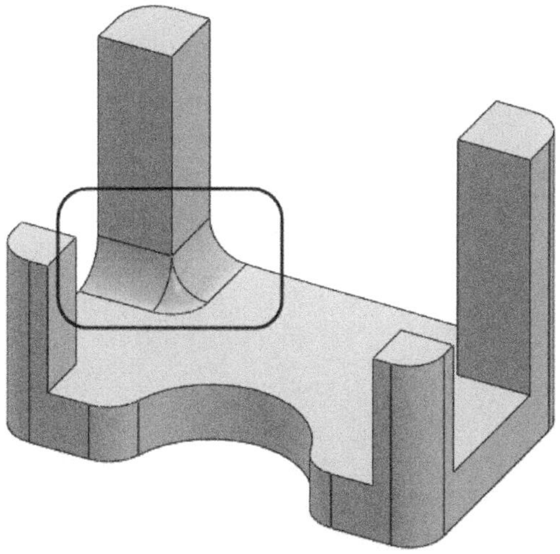

Setback Distance

If you create an edge fillet on three edges that come together at a corner, you have the option to control how these three fillets are blend together. Activate the **Fillet** command and select the three edges that meet together at a corner. On the **Fillet** PropertyManager, expand the **Setback Parameters** section, and then click in the **Setback Vertices** selection box. Select the corner point at which all the three fillets meet. One by one, select setback distances in the **Setback Distances** box, and then change the value in the **Distance** box. You can also click the **Set All** button to apply the same value for all three setbacks. Click **OK** to create the fillets with setback distance.

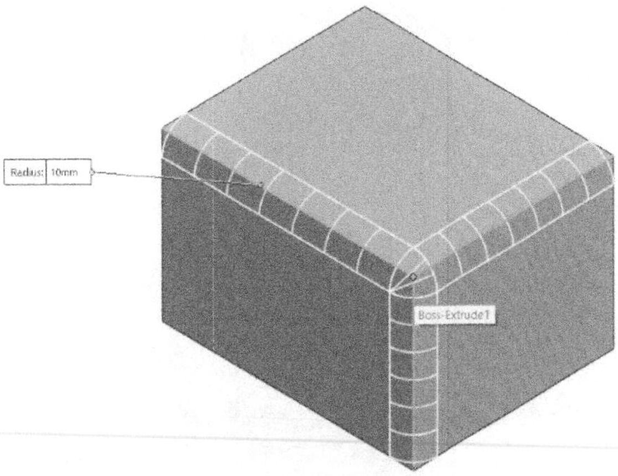

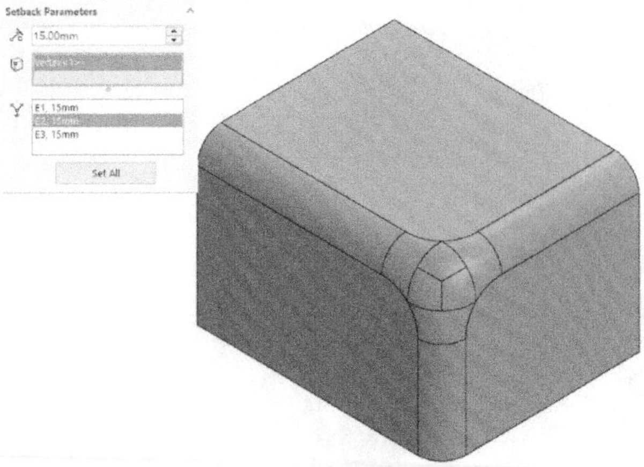

Variable Radius Fillet

SOLIDWORKS allows you to create a fillet with a varying radius along the selected edge. On the **Features** CommandManager, click **Fillets** drop-down > **Fillet** (or) click **Insert > Features > Fillet/Round** on the Menu bar. On the **Fillet** PropertyManager, click the **Variable Size Fillet** icon. Click on edge to fillet. In the **Variable Radius Parameters** section, change the value in the **Number of Instances** box. For example, if you enter 3 in the **Number of Instances** box, three points will be displayed between the selected edge vertices. On the selected edge, select points to define the variable radius. You will notice that the callouts appear on the selected points. The callouts display the radius and position. Also, the selected points are displayed in the **Attached Radii** box.

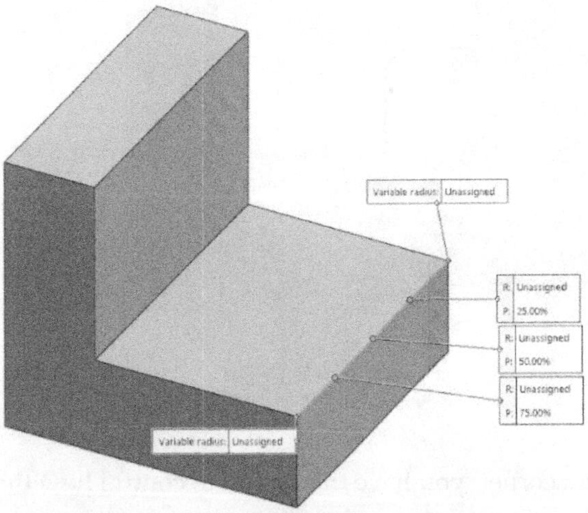

Double-click in the **Variable Radius** callout attached to the selected edge's start point, and then type in a value. Likewise, double-click in the **Variable Radius** callout attached to the selected edge's endpoint, and then change the value. Double-click in the **R** box of the callout with 25% P value, and then type in a value. Likewise, change the **R** values in the remaining callouts.

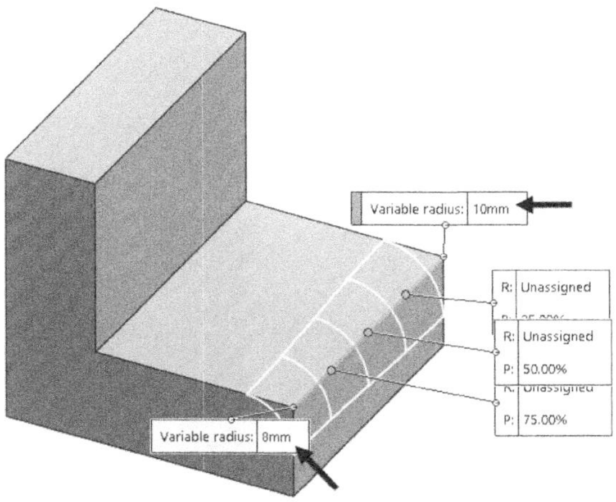

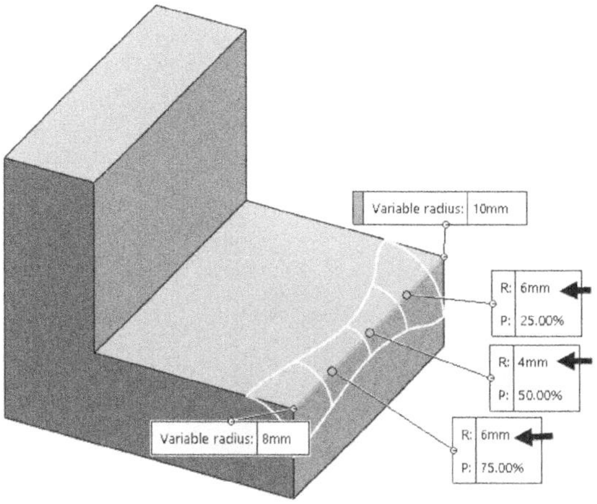

On the PropertyManager, select **Smooth transition** from the **Variable Radius Parameters** section to get a smooth fillet. Select **Straight transition** to get a straight transition fillet.

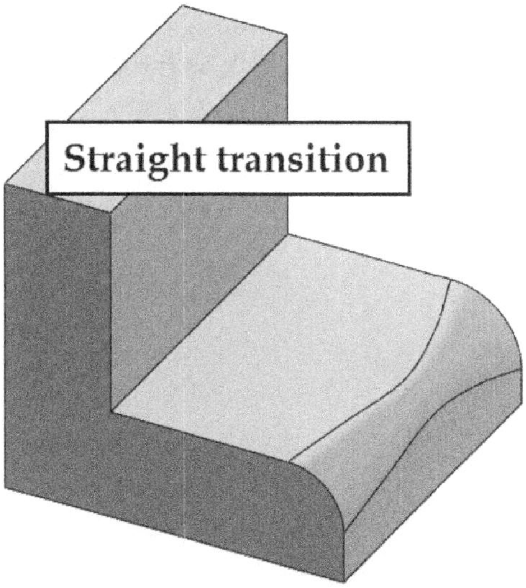

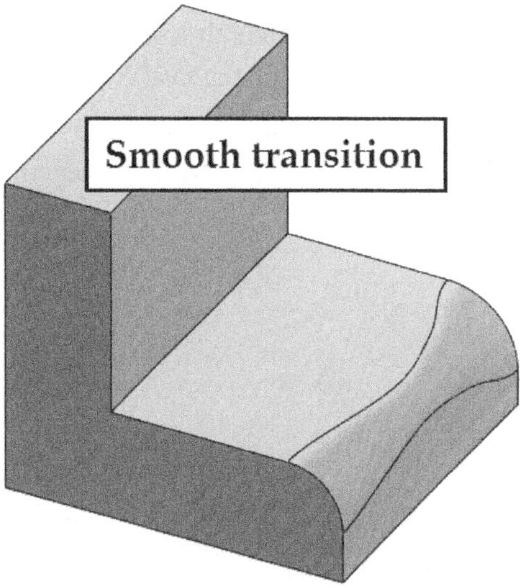

Click **OK** to create the variable radius fillet.

Face Fillet

This option creates a fillet between two faces. The faces are not required to be connected. On the **Features** CommandManager, click **Fillets** drop-down > **Fillet** (or) click **Insert > Features > Fillet/Round** on the Menu bar.

On the **Fillet** PropertyManager, click the **Face Fillet** icon. Click on the first side face, as shown. On the PropertyManager, click in the **Face Set 2** box and select the second side face, as shown. Under the **Fillet Parameters** section, select **Hold Line** from the **Fillet Method** drop-down. Click on edge, as shown. Click **OK** to create the face fillet.

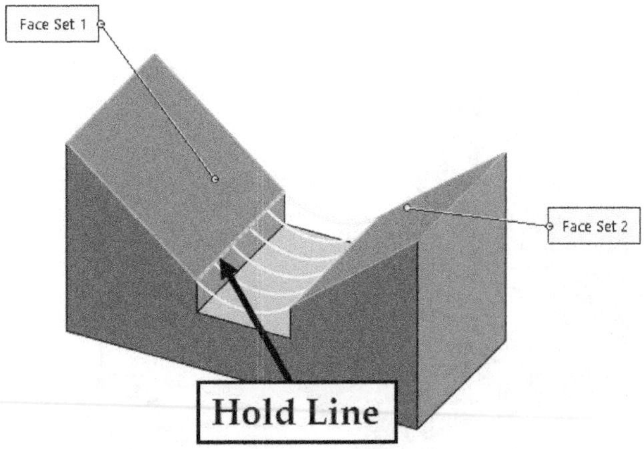

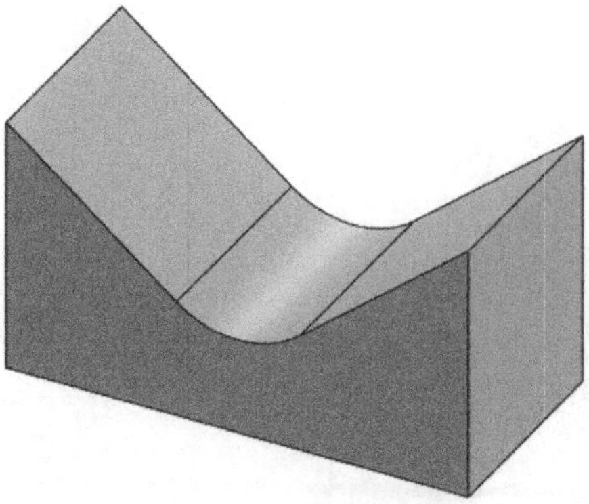

Full Round Fillet

This option creates a fillet between three faces. It replaces the center face with a fillet. On the **Features** CommandManager, click **Fillets** drop-down > **Fillet** (or) click **Insert > Features > Fillet/Round** on the Menu bar. On the PropertyManager, click the **Full Round** icon, and then select the first side face. Click on the **Center Face Set** box, and then select the face that needs to be replaced with a fillet. Click on the **Face Set 2** box and then select the second side face. Click **OK** to replace the center face with a fillet.

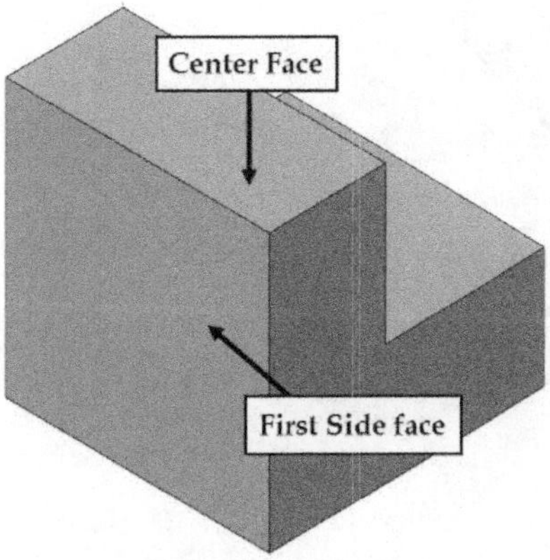

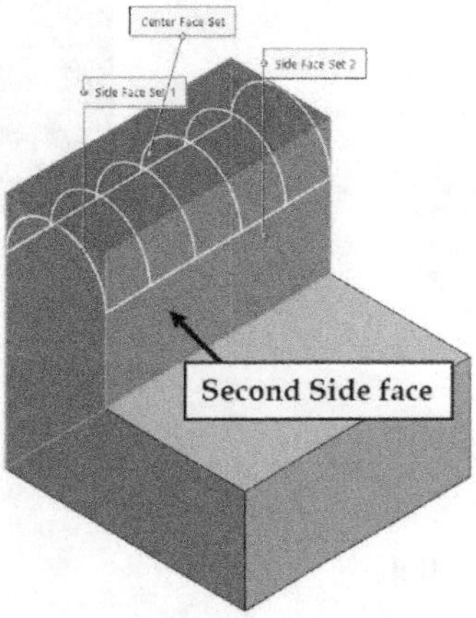

Using the FilletXpert

The **FilletXpert** allows users to quickly and easily create fillets in situation where it is difficult to create fillets using the **Manual** tab of the Fillet PropertyManager. For example, activate the **Fillet** command and try to add fillets of 4 mm radius to the edges, as shown. SOLIDWORKS displays an error message.

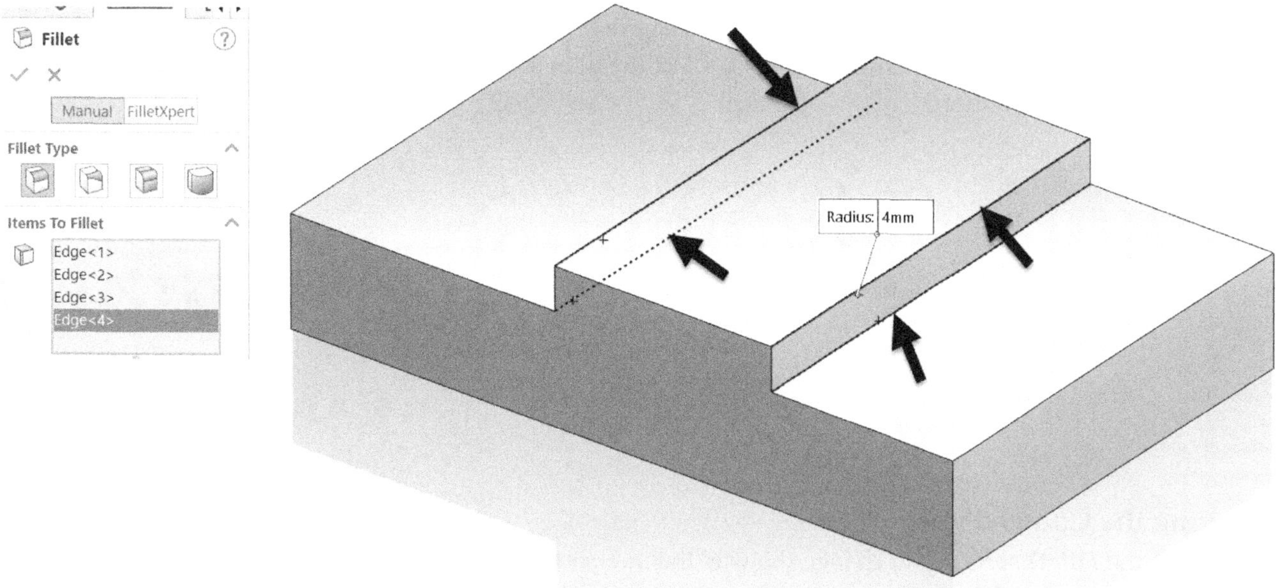

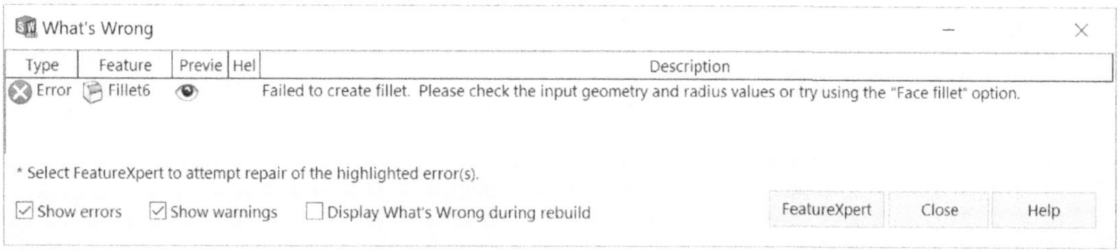

Close the **What's Wrong** dialog and click the FilletXpert tab on the Fillet PropertyManager. Next, select the edges to be filleted and type 4 in the **Radius** box. Click the **Apply** button on the PropertyManager and notice that the fillets are added to the selected edges. Click **OK** on the **FilletXpert** PropertyManager and notice that three Fillet features are added to the FeatureManager Design Tree. The FilletXpert option adds fillets to the selected edges without creating any error. In order to do so, it creates multiple fillet features and adjusts the sequence in which the fillets are added to the selected edges.

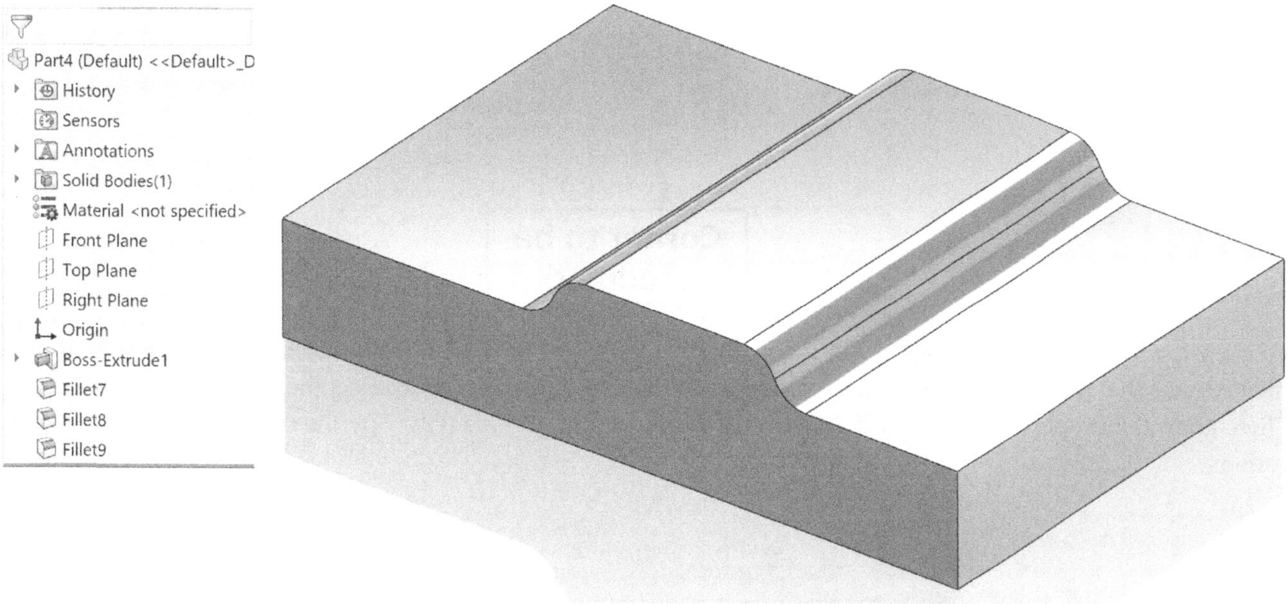

Changing the Fillet Size

You can use the **FilletXpert** tab to change the size of the existing fillets. To do this, click the **FilletXpert** tab on the PropertyManager and click the **Change** tab. Next, select the fillet to be resized. Type-in a new value in the Radius box and click the **Resize** button.

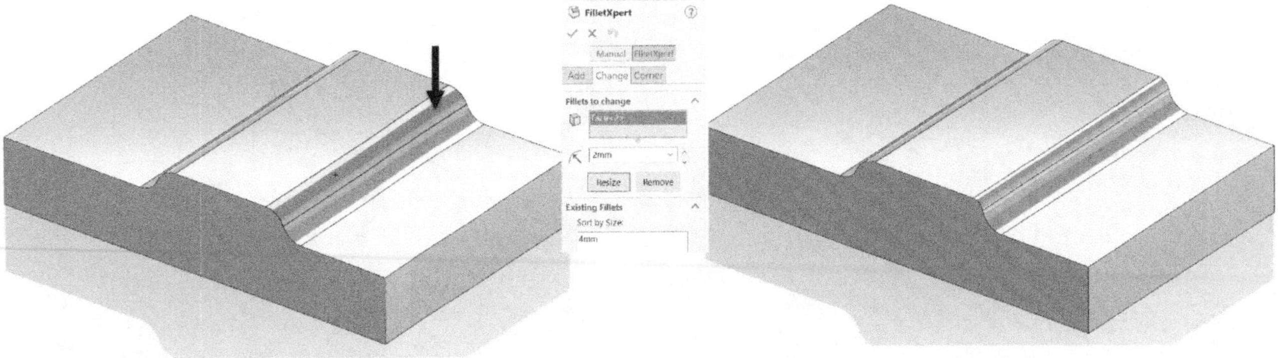

Changing the Corner Type

You can use the **FilletXperts** tab to change the way the intersection of three fillets is shown at a corner. Click the **Corner** tab on the FilletXpert Propertymanager and select the fillet corner to be modified. Click the Show Alternatives button on the PropertyManager; the **Select Alternatives** window appears. Select the alternative corner type from the **Select Alternatives** window.

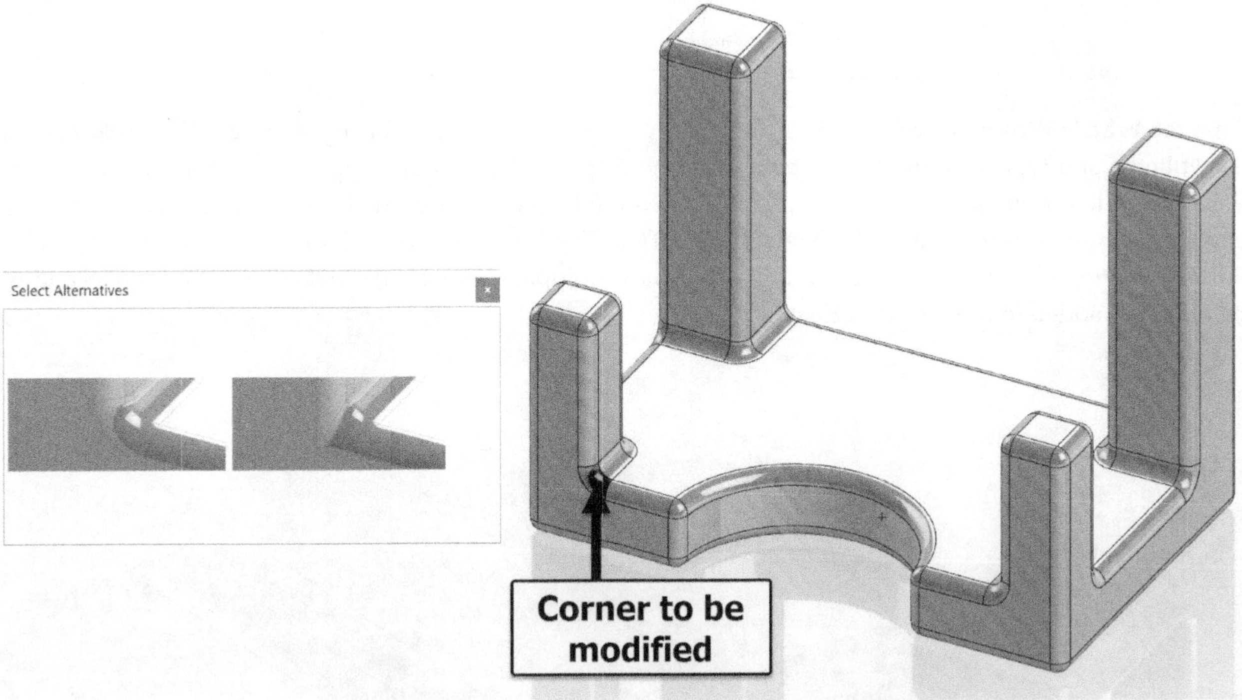

To copy the modified fillet corner, select it and click on the **Copy Targets** selection box. This will highlight other fillet corners that match with the selected corner. From the highlighted corners, select the ones you want to modify. Then, go to the PropertyManager and click on the **Copy to** button. This will copy the corner type to the selected corners.

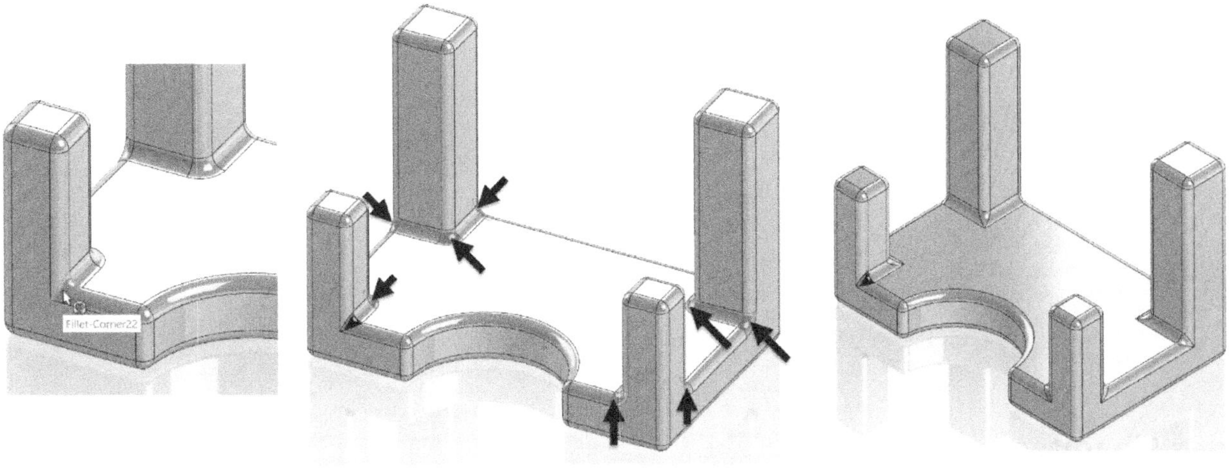

Chamfers

The **Chamfer** and **Fillet** commands are commonly used to break sharp edges. The difference is that the **Chamfer** command adds a bevel face to the model. A chamfer is also a placed feature. On the **Features** CommandManager, click **Fillet** drop-down > **Chamfer** button (or) click **Insert > Features > Chamfer** on the Menu bar. On the **Chamfer** PropertyManager, select chamfer **Chamfer Type**. You can select **Angle Distance, Distance Distance, Vertex, Offset Face** or **Face Face.** If you select **Chamfer Type > Angle Distance**, then type-in the distance and angle values of the chamfer in the **Chamfer Parameters** section. Click on edge (s) to chamfer. Click **OK**.

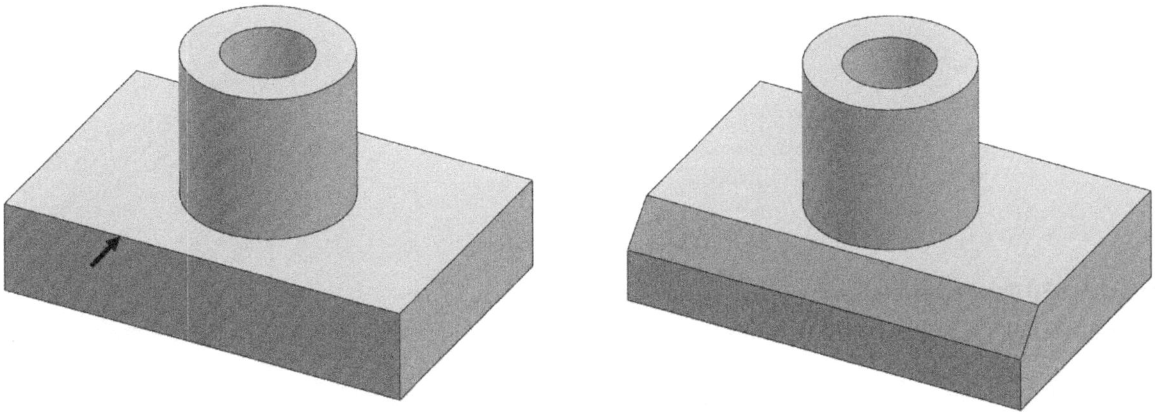

You can also convert a fillet into a chamfer. To do this, click on the fillet and select **Edit Feature**. On the PropertyManager, click the **Offset Face** icon. Next, specify the **Offset Distance** value in the **Chamfer Parameters** section. Click **OK**.

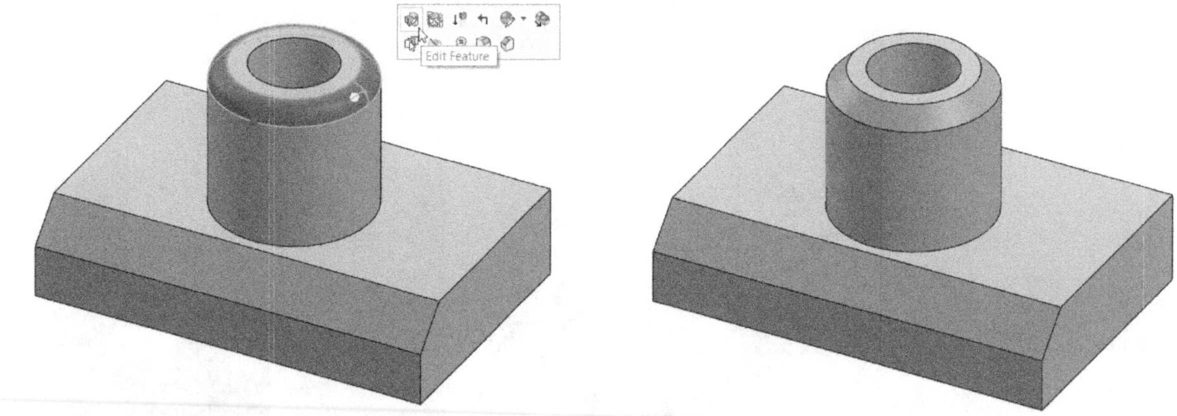

Vertex Chamfer

The **Vertex** option on the PropertyManager allows you to create a triangular shaped chamfer on a selected vertex. On the **Features** CommandManager, click **Fillets** drop-down > **Chamfer**. On the **Chamfer** PropertyManager, click the **Vertex** icon on the **Chamfer Type** section. Next, select a vertex from the model geometry. Note that you cannot select multiple vertices. Type-in values in the **D1**, **D2**, and **D3** boxes available in the **Chamfer Parameters** section. You can also check the Equal distance option to create a vertex chamfer with three equal edges.

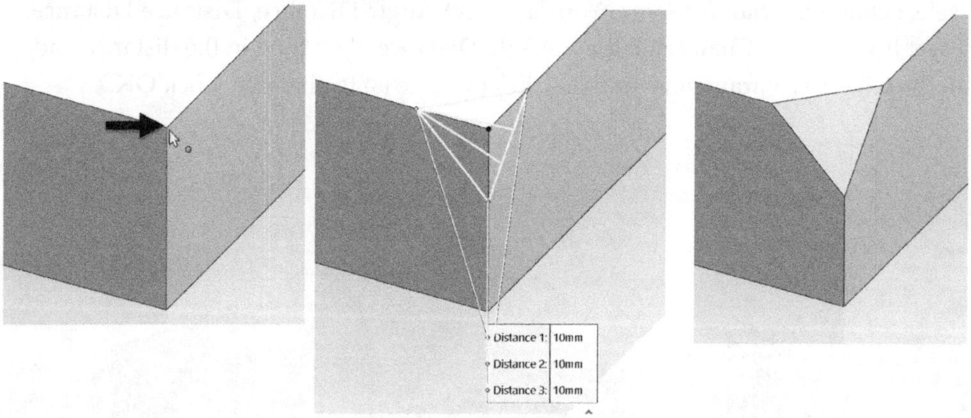

Multiple Distance Chamfer

The **Multiple Distance Chamfer** option allows you to create multiple distance chamfers under a single Chamfer feature. Activate the **Chamfer** command and click the **Offset Face** icon. Next, check the **Multiple Distance Chamfer** option in the **Chamfer Parameters** section; callouts are displayed on the selected edges. Change the **Distance** values of each callout. Click **OK** on the PropertyManagar.

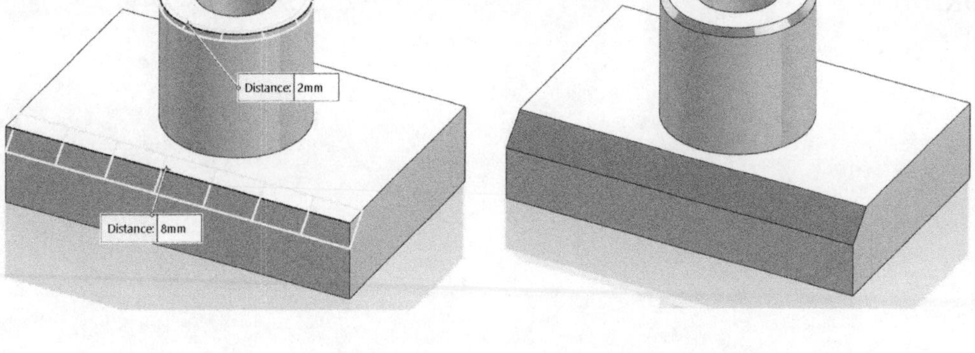

Draft

When creating cast or plastic parts, you are often required to add a draft on them so that they can be molded easily. A draft is an angle or taper applied to the components' faces to be removed from a mold easily. The following illustration shows a molded part with and without the draft.

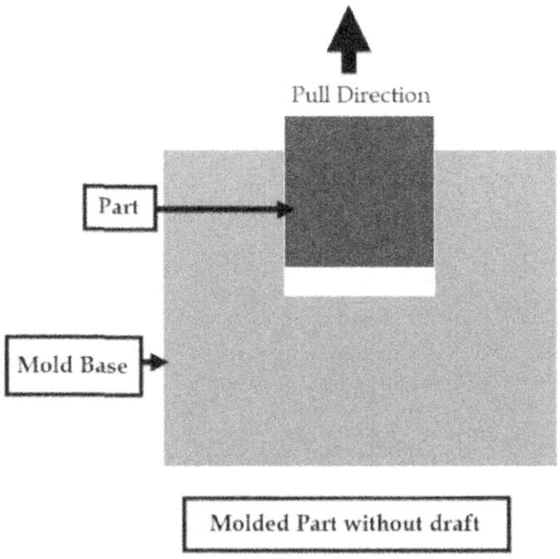

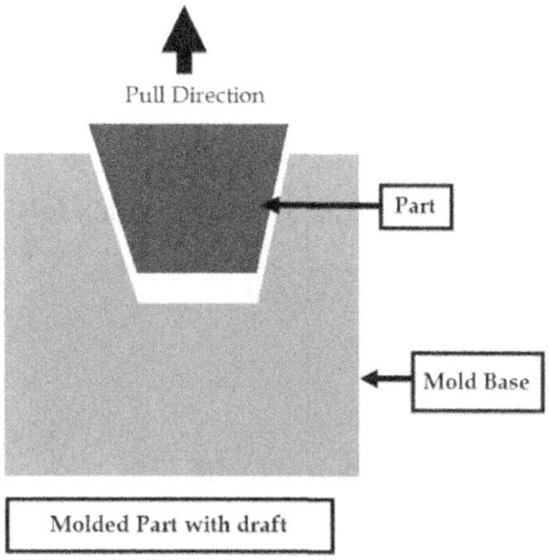

When creating *Extruded* features, you can predefine the draft. However, most of the time, it is easier to apply the draft after creating the features. On the **Features** CommandManager, click the **Draft** icon (or) click **Insert > Features > Draft** on the Menu bar. On the **DraftXpert** PropertyManager, click the **Manual** tab. Next, select **Types of Draft > Neutral plane**. Click in the **Neutral Plane** selection box and select a flat face or plane. This defines the neutral plane. The draft angle will be measured with reference to this face.

Click in the **Faces to draft** selection box and select the faces to draft. The **Face propagation** drop-down has five options: **None**, **Along Tangent**, **All Faces**, **Inner Faces**, and **Outer Faces**. The **None** option does not allow face propagation. The **Along Tangent** option allows you to select all the tangentially connected faces by selecting any one of them. The **All Faces** option selects all the faces normal to the neutral plane. The **Inner Faces** option selects the inner faces of the model that are normal to the neutral plane. The **Outer Faces** option selects the outer faces of the model.

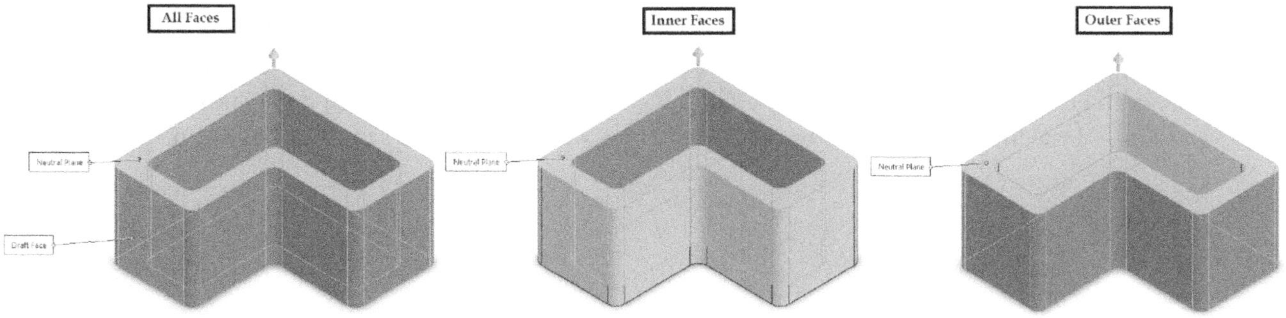

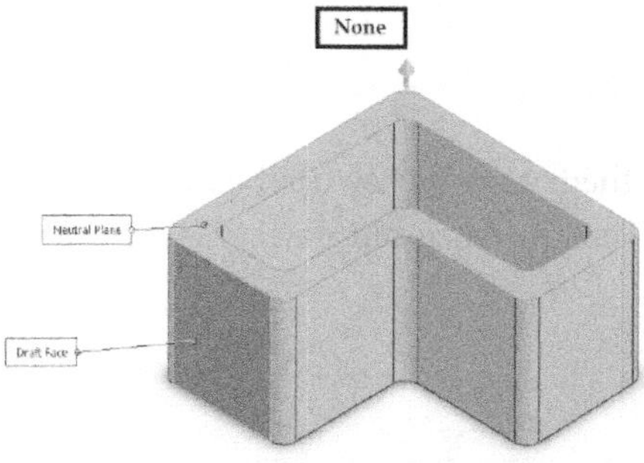

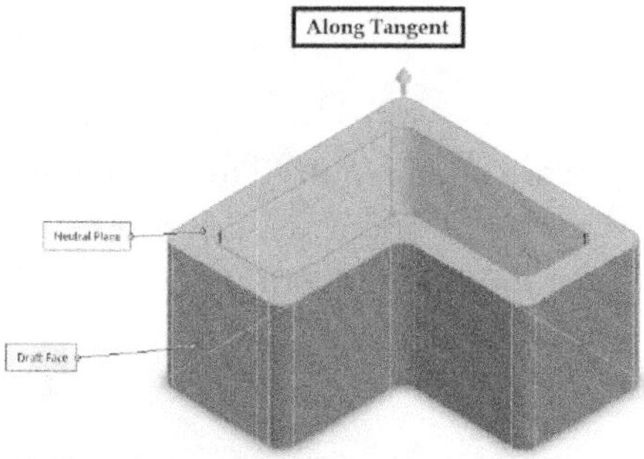

Type-in a value in the **Draft Angle** box to define the draft angle. If you want to reverse the draft direction, then click the arrow that appears on the geometry. Click **OK** to apply the draft.

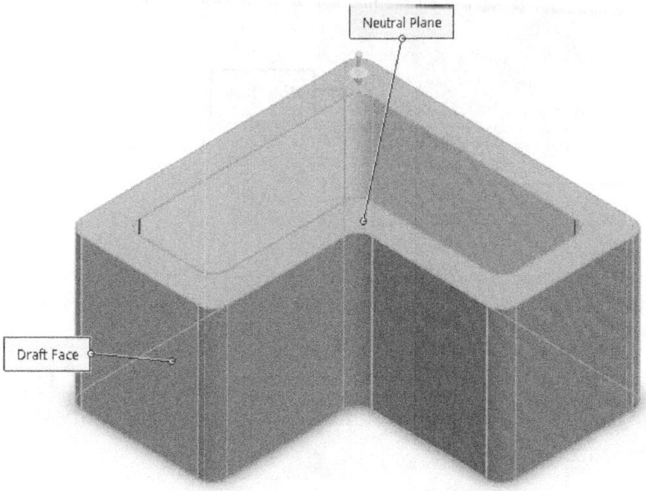

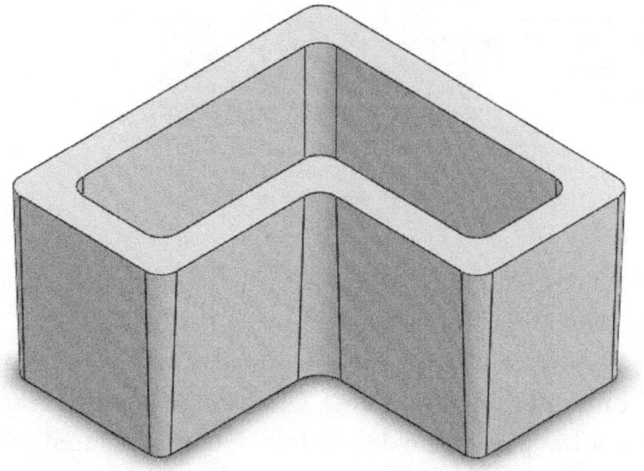

Shell

Shell is another useful feature that can be applied directly to a solid model. It allows you to take a solid geometry and make it hollow. This can be a powerful and time-saving technique when designing parts that call for thin walls such as bottles, tanks, and containers. This command is easy to use. You should have a solid part, and then activate this command from the **Features** CommandManager (or) click **Insert > Features > Shell** on the Menu bar. Select the faces to remove, and then type in a value in the **Thickness** box. If you want to add outside thickness, then Check the **Shell outside** option. If you want to add different thicknesses to some faces, click in the **Multi thickness faces** box, and then select the faces to add different thicknesses. You will notice that a thickness value appears on the selected face. Type in a new value in the **Multi thickness (es)** box to change it. Click **OK** to finish the feature.

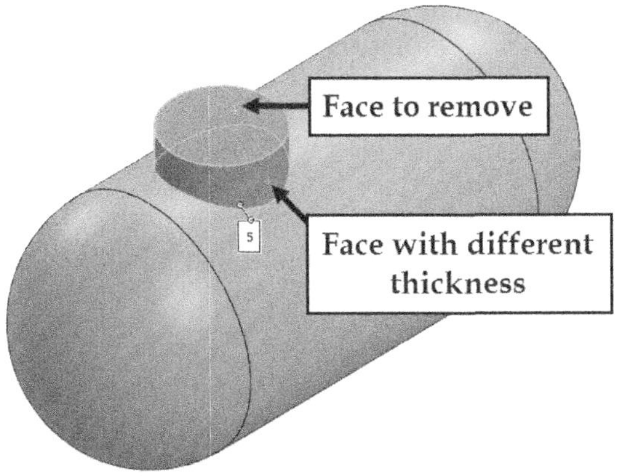

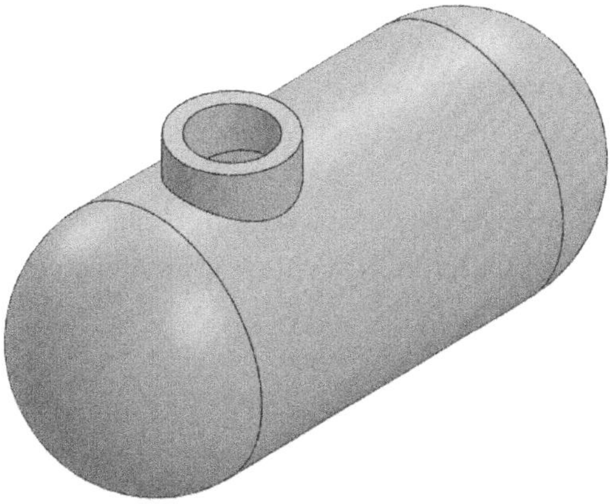

If you want to shell the solid body without removing any faces, just type in a value in the **Thickness** box and click **OK**. This creates the shell without removing the faces. Change the **Display Style** to **Wireframe** or **Hidden Lines** to view the shell.

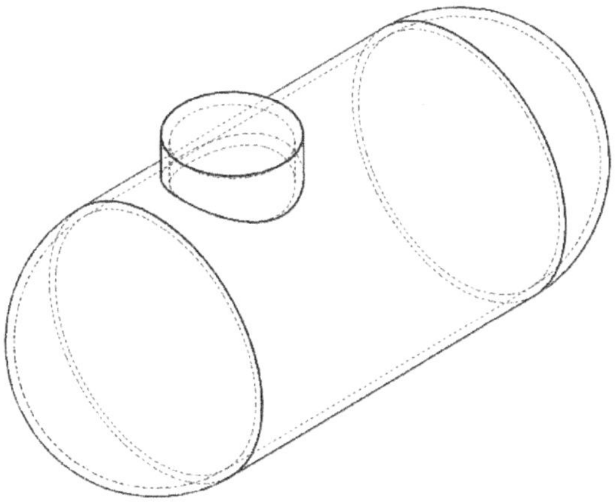

Best Practices

Proper Design Intent: Before creating a shell feature in SOLIDWORKS, it is essential to define the purpose and functional requirements of the part. Understanding why a shell is needed and how it impacts the part's behavior is crucial for a successful design.

Keep Wall Thickness Uniform: Maintaining a consistent wall thickness in your shell feature within SOLIDWORKS is vital. Uneven wall thickness can result in weak spots or defects in the final part, compromising its structural integrity.

Parametric Design: Utilize parametric modeling techniques in SOLIDWORKS to ensure that any changes made to the part design, such as dimensions or angles, automatically update the shell feature. This approach enhances design flexibility and efficiency.

Cautions

Wall Thickness Limits: When designing a shell feature in SOLIDWORKS, it is important to be aware of the minimum and maximum wall thickness limitations imposed by the chosen manufacturing process and material. Different processes have distinct requirements that must be considered.

Avoid Sharp Corners: Steer clear of creating sharp corners or sudden changes in wall thickness within your shell feature design in SOLIDWORKS. Such features can lead to stress concentration points and potential failure in the part.

Material Selection: The choice of material significantly influences the feasibility of a shell feature. Ensure that the selected material can be effectively molded, cast, forged, or machined with the desired wall thickness to achieve optimal results.

Examples

Example 1 (Millimetres)

In this example, you will create the part shown below.

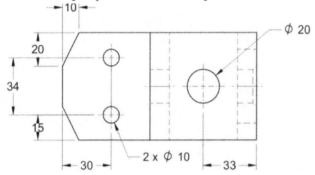

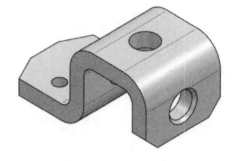

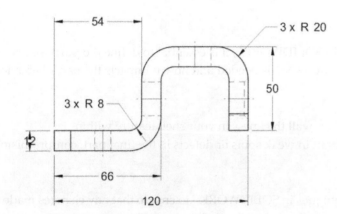

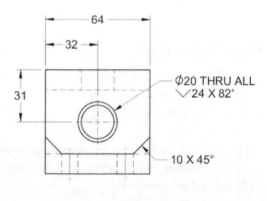

1. Start **SOLIDWORKS 2024**.
2. On the Menu bar, click **File > New**.
3. On the **New SOLIDWORKS Document** dialog, click the **Part** button, and then click **OK**; a new part file is opened.
4. On the CommandManager, click **Sketch** tab > **Sketch** drop-down > **Sketch**.
5. Select the **Front** plane and draw the sketch, as shown below (refer to Chapter 2 to learn how to draw sketches). Next, click **Exit Sketch** on the CommandManager.

6. Activate the **Extruded Boss/Base** command, and then select the sketch, if not selected.
7. On the **Extrude** PropertyManager, select **Direction 1** section > **End Condition** drop-down > **Mid Plane**.
8. Type 64 in the **Depth** box and click the **OK** icon on the PropertyManager.

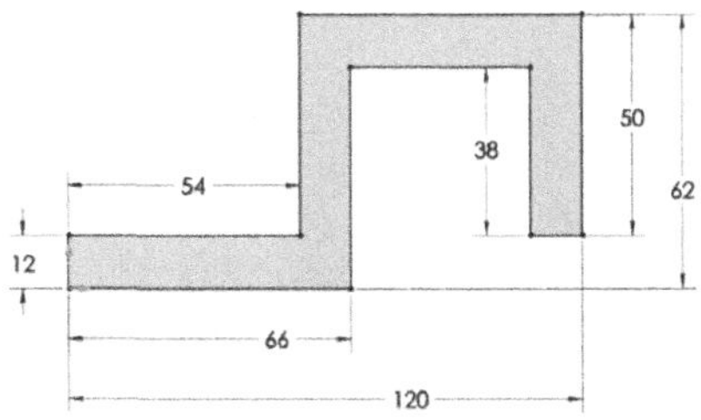

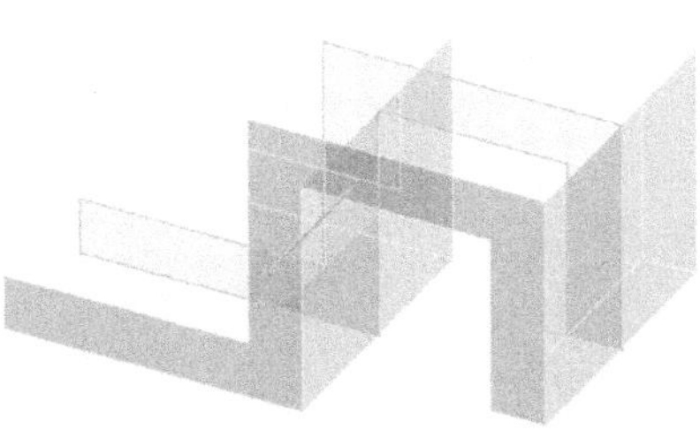

9. On the CommandManager, click **Features** tab > **Hole** drop-down > **Hole Wizard**; the **Hole Specification PropertyManager** appears.
10. On the PropertyManager, select the **Countersink** button and set the **Standard** to **ANSI Metric**.
11. Select **Type > Socket Countersunk Head Cap Screw**.
12. In the **Hole** section, select **Size > M20**, and then check the **Show custom sizing** option.
13. Type-in 20, 24, and 82 in the **Through Hole Diameter** , **Countersink Diameter** , and **Countersink Angle** boxes, respectively.
14. Set **End Condition** to **Through All**.
15. Click the **Positions** tab on the PropertyManager.
16. Click on the right-side face, and then click to place the hole.
17. On the CommandManager, click **Sketch** tab > **Smart Dimension**.
18. Select the center point of the hole and the right vertical edge of the placement face. Next, move the pointer up and click to place the dimension. Type 32 in the **Modify** box and click the green check.
19. Select the center point of the hole and the top edge of the placement face. Next, move the pointer toward the right and click to place the dimension. Type 31 in the **Modify** box and click the green check.
20. Click **OK** on the PropertyManager to create the countersink hole.

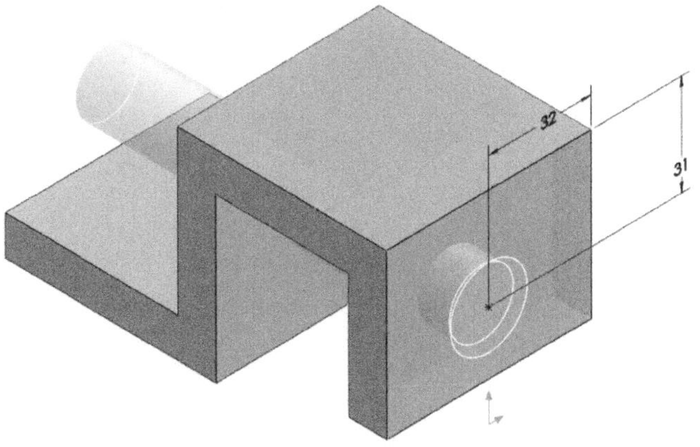

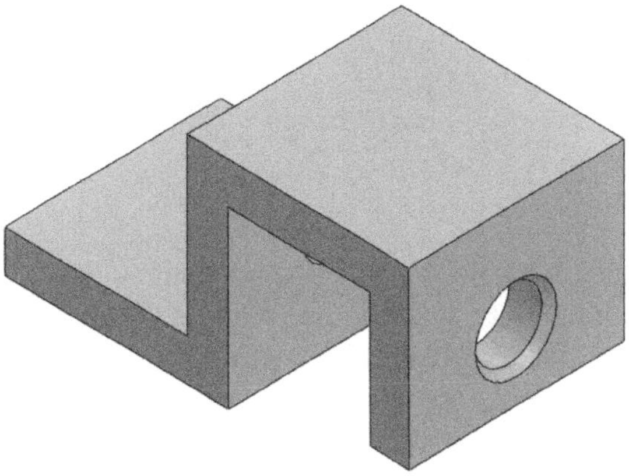

21. Activate the **Hole Wizard** command; the **Hole Specifications** PropertyManager appears.
22. On this PropertyManager, select the **Hole** icon; the **SOLIDWORKS** message box pops up on the screen. On this message box, select Reset the custom sizing values to their default values for the new hole type.

23. Select **Standard > ANSI Metric**, and then select **Size > Ø20**. Uncheck the **Show custom sizing** option.
24. Set **End Condition** to **Through All**.
25. Click the **Positions** tab on the PropertyManager, and then click on the top face of the model.
26. Click to locate the center point of the hole. Add dimensions to position the hole, as shown.

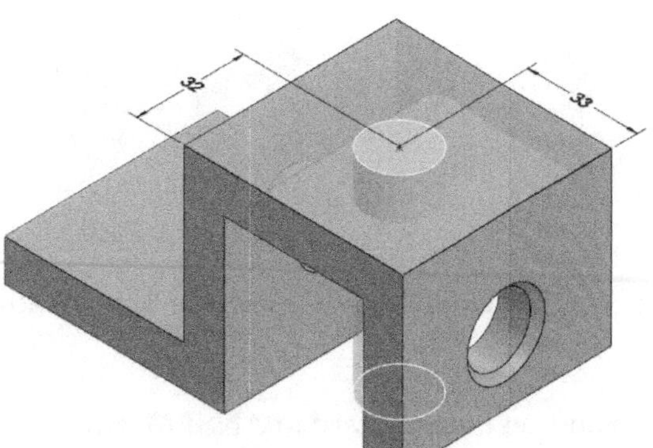

27. Click **OK** on the PropertyManager.
28. On the **View (Heads Up)** toolbar, click the **View Orientation** drop-down > **View Selector**; the **View Selector** appears.
29. On the **View Selector**, click on the top left corner; the view orientation of the model changes.

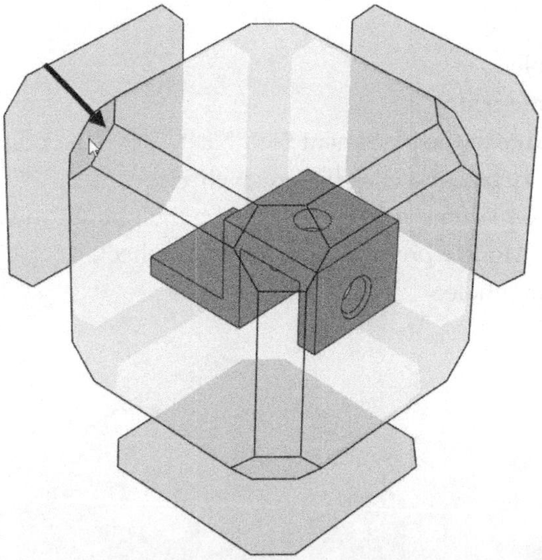

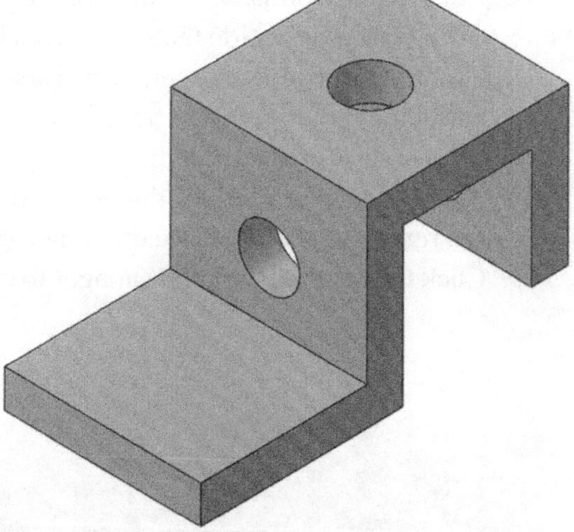

30. Activate the **Hole Wizard** command to open the **Hole Specifications** PropertyManager.
31. Select the **Hole** button and set the **Size** to 10.
32. Click the **Positions** tab, and then click on the lower top face of the model.
33. Click at two locations on the selected face to place holes.
34. Apply dimensions and relations to the center points of the holes, as shown.
35. Click **OK** on the PropertyManager to complete the hole feature.

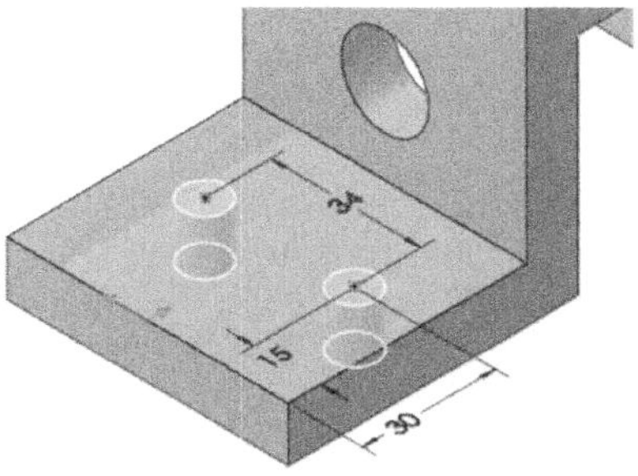

36. Click **Features** tab > **Fillet** drop-down > **Chamfer** on the CommandManager; the **Chamfer** PropertyManager appears.
37. On the PropertyManager, select the **Distance Distance** icon.
38. Click on the vertical edge of the model, as shown.
39. On the PropertyManager, select **Chamfer Parameters > Chamfer Method > Asymmetric**.
40. Set the **Distance 1** and **Distance 2** to **20** and **10**, respectively. Click **OK** on the PropertyManager.

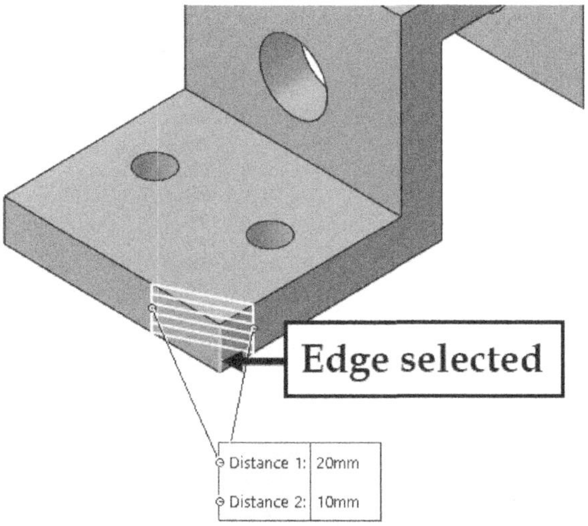

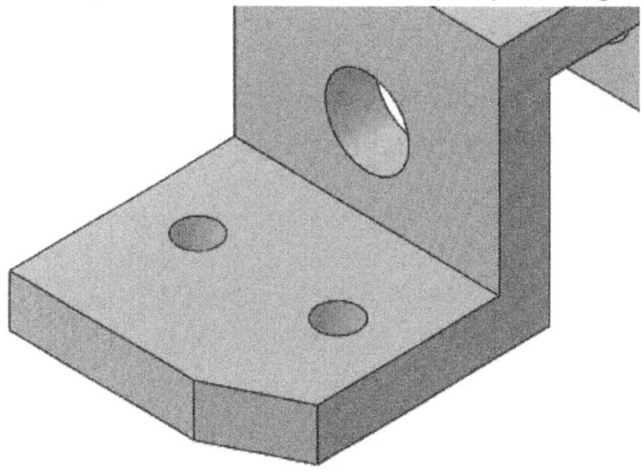

41. Likewise, create another chamfer, as shown.

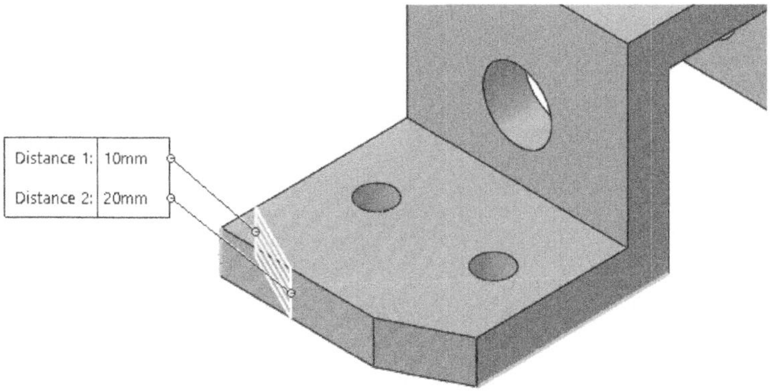

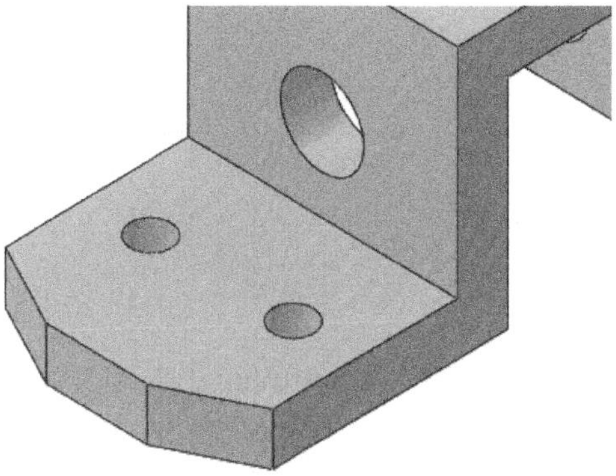

42. Click **Features** tab > **Fillet** drop-down > **Fillet** on the CommandManager.
43. On the **Fillet** PropertyManager, click the **Manual** tab, and then set the **Fillet type** to **Constant Size Fillet**.

44. Click on the horizontal edges of the geometry, as shown below.
45. Type-in **8** in the **Radius** callout attached to one of the selected edges and then press the Enter key.
46. Click the **OK** icon on the PropertyManager.

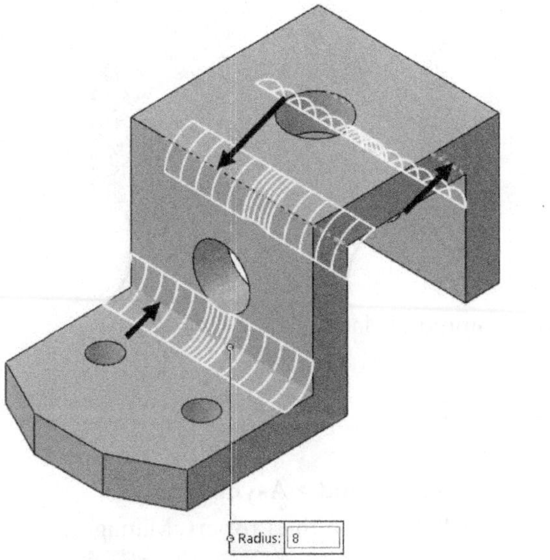

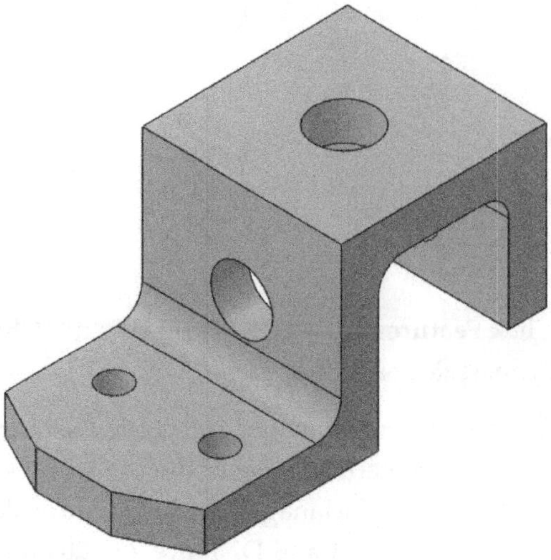

47. Activate the **Fillet** command and then click on the model's outer edges, as shown below.
48. Type-in **20** in the callout that is attached to the selected edges, and then press Enter. Click **OK** to complete the fillet feature.

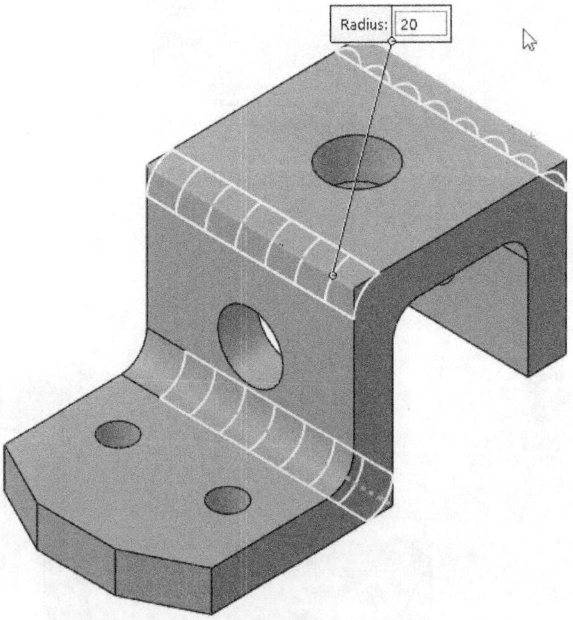

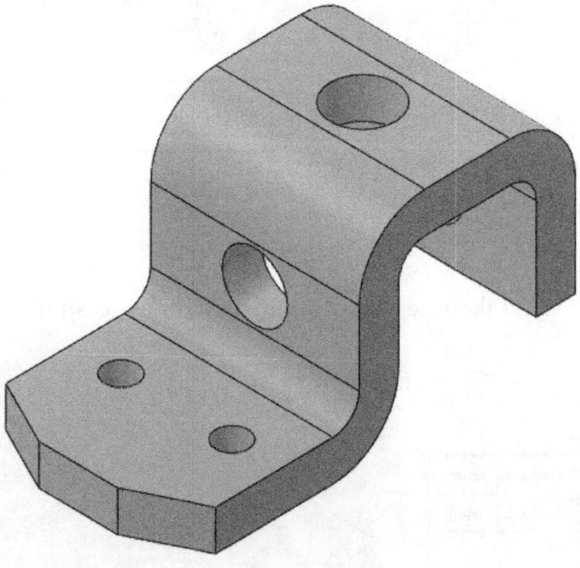

49. Change the orientation of the model view to Isometric by clicking the **View Orientation** icon on the **View (Heads Up)** toolbar and selecting the **Isometric** icon.
50. Click **Features** tab > **Fillet** drop-down > **Chamfer** on the CommandManager. Set the **Chamfer Type** to **Angle Distance**.
51. Click on the lower corners of the part geometry.

52. Type-in **10** and **45** in the **Distance** and **Angle** boxes of the callout attached to the selected edges. Click **OK** to chamfer the edges.

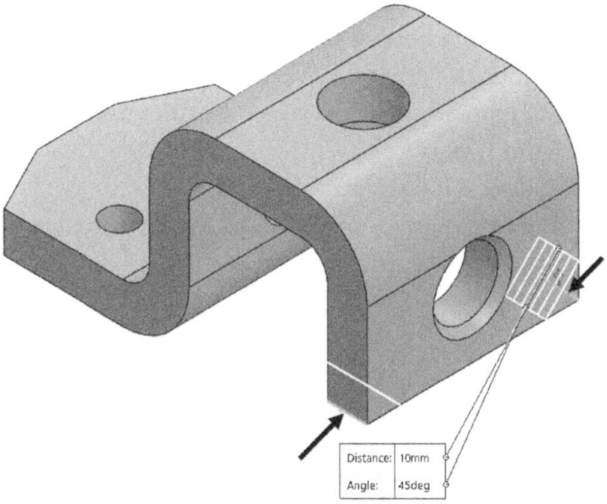

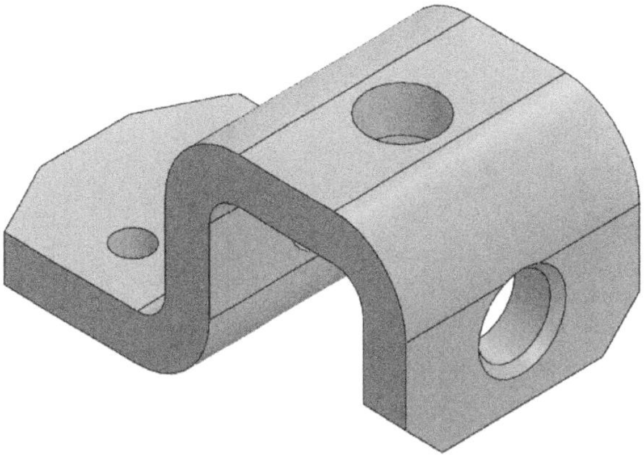

53. Save and close the file.

Example 2 (CSWP Mechanical Design)

In this tutorial, you will create the model given in the CSWP (Certified SolidWorks Associate) sample question. You can get more information about this exam by following the link given below:

https://www.solidworks.com/certifications/mechanical-design-cswp-mechanical-design

1. On the Quick Access Toolbar, click the **New** button.
2. On the **New SOLIDWORKS Document** dialog.
3. Click the **Part** button and click **OK**.
4. Select **MMGS** from the **Unit System** drop-down located at the bottom-right corner.
5. On the menu bar, click **Tools > Equations**.
6. On the **Equations, Global Variables, and Dimensions** dialog, click in the field under the **Global Variables** node, and then type **A**.

In SOLIDWORKS, you can define global variables, which are user-created names assigned numeric values. These variables can be applied in equations or directly in dimensions, making it easier for you to understand and modify equations. You create global variables in either the Equations dialog box or the Modify dialog box for dimensions.

7. Type **213** in the **Value/Equation** field next to the **A** field.
8. Likewise, create other global variables with the values, as shown.

Equations, Global Variables, and Dimensions

Filter All Fields

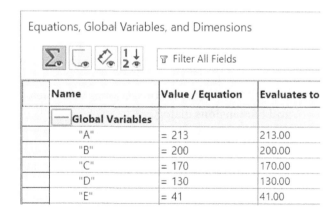

Name	Value / Equation	Evaluates to
Global Variables		
"A"	= 213	213.00
"B"	= 200	200.00
"C"	= 170	170.00
"D"	= 130	130.00
"E"	= 41	41.00

9. Click in the field below the **E** global variable under the **Global Variables** node, and then type **X**.
10. Click the **Value/Equation** field and select **Global Variable > A**. Next, type **/3**.

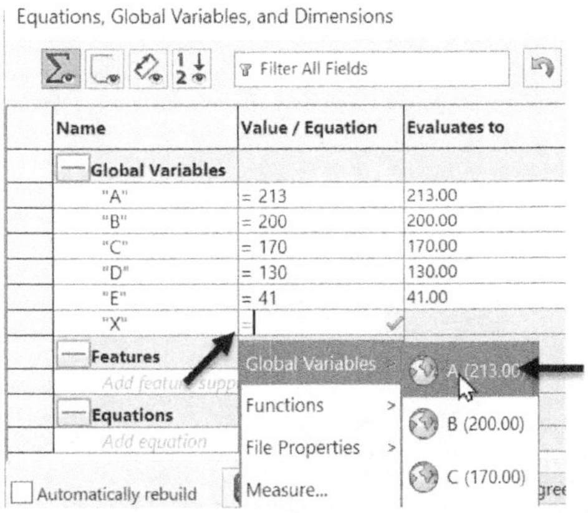

11. Click in the field below the **X** global variable under the **Global Variables** node, and then type **Y**.
12. Click the **Value/Equation** field and select **Global Variable > B**. Next, type **/3+10**.

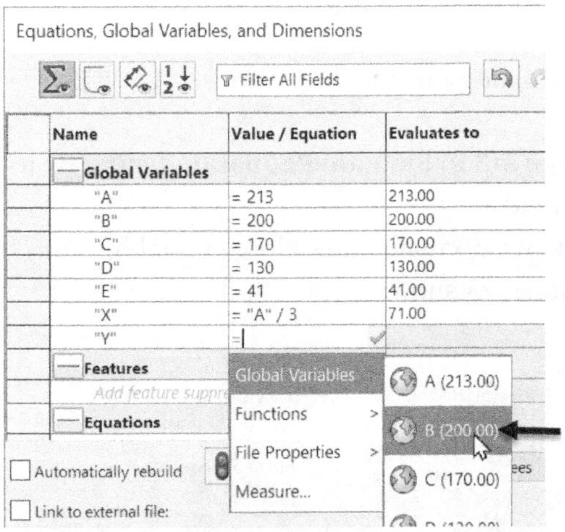

13. Click **OK** on the **Equations, Global Variables, and Dimensions** dialog.
14. On the CommandManager, click **Sketch > Sketch.**
15. Select the Top plane from the graphics window.
16. On the CommandManager, click **Sketch > Rectangle** drop-down **> Center Rectangle**.
17. Select the sketch origin to define the center of the rectangle. Move the pointer outward and click to create the rectangle.
18. On the CommandManager, click **Sketch > Smart Dimension.**
19. Select the vertical line. Next, move the pointer toward left.
20. Click to place the dimension.
21. On the **Modify** box, select the dimension value and type =.
22. Select **Global Variables > A**.
23. Click the green check on the **Modify** box.
24. With the **Smart Dimension** command active, select the bottom horizontal line.
25. Move the pointer downward and click.
26. On the **Modify** box, select the dimension value and type =.
27. Select **Global Variables > B**.
28. Click the green check on the **Modify** box.

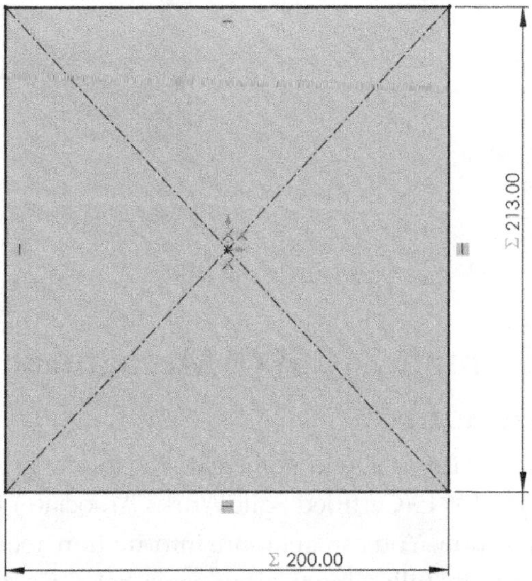

29. Click **Exit Sketch** on the CommandManager.
30. On the CommandManager, click **Features > Extruded Boss/Base**.
31. Select the sketch from the graphics window.
32. On the PropertyManager, type **25** in the **Depth** box available in the **Direction 1** section.
33. Click the **OK** icon on the PropertyManager.

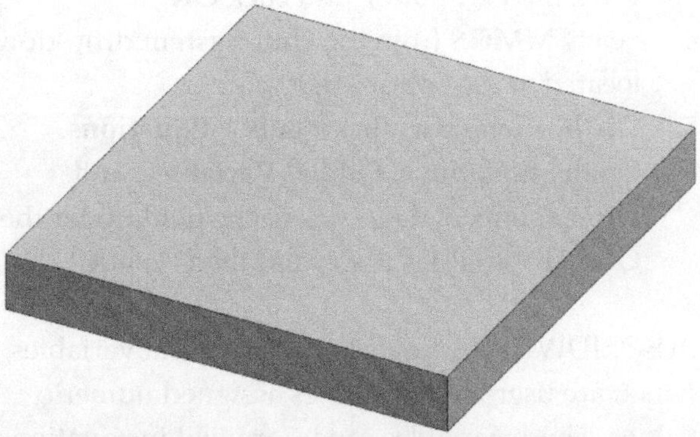

34. On the FeatureManager Design Tree, right-click on the Top plane, and then select **Sketch**.

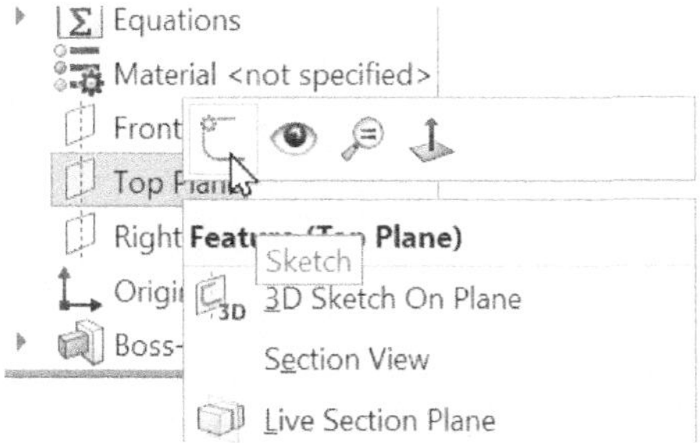

35. On the CommandManager, click **Sketch > Rectangle** drop-down **> Corner Rectangle**.
36. Select the lower right corner of the model. Next, move the pointer toward left and click.

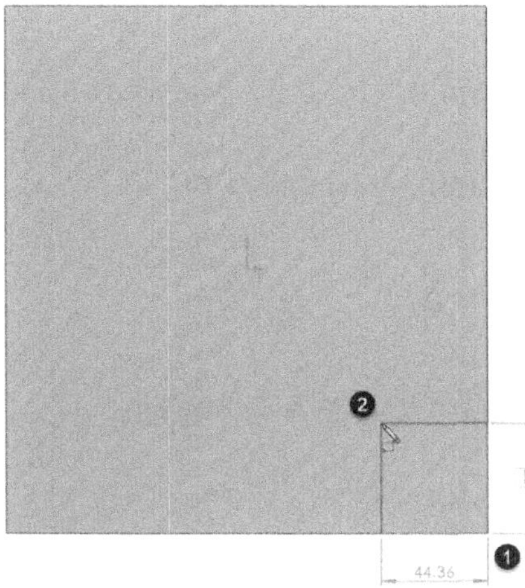

37. Add dimensions to the rectangle, as shown. Next, click **Exit Sketch** on the CommandManager.

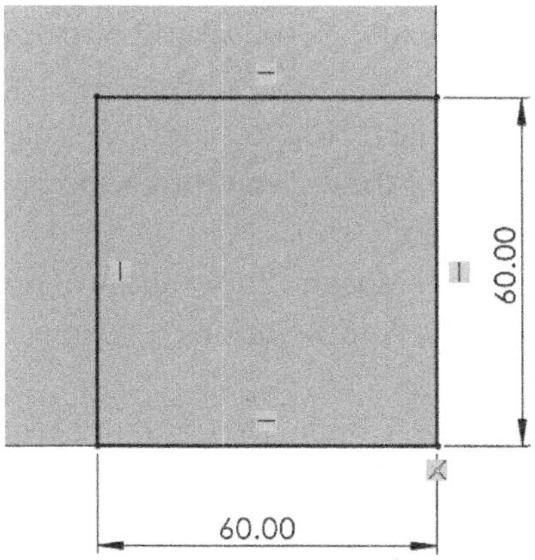

38. Type **35** in the **Depth** box and click **OK**.

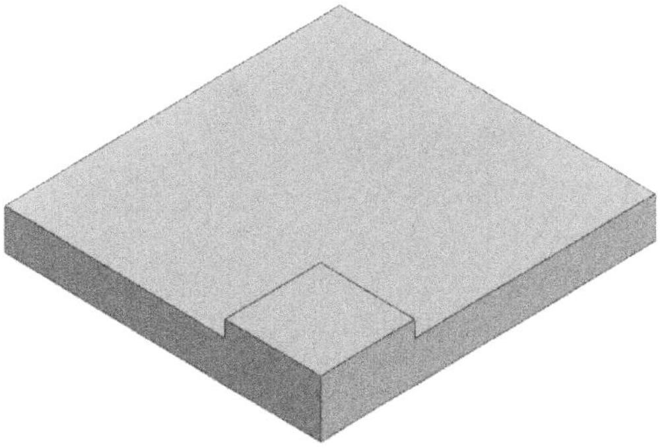

39. On the FeatureManager Design Tree, right-click on the Top plane, and then select **Sketch**.
40. On the CommandManager, click **Sketch > Line**.
41. Click on the left vertical edge of the model, as shown.
42. Move the pointer horizontally toward right and click.
43. Move the pointer vertically downward and click on the horizontal edge, as shown.

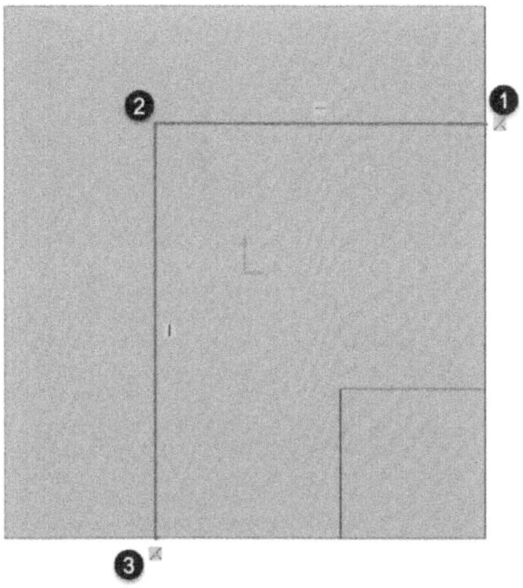

44. On the CommandManager, click **Sketch > Arc** drop-down > **Centerpoint Arc**.
45. Select the intersection point between the horizontal and vertical line.
46. Click on the vertical line to define the start point of the arc.
47. Move the pointer and click on the horizontal line to define the end point of the arc.

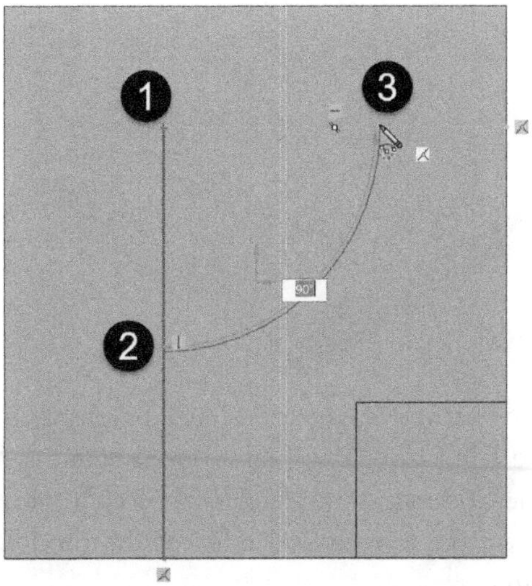

48. On the CommandManager, click **Sketch > Trim Entities**.
49. Click on the portions of the vertical and horizontal lines, as shown. Next, click **OK** on the PropertyManager.

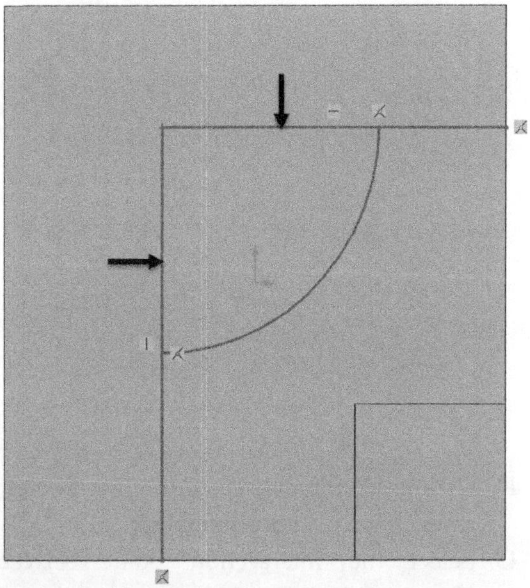

50. Activate the **Smart Dimension** command and select the horizontal edge. Next, select the horizontal line.
51. Move the pointer toward right and click.

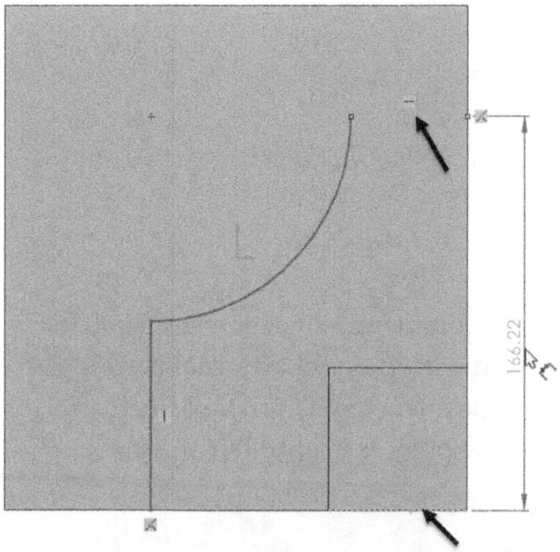

52. On the **Modify** box, select the dimension value and type =.
53. Select **Global Variables > C**.

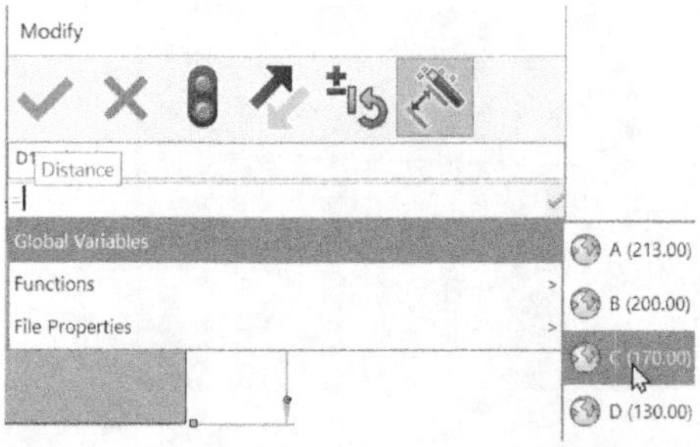

54. Click the green check on the **Modify** box.
55. Likewise, create a dimension between the vertical edge and the vertical line using the global variable C.
56. Add dimensions to the vertical and horizontal lines, as shown.
57. Press and hold the CTRL key, and then select the centerpoint of the arc and the endpoint of the horizontal line.
58. On the PropertyManager, click the **Horizontal** button under the **Add Relations** section.

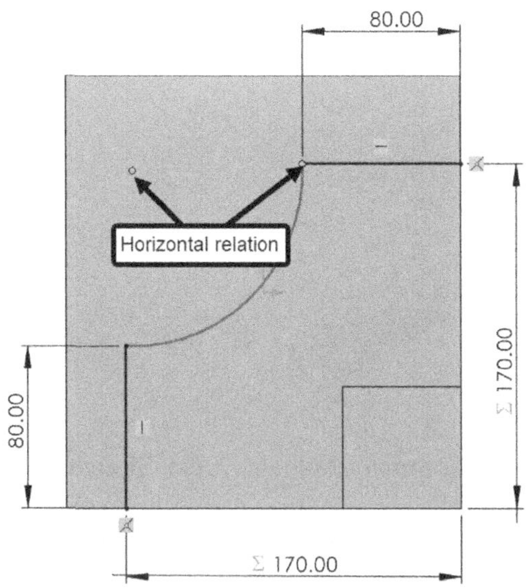

59. On the CommandManager, click **Sketch** tab > **Offset Entities** and select an element from the sketch.
60. Type-in **15** in the **Offset Distance** box available on the PropertyManager.
61. Check the **Reverse** option.
62. Check the **Cap ends** option and then select the **Lines** option.
63. Click **OK**.

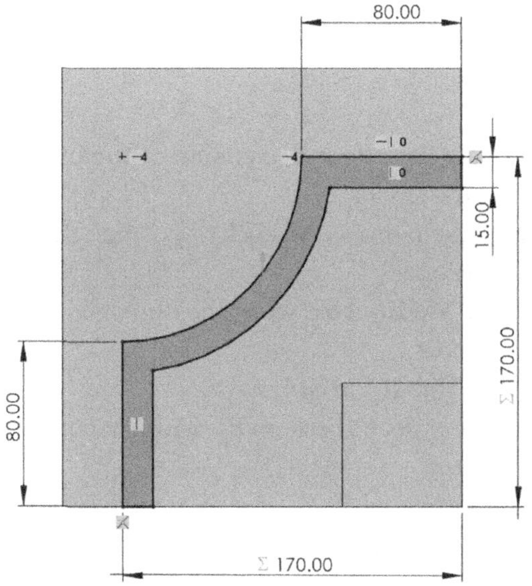

64. Click **Exit Sketch** on the CommandManager.
65. On the CommandManager, click **Features > Extruded Boss/Base**.
66. Select the sketch from the graphics window.
67. Type **95** in the **Depth** box and click **OK**.

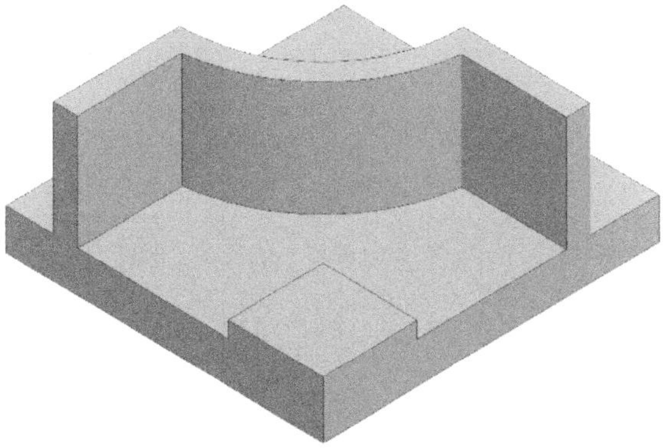

68. Click on the front face of the model.

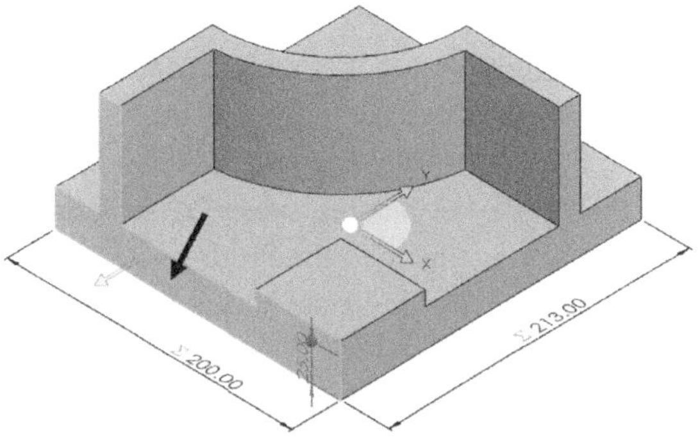

69. On the CommandManager, click **Sketch > Sketch**.
70. On the CommandManager, click **Sketch > Circle** drop-down **> Circle**.
71. Select the midpoint of the small horizontal edge, as shown.

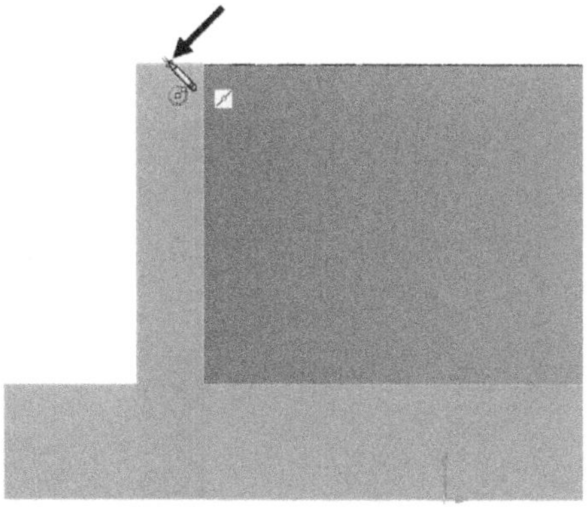

72. Activate the **Smart Dimension** command and select the circle.
73. Move the pointer and click to position the dimension.
74. On the **Modify** box, select the dimension value and type =.
75. Select **Global Variables > X**.

76. Click the green check on the **Modify** box.

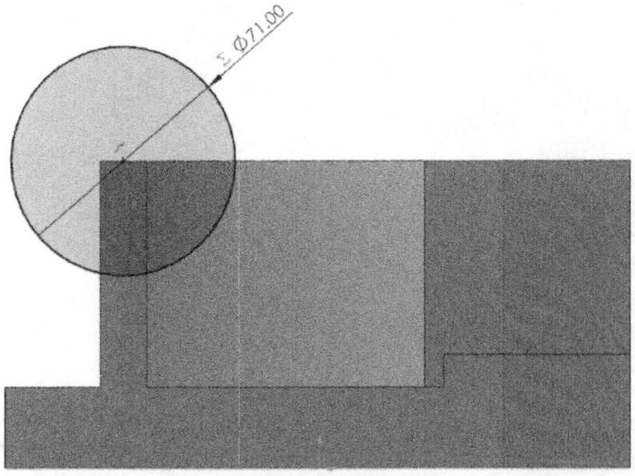

77. Click **Exit Sketch** on the CommandManager.
78. On the CommandManager, click **Features > Extruded Boss/Base**.
79. Select the sketch from the graphics window.
80. On the PropertyManager, select **Offset** from drop-down available in the **From** section.
81. Type **10** in the **Enter Offset Value** box available in the **From** section.
82. Select the **Blind** option from the **End Condition** drop-down.
83. Type **=** in the **Depth** box available in the **Direction 1** section.
84. Select **Global Variables > D**.
85. Click the **Reverse Direction** button next to the **End Condition** drop-down.

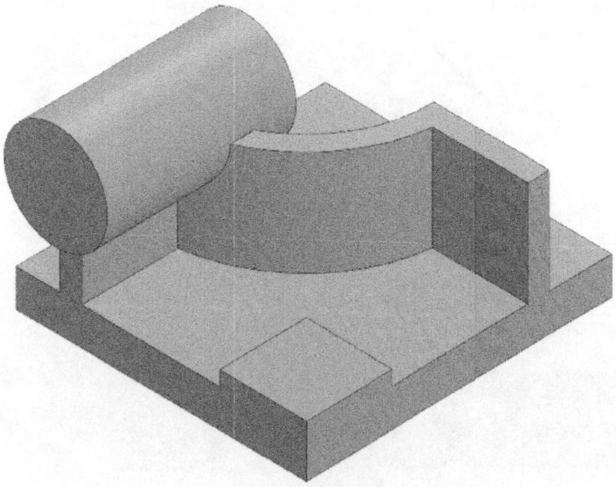

86. Click **OK** on the PropertyManager.
87. Click on the right face of the model.

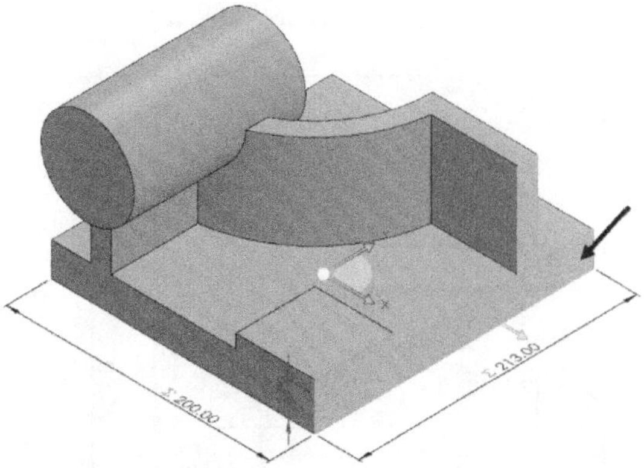

88. On the CommandManager, click **Sketch > Sketch**.
89. On the CommandManager, click **Sketch > Circle** drop-down **> Circle**.
90. Select the midpoint of the small horizontal edge, as shown.

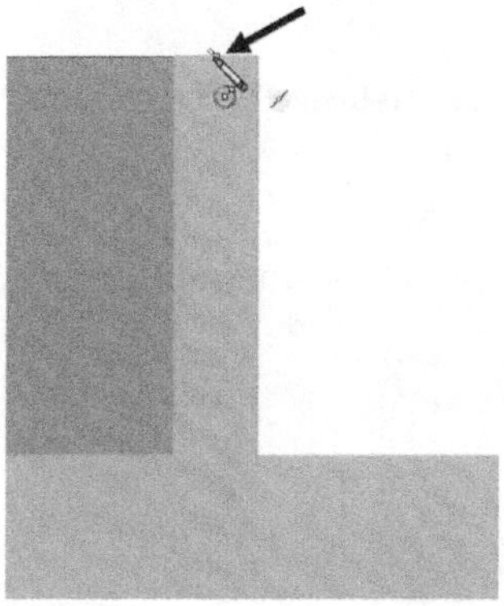

91. Activate the **Smart Dimension** command and select the circle.
92. Move the pointer and click to position the dimension.
93. On the **Modify** box, select the dimension value and type =.
94. Select **Global Variables > Y**.
95. Click the green check on the **Modify** box.

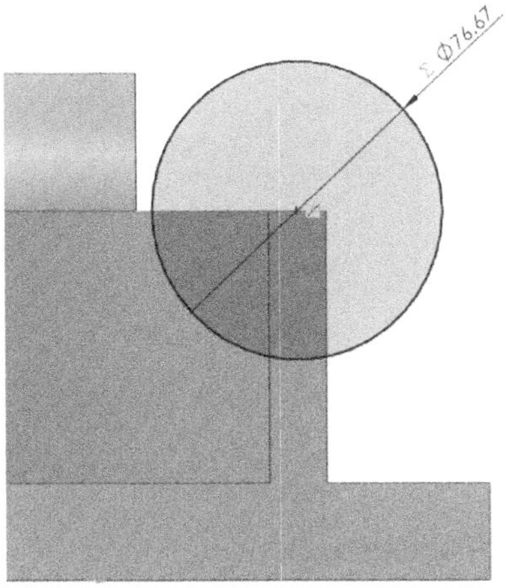

96. Click **Exit Sketch** on the CommandManager.
97. On the CommandManager, click **Features > Extruded Boss/Base**.
98. Select the sketch from the graphics window.
99. On the PropertyManager, select **Offset** from drop-down available in the **From** section.
100. Type **10** in the **Enter Offset Value** box available in the **From** section.
101. Select the **Blind** option from the **End Condition** drop-down.
102. Type **=** in the **Depth** box available in the **Direction 1** section.
103. Select **Global Variables > D**.
104. Click the **Reverse Direction** button next to the **End Condition** drop-down.

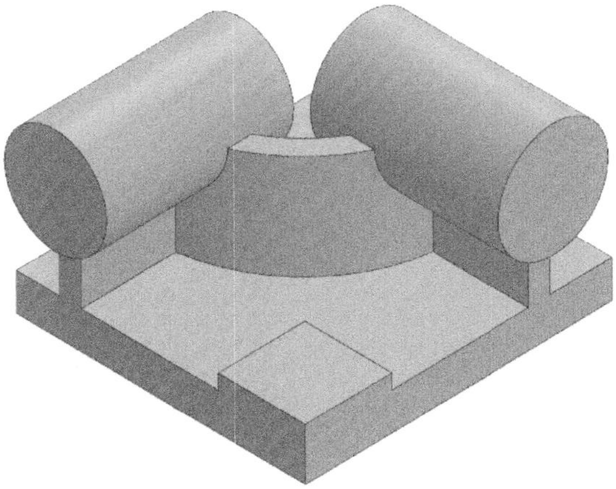

105. Click **OK**.

Creating the Fillet

1. Click **Features** tab > **Fillet** drop-down > **Fillet** on the CommandManager.
2. On the **Fillet** PropertyManager, click the **Manual** tab, and then set the **Fillet type** to **Constant Size Fillet**.
3. Click on the vertical edge of the geometry, as shown below.

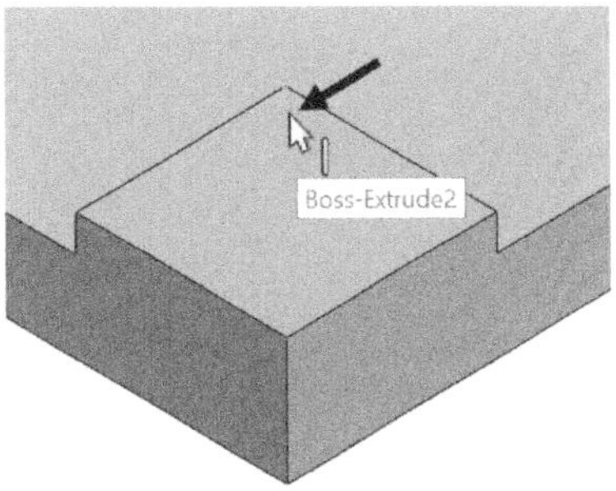

4. Type-in **9** in the **Radius** callout attached to one of the selected edges and then press the Enter key.
5. Click the **OK** icon on the PropertyManager.

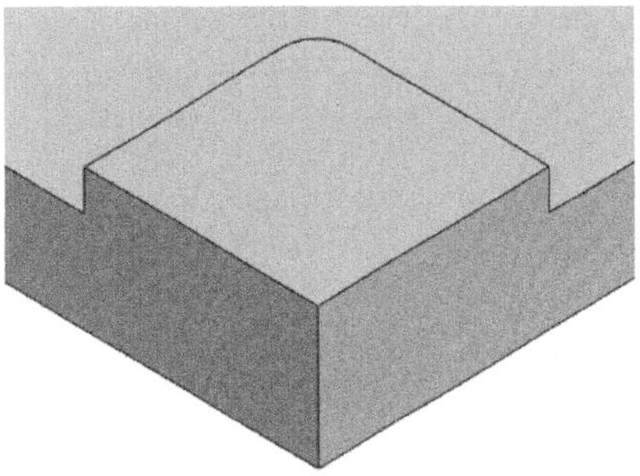

Creating the Extruded Cuts

44. On the **Features** CommandManager, click **Extruded Cut**.
45. Click on the top face of the first feature.

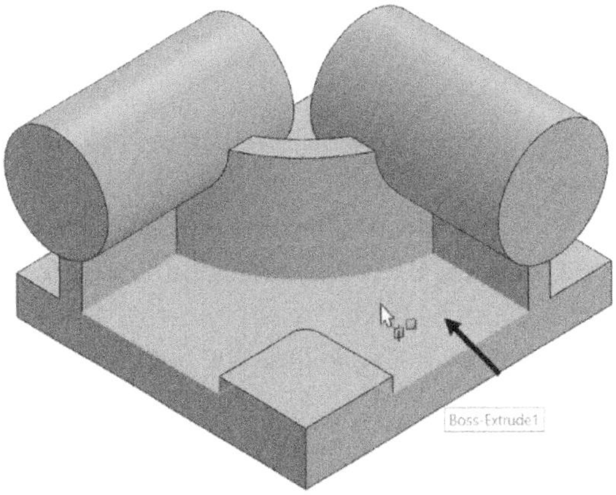

46. Again, select the same face.

47. On the CommandManager, click **Sketch** tab > **Offset Entities**.
48. Type-in **9** in the **Offset Distance** box available on the PropertyManager.
49. Check the **Reverse** option.

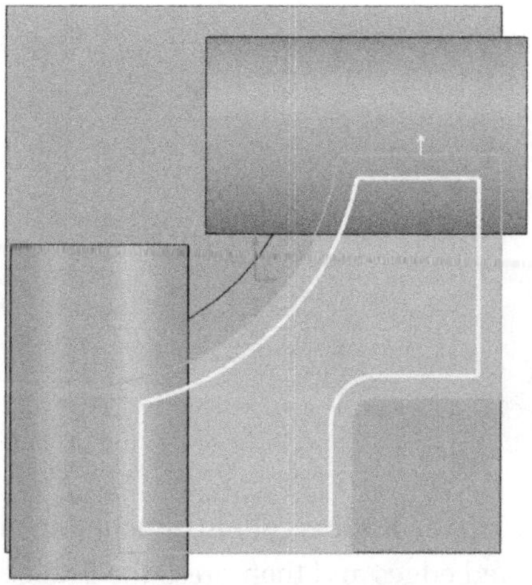

50. Click **OK** on the PropertyManager.

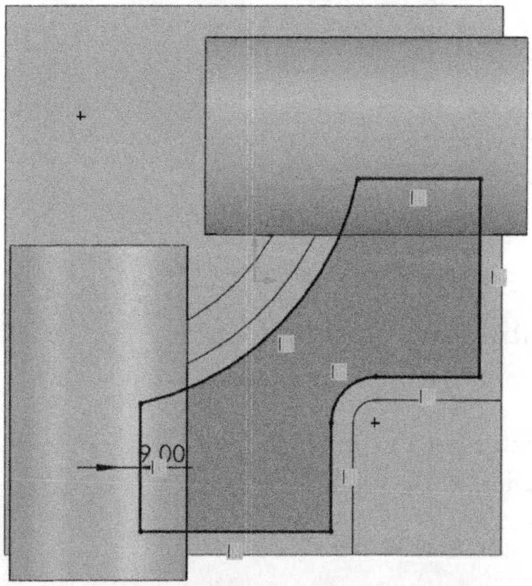

51. On the **Sketch** CommandManager, click **Exit Sketch**.
52. On the **Cut-Extrude PropertyManager**, under the **Direction 1** section, select **End Condition > Offset from Surface**.
53. Rotate the model and select the bottom face of it.

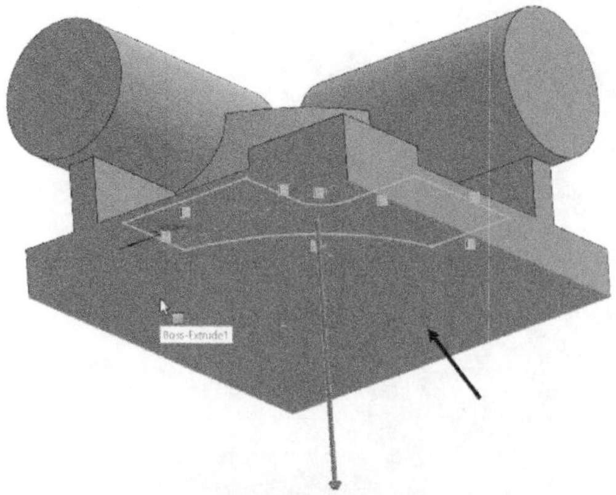

54. Type **5** in the **Offset Distance** box.
55. Click **OK** to create the cut offset from the bottom face.

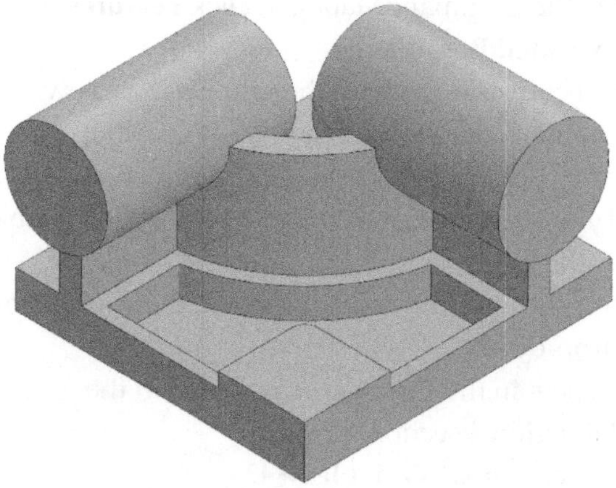

6. On the **Features** CommandManager, click **Extruded Cut** .
7. Click on the front face of the extruded feature, as shown.

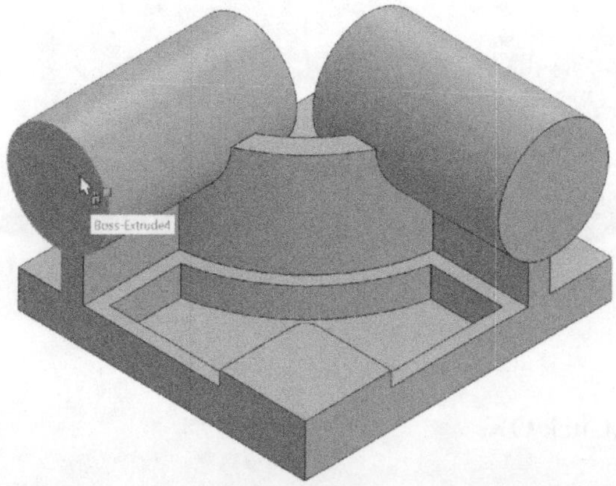

8. On the CommandManager, click **Sketch > Circle** drop-down **> Circle**.
9. Place the pointer on the circular edge, as shown.

The centerpoint of the circular edge is highlighted.

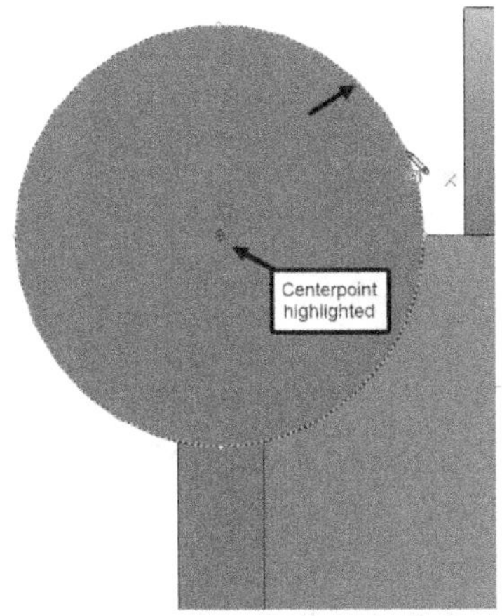

10. Select the centerpoint of the circular edge, move the pointer outward and click.
11. Activate the **Smart Dimension** command and select the circle.
12. Move the pointer and click to position the dimension.
13. On the **Modify** box, select the dimension value and type =.
14. Select **Global Variables > E.**
15. Click the green check on the **Modify** box.

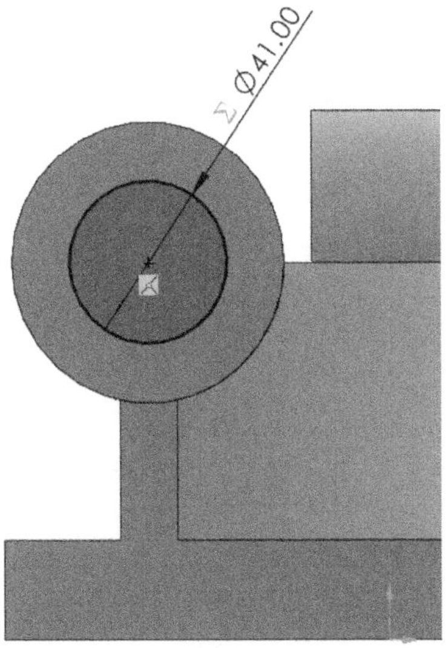

16. On the **Sketch** CommandManager, click **Exit Sketch**.
17. On the **Cut-Extrude PropertyManager**, under the **Direction 1** section, select **End Condition > Through All**.
18. Click **OK** to create the cut throughout the part geometry.

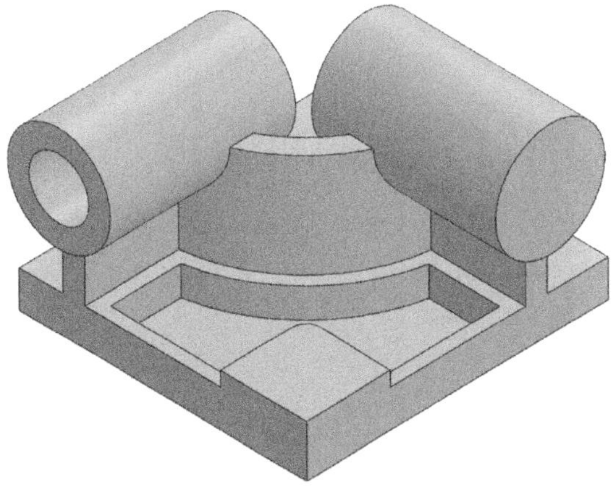

19. Likewise, create the other extruded cut feature, as shown.

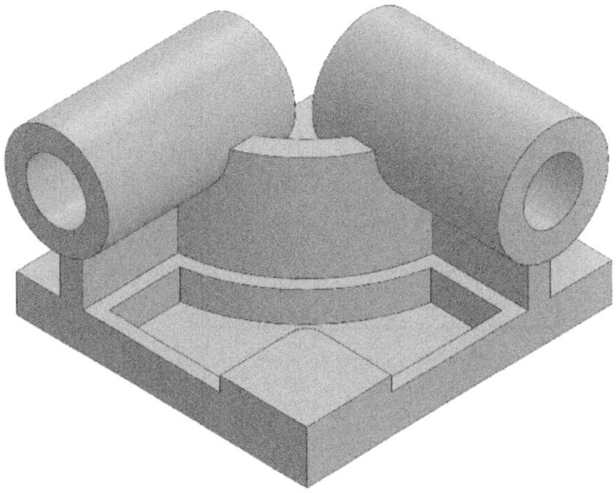

Creating the Hole

1. On the **Features** CommandManager, click **Hole Wizard**.
2. On the PropertyManager, select the **Counterbore** button and set the **Standard** to **ANSI Metric**.
3. Select **Type > Hex Bolt - ANSI B18.2.3.5M**.
4. Select **Size > M8**, and then check the **Show custom sizing** option.
5. Select **Fit > Close**.
6. Type-in 15, 30, and 10 in the **Through Hole Diameter**, **Counterbore Diameter**, and **Counterbore Depth** boxes, respectively.
7. Set **End Condition** to **Through All**.

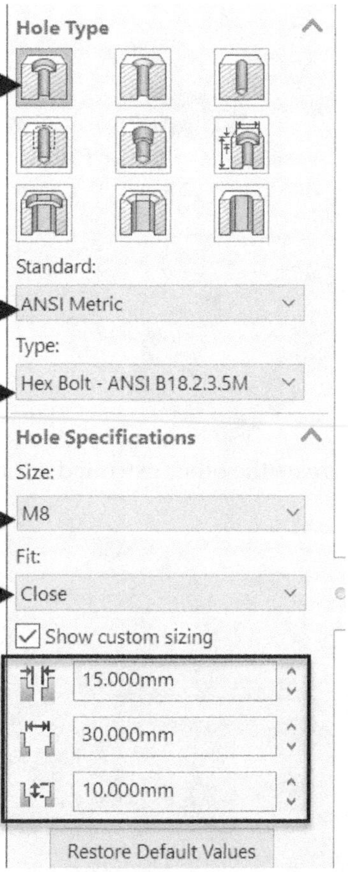

8. Uncheck all the options in the **Options** section.
9. Click the **Positions** tab on the PropertyManager.
10. Click on the top face of the second feature, and then click to place the hole.
11. On the CommandManager, click **Sketch** tab > **Smart Dimension**.
12. Select the center point of the hole and the right vertical edge of the placement face. Next, move the pointer up and click to place the dimension. Type **30** in the **Modify** box and click the green check.
13. Select the center point of the hole and the top edge of the placement face. Next, move the pointer toward the right and click to place the dimension. Type 30 in the **Modify** box and click the green check.
14. Click **OK** on the PropertyManager to create the counterbore hole.

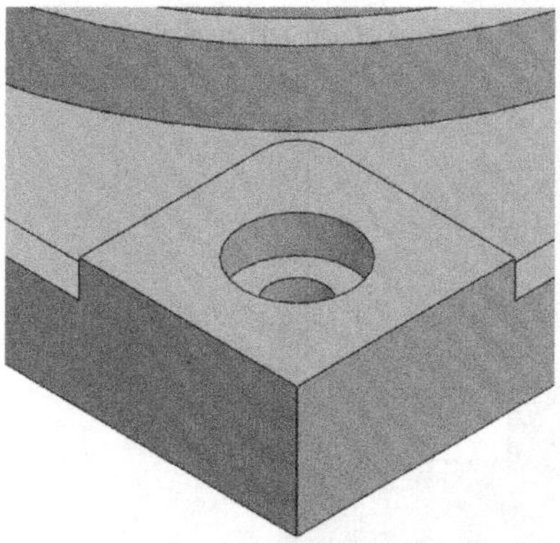

Creating the Fillets and Chamfers

1. Click **Features** tab > **Fillet** drop-down > **Fillet** on the CommandManager.
2. On the **Fillet** PropertyManager, click the **Manual** tab, and then set the **Fillet type** to **Constant Size Fillet**.
3. Click on the vertical edges of the geometry, as shown below.

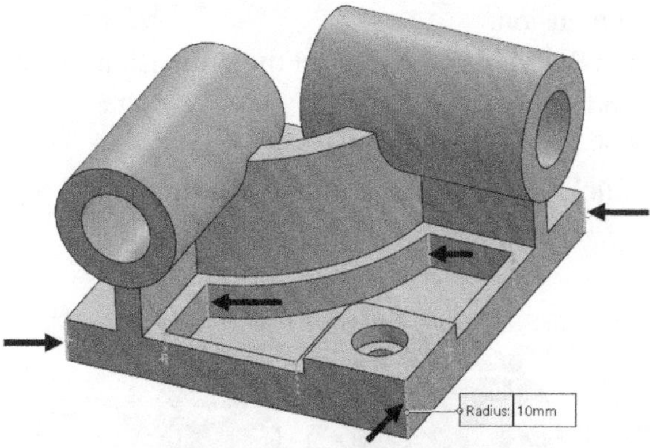

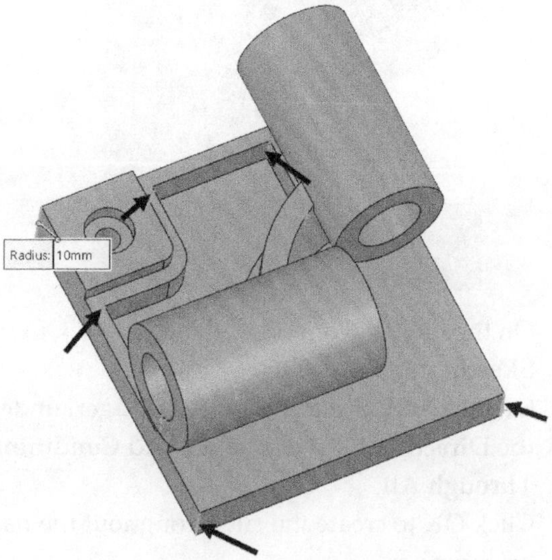

4. Type-in **10** in the **Radius** callout attached to one of the selected edges and then press the Enter key.
5. Click the **OK** icon on the PropertyManager.

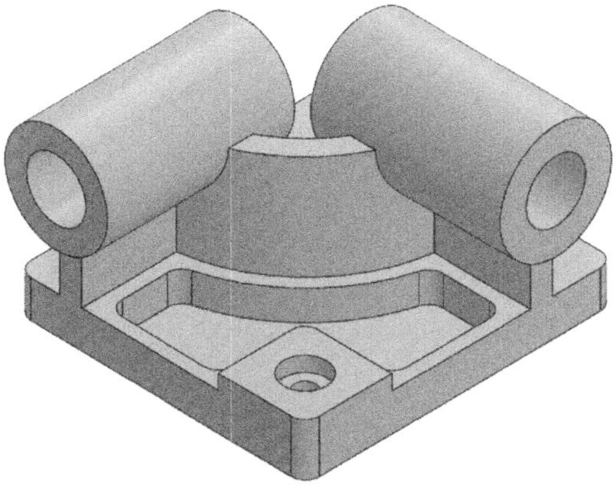

6. Click **Features** tab > **Fillet** drop-down > **Chamfer** on the CommandManager; the **Chamfer** PropertyManager appears.
7. On the PropertyManager, select the **Angle Distance** icon.
8. Set the **Distance** and **Angle** to **2** and **45**, respectively.
9. Click on the circular edges of the model, as shown.

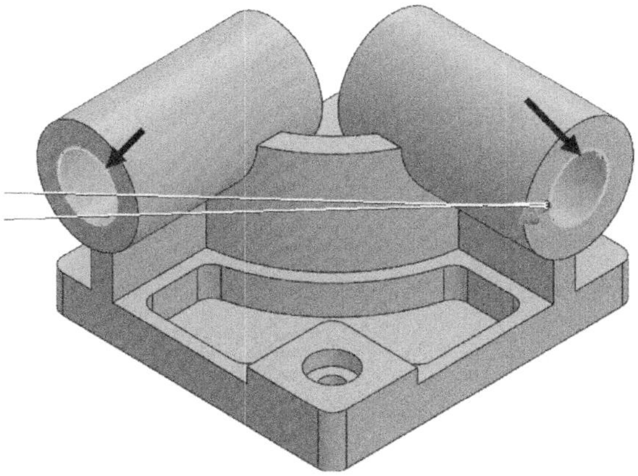

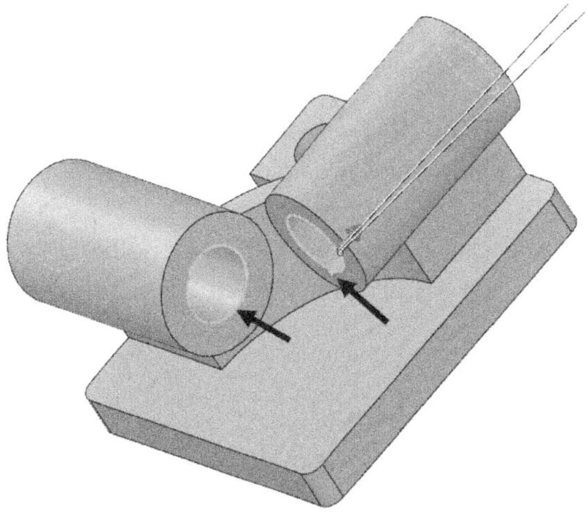

10. Click **OK** on the PropertyManager.

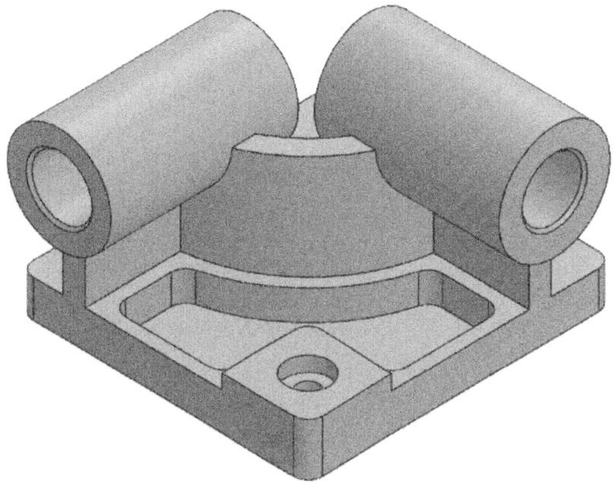

Adding Material to the Model

1. In the FeatureManager Design Tree, right-click on the **Material** node and select **Edit Material**.

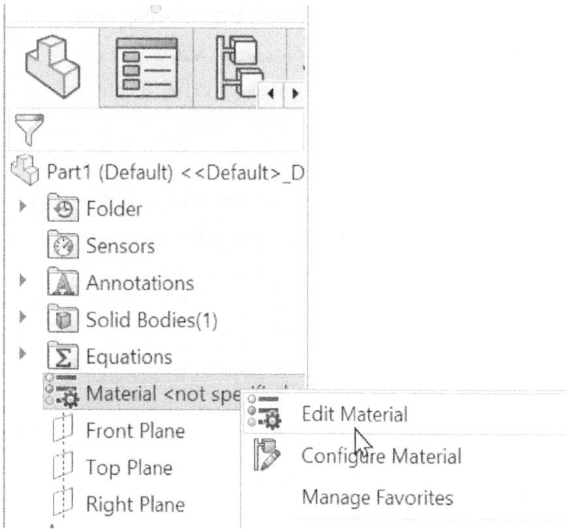

2. On the **Material** dialog, expand the **Steel** folder and select the **Alloy Steel** material.

3. Click **Apply** and **Close** on the **Material** dialog.
4. On the CommandManager, click **Evaluate > Mass Properties**  ; on the **Mass Properties** dialog, the **Mass** is displayed.
5. Close the **Mass Properties** dialog.
6. Click **Save** on the Quick Access Toolbar.
7. Type **C4_Tutorial2** in the **File name** box and click **Save**.

Updating the Model

1. In the FeatureManager Design Tree, right-click on the **Equations** node and select **Manage Equations**.

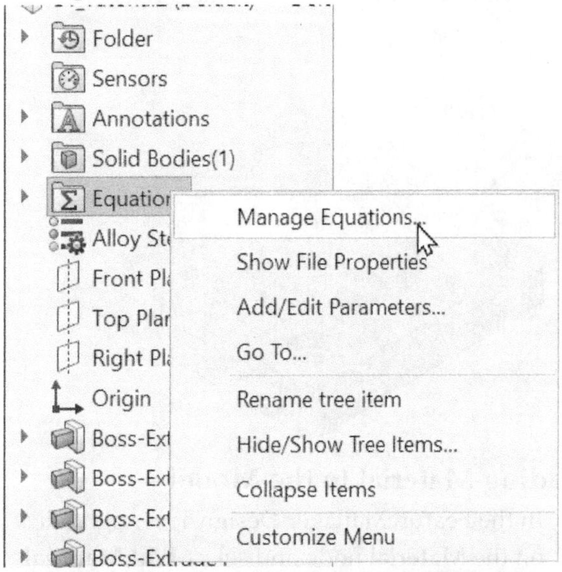

2. On the **Equations, Global Variables, and Dimensions** dialog, select the value in the **Value/ Evaluation** box next to the **A** global variable.
3. Type **225.**
4. Likewise, change the **B, C, D,** and **E** values to **210, 176,137** and **39**, respectively.

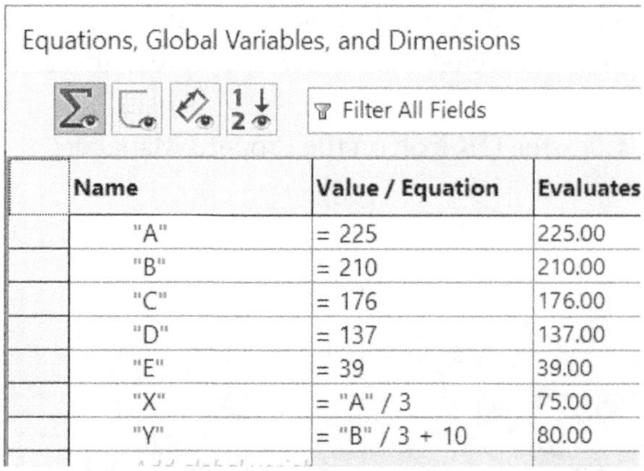

Equations, Global Variables, and Dimensions
Filter All Fields
Name | Value / Equation | Evaluates
"A" | = 225 | 225.00
"B" | = 210 | 210.00
"C" | = 176 | 176.00
"D" | = 137 | 137.00
"E" | = 39 | 39.00
"X" | = "A" / 3 | 75.00
"Y" | = "B" / 3 + 10 | 80.00

5. Click **OK**; the model is updated.

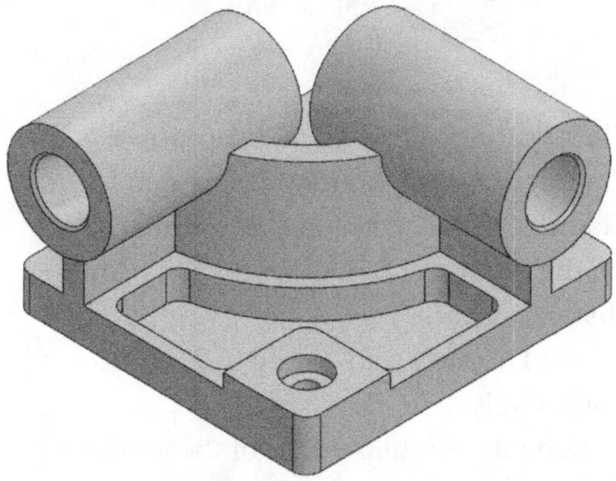

6. Click **Evaluate > Mass Properties** on the CommandManager. The **Mass** value is updated.
7. Close the **Mass Properties** dialog.
8. In the FeatureManager Design Tree, right-click on the **Equations** node and select **Manage Equations**.
9. On the **Equations, Global Variables, and Dimensions** dialog, select the value in the **Value/ Evaluation** box next to the **A** global variable.
10. Type **209.**
11. Likewise, change the **B, C, D,** and **E** values to **218, 169,125** and **41**, respectively.

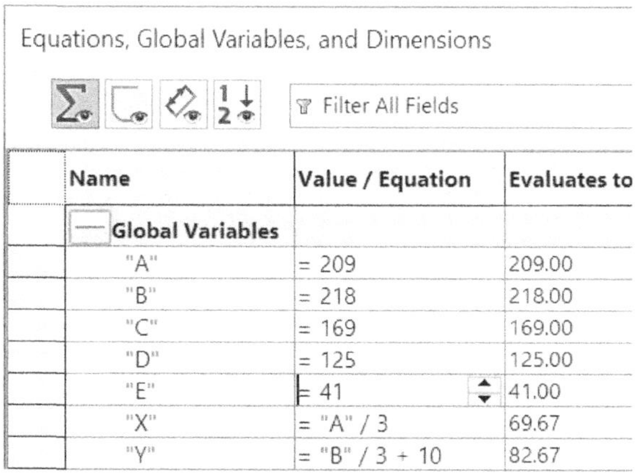
Equations, Global Variables, and Dimensions

Filter All Fields

Name	Value / Equation	Evaluates to
Global Variables		
"A"	= 209	209.00
"B"	= 218	218.00
"C"	= 169	169.00
"D"	= 125	125.00
"E"	= 41	41.00
"X"	= "A" / 3	69.67
"Y"	= "B" / 3 + 10	82.67

12. Click **OK**; the model is updated.

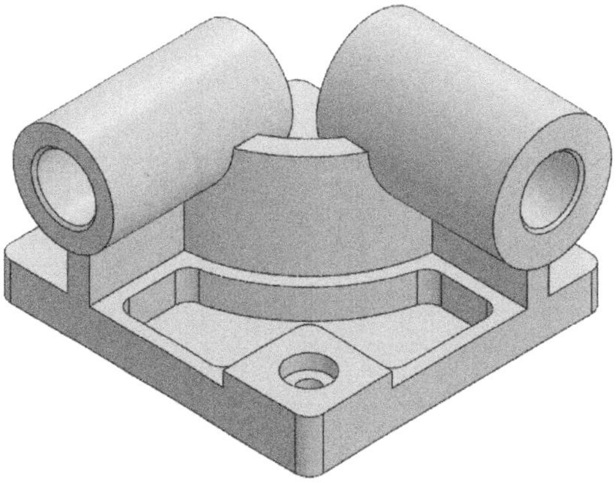

13. Click **Evaluate > Mass Properties** on the CommandManager. The **Mass** value is updated.
14. Close the **Mass Properties** dialog.
15. Save the file as C4_example2 and close it.

Questions

1. What are placed features'?
2. Which option allows you to create chamfer with unequal setbacks?
3. Which option allows you to create a variable radius fillet?
4. When you create a thread on a cylindrical face, will the cylinder's diameter remain the same or not?
5. How to create a shell feature without removing any face?

Exercises

Exercise 1 (Millimetres)

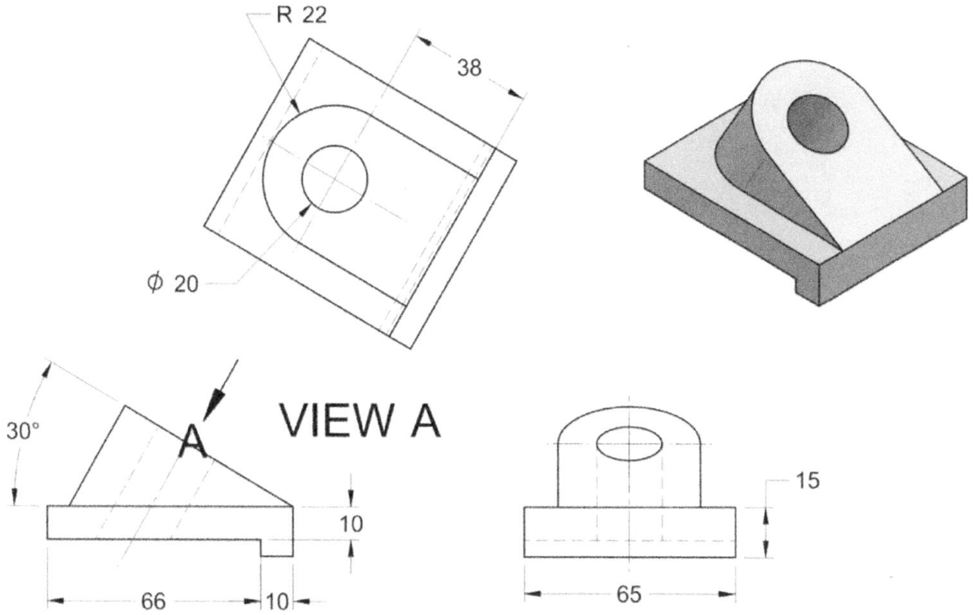

Exercise 2 (Inches)

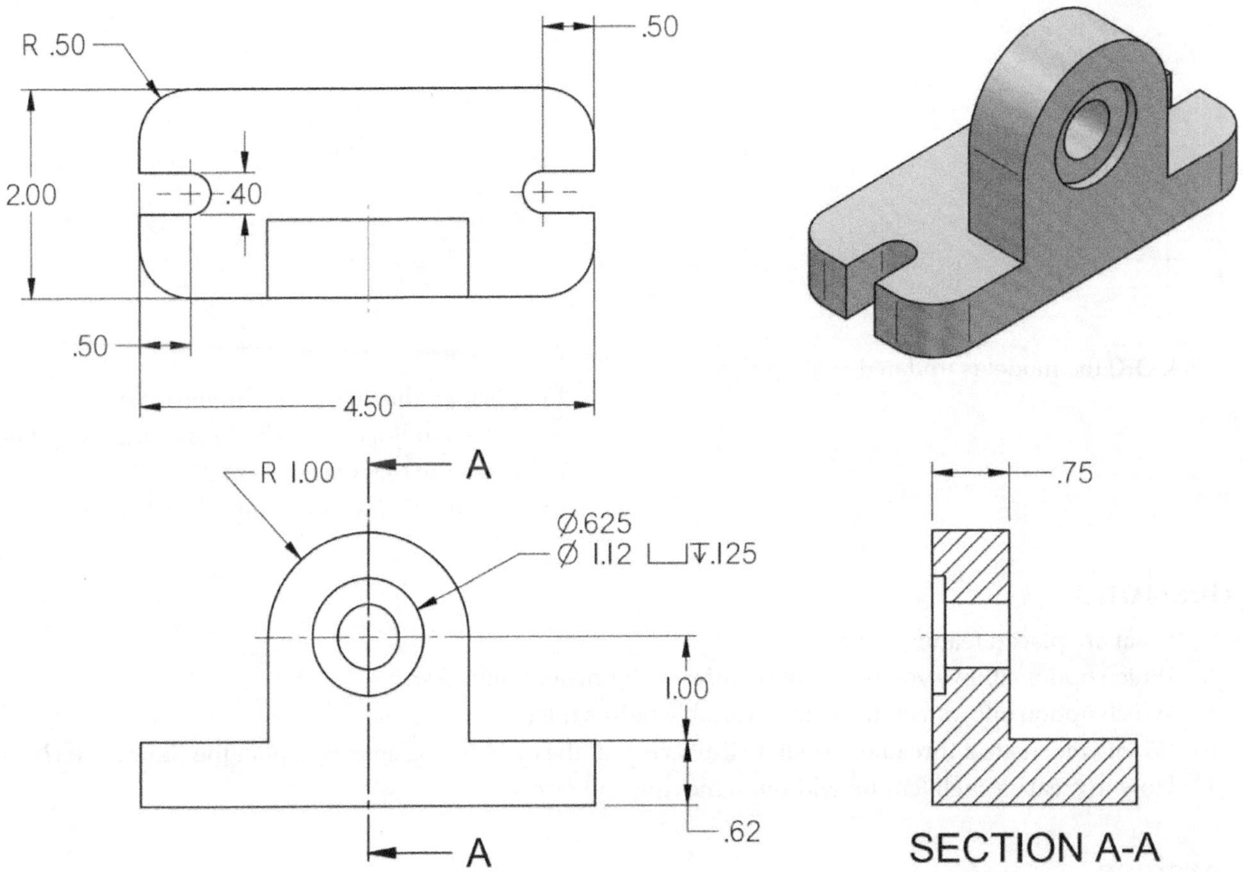

Chapter 5: Patterned Geometry

When designing a part geometry, there are often elements of symmetry in each part, or there are at least a few features repeated multiple times. In these situations, SOLIDWORKS offers you some commands that save you time. For example, you can use mirror features to design symmetric parts, which makes designing the part quicker. This is because you only have to design a portion of the part and use the mirror feature to create the remaining geometry.

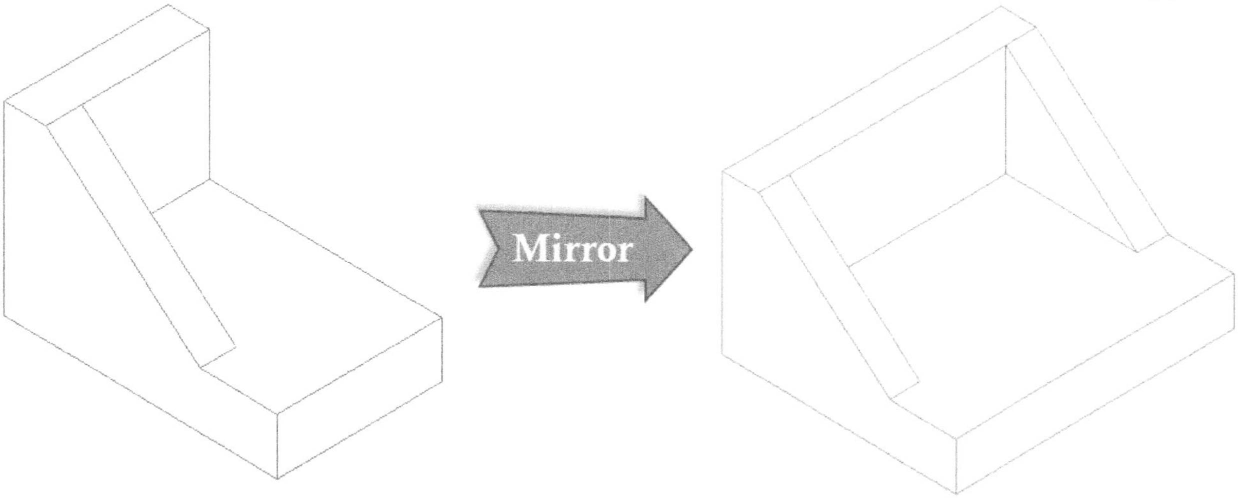

Also, there are some pattern commands to replicate a feature throughout the part quickly. They save you time from creating additional features individually and help you modify the design easily. If the design changes, you only need to change the first feature, and the rest of the pattern features will update automatically. In this chapter, you will learn to create the mirrored and pattern geometries using the commands available in SOLIDWORKS.

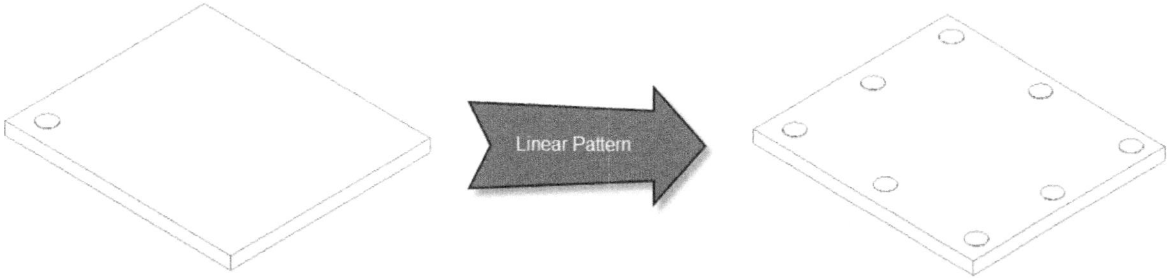

The topics covered in this chapter are:

- *Mirror* features
- *Linear Patterns*
- *Circular Patterns*
- *Curve Driven Patterns*
- *Sketch Driven Patterns*
- *Fill Patterns*

Mirror

If you are designing a symmetric part, you can save time by using the **Mirror** command. Using this command, you can replicate the individual features of the entire body. To mirror features (3D geometry), you need to have a face or plane to use as a reference. You can use a model face, default plane, or create a new plane if it does not exist where it is needed.

Activate the **Mirror** command (click **Features > Mirror** on the CommandManager). On the **Mirror** PropertyManager, click in the **Mirror Face/Plane** selection box and select the reference plane about which the features are to be mirrored. On the part geometry, click on the features to mirror. Next, click **OK** on the PropertyManager.

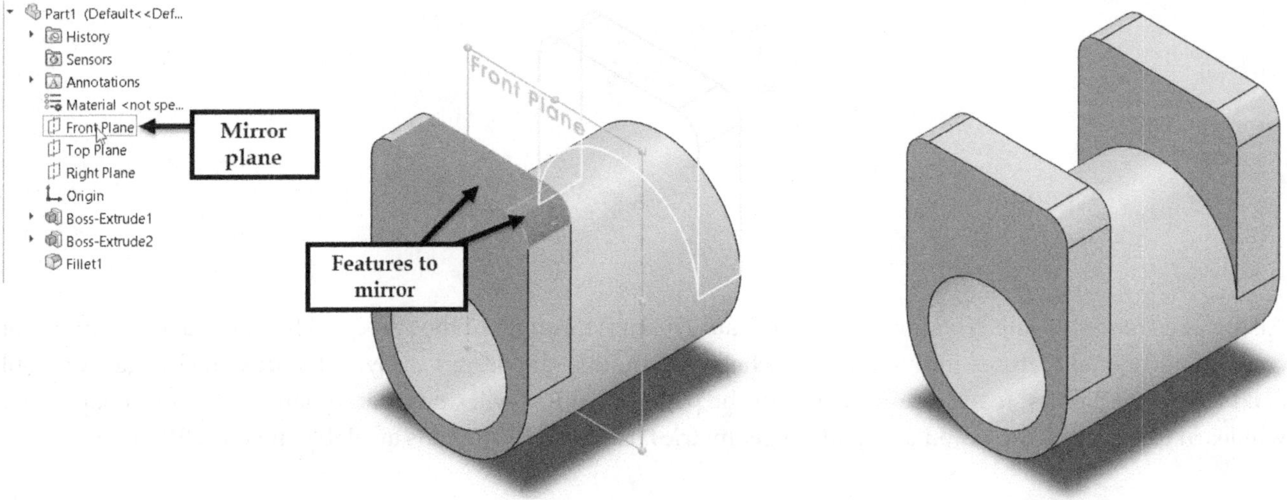

Now, if you make changes to the original feature, the mirrored feature will be updated automatically.

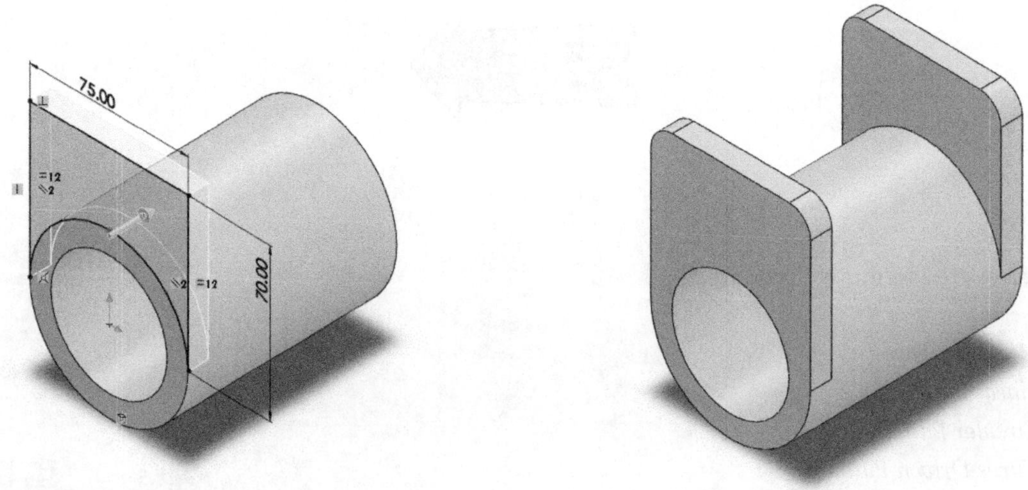

Mirror Bodies

If the part you are creating is entirely symmetric, you can save more time by creating half of it and mirroring the entire geometry rather than individual features. Activate the **Mirror** command (On the CommandManager, click **Features > Mirror**) and click in the **Mirror Face/Plane** selection box. Select the face or plane about which the geometry is to be mirrored. Next, expand the **Bodies to Mirror** section on the PropertyManager and select the model geometry. The **Merge solids** option in the **Options** section allows you to specify whether the mirrored body will be joined with the source body or a separate body will be created. Click **OK** to complete the mirror geometry.

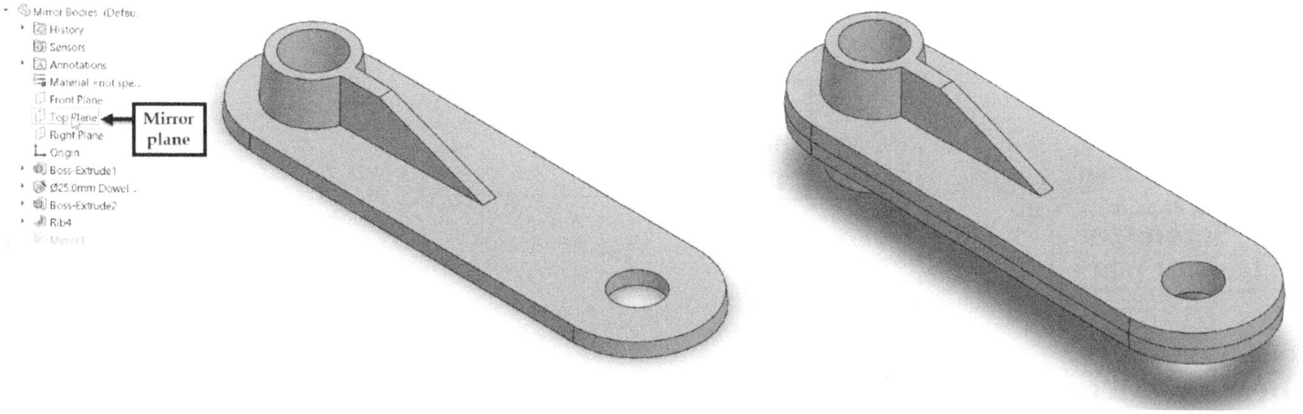

Mirroring about Two Planes

SOLIDWORKS allows you to mirror features about two planes. To do this, activate the **Mirror** command and select the mirror plane from the FeatureManager Design Tree or graphics window. Next, select the secondary mirror plane from the FeatureManager Design Tree or graphics window. Select the features to be mirrored from the graphics window and click **OK**.

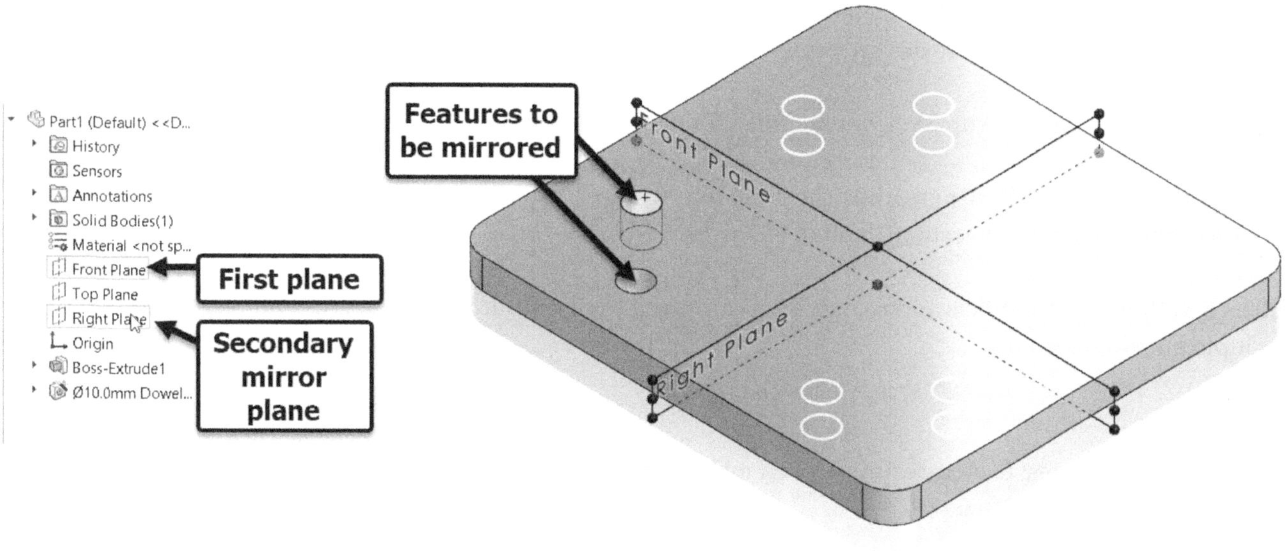

Likewise, you can mirror the entire part body about two planes. Activate the **Mirror** command and click in the **Mirror Face/Plane** selection box. Select the face or plane about which the geometry is to be mirrored. Next, select the secondary mirror face from the model. Next, expand the **Bodies to Mirror** section on the PropertyManager and select the model geometry. The **Merge solids** option in the **Options** section allows you to specify whether the mirrored body will be joined with the source body or a separate body will be created. Click **OK** to complete the mirror geometry.

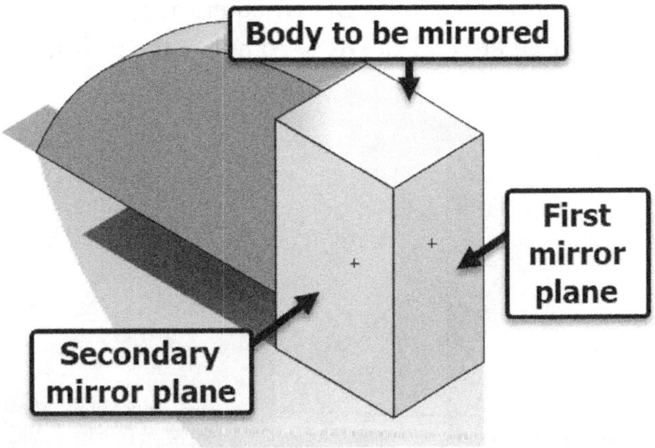

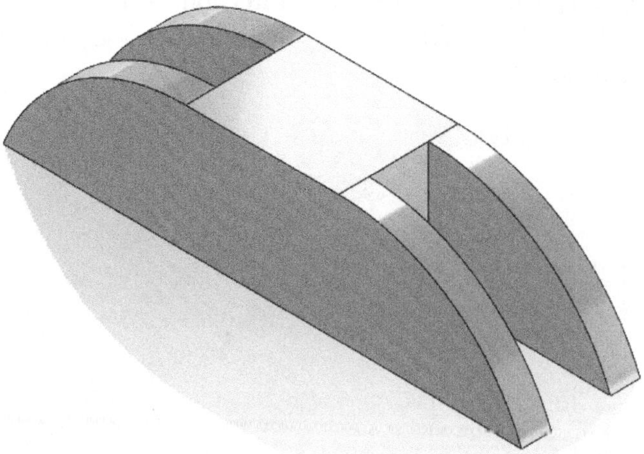

Best Practices

Symmetry: Ensure your part displays clear symmetry by placing features intended for mirroring appropriately relative to the axis or plane chosen as the mirror reference.

Part Modeling Order: Construct the initial features in the correct sequence, typically creating primary features before applying the mirror feature.

Reference Planes or Axes: Establish reference planes or axes defining symmetry and mirror planes to ensure consistency and simplify future design modifications.

Fully Defined Sketches: Ensure sketches are fully defined to prevent unintended changes during mirroring.

Check for Interferences: Post-mirroring, check for interferences, especially in complex designs where mirrored features may overlap or interfere with other elements.

Fillets and Rounds: Add fillets and rounds after mirroring to ease future modifications without adjusting multiple mirrored features.

Cautions

Asymmetrical Features: Exercise caution when mirroring features lacking perfect symmetry to avoid unexpected results and design issues.

Internal Features in Multi-body Parts: Hide or exclude internal bodies or features that shouldn't be mirrored in multi-body parts.

Revisions: When modifying one side of a mirrored part, adjust mirrored features accordingly.

Tips for Specific Part Types

Plastic Parts: Consider draft angles and undercuts when mirroring complex shapes. Utilize the draft feature and conduct draft analysis to identify issues.

Casted Parts: Be mindful of parting lines and draft angles in casted parts. Note that mirrored features can complicate design and affect mold creation.

Forged Parts: Account for extensive machining after forging when mirroring. Allow for machining allowances in your design.

Machined Parts: Mirroring machined parts is more flexible. Ensure mirrored features do not result in excess material requiring additional machining.

Create Patterns

SOLIDWORKS allows you to replicate a feature using the pattern commands (**Linear Pattern, Circular Pattern, Curve Driven Pattern, Sketch Driven Pattern, Table Driven Pattern, Fill Pattern** and **Variable Pattern)**. The following sections explain the different patterns that can be created using these pattern commands.

Linear Pattern

To create a linear pattern, you must first activate the **Linear Pattern** command (On the CommandManager, click **Features > Pattern** drop-down **> Linear Pattern**). Next, select an edge, face, or axis to define **Direction 1** of the pattern. Likewise, select an edge, face or axis to define the second direction of the pattern. Next, select the feature to pattern from the model geometry. You will notice that a pattern preview appears on the model.

Type-in values in the **Spacing** and **Number of Instances** boxes available in the **Direction 1** section. Click the **Reverse Direction** icon if you want to reverse the pattern direction. Likewise, set the parameters (**Spacing** and **Number of Instances**) of the pattern in direction 2.

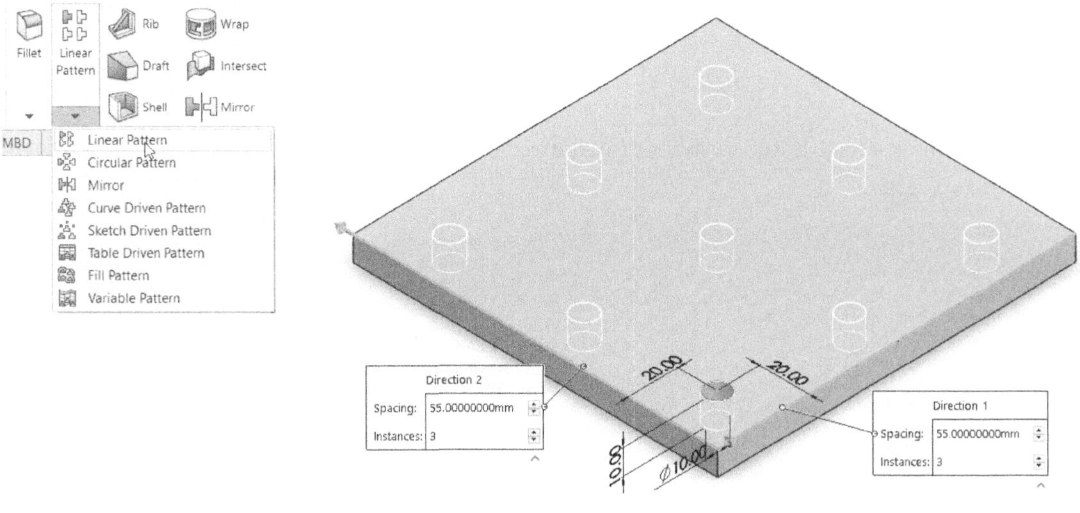

Up to reference

The **Up to reference** option allows you to specify the spacing between the instances by selecting a reference element. This option is helpful while creating a linear pattern of holes on the rectangular block, as shown. When you change the rectangular block's length, the distance between the end holes of the pattern and the boundary edge is not maintained. The **Up to reference** option helps you to solve this problem.

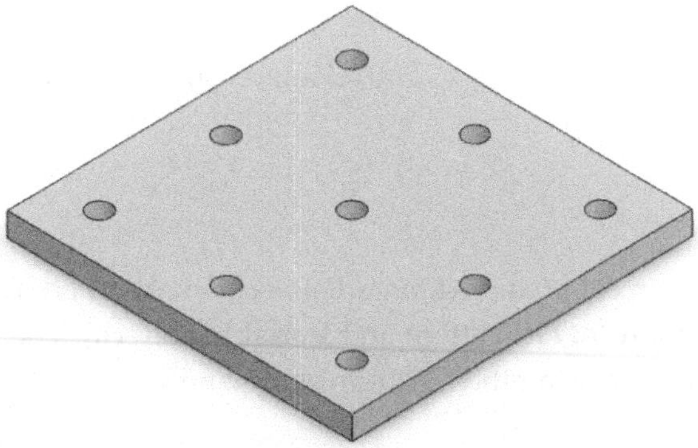

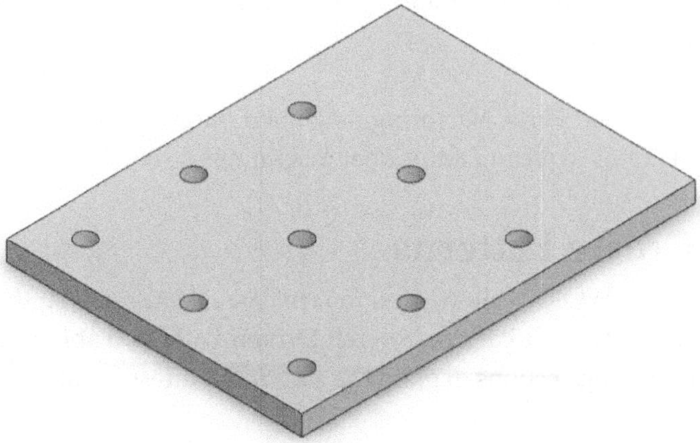

Select the **Up to reference** option in the **Direction 1** section, and then select the rectangular block's vertex, as shown. You will notice that another instance of the hole is added to the linear pattern.

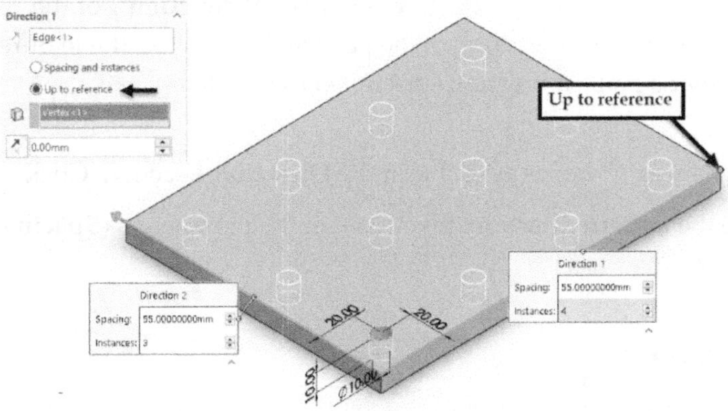

If you change the rectangular block's length, the number of instances is adjusted to maintain the distance between the end holes and the boundary edge.

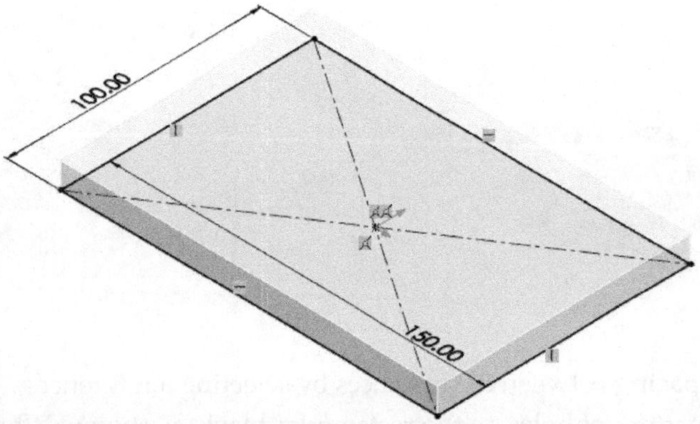

The **Up to reference** option allows you to maintain the distance between the end holes and the reference element even if you want to specify the number of instances. To do this, click the **Set Number of Instances** icon and click in the **Offset distance** box displayed below the **Reference Geometry** selection box. Next, select the dimension between the seed instance and the edge, as shown.

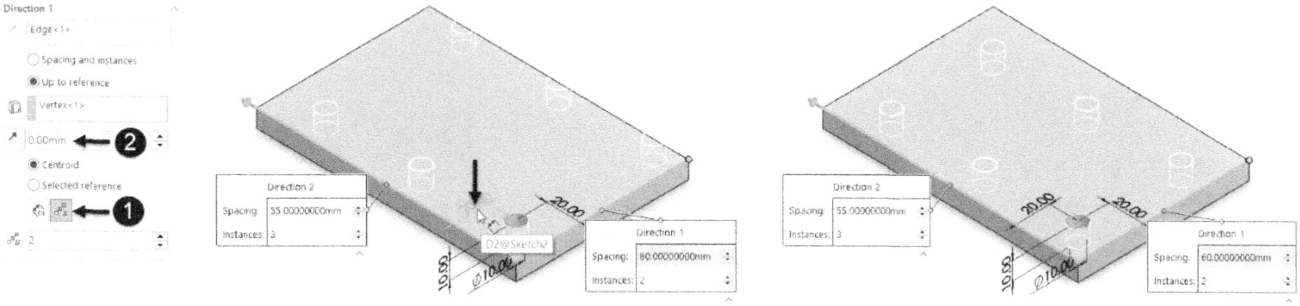

Now, change the **Number of Instances** value or length of the rectangular block; the linear pattern is updated.

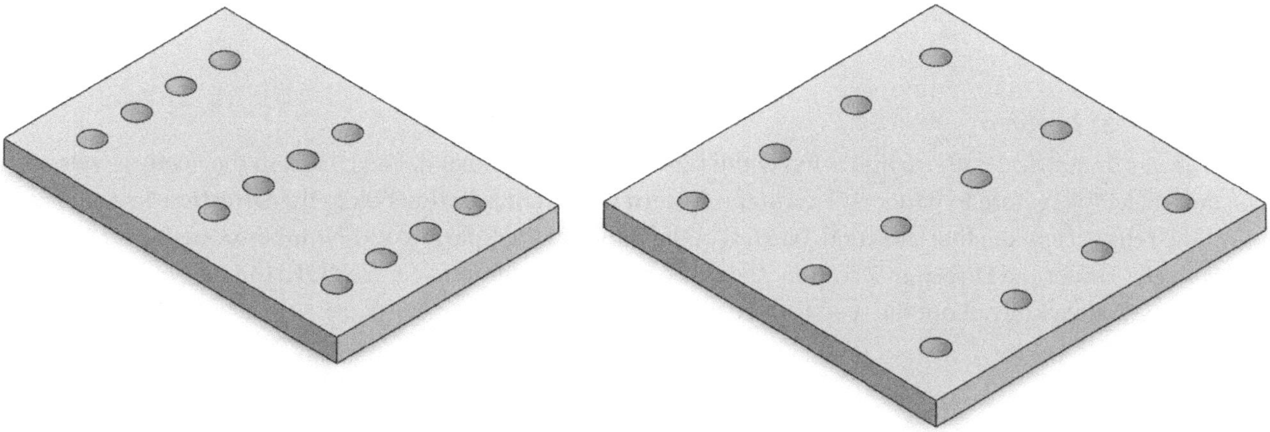

If you want to suppress an occurrence of the pattern, then expand the **Instance to skip** section in the **Linear Pattern** PropertyManager. Next, select the pink dots from the pattern preview.

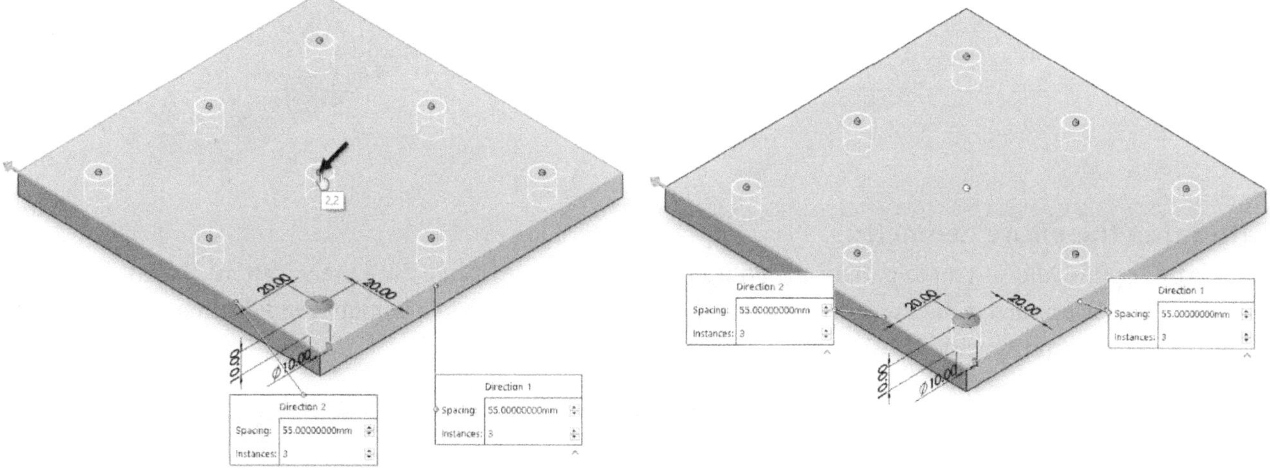

Check the **Pattern seed only** option in the **Direction 2** section if you want to pattern only the seed instance along the second direction.

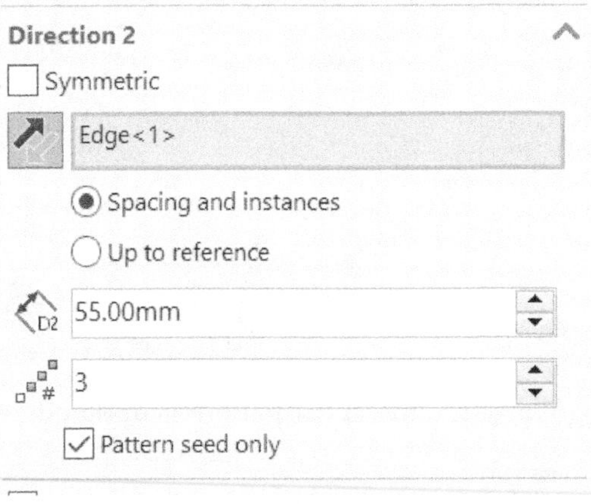

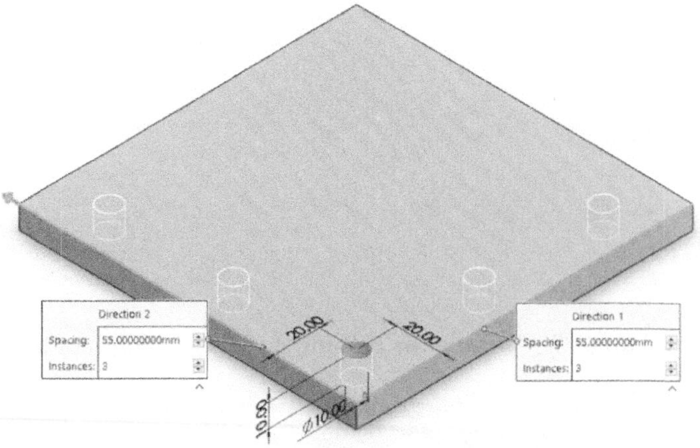

Symmetrical Pattern

To create a symmetrical pattern, first activate the **Linear Pattern** command. Next, click in the Features selection box and select the feature to pattern. Click in the **Pattern Direction** selection box in the **Direction 1** section and select the reference of the first direction. Next, type-in values in the **Spacing** and **Number of Instances** boxes, respectively. Under the **Direction 2** section, check the **Symmetric** option and click **OK**; the selected feature is patterned symmetrically about the seed feature.

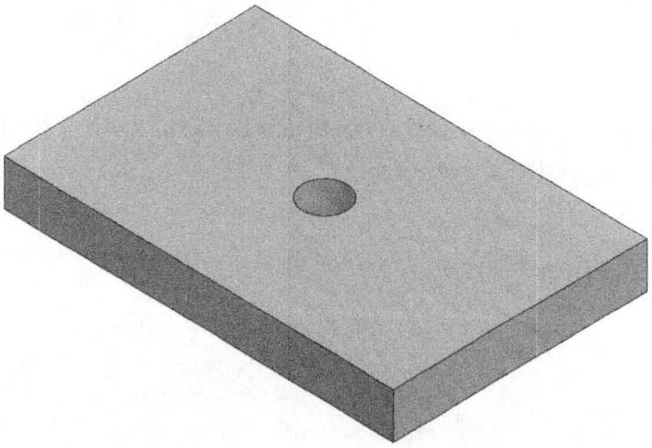

Patterning the entire geometry

The **Bodies** option allows you to pattern the entire part geometry. Activate the **Linear Pattern** command and check the **Bodies** option on the PropertyManager. Next, select the bodies to be patterned. Define the direction, occurrence count, and spacing between the instances.

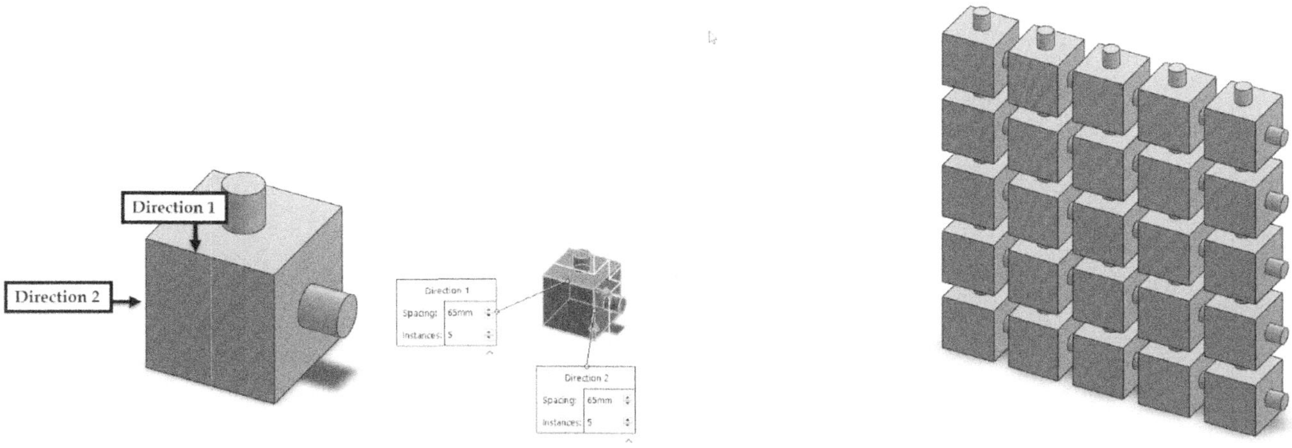

Best Practices

Design Intent: Begin by comprehensively understanding your design intent. Determine where linear patterns are essential to streamline processes and ensure that modifications to the original feature are automatically applied to the pattern.

Efficient Geometry: It is crucial to ensure that the geometry intended for patterning is well-constructed. Avoid intricate or overly detailed features, as patterns can significantly inflate file sizes and potentially impede system performance.

Standard Units: Always operate within standard units and accurately configure your units in the document properties. This practice is fundamental in guaranteeing the precision and consistency of your patterns.
Use Patterns Sparingly: While patterns are a potent tool, excessive use can overly complicate models. Utilize them judiciously, as an abundance of patterns can render your model more challenging to manage and modify.

Cautions:

Overdefinition: Exercise caution regarding over-defining sketches or features. Excessive dimensions or relations in the original feature can lead to complications when creating patterns.

Complex Patterns: Refrain from generating excessively intricate patterns, particularly in large assembly files. The presence of numerous components and intricate patterns escalates the likelihood of performance issues.

Maintain Modifiers: When updating patterns, ensure that any modifiers or dependent features are also updated accordingly. Neglecting this step may result in unintended consequences within your design.

Circular Pattern

The circular pattern is used to pattern the selected features circularly. Activate the circular pattern command (click **Features > Pattern** drop-down > **Circular Pattern** on the CommandManager) and select the feature to pattern from the model geometry. Click in the **Pattern Axis** selection box in the **Direction 1** section, select the axis from the model geometry or click on a cylindrical face; the rotation axis is defined. Usually, the rotation axis is perpendicular to the plane/face on which the selected feature is placed.

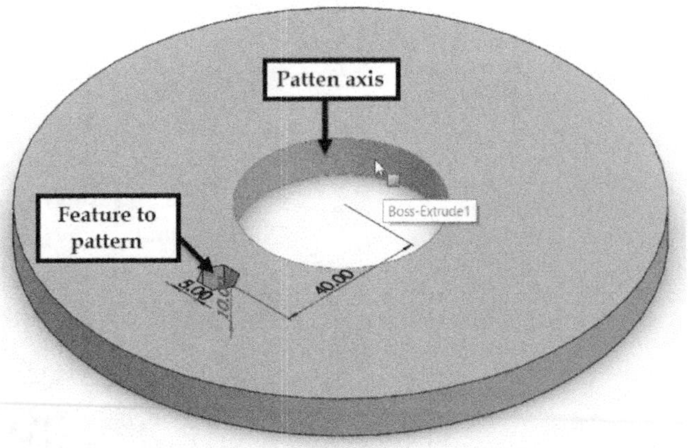

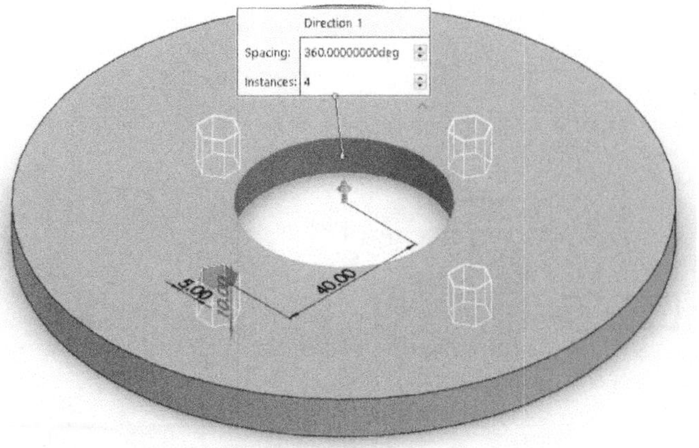

Under the **Direction 1** section, select the **Equal Spacing** option. Type-in values in the **Angle** and **Number of Instances** boxes of the **Direction 1** section. The total number of instances that you specify will be fitted in the occurrence angle value.

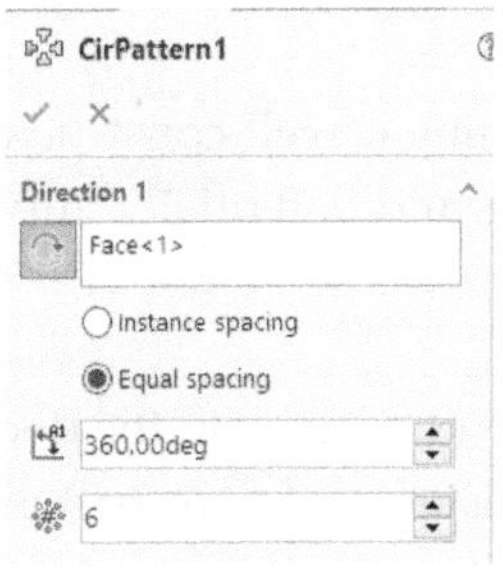

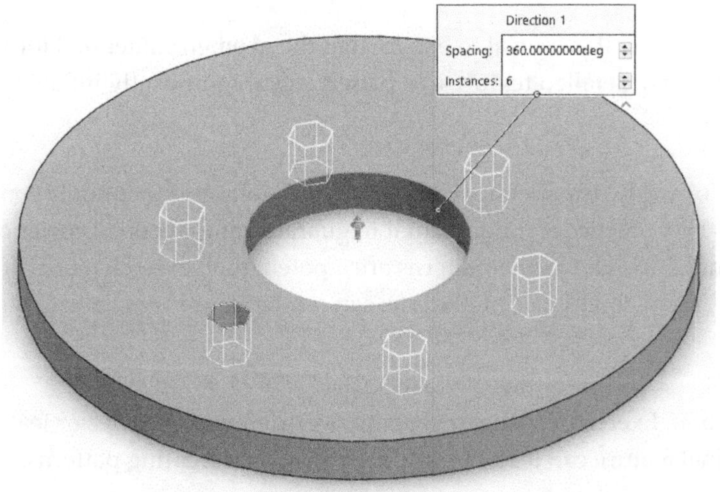

Select **Instance Spacing** if you want to type in the instance number and the angle between individual instances.

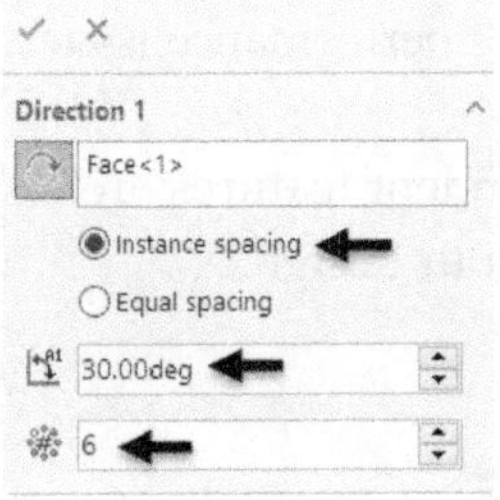

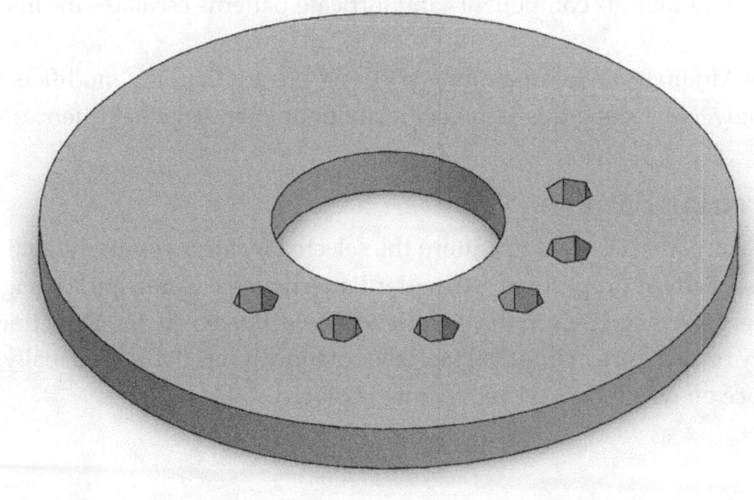

Instances to Vary

The **Instances to Vary** option allows you to specify the increments in parameters of the pattern instances. Check the **Instances to Vary** option on the **Circular Pattern** PropertyManager. Next, type in a value in the **Spacing Increment** box to specify the increments in instance spacing.

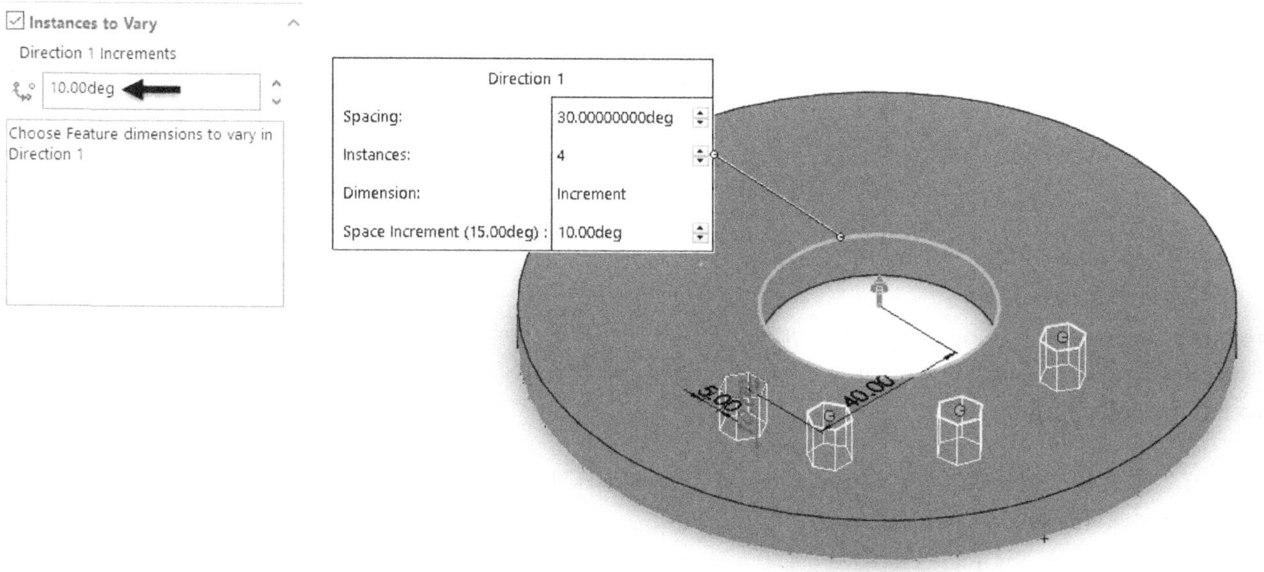

Click in the selection box displayed below the **Spacing Increment** box. Next, select any one of the dimensions of the patterned feature.

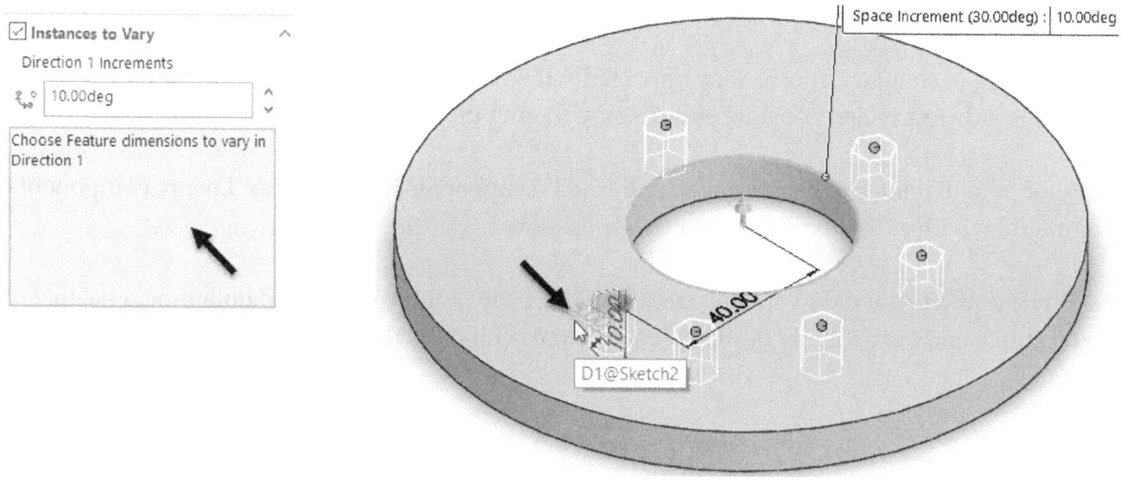

Click in the **Increment** box and type in a value. Click **OK** on the PropertyManager to update the pattern.

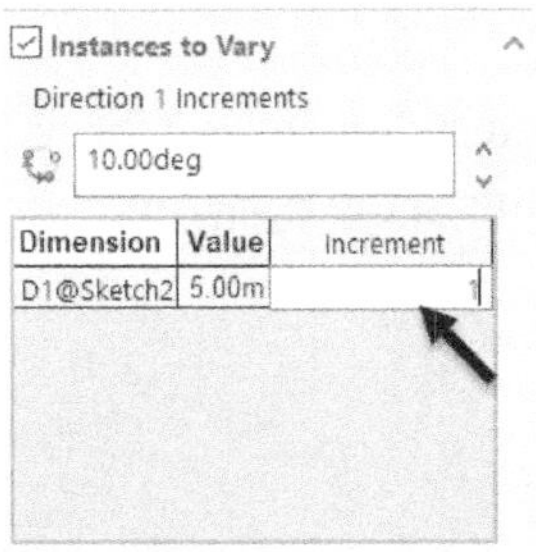

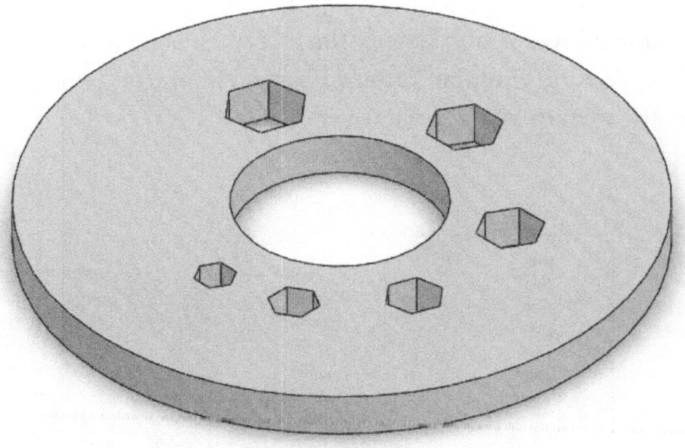

Best Practices

Modeling Intent: Before using circular patterns in SOLIDWORKS, ensure the accurate positioning and design of the original feature or component to avoid issues in the final design caused by errors or discrepancies propagated through the pattern.

Symmetry: Circular patterns are ideal for symmetric parts in SOLIDWORKS, especially when working with regularly spaced features like bolts or screws.

Feature Order: In SOLIDWORKS, create the central feature before applying the circular pattern as the order of features matters and changing it can affect the pattern's outcome.

Cautions

Performance: Excessive use of circular patterns may impact SOLIDWORKS' performance, particularly in complex assemblies. Use patterns judiciously and consider suppressing them if performance becomes an issue.

Interference: Watch for potential interference between patterned components in assemblies. Ensure components do not intersect or interfere with each other post-pattern application.

Pattern Instances: While powerful, avoid overusing circular patterns in SOLIDWORKS. Maintaining a balance between patterned features and individually designed components enhances flexibility.

For Sheet Metal Parts

- Bend Features: Circular patterns can efficiently create symmetric bend features in sheet metal parts, ensuring uniformity and design effectiveness.
- Flange Patterns: Utilize circular patterns to replicate flanges in sheet metal components, streamlining the design process and ensuring consistency.

For Weldment Structures

- Structural Members: Employ circular patterns to duplicate structural members in weldment structures, facilitating efficient design and assembly processes.
- Gusset Plates: Use circular patterns for gusset plates in weldments to maintain structural integrity and uniformity across the design.

Curve Driven Pattern

You can create a pattern along a selected curve or edge using the **Curve Driven Pattern** command. Activate the **Curve Driven Pattern** command (click **Features > Pattern** drop-down > **Curve Driven Pattern**) and select a curve, edge or sketched path. On the **Curve Driven Pattern** PropertyManager, click on the **Features to Pattern** selection button and click on the feature to pattern. Next, specify the **Number of Instances** . Select an option from the **Curve Method** section, and then specify the **Spacing** value.

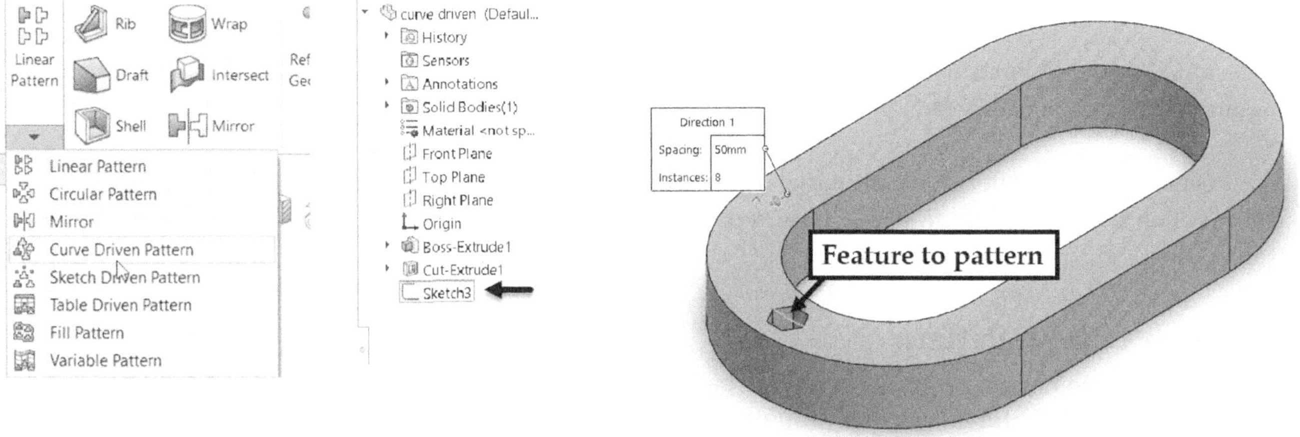

Specify the orientation using the **Alignment** section, which has two options: **Tangent to curve**, and **Align to seed**. The options in this section are explained in the figure, as shown. Click **OK** to create the pattern along the curve.

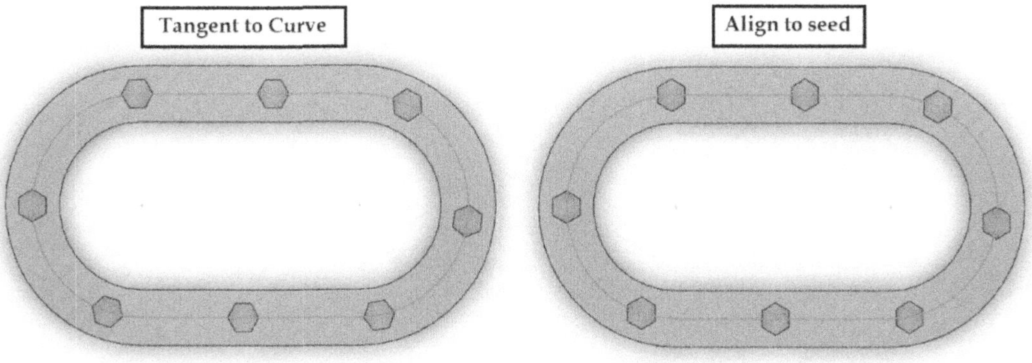

Best Practices:

Design Intent: When employing Curve Driven Patterns in SOLIDWORKS, it is essential to ensure clarity in design intent. Understand how modifications to dimensions will affect the model to uphold consistency and accuracy.

Curve Selection: Choose curves that accurately depict the desired pattern path. Thoughtful curve selection is pivotal for achieving the intended design outcome.

Feature Order: Similar to other patterns, the sequence of features holds significance in Curve Driven Patterns. Establish the base feature prior to pattern application to ensure a seamless and predictable outcome.

Cautions:

Complexity: Steer clear of excessively intricate curves that could pose challenges in pattern creation or yield unexpected outcomes. Opt for simple and well-defined curves for optimal performance.

Interference Check: Following the application of Curve Driven Patterns, assess for potential interference among patterned components. Confirm that patterned features do not intersect or interfere with each other.

Tips:

Curve Editing: Modify curves as necessary to maintain the desired shape and path for the pattern. Adjustments to curves can significantly influence the final design outcome.

Pattern Validation: Validate the pattern by generating a sample component to verify alignment with design requirements before widespread application.

Documentation: Clearly document the curve-driven pattern process within SOLIDWORKS to facilitate collaboration and future modifications. Thorough documentation enhances design clarity and aids in troubleshooting.

Sketch Driven Pattern

The **Sketch Driven Pattern** command is used to pattern the feature or body by using the sketch points. Activate this command (click **Features > Pattern** drop-down **> Sketch Driven Pattern** on the CommandManager) and select any one of the points; the entire sketch is selected. Next, select the feature to pattern; the preview of the pattern is displayed. Click **OK** to create the pattern.

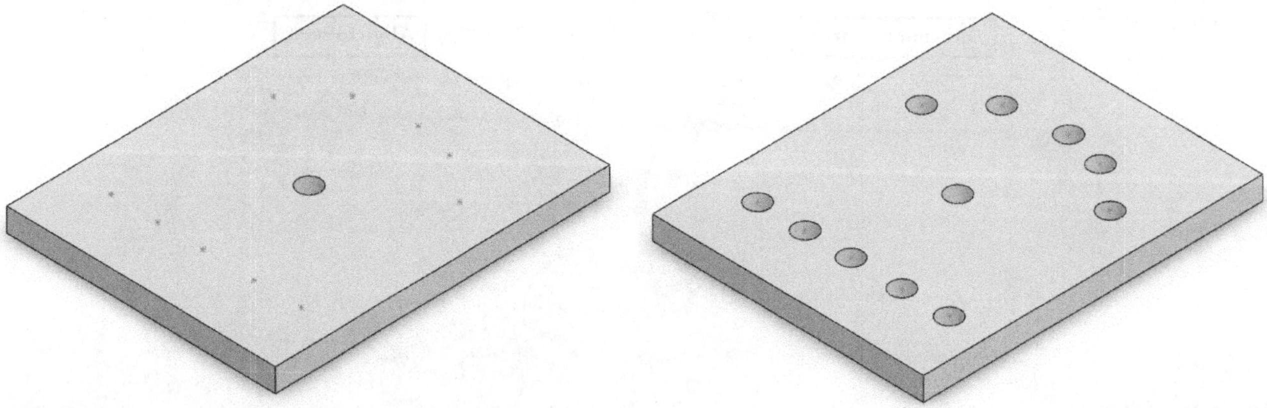

Fill Pattern

This command creates a pattern of a feature by filling it in a defined region. You can create four different types of fill patterns: **Perforation, Circular, Square**, and **Polygon.**

Perforation Fill Pattern

In this type of pattern, the features are arranged in a staggered fashion. Activate the **Fill Pattern** command (click **Features > Pattern drop-down > Fill Pattern** on the CommandManager) and select the face or sketch to define the fill boundary.

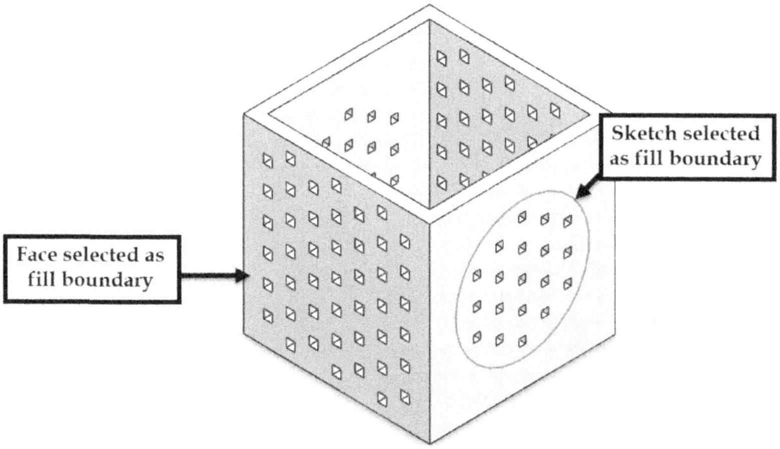

Next, set the **Pattern Layout** to **Perforation** on the PropertyManager. Next, click in the **Features to Pattern** selection box, and then select the feature to pattern. You can also use the **Create seed cut** option to create a fill pattern of predefined cuts such as **Circle**, **Square**, **Diamond**, and **Polygon**.

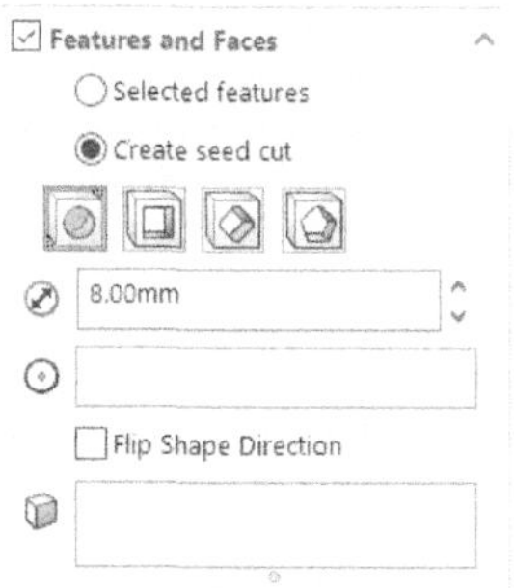

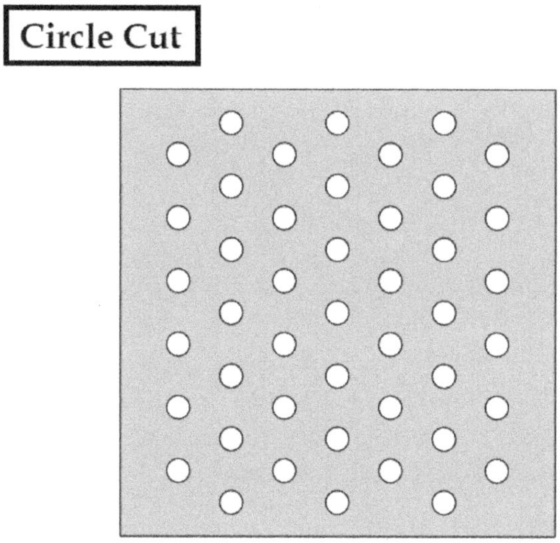

Square Cut

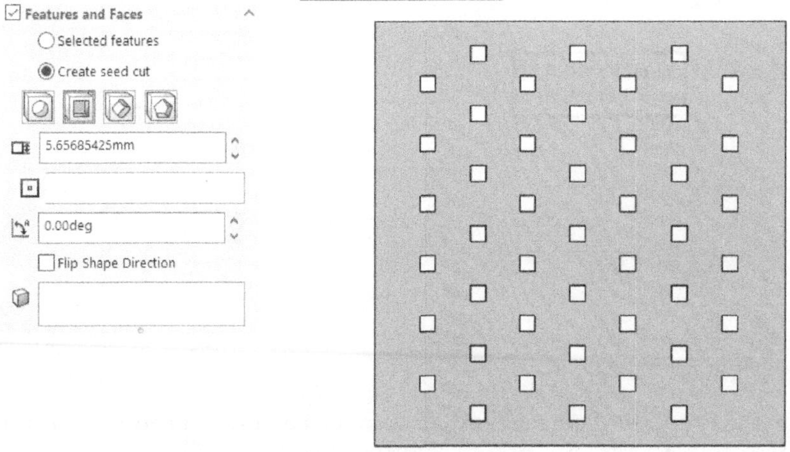

Diamond Cut

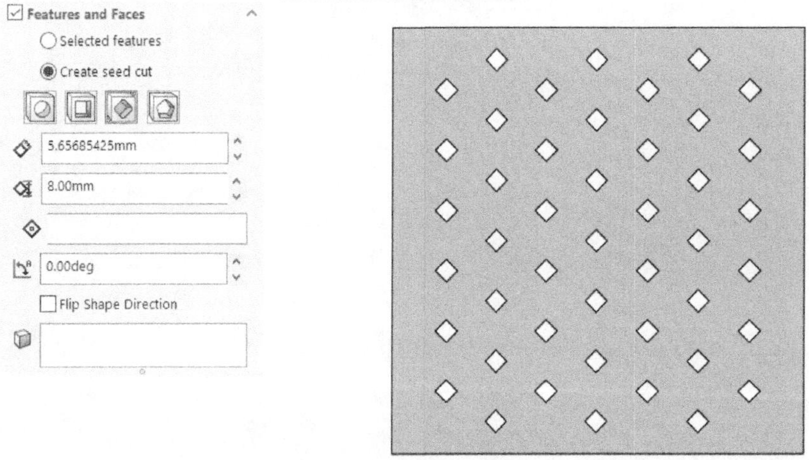

Polygon Cut

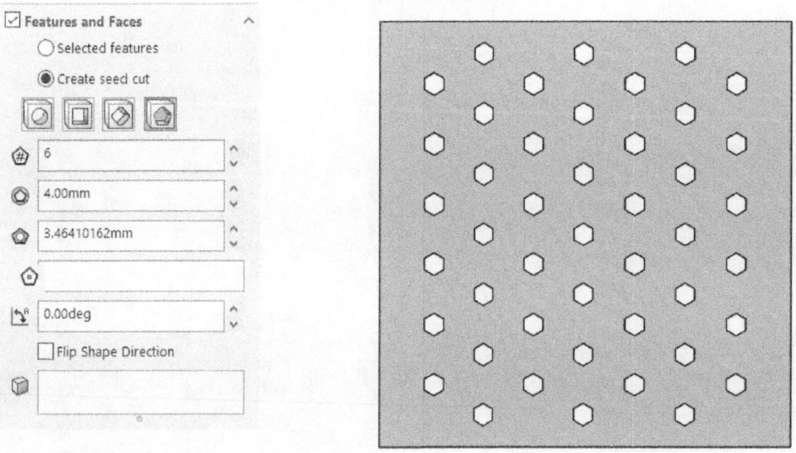

Next, define the spacing between the instances. Specify the **Instance Spacing** and **Stagger Angle** values. If you want to offset the pattern inward from the fill boundary, then type in a value in the **Margins** box. Next, click in the **Pattern Direction** box and select an edge or line to define the stagger direction.

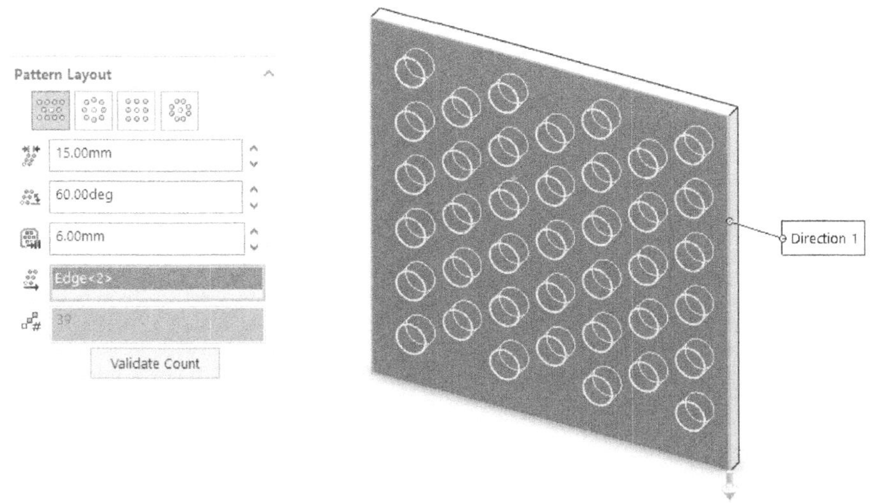

Click **OK** to create the perforated pattern.

Square Fill Pattern

The **Square** option in the **Pattern Layout** section creates a fill pattern in the form of square loops. Activate the **Fill Pattern** command and select the fill boundary. Set the **Pattern Layout** to **Square** and select the spacing method: **Target spacing** or **Instances per side.** If you select the **Target spacing** method, you need to specify the **Loop spacing** and **Instance spacing** values.

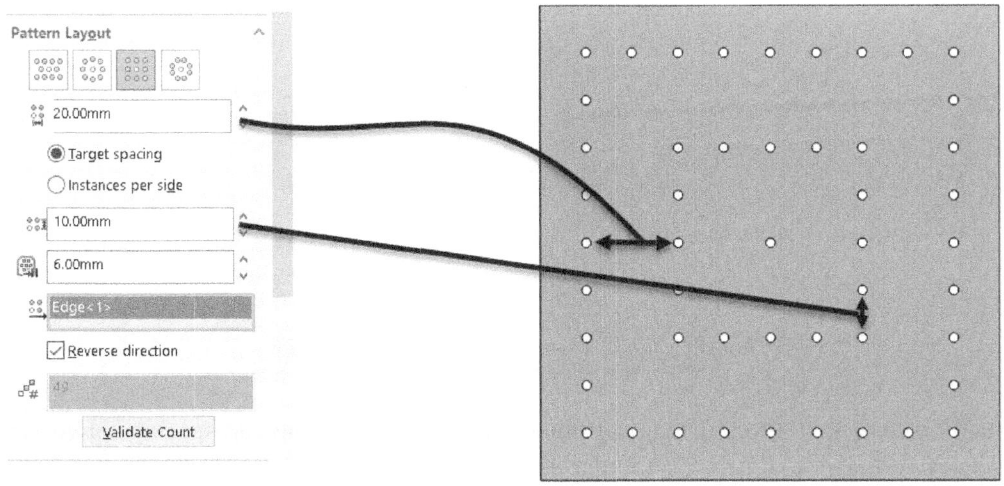

If you select the **Instances per side** method, then you need to specify the **Loop spacing** and **Number of instances** values. Specify the feature to pattern and click **OK** on the PropertyManager to create the square fill pattern.

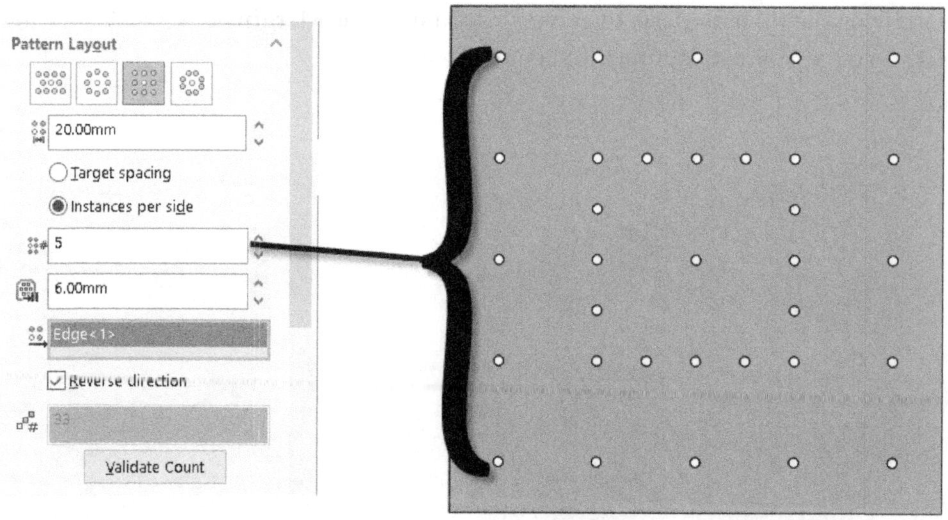

Circular Fill Pattern

In this type of pattern, the features are filled in a radial fashion inside the selected fill boundary. To create a circular fill pattern, set the **Pattern Layout** to **Circular** on the PropertyManager. Next, define the spacing between the instances. This can be done by using the spacing methods. There are two methods to define the spacing between the instances: **Target Spacing** and **Instances per loop**. The **Target Spacing** method creates a pattern using the spacing between the rings and spacing between the instances.

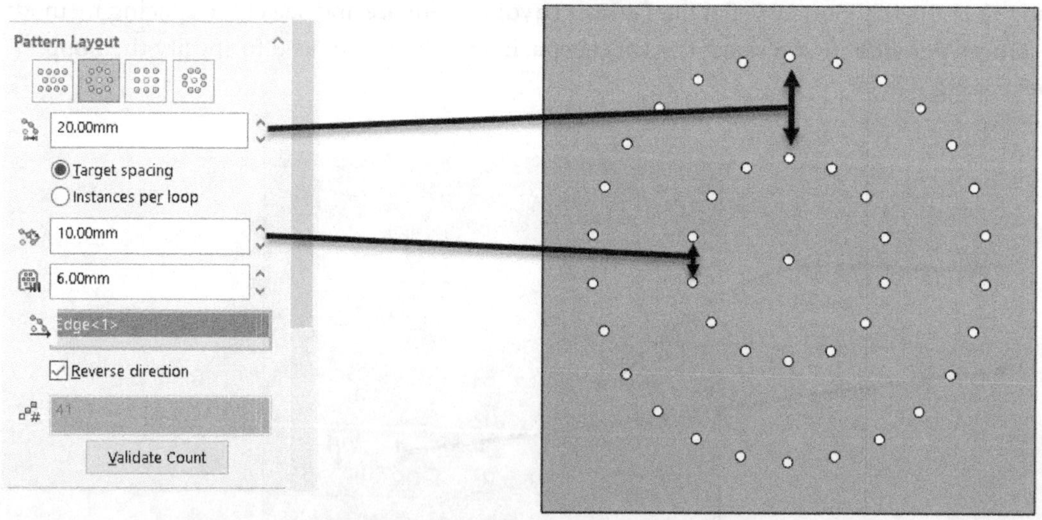

The **Instances per loop** method creates a pattern using the number of instances per ring and spacing between the rings. Click **OK** on the PropertyManager to create the circular fill pattern.

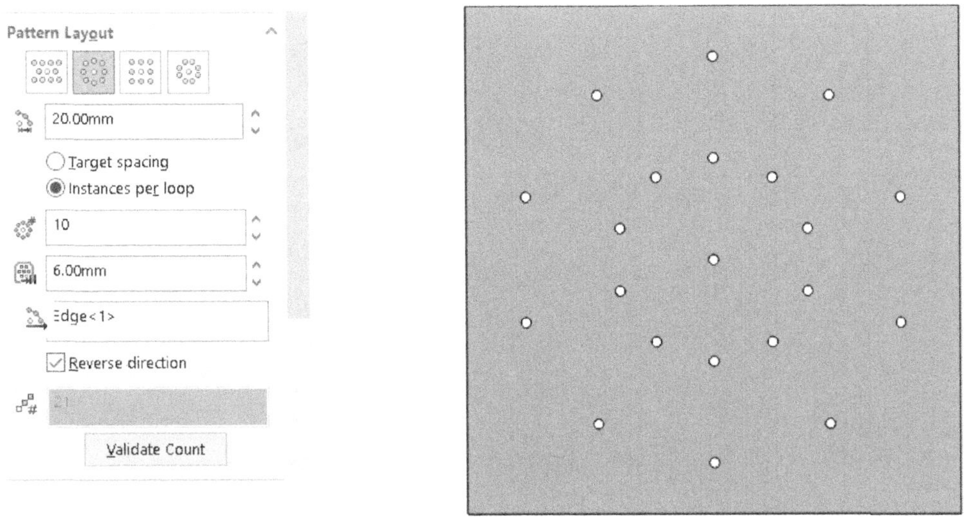

Polygon Fill Pattern

In this type of pattern, the features are filled in a polygonal fashion inside the selected fill boundary. To create this pattern, click the **Polygon** icon in the **Pattern Layout** section. Next, select the fill boundary and feature to pattern. In the **Pattern Layout** section, specify the spacing options: **Target spacing** and **Instances per side**. If you select the **Target spacing** option, you need to specify the **Loop spacing**, **Polygon sides**, and **Instance spacing**.

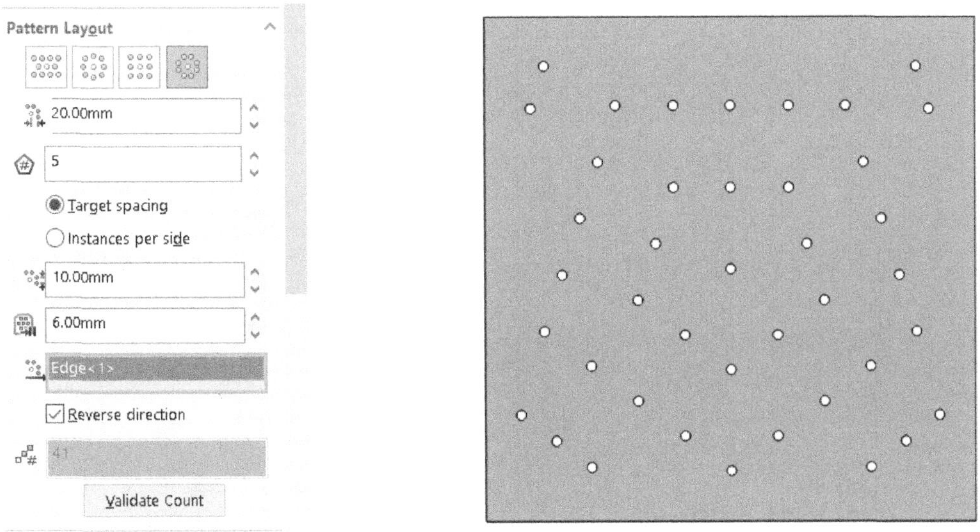

If you select the **Instances per side** option, you need to specify the **Loop spacing**, **Polygon sides**, and **Number of Instances**.

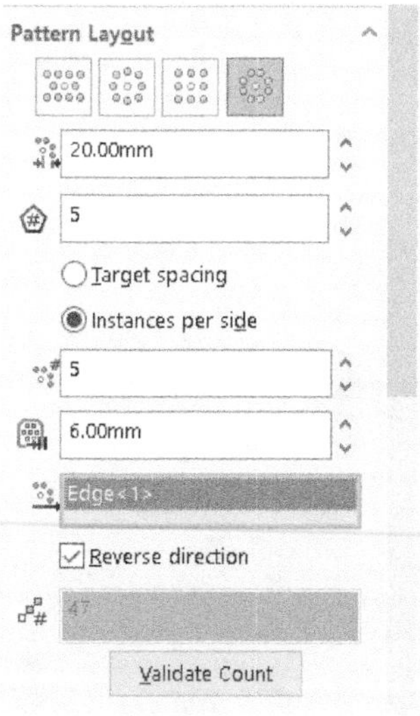

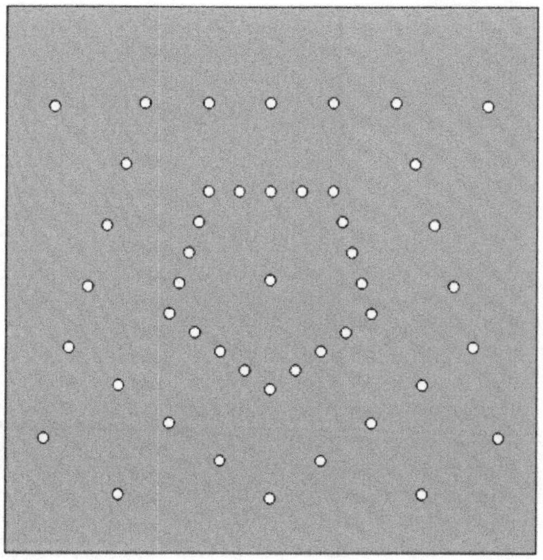

Example 1

In this example, you will create the part shown next.

.

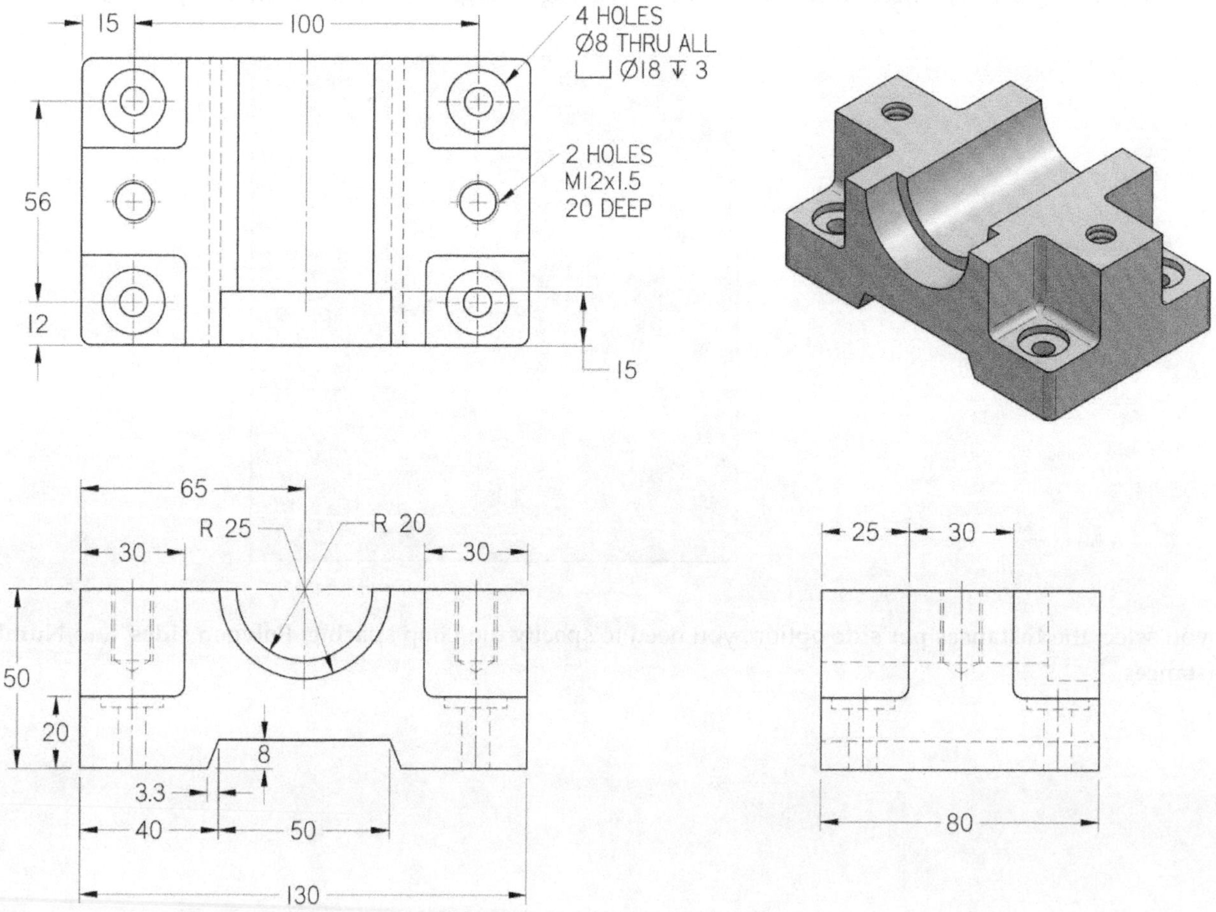

1. Start **SOLIDWORKS 2024**.
2. Open a new part file.
3. Activate the **Extruded Boss/Base** command.
4. Click on the Front plane.
5. Create a rectangular sketch, and then click **Exit Sketch** on the CommandManager.

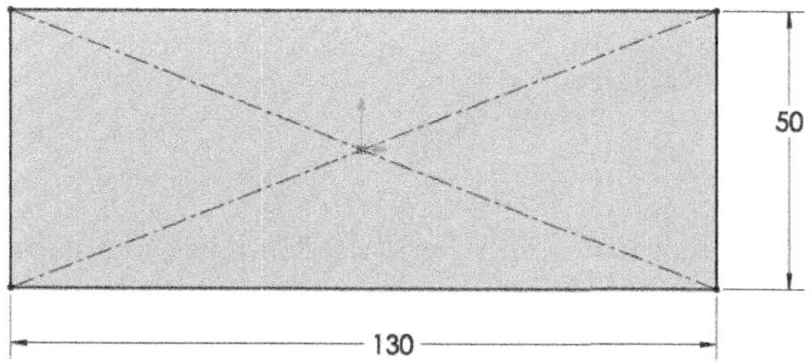

6. On the **Boss-Extrude** PropertyManager, select **End Condition > Mid Plane**.
7. Type in **80** in the **Depth** box.
8. Click **OK** to complete the *Extruded Boss* feature.

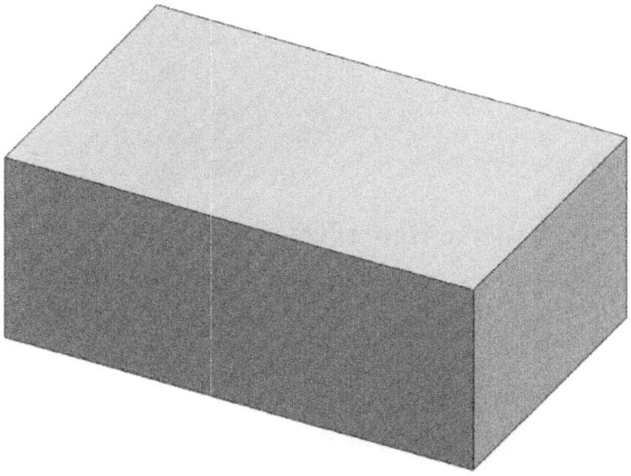

9. On the **Sketch** CommandManager, click the **Sketch** icon.
10. Click on the top face of the part geometry.
11. On the **Sketch** CommandManager, click **Rectangle** drop-down > **Corner Rectangle**.
12. Click on the corner point of the top face, as shown.

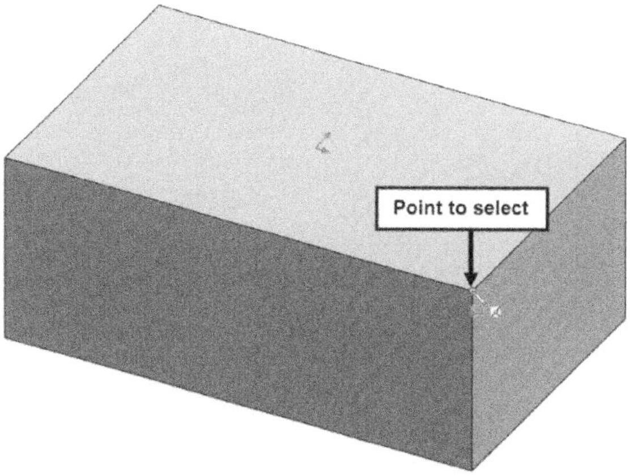

13. Move the pointer in the top-left direction, and then click.

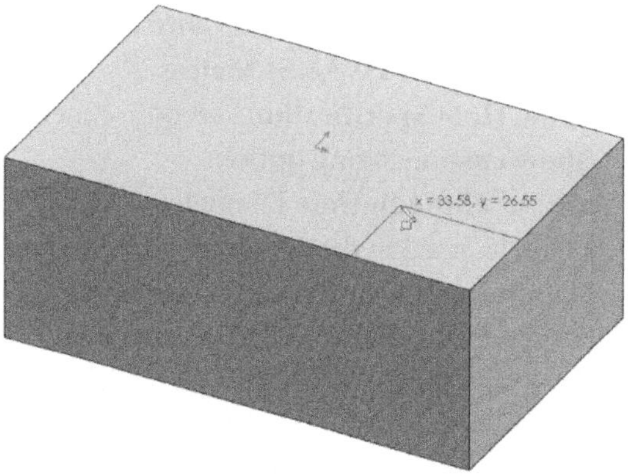

14. Add dimensions to the sketch.

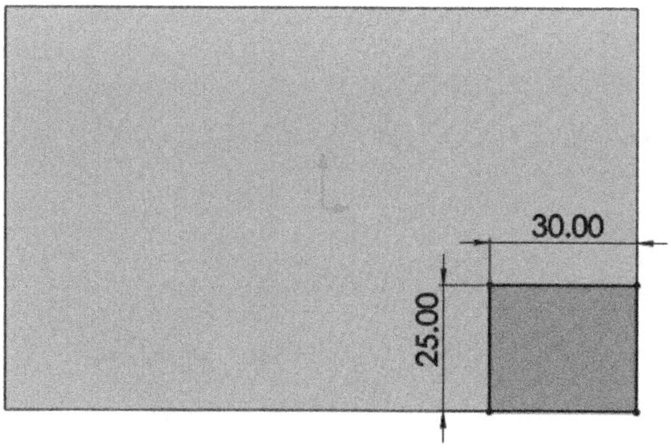

15. Click the **Exit Sketch** icon on the CommandManager.
16. Activate the **Extruded Cut** command.
17. Select the sketch, if not already selected.
18. Type **30** in the **Depth** box.
19. Click **OK** on the PropertyManager.

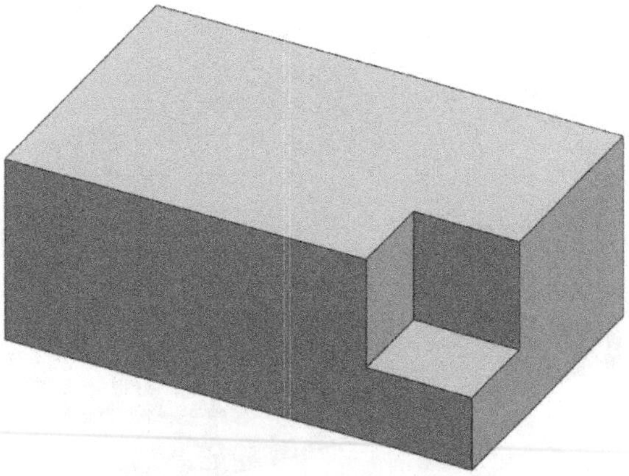

20. Activate the **Hole Wizard** command.
21. On the **Hole Specification** PropertyManager, select **Hole Type > Counterbore**.
22. Select **Standard > ANSI Metric**.
23. In the **Hole Specifications** section, check the **Show custom sizing** option.
24. Set the **Through Hole Diameter** value to 8.
25. Enter **18** and **3** in the **Counterbore Diameter** and **Counterbore Depth** boxes, respectively.
26. Set the **End Condition** to **Through All**.
27. Click the **Positions** tab, and then click on the bottom face of the **Extruded Cut** feature.
28. Specify the hole location on the selected face.

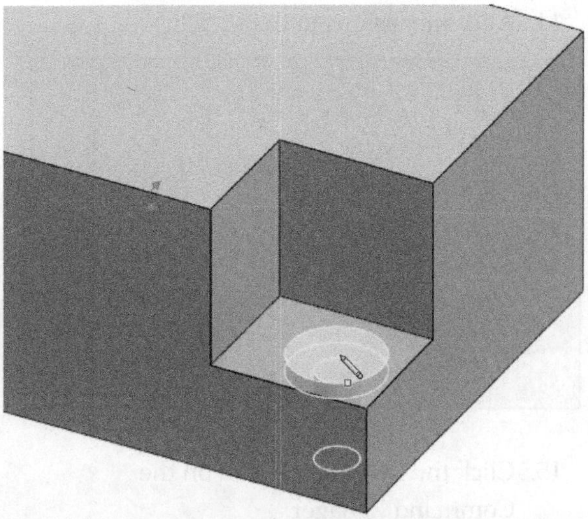

29. Activate the **Smart Dimension** command.
30. Add dimensions between the hole position point and the adjacent edges, as shown.

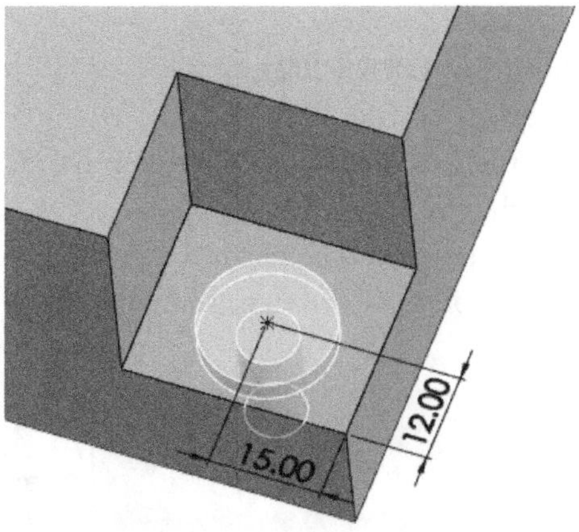

31. Click **OK** on the PropertyManager to create the counterbore hole.

32. Activate the **Hole Wizard** command.
33. Click the **Type** tab on the **Hole Specification** PropertyManager, and then select **Hole Type > Straight Tap**.
34. Select the **Reset the custom sizing values to their default values for the new hole type** option from the **SOLIDWORKS** dialog (Leave this step if the **SOLIDWORKS** dialog does not appear).
35. Select **Standard > ANSI Metric**.
36. Select **Type > Tapped Hole**.
37. In the **Hole Specifications** section, select **Size > M12x1.5**.
38. Select **End Condition > Blind**.
39. Enter 20 in the **Blind Hole Depth** and **Tap Thread Depth** boxes, respectively.
40. In the **Options** section, click the **Cosmetic Thread** icon.

41. Uncheck the **Near side countersink** option.
42. Click the **Positions** tab, and then click on the top face of the model.
43. Specify the position of the hole.
44. Add dimensions to define the hole location.

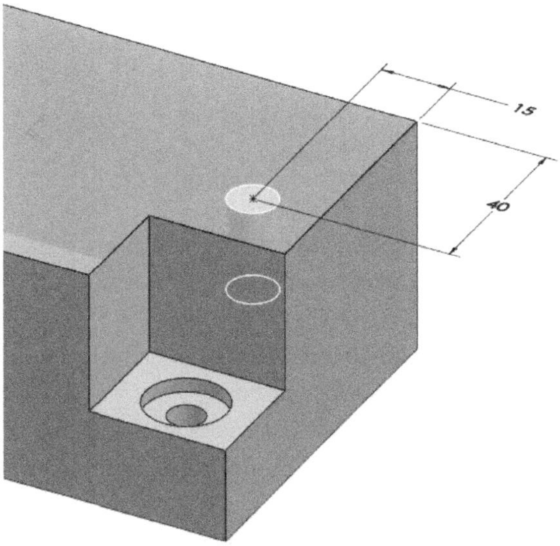

45. Click **OK** on the PropertyManager to create the threaded hole.
46. On the **Features** CommandManager, click **Pattern** drop-down > **Linear Pattern** (or) click **Insert > Pattern/Mirror > Linear Pattern** on the Menu.
47. On the **Linear Pattern** PropertyManager, under the **Direction 1** section, select **Spacing and instances**.
48. Type in **100** and **2** in the **Spacing** and **Number of Instances** boxes, respectively.
49. Under the **Direction 1** section, click in the **Pattern Direction** selection box and then click on the part geometry's top front edge.

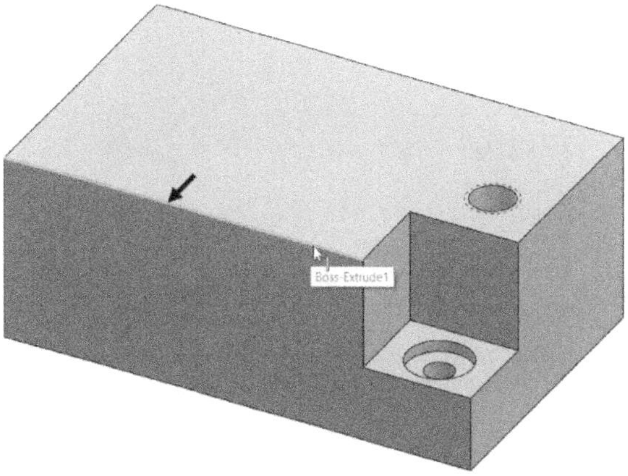

50. Under the **Features and Faces** section, click in the **Features to Pattern** selection box and select the *Extruded Cut* feature.
51. Expand the **Direction 2** section.
52. Under the **Direction 2** section, type in **55** and **2** in the **Spacing** and **Number of Instances** boxes.
53. Under the **Direction 2** section, click in the **Pattern Direction** box and click on the part geometry's top side edge.

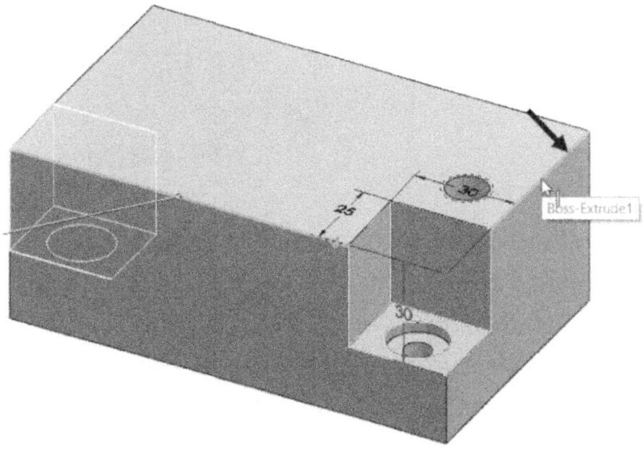

54. Click the **Reverse Direction** icon in the **Direction 2** section if the pattern preview is displayed outside the model.
55. Uncheck the **Geometry pattern** option in the **Options** section.
56. Click **OK** to pattern the *Extruded Cut* feature.

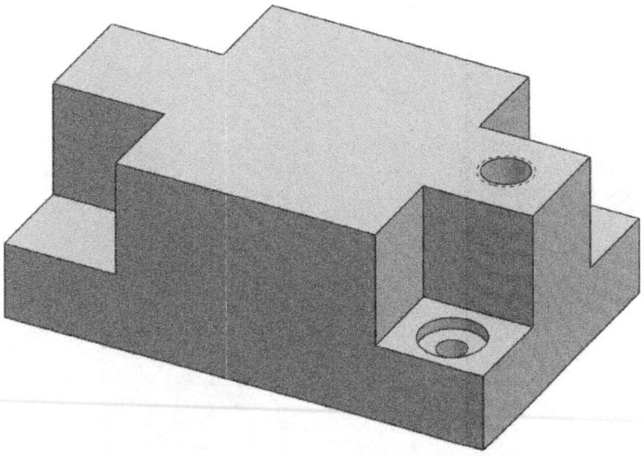

57. Likewise, pattern the counterbore hole. The pattern parameters are the same.

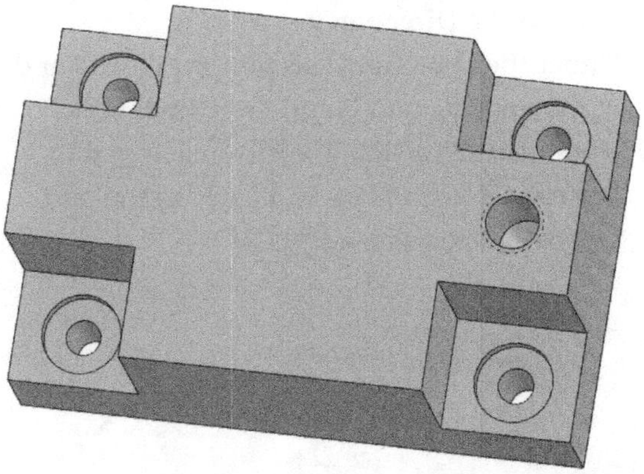

58. On the **Features** CommandManager, click **Linear Pattern > Mirror** (or) click **Insert > Pattern/Mirror > Mirror** on the Menu.
59. In the FeatureManager Design Tree, click the **Right Plane** to define the mirror plane.

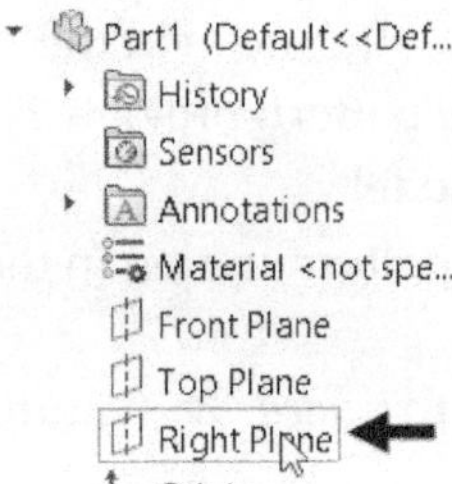

60. On the **Mirror** PropertyManager, click in the **Features to mirror** selection box and select the threaded hole feature.
61. Click **OK** to mirror the selected features.

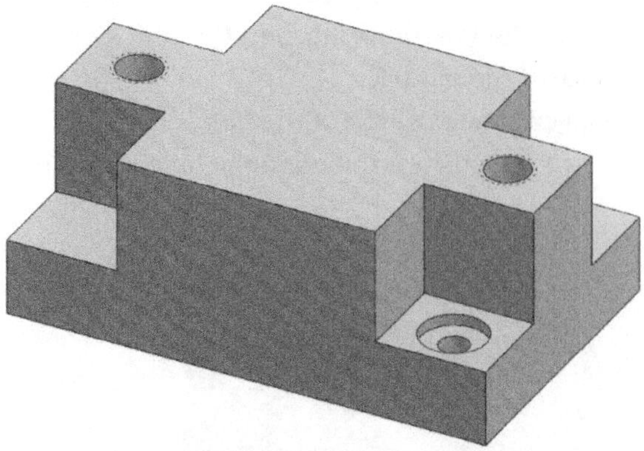

62. Activate the **Hole Wizard** command.
63. On the **Hole Specification** PropertyManager, select **Hole Type > Counterbore**.
64. In the **Hole Specifications** section, check the **Show custom sizing** option.
65. Set the **Through Hole Diameter** value to 40.
66. Type in **50** and **15** in **Counterbore Diameter** and **Counterbore Depth** boxes, respectively.
67. Set the **End Condition** to **Through All**.
68. Click the **Positions** tab, and then click on the front face of the model.
69. Place the pointer on the top edge of the front edge, and then select its midpoint, as shown.

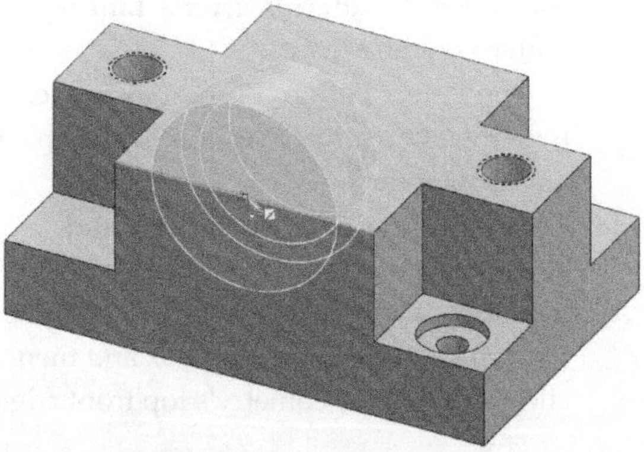

70. Click **OK** on the PropertyManager to create the counterbore hole.

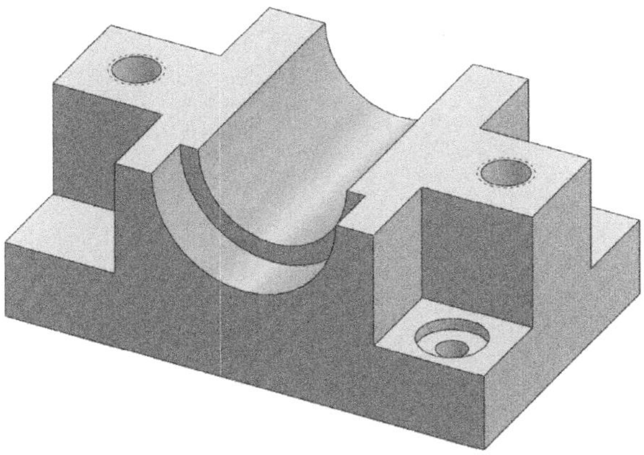

71. Draw a sketch on the part geometry's front face and create an *Extruded Cut* throughout the geometry.

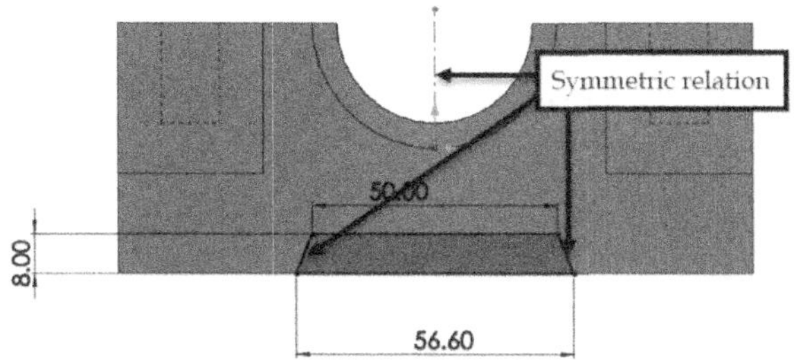

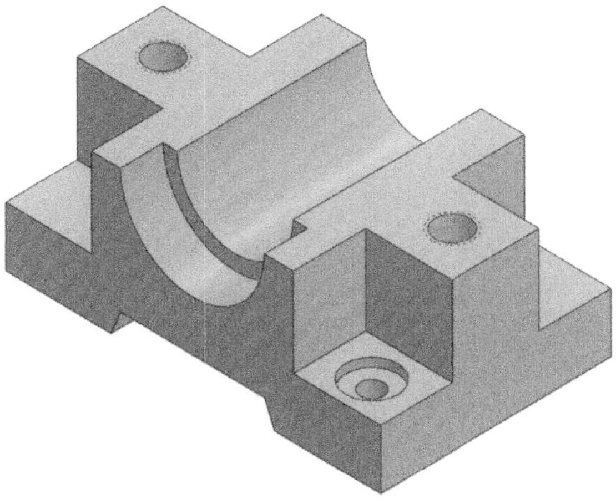

72. Fillet the sharp edges of the *Pocket* features. The fillet radius is 2 mm.

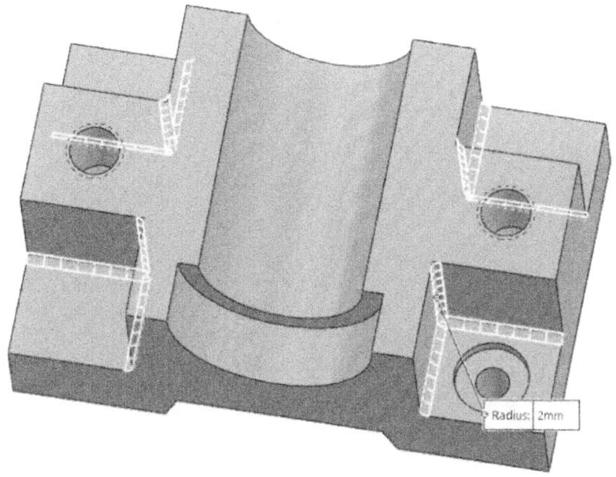

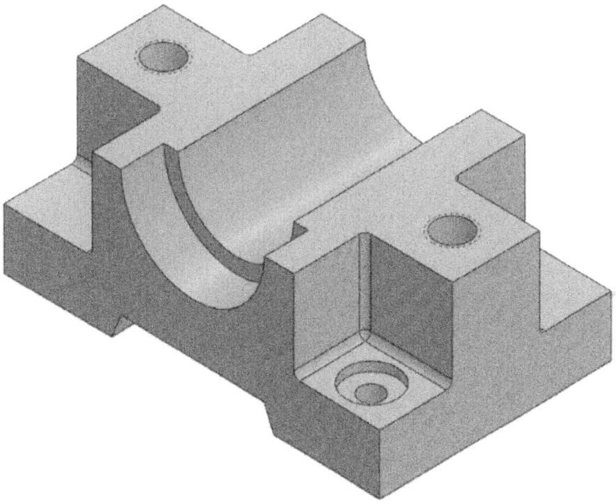

73. Save and close the part file.

Questions

1. Describe the procedure to create a mirror feature.
2. List any two types of patterns.
3. List the types of fill patterns.
4. What is the use of the **Up to reference** option in the Linear pattern?

Exercises

Exercise 1

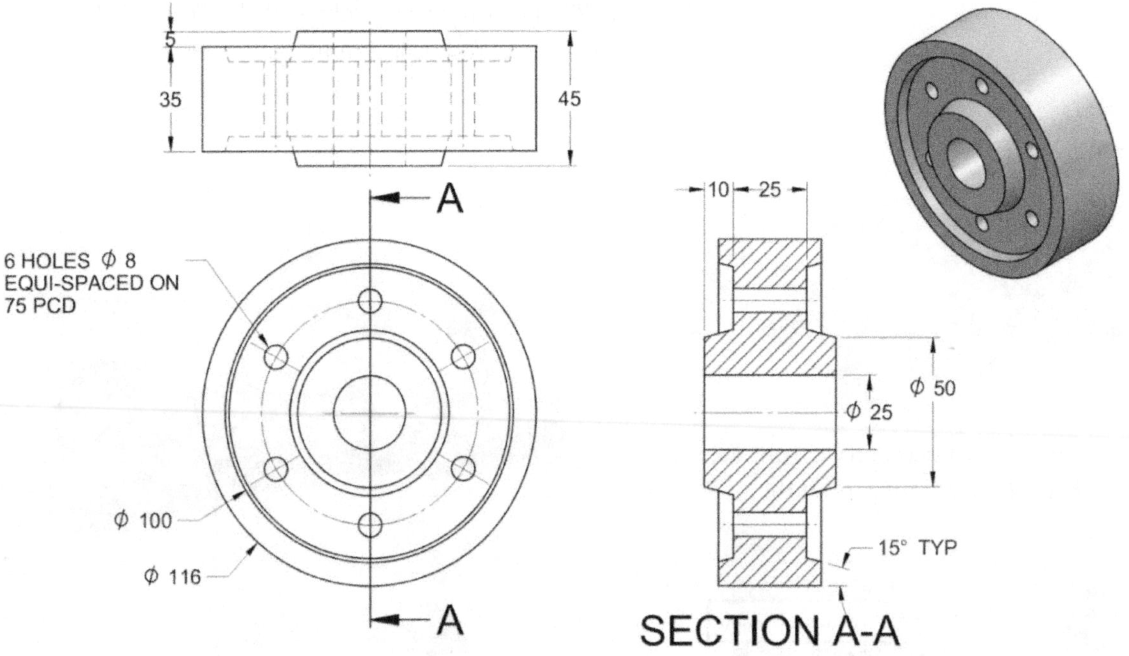
5
35
45
A
6 HOLES Φ 8
EQUI-SPACED ON
75 PCD
Φ 100
Φ 116
A
10
25
Φ 50
Φ 25
15° TYP
SECTION A-A

Exercise 2

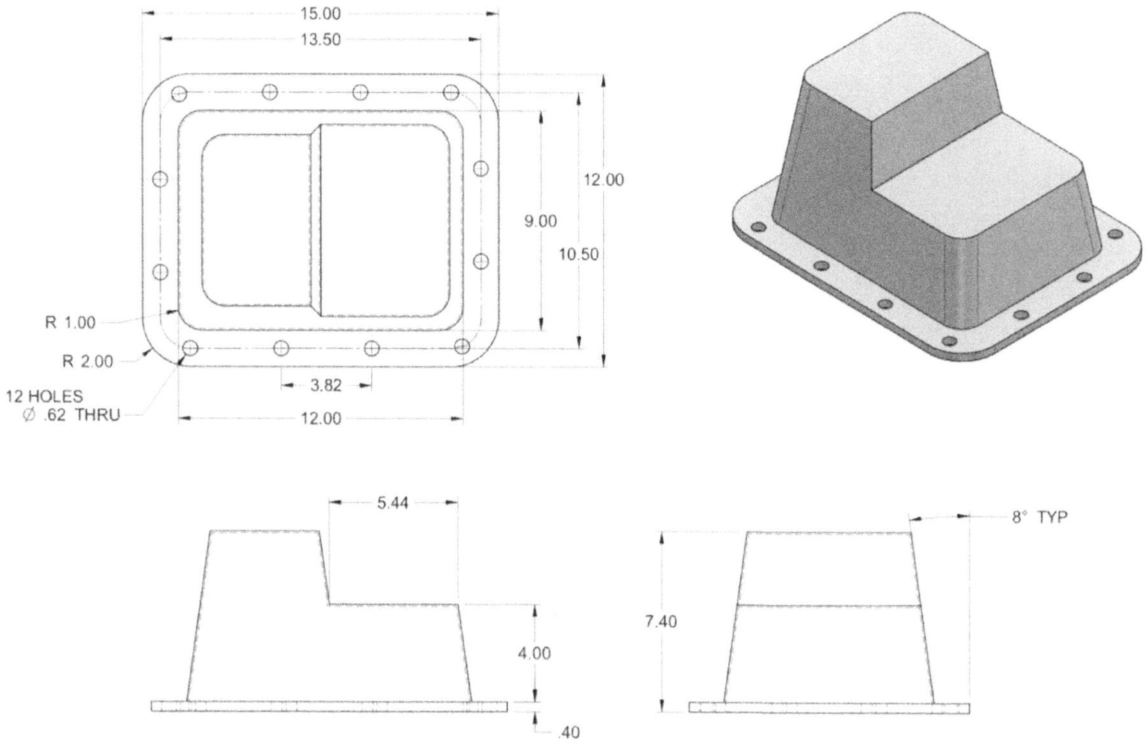

SHEET THICKNESS = 0.079 in

Chapter 6: Sweep Features

The Swept Boss/Base feature

The **Swept Boss/Base** command is one of the basic commands available in SOLIDWORKS that allow you to generate solid geometry. It can be used to create simple geometry as well as complex shapes. A sweep is composed of two items: a cross-section and a path. The cross-section controls the shape of the sweep while the path controls its direction. For example, take a look at the angled sweep feature shown in the figure. This is created using a simple sweep with the ellipse as the profile and an angled line as the path.

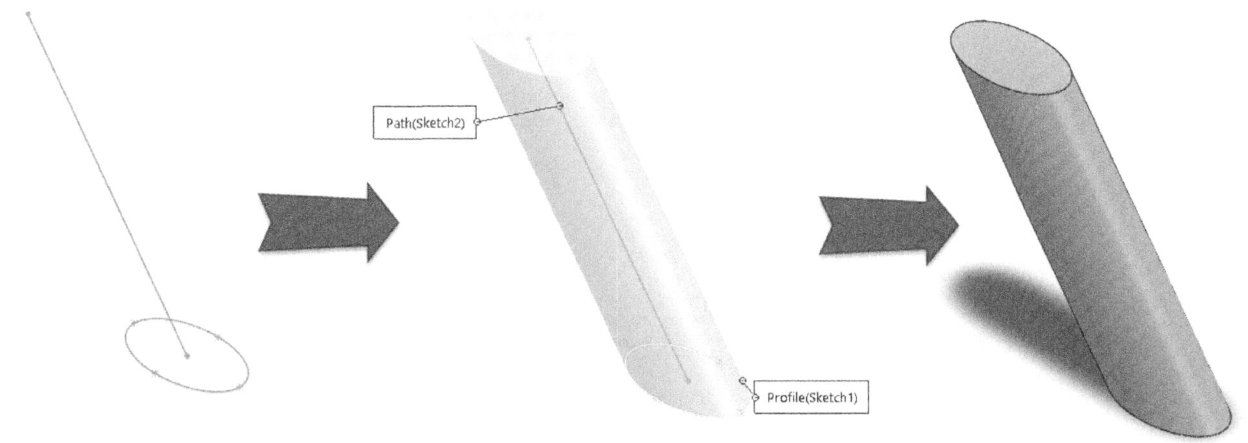

By making the path a bit more complicated, you can see that a sweep allows you to create shapes you would not be able to create using commands such as Extrude or Revolve.

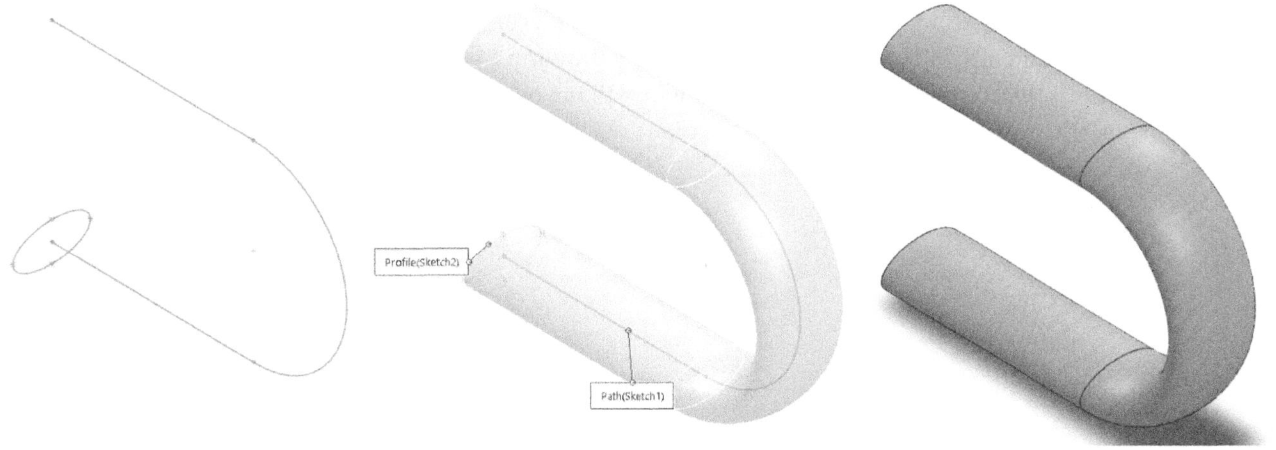

To take the sweep feature to the next level of complexity, you can add guide curves to it. By doing so, the shape of the geometry is controlled by guide curves and paths. For example, the elliptical cross-section in the figure varies in size along the path because a guide curve controls it.

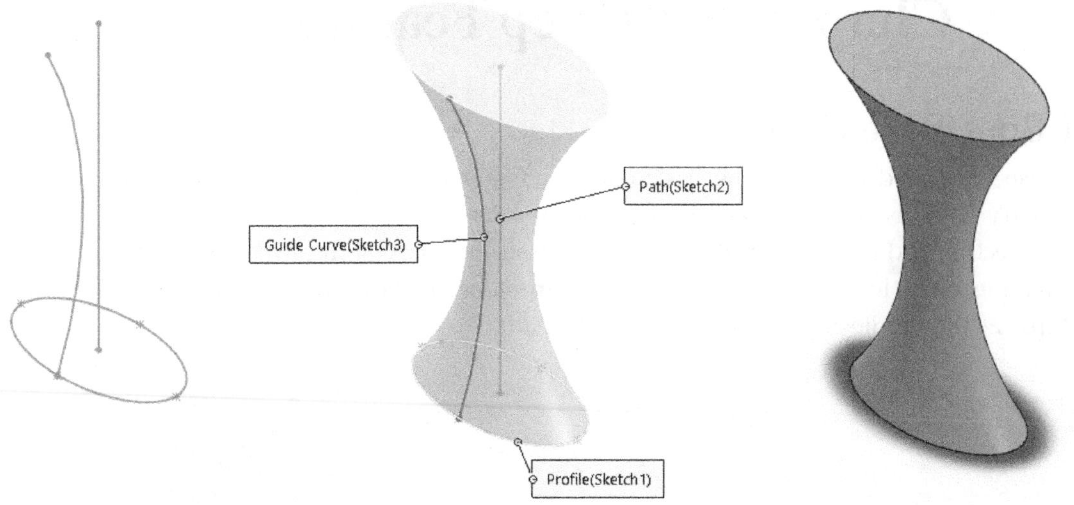

The topics covered in this chapter are:

- *Path sweeps*
- *Scaling and twisting the profile along the path*
- *Path and Guide curve sweeps*
- *Swept Cutouts*

Path sweeps

This type of sweep requires two elements: a path and profile. The profile defines the shape of the sweep along the path. A path is used to control the direction of the profile. A path can be a sketch or an edge. To create a sweep, you must first create a path and a profile. Create a path by drawing a sketch. It can be an open or closed sketch. Next, click **Features > Reference Geometry** drop-down **> Plane** on the CommandManager, and then create a plane normal to the path. Sketch the profile on the plane normal to the path.

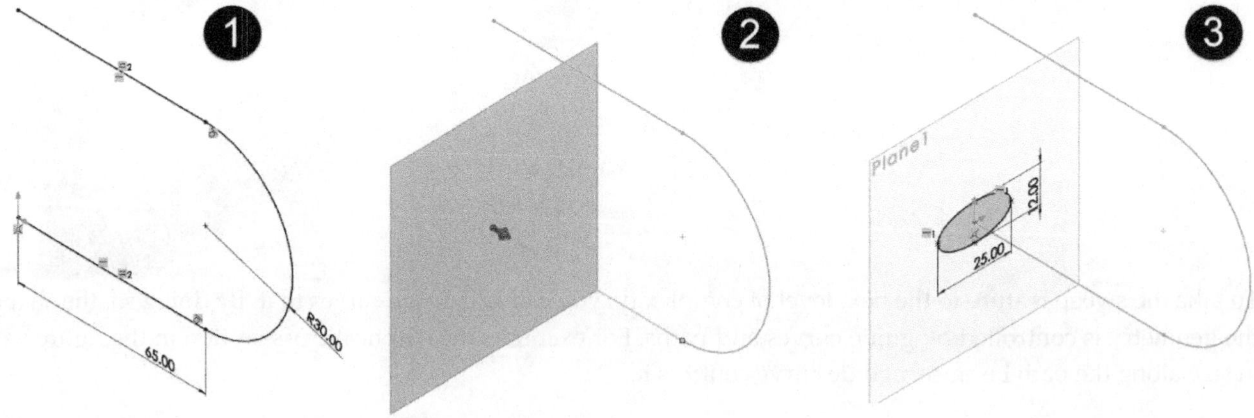

Activate the **Swept Boss/Base** command (click **Features > Swept Boss/Base** on the CommandManager). As you activate this command, the PropertyManager appears, showing two options to create the sweep: **Sketch Profile** and **Circular Profile**. Select the **Sketch Profile** option from the **Profile and Path** section on the PropertyManager. Click in the **Profile** selection box, and then select the profile from the graphics window. Next, select the path and click **OK**.

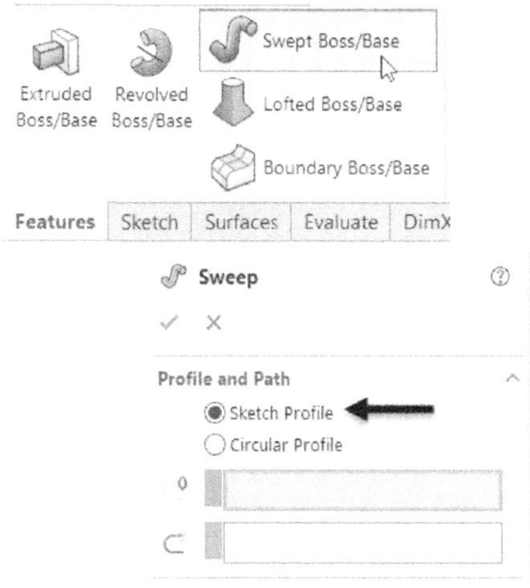

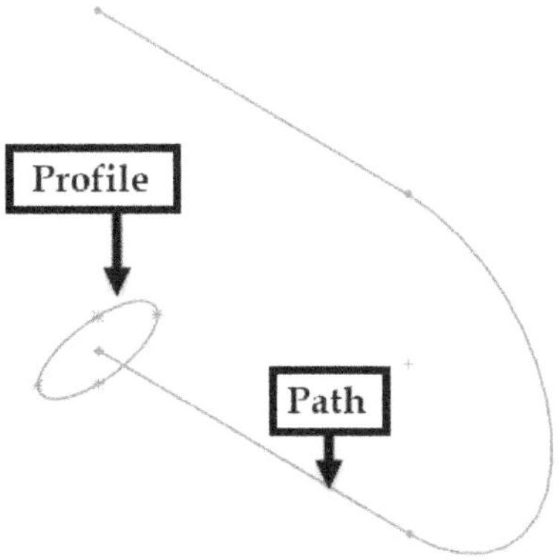

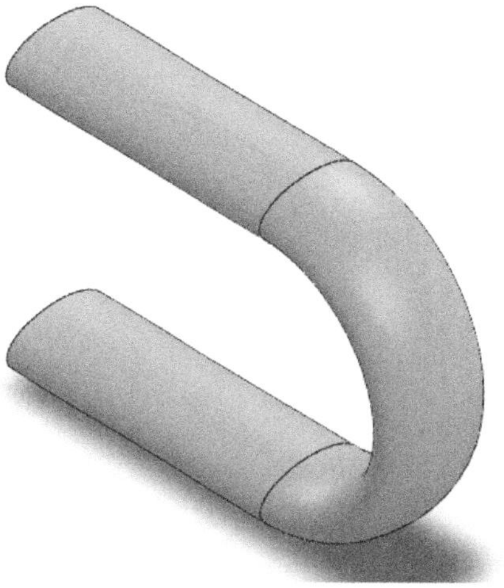

Sweeping the Circular Profile

The **Circular Profile** option in the **Profile and Path** section helps you create a swept boss without drawing a profile sketch. Select this option from the PropertyManager, and then select the path from the graphics window. Next, specify the diameter of the circular profile in the **Diameter** box available on the PropertyManager. You can also check the **Thin feature** option if you want to create a hollow tube.

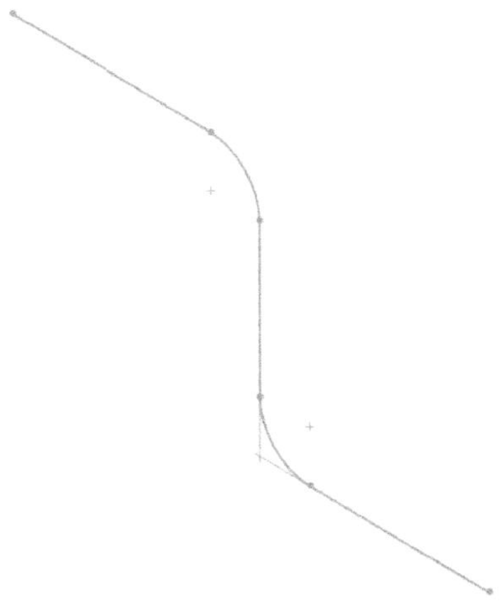

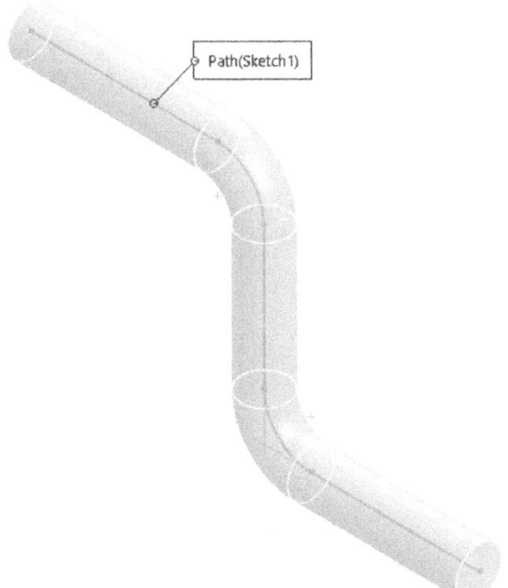

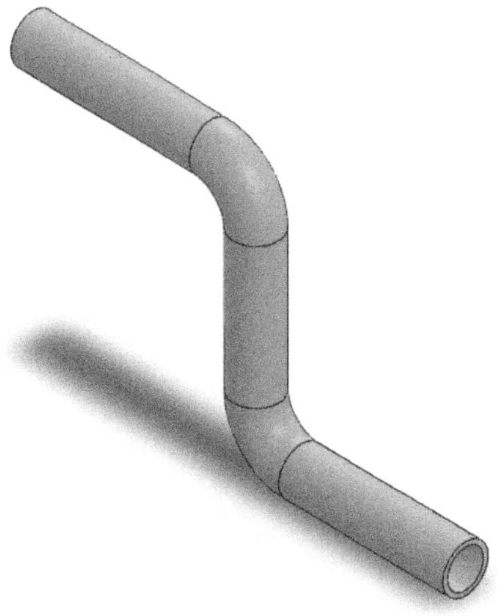

SOLIDWORKS will not allow the sweep to result in a self-intersecting geometry. As the profile is swept along a path, it cannot come back and cross itself. For example, if the sweep profile is larger than the curves on the path, the resulting geometry will intersect, and the sweep will fail.

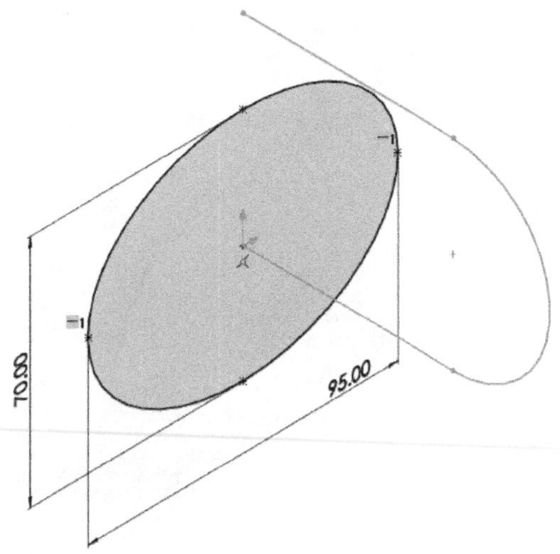

Type	Feature	Preview	Help	Description
Error	Sweep1			Sweep operation failed to complete.

Show errors   Show warnings   Display What's Wrong during rebuild   Close   Help

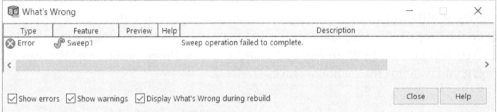

A profile must be created as a sketch. However, a path can be a sketch, curve or an edge. The following illustrations show various types of paths and resultant sweep features.

Sketch Path

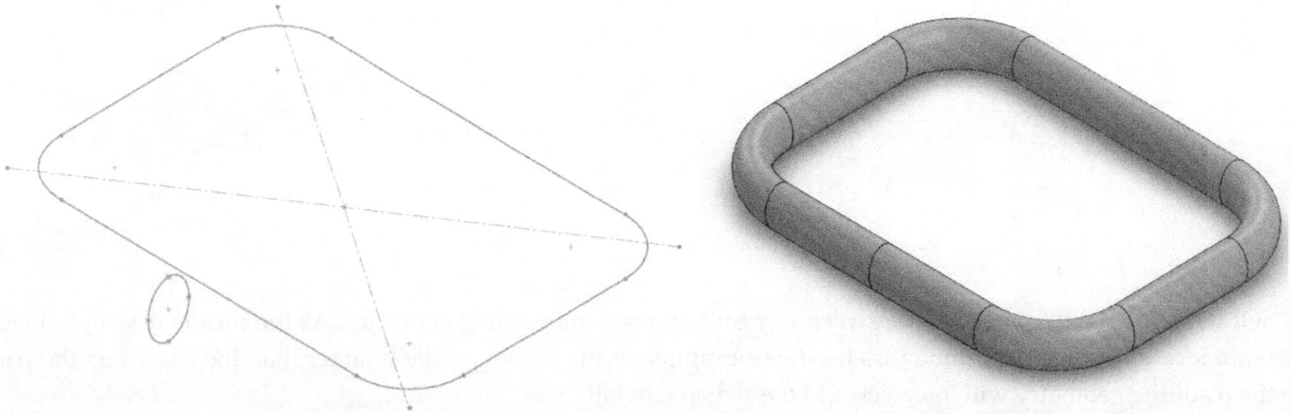

Curve Path

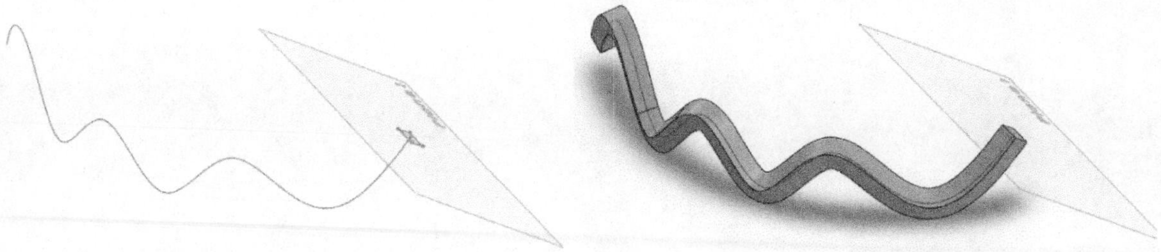

Edge Path

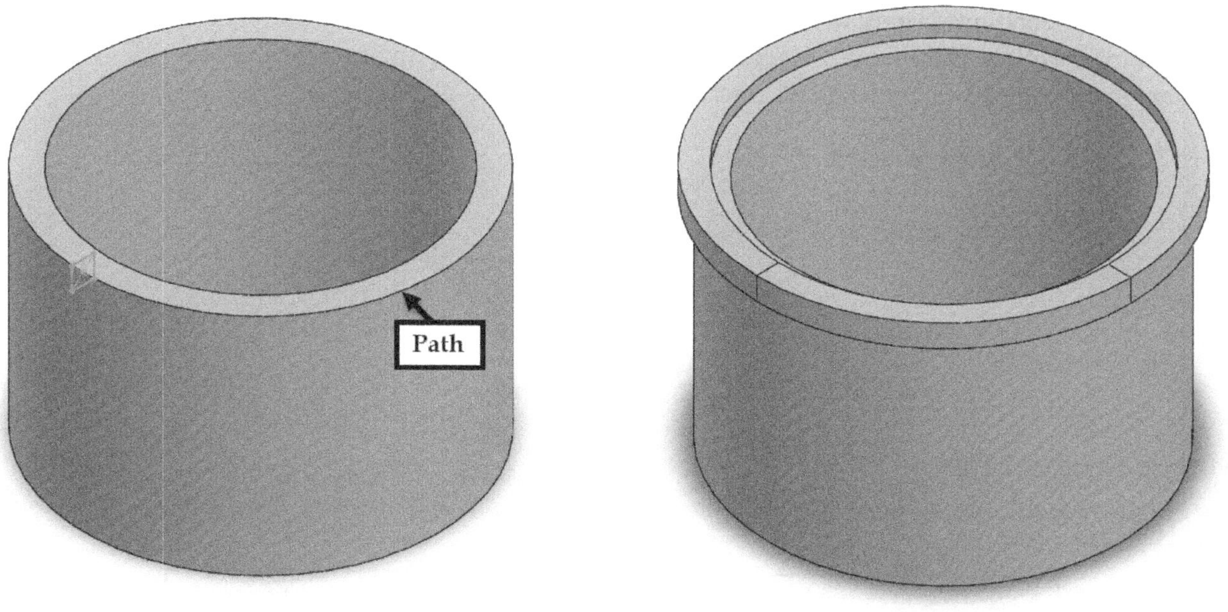

SOLIDWORKS has three options to specify the swept boss's direction if the profile is not located on the path's endpoint. These options are **Direction 1**, **Bidirectional**, and **Direction 2**.

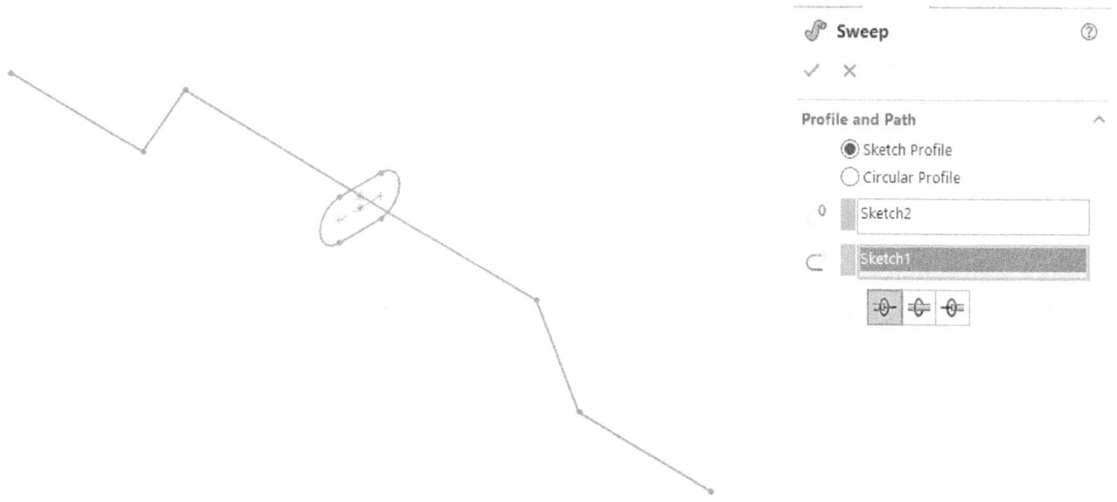

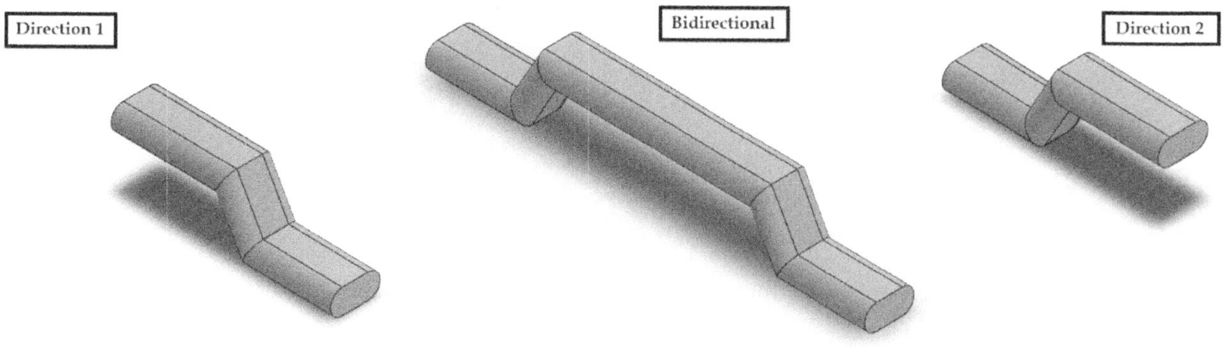

Profile Orientation

The **Profile Orientation** drop-down available in the **Options** section defines the orientation of the resulting geometry. The **Follow Path** option sweeps the profile in the direction normal to the path. The **Keep Normal Constant** option sweeps the profile in the direction parallel to itself.

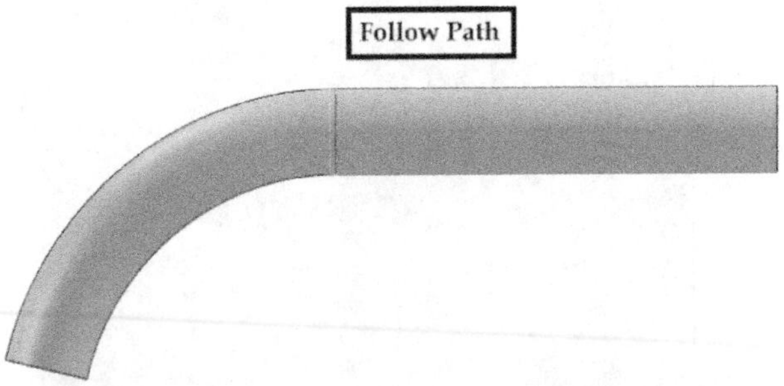

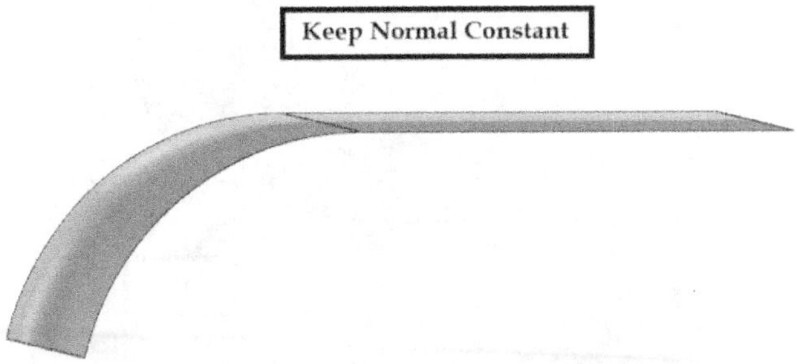

Merge tangent faces

The **Merge tangent faces** option available in the **Options** section merges the side faces of the swept boss feature tangent to each other.

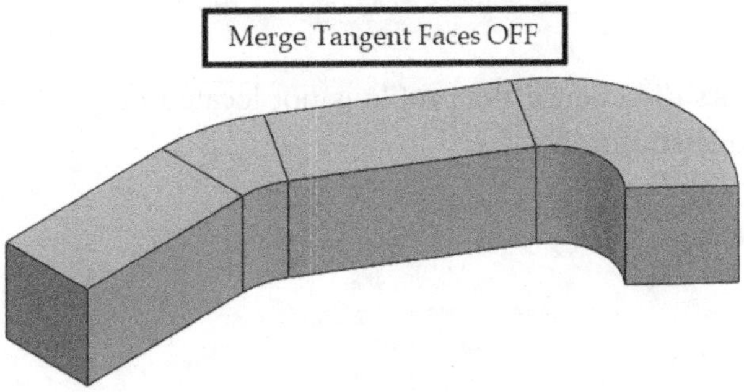

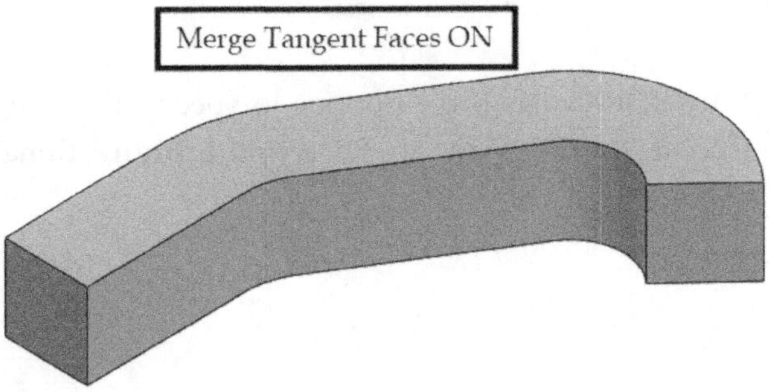

Profile Twist

SOLIDWORKS allows you to twist the profile along the path. Define the path and profile and select the **Specify twist value** option from the **Profile Twist** drop-down in the **Options** section. The **Twist Control** options in the **Options** section helps you to apply a twist to the profile.

The **Revolutions** option turns the cross-section by the value that you enter in the **Direction 1** box.

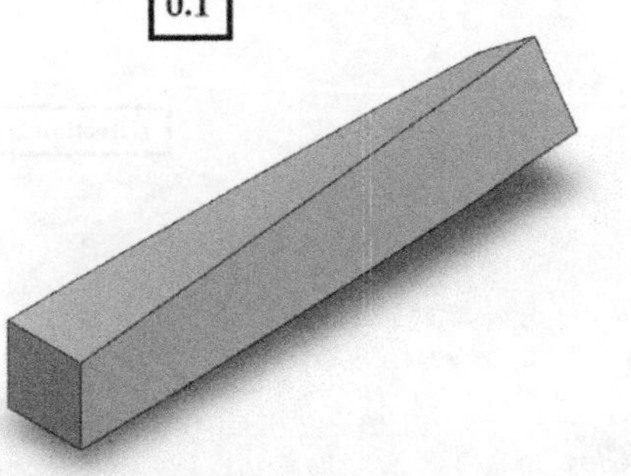

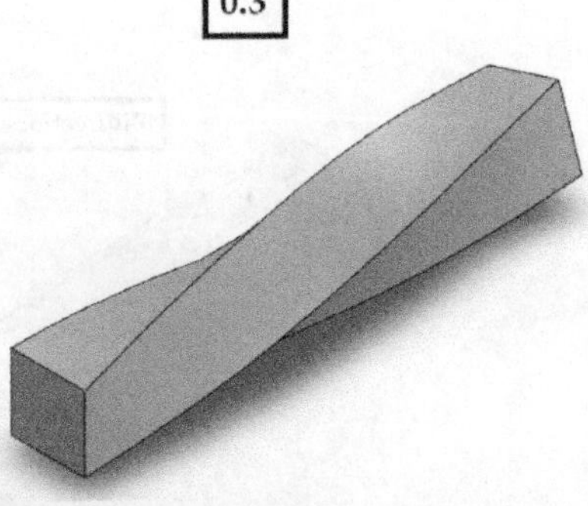

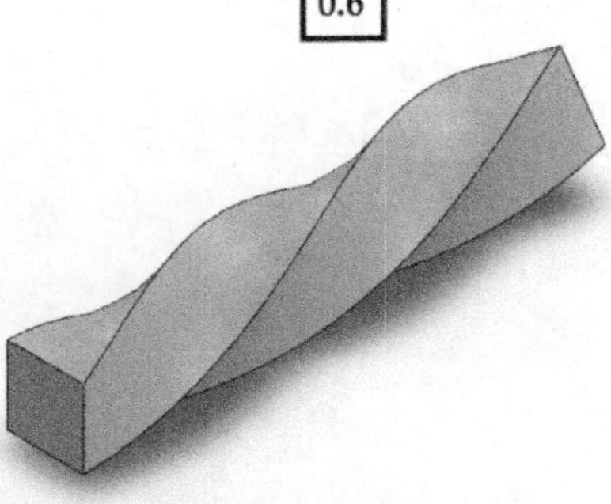

The **Radians** option twists the profile by the angle value that you specify in radians. 1 is equal to 57.3 degrees.

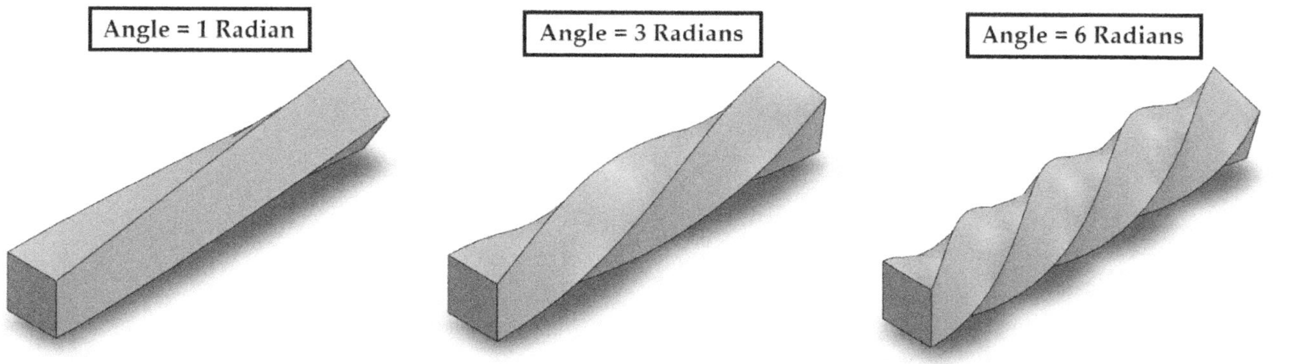

The **Degrees** option twists the profile by an angle that you specify in degrees.

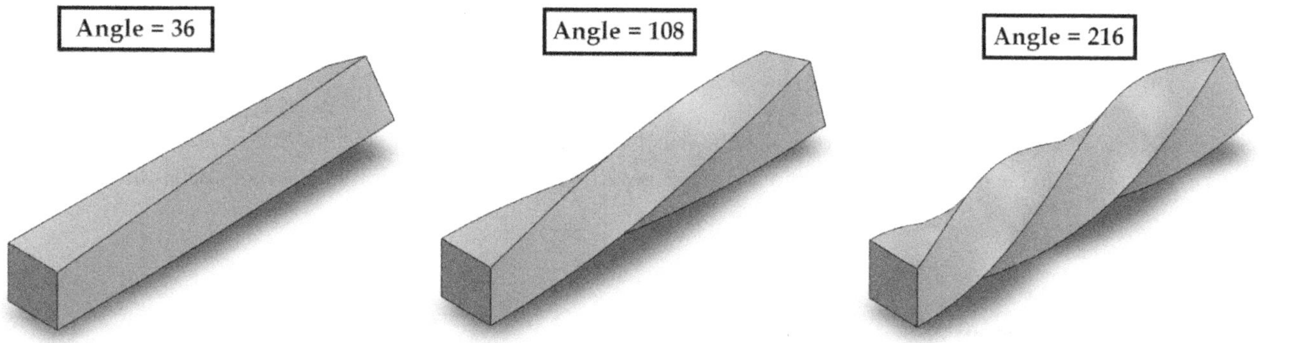

The **Profile Twist** drop-down has other options that will be useful while sweeping a profile along a non-planar path. For example, define a path and profile similar to the one shown in the figure. Next, select the options from the **Profile Twist** drop-down and view the result.

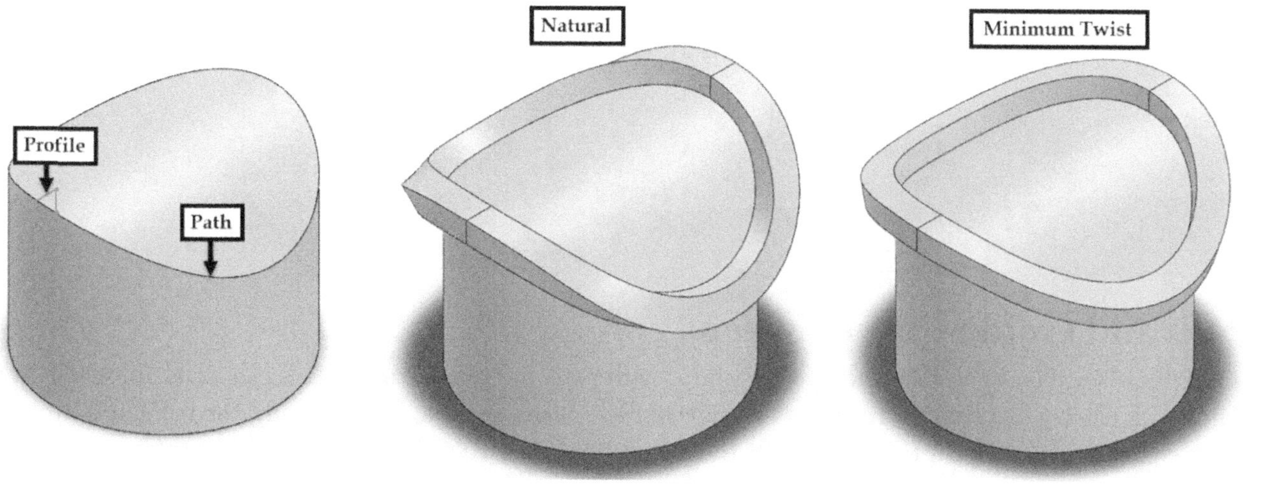

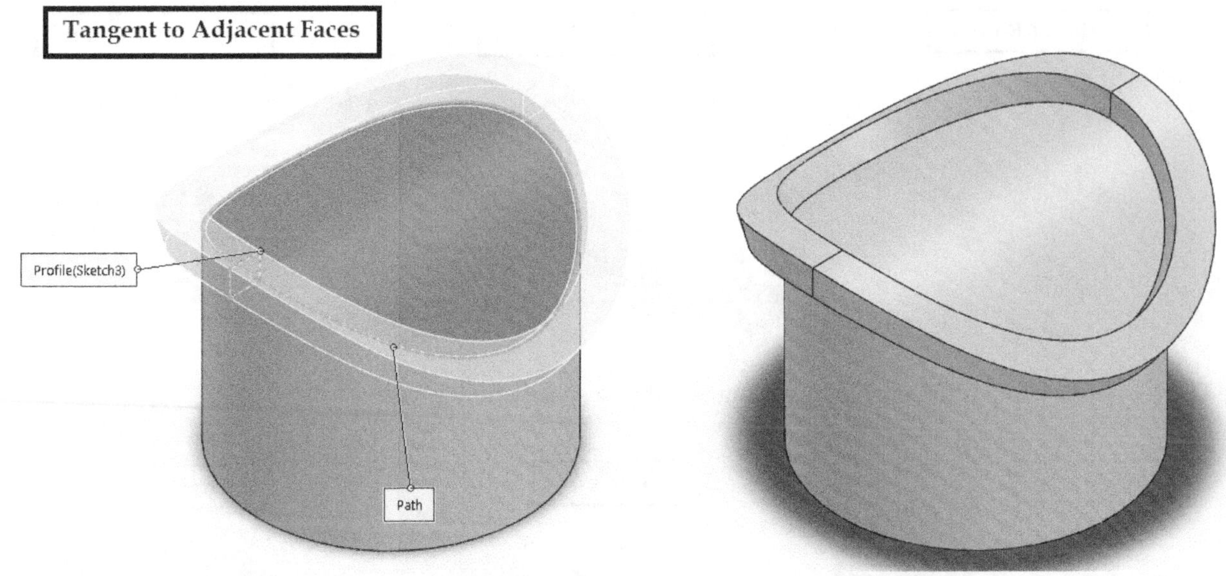

Select the **Specify Direction Vector** from the **Profile Twist** drop-down and select a line or axis. The orientation of the profile and axis becomes the same, and the profile will be swept, maintaining the axis's orientation.

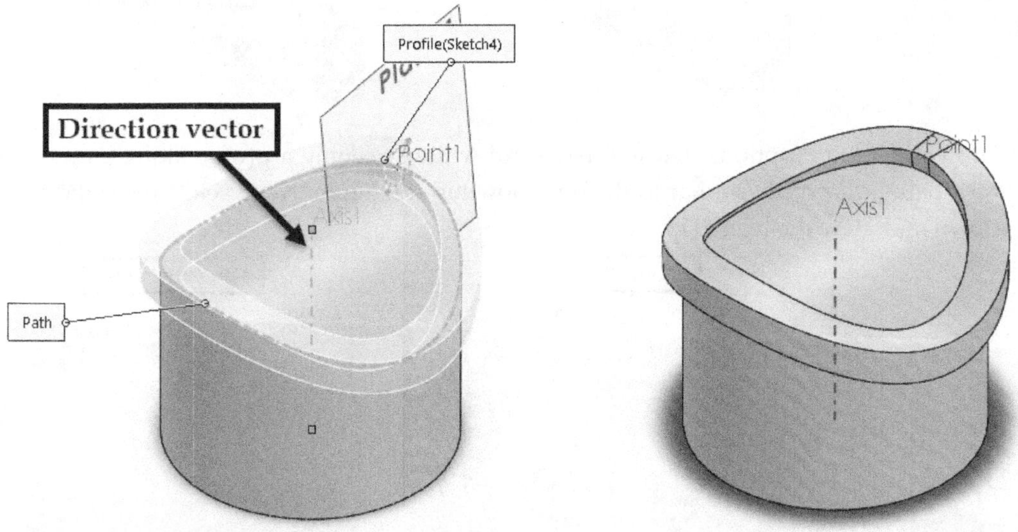

Path and Guide Curve Sweeps

SOLIDWORKS allows you to create sweep features with path and guide curves. This can be useful while creating complex geometry and shapes. To create this type of sweep feature, first, create a path. Next, create a profile and guide curve, as shown in the figure. Activate the **Swept Boss/Base** command and select the profile.

Click on the **Path** selection box and select the path. Next, expand the **Guide Curves** section and select the guide curve. A preview of the geometry will appear. Click **OK** to complete the feature.

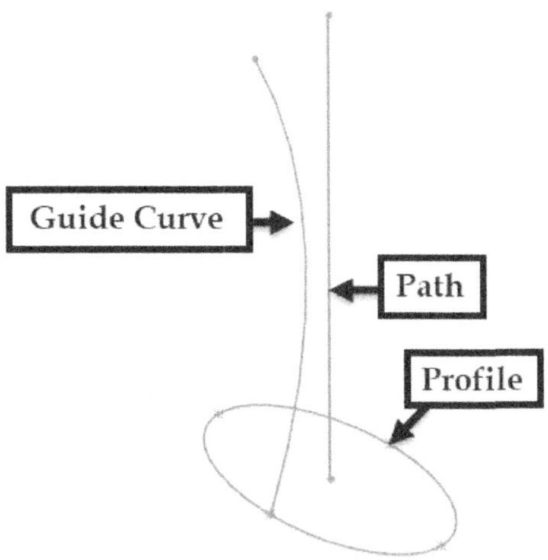

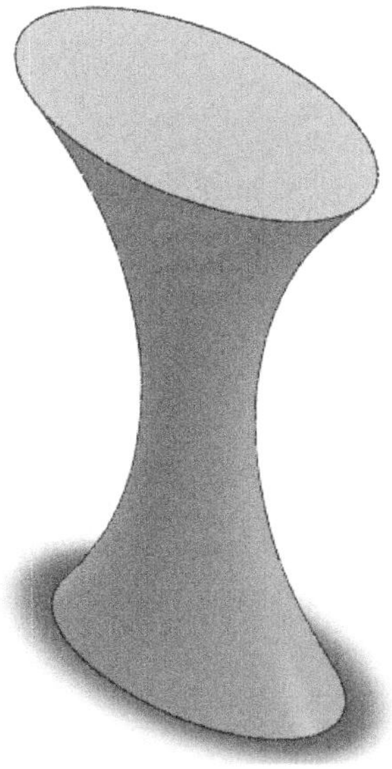

Swept Cut

In addition to adding swept features, SOLIDWORKS allows you to remove geometry using the swept feature. To do this, activate the **Swept Cut** command (click **Features > Swept Cut** on the CommandManager) and select the sweep type from the **Profile and Path** section. Select the profile and path. Next, expand the **Options** section and check the **Align with end faces** option to create the cut throughout the geometry. Click **OK** to create the swept cut.

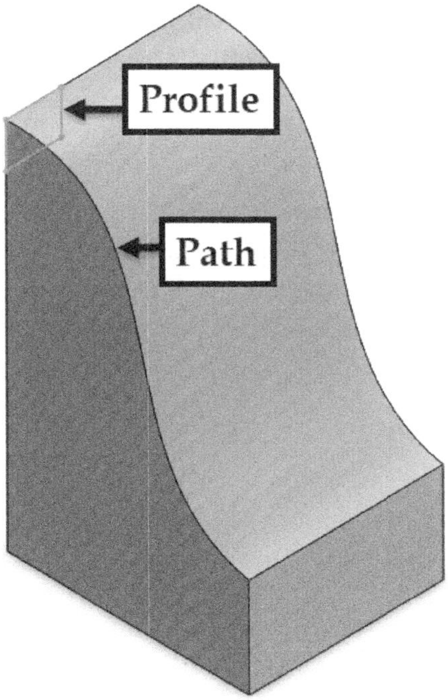

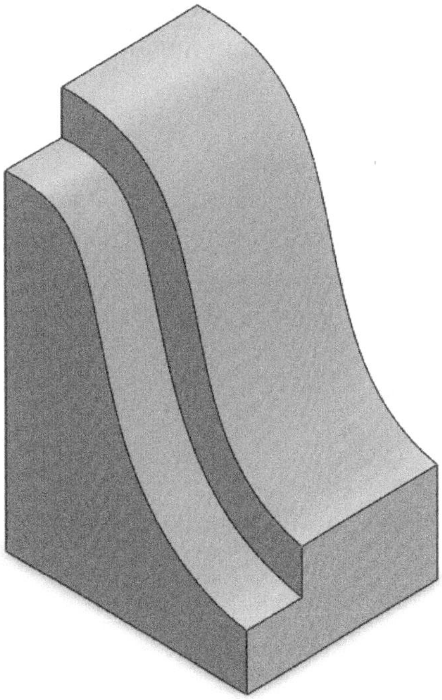

Best Practices

General Swept Features:

Before creating a swept feature, meticulous planning of your design is crucial. Understand the path and profile you will work with. Note that intricate paths or profiles might impede system performance or result in errors. It is recommended to commence with simple sweeps and progressively increase complexity as you gain familiarity with the tool.

Plastic Parts:
When designing for plastic parts, ensure that the path and profile align with the plastic manufacturing process. Consider draft angles and wall thickness. Avoid overly complex or tight sweeps, as they could complicate molding or lead to defects. For 3D printing, bear in mind that support structures may be essential, particularly for overhangs or intricate geometries.

Casted Parts:
In the context of casted parts, emphasize draft angles and steer clear of undercuts in your swept features. This approach facilitates easy removal from the mold. Sharp internal corners can pose issues in casted parts, potentially causing defects or casting challenges. Incorporate fillets and employ gradual transitions in your swept features to streamline the casting process.

Forged Parts:
When designing for forged parts, ensure that your swept features exhibit smooth transitions to mitigate stress concentrations and material inconsistencies. Exercise caution with intricate shapes, as they might pose forging difficulties or increase costs. Combining lofted features with sweeps can aid in crafting intricate, seamless transitions conducive to forging processes.

Machined Parts:
For machined parts, craft swept features with seamless transitions to prevent internal stresses or machining complexities. Verify that the geometry avoids deep pockets or sharp angles that could challenge cutting tool access. Design parts with appropriate clearances and assess the toolpath to guarantee efficient machining of the swept features.

3D Printed Parts for Makers:
When developing swept features for 3D printed parts, factor in the layering process during design. Pay attention to overhangs and contemplate incorporating support structures if necessary. Employ features like fillets and chamfers to reduce sharp angles, which can be problematic for 3D printing. Monitor wall thickness to uphold structural integrity in the printed part.

Examples

Example 1

In this example, you will create the part shown below.

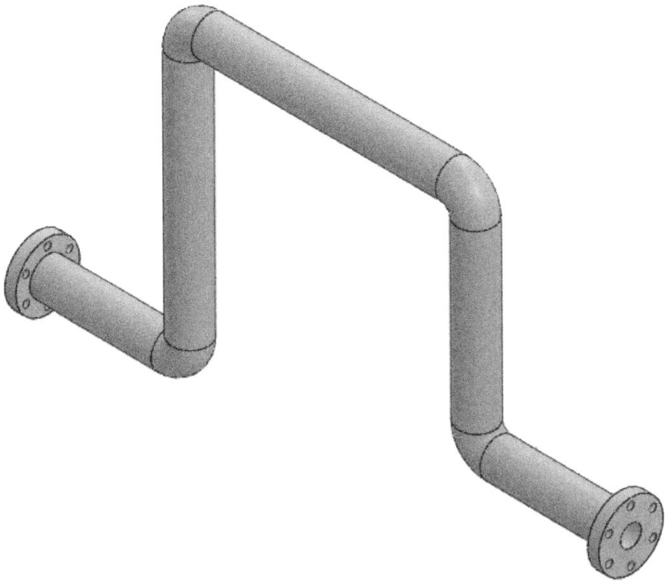

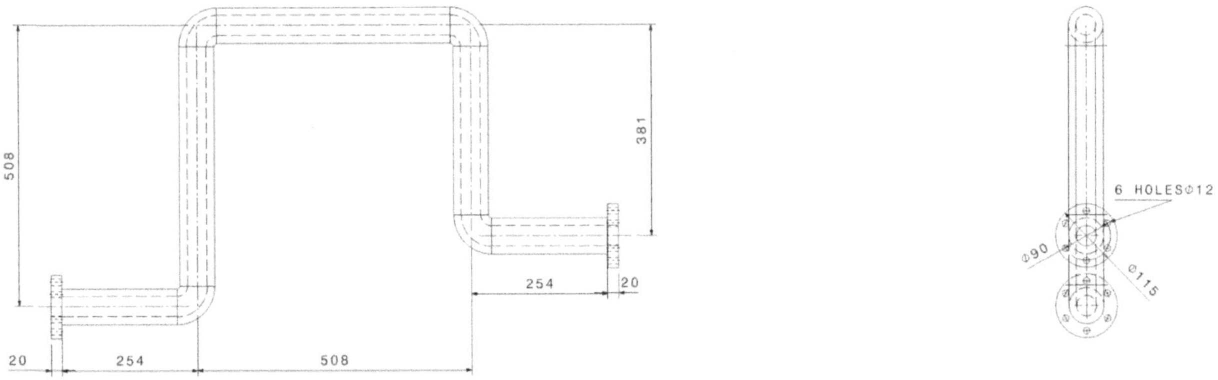

PIPE I.D. 51

PIPE O.D. 65

1. Start **SOLIDWORKS 2024.**
2. Open a new part file.
3. Draw the sketch on the Front plane, as shown in the figure.
4. Click **Exit Sketch** to complete the sketch.

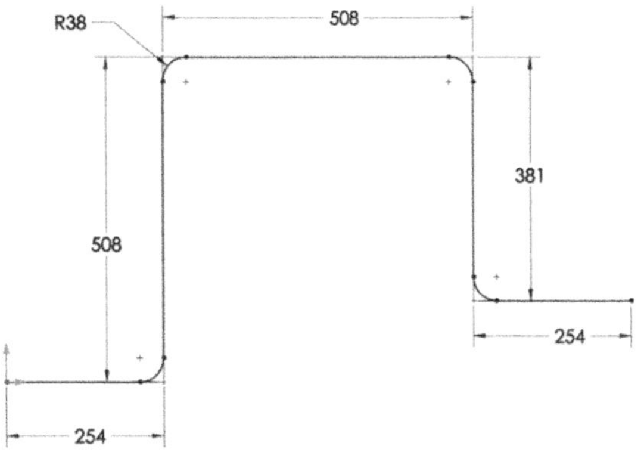

5. Change the **View orientation** to **Isometric** view.
6. On the **Features** CommandManager, click **Reference Geometry** drop-down > **Plane** icon.
7. Select the lower horizontal line of the sketch to define the first reference.
8. Click on the end-point of the sketch to define the plane location.

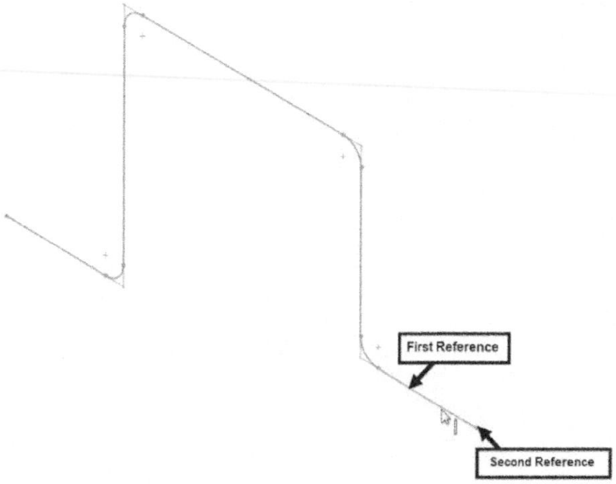

9. Click **OK** to create a new plane.

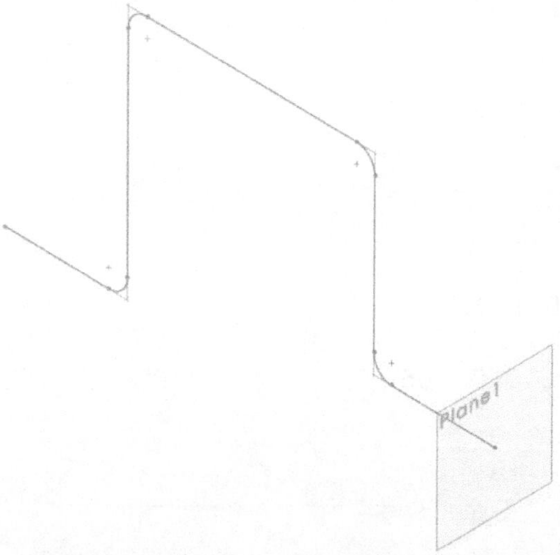

10. Start a sketch on the new plane (refer to **Chapter 2: Sketch Techniques** to learn about starting a sketch).
11. Draw a circle and make it's center coincident with the endpoint of the previous sketch.
12. Apply 65 mm diameter to the circle.

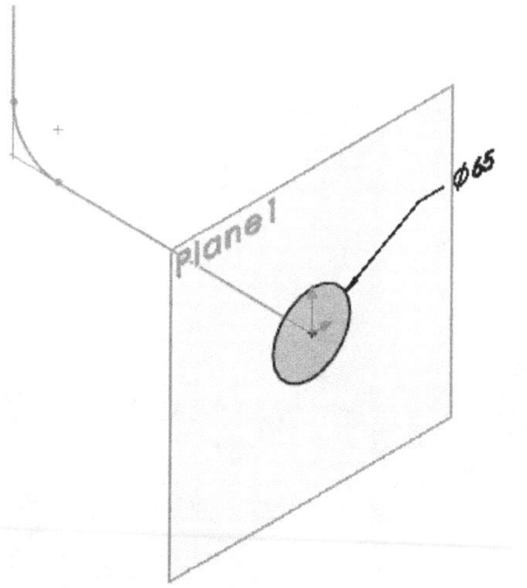

13. Click the **Exit Sketch** icon.
14. On the **Features** CommandManager, click the **Swept Boss/Base** icon (or) click **Insert > Boss/Base > Swept** on the Menu.
15. Select the circle to define the profile.
16. Select the first sketch to define the path.
17. On the **Sweep** PropertyManager, check the **Thin Feature** option and type in 14 in the **Thickness** box.
18. Click the **Reverse Direction** icon if the thin feature's preview is displayed outside the circle.
19. Click **OK** to create the *Swept Boss/Base* feature.

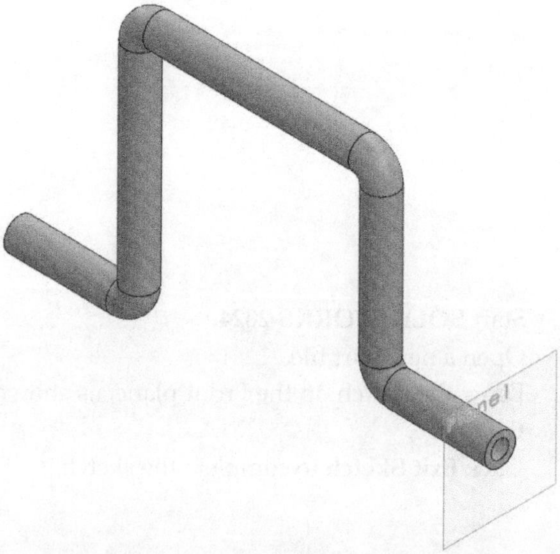

20. Activate the **Extrude Boss/Base** command.
21. Click on the front-end face of the *Swept Boss/Base* feature.

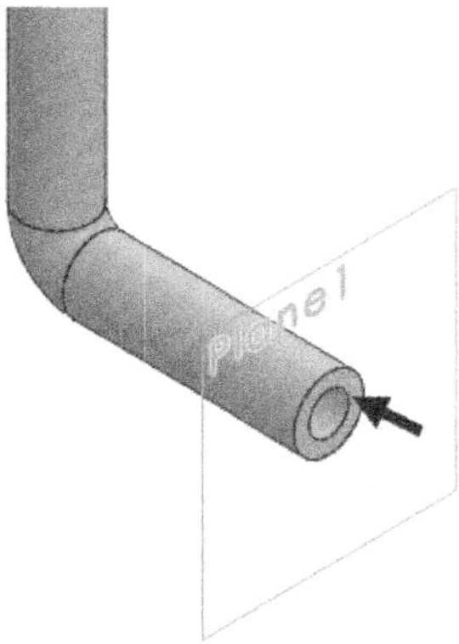

22. On the **Sketch** CommandManager, click **Convert Entities** and click on the inner circular edge.

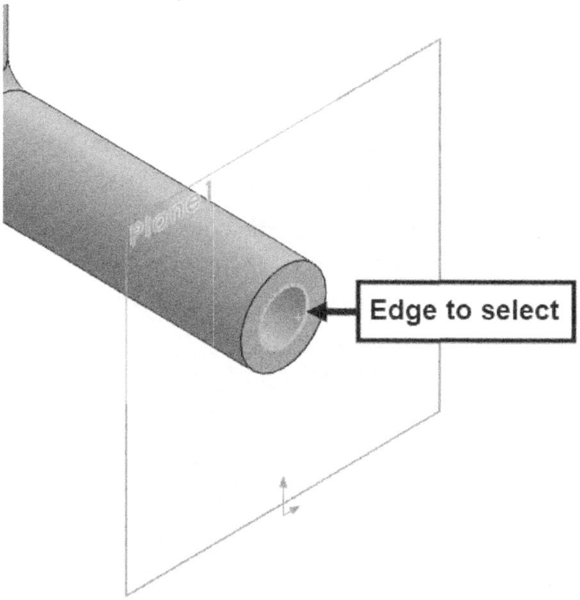

23. On the **Convert Entities** PropertyManager, click **OK** to project the curve onto the sketch plane.

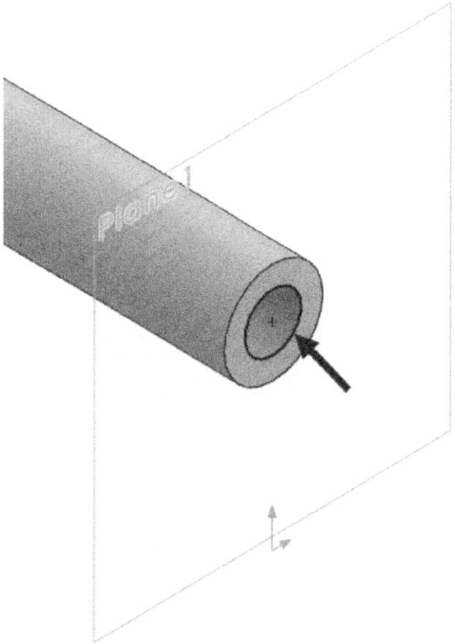

24. Draw a circle of 115 mm in diameter.

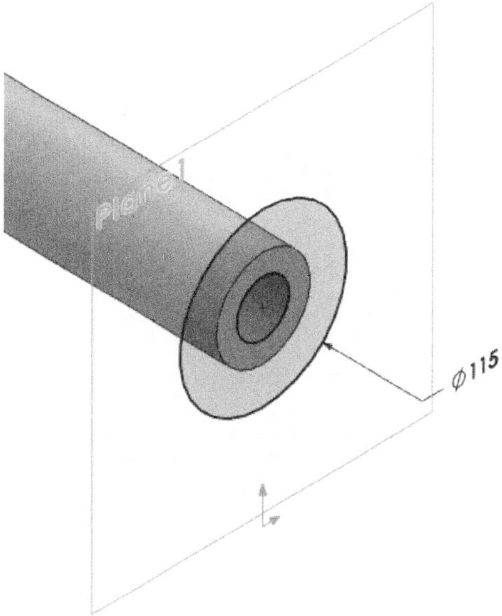

25. Click **Exit Sketch** to complete the sketch.
26. Type-in **20** in the **Depth** box. Click **OK** to complete the *Extrude Boss/Base* feature.

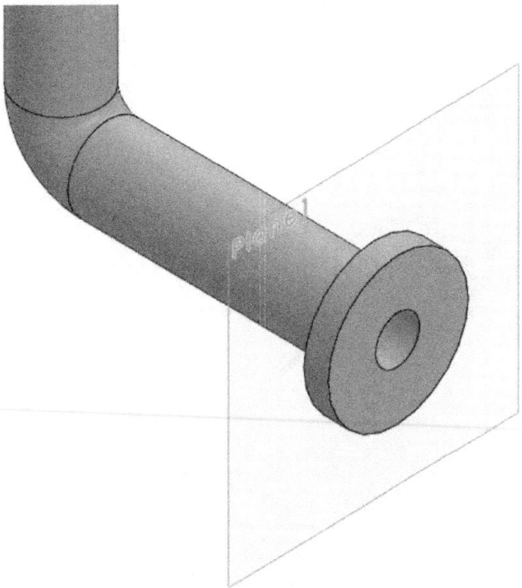

27. Create a hole of 12 mm in diameter on the *Extruded Boss/Base* feature (refer to Chapter 4: Placed Features to learn how to create holes).

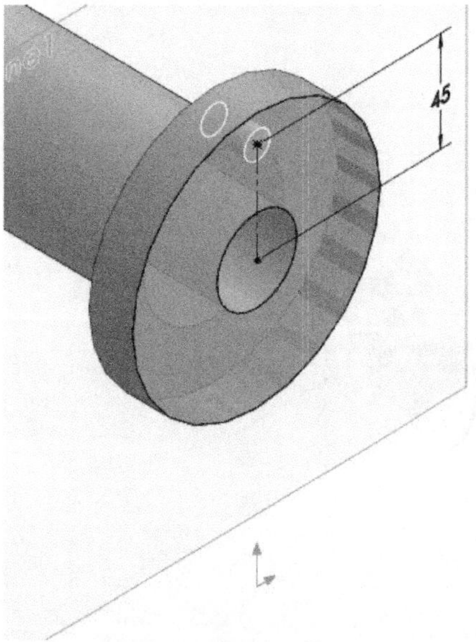

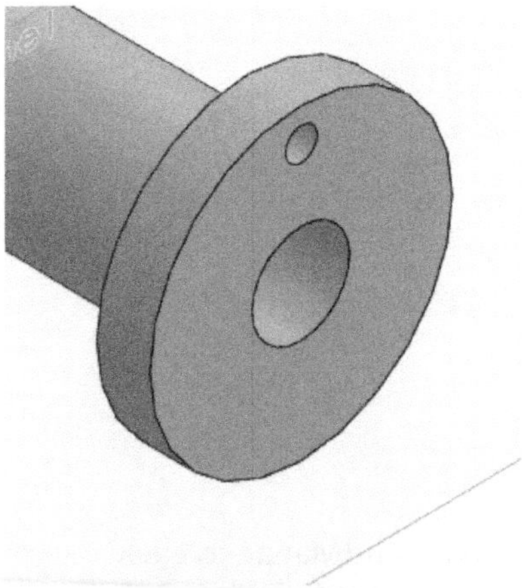

28. On the **Features** toolbar, click **Pattern** drop-down > **Circular Pattern**.
29. On the **CirPattern1** PropertyManager, select **Instance spacing**.
30. Under the **Features and Faces** section, click in the **Features to Pattern** box and select the hole.
31. Under the **Direction 1** section, click in the **Pattern Axis** box and select the outer cylindrical face of the *Extruded Boss/Base* feature.
32. Type-in **60** and **6** in the **Angle** and **Number of Instances** boxes, respectively.
33. Click **OK** to pattern the hole.

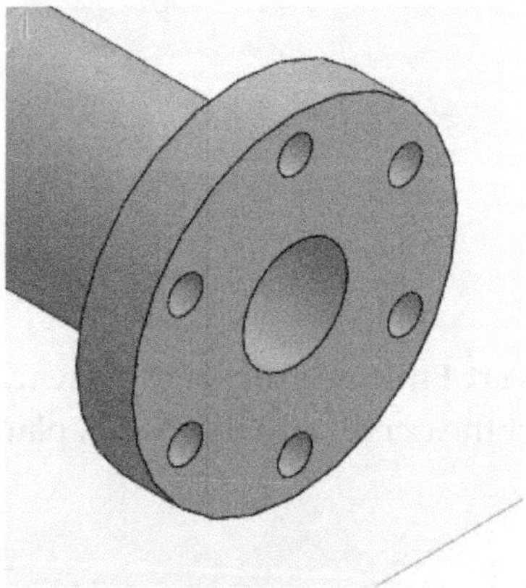

34. Create the *Extruded Boss/Base, Hole,* and *Circular Pattern* features on the other end of the model.

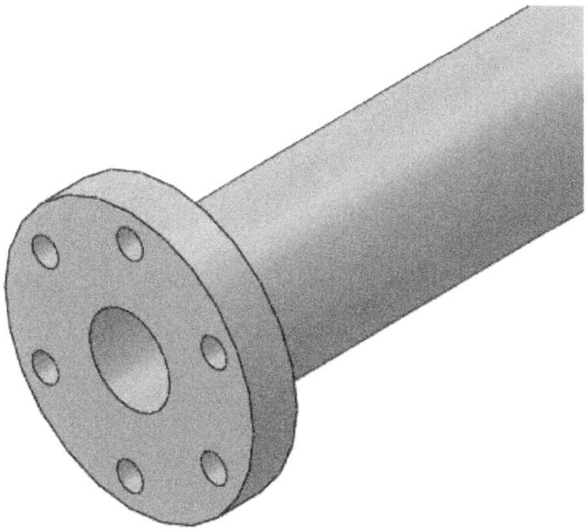

35. Save and close the part file.

Questions

1. List the types of paths that can be used to create *Swept Boss/Base* features.
2. What is the use of the **Align with end faces** option?
3. Explain the use of the **Profile Twist** option in the **Swept Boss/Base** command?
4. What is the use of the **Profile Orientation** option?

Exercises

Exercise1

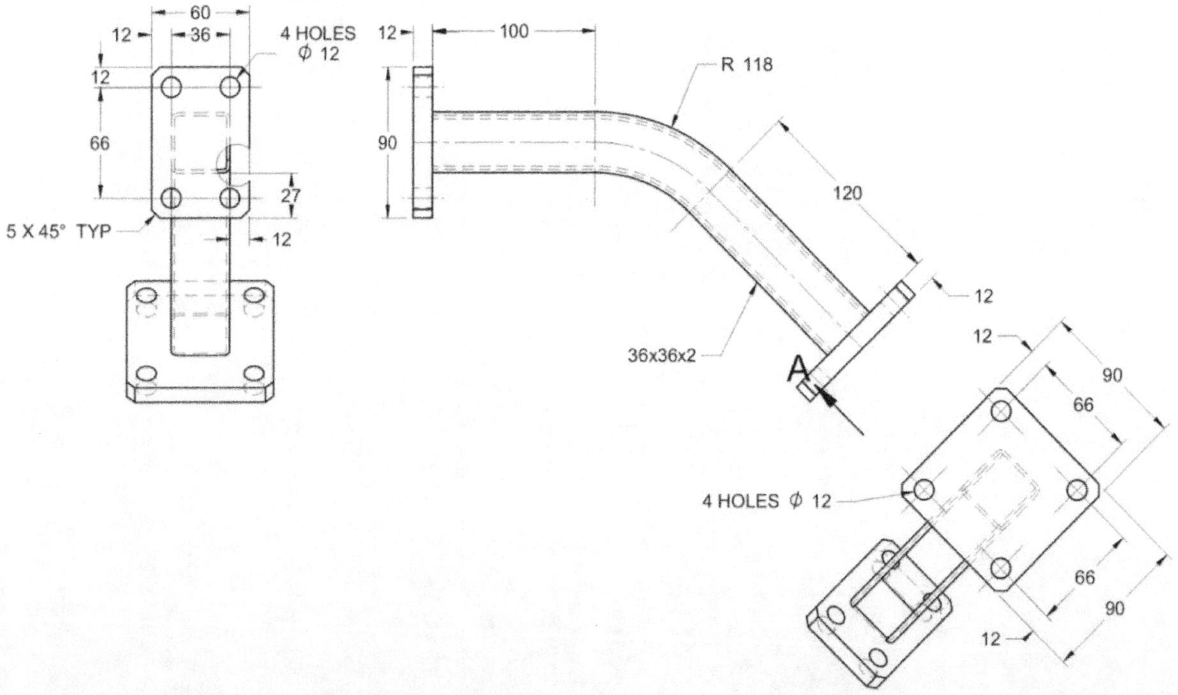

Chapter 7: Loft Features

The **Lofted Boss/Base** command is one of the advanced commands available in SOLIDWORKS that allows you to create simple and complex shapes. A basic loft is created by defining two profiles and joining them together. For example, if you create a loft feature between a circle and a square, you can quickly change the solid's cross-sectional shape. This ability is what separates the loft feature from the sweep feature.

The topics covered in this chapter are:

- *Basic Lofts*
- *Loft profiles*
- *Profile geometry*
- *Start-end constraints*
- *Guide Curves*
- *Center Line Loft*
- *Closed loft*
- *Adding loft sections*
- *Loft Cutouts*

Lofted Boss/Base

This command creates a loft feature between different profiles. To create a loft, first, create two or more profiles on different planes. The planes can be parallel or perpendicular to each other. Activate the **Lofted Boss/Base** command (click **Features > Lofted Boss/Base** on the CommandManager). Next, click in the **Profile** selection box under the **Profiles** section on the PropertyManager and select two or more profiles. Click **OK** to create a basic loft feature.

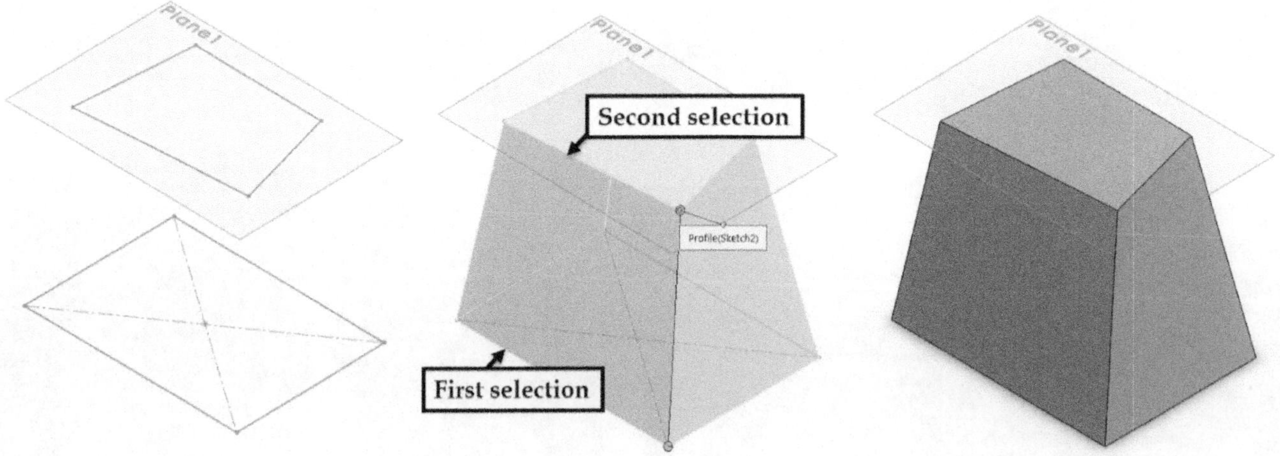

While selecting the profiles, you need to select the elements on the same side of both the profiles. By mistake, if you have selected elements on a different side, the lofted boss will be created with a twist.

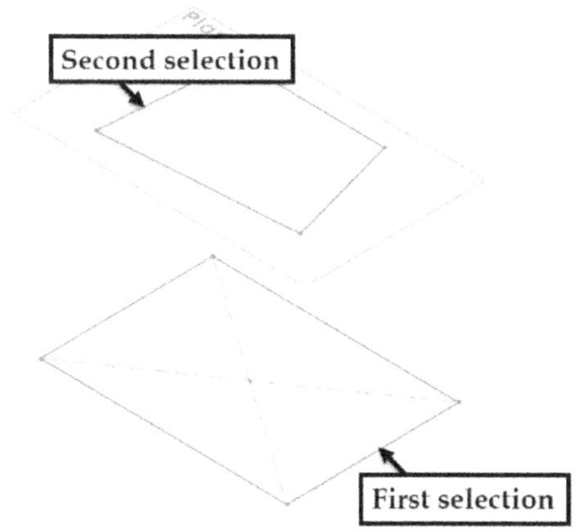

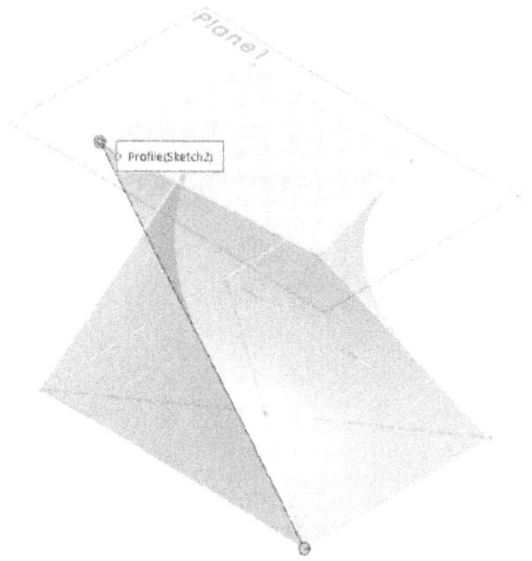

You can fix this unwanted twisting by dragging the handles displayed on the profiles. To do this, press and hold the left mouse button on the handle displayed on the second profile. Next, drag and release the handle on the vertex, which is on the same side of the first profile handle.

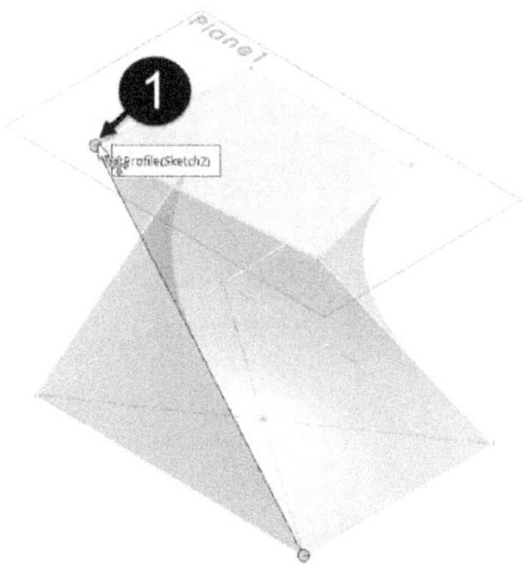

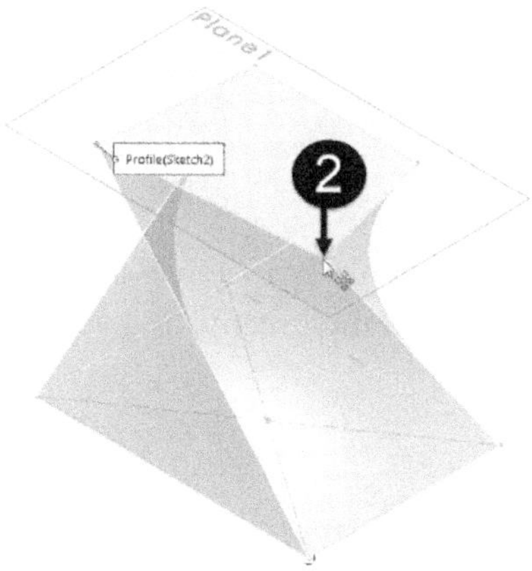

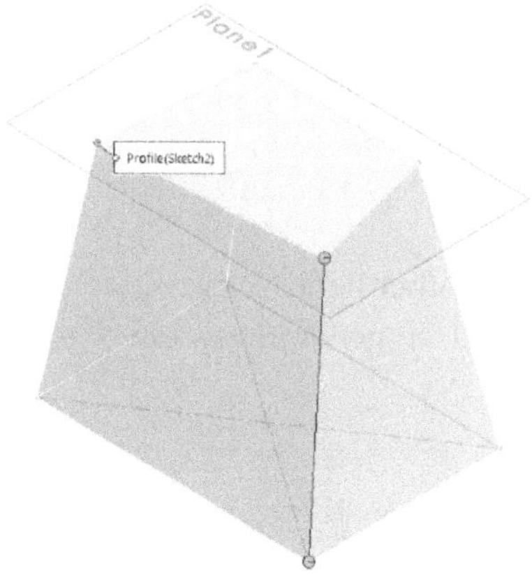

Loft Profiles

In addition to 2D sketches, you can also define loft profiles by using different element types. For instance, you can use existing model faces, surfaces, curves, and points. The only restriction is that the points can be used at the beginning or end of a loft.

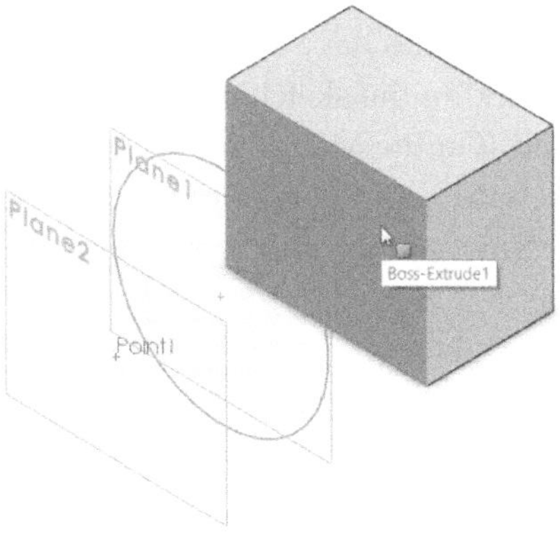

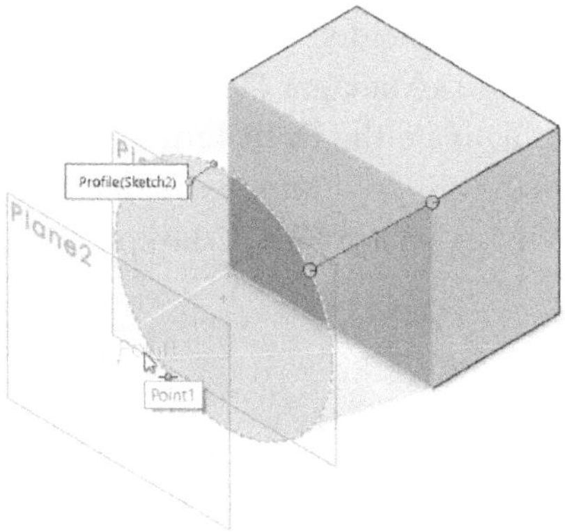

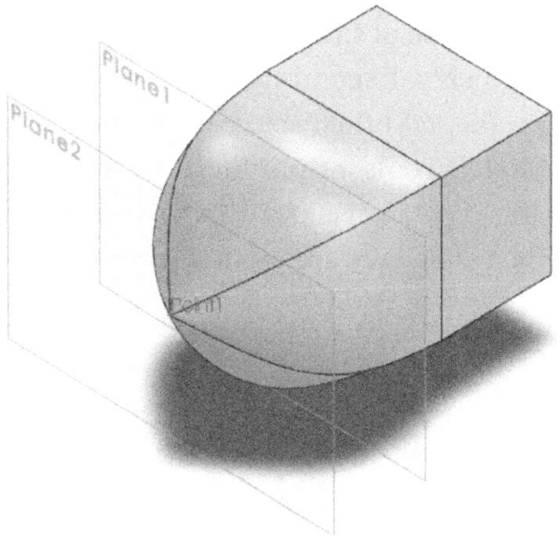

Profile Geometry

Profiles used for creating lofts should have a matching number of segments. For example, a four-sided profile will loft nicely to another four-sided profile despite the differences in the individual segments' shape. The **Lofted Boss/Base** command generates smooth faces to join them.

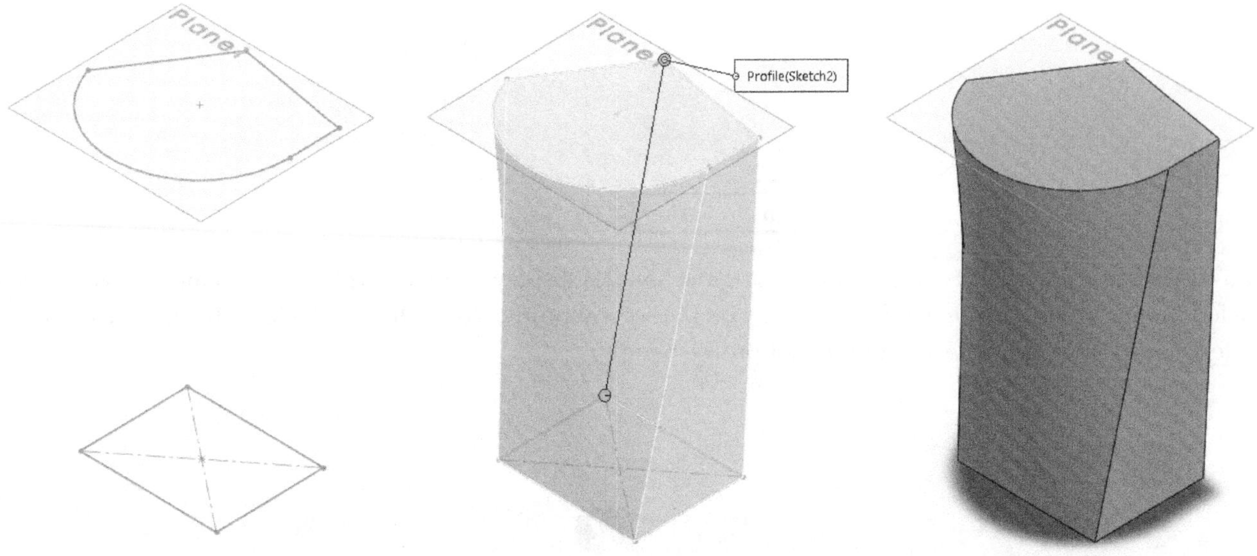

On the other hand, a five-sided profile will not loft nicely to a two-sided profile. Although SOLIDWORKS succeeds in generating a loft, it maps the endpoints incorrectly, and you may not get the desired result.

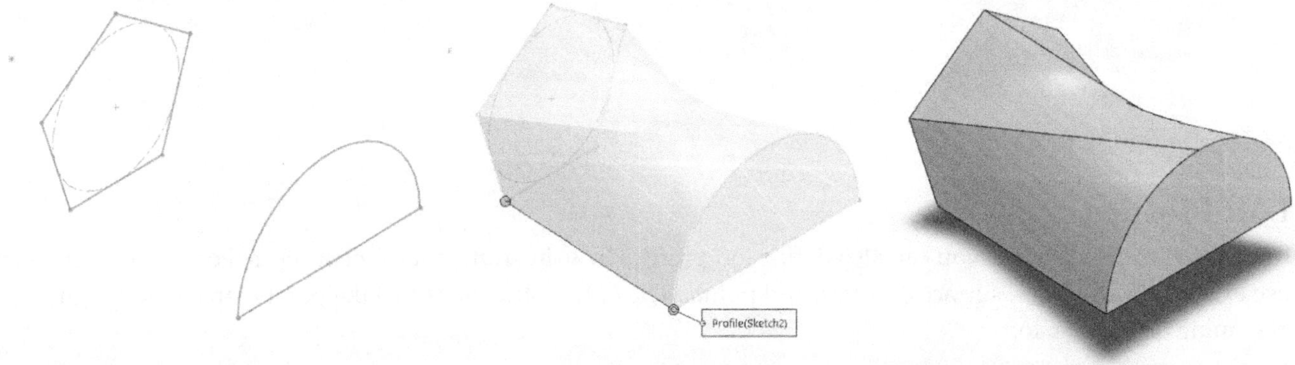

To get the desired result, you have to split the two-sided profile so that both the profiles have an equal number of segments. Expand the **Loft** feature in the FeatureManager Design Tree, right click on the sketch of the two-sided profile, and then select **Edit Sketch**. Next, activate the **Split Entities** command (On the Menu bar, click **Tools > Sketch Tools > Split Entities**) and click on the arc, split into two elements. Likewise, select two more points on the arc. You can also use dimensions to define the exact location of the split points. Now, exit the sketch and notice that the connectors are plotted nicely.

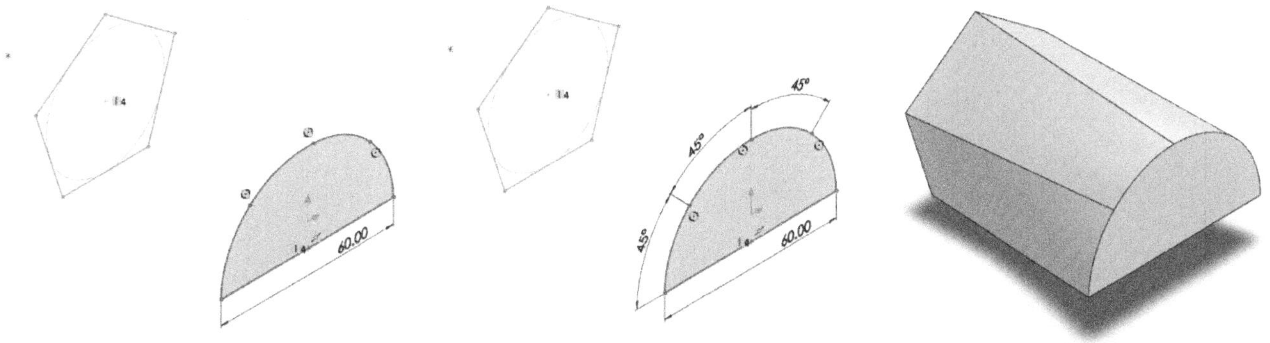

Start-End Constraints

The shape of a loft is controlled by the profile sketches and the plane location. However, the **Start/End Constraints** connected to the profile sketches can control the side faces' behavior. If you would like to change the side faces' appearance, you can use the **Start-End Constraints** section either at the beginning of the loft, or at the end of the lofts, or both.

Normal to Profile

Click the **Start-End Constraints** section on the PropertyManager and select **Start Constraint** drop-down **> Normal to Profile**. Next, enter a value in the **Draft Angle** box; you can notice that the loft's beginning starts at an angle to the profile. You can control how much influence the angle will have by adjusting the parameter in the **Start Tangent Length** box. A lower value will have a lesser effect on the feature. As you increase the value, the more noticeable the effect will be, eventually. If you increase the number high enough, the direction angle will lead to self-intersecting results.

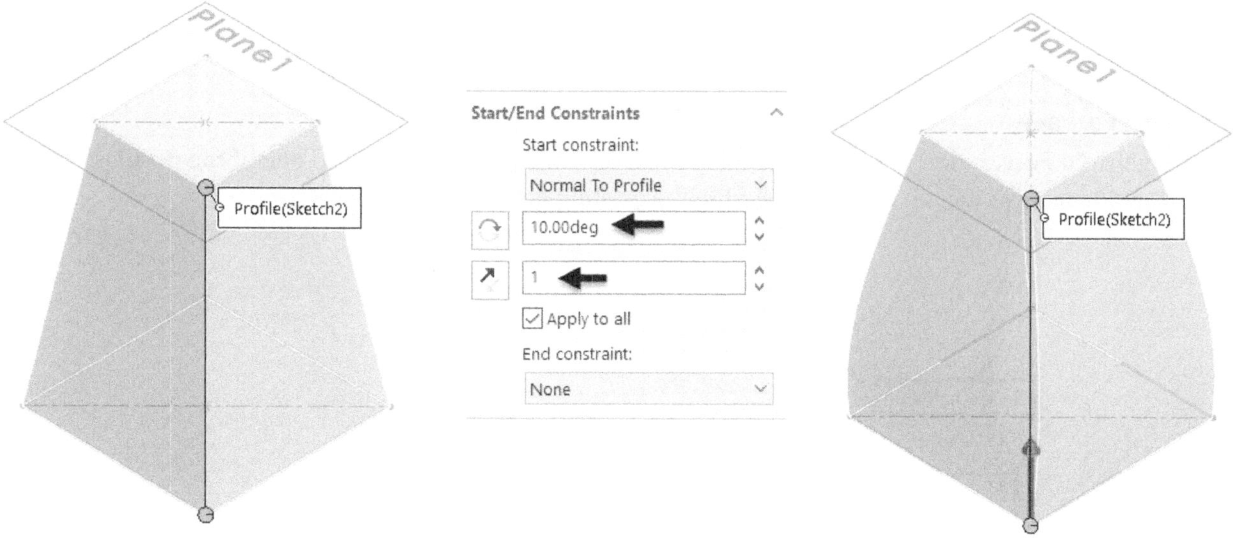

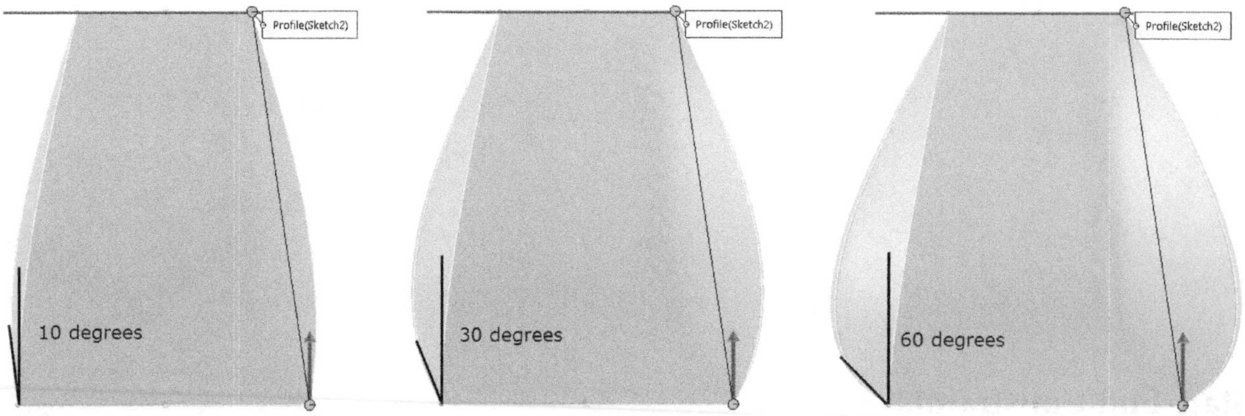

Likewise, you can also apply the **Normal to Profile** constraint to the second profile of the loft.

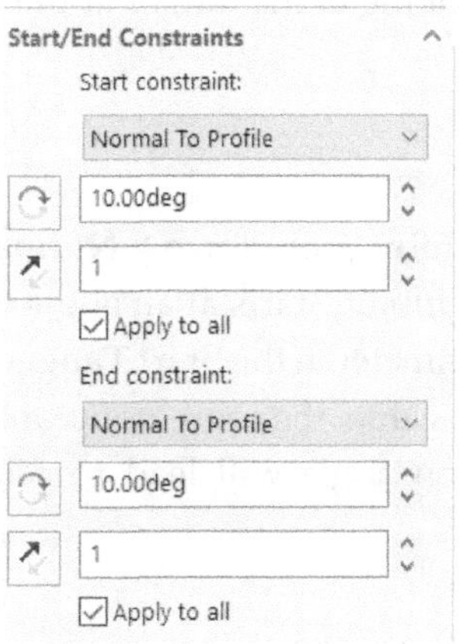

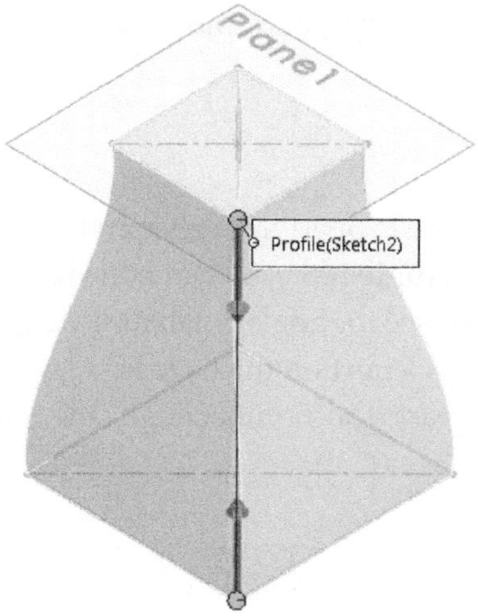

Tangent To Face

The **Tangent To Face** option is available when you select an existing face loop as one of the profiles. This option makes the side faces of the loft feature tangent to the existing geometry's side faces.

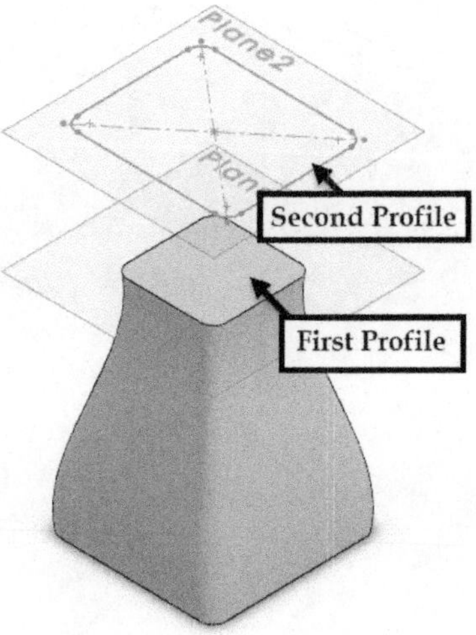

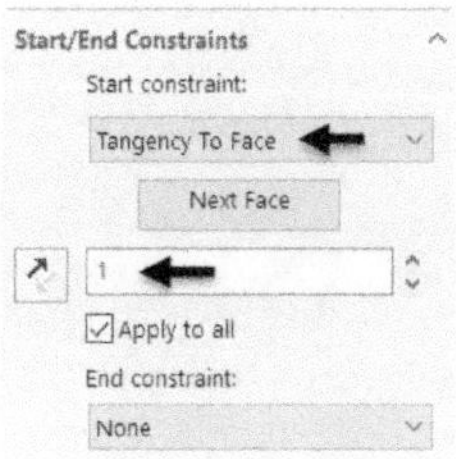

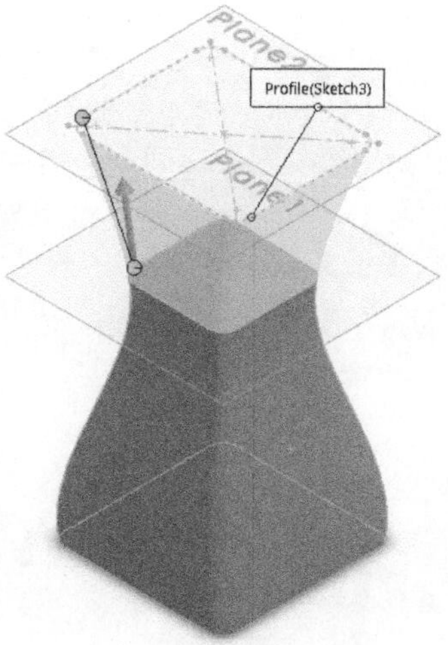

Curvature To Face

The **Curvature To Face** option is available when you select an existing face loop as one of the profiles. This option makes the loft feature curvature's side faces continuous with the existing geometry's side faces. You can change the curvature magnitude using the Start Tangent length and End Tangent length boxes, respectively. You can also reverse the curvature using the **Reverse Direction** icons located next to the **Start Tengent length** and **End Tangent length** boxes.

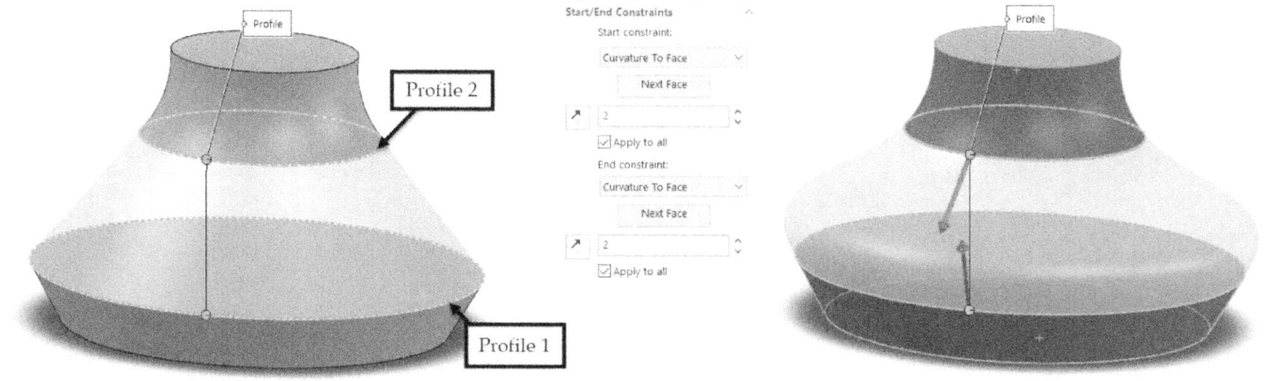

Direction Vector

This option is used to control the behavior of the side faces using a direction vector. The direction vector can be defined by selecting a sketch, edge, face, or plane.

After selecting the loft feature's start and end profiles, click the **Start-End Constraints** section on the PropertyManager and select **Start Constraint** drop-down **> Direction Vector**. Select a line, edge, or plane to define the direction vector and enter a **Draft Angle** box value. You can control how much influence the angle will have by adjusting the parameter in the **Start Tangent Length** box. Select **Apply to all** option to apply the same for all vertices. Click **OK**.

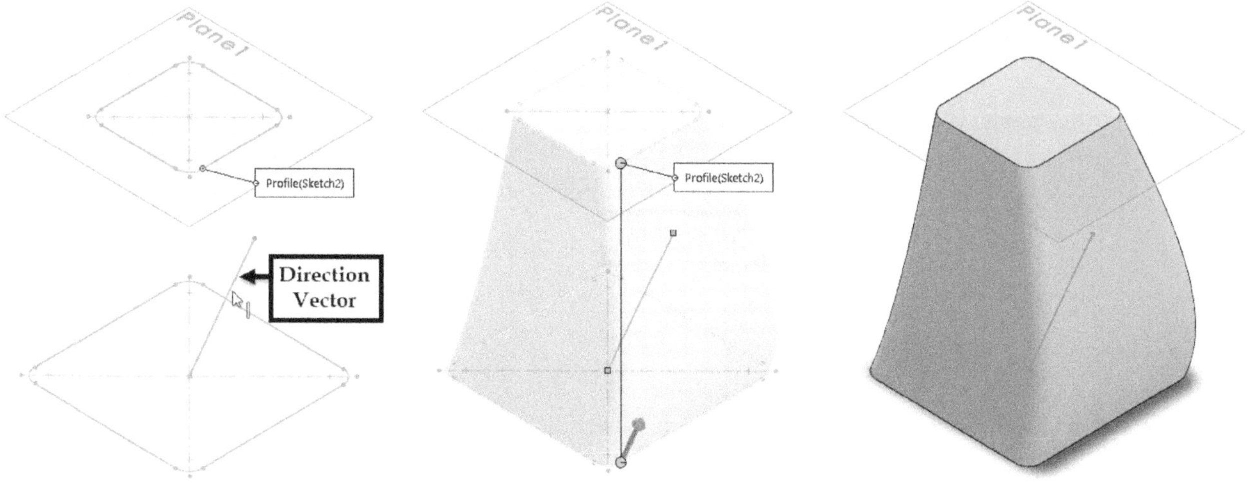

Guide Curves

Similar to the **Start/End Constraints** options, guide curves allow you to control a loft's behavior between profiles. You can create curves by using 2D or 3D sketches. For example, start a sketch on the plane intersecting with the profiles, and then create a spline, as shown.

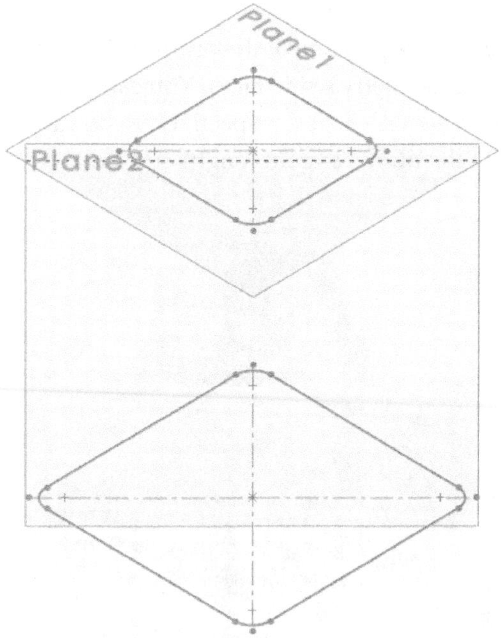

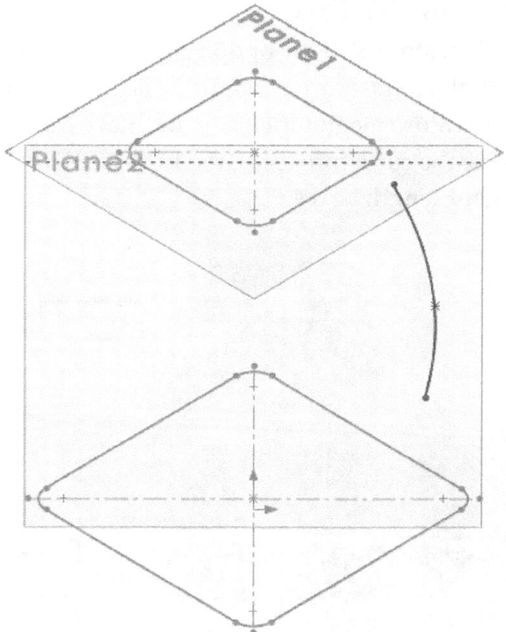

Next, you need to connect the endpoints of the spline with the profiles. To do this, apply the **Coincident** relation between the endpoints of the spline and the profiles. Click **Exit Sketch** on the CommandManager.

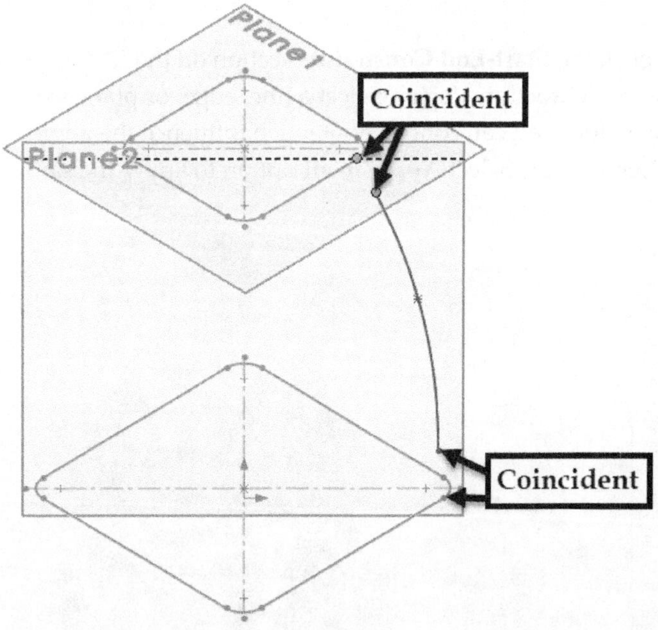

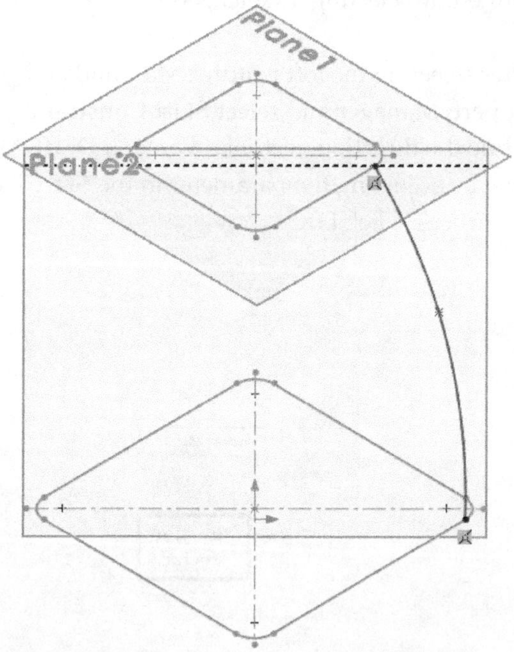

Likewise, create a sketch on the other plane, as shown. Next, click **Exit Sketch**.

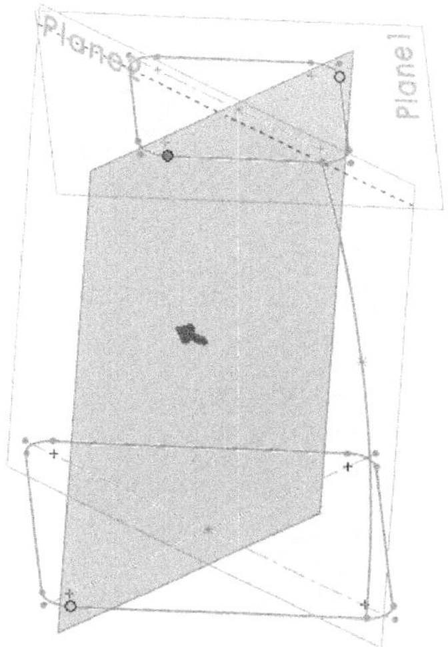

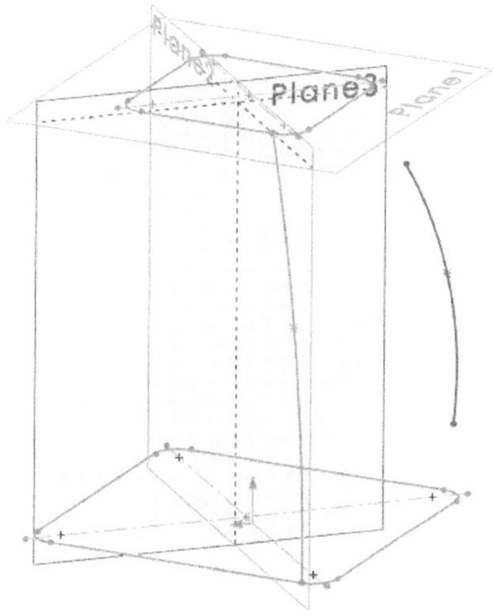

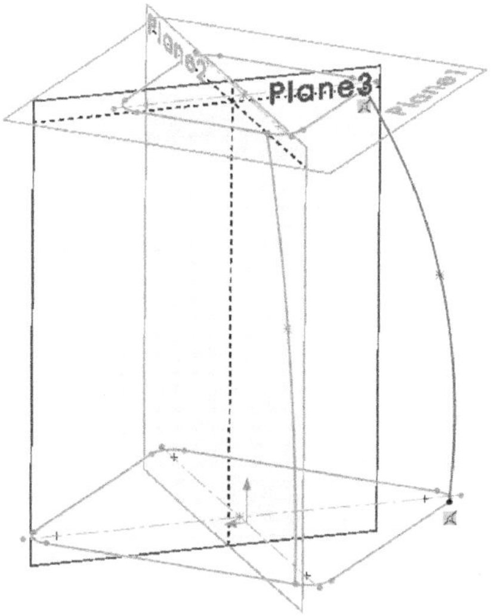

Now, activate the **Loft Boss/Base** command and select the profiles. To select guide curves, click in the **Guide Curves** section and select the guide curves; you will see the preview updates. Notice that the edges with rails are affected.

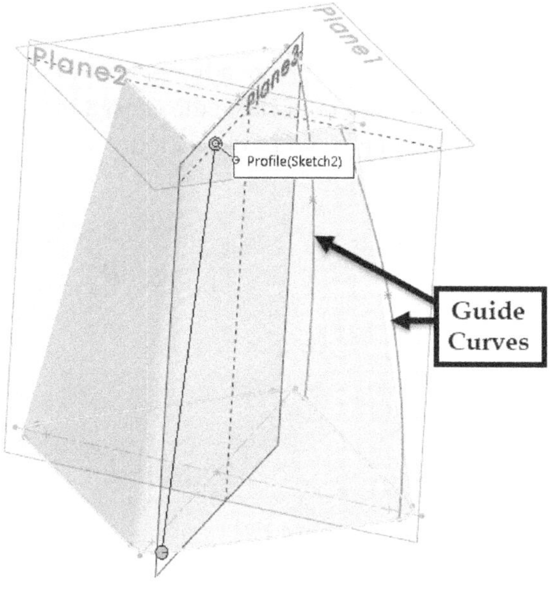

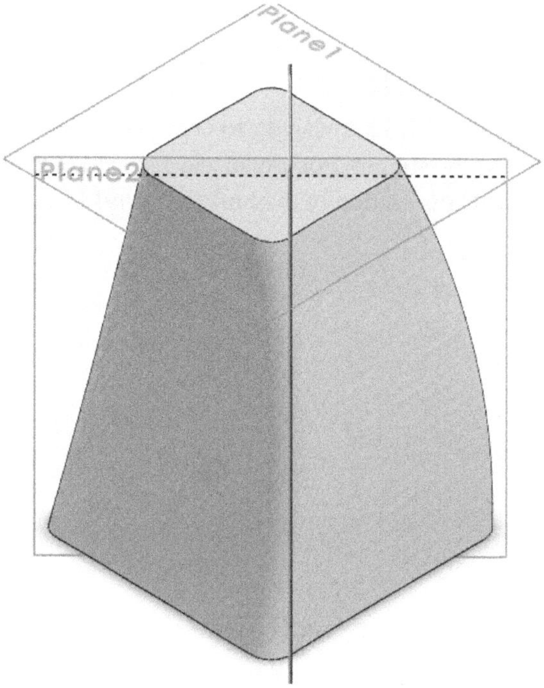

Centerline Parameters

The **Centerline Parameters** section helps you create a loft feature with the help of a centerline passing through the selected profiles. First, create a centerline passing through all the profiles, as shown. Next, activate the **Lofted**

Boss/Base command and select the profiles. Next, expand the **Centerline Parameters** section, and then select the centerline. Click **OK** to create the loft feature.

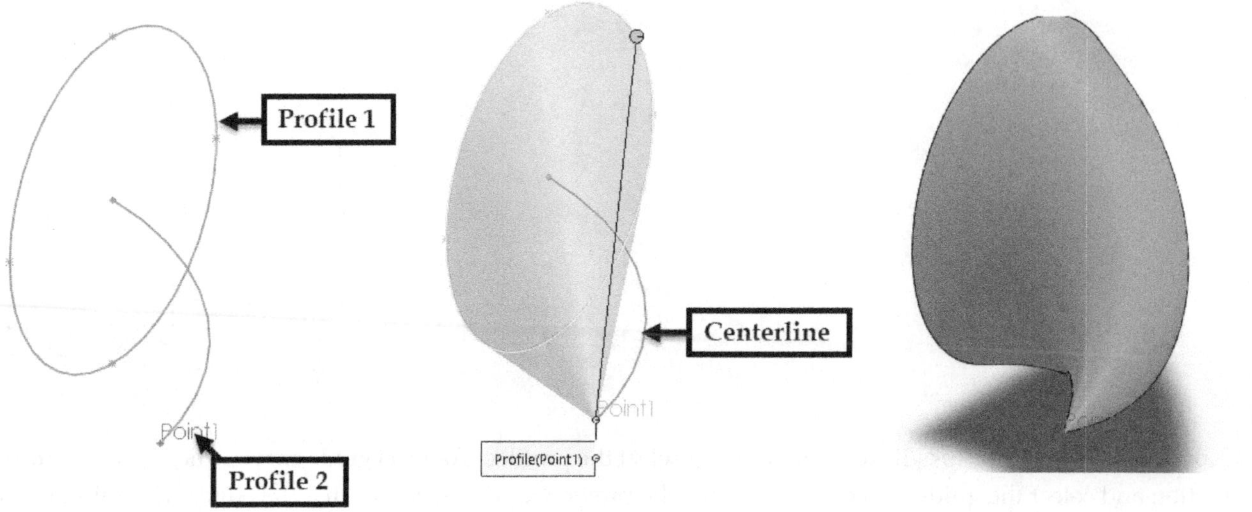

Closed Loft

SOLIDWORKS allows you to create a loft that closes on itself. For example, to create a model that lofts between each of the shapes, you must select four sketches, as shown in the figure, and then check the **Closed loft** option in the **Options** section of the **Loft** PropertyManager. Next, click **OK**; this will give you a closed loft.

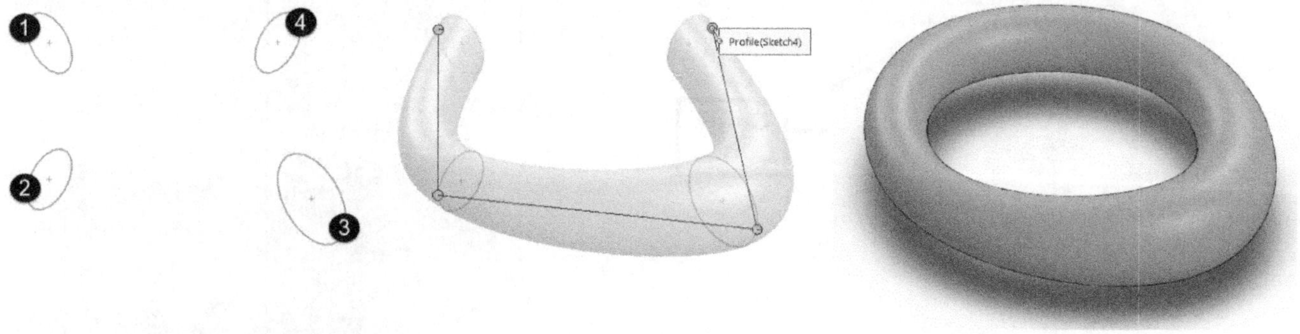

Adding Loft Sections

You can add sections to the created loft feature. Adding sections to the loft can change the loft feature. For instance, you can add a new section between the loft feature profiles, as shown.

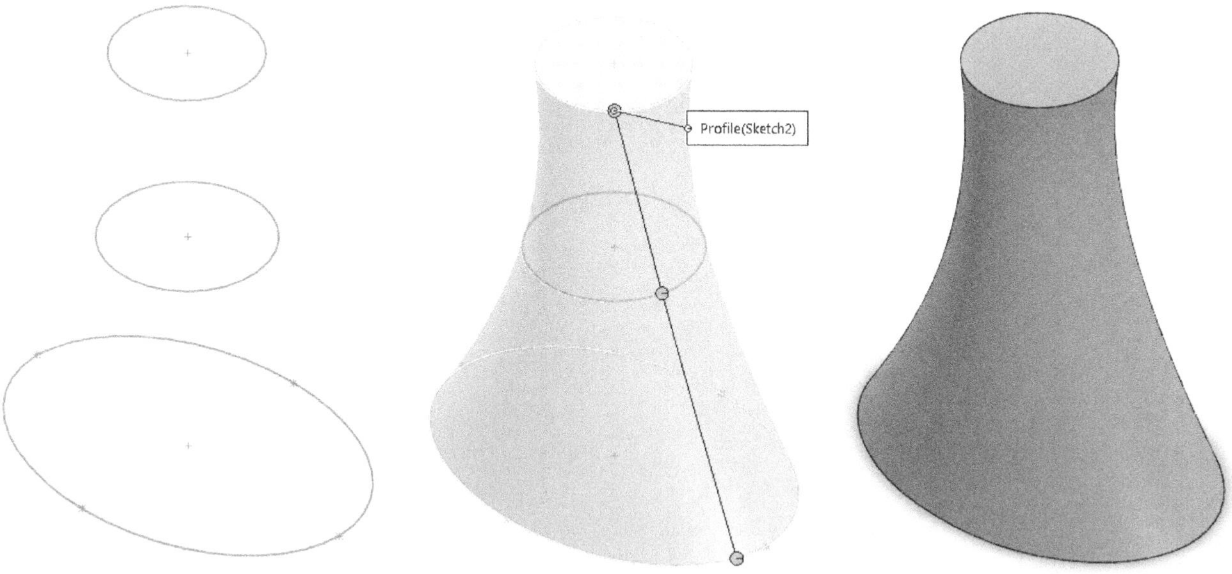

To add a loft section, right-click on the loft feature and select **Add Loft Section**; a temporary plane appears. You can click and drag the plane to position the new loft section. Next, click **OK** to add a loft section.

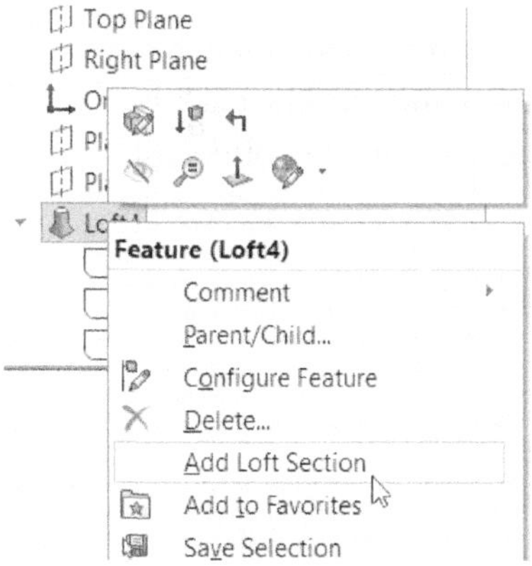

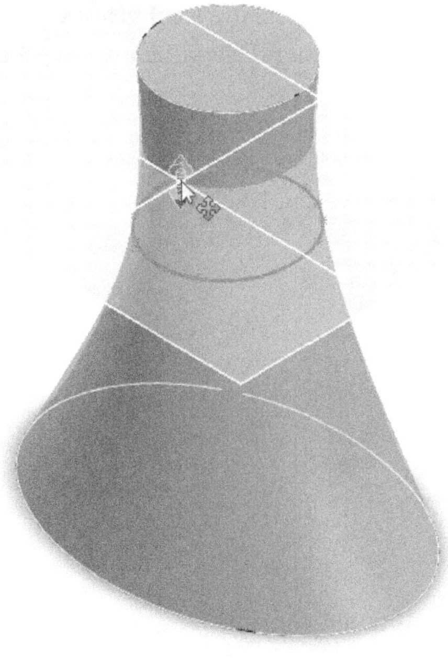

To edit the section, expand the **Loft** feature, right click on the sketch on the newly added section, and then select **Edit Sketch**; you will notice that a closed spline is created. You can edit the spline (or) delete it and create another sketch. Next, click **Exit Sketch** to update the model.

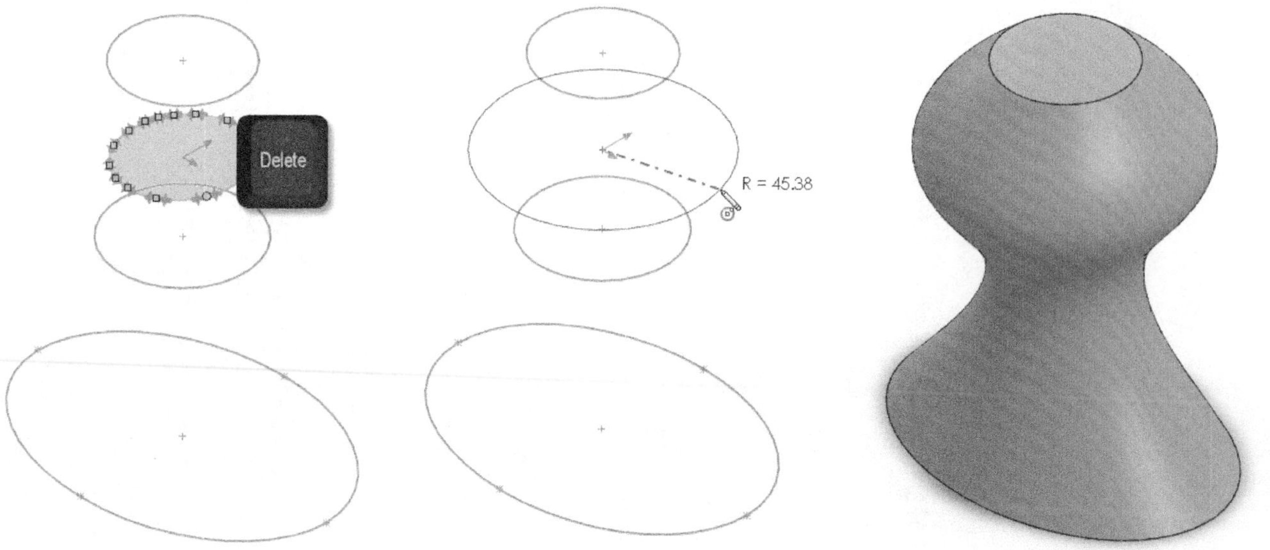

Lofted Cut

Like other standard features such as extrusion, revolve, and sweep, the loft feature can add material or remove material. You can remove material by using the **Lofted Cut** command. Activate this command by clicking **Features > Lofted Cut** on the CommandManager and select the profiles. Click **OK** to create the lofted cut.

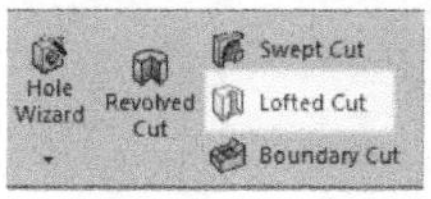

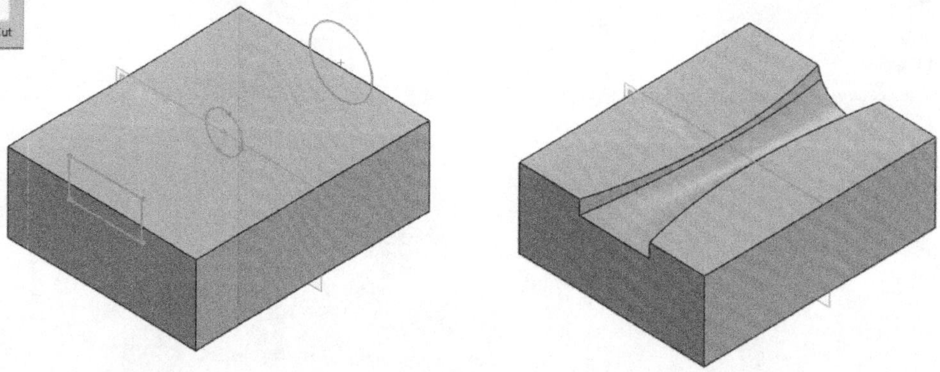

Best Practices

Plan Your Design: Before commencing, ensure a clear design plan is in place. Sketch the profiles intended for lofting to save time and prevent design errors.

Profile Sketching: Ensure profile sketches have the same number of vertices and consistent orientation to avoid twisting during lofting.

Maintain Tangency: Achieve smooth transitions by ensuring tangency or curvature continuity between profiles and guide curves to minimize shape changes.

Simplify Sketches: Keep sketches simple to streamline the lofting process; complex elements can complicate and introduce issues.

Cautions

Profile Compatibility: Exercise caution when lofting between profiles with varying shapes or dimensions to prevent quality issues.

Additional Tips for Specific Part Types

Plastic Parts:
Incorporate draft angles to facilitate part ejection from molds.

Casted Parts:
Account for casting shrinkage in lofted features design. Ensure geometry supports casting processes without trapping molten material.

Forged Parts:
Consider deformation during forging that can impact lofted features' shape.

Machined Parts:
Ideal for parts with intricate contours like aircraft components. Note that machining complex designs can be time-intensive.

3D Printed Parts:
Understand 3D printer limitations like minimum feature sizes and support needs. Utilize lofted features for unique shapes suitable for 3D printing, ensuring compatibility with your printer setup.

Examples

Example 1

In this example, you will create the part shown below.

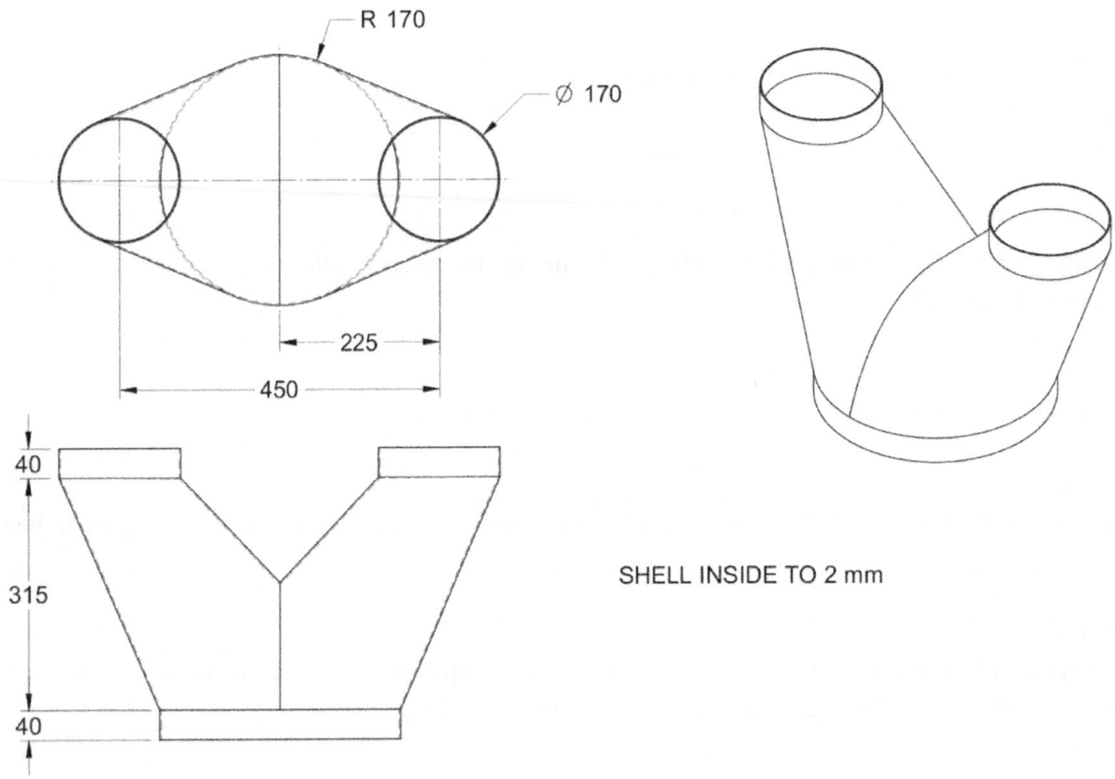

1. Start **SOLIDWORKS 2024**
2. Open a new part file.
3. Start a new sketch on the Top plane and draw a circle of 340 mm in diameter.

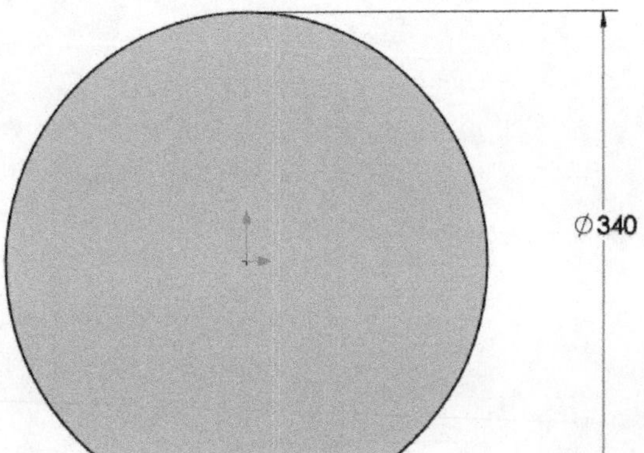

4. Exit the sketch.
5. Create the *Extruded Boss/Base* feature with 40 mm depth.

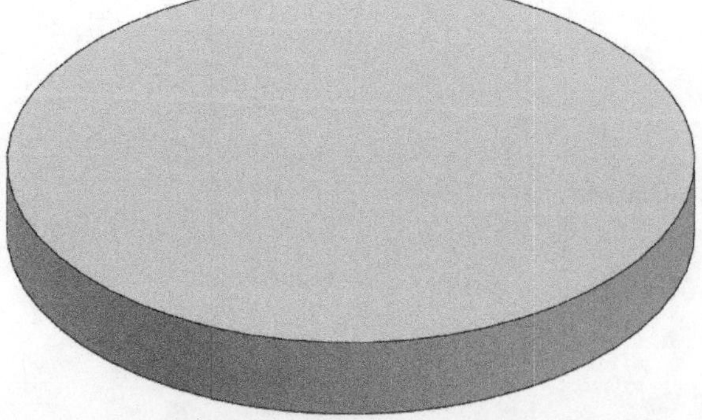

6. On the **Features** CommandManager, click the **Reference Geometry** drop-down > **Plane** icon.

7. Click on the top face of the geometry and type-in 315 in the **Offset distance** box.
8. Click **OK** to create an offset plane.
9. Start a sketch on the offset plane.
10. Draw a circle of 170 mm diameter and add dimensions to it.
11. Exit the sketch.

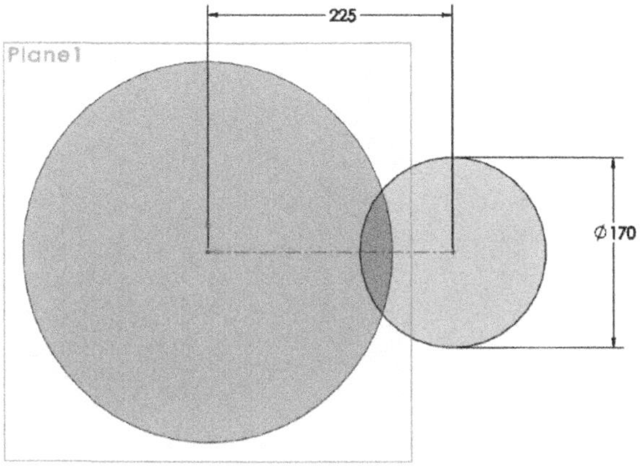

12. Deselect the sketch, if selected.
13. On the **Features** CommandManager, click the **Lofted Boss/Base** icon.
14. Click on the circle at the location, as shown in the figure.
15. Click on the circular edge of the base to define the second section. The location at which you click on the circular sketch should be the same as that of the first section.

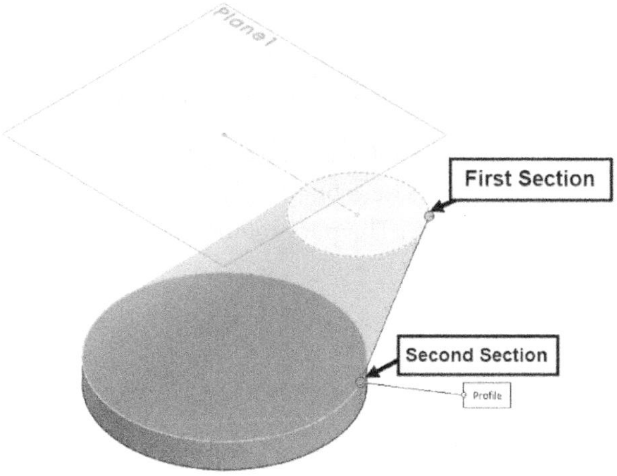

16. Click **OK** to create the *Loft* feature.

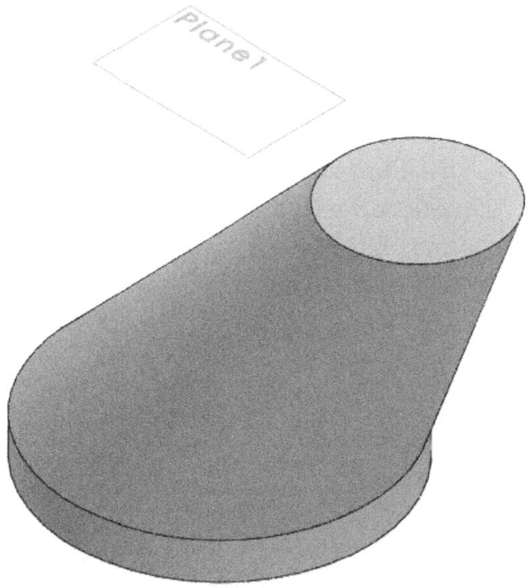

17. Activate the **Extruded Boss/Base** command and click on the top face of the *Loft* feature.
18. On the **Features** CommandManager, click on the **Convert Entities** icon.
19. Select the top circular edge of the loft feature.
20. Click **OK** on the PropertyManager.
21. Click **Exit Sketch** on the CommandManager.
22. Type-in 40 in the **Depth** box, and then click **OK** to create the *Extruded Boss/Base* feature.

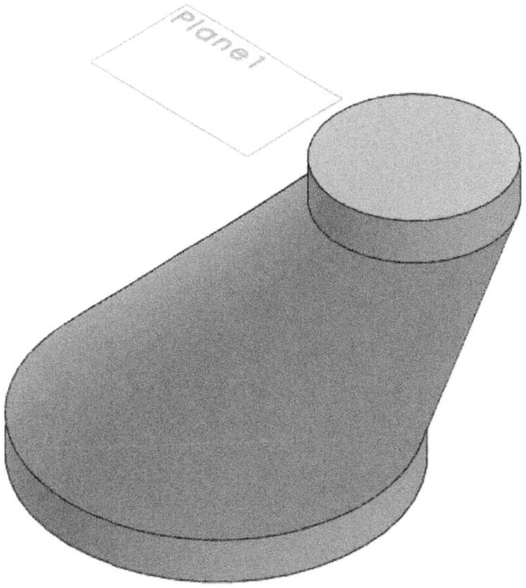

23. On the **Features** CommandManager, click the **Mirror** icon, and then select the Right plane from the FeatureManager Design Tree to define the mirroring element.
24. Select the Loft1 and Boss-Extrude2 from the FeatureManager Design Tree.
25. Check the **Geometry Pattern** option under the **Options** section.
26. Click **OK** to mirror the entire solid body.

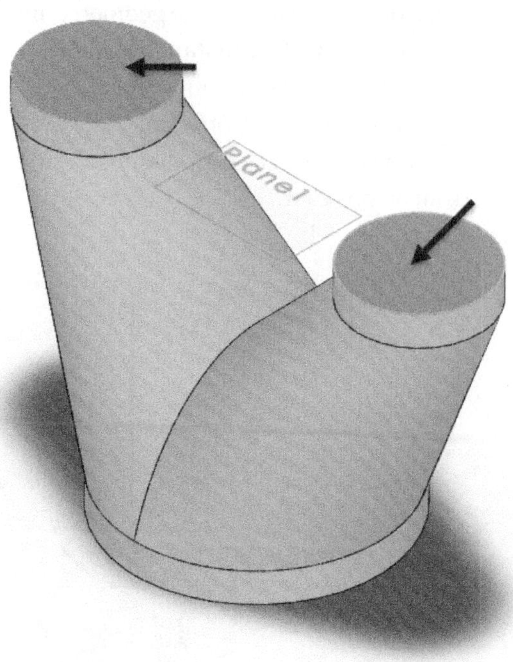

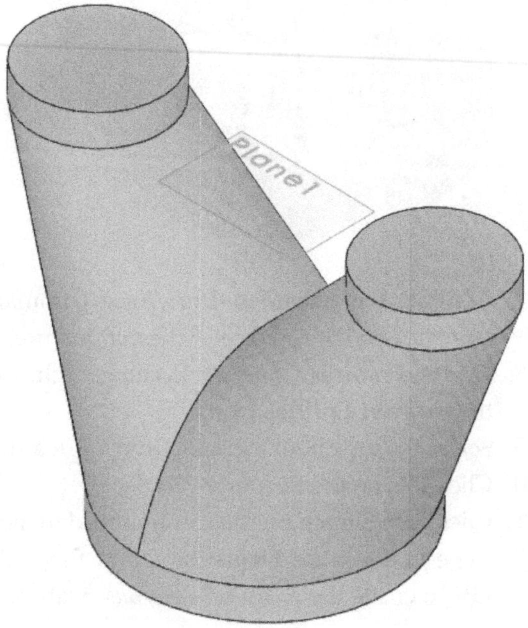

27. On the **Features** CommandManager, click the **Shell** icon.
28. Click on the flat faces of the model geometry.

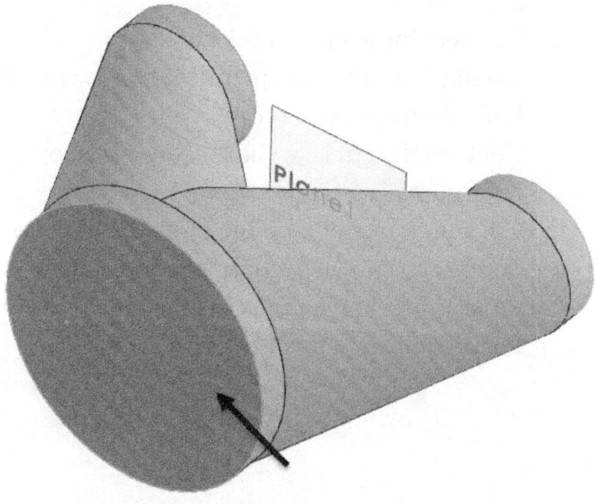

29. On the **Shell1** PropertyManager, type-in 2 in the **Thickness** box.
30. Click **OK**. The part geometry is shelled.

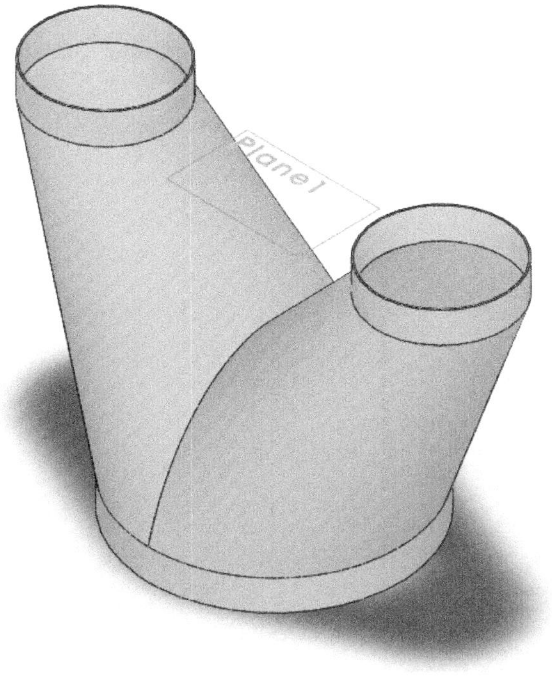

31. Save and close the part file.

Example 2 (Inches)

In this example, you will create the part shown below.

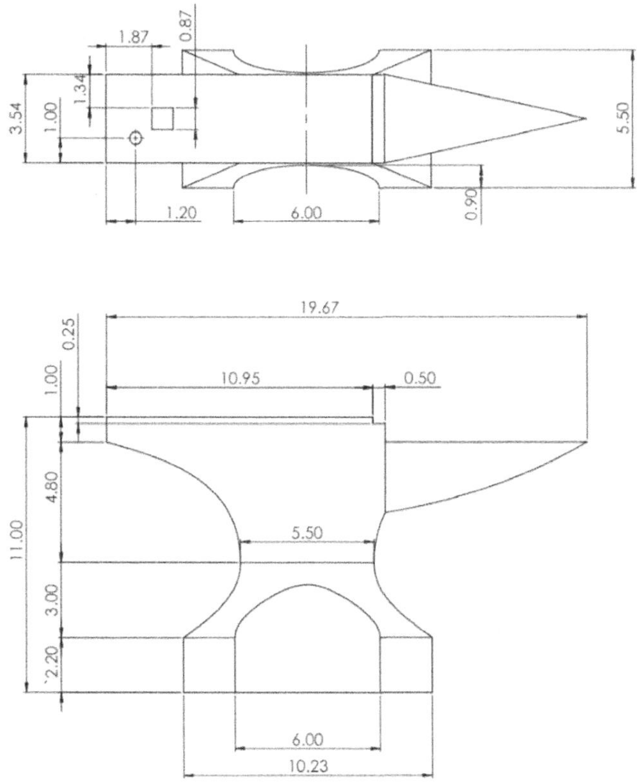

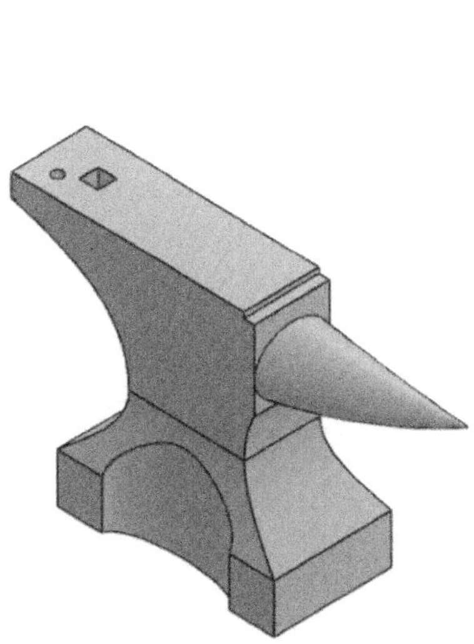

1. Open a new file in the **Part** Environment.
2. On the Status bar, click on the MMGS drop-down and select IPS (inch-pound-second).
3. On the CommandManager, click **Sketch > Sketch.** Next, select the top plane to start the sketch.

4. On the CommandManager, click **Sketch > Rectangle > Center Rectangle.**
5. Click on the origin point to define the center point of the rectangle. Move the mouse pointer diagonally upward and click to draw a rectangle.
6. Activate the **Smart Dimension** command and apply dimensions to the sketch, as shown below. Click **Exit Sketch** on the CommandManager.

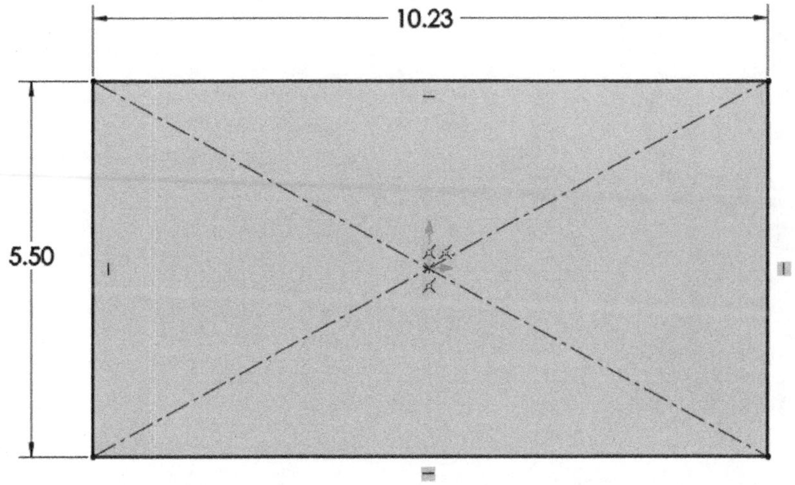

7. On the **View (Heads-Up)** toolbar, click **View Orientation > Isometric**.
8. Click **Features > Extruded Boss/Base** on the CommandManager and select the sketch (if not already selected).
9. On the **Boss-Extrude** PropertyManager, select **End Condition > Blind** and type-in 2.2 in the **Depth** box. Next, click **OK** to create the *Extrude* feature.

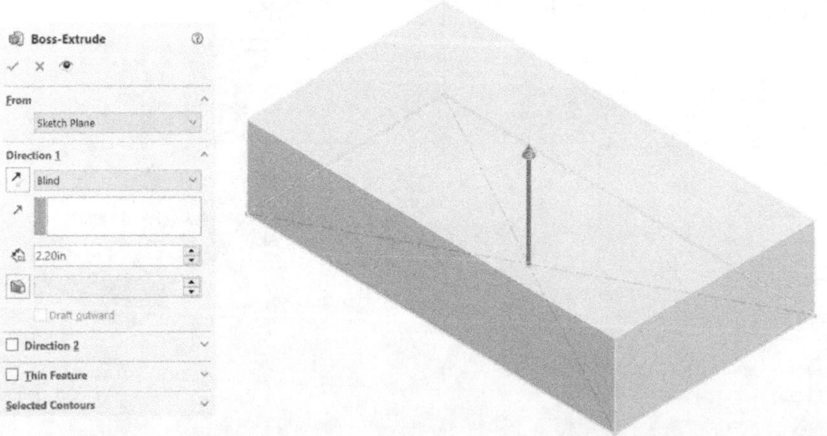

10. Click **Features** tab > **Reference Geometry > Plane** on the CommandManager and click on the top face of the model.
11. On the **Plane** PropertyManager, type-in 3 in the **Offset distance** box and click **OK** to create an offset plane.

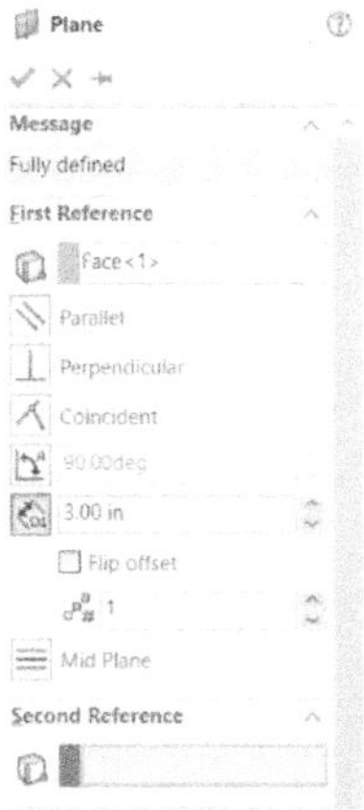

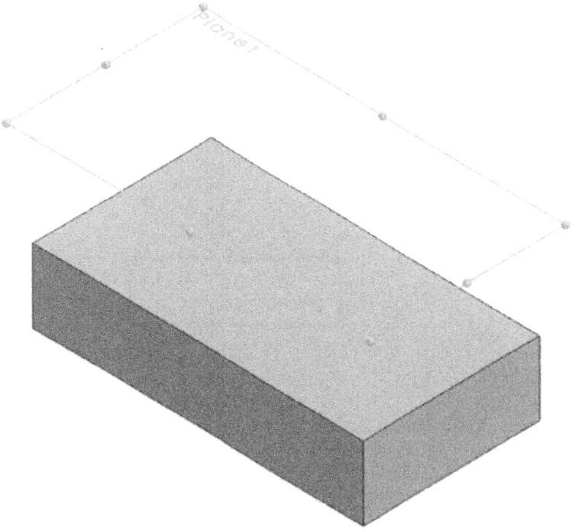

12. Click **Sketch > Sketch** on the CommandManager and click on the newly created plane.
13. On the **View (Heads-Up)** toolbar, click **View Orientation > Normal To** to change the view orientation normal to the sketch plane.
14. Click **Sketch > Rectangle > Center Rectangle** on the CommandManager.
15. Click on the origin point to define the center point of the rectangle. Move the mouse pointer upward and click to draw a rectangle.
16. Activate the **Smart Dimension** command and apply dimensions to the sketch, as shown in the figure.
17. Click **Exit Sketch** on the CommandManager and change the view orientation to Isometric.

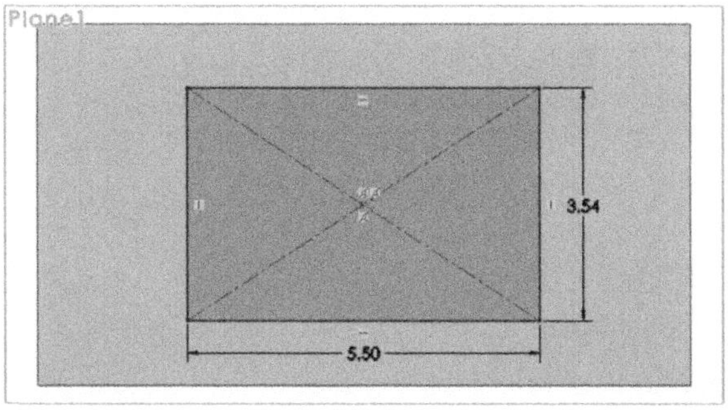

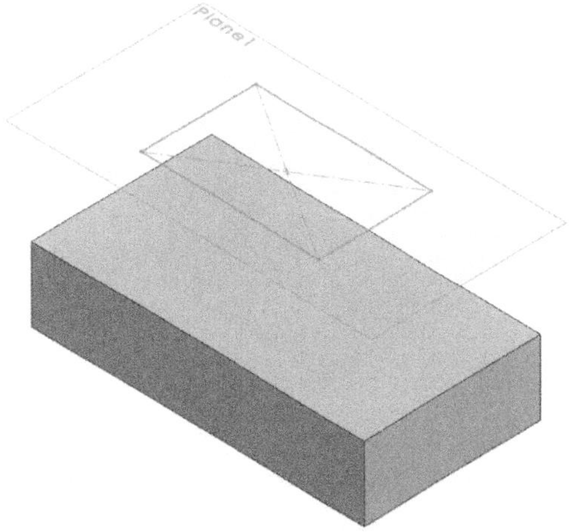

18. Click **Features > Loft Boss/Base** on the CommandManager; the **Loft** PropertyManager appears on the screen.
19. Click the top face of the first feature at the right corner, as shown.
20. Select the rectangular sketch by clicking on the right corner, as shown.

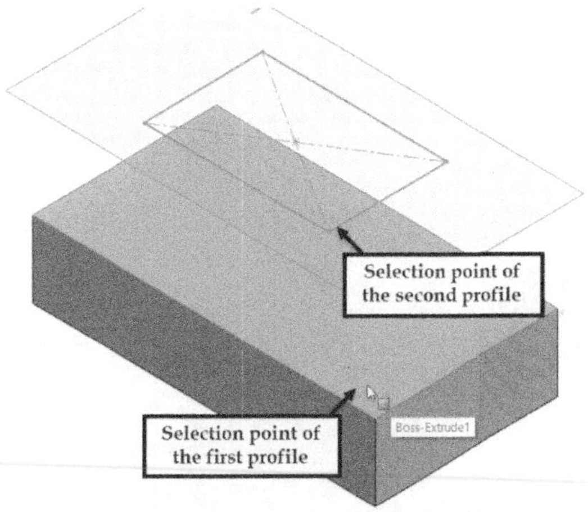

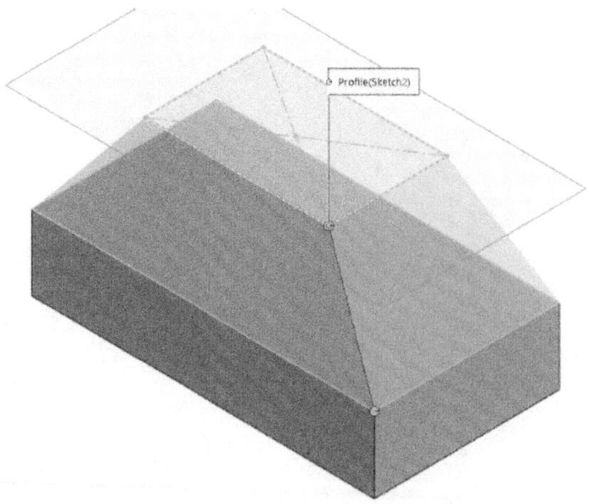

21. On the PropertyManager, expand the **Start/End Constraints** section.
22. Click **Start Constraint > None** and **End Constraint > Normal To Profile**.
23. Type-in 0 and 1 in the **Draft Angle** and **End Tangent Length** boxes, respectively. Click **OK** on the PropertyManager to create the **Loft** feature.

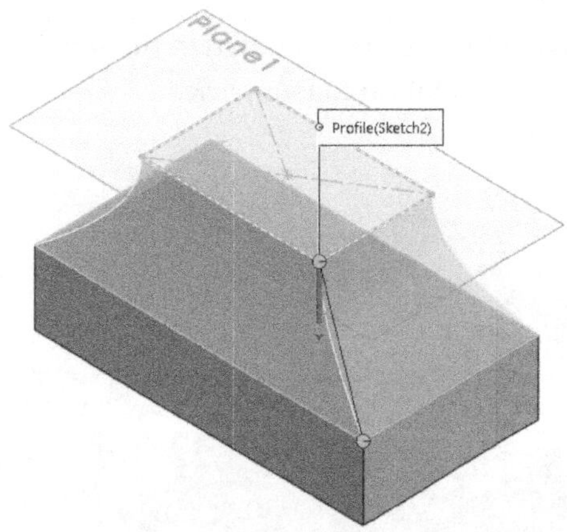

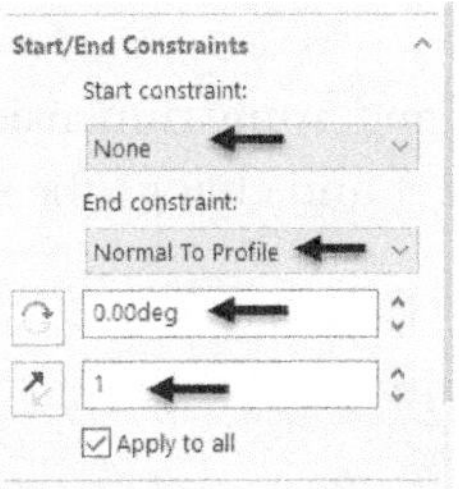

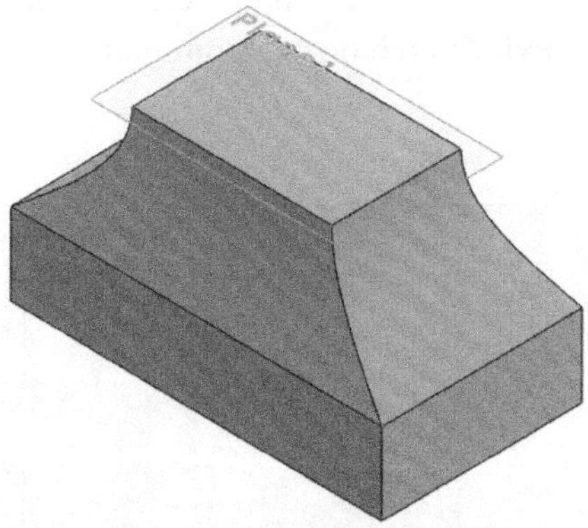

24. Click **Sketch > Sketch** on the CommandManager, and then select the top face of the loft feature.
25. On the **View(Heads-Up)** toolbar, click **View Orientation** drop-down > **Normal To**.
26. On the CommandManager, click **Sketch** tab > **Ellipse**. Next, check the **Add construction lines** option on the **Options** section of the PropertyManager.
27. Place the pointer on the bottom horizontal edge and select the midpoint. Move the mouse pointer upward and click to specify the first axis of the ellipse. Next, move the pointer toward the right and click to create the second axis of the ellipse. Press Esc to select the ellipse.

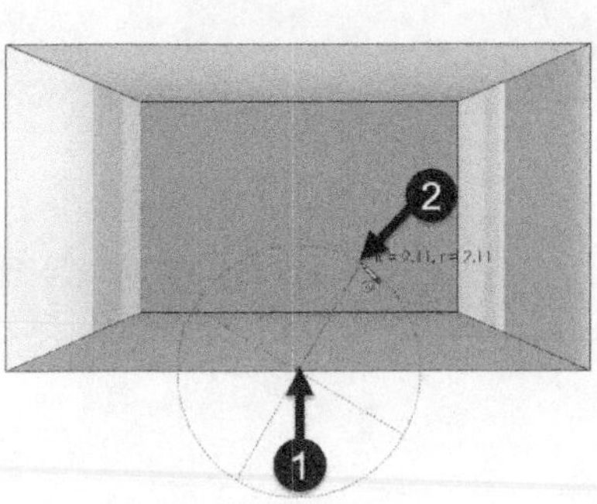

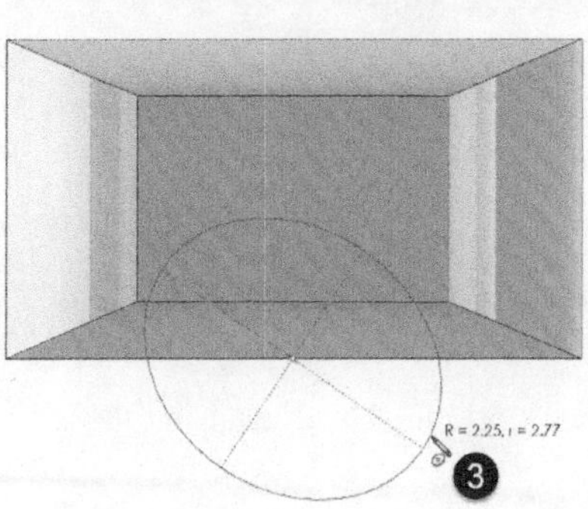

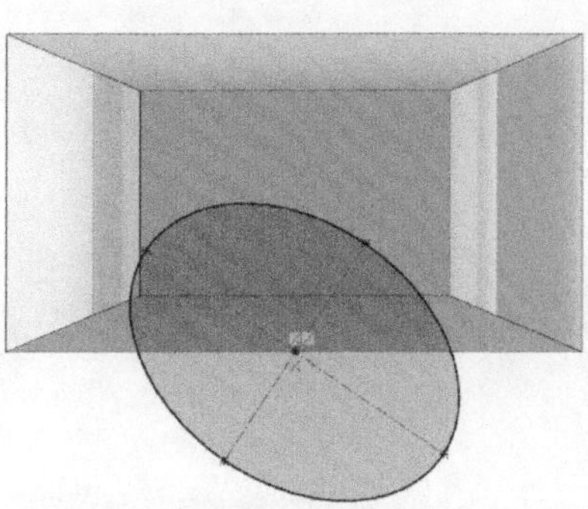

28. Click **Display/Delete Relations > Add Relations** on the CommandManager.
29. Select the top and bottom quadrant points of the ellipse, as shown.
30. On the PropertyManager, click the **Vertica**l icon, and then click **OK**.

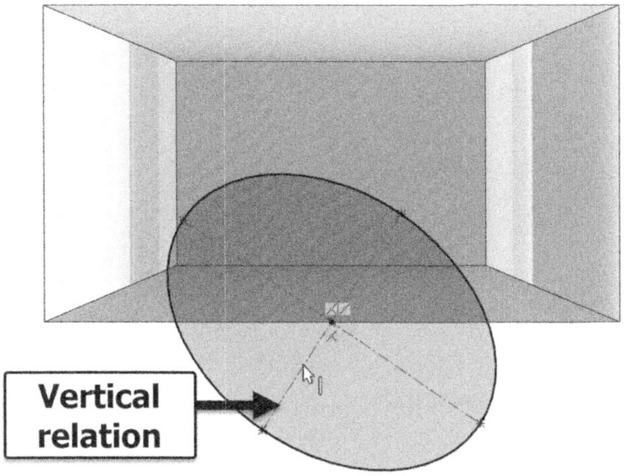

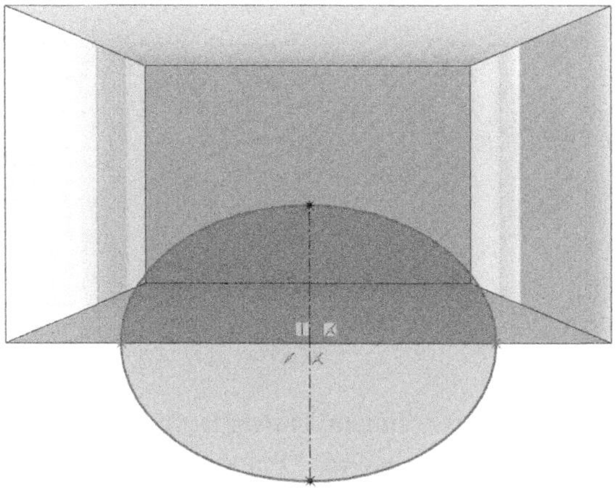

31. Click **Sketch > Smart Dimension** on the CommandManager and select the horizontal axis, as shown. Place the dimension, type 6 in the dimension box, and press Enter.

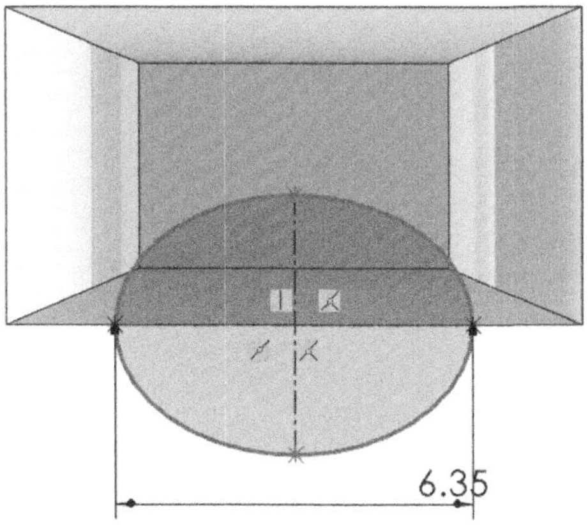

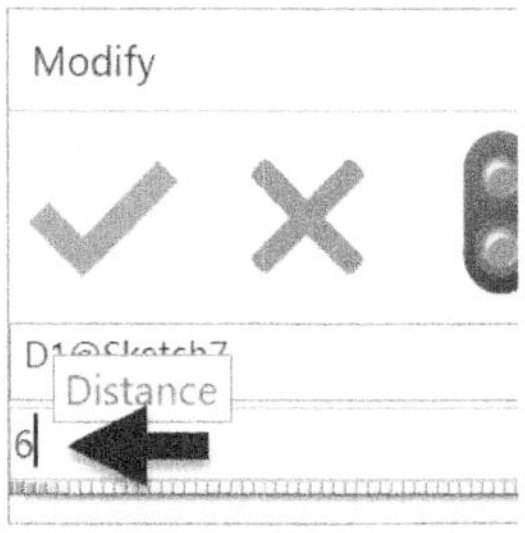

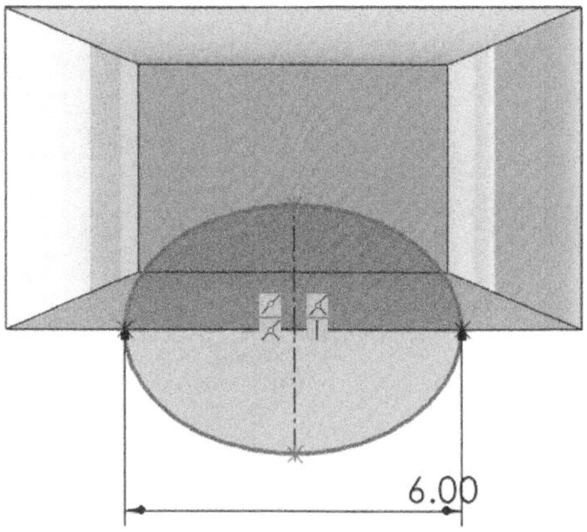

32. Select the remaining two vertical quadrant points and place the dimension. Type 1.8 in the dimension box and press Enter.

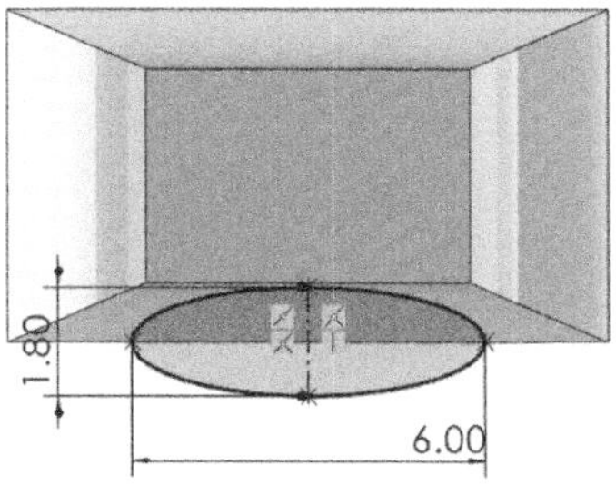

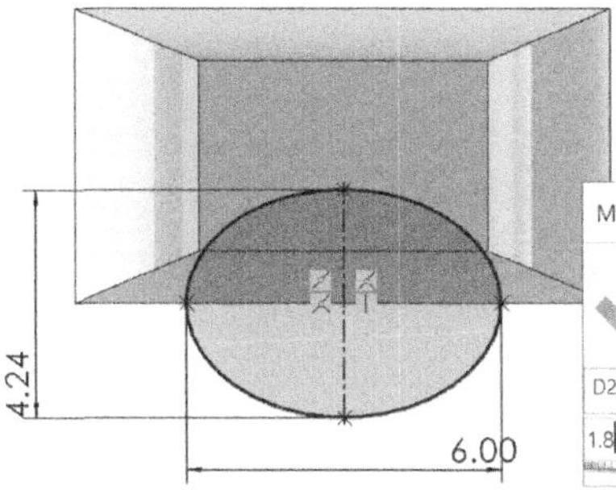

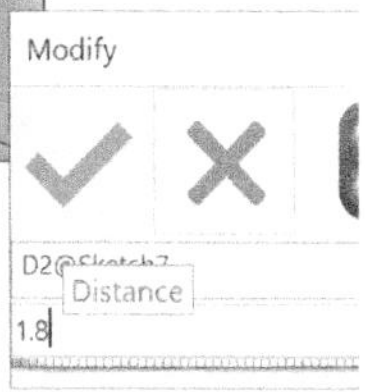

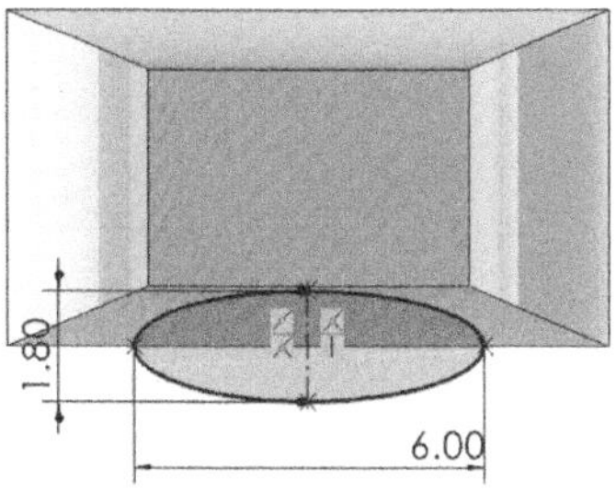

33. Make sure that the midpoint of the *Ellipse* is coincident with the horizontal edge of the *Extrude* feature. Click **Exit Sketch** on the CommandManager.
34. On the **View (Heads-Up)** toolbar, click **View Orientation > Isometric**.

35. Click **Features > Extruded Cut** on the CommandManager and select the sketch.
36. On the **Cut-Extrude** PropertyManager, select **End Condition > Through All**. Click **OK** on the PropertyManager to create the extruded cut.

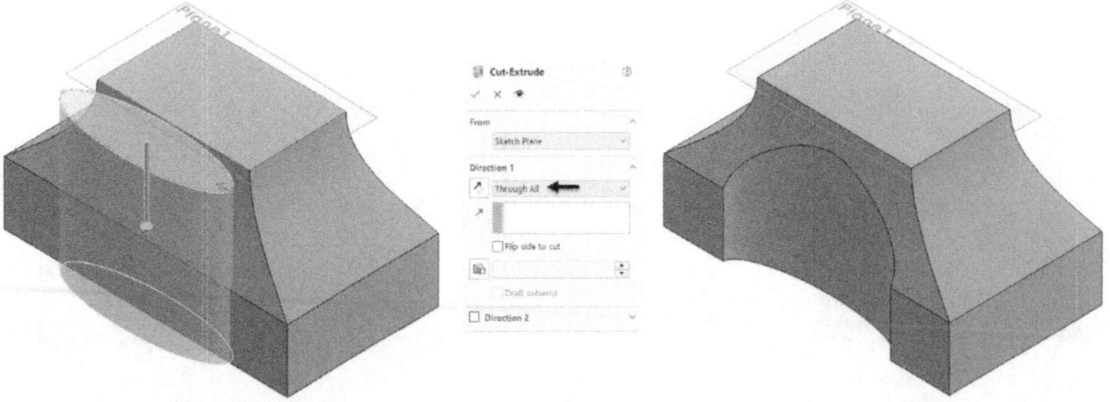

37. On the CommandManager, click **Features** tab > **Mirror**. Next, expand the FeatureManager Design Tree and select the Front Plane.
38. Click in the **Features to Mirror** selection box, and then select the *Extruded Cut* feature. Click **OK** to create the mirror feature.

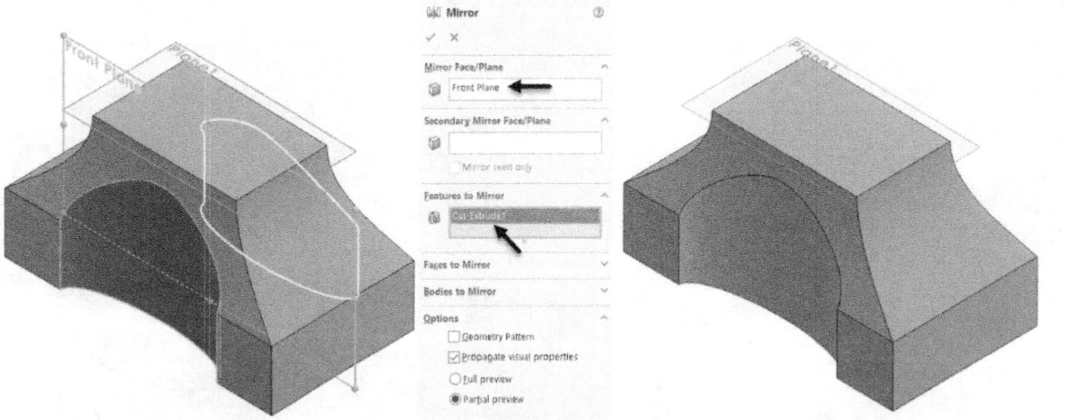

39. Click on **Plane1** displayed on the top face of the model and select the **Hide** icon.
40. Click **Features > Reference Geometry > Plane** on the CommandManager and click on the top face of the loft feature. Enter 4.8 in the **Offset distance** box and click **OK**.

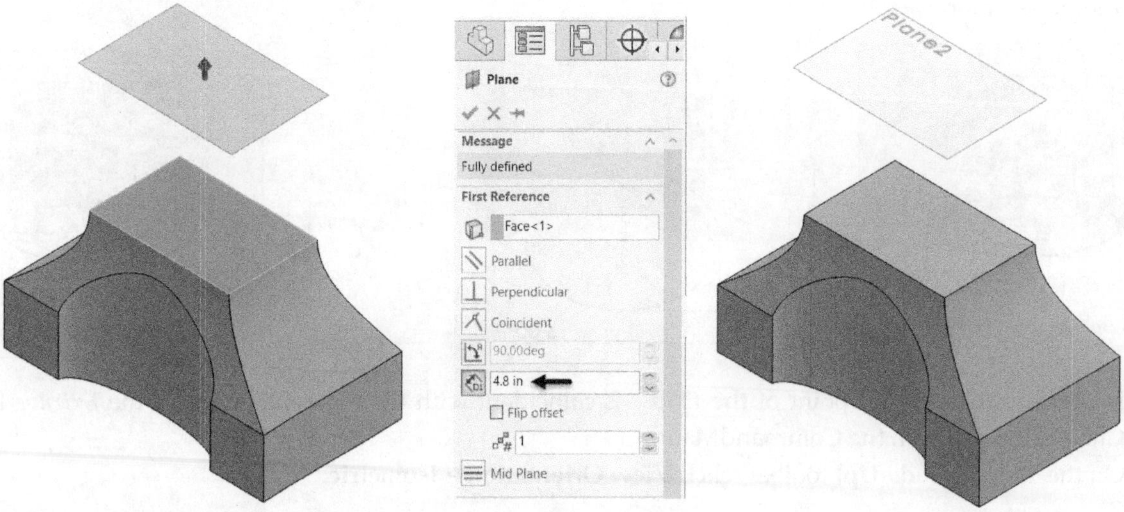

41. Click **Sketch > Rectangle > Center Rectangle** on the CommandManager and click on the newly created plane.
42. On the **View (Heads-Up)** toolbar, click **View Orientation > Normal To** to change the view orientation normal to sketch.
43. Create a rectangle and add dimensions to it, as shown.

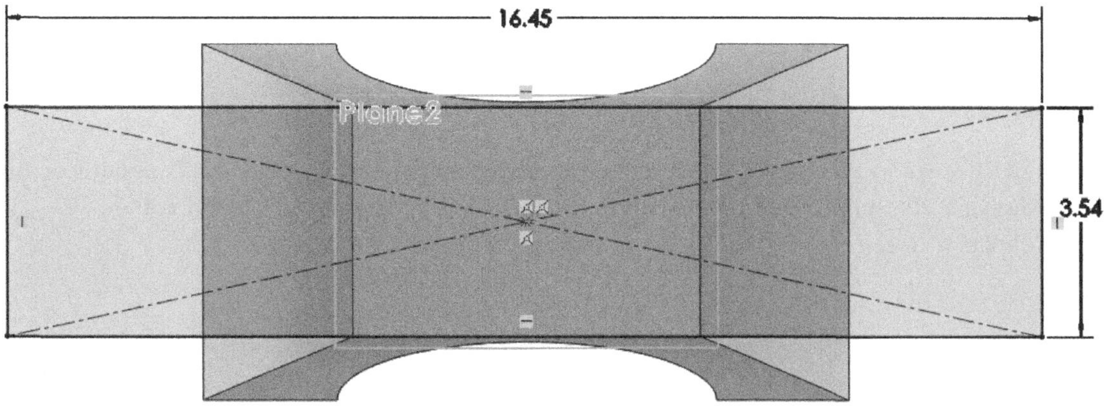

44. Click **Exit Sketch** on the CommandManager.
45. On the **View (Heads-Up)** toolbar, click **View Orientation > Isometric**.
46. Click **Features > Reference Geometry > Plane** on the CommandManager. Next, expand the FeatureManager Design Tree and select the Front Plane.
47. Click on the top right corner of the loft feature, as shown.
48. Likewise, create another plane at the backside.

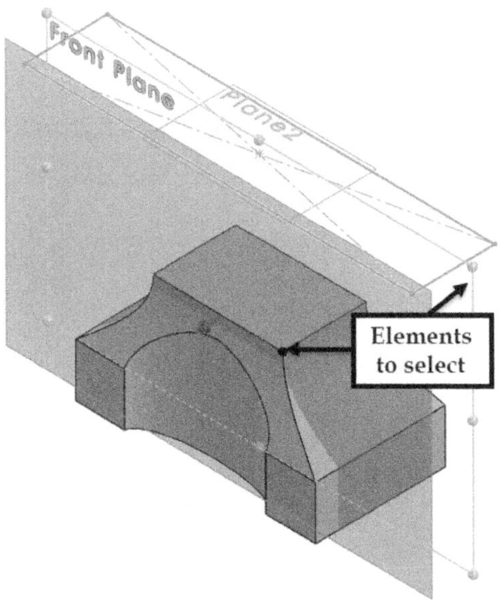

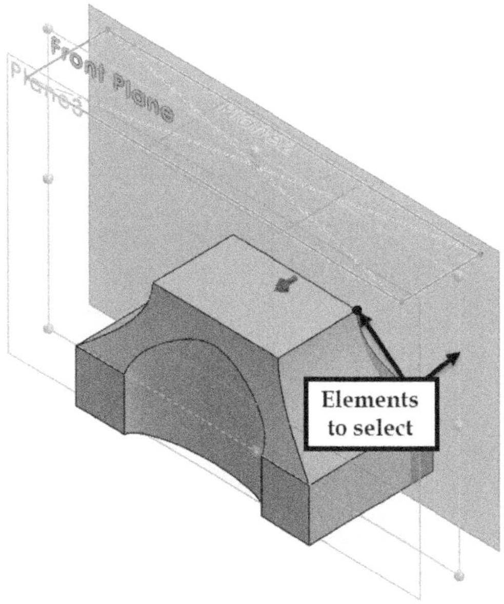

49. Click **Sketch > Sketch** on the CommandManager and select the plane created on the front side.
50. On the **View (Heads-Up)** toolbar, click **View Orientation > Normal To** to change the view orientation normal to sketch.
51. Click **Sketch > Convert Entities** on the CommandManager and click the edge, as shown. Click **OK** on the PropertyManager.

52. Click **Sketch > Spline > Spline** on the CommandManager and specify the points, as shown. Next, right-click and select **Select**.

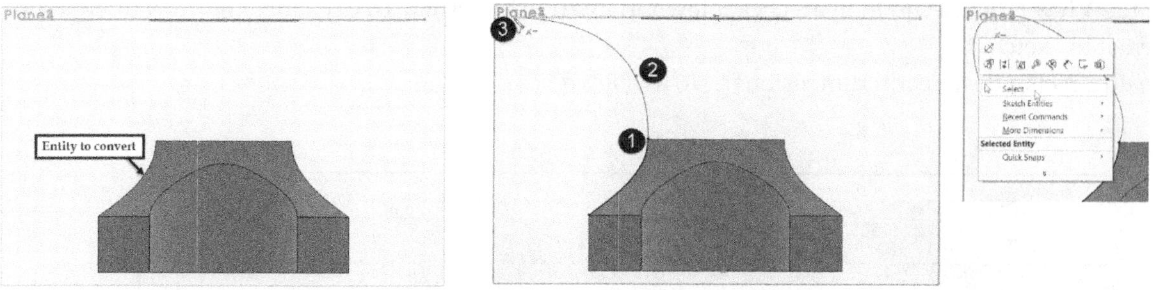

53. Press the CTRL key and click on the converted entity and the spline. Next, click the **Tangent** icon on the PropertyManager; the spline will become tangent to the converted entity. Click **OK** on the PropertyManager.

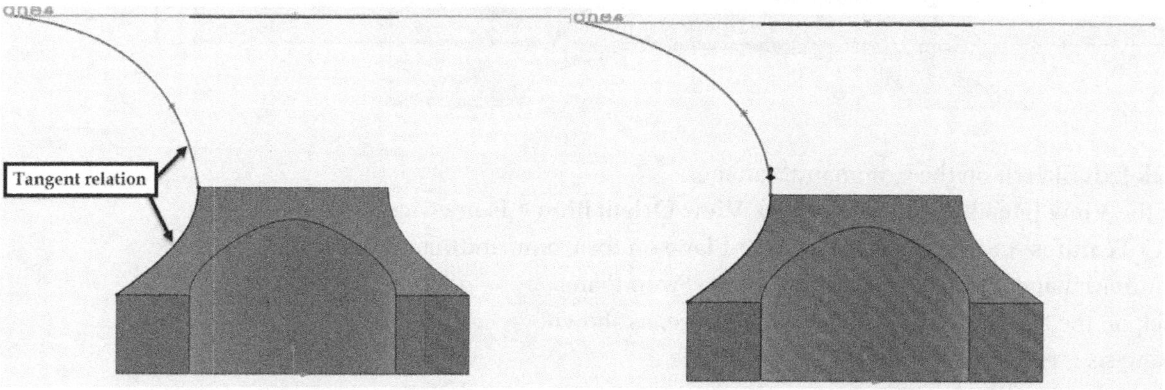

54. Zoom-in to the top-end point of the spline and select it. Next, press the Ctrl key and select the rectangular sketch.
55. Click the **Pierce** icon on the PropertyManager, and then click **OK**.
56. On the CommandManager, click **Sketch > Smart Dimension**, and then apply the linear dimensions to the middle point of the spline, as shown.

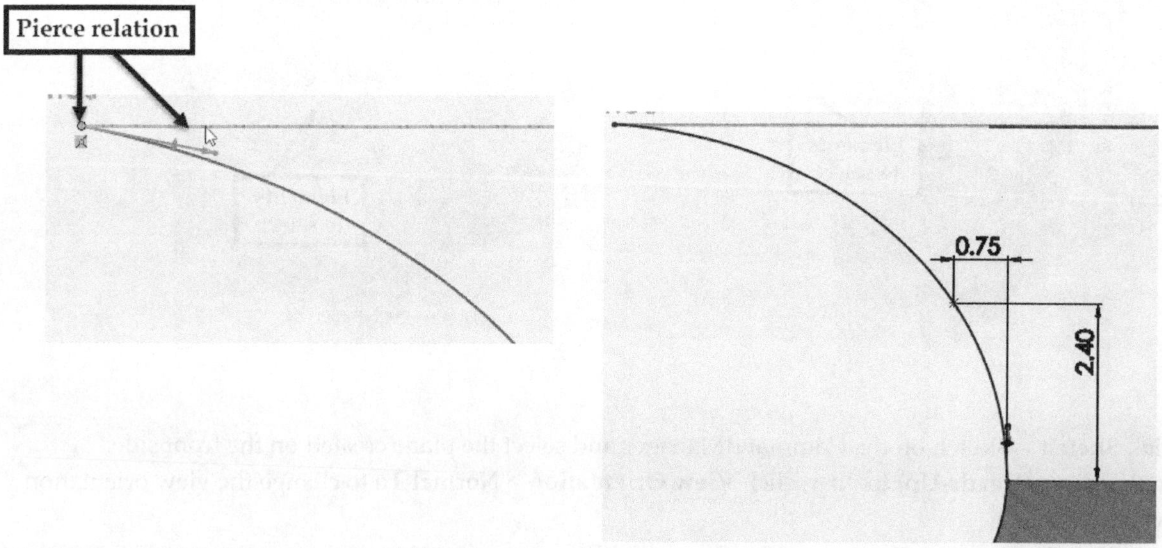

57. Zoom-in to the spline handle displayed at the bottom. Next, activate the **Smart Dimension** command and select the Tangency Magnitude Handle.

58. Move the pointer and place the dimension. Type-in 3 in the **Modify** box and press Enter.

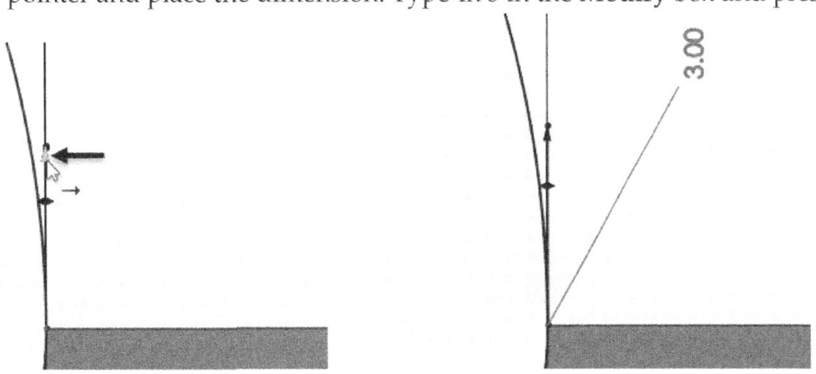

59. Likewise, convert the right edge and then draw a spline connected to it. Next, apply the **Tangent** relation between the converted entity and the spline. In addition to that, create the Pierce relation between the spline's endpoint and the rectangular sketch.
60. Apply dimensions to the spline and its Tangency Magnitude Handle. Click **Exit Sketch** on the CommandManager.

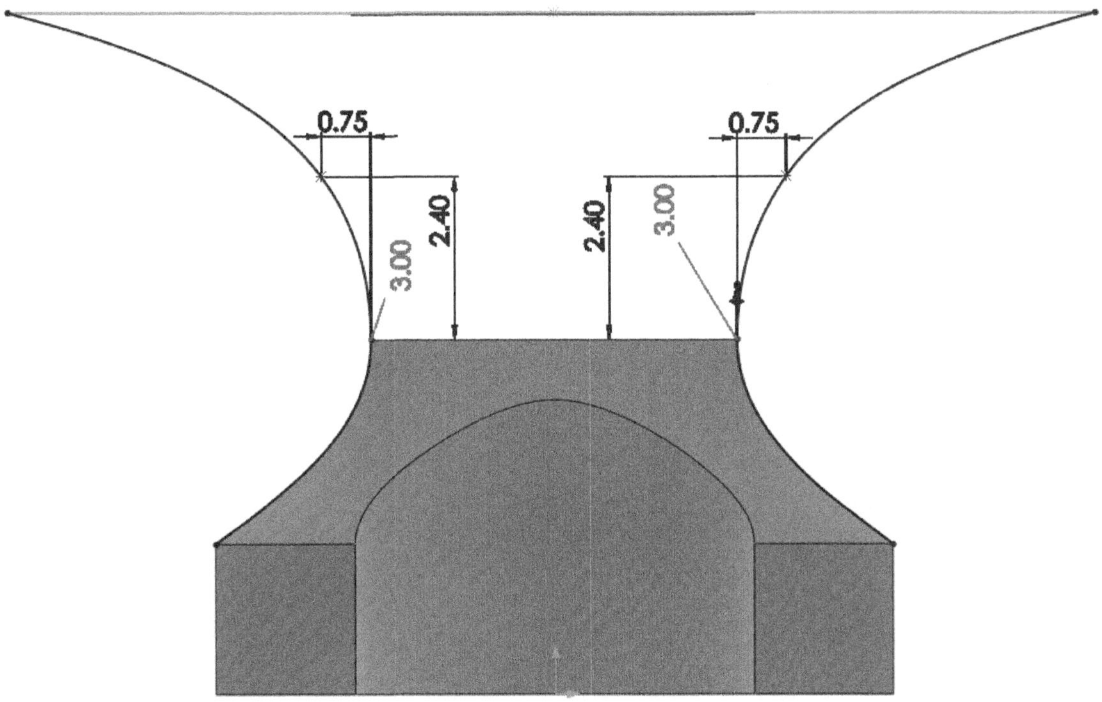

61. On the **View (Heads-Up)** toolbar, click **View Orientation > Isometric**.
62. Click **Sketch > Sketch** on the CommandManager and select the plane displayed on the backside.
63. Click **Sketch > Convert Entities** on the CommandManager and select the entities of the last sketch, as shown. Click **Exit Sketch** on the CommandManager.

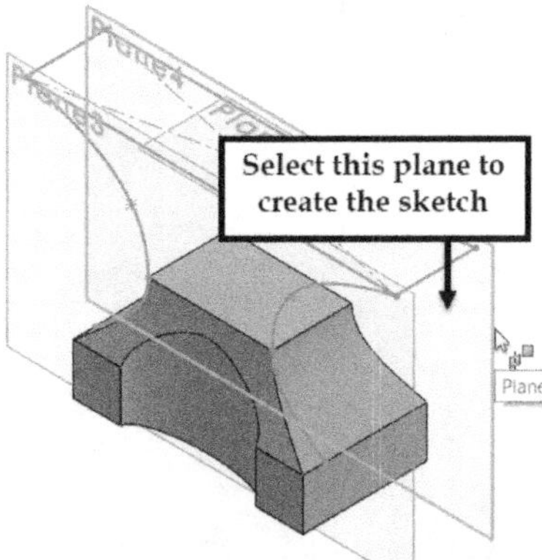

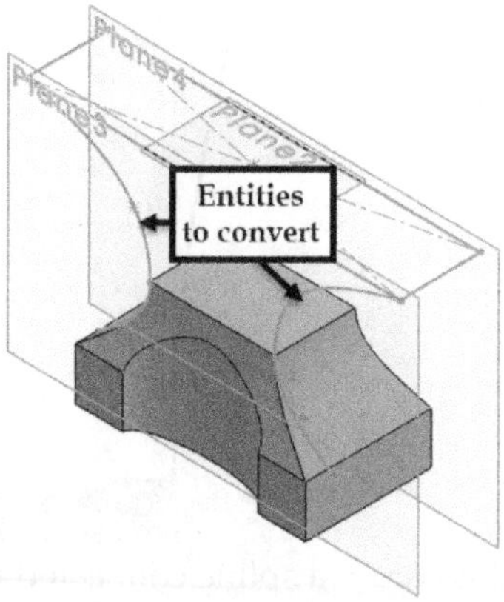

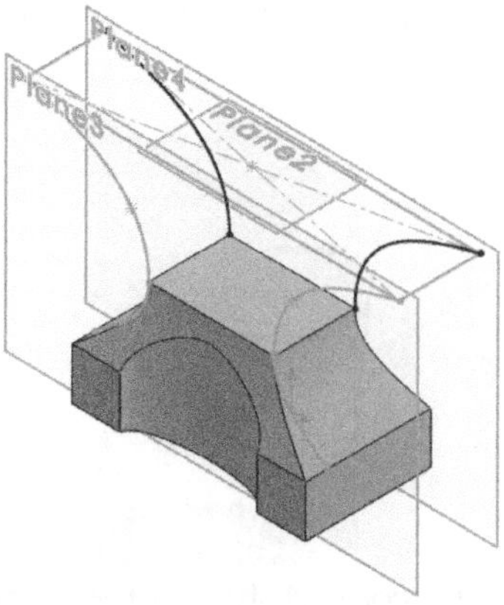

64. Click **Features > Lofted Boss/Base** on the CommandManager; the **Loft** PropertyManager appears on the screen.
65. Select the first profile by clicking on the model's top face at the location, as shown.
66. Select the second profile by clicking on the corner point of the rectangle, as shown.
67. On the **Loft** PropertyManager, click in the **Guide Curves** selection box and select the first spline.
68. On the Pop-up toolbar, click the **Select Open Loop** icon, and then click **OK**.

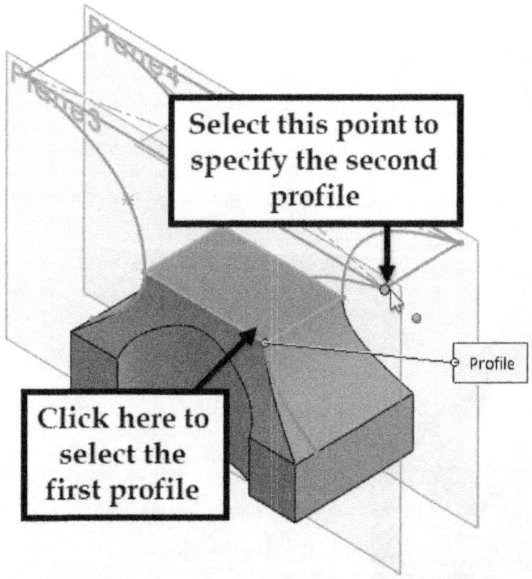

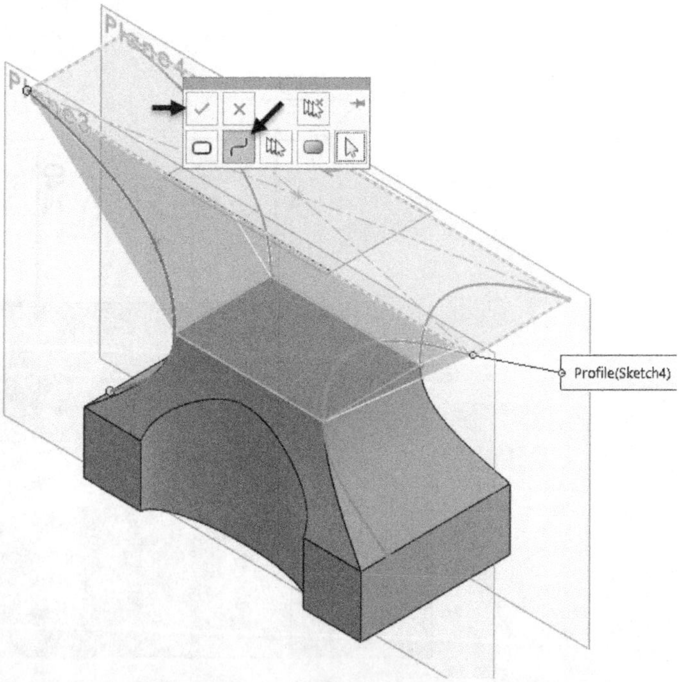

69. Likewise, select the other splines and click **OK** on the **Loft** PropertyManager to create the loft.

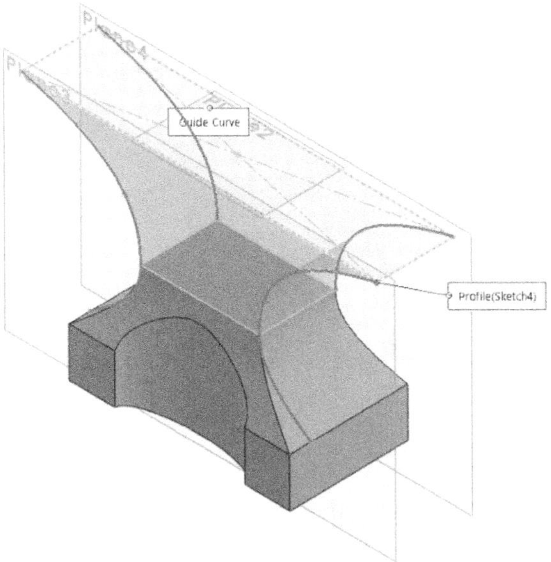

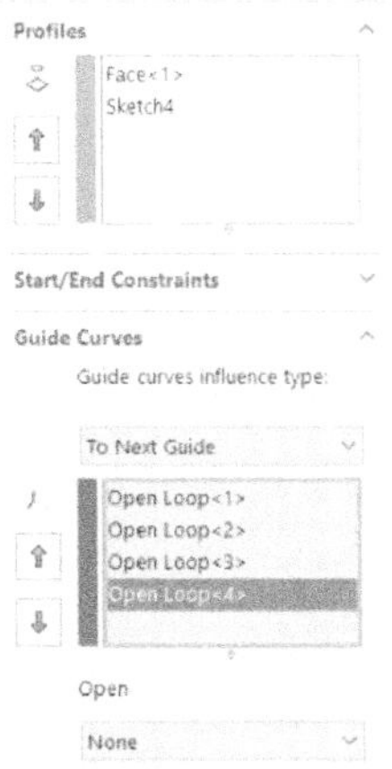

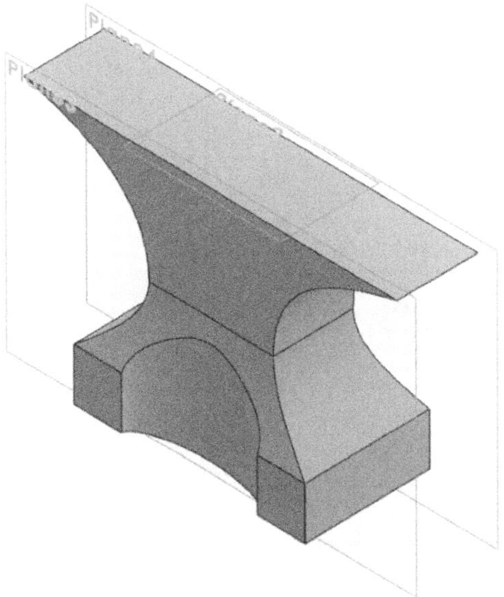

70. Click **Sketch > Center Rectangle > Corner Rectangle** on the CommandManager and click on the top face of the model.
71. Select the corner point of the model, as shown. Next, move the pointer, and then click on edge, as shown.
72. Activate the **Smart Dimension** command and apply the dimension to the rectangle, as shown.
73. Click **Exit Sketch** on the CommandManager.

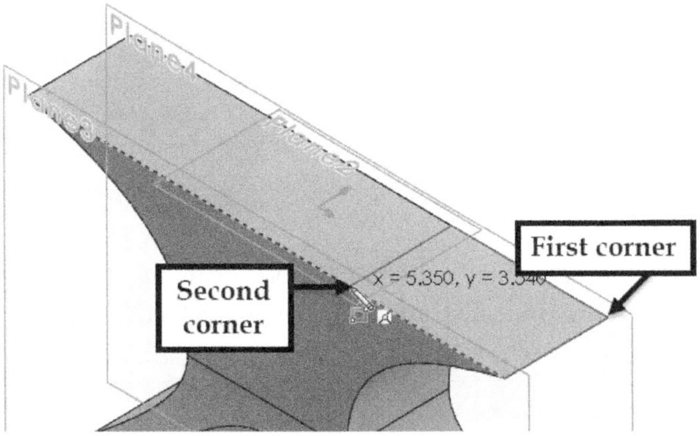

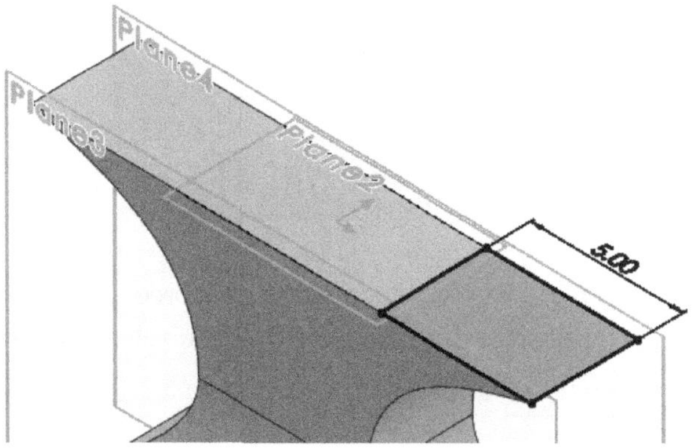

74. Activate the **Extruded Cut** command (on the CommandManager, click **Features** tab > **Extruded Cut**) and select the sketch.
75. On the **Cut-Extrude** PropertyManager, under the **Direction 1** section, select **End Condition > Blind** and type-in 3.5 in the **Depth** box. Next, click **OK** to create the extruded cut feature.

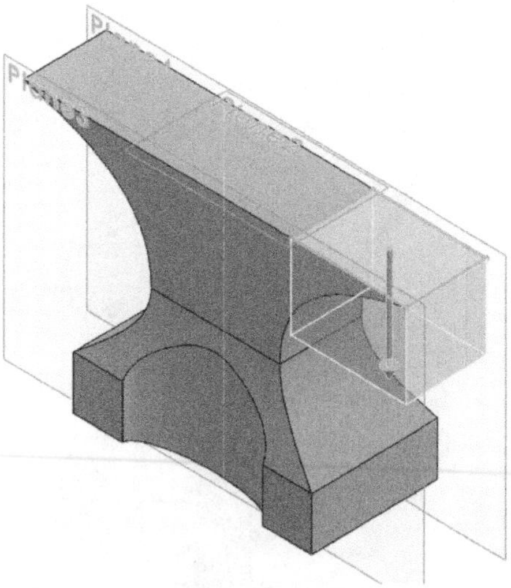

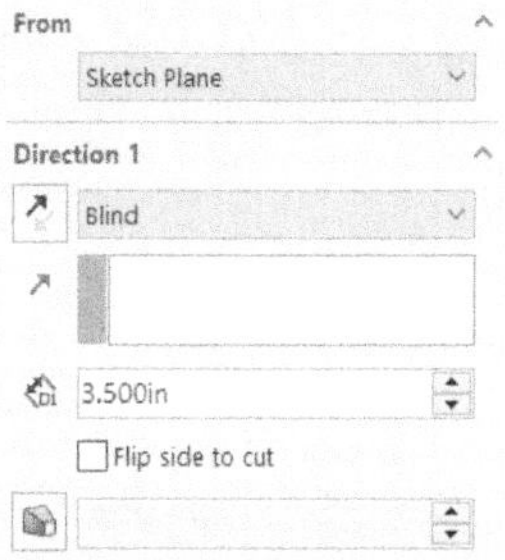

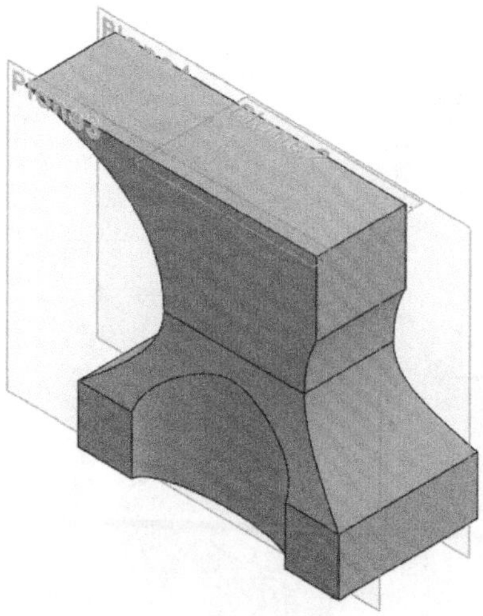

76. Click **Features > Extruded Boss-Base** on the CommandManager, and then click on the top face of the model.
77. On the CommandManager, click **Sketch > Rectangle** drop-down > **Corner Rectangle**.
78. Specify the first and second corners of the rectangle, as shown. Next, click **Exit Sketch**.
79. On the **Boss-Extrude** PropertyManager, type 1 in the **Depth** box, and then click **OK**.

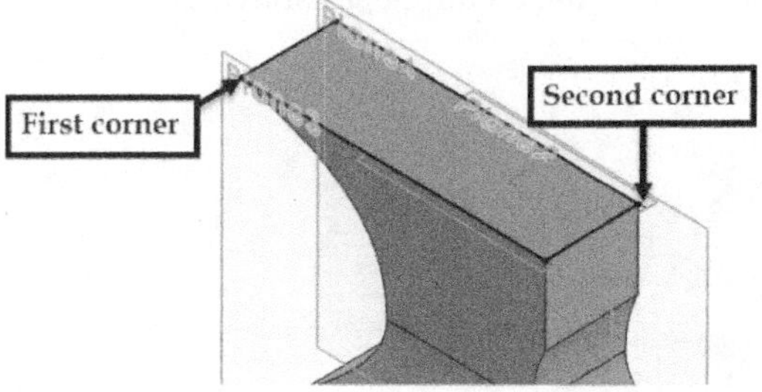

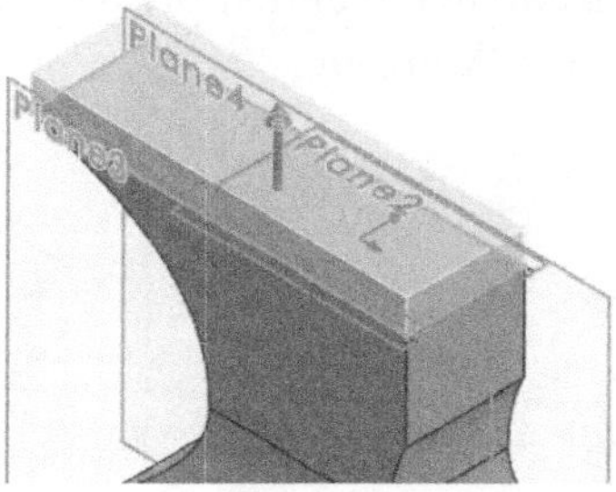

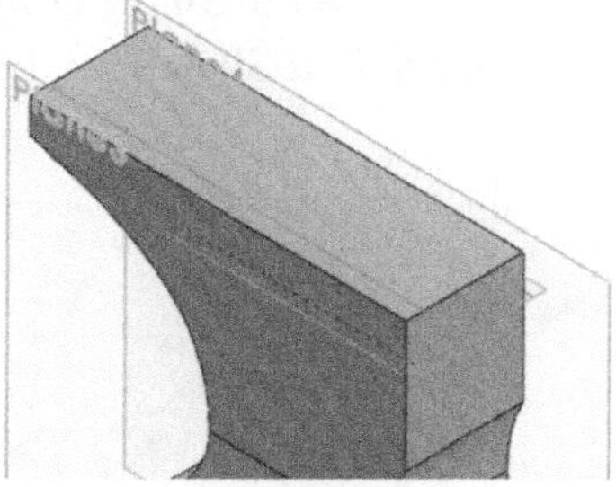

80. In the FeatureManager Design Tree, press and hold the left mouse button on the bar located at the bottom, and then drag it upward. Next, release the bar above the **Boss-Extrude2** feature; the extruded feature is suppressed.

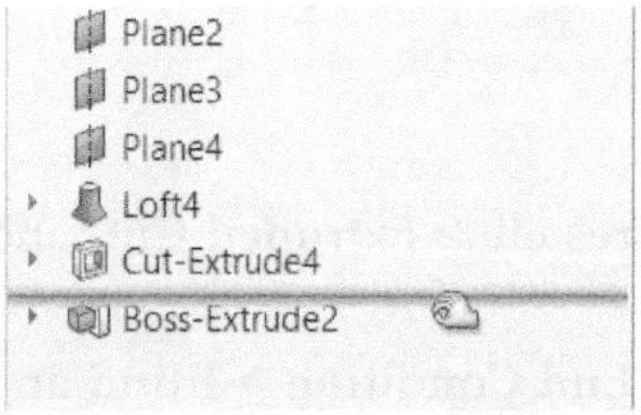

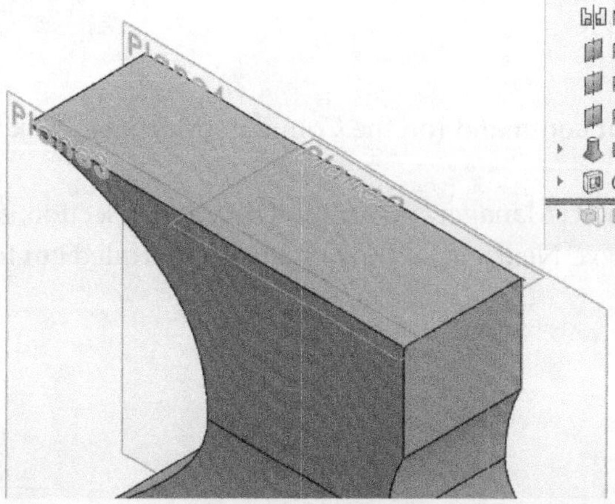

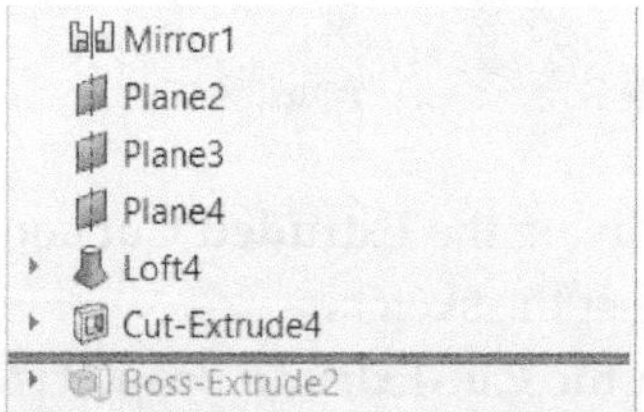

81. Click **Sketch > Ellipse** on the CommandManager and click on the right face, as shown.
82. Create the ellipse by specifying its center and quadrant points, as shown.

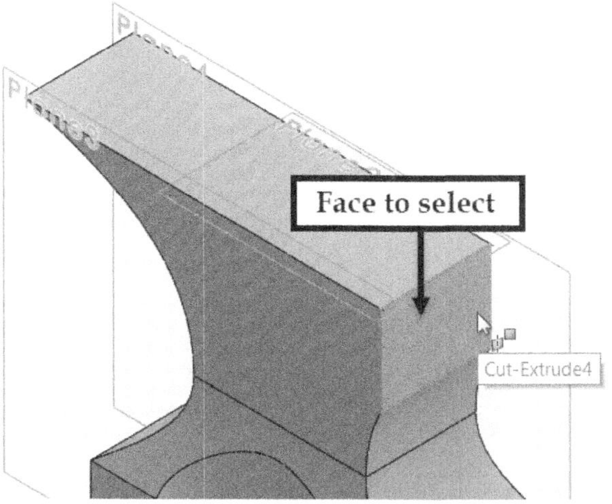

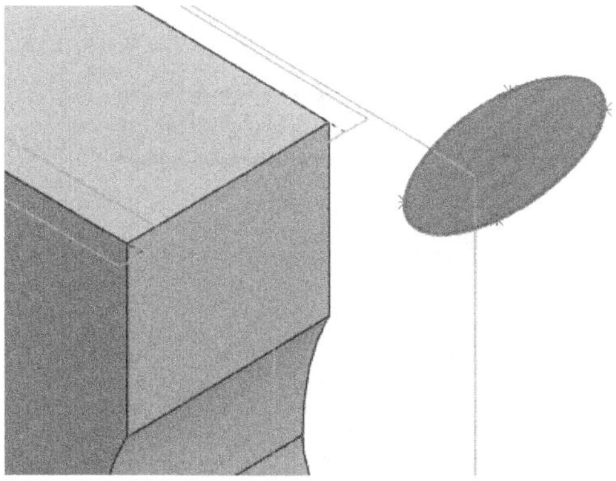

83. Press the Ctrl key and select the lower quadrant point and the sketch plane's lower horizontal edge, as shown.
84. On the PropertyManager, click the **Midpoint** icon, and then click **OK**.

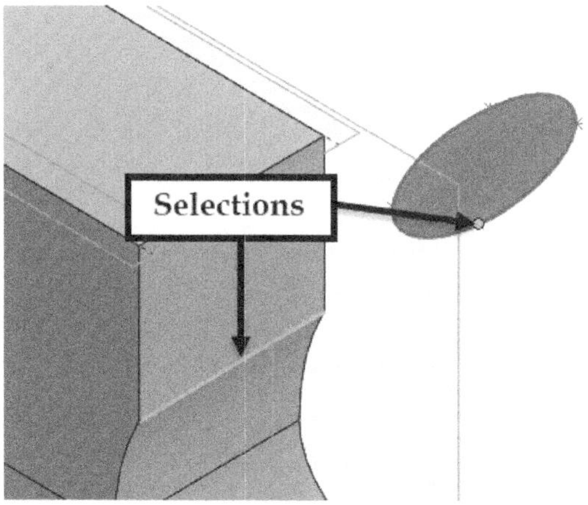

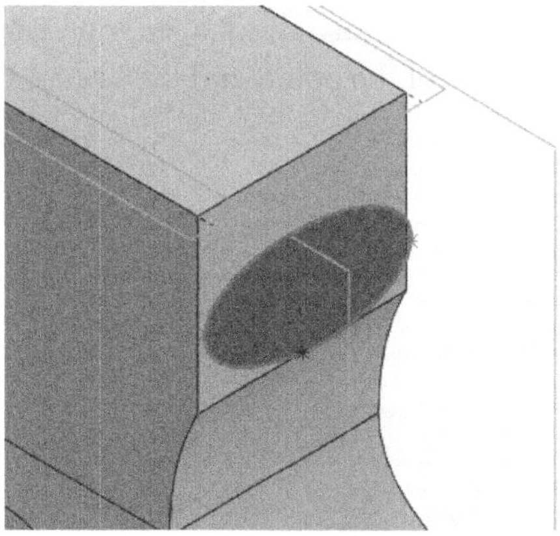

85. Likewise, apply the **Midpoint** relation between the right quadrant point of the ellipse and the model's right vertical edge, as shown.

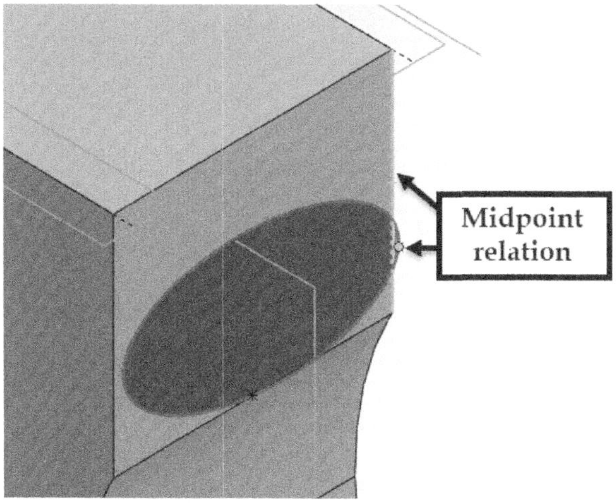

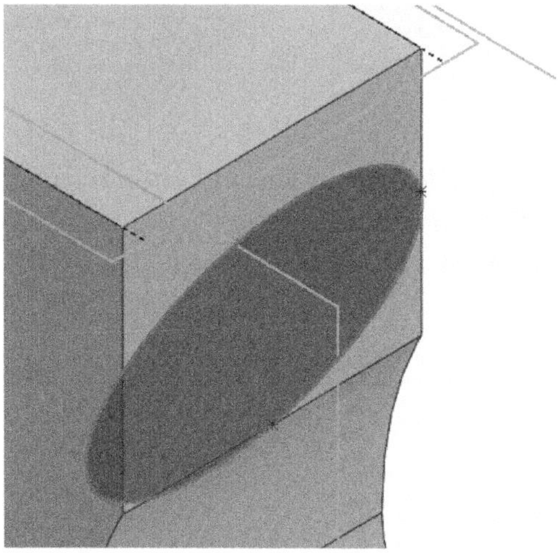

86. Press and hold the Ctrl key and then select the top horizontal edge of the sketch plane and the ellipse's top quadrant. Next, click the **Midpoint** icon on the PropertyManager, and then click **OK**.

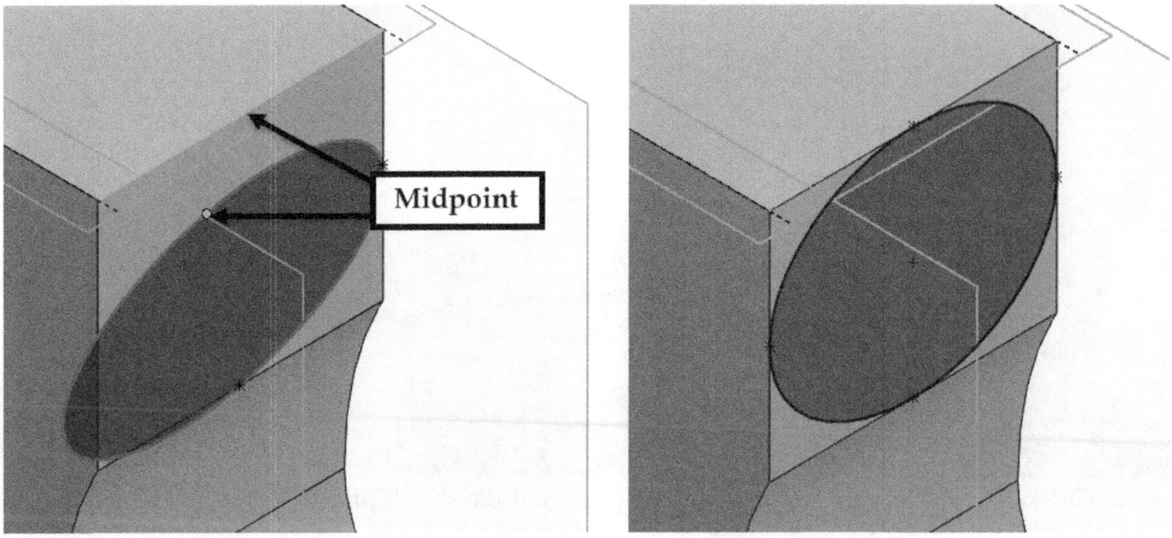

87. Click **Exit Sketch** on the CommandManager.
88. Create a new plane offset to the right flat face. The Offset distance is 8.22.
89. Click **Sketch > Point** on the CommandManager and click on the newly created plane.
90. Click to specify the location of the point, and then press **Esc** to deactivate the tool.

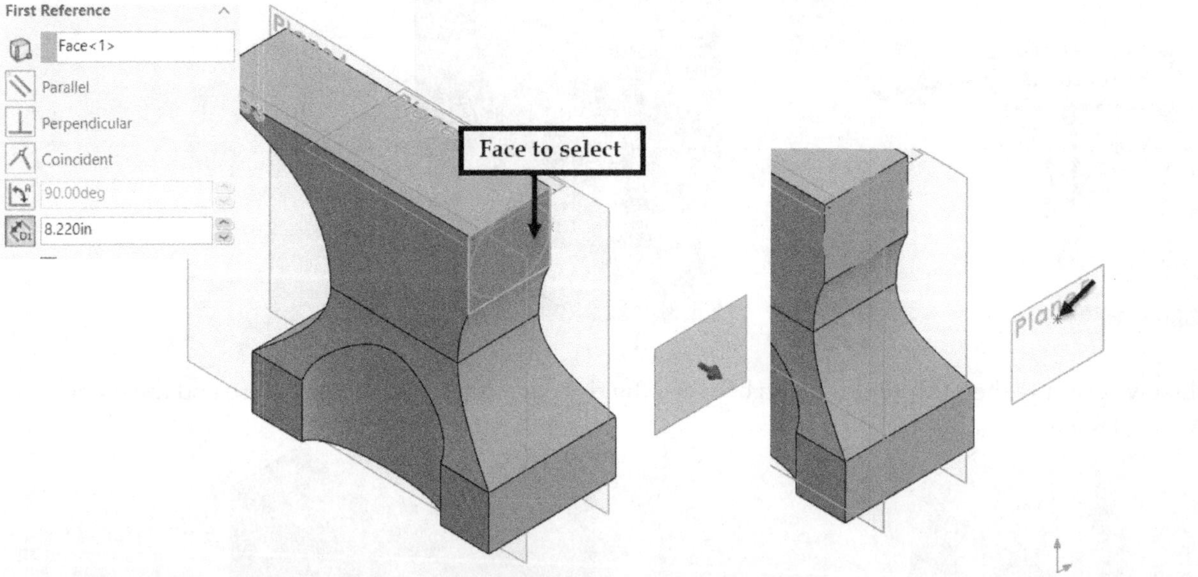

91. Press and hold the Ctrl key and select the newly created point and the ellipse's top quadrant point.
92. Click the **Coincident** icon on the PropertyManager. Next, click **OK**.
93. Click **Exit Sketch** on the CommandManager.

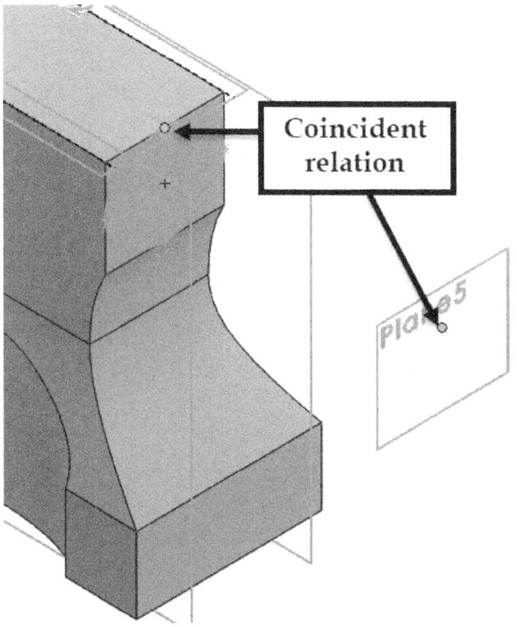

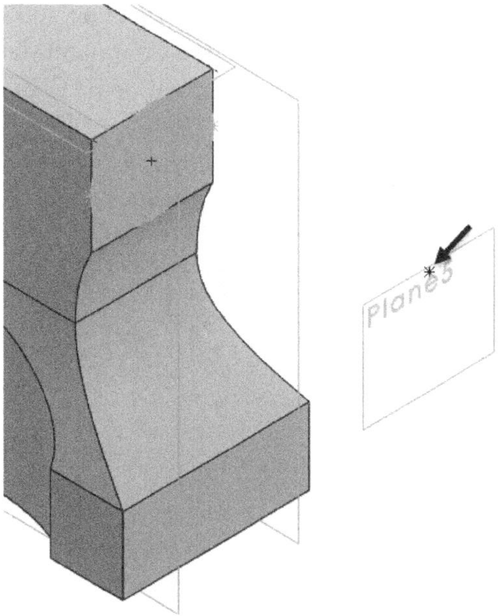

94. Click **Sketch > Sketch** on the CommandManager and click on the Front plane, as shown.
95. On the **View (Heads-Up)** toolbar, click **View Orientation > Normal To**.
96. Click **Sketch > Spline** on the CommandManager.
97. Specify the first two points of the spline, as shown. Next, select the sketch point displayed on the plane to specify the third point. Right-click and select **Select** to create the spline.

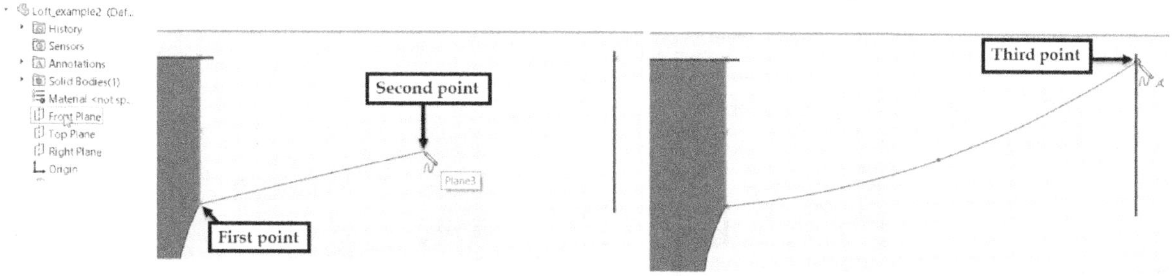

98. Activate the **Smart Dimension** command (click **Sketch > Smart Dimension** on the CommandManager), and then apply dimensions to the spline's midpoint.

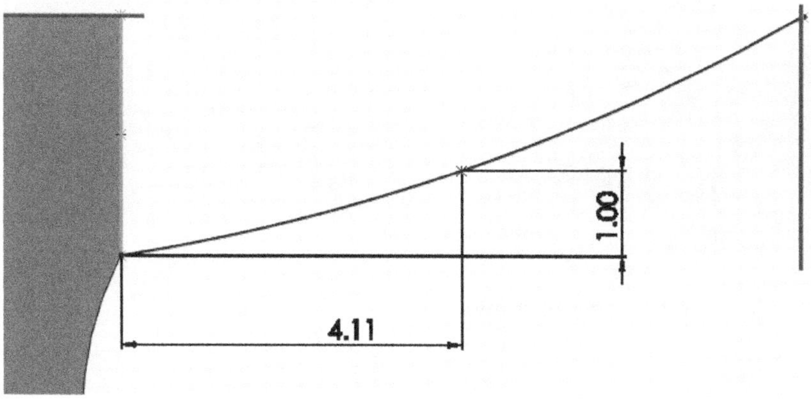

99. Click **Exit Sketch** on the CommandManager.

100. Click **Sketch > Sketch** on the CommandManager and select the Front Plane from the FeatureManager Design Tree. Click **Sketch > Line** on the CommandManager.
101. Create a line connecting the top quadrant point of the ellipse and the sketch point. Next, right-click and select **Select**.
102. Click **Exit Sketch** on the CommandManager.
103. On the **View (Heads-Up)** toolbar, click **View Orientation > Isometric**.

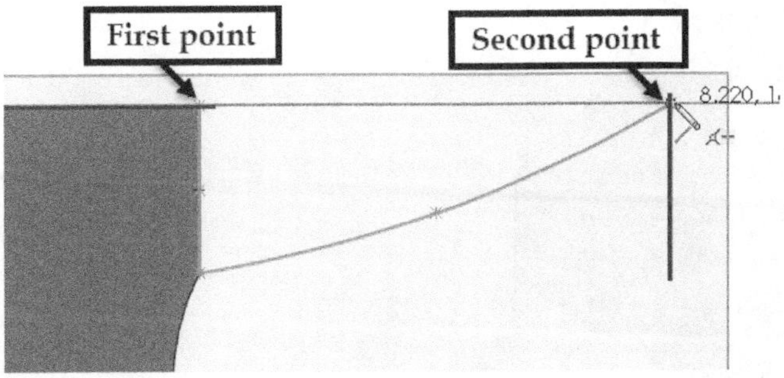

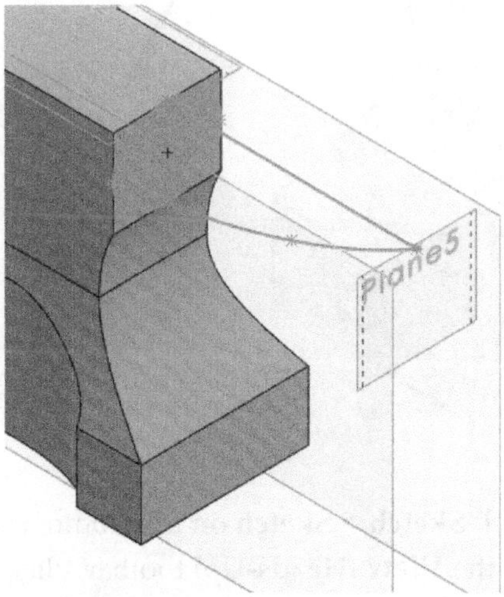

104. Click **Features > Lofted Boss-Base** on the CommandManager. Select the ellipse and the sketch point.
105. On the **Loft** PropertyManager, click in the **Guide Curves** selection box. Select the spline (guide curve 1) and line (guide curve 2), as shown. Next, click **OK** to create the loft feature.

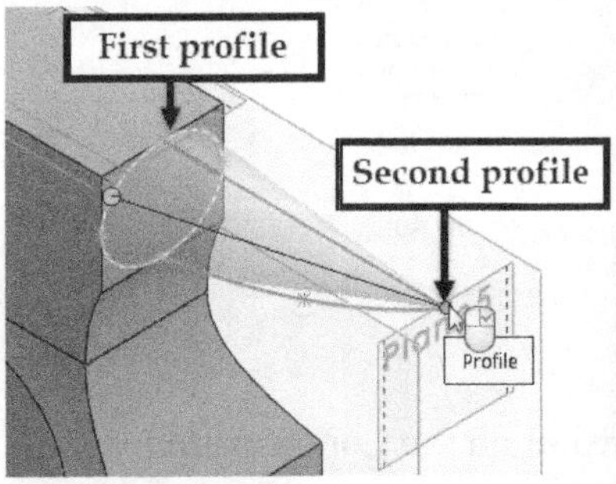

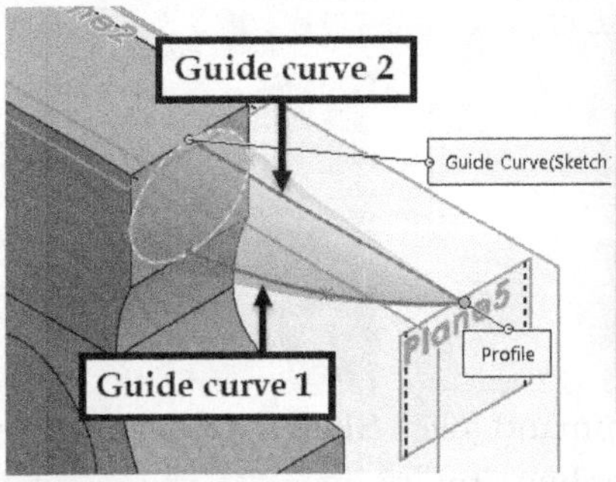

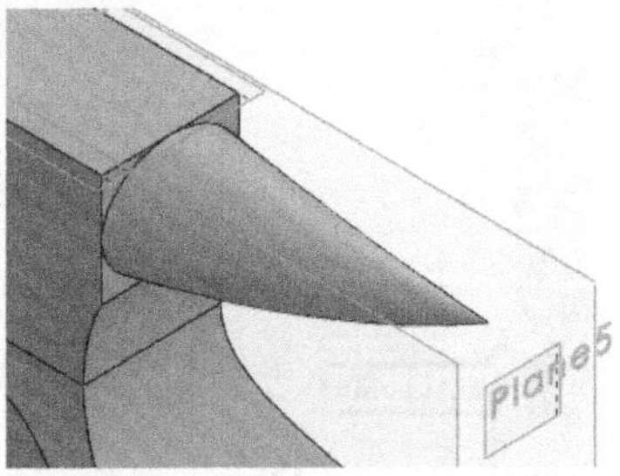

106. Drag the bar in the FeatureManager Design Tree downwards; the **Boss-Extrude2** feature is unsuppressed.

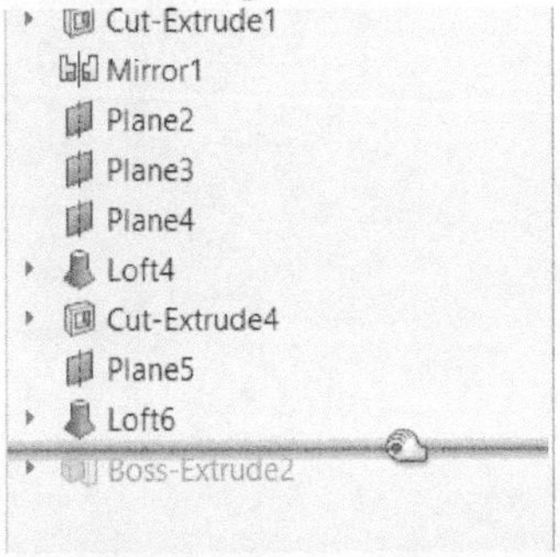

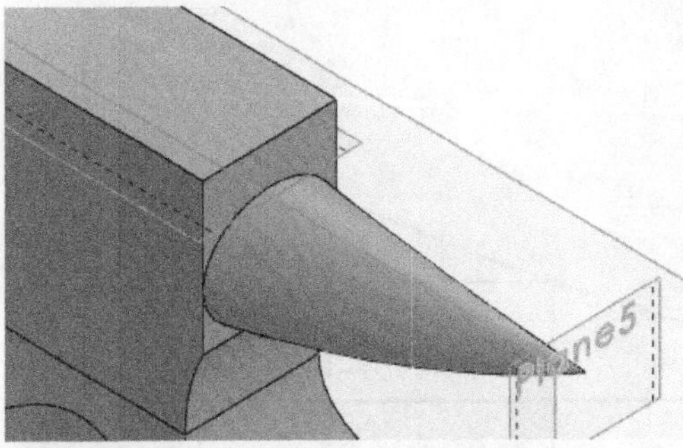

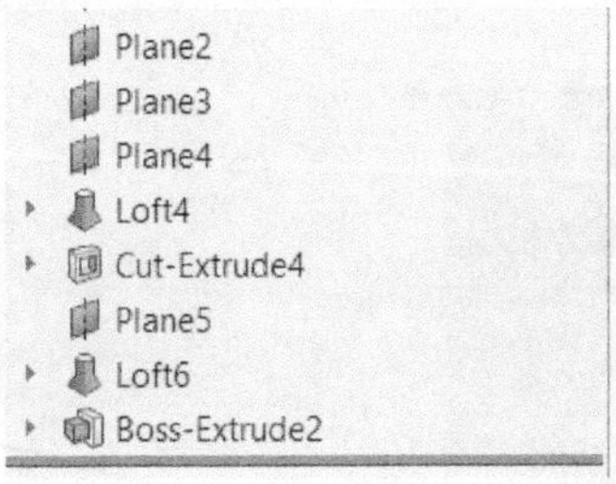

107. Create an extruded cut feature on the side face, as shown.

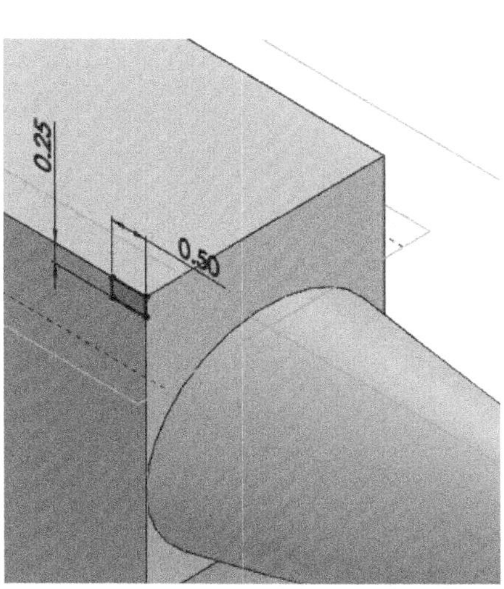

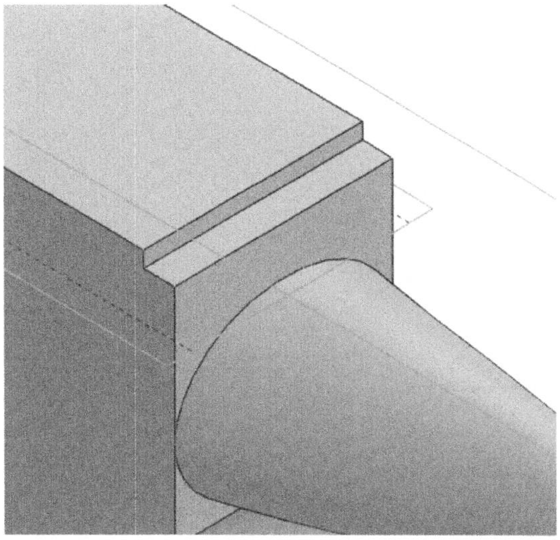

108.Create the circular and square-cut features on the top face, as shown.

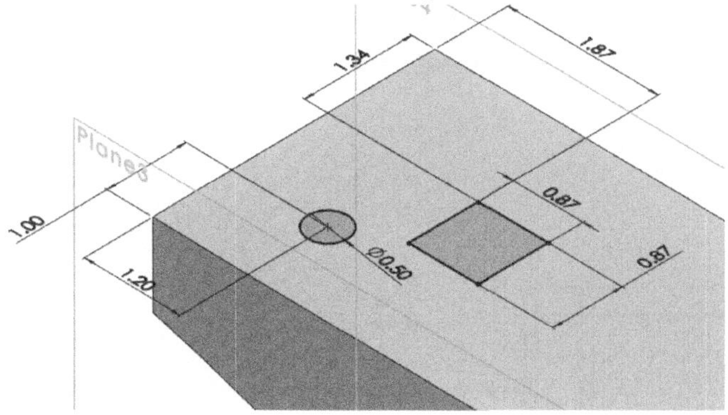

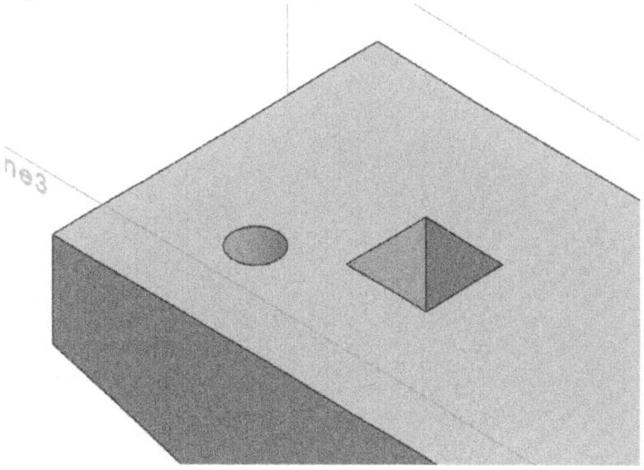

109.Save and close the file.

Questions

1. Describe the procedure to create a *Lofted Boss/Base* feature.
2. List the **Start-End Constraint** options.
3. List the type of elements that can be selected to create a *Lofted Boss/Base* feature.

Exercises

Exercise 1

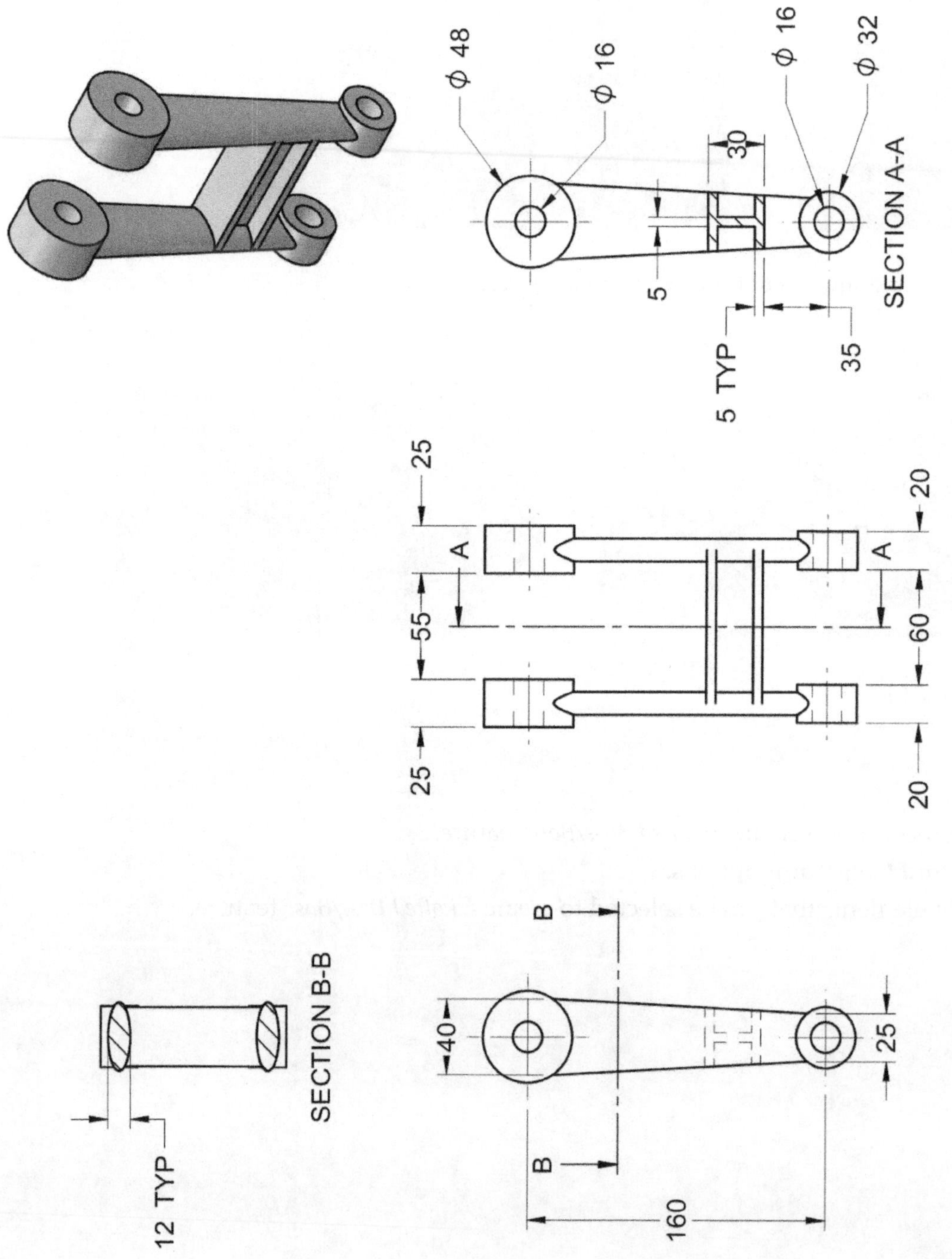

Chapter 8: Additional Features and Multibody Parts

SOLIDWORKS offers you some additional commands and features which will help you to create complex models. These commands are explained in this chapter.

The topics covered in this chapter are:

- *Ribs*
- *Vents*
- *Dome*
- *Wrap*
- *Multi-body parts*
- *Split bodies*
- *Boolean Operations*
- *Create cut features in Multi-body parts*
- *Lip/Groove*
- *Mounting Bosses*
- *Indent*

Rib

This command creates a rib feature to add structural stability, strength, and support to your designs. Just like any other sketch-based feature, a rib requires a two-dimensional sketch. Create a sketch, as shown in the figure and activate the **Rib** command (click **Features > Rib** on the CommandManager). Select the sketch; two yellow lines appear, showing the thickness side of the rib. You can add the rib material to either side of the sketch line or evenly to both sides. Set the **Thickness** type to **Both Sides** to add material to both sides of the sketch line. Next, type in the rib feature's value in the **Rib thickness** box displayed on the PropertyManager.

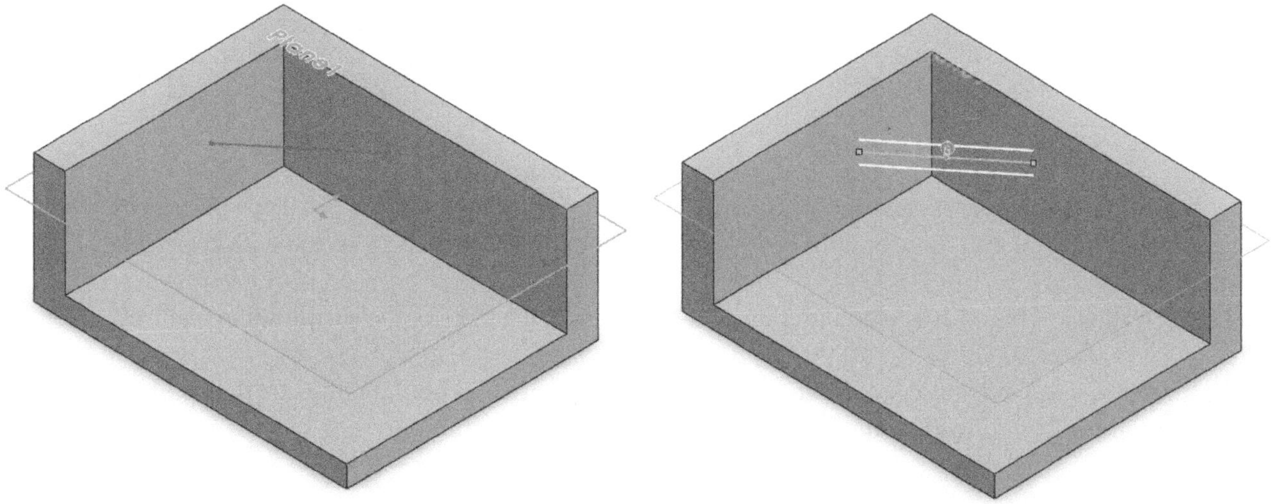

You can define the rib feature's direction by using the **Parallel to Sketch** or **Normal to Sketch** icon.

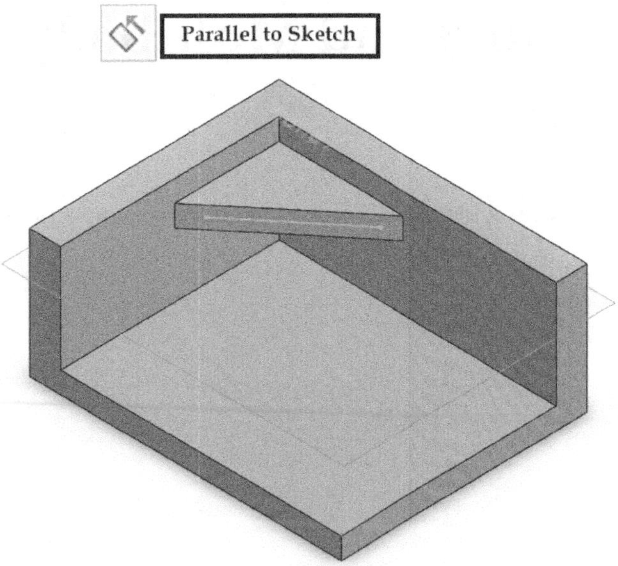

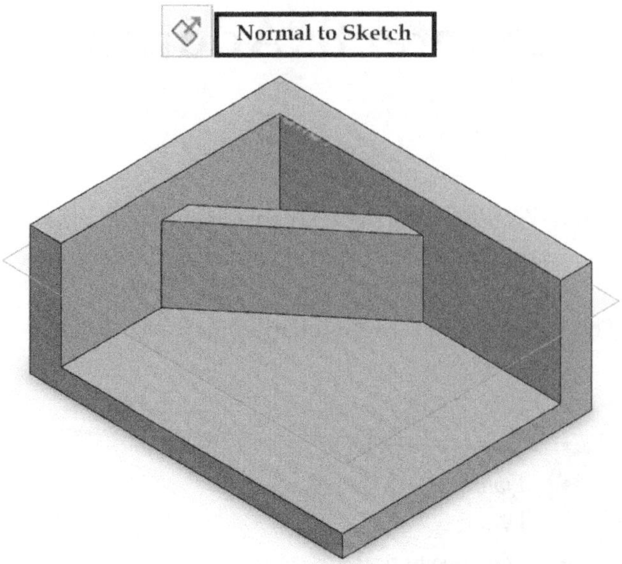

To apply draft to the rib feature, click the **Draft On/Off** icon, and then type in a value in the **Draft Angle** box. You can apply the draft to the rib feature either from the sketch plane or wall interface by selecting the **At sketch plane** or **At wall interface** options, respectively. Check/uncheck the **Draft outward** option to change the direction of the draft.

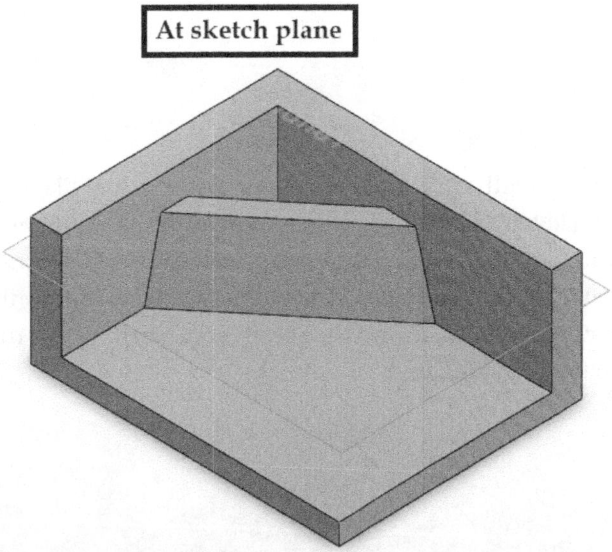

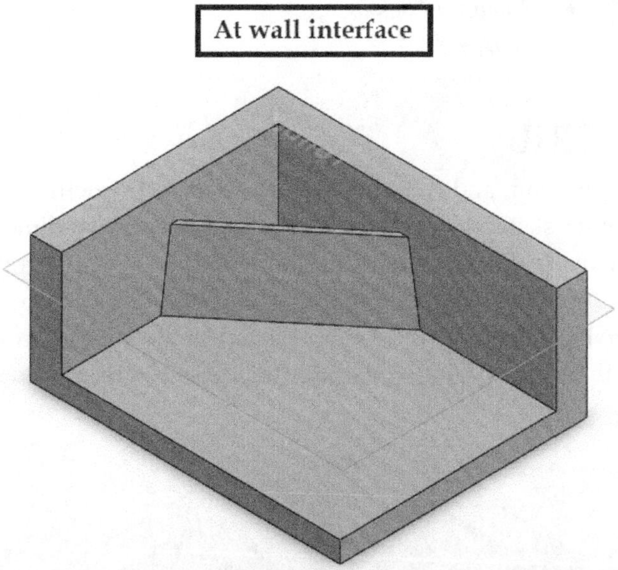

If you select the **Normal to sketch** option from the **Extrusion direction** section, the **Type** section appears with Linear and Natural options. These options are useful while creating a rib using a sketch with curve edges.

The **Linear** option creates a rib by extending the material to meet the faces of the surrounding features.

The **Natural** option extends the curve by continuing its contour equation until it will meet the boundary. For example, the arc will be converted into a circle.

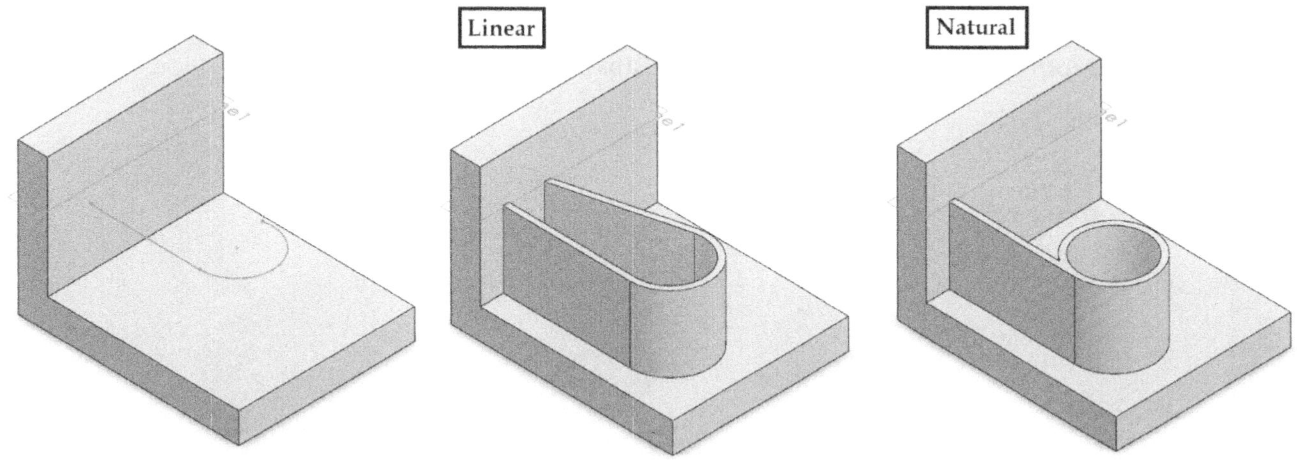

The **Selected Contours** section is used while creating a rib feature using a sketch that has multiple contours. Expand this section and select the contours to be included in the rib feature.

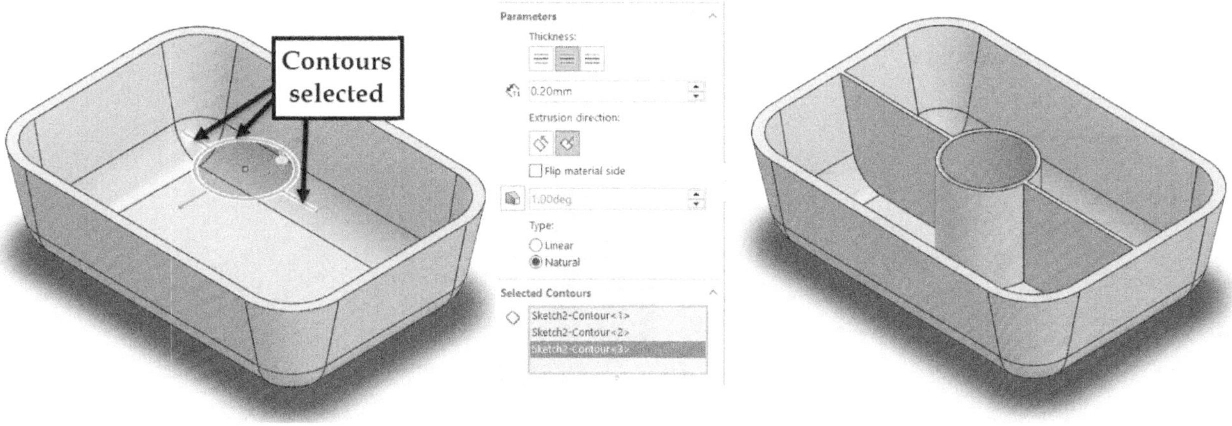

Vent

This command allows you to take a two-dimensional sketch of a vent and convert it into a 3D cutout. To create a vent feature, first, create a 2D sketch and activate the **Vent** command (click **Insert > Fastening Features > Vent** on the Menu bar or click the **Vent** icon on the **Sheet Metal** CommandManager). As you activate this command, the **Vent** PropertyManager appears on the screen.

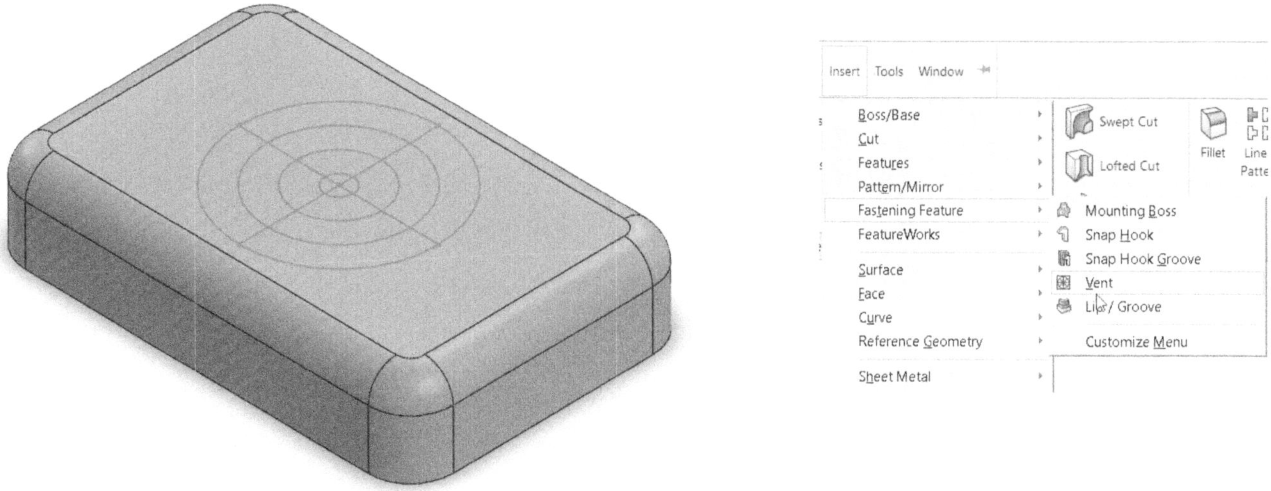

Select the boundary of the vent. Next, click in the **Ribs** selection box and select the ribs from the sketch. Specify the parameters (**Depth**, **Width**, and **Offset**) in the **Ribs** section.

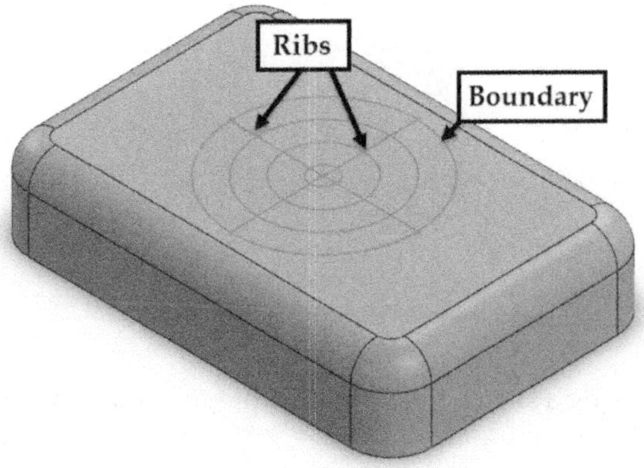

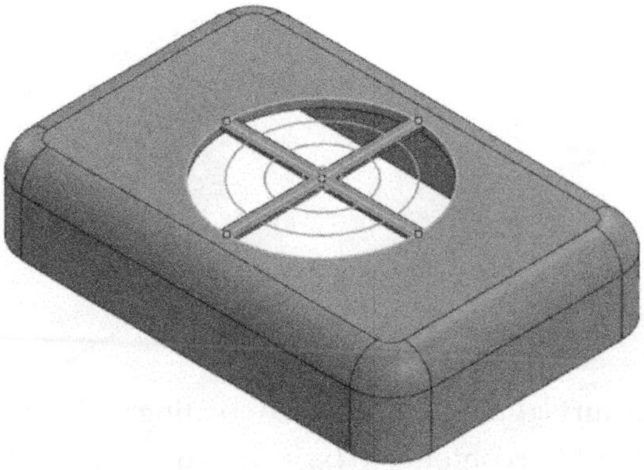

Click in the **Spars** selection box and select the spars, as shown. Likewise, click in the **Fill-In Boundary** selection box and select the fill-in boundary.

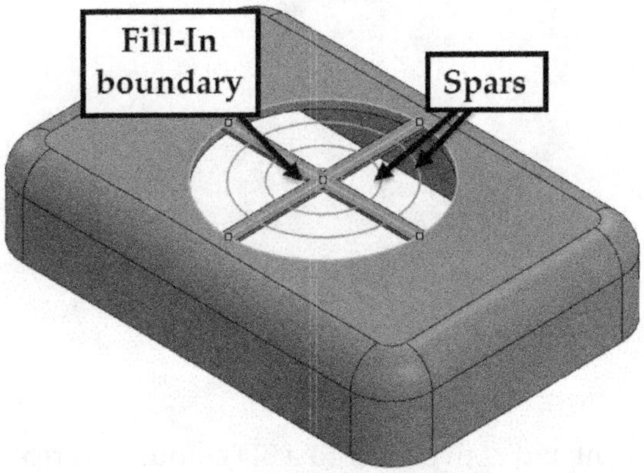

In the **Geometry Properties** section, click the **Draft On/Off** icon and type in a value in the **Draft Angle** box. Also, type in a value in the **Radius for fillets** box if you want to fillet the corners. Click **OK** to create the vent.

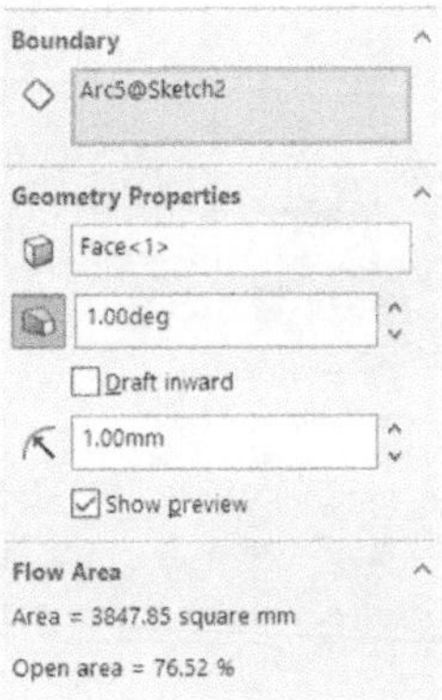

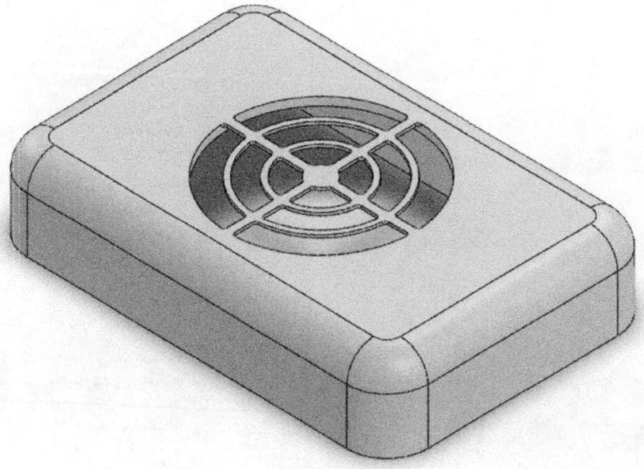

Dome

This command is used to add one or more domes to the planar or non-planar faces. Activate the **Dome** command (on the Menu bar, click **Insert > Features > Dome**), and then click on the **Face to Dome** selection box. Select the face that you want to dome and specify a value in the **Distance** box. Click the **Reverse Direction** button to reverse the dome feature. Click the **OK** button to create the dome.

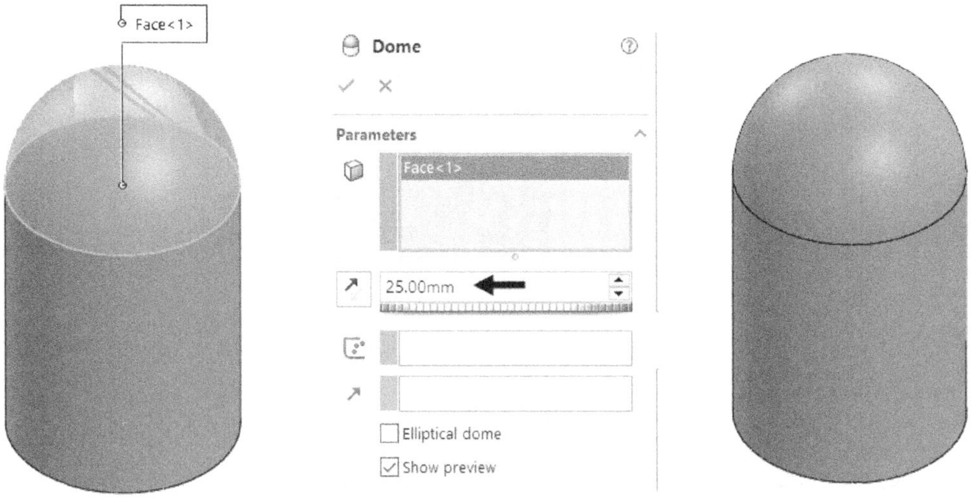

You can create an elliptical dome by selecting the **Elliptical dome** option on the **Dome** PropertyManager.

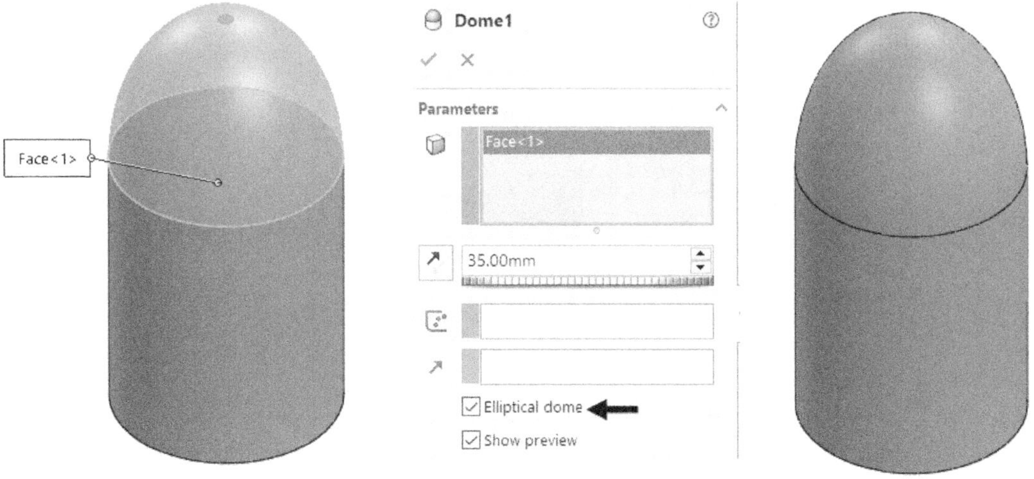

Check the **Continuous dome** option to create a dome continuous to all the boundary edges.

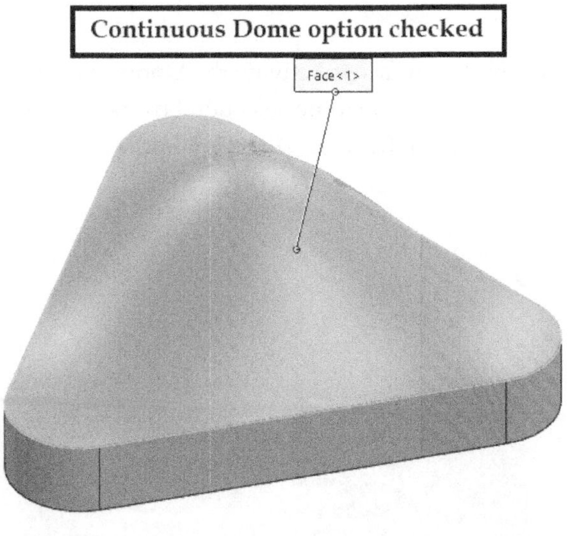

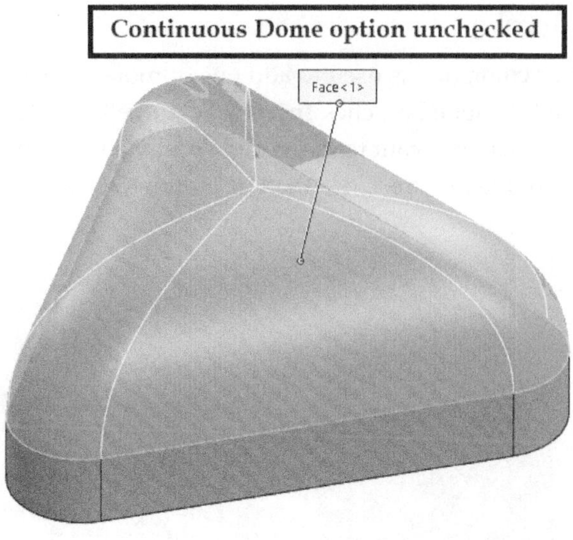

You can also define the height of the dome by constraining it to a point. To do this, click in the **Constraint Point or sketch** selection box, and then select a point.

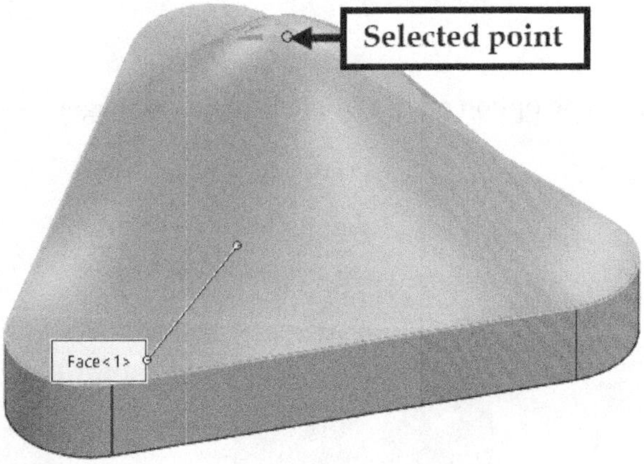

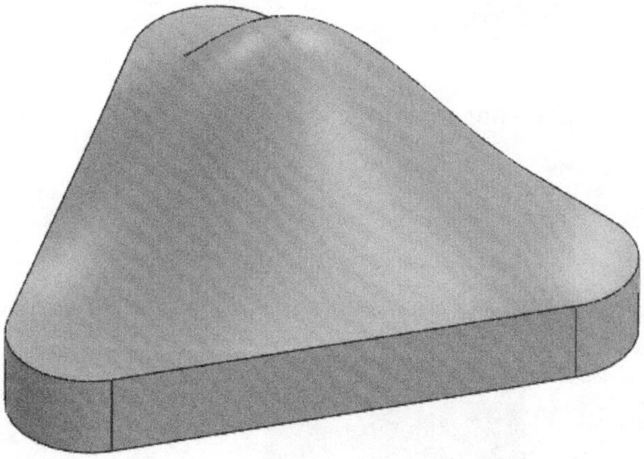

Wrap

This command helps you to emboss or deboss or scribe a sketch on any planar or non-planar face. First, you need to have a solid body and a sketched profile. Activate the **Wrap** command (On the CommandManager, click **Features > Wrap**), and then select the sketch profile that you want to emboss or deboss or scribe on the planar or non-planar face. On the **Wrap** PropertyManager, select the **Emboss** or **Deboss**, or **Scribe** icon from the **Wrap Type** section.

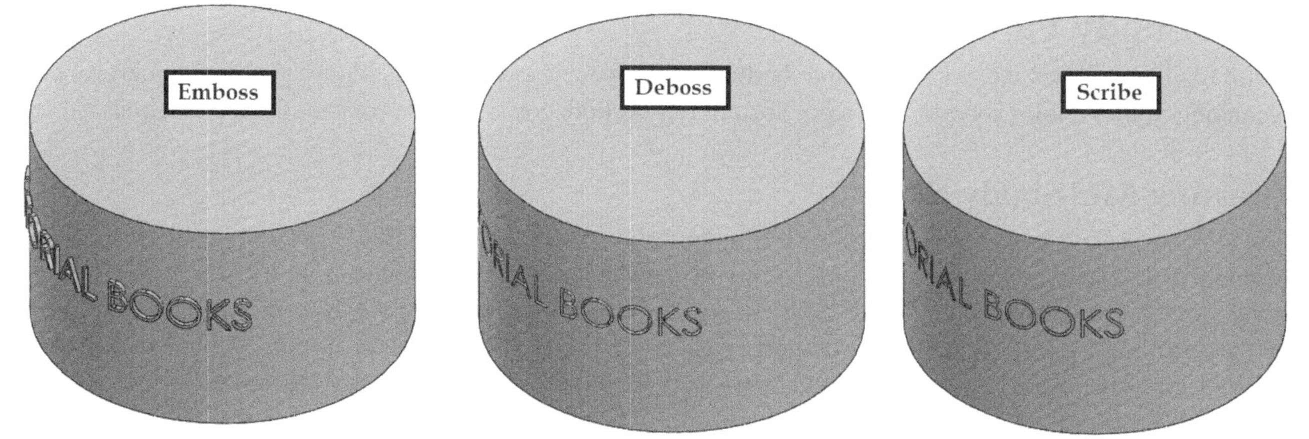

Click the **Analytical** under the **Wrap Method** section. Click in the **Face for Wrap** selection box and select the face to be wrapped. Next, type in a value in the **Thickness** box. Click **OK** to wrap the sketch.

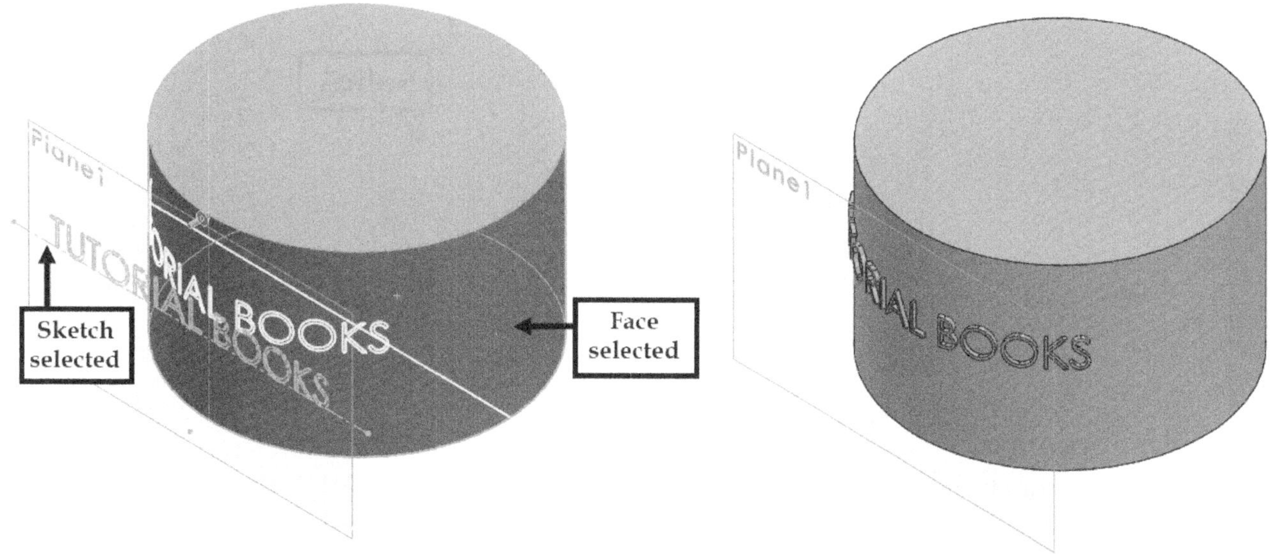

Activate the Wrap command and select the sketch to be wrapped around the model geometry. Next, select a non-planar spline surface. Select an option from the **Wrap Type** section and click the **Spline Surface** icon under the **Wrap Method** section. Next, click **OK** to wrap the sketch around the spline surface.

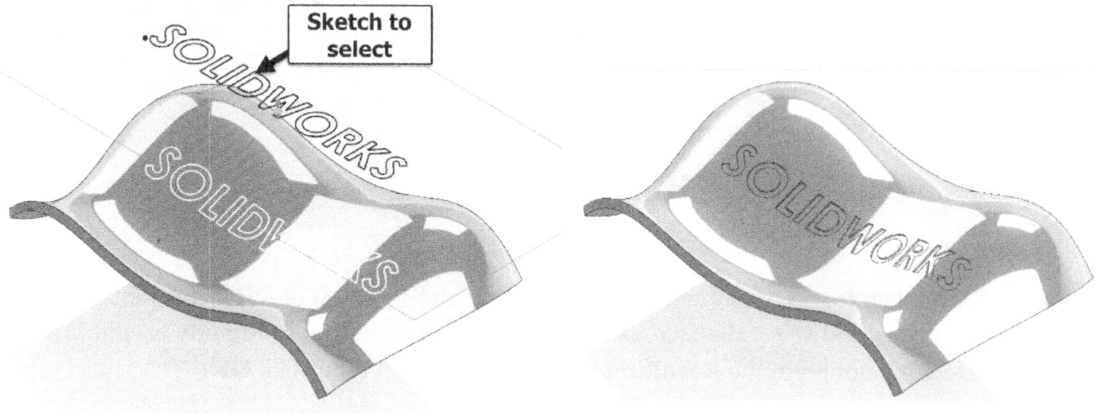

Multi-body Parts

SOLIDWORKS allows the use of multiple bodies when designing parts. This allows you to use several design techniques that would otherwise not be possible. In this section, you will learn some of these techniques.

Creating Multibody Parts

The number of bodies in a part can change throughout the design process. SOLIDWORKS makes it easy to create separate bodies inside a part geometry. Also, you can combine multiple bodies into a single body. To create multiple bodies in part, first, create a solid body, and then create any sketch-based feature such as extruded, revolved, swept, or loft feature. While creating the feature, ensure that the **Merge result** option is unchecked on the PropertyManager. Next, expand the **Solid Bodies** folder in the **FeatureManager Design Tree** and notice the multiple solids.

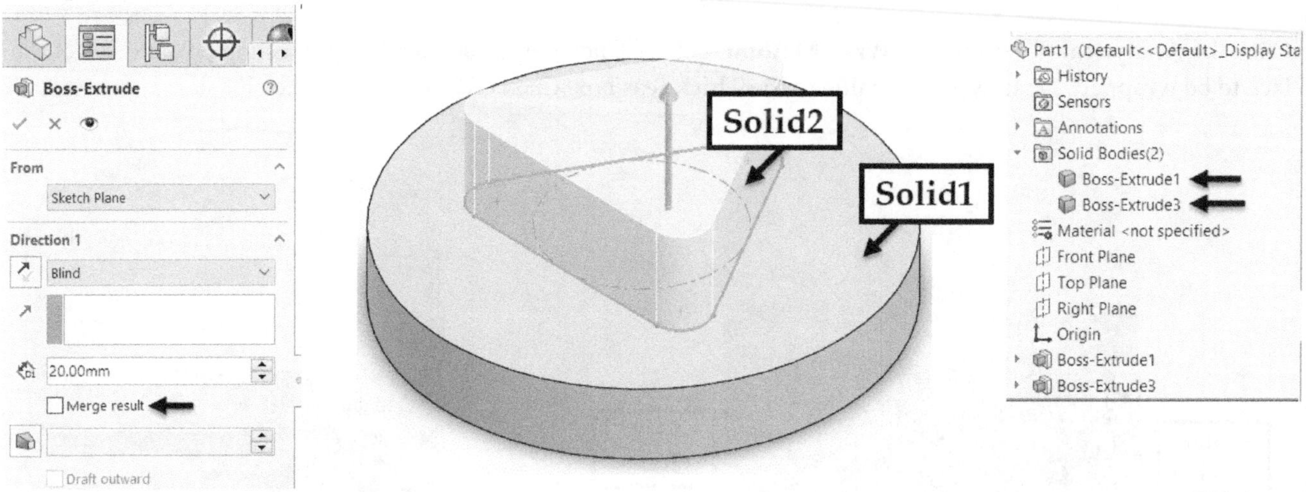

The Split command

The **Split** command can be used to separate single bodies into multiple bodies. This command can be used to perform local operations. For example, if you apply the shell feature to the front portion of the model shown in the figure, the whole model will be shelled. To solve this problem, you must split the solid body into multiple bodies (In this case, separate the front portion of the model from the rest).

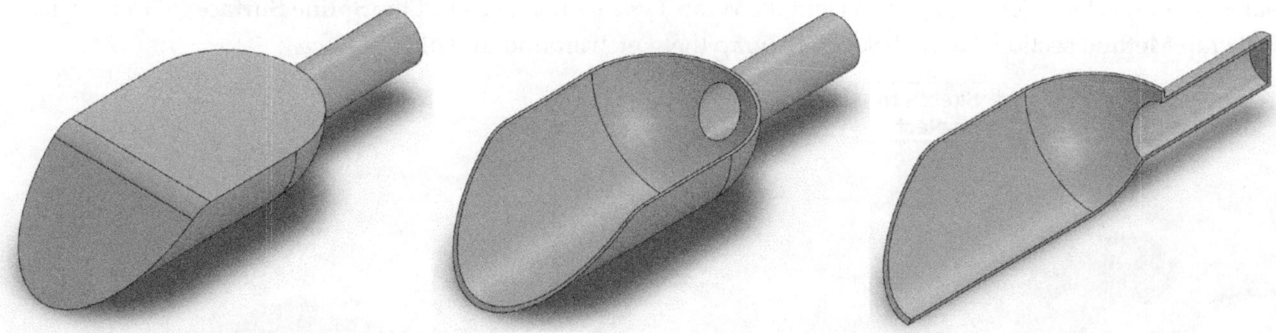

Activate the **Split** command (click **Insert > Features > Split** on the Menu bar). On the **Split** PropertyManager, click in the **Trim Tools** selection box and select the face, as shown. Click **Cut Part**; the solid body is split into two separate bodies. Click the **Select All** option in the **Resulting Bodies** section, and then click **OK**.

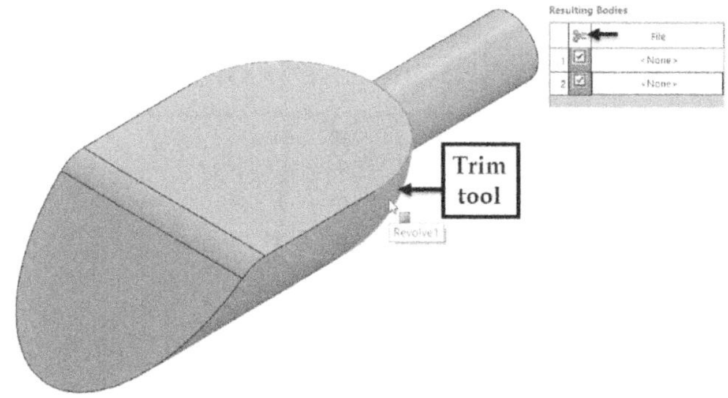

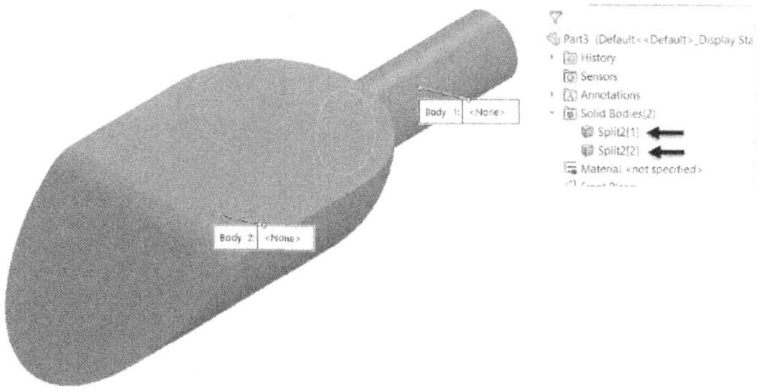

Now, create the shell feature on the split body.

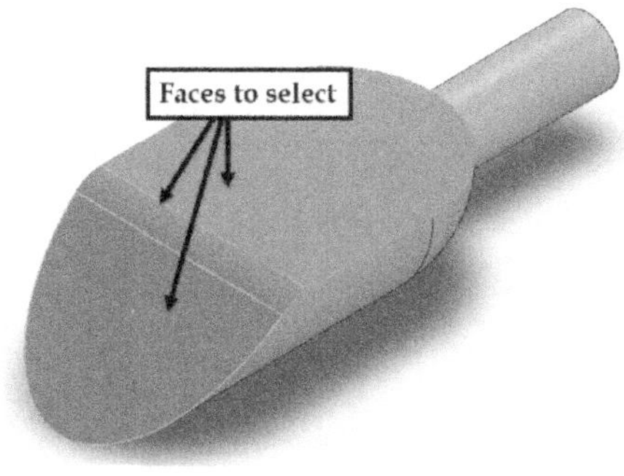

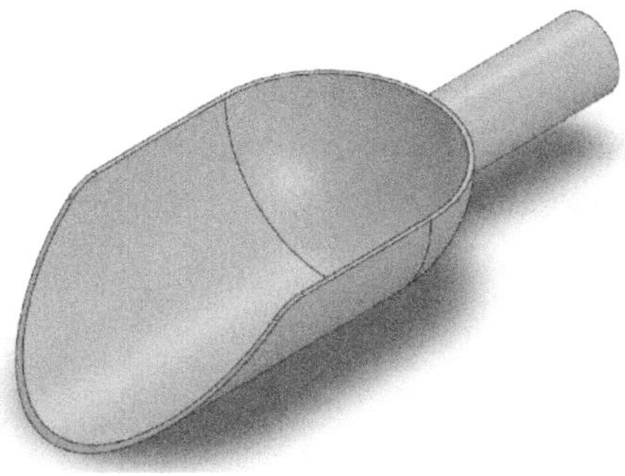

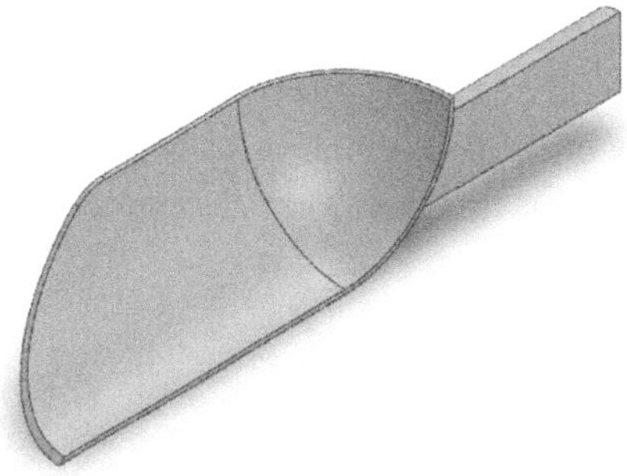

Add

If you apply fillets to the edges between two bodies, it will show a different result, as shown in the figure. In order to solve this problem, you must combine the two bodies.

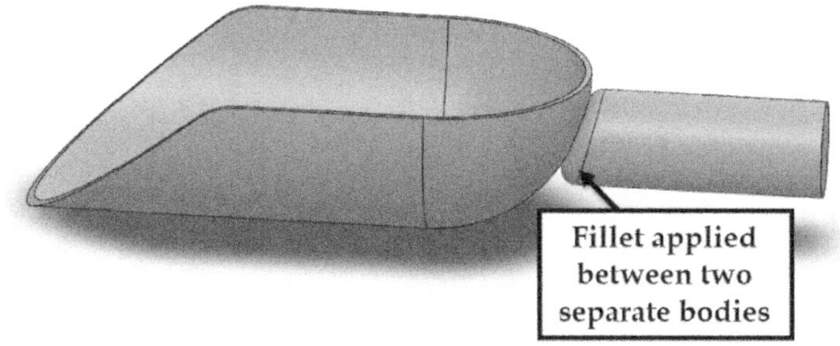

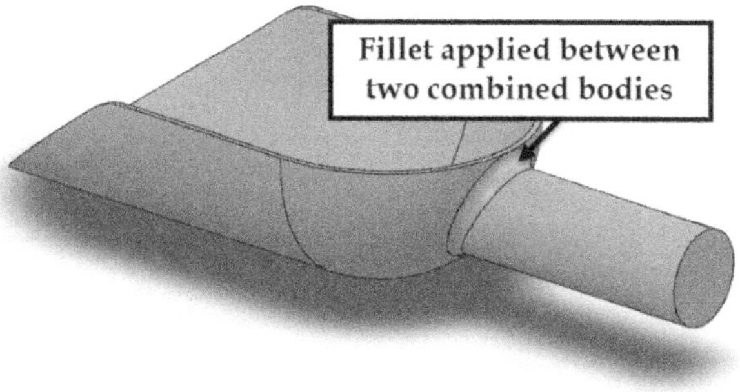

Activate the **Combine** command (on the Menu bar, click **Insert > Features > Combine**) and click the **Add** option from the **Operation Type** section. Next, select the two bodies. Click **OK** to join the bodies. Now, apply fillets to the edges.

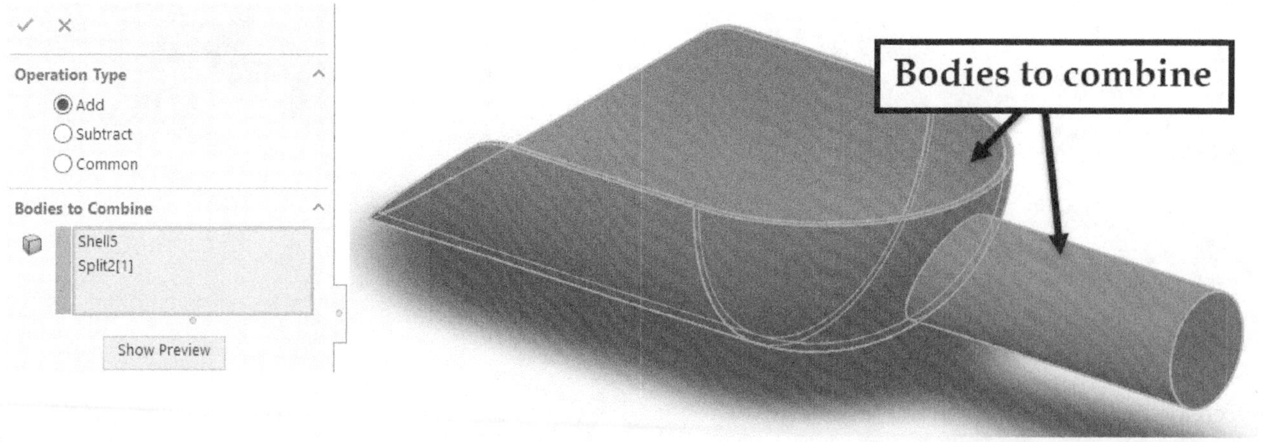

Common

By using the **Common** option, you can generate bodies defined by the intersecting volume of two bodies. Activate **Combine** command (click **Insert > Features > Combine** on the Menu bar). On the **Combine** PropertyManager, select the **Common** option and select two bodies. Click **OK** to see the resultant single solid body.

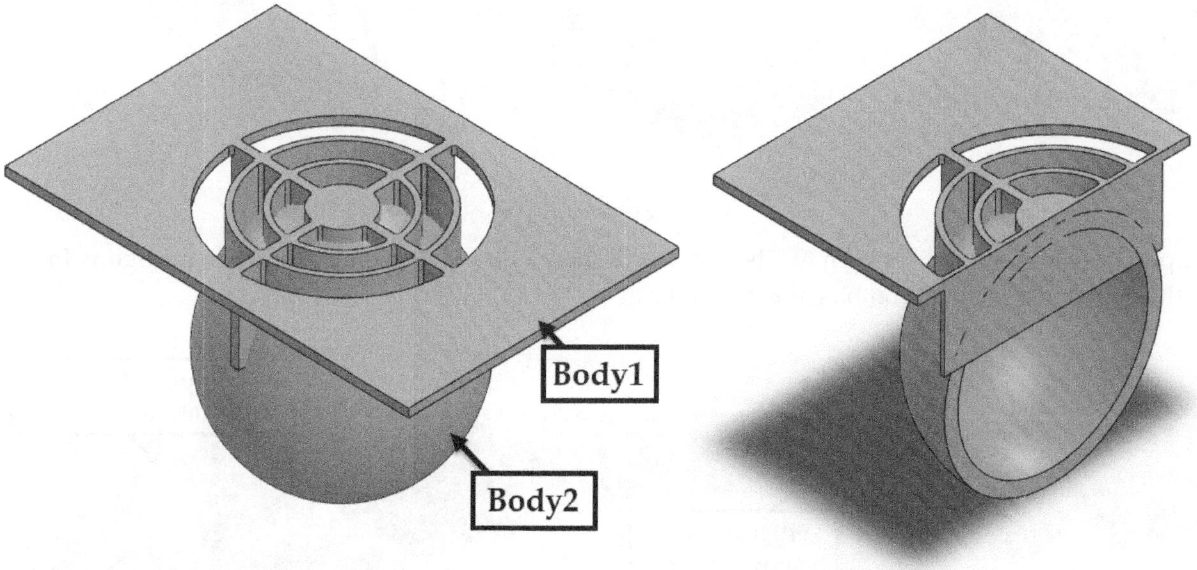

Subtract

This option performs the function of subtracting one solid body from another. Activate the **Combine** command (click **Insert > Features > Combine** on the Menu bar) and select the **Subtract** option from the **Operation Type** section on the **Combine** PropertyManager. Next, select the main and tool body. Click **OK** to subtract the tool body from the main body.

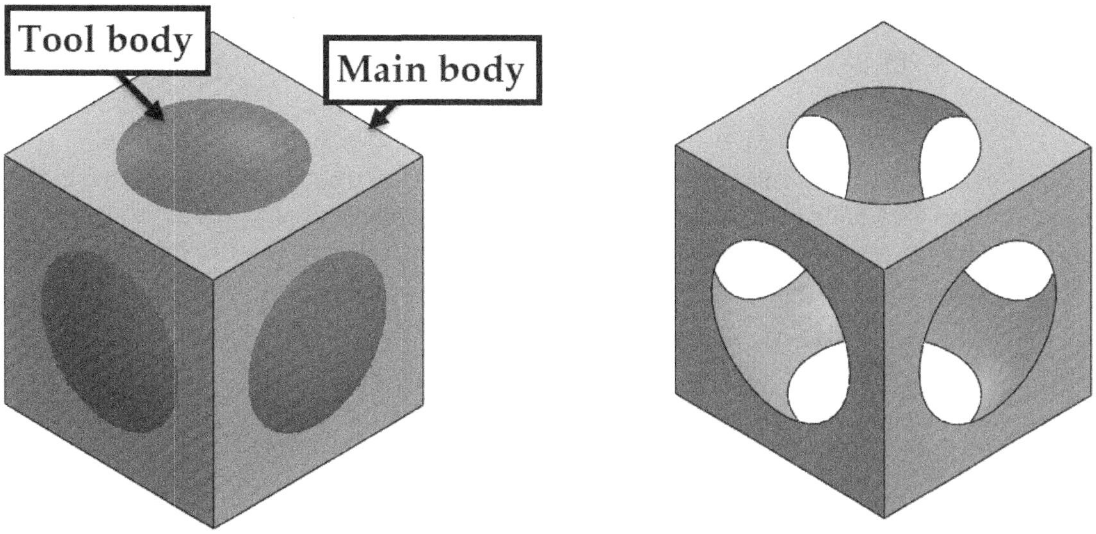

While selecting the main body, you can check the **Make main body transparent** option to make it transparent. This allows you to select bodies that are relatively smaller and are completely submerged within the main body.

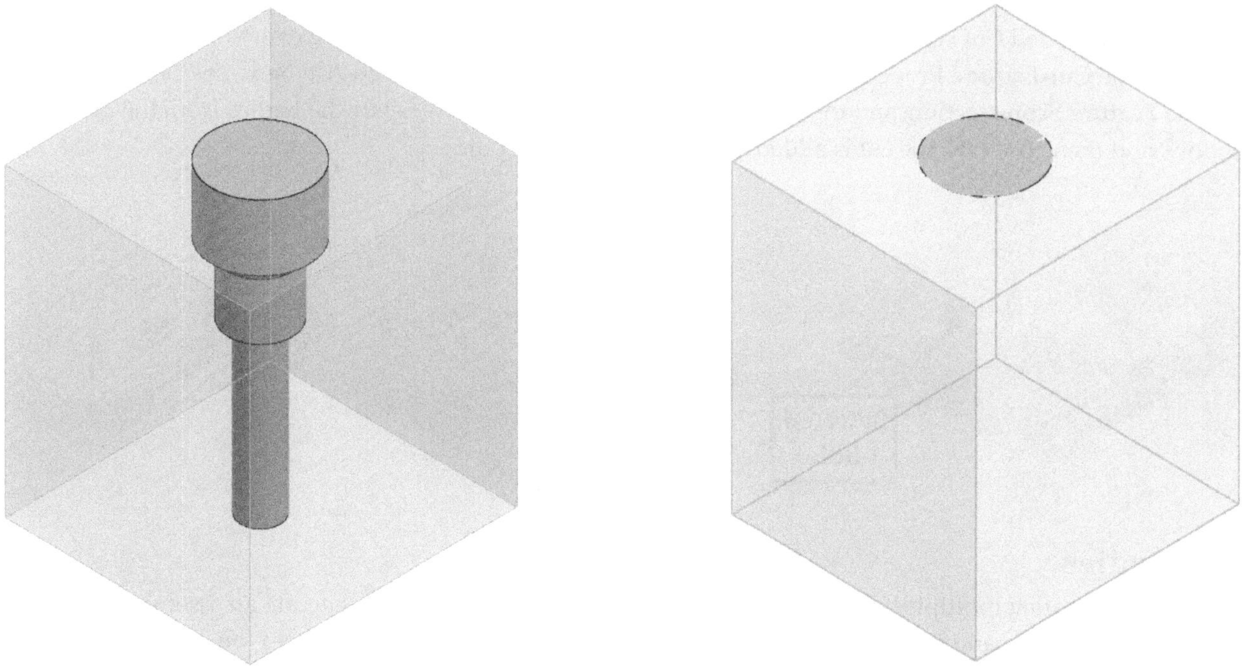

Make main body transparent checked

Make main body transparent unchecked

Creating Cut Features in Multi-body Parts

SOLIDWORKS has some additional options (**All Bodies** and **Selected Bodies**) while creating cuts in Multi-body parts. These options are available on all the Boss/Base or Cut PropertyManagers in the **Feature Scope** section. The options in the **Feature Scope** section are explained next.

All bodies

Create three separate bodies in a part file, and then create a sketch on the active body, as shown. Activate the **Extruded Cut** command (on the CommandManager, click **Features > Extruded Cut**) and select the sketch. On the **Cut-Extrude** PropertyManager, select **End Condition > Through All**. Next, select **All bodies** from the **Feature Scope** section. Click **OK** and notice that the cut is added to all the bodies of the part.

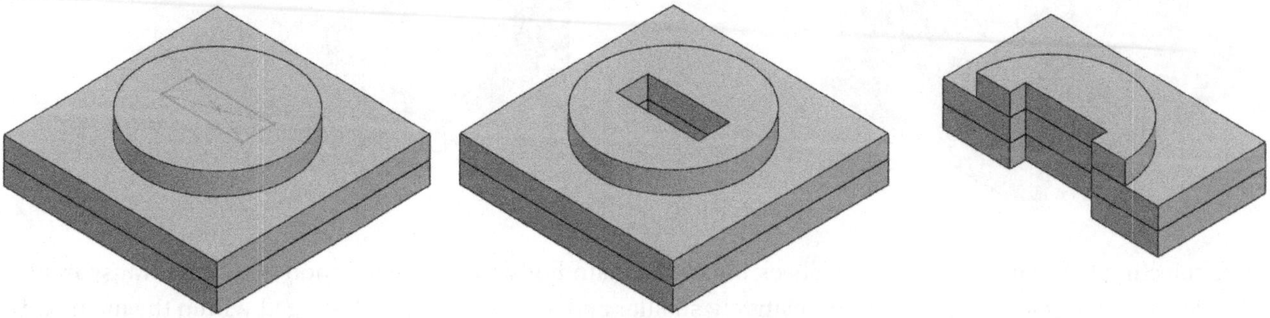

Selected Bodies

Activate the **Extruded Cut** command (on the CommandManager, click **Features > Extruded Cut**) and select the sketch. On the **Cut-Extrude** PropertyManager, select **End Condition > Through All**. Next, select **Selected bodies** from the **Feature Scope** section, and then uncheck the **Auto-select** option. Select the bodies from the graphics window, and then click **OK**; the cut is added only to the selected bodies.

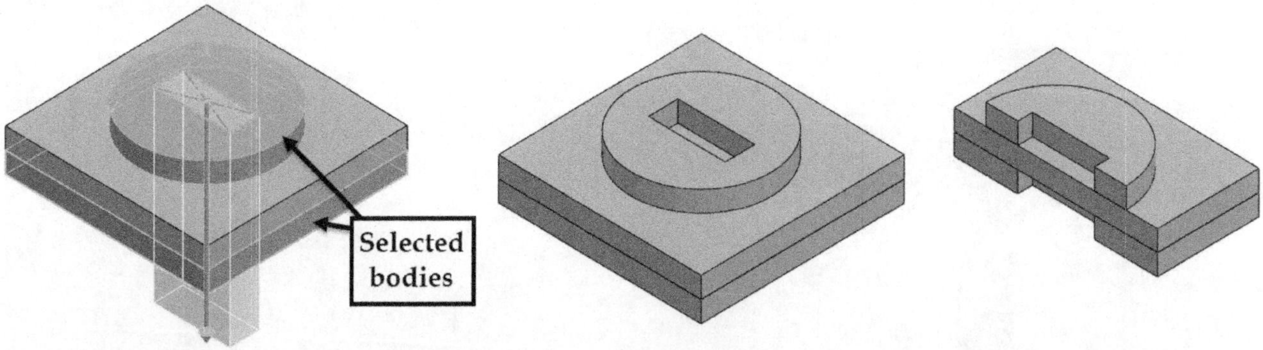

Save Bodies

In addition to creating multiple bodies, SOLIDWORKS also offers an option to generate an assembly from the resulting bodies. For example, create the model shown in the figure and split it into two separate bodies.

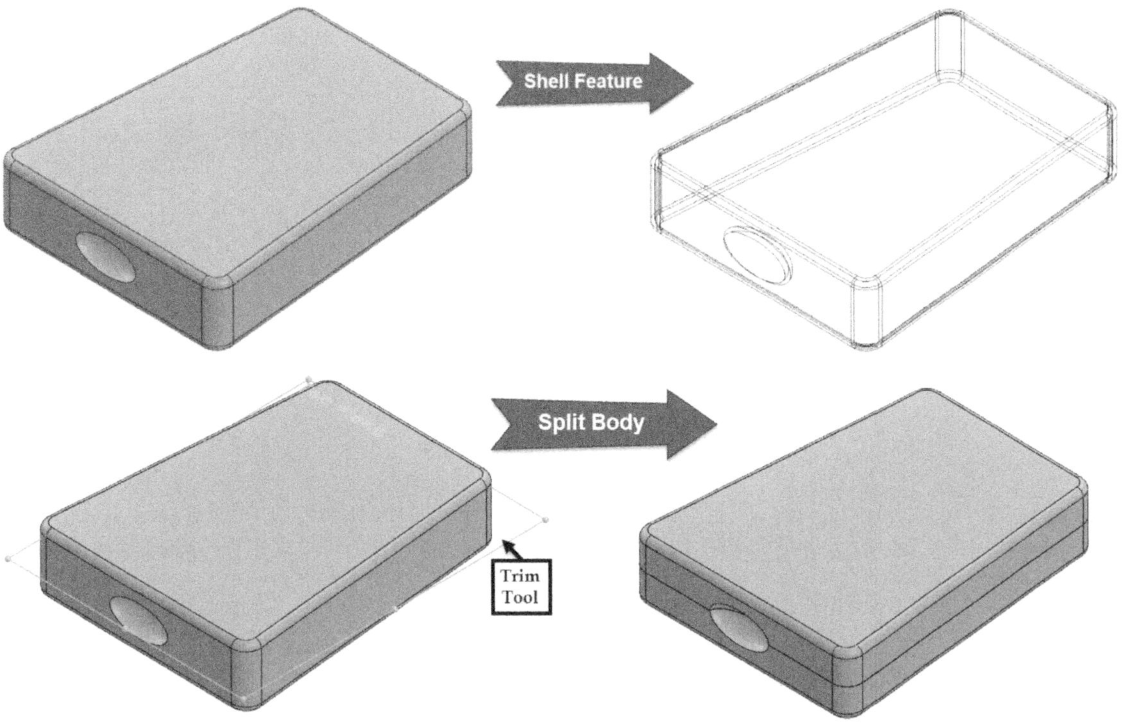

Activate the **Save Bodies** command (click **Insert > Features > Save Bodies** on the Menu bar); the **Save Bodies** PropertyManager appears. Click **Select All** in the **Resulting bodies** section. Next, click the **Browse** button in the **Create Assembly** section and then specify the assembly file's name and location. Click the **Save** button, and then click **OK** on the **Save bodies** PropertyManager.

Make Multibody Part

The **Make Multibody Part** command allows you to convert an entire assembly into multibody part. To do this, open the assembly that you want to convert into a multibody part. Next, click **Tools > Make Multibody Part** from menu bar. On the PropertyManager, check the options to decide which other assembly entities to transfer, such as surface bodies, reference geometry, and materials. By default, the multi-body part will be linked to the assembly from which it is created. However, you check the Break link to original assembly option in the Link section of the PropertyManager to break the link. Next, click **OK** on the PropertyManager; the Make Multibody Part feature will appear in the FeatureManager design tree on the left-hand side of the screen.

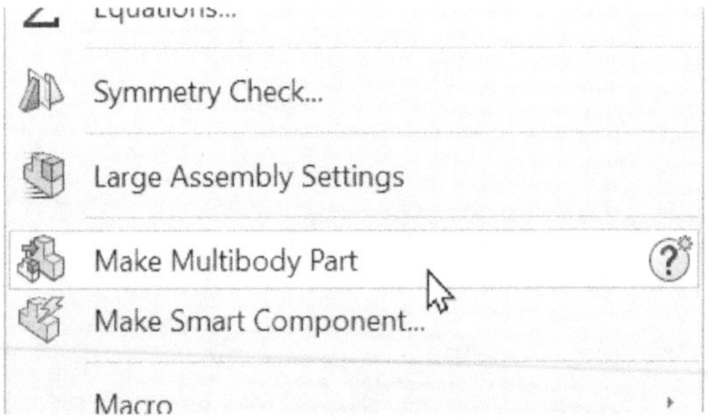

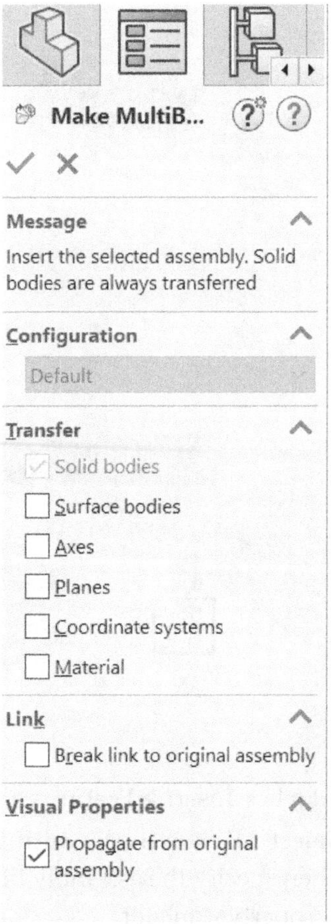

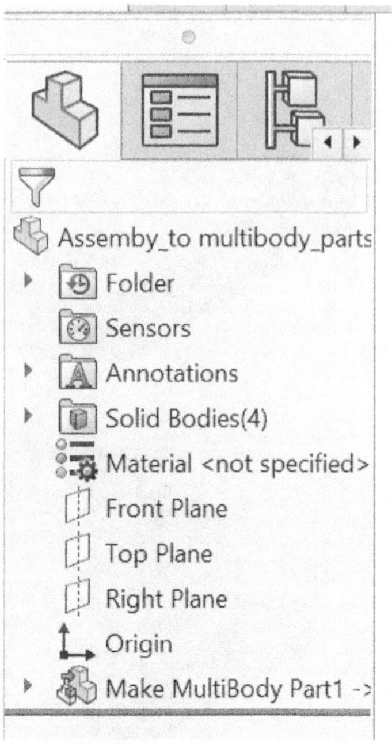

Advantages of Multi-Body Parts

Single Part Number: Simplifies bill of materials (BOM) management by representing multiple components with a single part number.

Use Cases: Beneficial for purchased parts, improving file loading speed and simplifying assembly processes.

File Sharing: Simplifies file sharing and reduces file size, especially useful for fit checks or simplifying complex assemblies.

Complexity Reduction: Simplifies complex assemblies by reducing the overall number of parts required, thus improving file management.

Assembly Replacement: Can replace assemblies in certain cases, particularly when dealing with purchased parts or inserting parts into assemblies.

Configuration Handling: Efficiently handles configuration settings, reducing the need to add configurations for each part at the assembly level.

File Management: Helps in managing fewer files, reducing potential downstream problems associated with large assemblies.

Disadvantages of Multi-Body Parts:

Graphical Performance: May have slower graphical performance compared to assemblies, depending on project requirements.

Dynamic Motion: Limited access to dynamic motion and flexible assemblies, which may be essential for certain designs.

Performance Considerations: While offering faster file loading, graphical performance may not be as efficient as assemblies.

File Management: While reducing file count, may still pose challenges in managing complexity, particularly for designs requiring extensive configurations or dynamic motion.

Evaluation Complexity: Requires careful evaluation of design complexity to determine whether multi-body parts or assemblies are more suitable, adding an extra step in design decision-making.

Lip/Groove

This command allows you to create lips and grooves on the edges of parts, saving you time by not having to create a series of manual cuts. You can add a lip/groove feature to the multi-body part or assembly. For example, the following figure shows a multi-body part containing two bodies. Click the **Section View** icon on the **View HeadsUp** toolbar and notice that both the bodies are shelled out with some features added on the inside. You can add the lip/groove feature at the location where the two bodies meet each other.

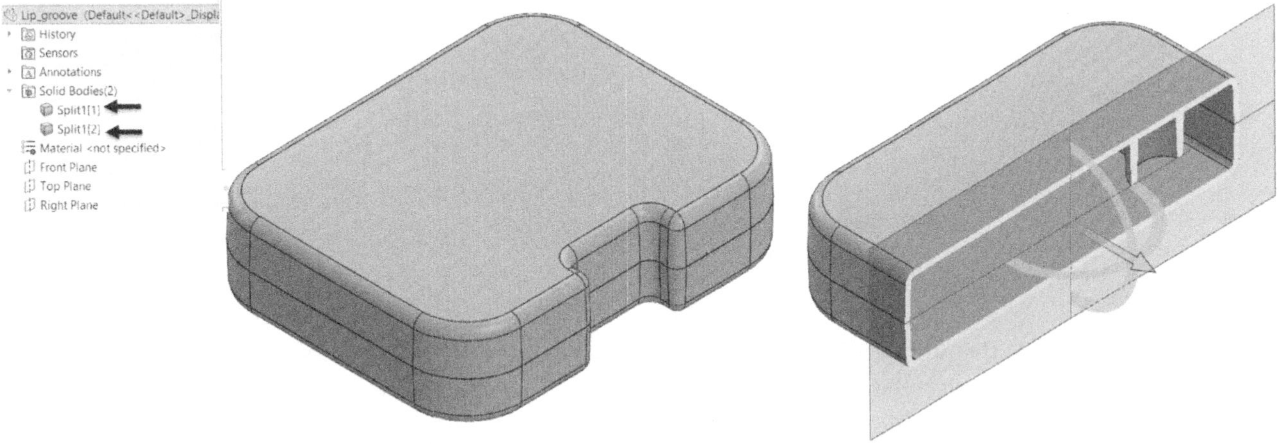

Activate this command (click **Insert > Fastening Feature > Lip/Groove** on the menu bar) and select the body that will have the groove. Next, select the body that will have the lip.

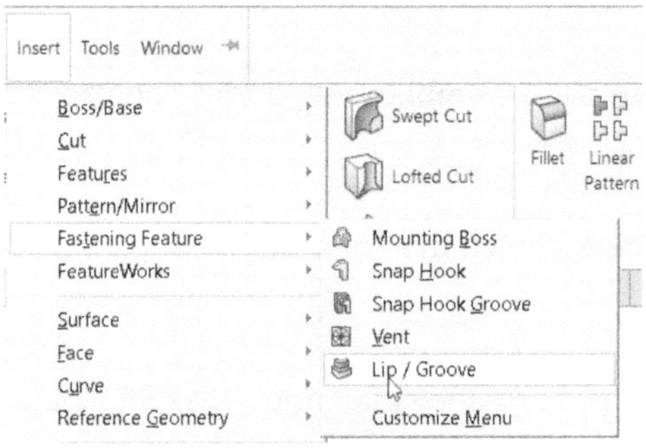

Expand the FeatureManager Design Tree and select the Top plane. This defines the direction of the lip/groove feature. In general, the plane should be parallel to the parting line.

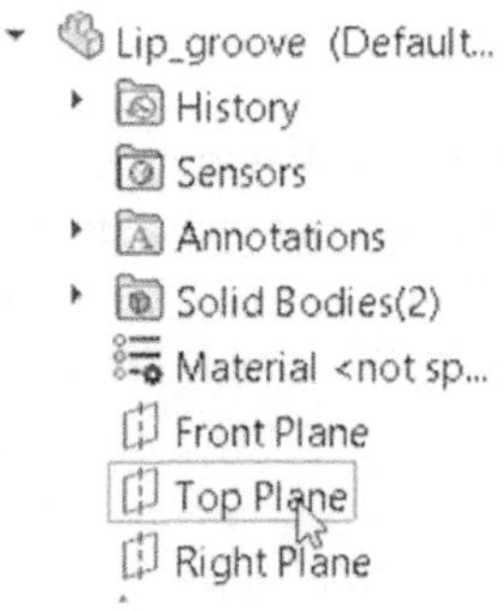

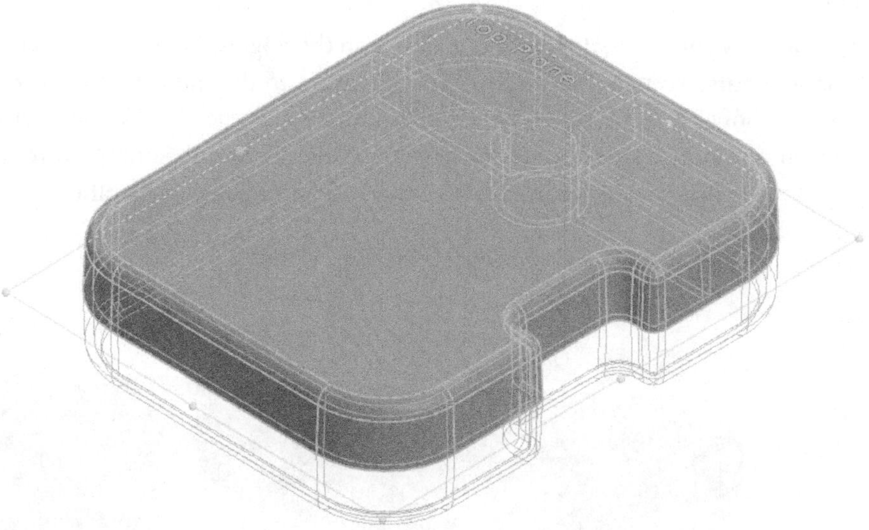

Select the faces on which the groove will be created. Next, click in the edge selection box available in the Groove Selection section, and then select the edges along which the groove will be created.

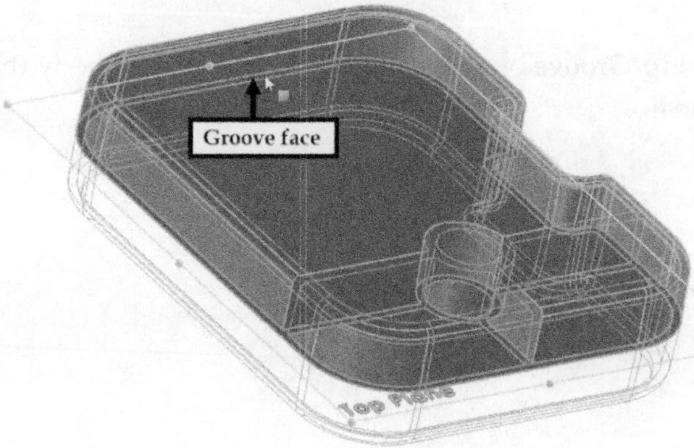

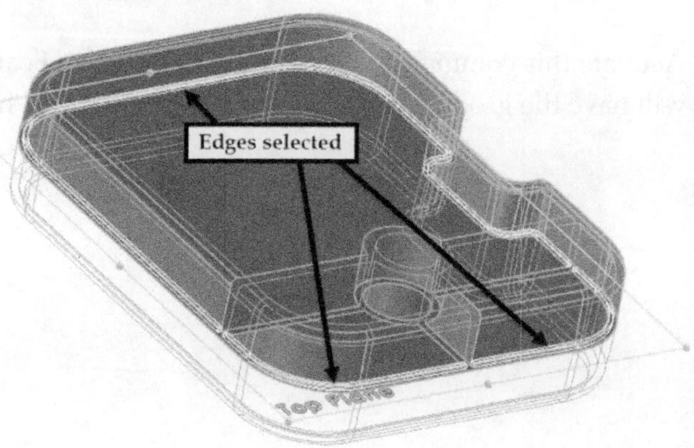

Zoom-in to the rib feature and notice that gaps are displayed at the intersections. Check the **Jump gaps** option in the **Groove Selection** section.

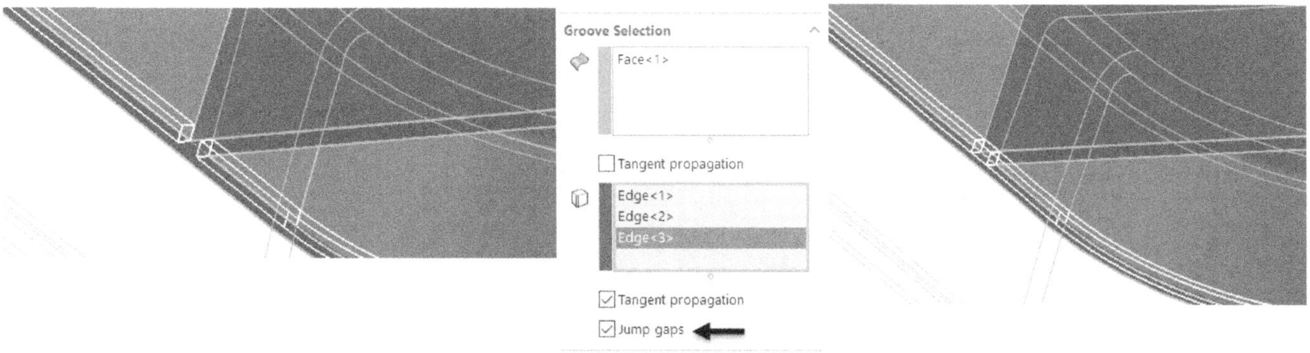

Click in the face selection box in the **Lip Selection** section, and then select the face on which the lip will be created. Next, click in the edge selection box and select the lip body's inner edge, as shown.

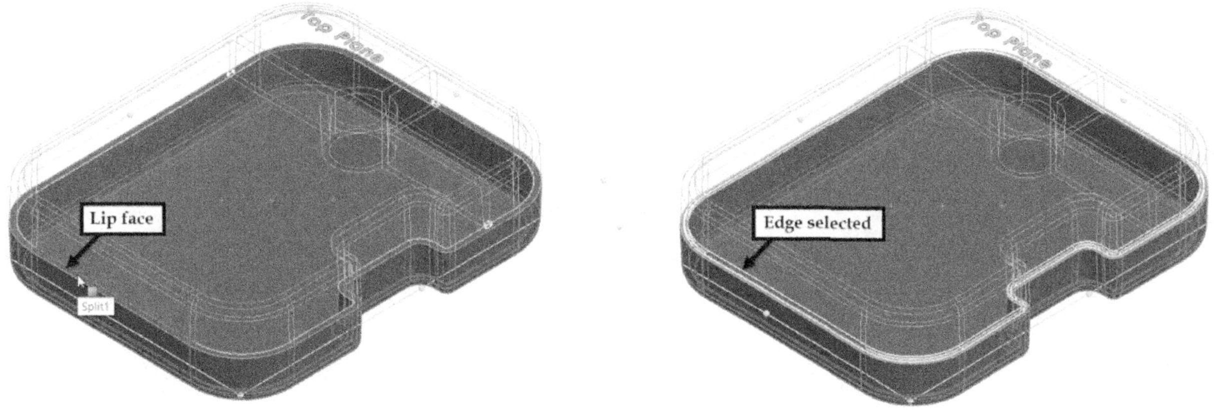

Next, specify the parameters of the lip/groove feature in the **Parameters** section. The parameters are self-explanatory. The **Link matched values** option specifies all the values by using any single value that you enter. The **Maintain existing wall faces** option creates the lip/groove feature by maintaining the draft or curvature of the wall faces. If you uncheck this option, the lip/groove feature will be created vertically.

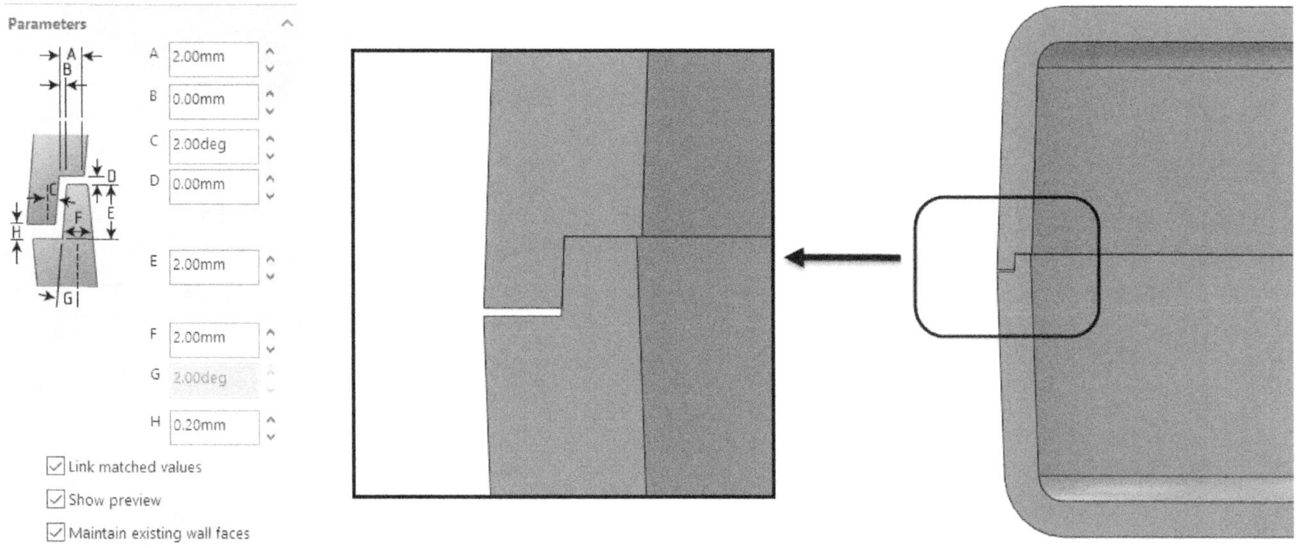

Mounting Boss

The process of creating mounting bosses can be automated using the **Mounting Boss** command. Activate the **Mounting Boss** command (click **Insert > Fastening Feature > Mounting Boss** on the menu bar) and select the top face of the model. Next, click in the Circular edge selection box and select a circular edge to define the mounting boss's location. Select the boss-type from the **Boss Type** section.

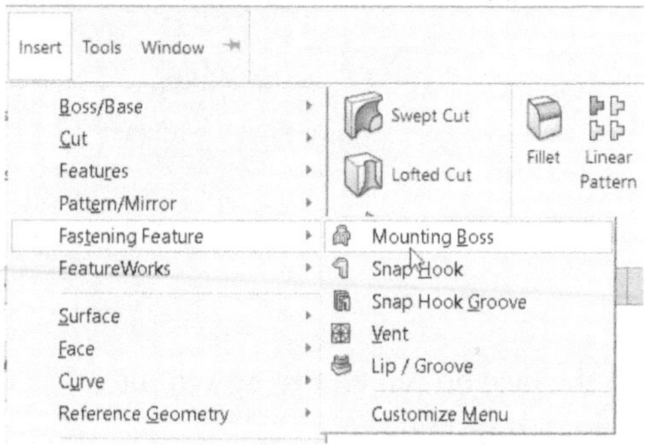

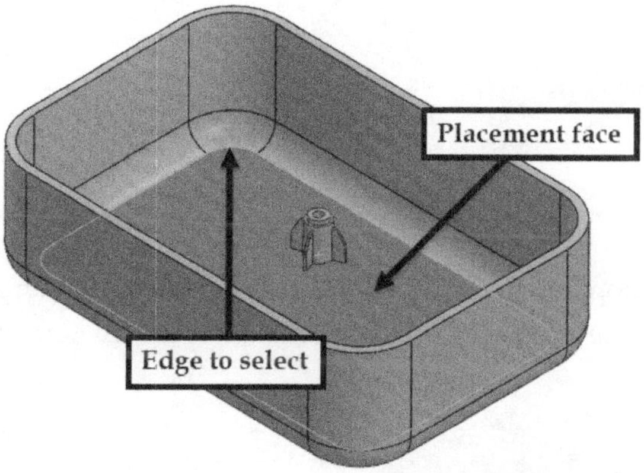

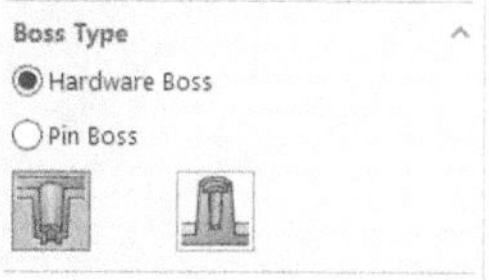

Define the parameters of the mounting boss in the **Boss** and **Fins** section. The parameters are self-explanatory. Click **OK** to complete the feature.

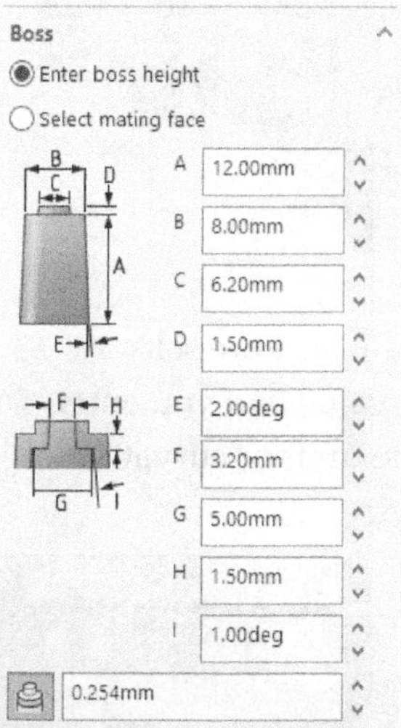

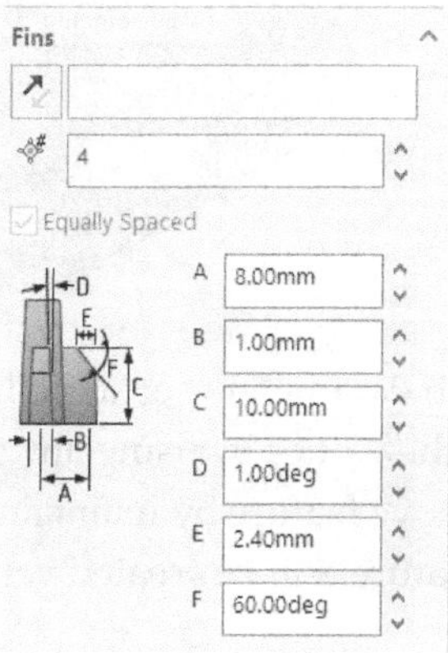

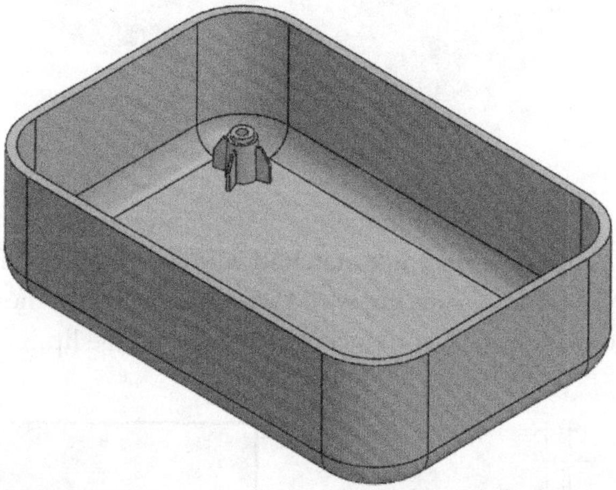

Indent

This command allows you to change the shape of a solid body by using another solid body. The solid body that is changed is called the target body, and the solid body that causes the changes is called the tool body. To create an emboss feature, you must have two solid bodies in a part. Activate the **Indent** command (click **Insert > Features > Indent** on the menu bar) and select the target and tool bodies. Type-in values in the **Thickness** and **Clearance** boxes.

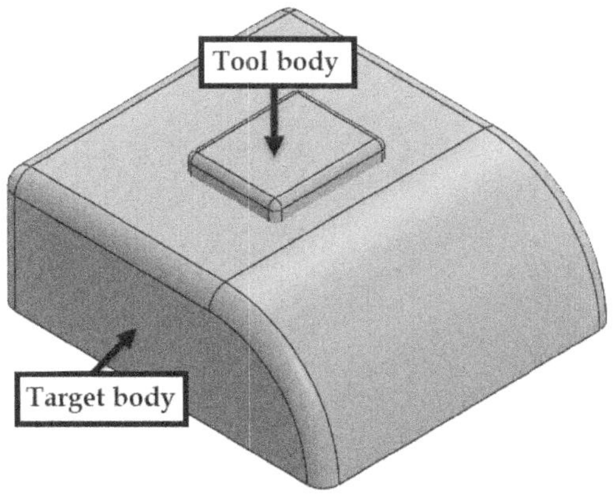

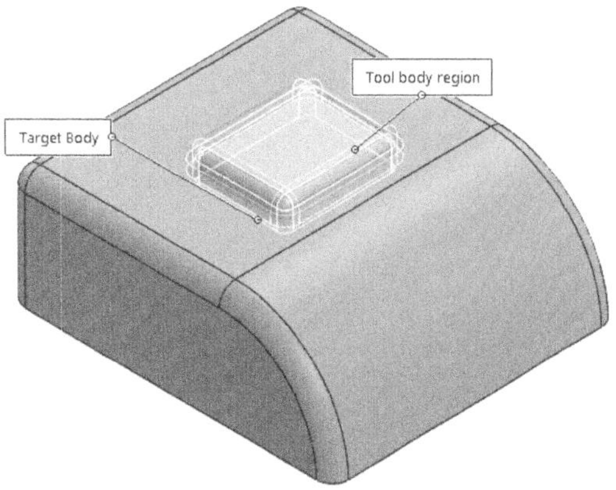

Use the **Keep Selections** or **Remove Selections** option on the PropertyManager to define the side on which the body is indented. Click the green check on the PropertyManager to complete the indent feature.

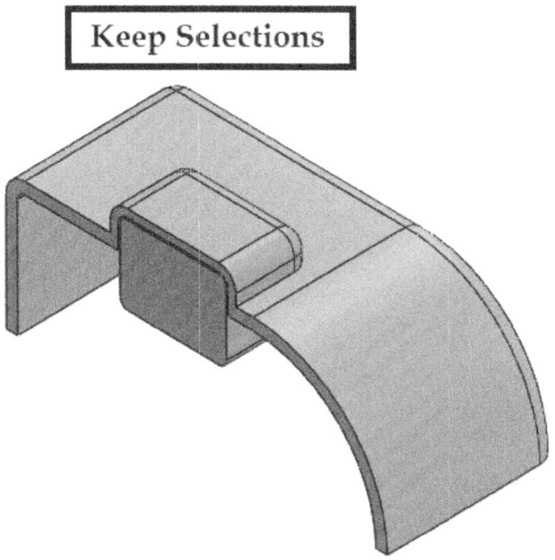

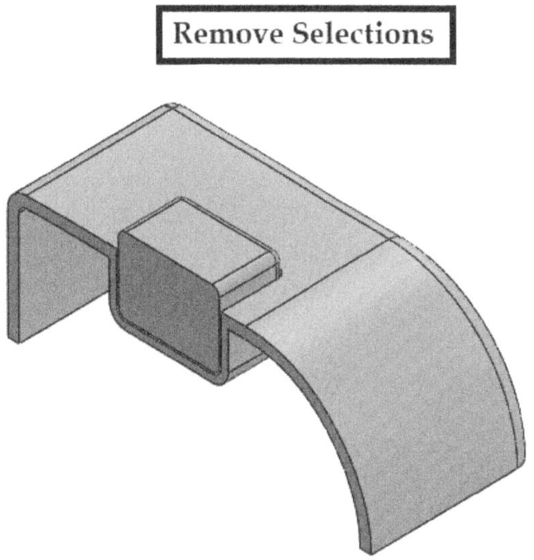

Bounding Box

You can utilize the **Bounding Box** command to encase a 3D geometry within a minimum volume. It serves various purposes such as determining the space required for shipping and packaging a product, as well as finding the length, width, and height of the stock required for a body. On the CommandManager, click **Features** > **Reference Geometry** drop-down > **Bounding Box**. On the PropertyManager, select the Type of Bounding Box. There are two types of bounding boxes that you can create: **Rectangular** and **Cylinder**. The Rectangular bounding box is useful for parts and assemblies of various shapes. The Cylinder bounding box is useful for parts and assemblies that are cylindrical in shape. Next, you need to select the **Reference Face/Plane** option. There are two options: **Best Fit** and **Custom Plane**. The **Best Fit** option is selected by default. It selects the XY plane as the reference plane. You can also select the **Custom Plane** option and define a reference plane based on your requirement. Next, examine the options in the **Options** section (**Include hidden bodies**, **Include surfaces**, and **Show preview**). Click **OK** to create the bounding box.

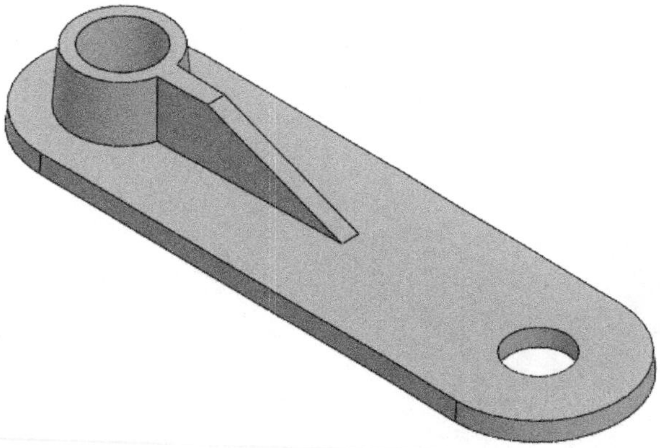

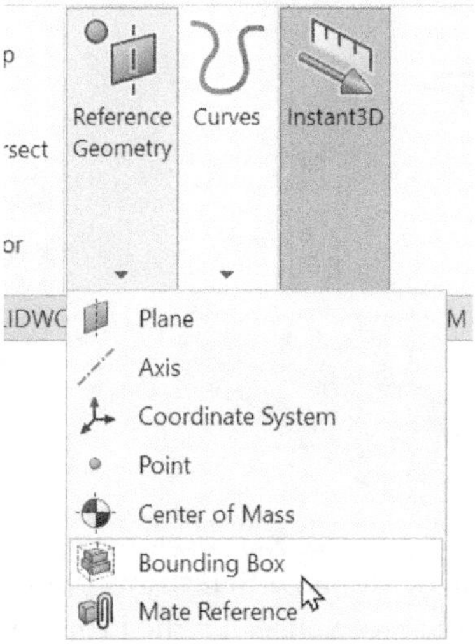

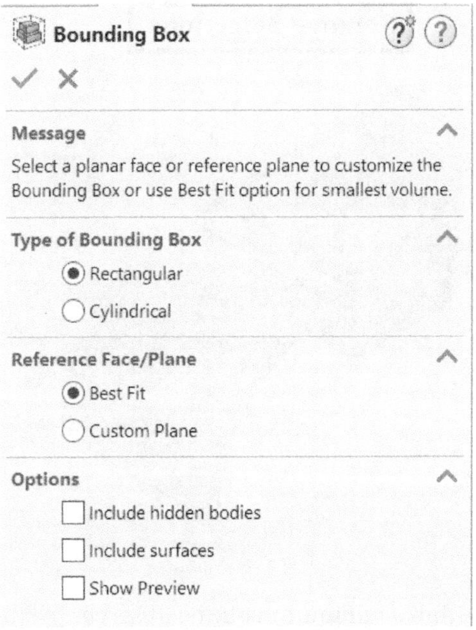

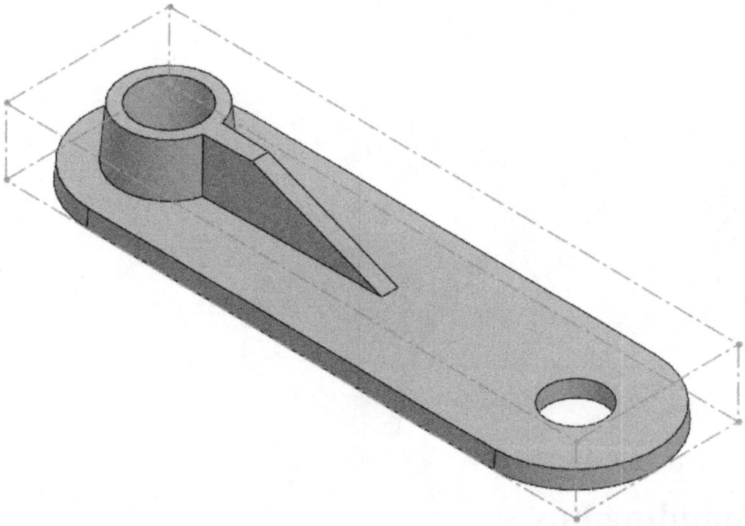

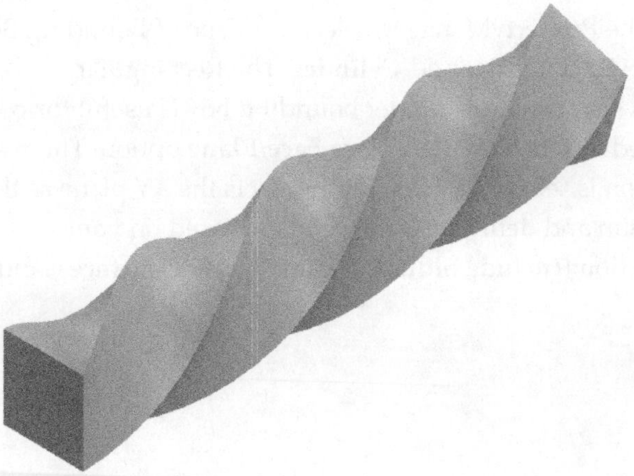

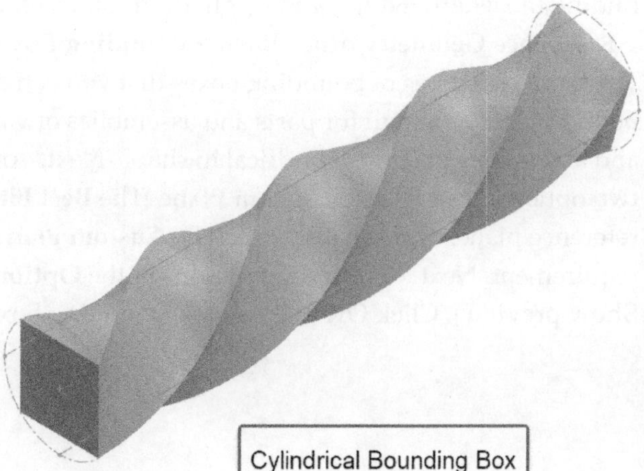

Cylindrical Bounding Box

Examples

Example 1 (Millimetres)

In this example, you will create the multi-body part shown next.

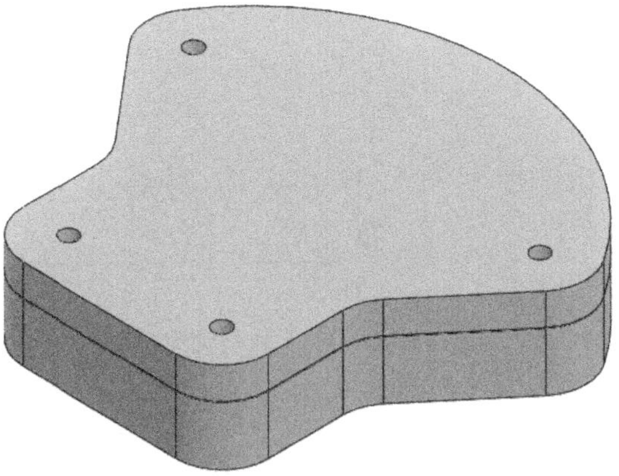

Lower Body

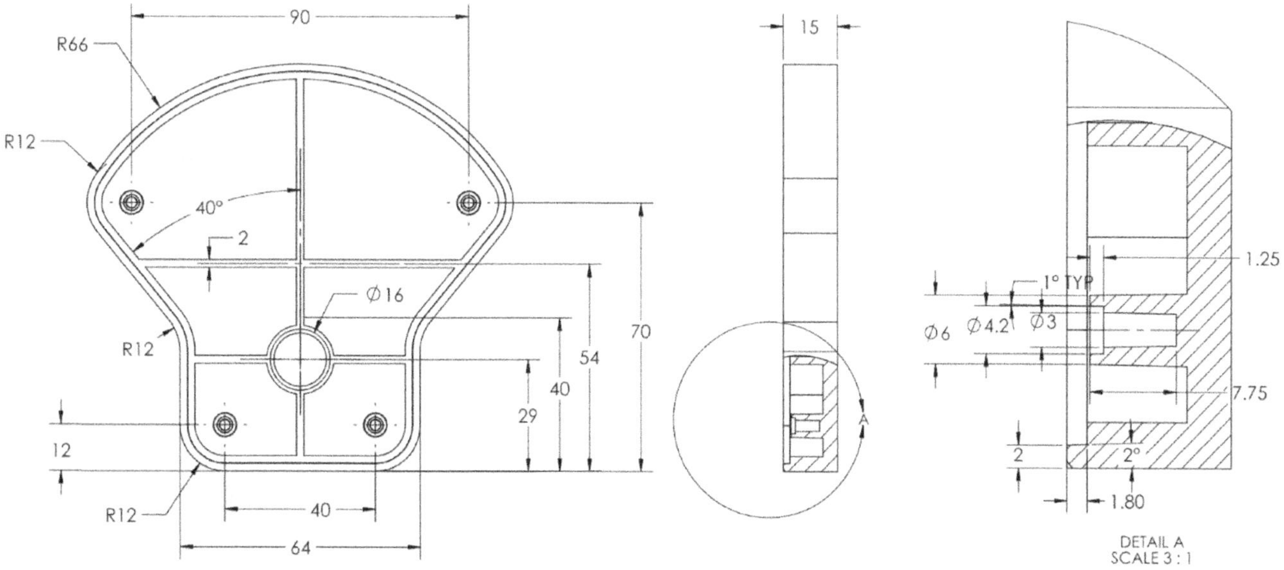

Upper Body

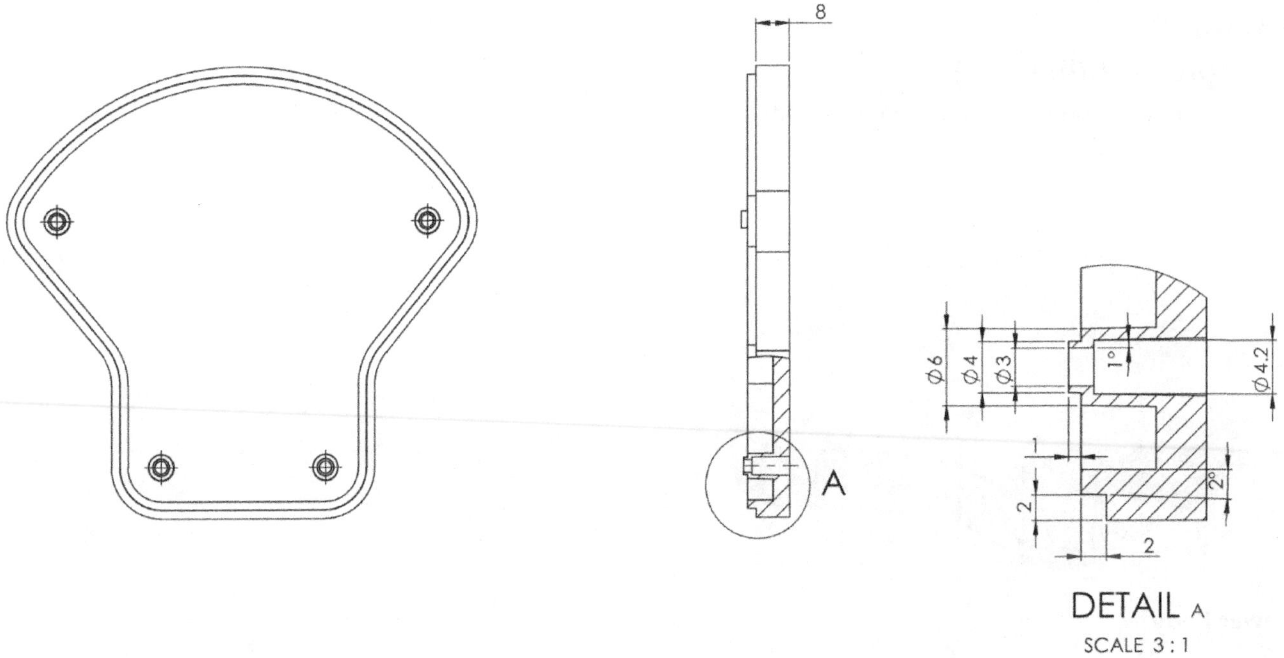

1. Start **SOLIDWORKS 2024**.
2. Click the **New** icon on the Quick Access Toolbar. Next, click the **Part** button, and then click **OK**; a new part file is opened.
3. To start a new sketch, click **Sketch > Sketch** on the CommandManager, and then select the Top Plane.
4. On the CommandManager, click **Sketch > Line** and draw the sketch, as shown in the figure below. Also, create a vertical centerline passing through the origin.
5. On the CommandManager, click **Sketch > Mirror Entities**, and then mirror the vertical and inclined line about the centerline. Next, apply the **Merge** relation between the endpoints, as shown.

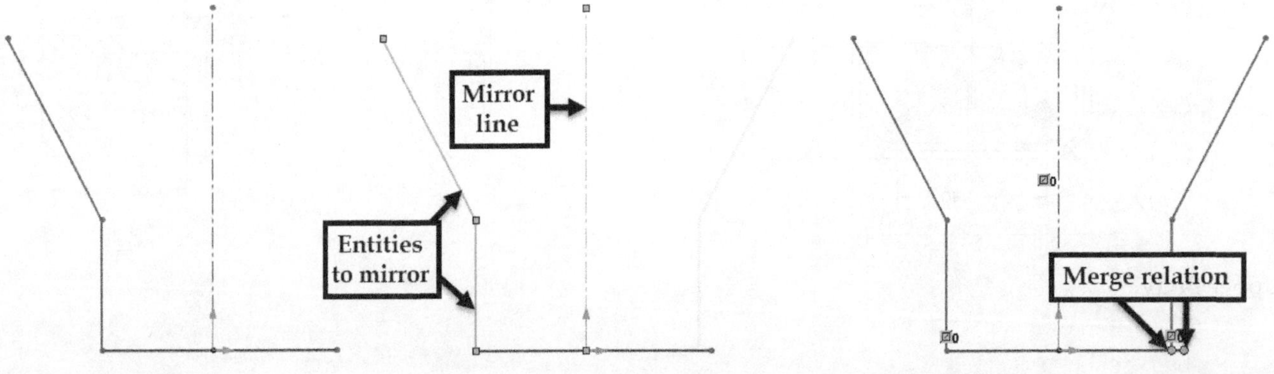

6. On the CommandManager, click **Sketch > Arc drop-down > 3 Point Arc**, and then create an arc by specifying the points in the sequence, as shown.
7. Apply dimensions to the sketch.

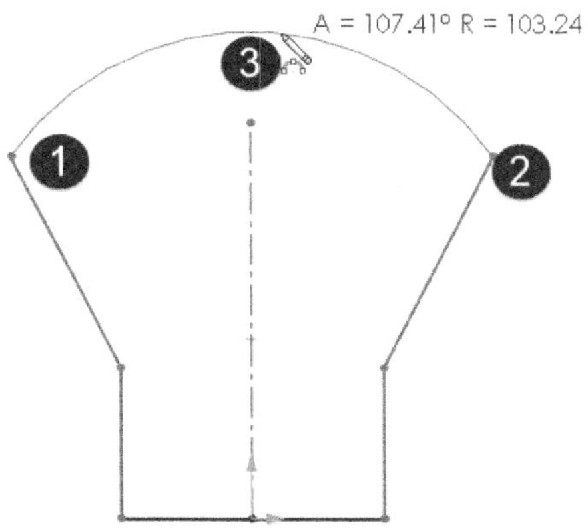

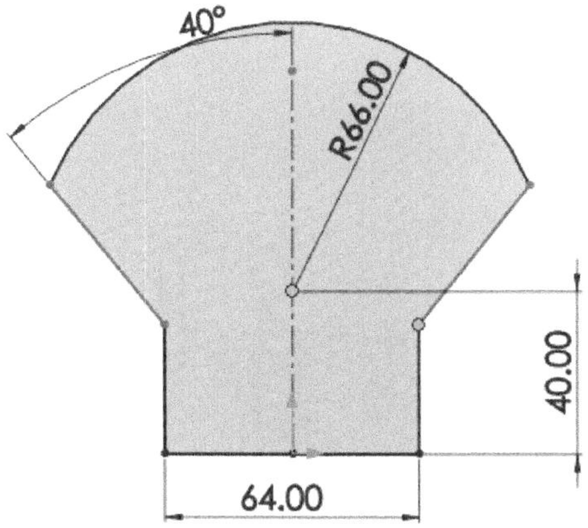

8. On the CommandManager, click **Sketch > Fillet drop-down > Sketch Fillet**, and then type **12** in the **Fillet Radius** box. Create the fillet at the sharp corners, as shown.
9. On the CommandManager, click **Sketch > Smart Dimension**. Add a linear dimension between the center points of the fillets, as shown.

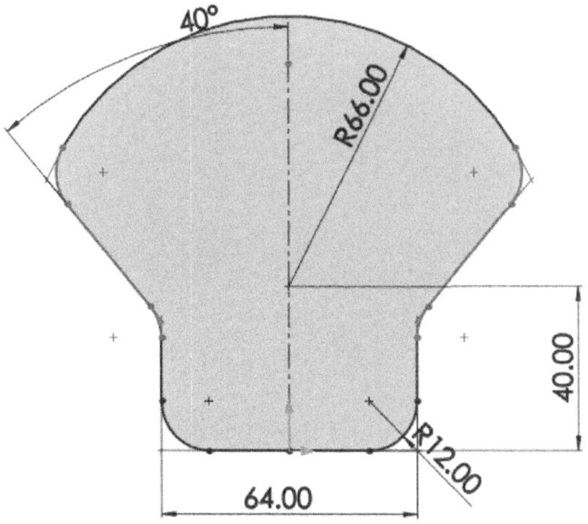

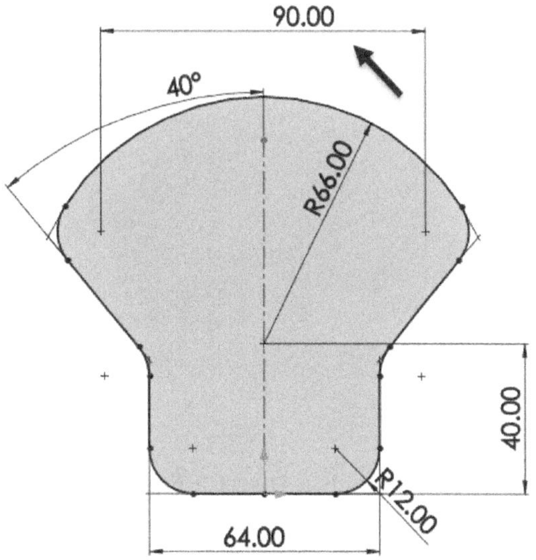

10. Click **Exit Sketch** on the CommandManager, and then create the *Extrude Boss/Base* feature of 23 mm depth.

11. On the CommandManager, click **Features** tab > **Shell**. Next, type 4 in the **Thickness** box and click **OK**.

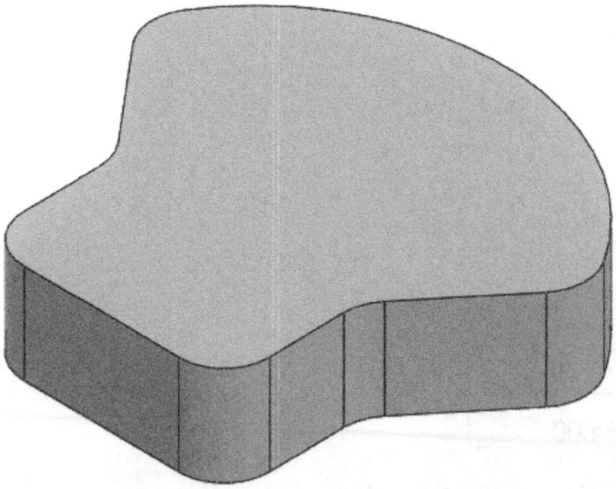

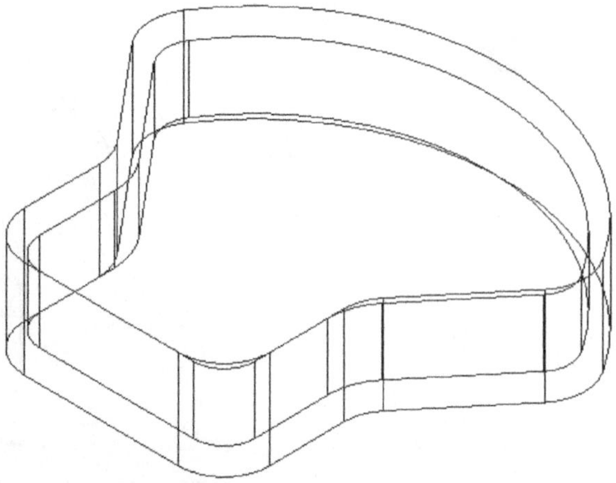

12. Create a reference plane offset to the top face in the downward direction. The offset distance is **8** mm.
13. Activate the **Split** command (on the menu bar, click **Insert > Features > Split**). Select the reference plane and click the **Cut Part** button.
14. Click the **Select all** option in the **Resulting Bodies** section. Click **OK** to split the model into two bodies.

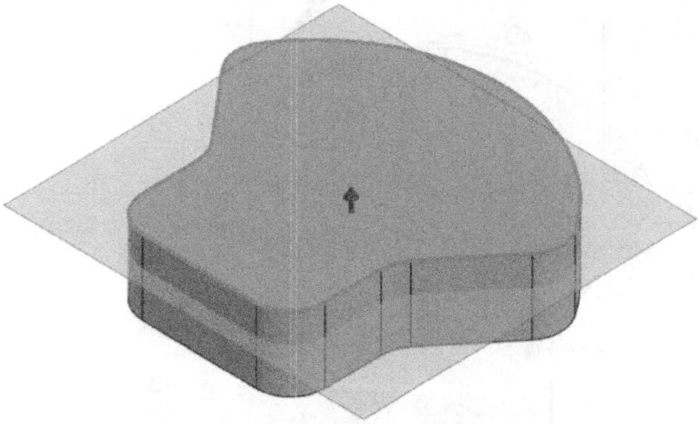

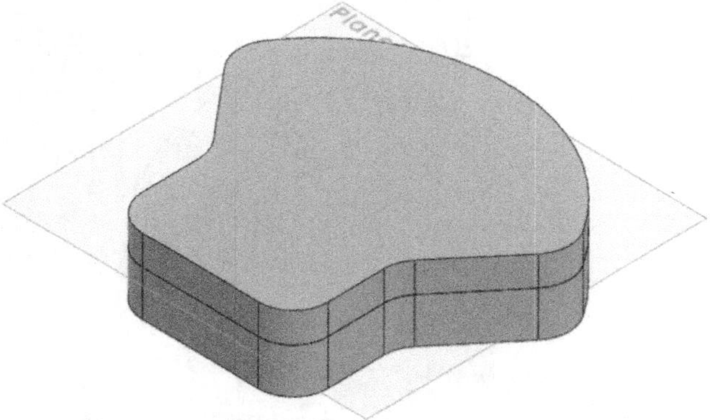

15. On the menu bar, click **Insert > Fastening Feature > Lip /Groove** . Select the groove and lip bodies, as shown. Next, select the offset plane to define the direction of the lip/groove feature.
16. Select the top face of the groove body. Next, click in the edge selection box in the **Groove Selection** section and select the Shell feature's inner edge to specify the path edges.

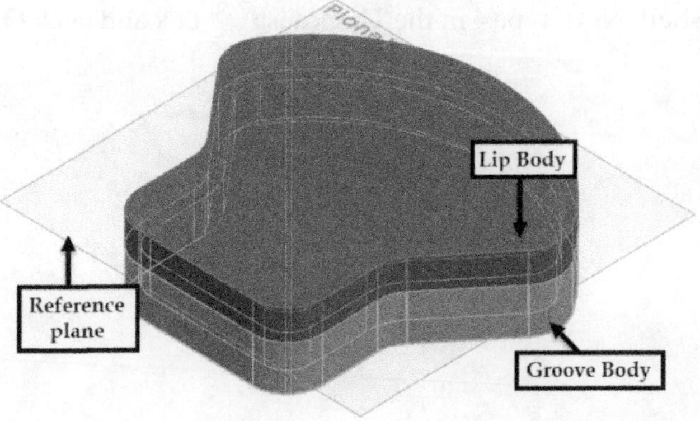

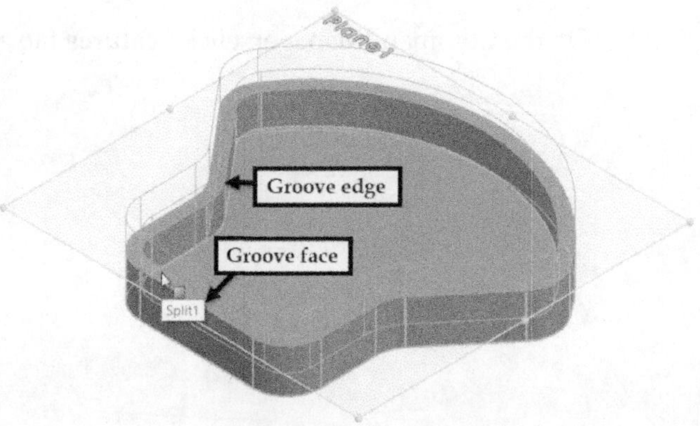

17. Click in the face selection box in the **Lip Selection** section, and then select the lip body's bottom face.

18. Click in the edge selection box in the **Lip Selection** section, and then select the lip body's inner edge.

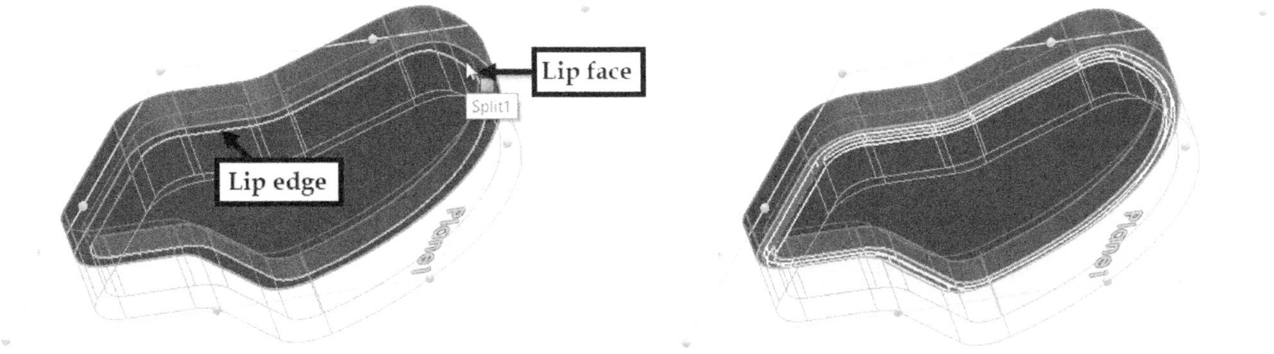

19. In the **Parameters** section, set the **E** and **F** values to 2 and the **H** value to 0.2. Make sure that the **Link matched values** option is checked. Click **OK** to create the lip/groove feature.
20. Click on the top solid body and select the **Hide** option.

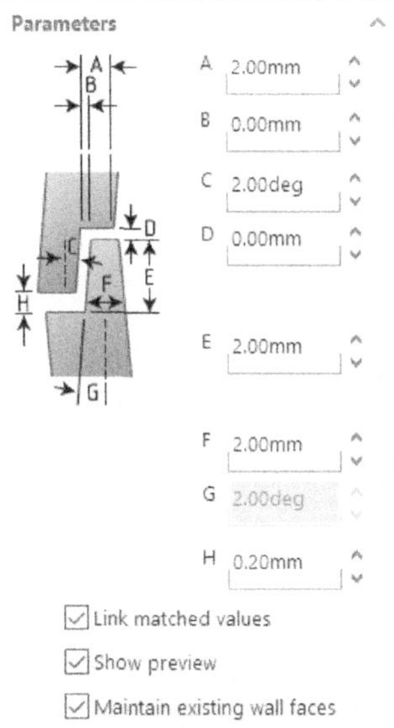

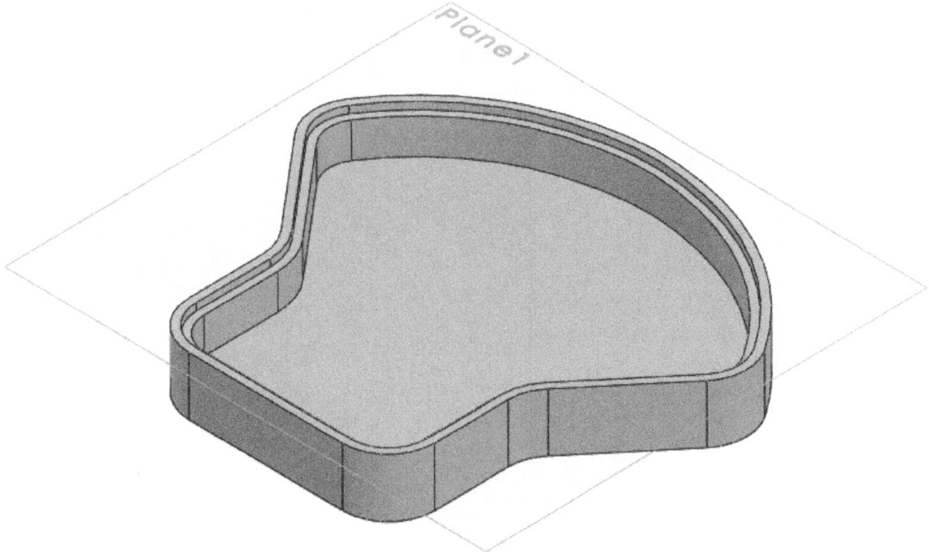

21. Activate the **Mounting Boss** command (click **Insert > Fastening Feature > Mounting Boss** on the menu bar) and click on the flat face, as shown.
22. Click in the Circular edge selection box, and then select the edge, as shown.

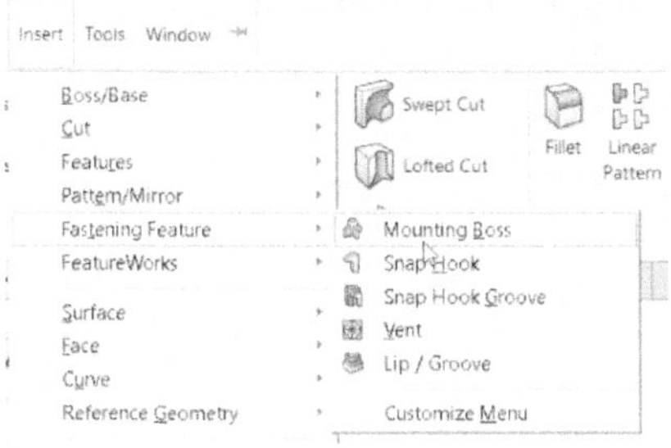

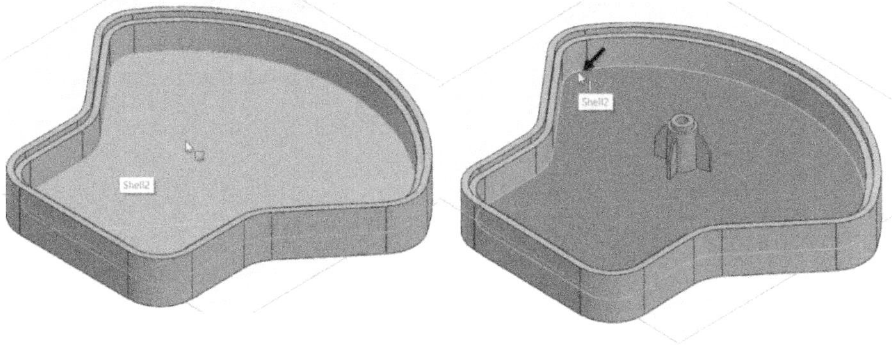

23. Set the **Boss Type** to **Hardware Boss**, and then click the **Thread** icon.
24. Go to the **Fins** section, and type **0** in the **Enter number of fins** box.

25. Select the **Select mating face** option from the **Boss** section. Next, click the groove face.
26. Type-in values in the **Boss** section, as shown. Click **OK** to create the mounting boss.

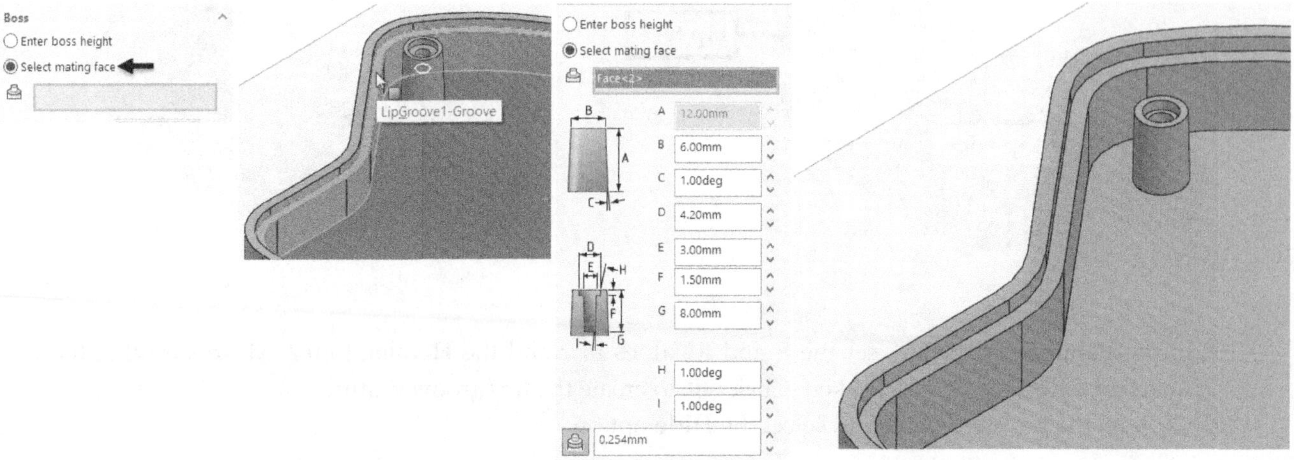

27. Likewise, create another mounting boss, as shown.

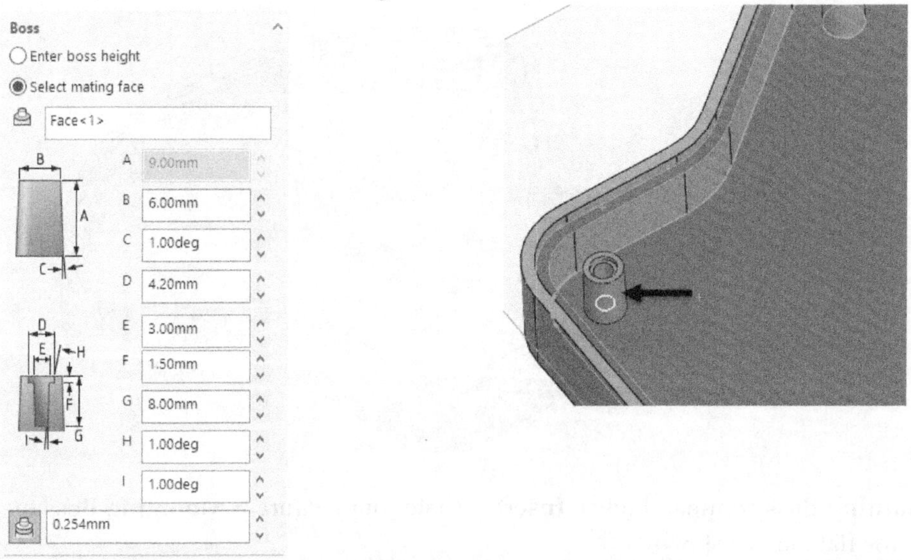

28. On the CommandManager, click **Features > Mirror** and select the **Right Plane** from the FeatureManager Design Tree.
29. Select the mounting boss features from the model and click **OK** to mirror them.

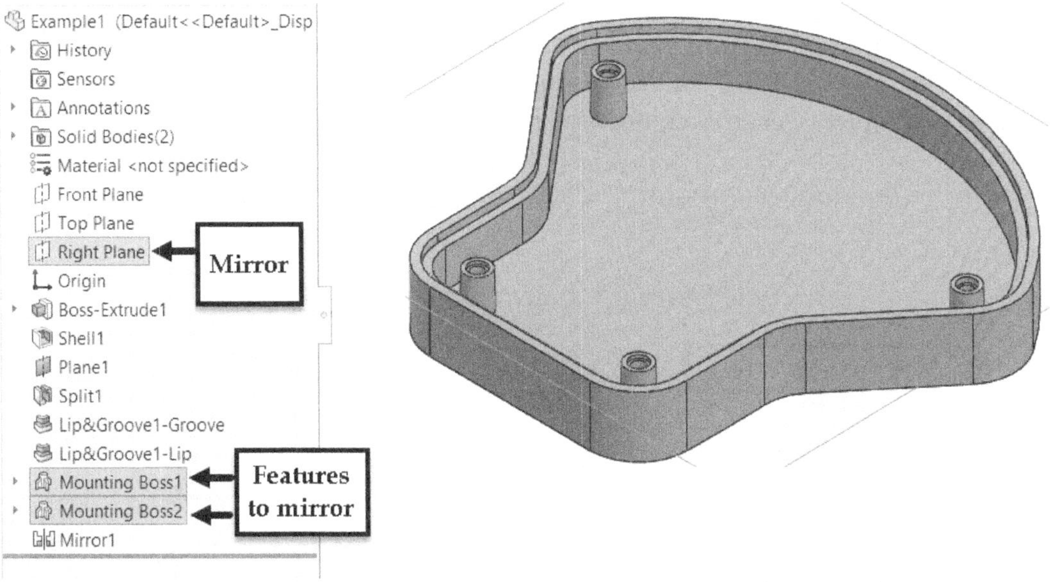

30. In the FeatureManager Design Tree, expand the **Solid Bodies** folder, click on the **Lip&Groove1-Lip** body, and then select **Show**.

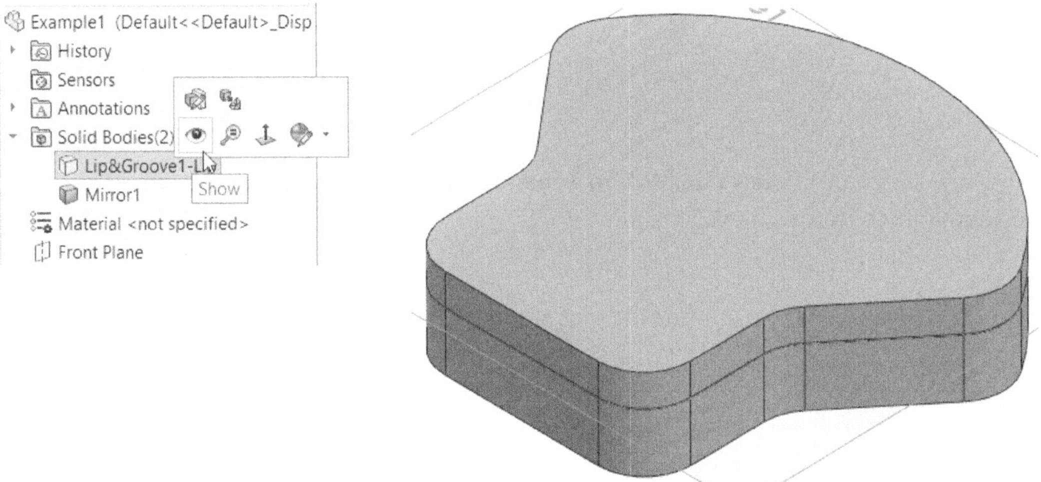

31. Click the **Section View** icon on the **View Heads Up** toolbar.
32. Click the **Right Plane** icon in the **Section 1** section, and then drag the section plane to the location, as shown. Click **OK** on the PropertyManager.

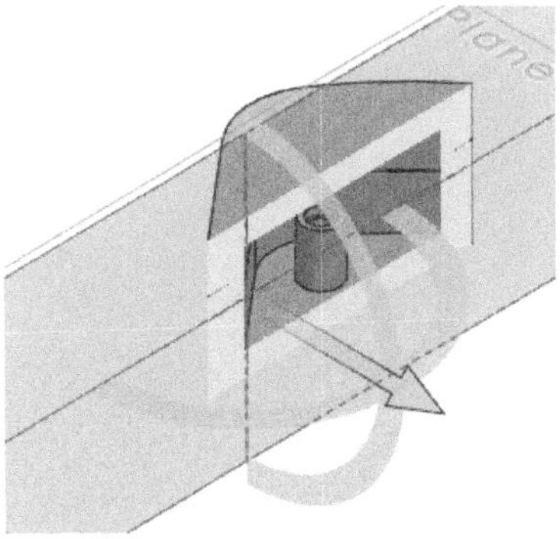

33. On the Menu bar, click **Insert > Fastening Feature > Mounting Boss** and select the flat face.
34. Click in the Circular edge selection box, and then select the edge, as shown.

35. Set the **Boss Type** to **Hardware Boss**, and then click the **Head** icon.
36. Go to the **Fins** section, and type **0** in the **Enter number of fins** box.
37. Select the **Select mating face** option from the **Boss** section. Select the top face of the mounting boss, as shown.
38. Type-in values in the **Boss** section, as shown. Click **OK** to create the mounting boss feature.

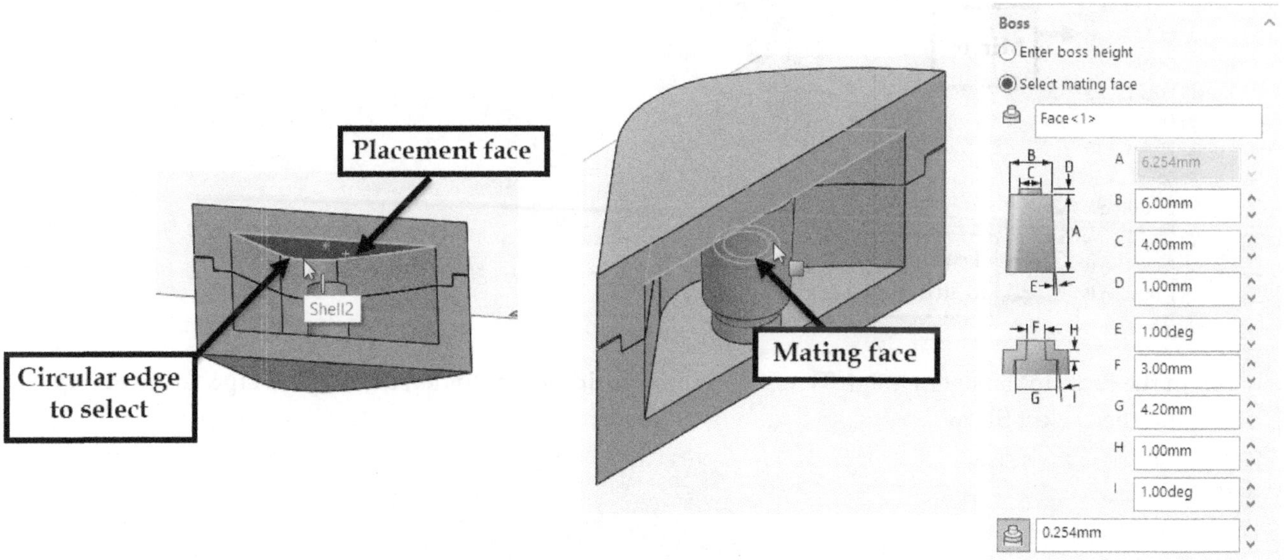

39. Click the **Section View** icon on the **View (Heads Up)** Toolbar to deactivate the section view.
40. Likewise, create another boss feature, as shown.

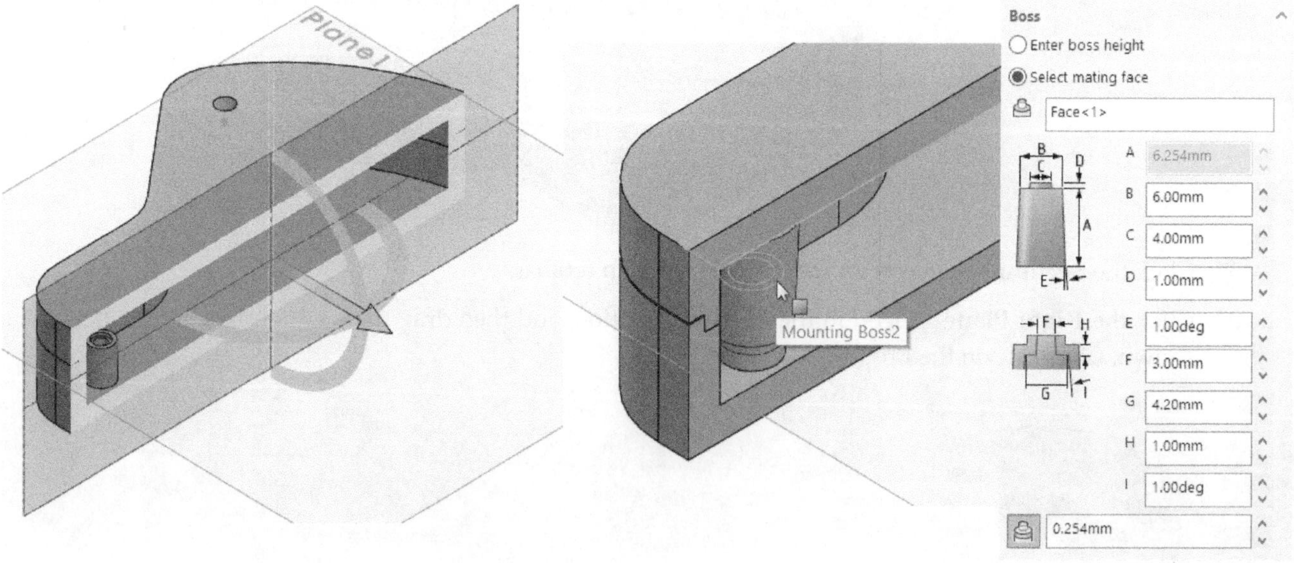

41. Mirror the mounting bosses about the right plane.

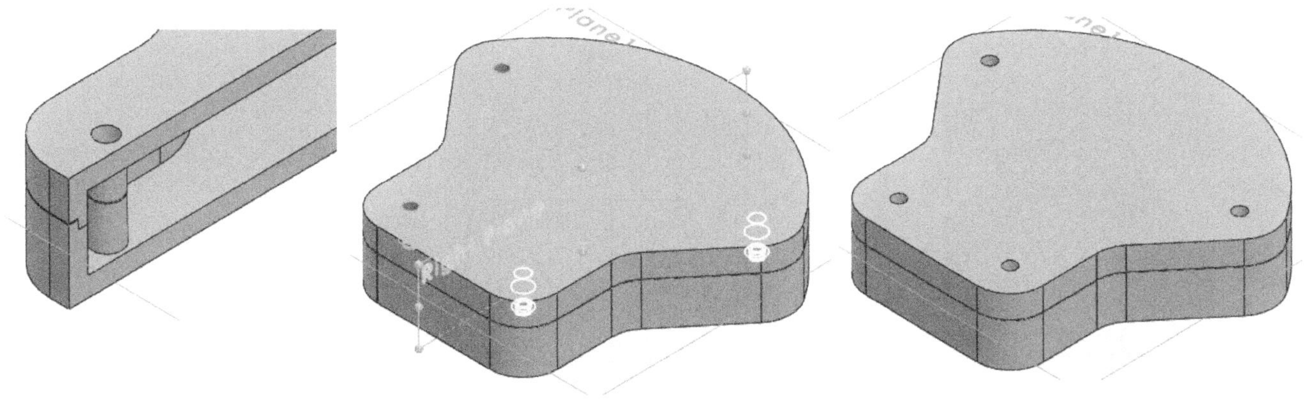

42. Hide the solid upper body.
43. On the CommandManager, click **Sketch > Sketch** and select the groove face.
44. Create a sketch using the **Line** and **Circle** commands, and then add dimensions to it. Click **Exit Sketch** on the CommandManager.
45. On the CommandManager, click **Features > Rib** and select the horizontal lines of the sketch.
46. On the **Rib** PropertyManager, type **2** in the **Rib Thickness** box and click the **Both Sides** icon.
47. Select the lower body and click **OK** to complete the rib feature.

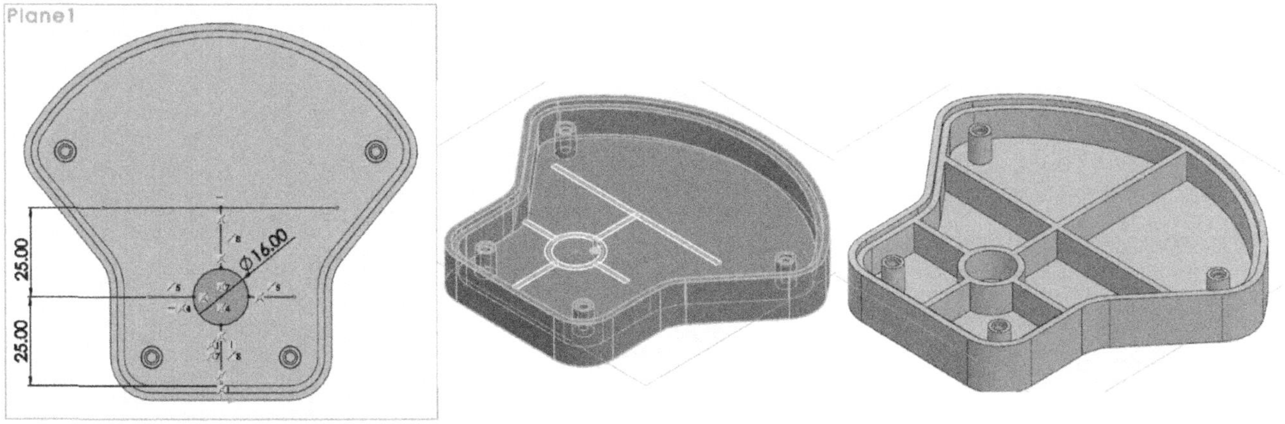

48. Save and close the file.

Questions

1. What is the use of the **Vent** command?
2. How many types of ribs can be created in SOLIDWORKS?
3. Why do we create multi-body parts?
4. Describe the terms 'Rib' and 'Spar' in the *Vent* feature.
5. How to save bodies as individual parts?

Exercises

Exercise 1

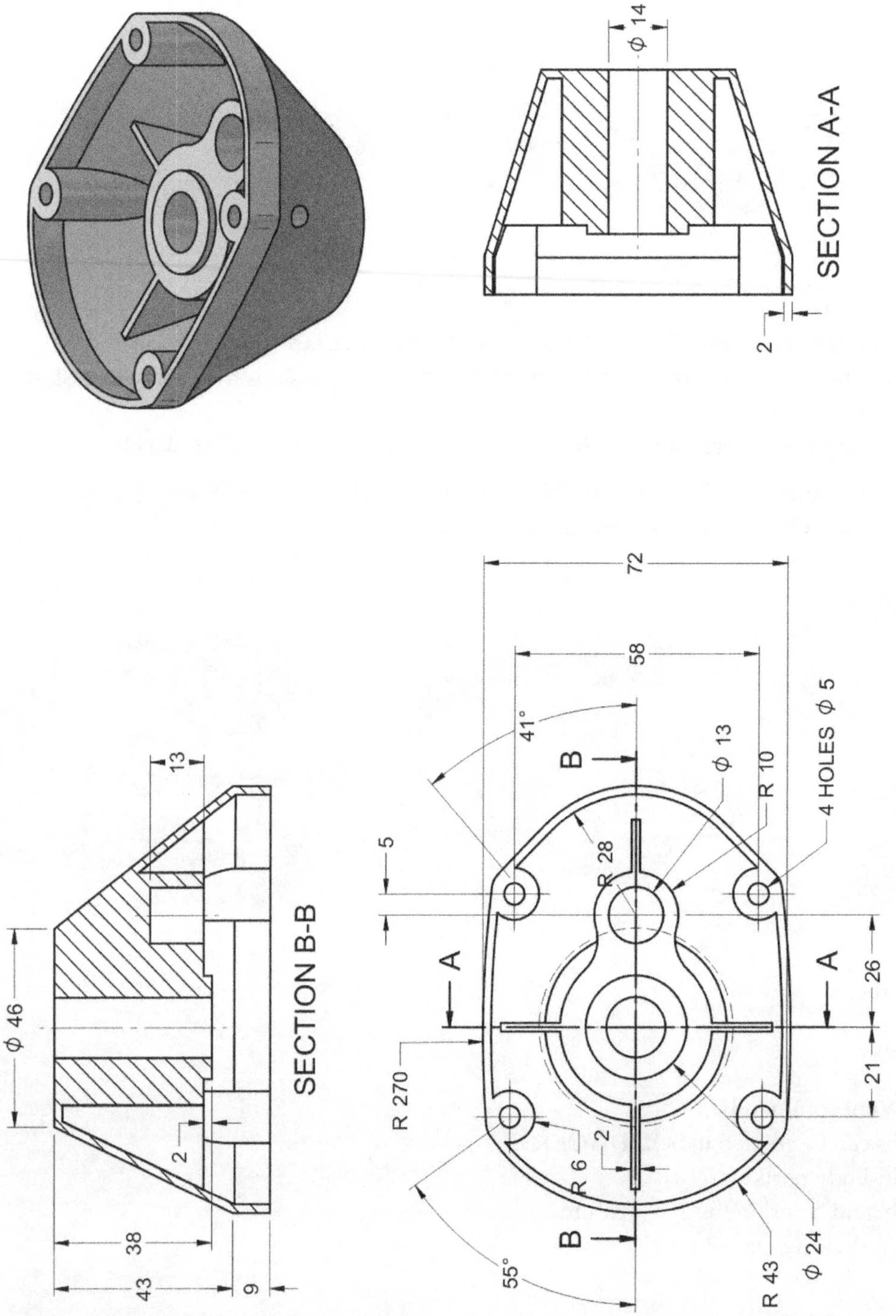

Exercise 2

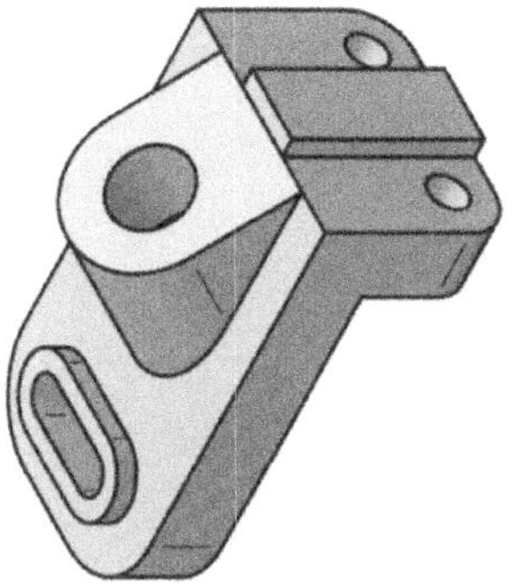

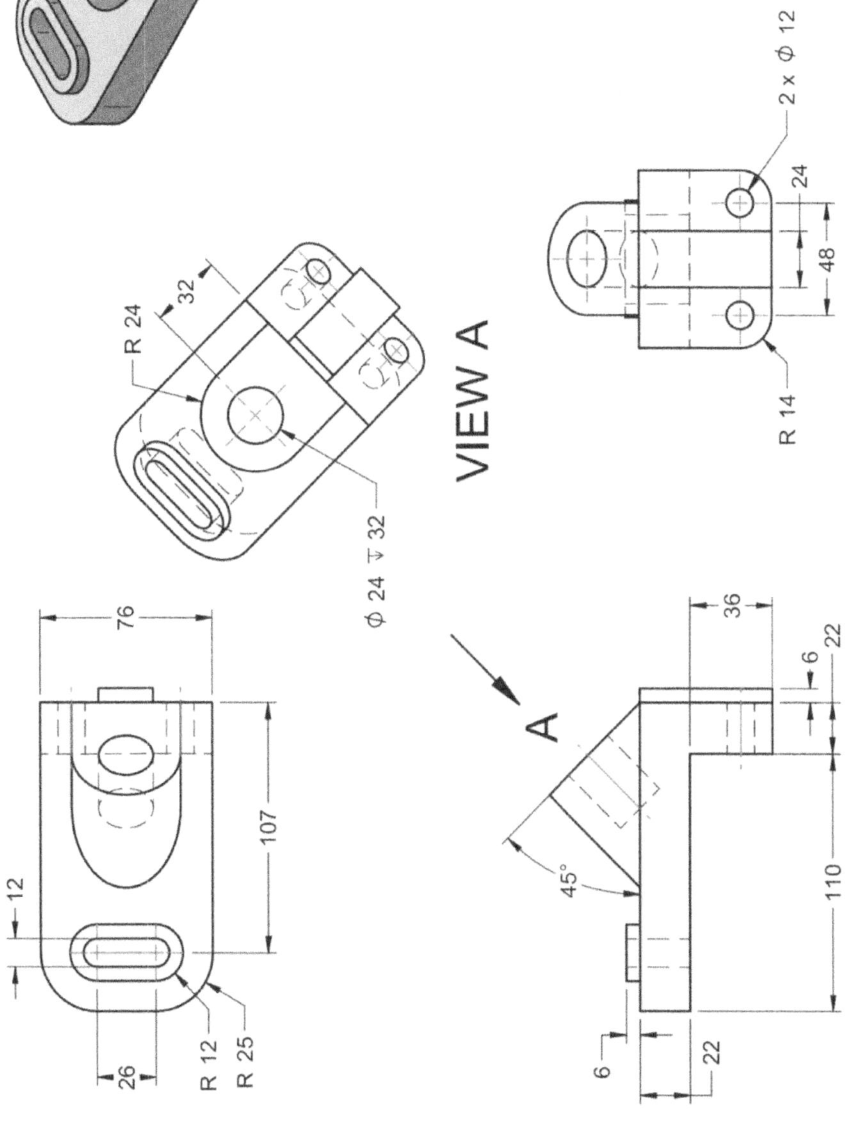
VIEW A
A
2 x Φ 12
R 14
24
48
R 24
32
Φ 24 ↧ 32
76
107
12
26
R 12
R 25
45°
36
6
22
110
6
22

Exercise 3 (Inches)

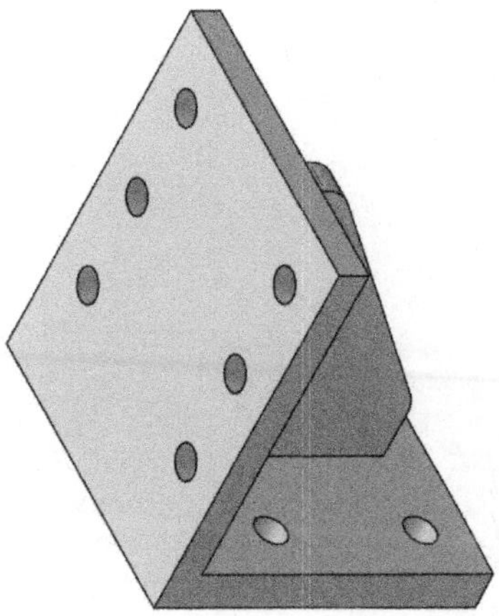

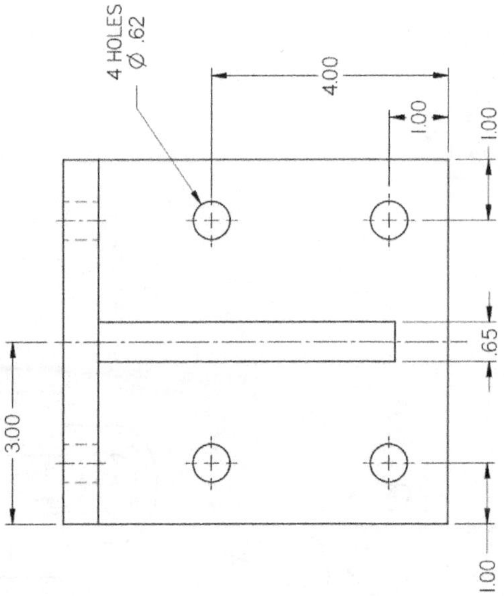

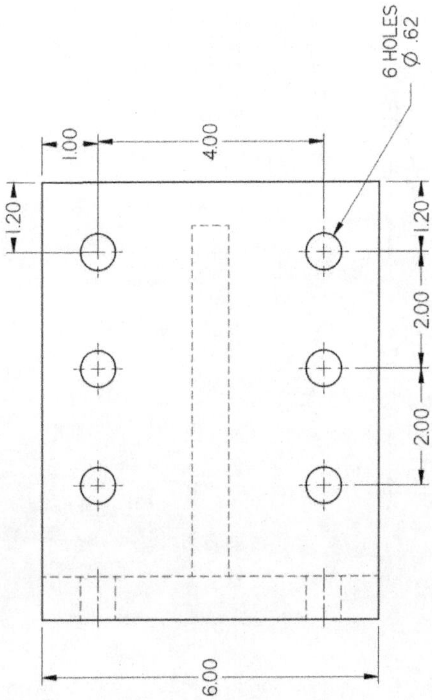

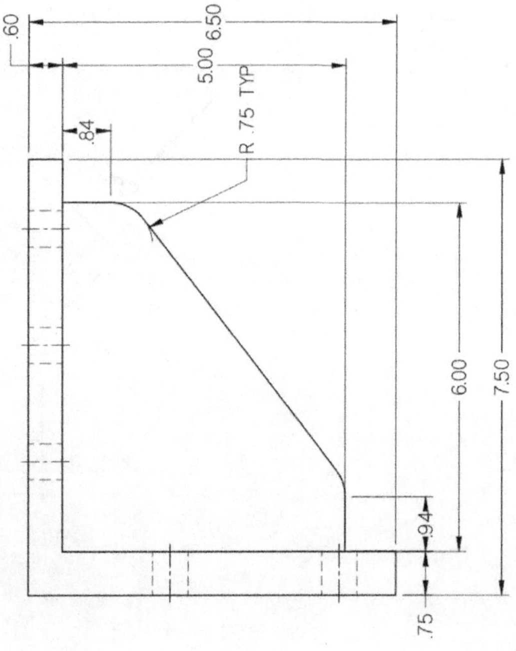

Chapter 9: Modifying Parts

In the design process, it is not required to achieve the final model in the first attempt. There is always a need to modify the existing parts to get the desired part geometry. In this chapter, you will learn various commands and techniques to make changes to a part.

The topics covered in this chapter are:

- *Edit Sketches*
- *Edit Features*
- *Suppress Features*
- *Move Faces*

Edit Sketches

Sketches form the base of a 3D geometry. They control the size and shape of the geometry. If you want to modify the 3D geometry, most of the time, you are required to edit sketches. To do this, click on the feature and select **Edit Sketch**. Now, modify the sketch and click **Exit Sketch** on the CommandManager. You will notice that the part geometry updates immediately.

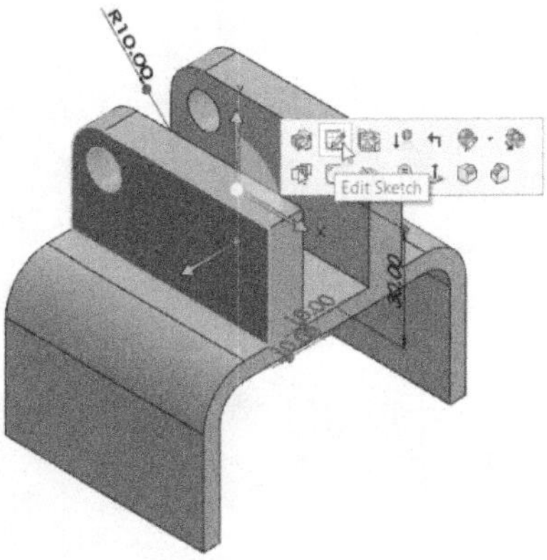

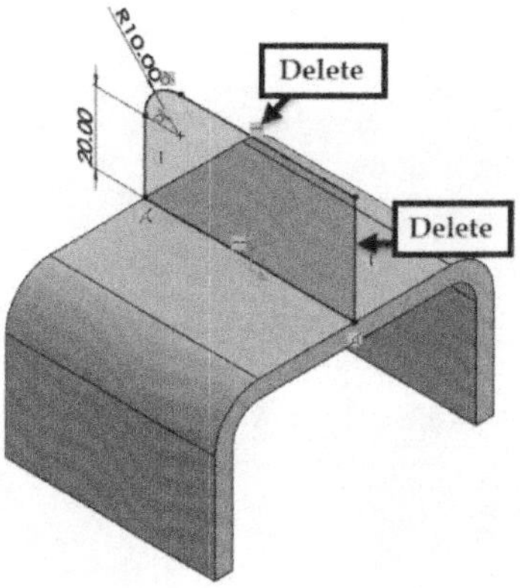

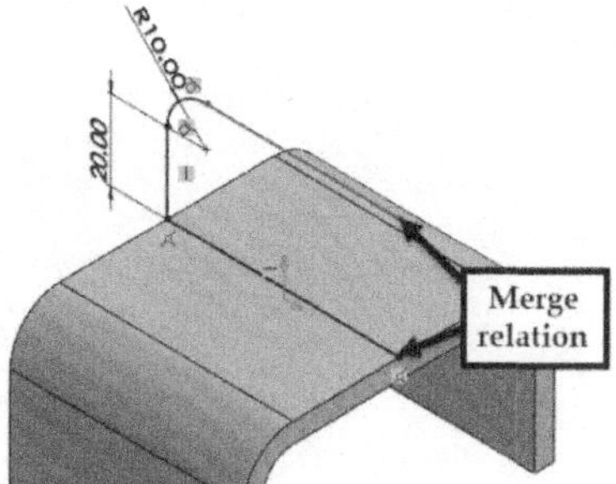

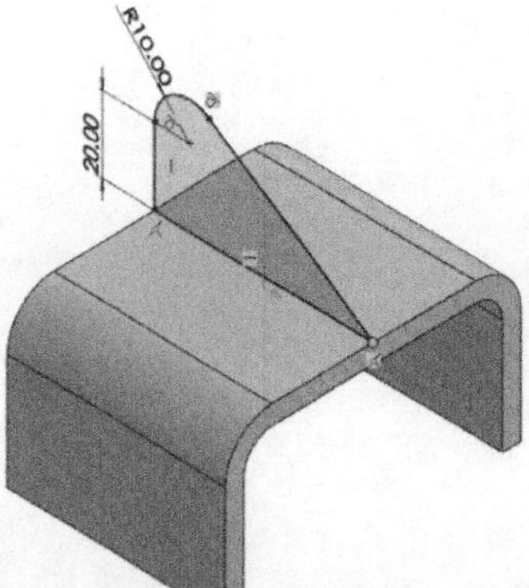

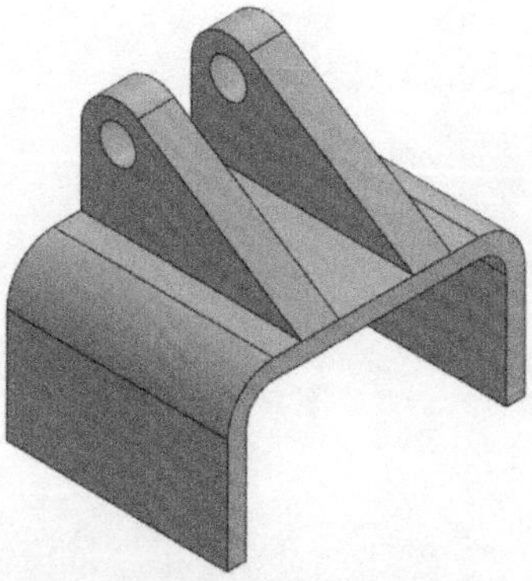

Managing sketches requires care and consideration. Deleting sketches must be approached cautiously to prevent complications. Here are some key points to consider when dealing with sketches:

Loss of References: SOLIDWORKS can lose references easily when deleting a sketch serving as a reference for others. This can lead to dependent sketches becoming overdefined or unsolvable, causing errors and difficulties in the model.

Difficulty in Editing: Removing a sketch makes it harder to edit the model later on. Keeping the original sketch intact and making changes gradually is generally more practical than starting from scratch. This maintains control over the model and allows for easier adjustments.

Unintended Consequences: Eliminating a sketch can affect different parts of the model unexpectedly. For instance, changing a line in a parent sketch could create issues for child features reliant on that line. These features might experience rebuild errors since they continue referring to the old line's faces or edges, not the new ones.

To handle sketches efficiently, you should exercise restraint while employing the Delete function. Preserve sketches and their constraints to ensure stability and predictability during the modeling process. This approach saves time and prevents unwanted complications.

Edit Features

Features are the building blocks of model geometry. To modify a feature, click on it and select **Edit Feature**. The PropertyManager related to the feature appears. On the PropertyManager, modify the parameters of the feature and click **OK**. The changes take place immediately.

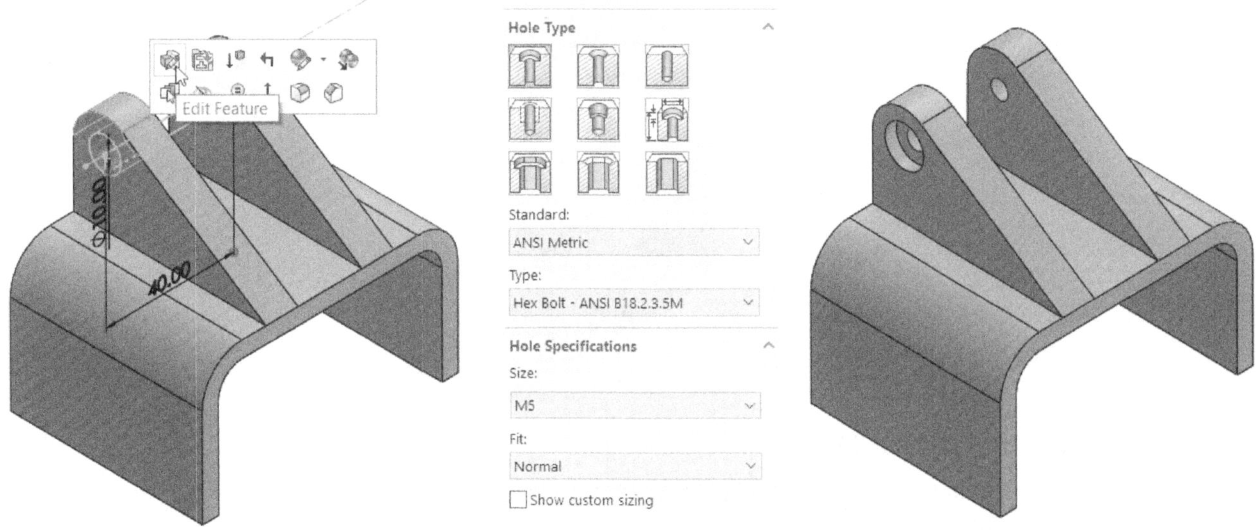

Suppress Features

Sometimes you may need to suppress some features of model geometry. Click on the feature to suppress, and then select **Suppress**.

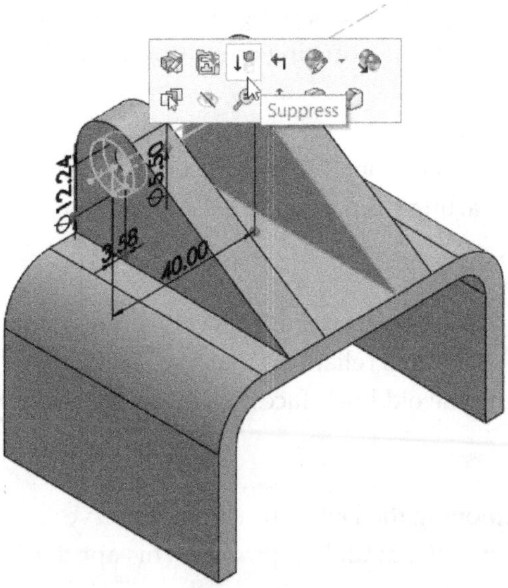

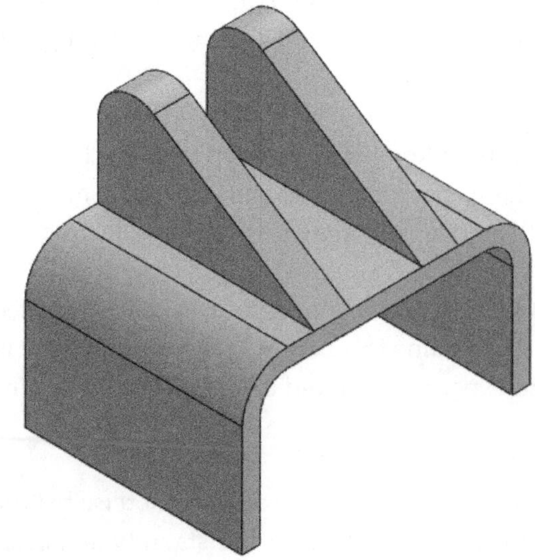

Resume Suppressed Features

If you want to resume the suppressed features, then right click on the suppressed feature in the FeatureManager Design Tree, and then select **Unsuppress**; the feature is resumed.

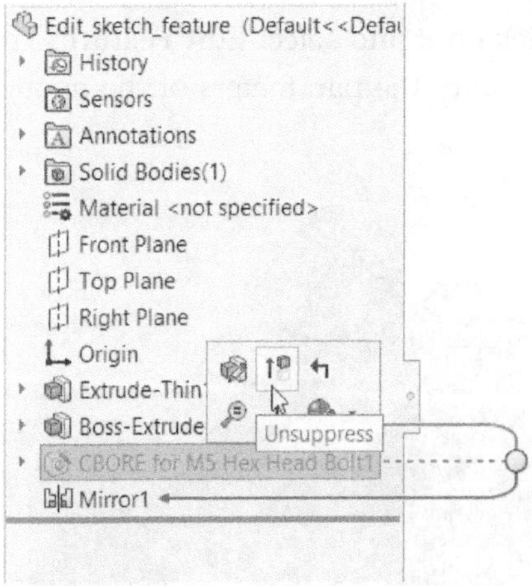

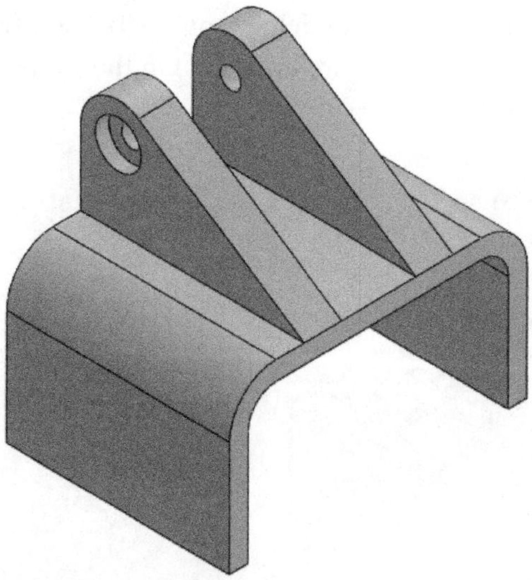

The Move Face command

SOLIDWORKS provides you with a special tool called **Move Face** (on the Menu bar, click **Insert > Face > Move**) to modify faces and planes of part geometry. You can perform three operations using this tool: **Offset**, **Translate** and **Rotate** faces.

Offset Faces

The Offset option moves the selected face in the direction perpendicular to it. To offset a face, select the **Offset** option from the **Move Face** section of the PropertyManager. Next, click on the face to offset, and type in a value in the **Distance** box available in the **Parameters** section. You can change the offset direction by checking the **Flip direction** option.

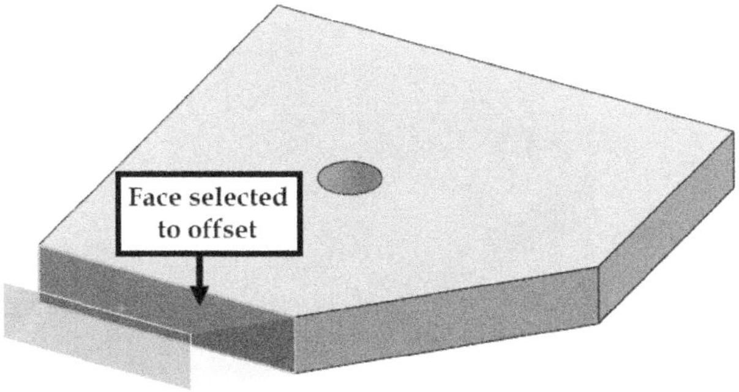

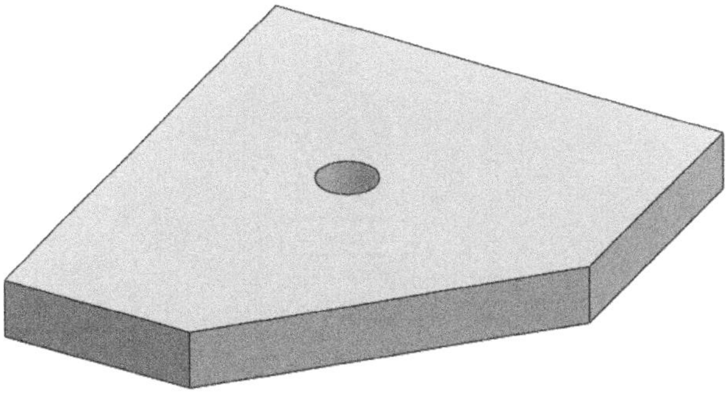

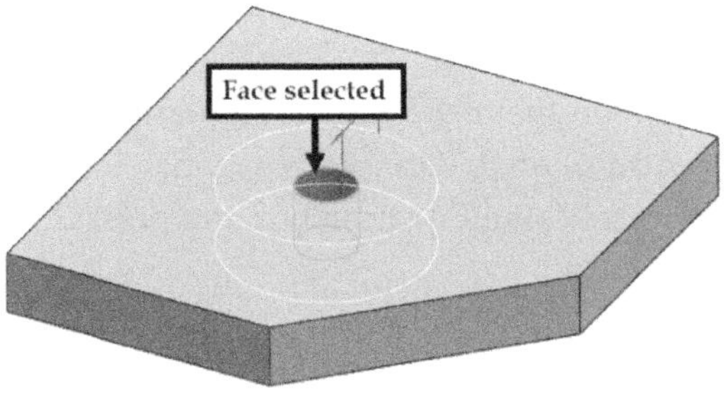

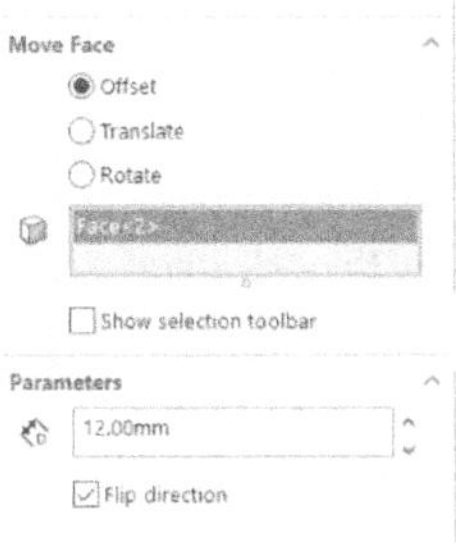

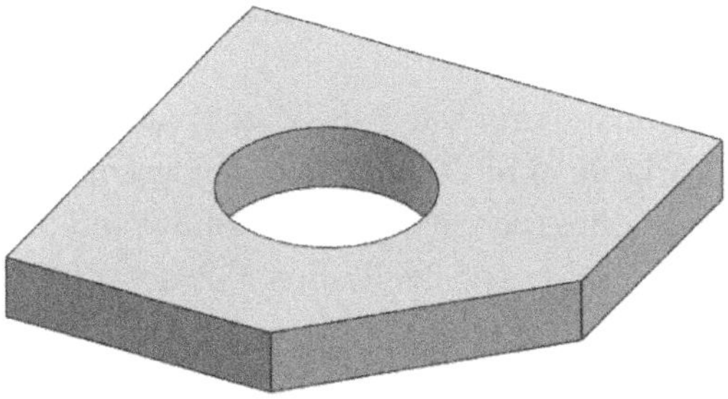

Translate Faces

The **Translate** option allows you to move the selected face(s) along X, Y, or Z directions. To translate faces in a particular direction, select the **Translate** option from the **Move Faces** section of the PropertyManager. Next, click on the face to translate, and then select any of the arrows that appear on the screen. Press and hold the mouse button, and then drag the pointer. You can specify the translation distance using the arrows or the **Distance** box available in the **Parameters** section of the PropertyManager.

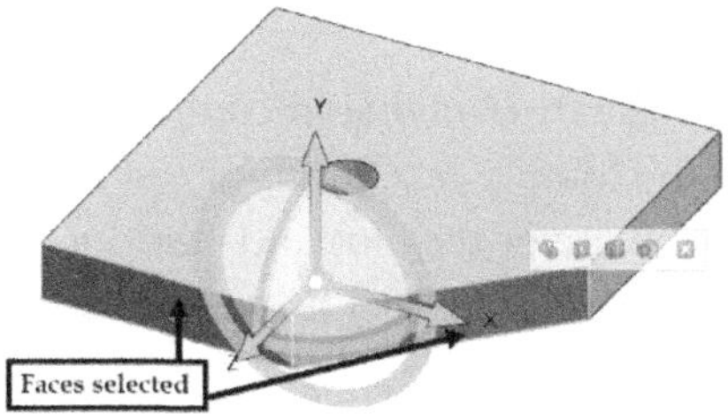

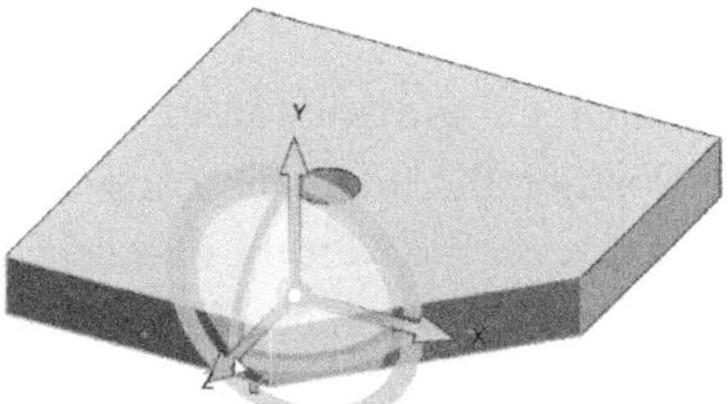

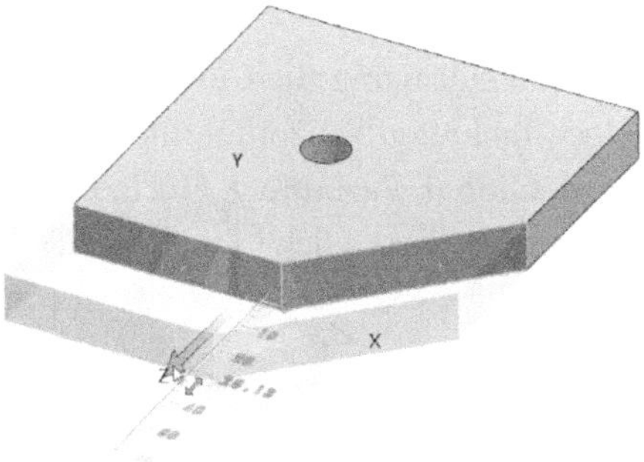

Rotate faces

The **Rotate** option helps you to rotate the selected faces about an axis. To rotate a face, you must click on it to display the manipulator. Click in the **Axis reference** box in the **Parameters** section, and then select an edge. This defines the axis of rotation. Now, type in a value in the **Rotation angle** box displayed in the **Parameters** section.

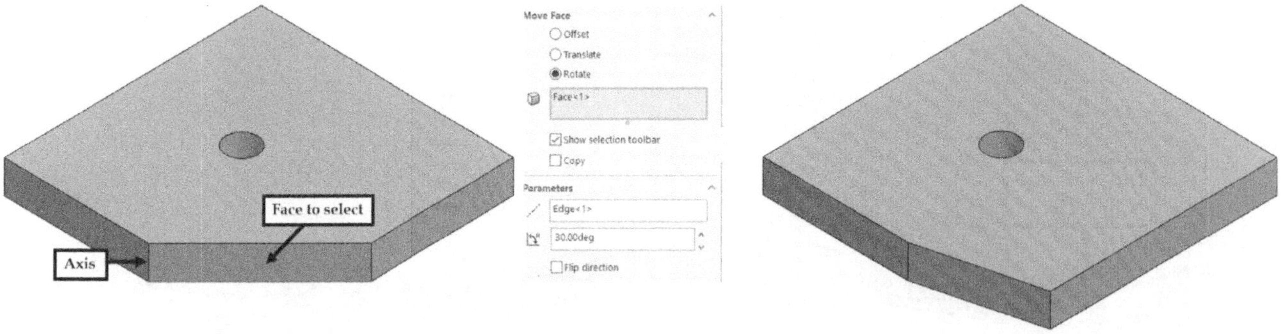

The Move/Copy Body command

The **Move/Copy Body** command can be used to translate or rotate bodies. For example, activate the Create Sketch command and select the Right plane. Next, click **Tools > Equations** on the menu bar; the **Equations, Global Variable, and Dimensions** dialog appears. On this dialog, click in **Add Global Variable** box under the **Global Variables** section. Type **Horizontal** and press ENTER. Next, type 50 in the **Value/Equation** box and press ENTER. Likewise, create the **Radius** global variable and set its **Value/Equation** to 20.

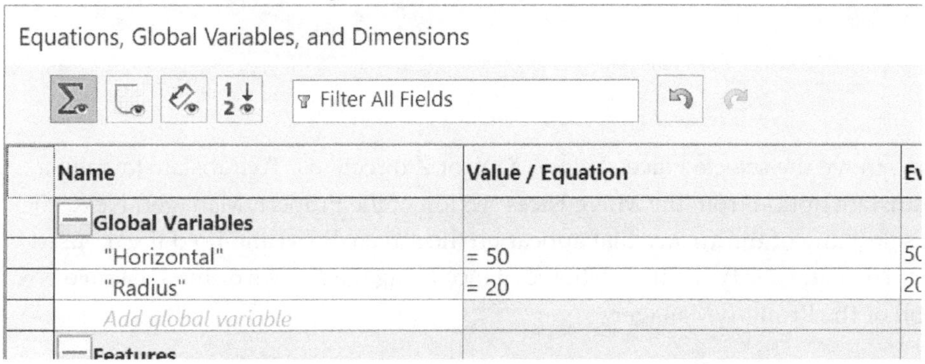

Create a slot using the **Centerpoint Straight Slot** command. Next, activate the **Smart Dimension** command and select the lower horizontal line of the slot. Position the dimension and click in the Modify box and type "=". Next, select **Global Variable > Horizontal**. Click **OK** on the Modify dialog. Select the arc of the slot and position the dimension. Next, click in the **Modify** box and type "=". Select **Global Variable > Radius**. Click **OK** on the **Modify** dialog. Click **Exit Sketch** on the CommandManager.

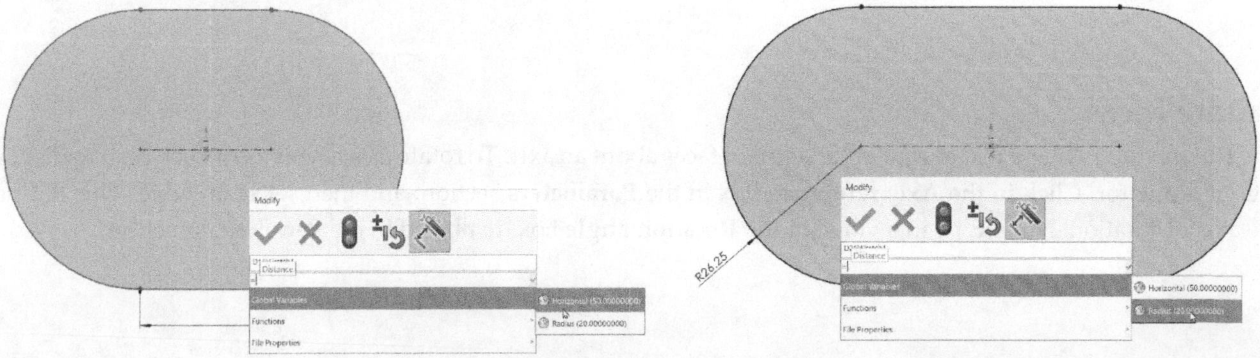

On the CommandManager, click **Features > Reference Geometry > Plane**. Select the newly created slot and the endpoint of the horizontal line of the slot. Next, click **OK** to create a plane normal to the slot. Activate the **Create Sketch** command and select the newly created plane.

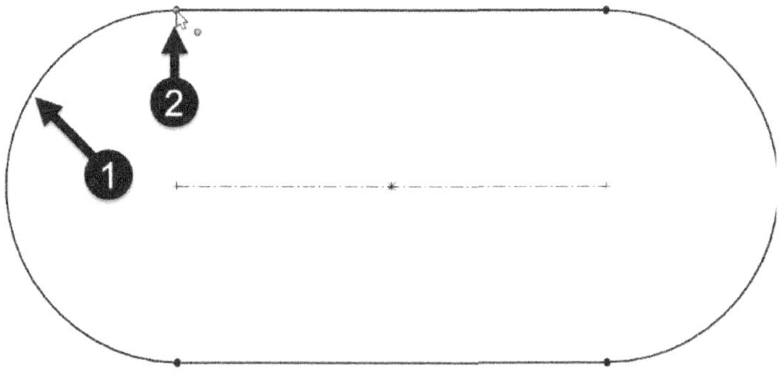

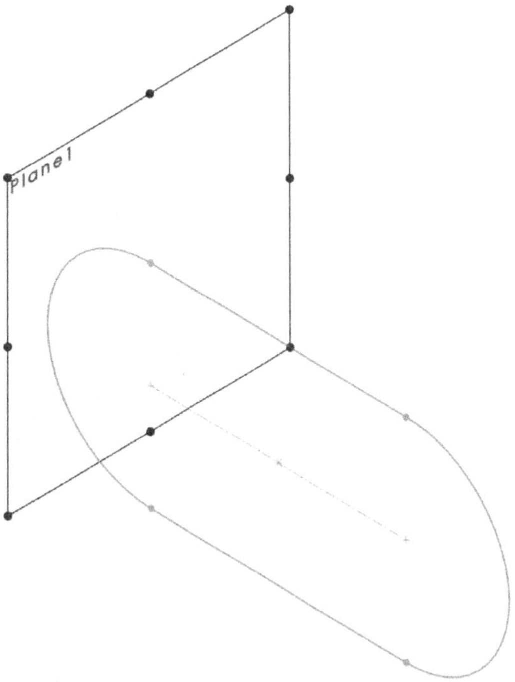

Create a circle coincident to the sketch origin and click **Exit Sketch**. Next, create a solid body using the **Swept Boss/Base** command, as shown.

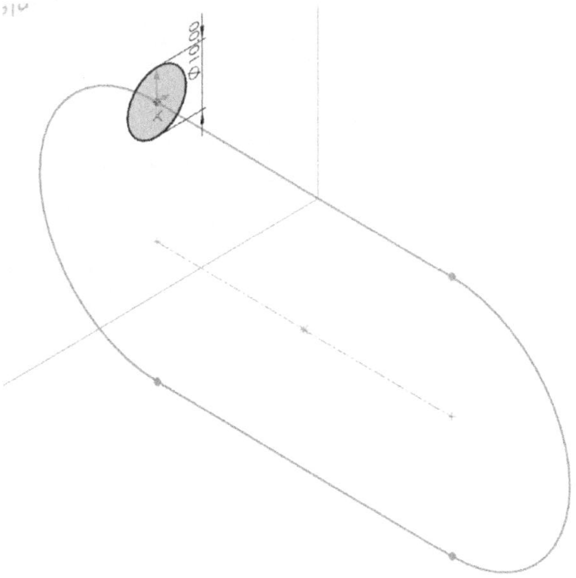

On the menu bar, click **Insert > Features > Move/Copy Body**. Select the solid body from the graphics window and check the **Copy** option on the PropertyManager. Type **1** in the **Number of Copies** box. Next, expand the **Rotate** section and type **90** in the **X Rotation Angle** box. Click **OK** to create a rotated copy of solid body.

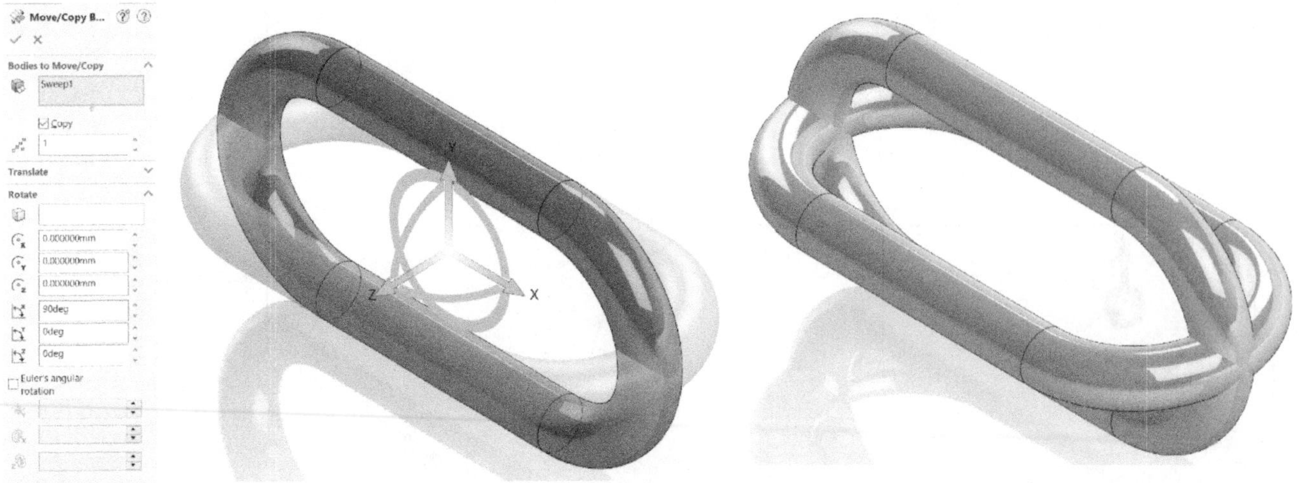

On the menu bar, click **Insert > Features > Move/Copy Body**. Select the rotated solid body from the graphics window and uncheck the **Copy** option on the PropertyManager. Next, expand the **Transition** section and click in the Delta X box. Next, type "=" in the **Delta X** box and select **Global Variables > Horizontal**. Next, Type "+" and select **Global Variable > Radius**; the **Delta X** value is calculated by adding the Horizontal and Radius values. Also, the solid body is moved along the X direction up to the value specified in the **Delta X** box. Click **OK** to complete the **Move/Copy** operation.

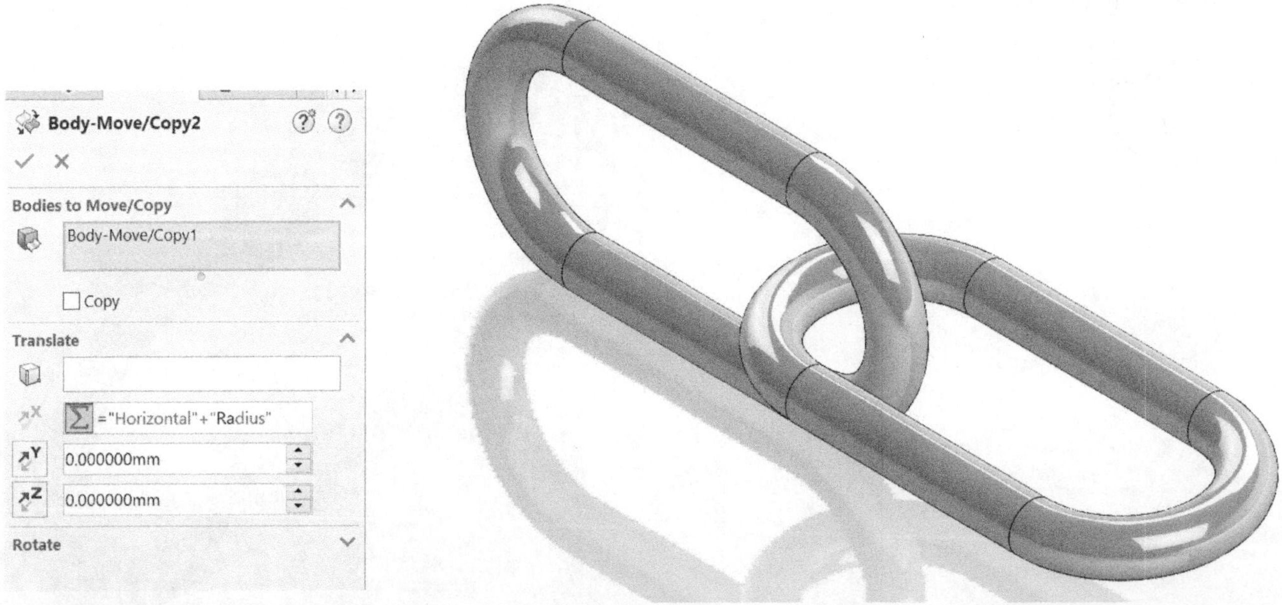

In the FeatureManager Design Tree, right-click on the **Equations** node and select **Manage Equations**. Next, type **Delta X** in the **Equation** box of the **Horizontal+Radius** equation.

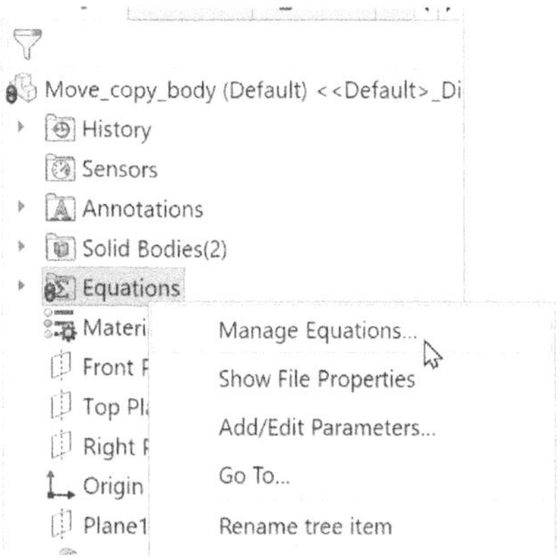

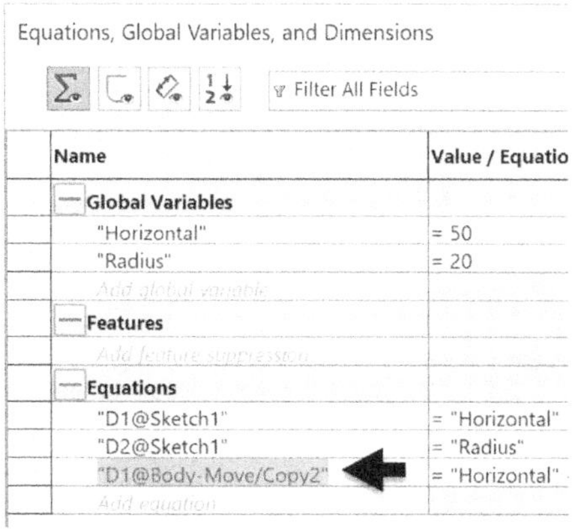

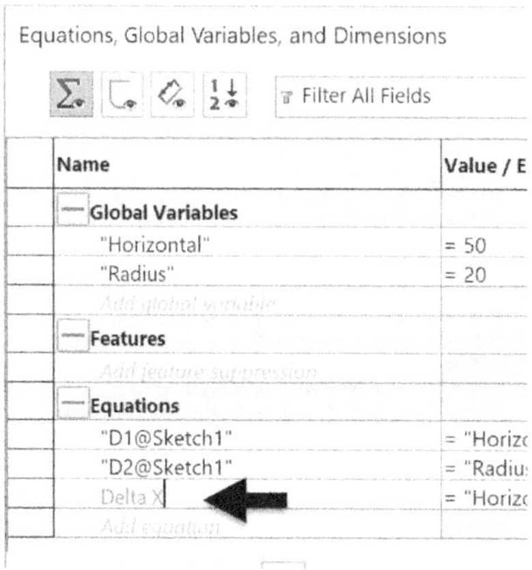

On the CommandManager, click **Features > Linear Pattern**. Next, check the **Bodies** option on the PropertyManager and select the two bodies from the graphics window. In the **Direction 1** section, click in the **Spacing** box and type "=". Next, select **Global Variables > Delta X**. Click before **Delta X** and type "2*". Type **3** in the **Number of Instances** box. Click in the **Pattern Direction** box in the **Direction 1** section and select the Right Plane from the FeatureManager Design Tree available in the graphics window. Click **OK** to complete the linear pattern.

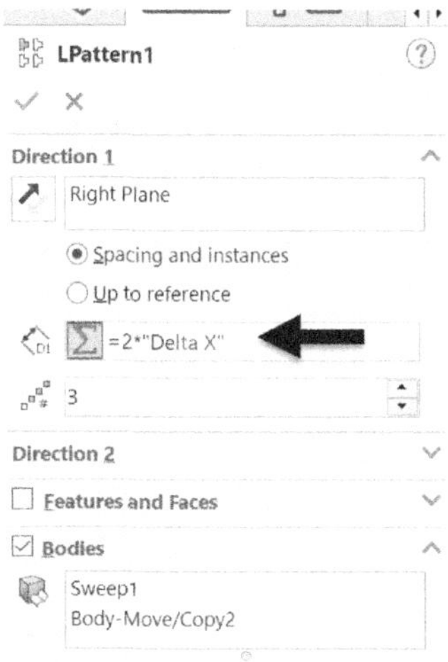

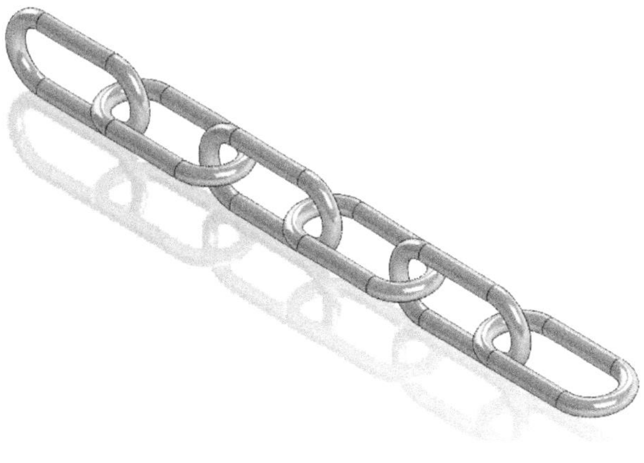

Examples

Example 1 (Inches)

In this example, you will create the part shown below and then modify it.

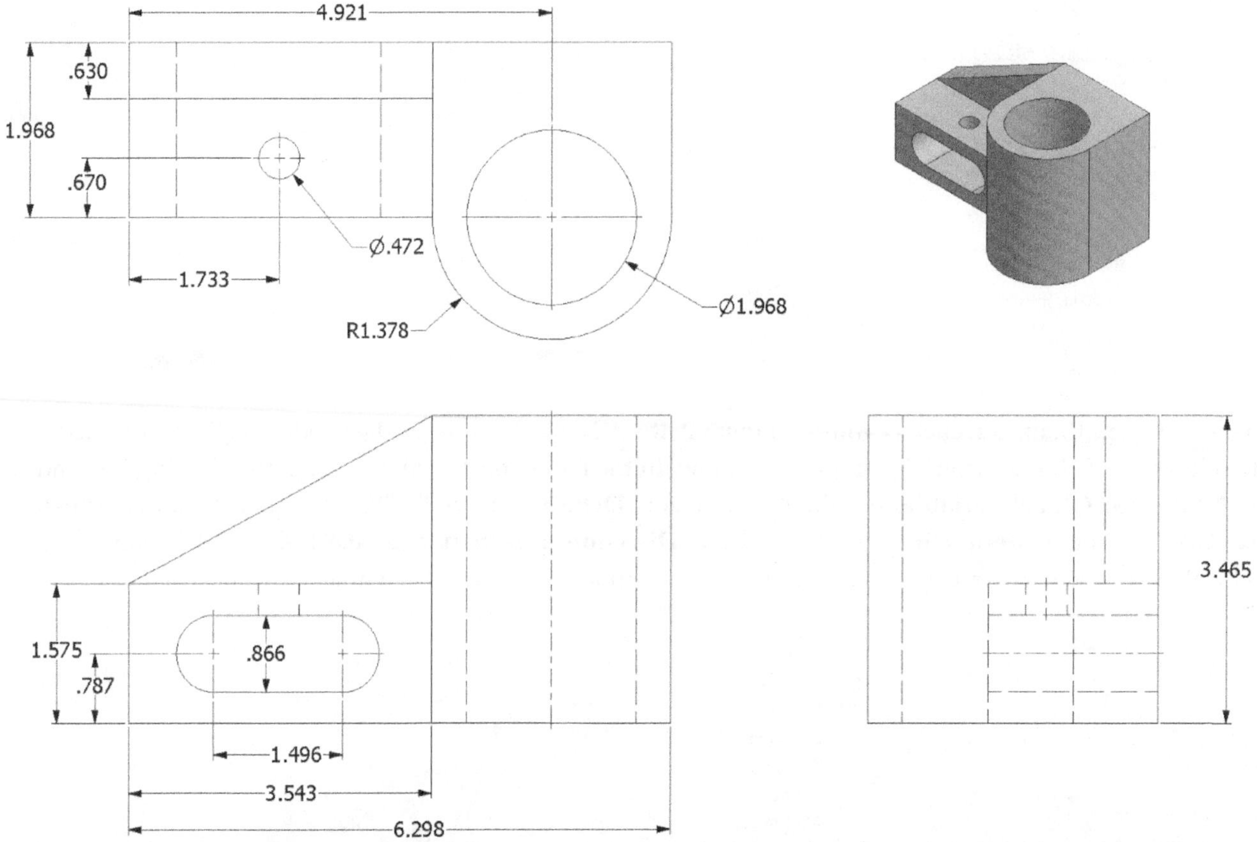

1. Start **SOLIDWORKS 2024** and open a part file and create the part using the tools and commands available in SOLIDWORKS. You can also download this part file from the companion website.

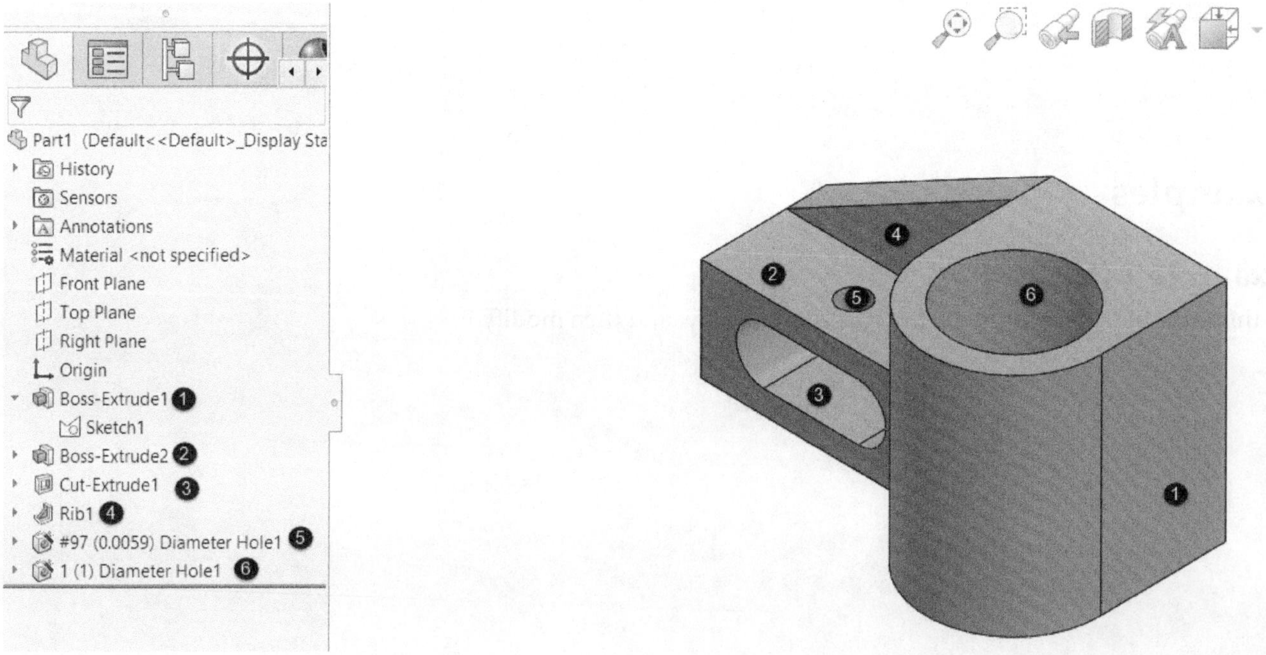

2. Click on the large hole and select **Edit Feature**; the **PropertyManager** appears. On the **PropertyManager**, select **Hole Type > Counterbore**. Next, click the **Reset custom sizing values to the default values to the new hole type** option on the **SOLIDWORKS** message box.
3. Check the **Show custom sizing** option, and enter **1.378, 1.968**, and **0.787** in the **Through Hole Diameter**, **Counterbore Diameter**, and **Counterbore Depth** boxes, respectively. Click **OK**.

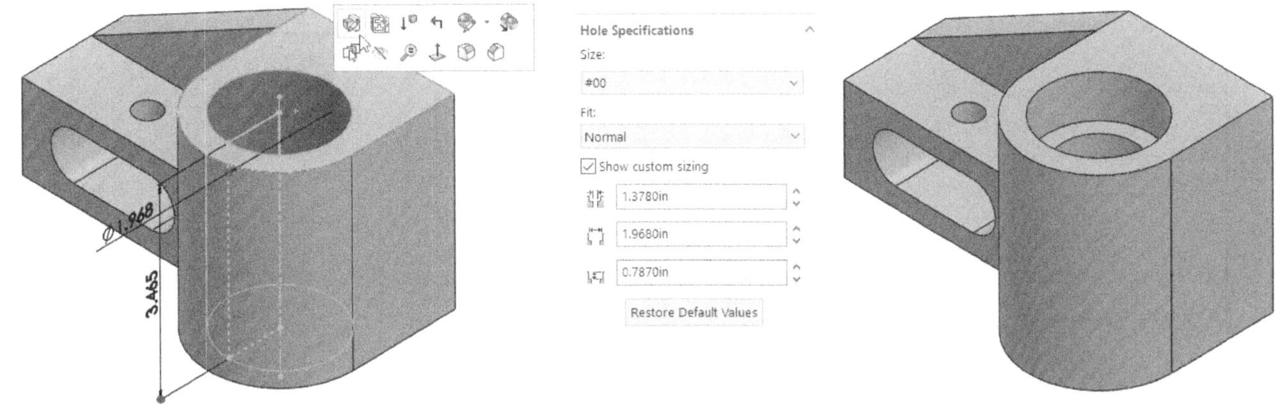

4. Click on the rectangular *Extruded Boss/Base* feature and select **Edit Sketch**. Modify the sketch, as shown. Click **Exit Sketch**.

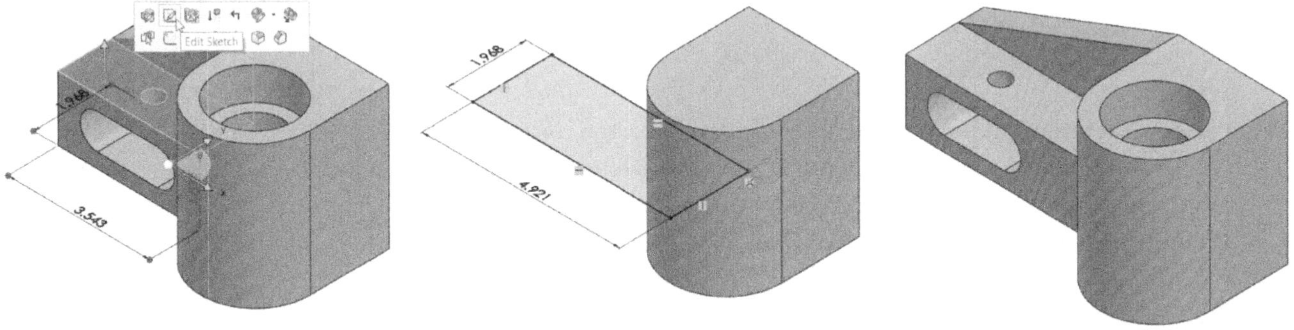

5. Click on the slot and select **Edit Sketch**.
6. Delete the slot's length dimension and add a new dimension between the right-side arc and the right vertical edge.

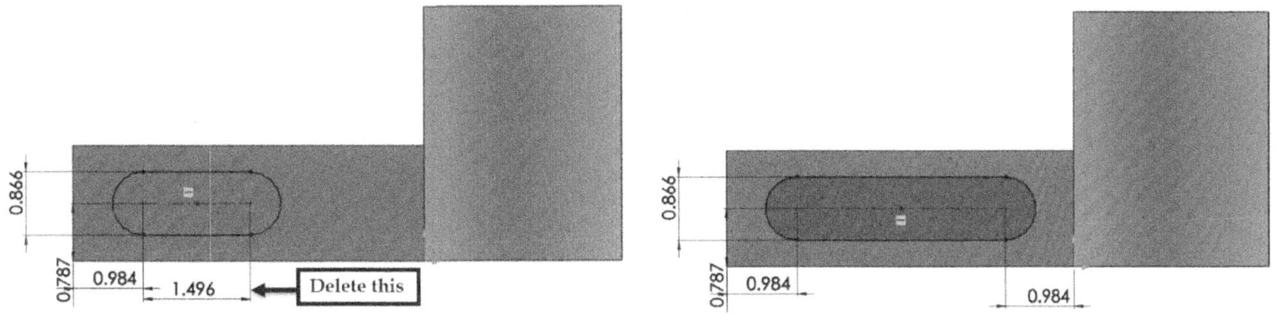

7. Delete the dimension between the centerline of the slot and the horizontal edge.
8. Apply the **Horizontal** relation between the endpoint of the slot and the midpoint of the left vertical edge. Click **Exit Sketch** on the CommandManager.

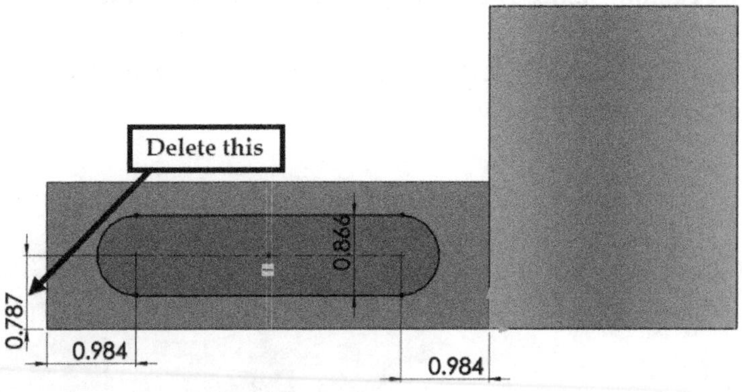

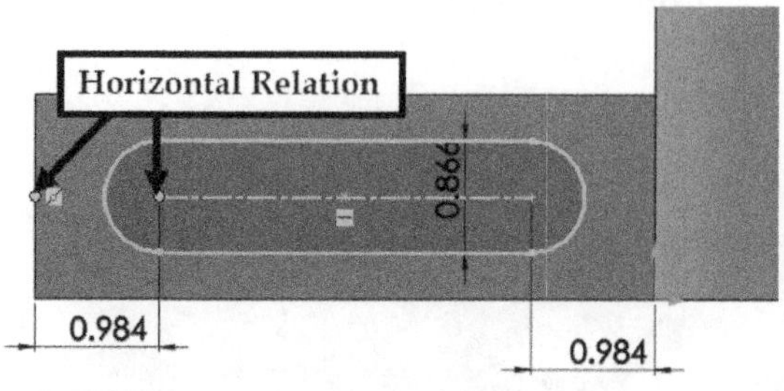

9. Click on the small hole, and then select **Sketch 5** from the breadcrumbs. Next, click **Edit Sketch**. Delete the positioning dimensions.

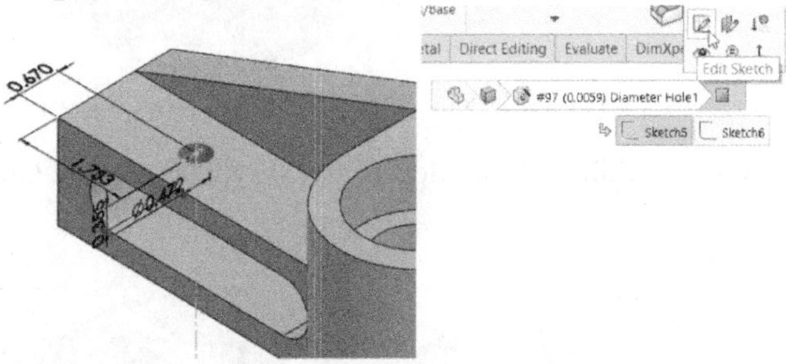

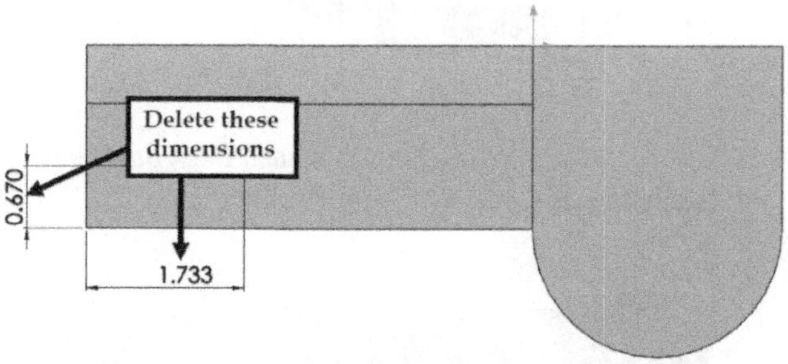

10. Create a centerline and make its ends coincident with the corners, as shown below.
11. On the CommandManager, click **Sketch > Display/Delete Relations > Add Relation**, and select the hole point. Next, select the centerline, and then click **Midpoint** on the PropertyManager. Click **OK** on the PropertyManager, and then click **Exit Sketch**.

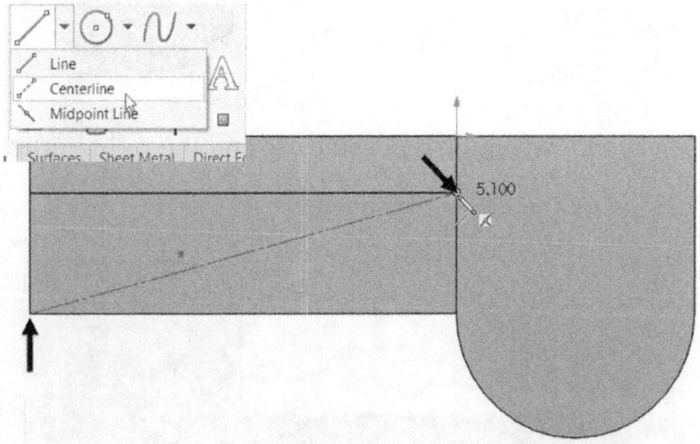

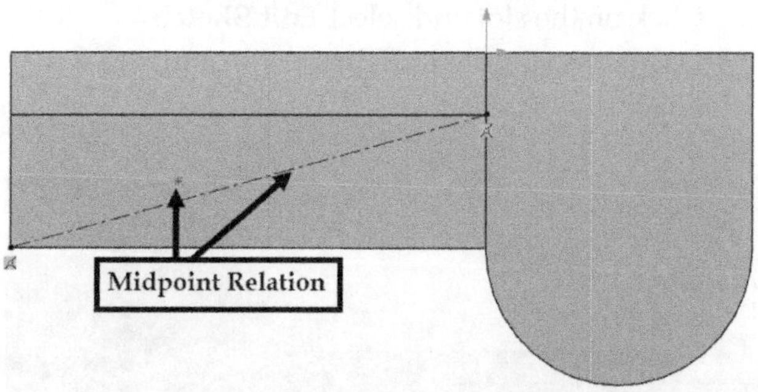

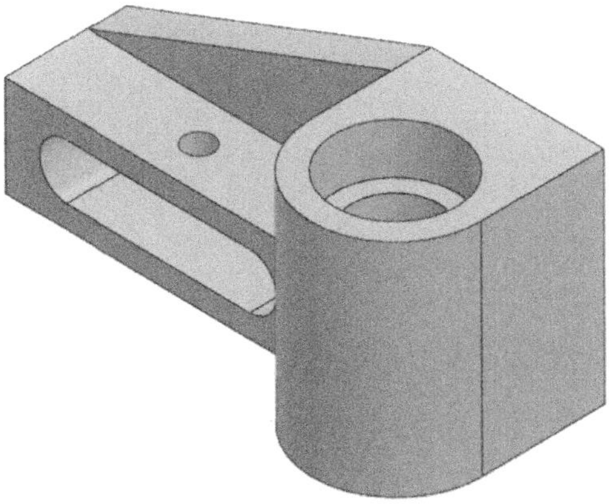

12. Now, change the size of the rectangular extruded feature. You will notice that the slot and hole are adjusted automatically.

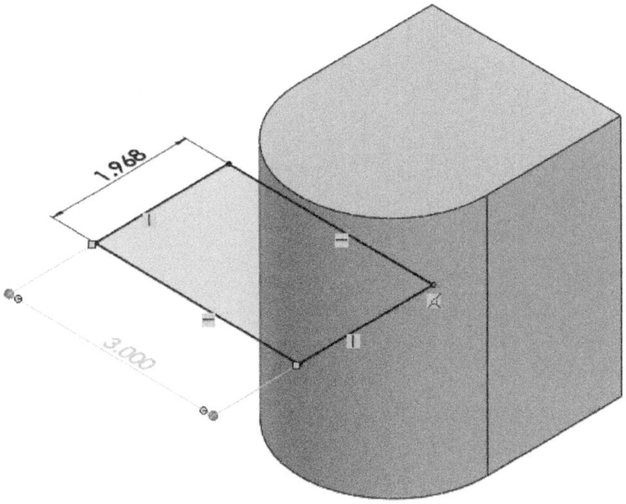

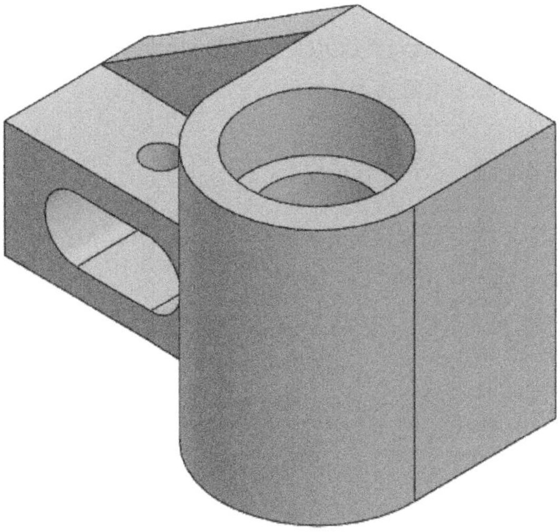

13. Save and close the file.

Example 2 (Millimetres)

In this example, you will create the part shown below and then modify it using the editing tools.

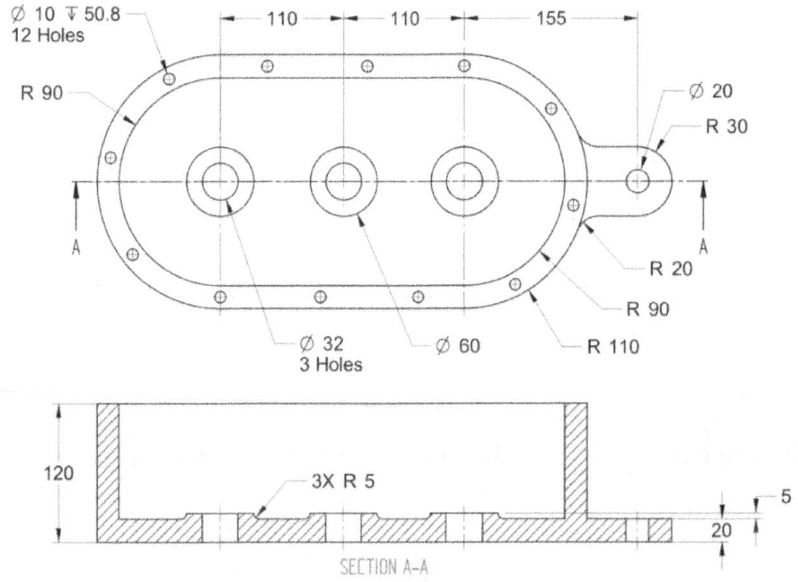

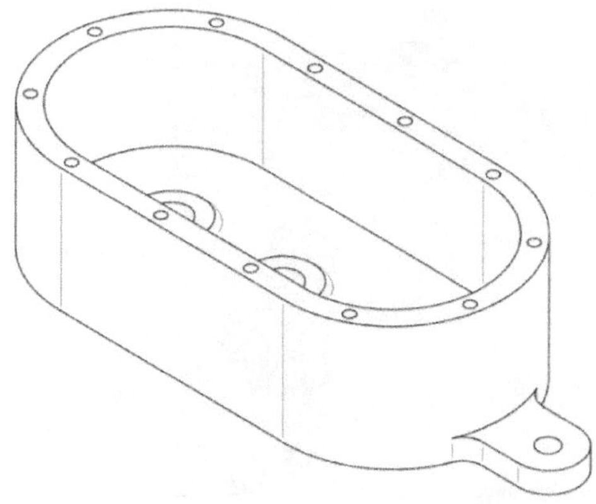

1. Start **SOLIDWORKS 2024**.
2. Create the part using the tools and commands in SOLIDWORKS.

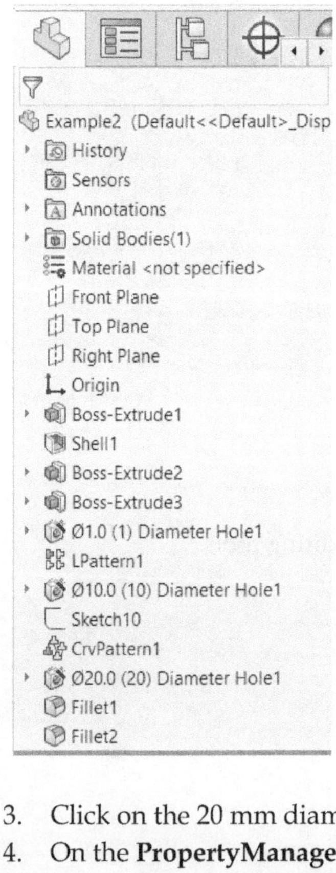

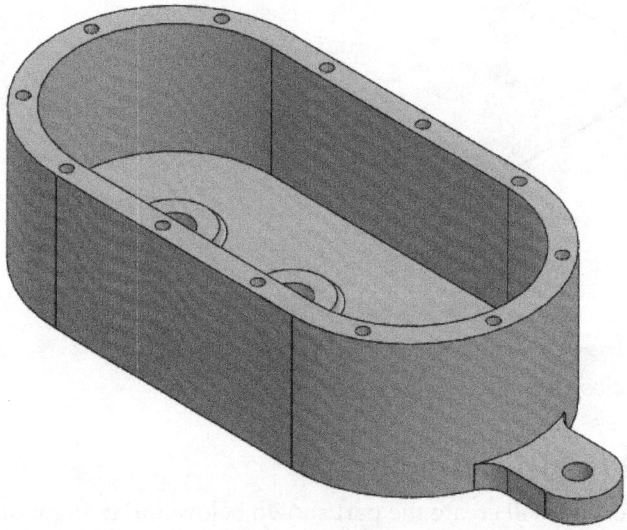

3. Click on the 20 mm diameter hole, and then click **Edit Feature**; the **PropertyManager** appears.
4. On the **PropertyManager**, select **Hole Type > Counterbore**.
5. In the **Hole Specifications** section, select **Size > M20**. Click **OK** to close the PropertyManager.

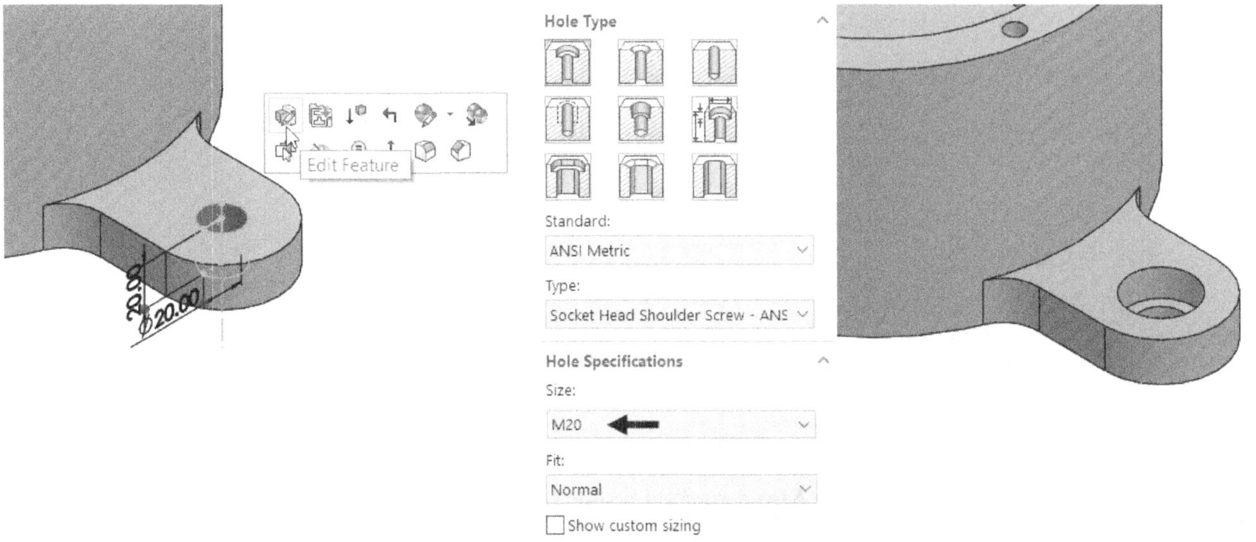

6. On the Menu bar, click **Insert > Face > Move**, and then select **Translate** from the PropertyManager.
7. Click on the counterbore hole and click the **Internal to feature** icon on the toolbar that appears on the screen. Next, select the cylindrical face concentric to it.
8. Select the arrow pointing toward the right. Next, press and hold the left mouse button on the selected arrow, drag, and then release it.

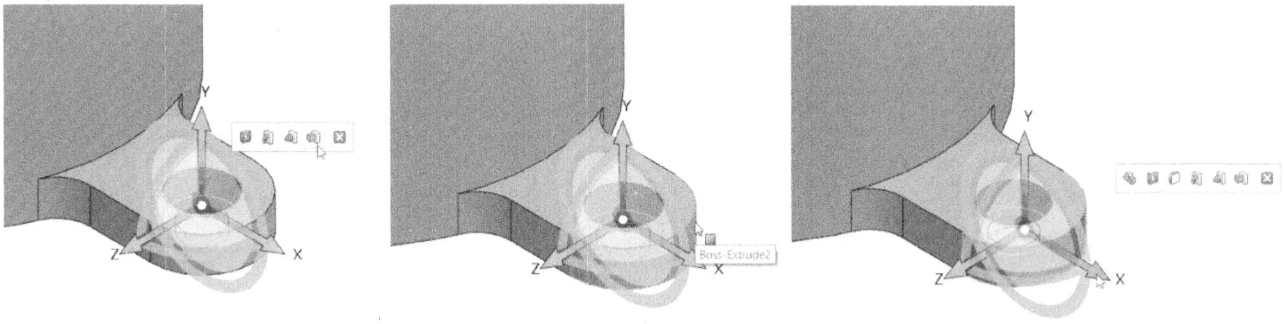

9. Type **20** in the **Distance** box available in the **Parameters** section of the PropertyManager and press **OK**.

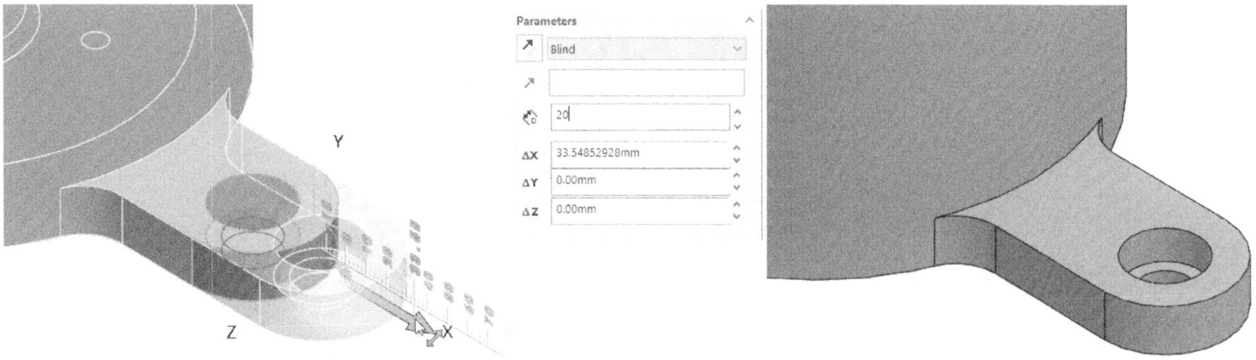

10. Click on any one of the holes of the curve driven pattern, and then select **Edit Feature**.
11. Type **14** in the **Number of Instances** box and click **OK** to update the pattern.

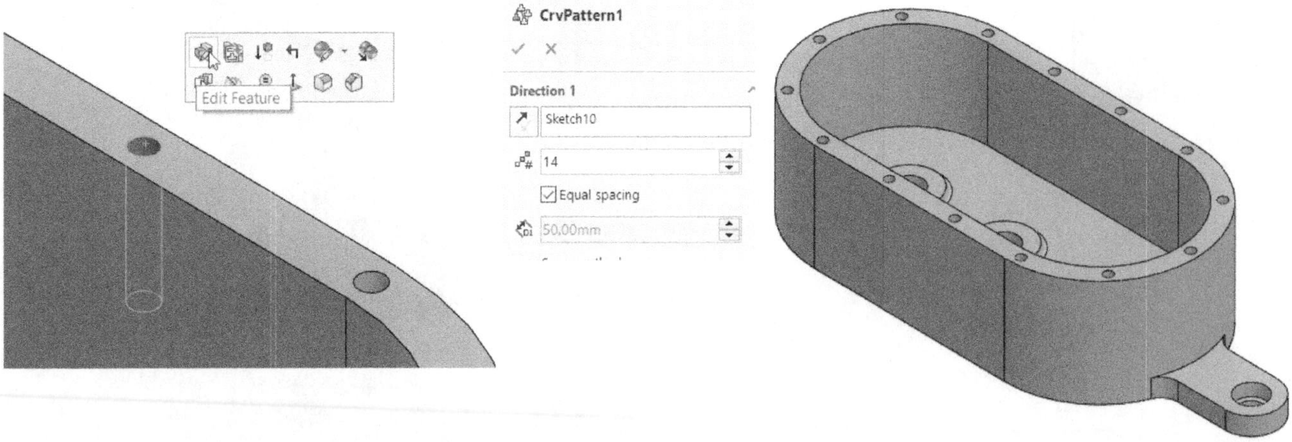

12. On the Menu bar, click **Insert > Face > Move**, and then select the **Translate** option from the PropertyManager.
13. Select any one of the holes of the curve driven pattern, and then select the **Internal to Features** option from the toolbar.
14. Likewise, select the base hole, and then select the **Internal to Feature** option.
15. Click on the top face of the geometry.

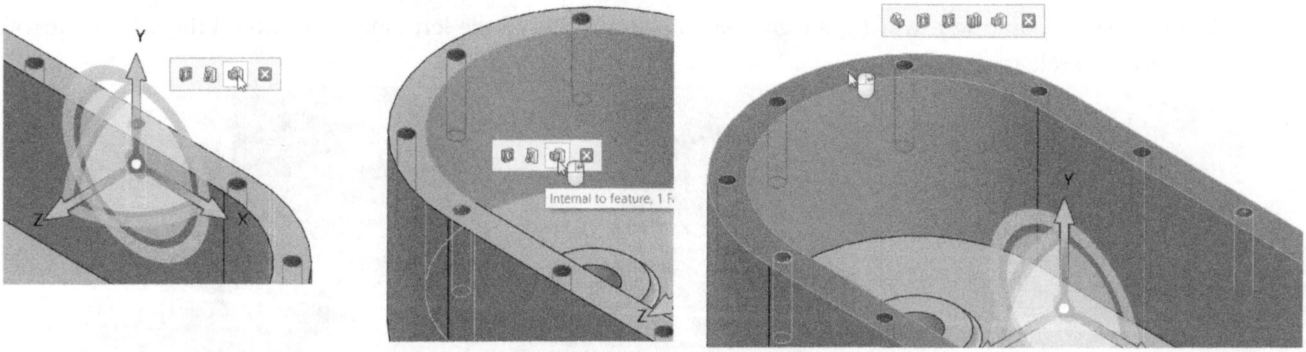

16. Click on the arrow pointing upwards. Press and hold the left mouse button and drag the mouse pointer down. Type **40** in the **Distance** box and click **OK** to update the model.

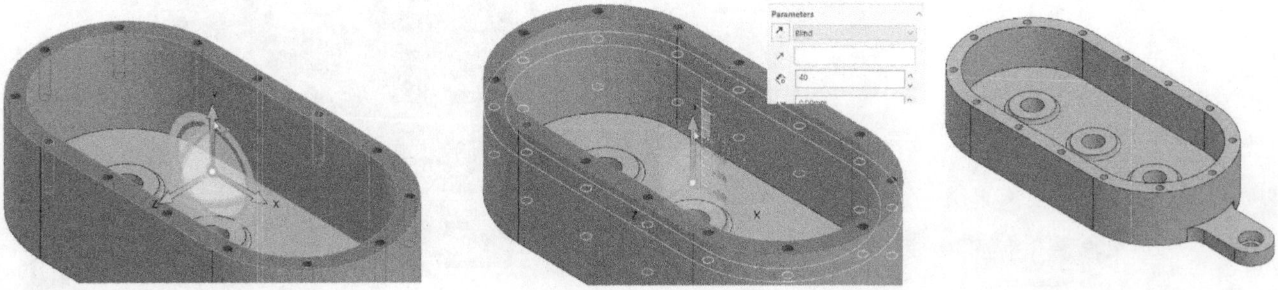

17. Save and close the file.

Example 3 (CSWP Mechanical Design Step 4 and 5)

In this tutorial, you will complete Step 4 and Step 5 of the CSWP sample question. These steps involve editing the model created in Example 2 of Chapter 4.

1. Open the file created in Example 2 of Chapter 4.

2. In the FeatureManager Design Tree, right-click on the **Equations** node and select **Manage Equations**.
3. On the **Equations, Global Variables, and Dimensions** dialog, select the value in the **Value/ Evaluation** box next to the **A** global variable.
4. Type **221**.
5. Likewise, change the **B**, **C**, **D**, **E**, and **Y** values to **211,165,121**, **37**, respectively.

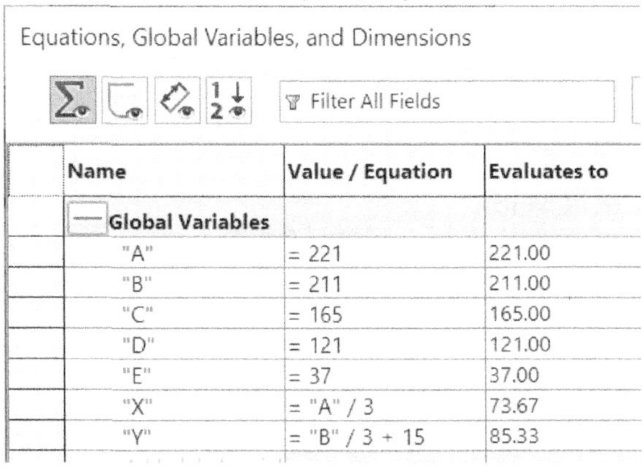

Name	Value / Equation	Evaluates to
Global Variables		
"A"	= 221	221.00
"B"	= 211	211.00
"C"	= 165	165.00
"D"	= 121	121.00
"E"	= 37	37.00
"X"	= "A" / 3	73.67
"Y"	= "B" / 3 + 15	85.33

6. Click **OK**.

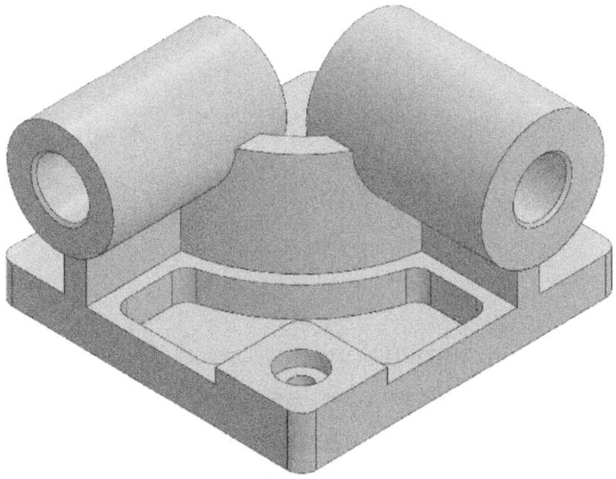

7. In the FeatureManager Design Tree, right-click on the **CBORE for M8 Hex Head Bolt1** feature and select **Delete**.

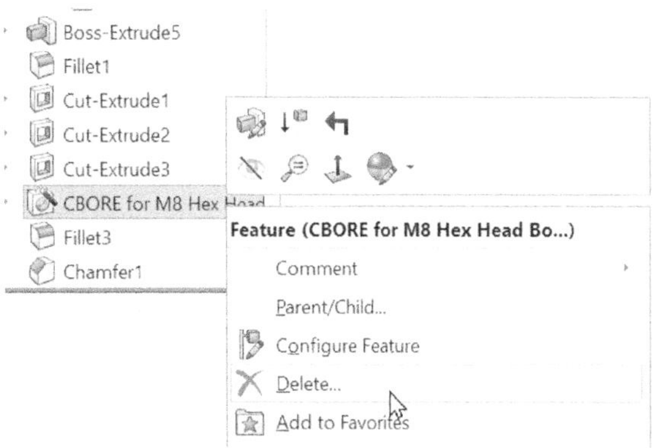

8. Click **Yes** on the **Confirm Delete** dialog.

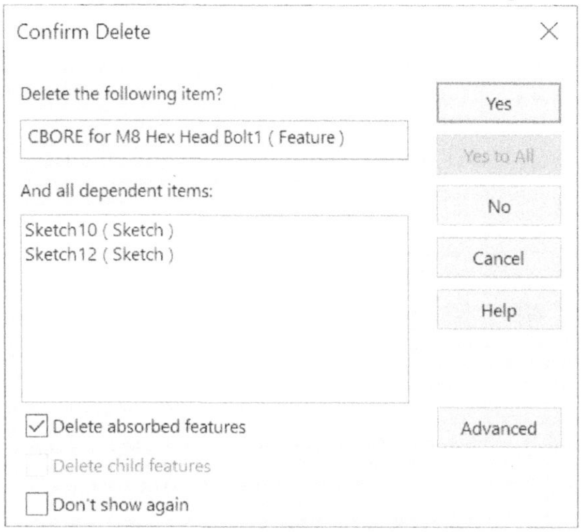

9. Right-click on the second Boss-Extrude feature in the graphics window or FeaturManager Design Tree. Next, select the **Delete** option.

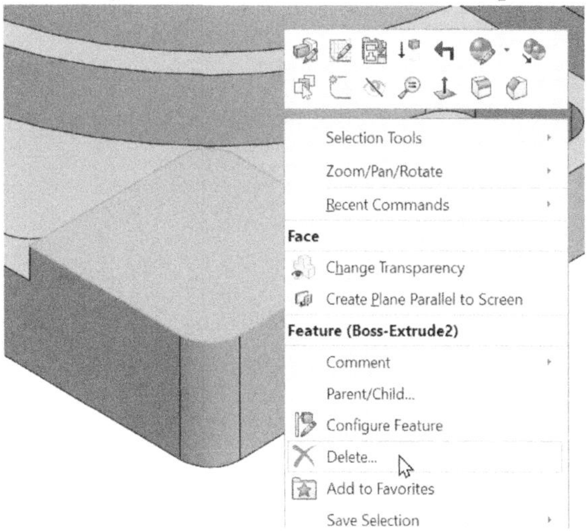

10. On the **Confirm Delete** dialog, check the **Delete absorbed features** option and click **Yes**.

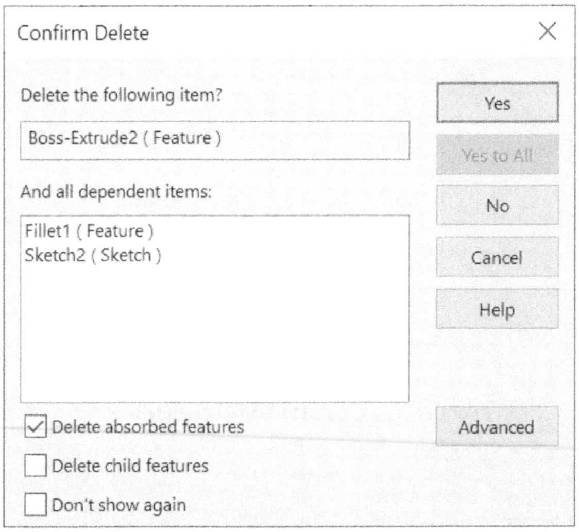

11. Click **Continue (Ignore Error)** on the **SOLIDWORKS** dialog.

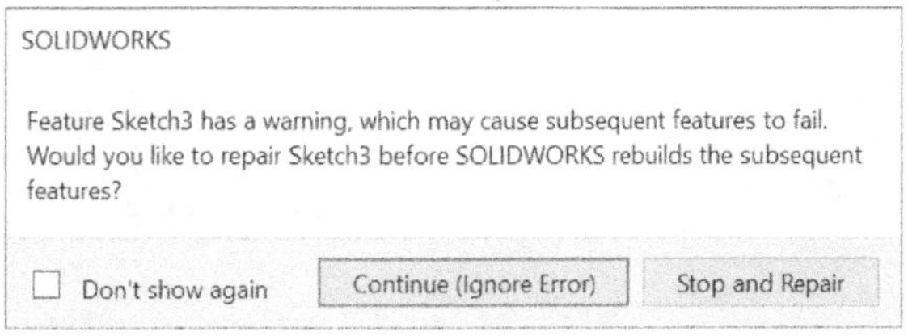

When you delete a feature in SOLIDWORKS, it can cause errors elsewhere in the design. In this case, you may notice an error symbol displayed next to the Boss_Extrude feature that is linked to the deleted feature. To fix this issue, you will need to edit the sketch associated with the Boss_Extrude feature to remove the error. Once you have made the necessary changes to the sketch, you can rebuild the feature and the error should be resolved.

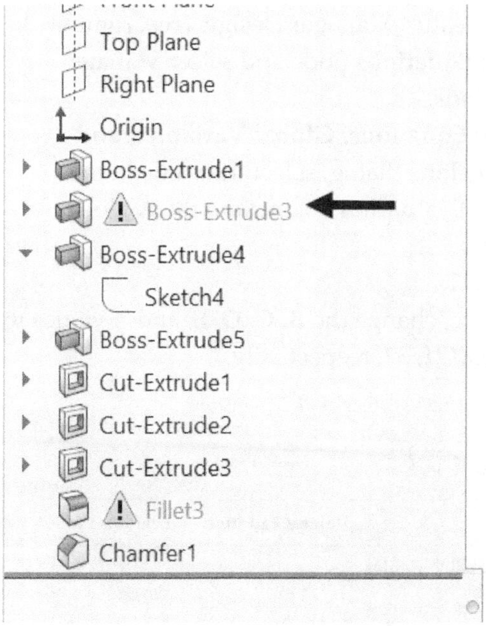

12. Right-click on the Boss-Extrude feature with the error symbol. Next, select Edit Sketch.

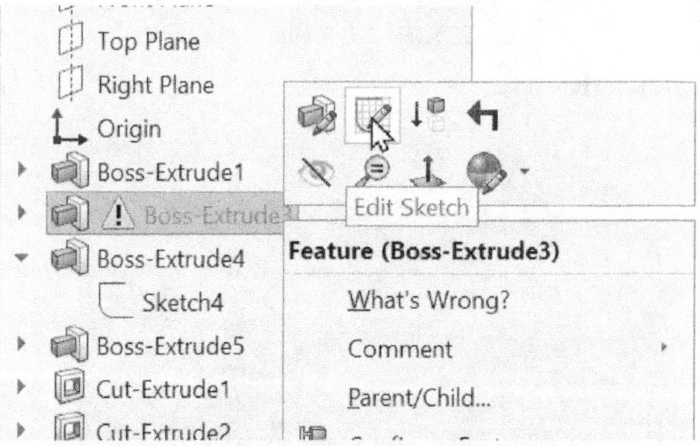

Notice that two dimensions of the sketch are highlighted in golden color. As they are attached to the deleted feature. you need to reattach them to the existing edges of the model. It appears that two dimensions within the sketch have been highlighted in a golden color. However, these dimensions were previously attached to a feature that has since been deleted. In order to proceed further, you will need to reattach these dimensions to the existing edges of the model. This will ensure that your sketch remains accurate and aligned with the rest of your design.

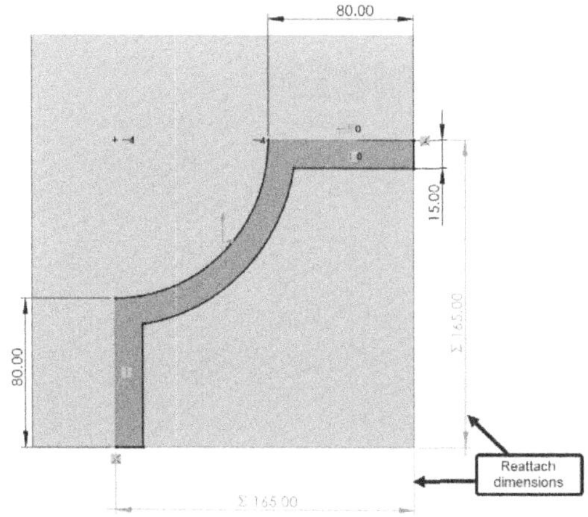

13. Select the vertical dimension and notice a square dot on the dimension extension line.
14. Press and hold the left mouse button and drag the pointer downward.

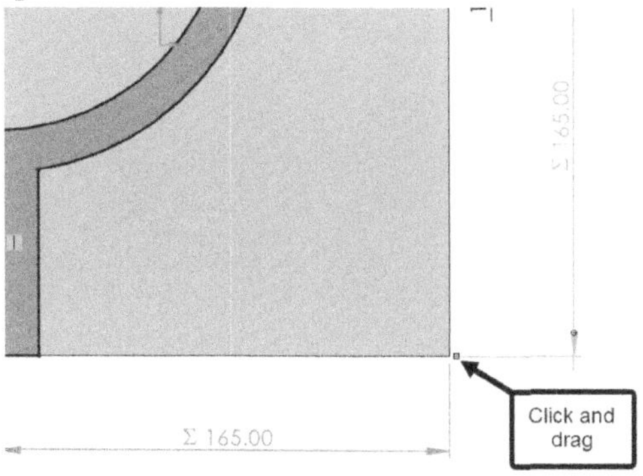

15. Place the pointer on the lower horizontal edge; the edge is highlighted in blue.
16. Release the pointer to attach the dimension to the horizontal edge.

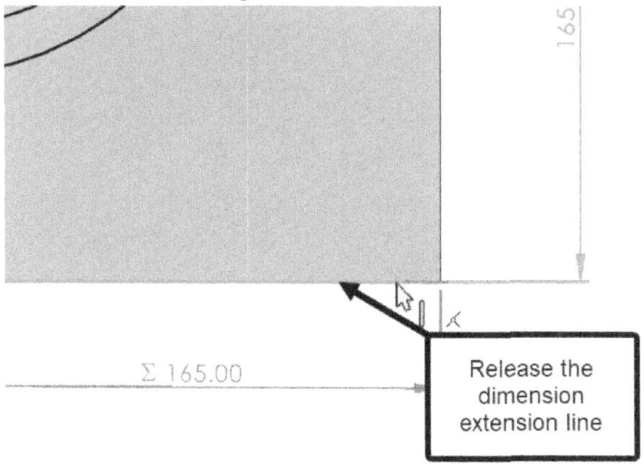

17. Likewise, attach the horizontal dimension to the vertical edge.

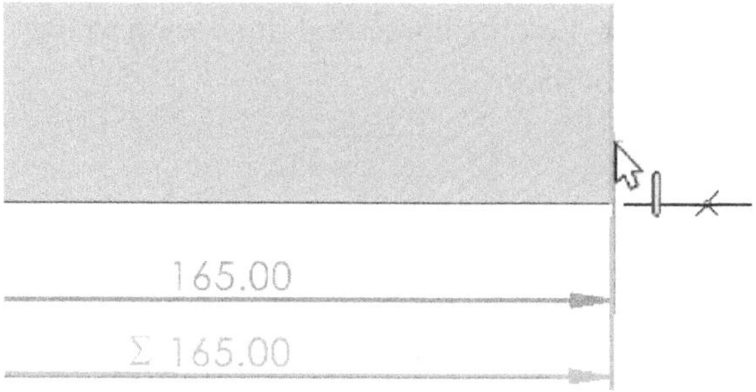

Notice that the equation symbol disappears from the reattached dimensions. You need to link the reattached dimensions to the global variables to avoid future errors.

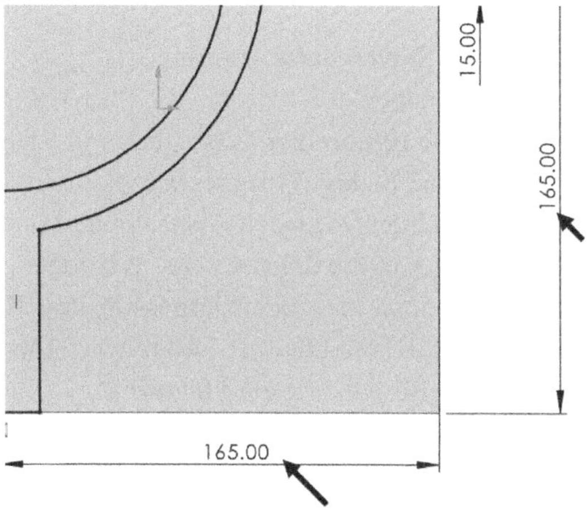

18. Double-click on the vertical dimension.
19. On the **Modify** box, select the dimension value and type =.
20. Select **Global Variables > C**.

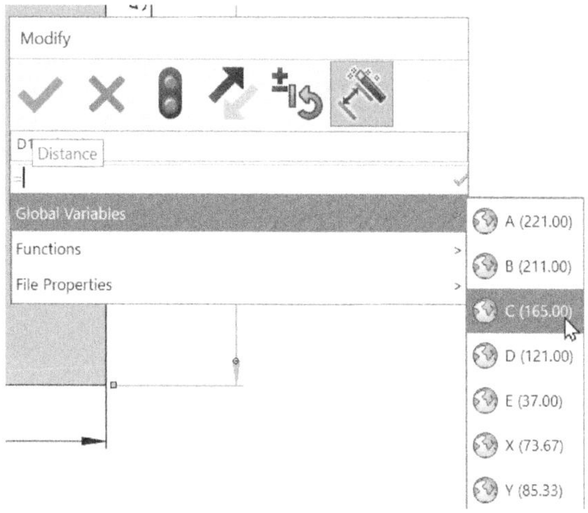

21. Click the green check on the **Modify** box.

22. Likewise, link the horizontal dimension to the global variable **C**.

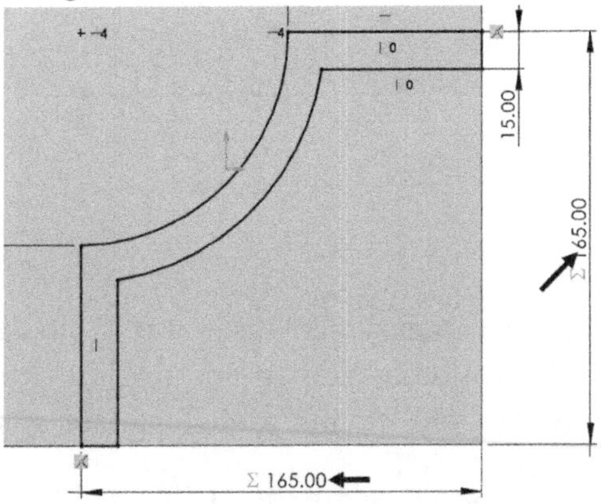

23. Click the **Exit Sketch** button on the CommandManager.
24. Click **Continue (Ignore Error)** on the **SOLIDWORKS** dialog. This message appears because of the fillet feature that was created on one of the edges of the deleted Boss-Extrude feature. To address the underlying issue, you need to revisit the fillet feature and remove the references pointing the deleted feature.
25. In the FeatureManager, right-click on the **Fillet** feature with the error symbol. Next, select **Edit Feature**.

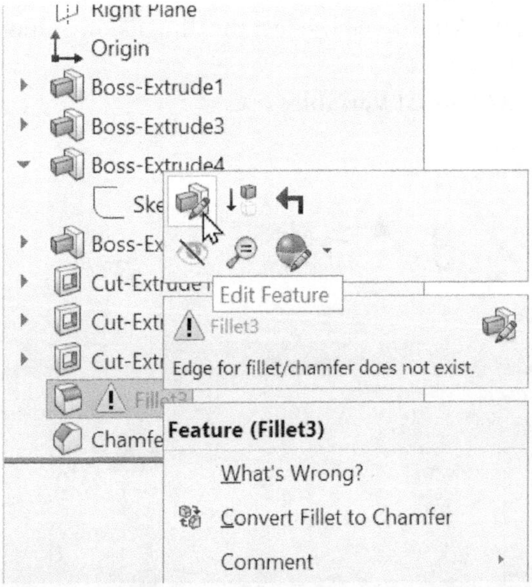

26. On the **Fillet** PropertyManager, in the **Items to Fillet** selection box, press and hold CTRL key and select the two missing edges.

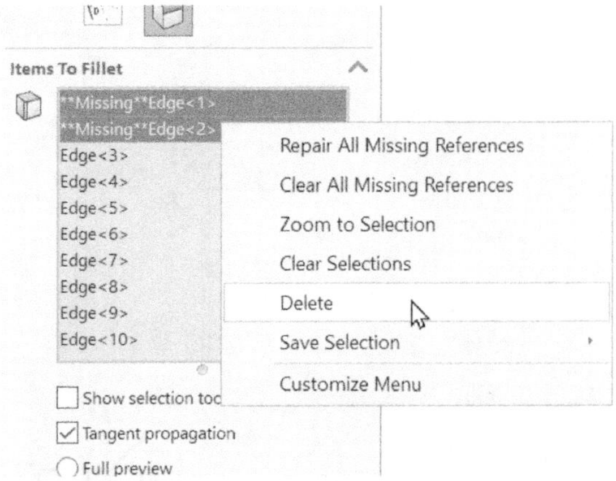

27. Select the corner edge of the cut feature, as shown.

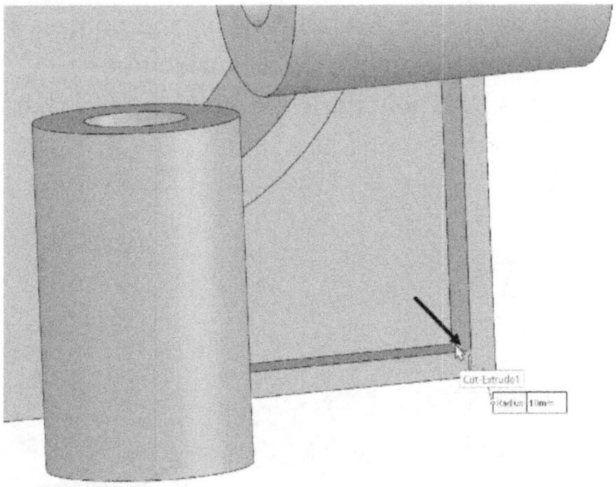

28. Click **OK** on the PropertyManager; the error symbol disappears.

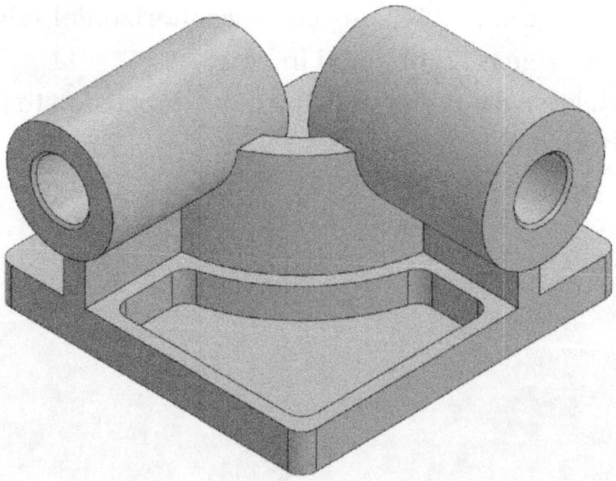

29. On the **Features** CommandManager, click **Extruded Cut** .
30. Click on the top face of the first feature.

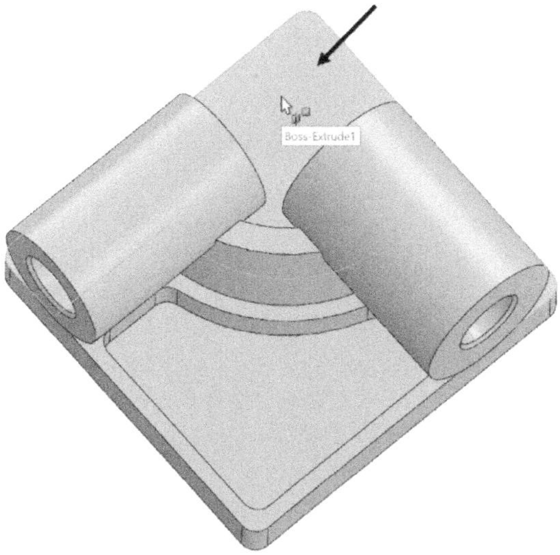

31. On the CommandManager, click **Sketch** tab > **Offset Entities**.
32. Select the horizontal edge of the model, as shown.

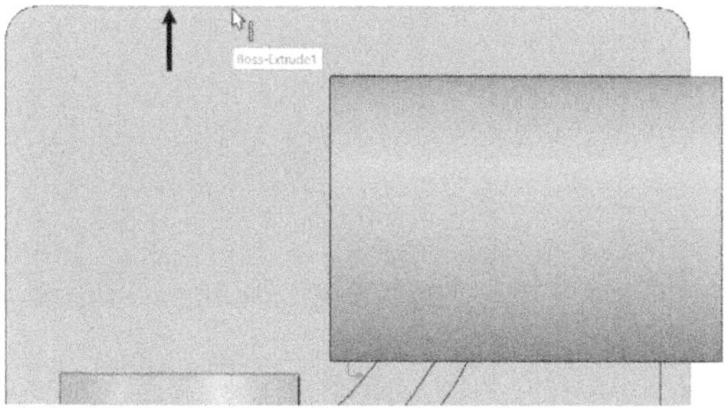

33. Type-in **9** in the **Offset Distance** box available on the PropertyManager.
34. Check the **Reverse** option.
35. Click the **Keep Visible** icon on the PropertyManager.

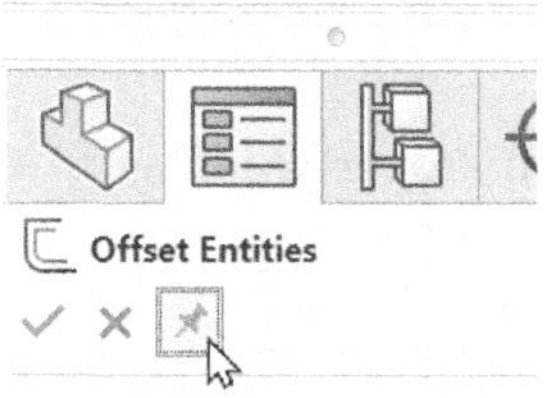

36. Click **OK** on the PropertyManager.
37. Click on the left vertical edge.
38. Click **OK**.
39. Click on the curved edge and click **OK** twice.

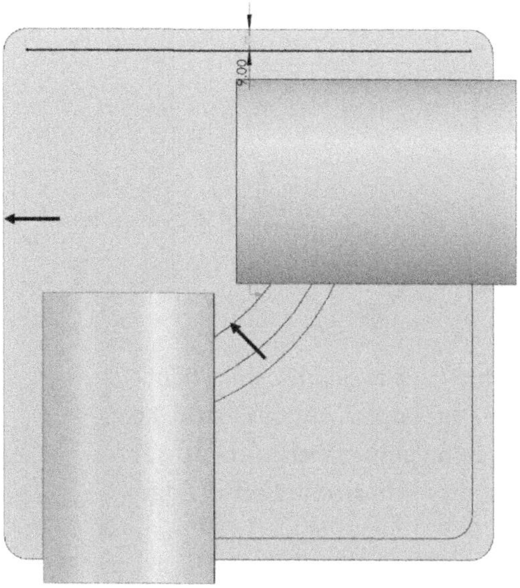

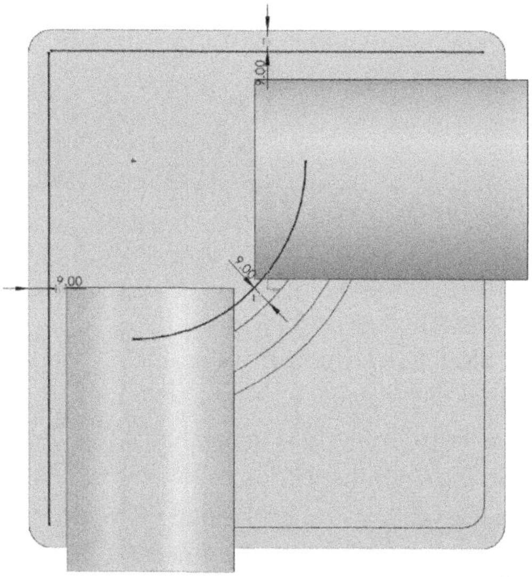

40. Click the **Line** icon on the CommandManager.
41. Select the lower endpoint of the arc.
42. Move the pointer horizontally toward left and click on the vertical offset line.

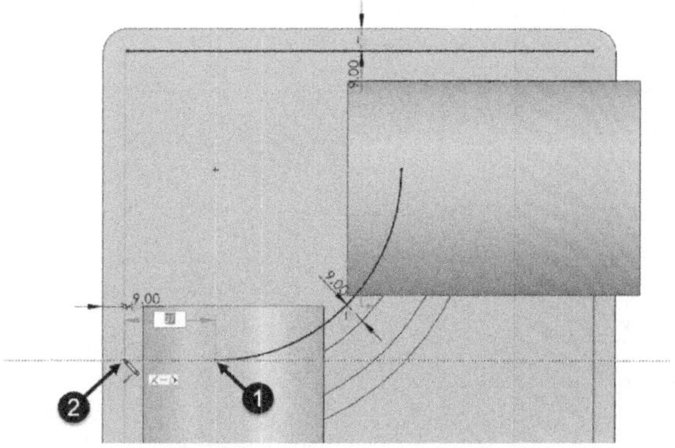

43. Double-click to end the line chain.
44. Select the upper endpoint of the arc.
45. Move the pointer vertically upward.
46. Click on the horizontal offset edge.

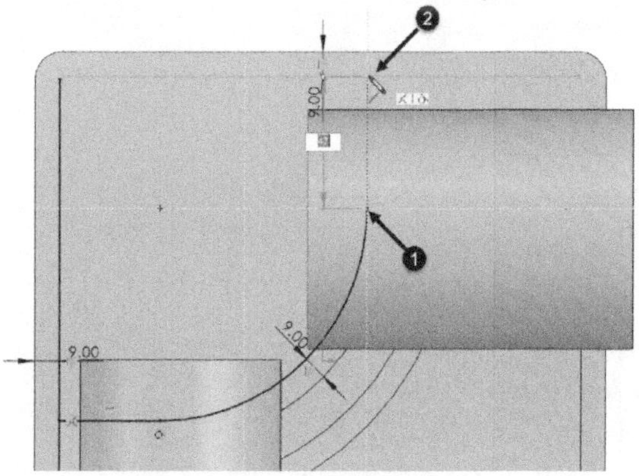

47. Right-click and select **Select**.
48. Click **Sketch > Trim Entities** on the CommandManager.
49. On the **Trim** PropertyManager, click the **Corner** icon in the **Options** section.
50. Select the vertical and horizontal offset lines, as shown. A corner is formed connecting the two lines.

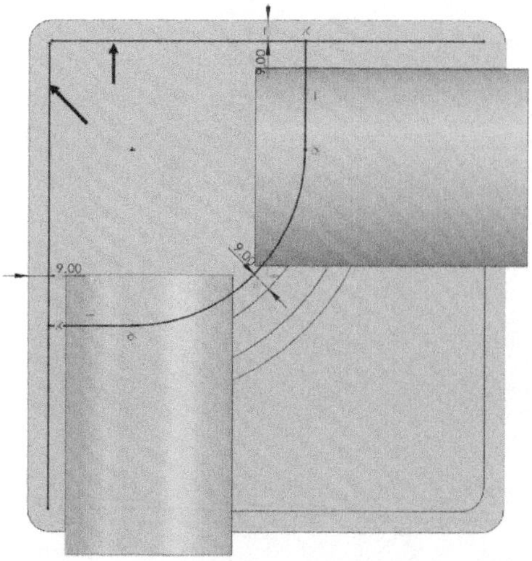

51. On the **Trim** PropertyManager, click the **Trim to closest** icon in the **Options** section.
52. Select the portions of the vertical and horizontal lines, as shown.

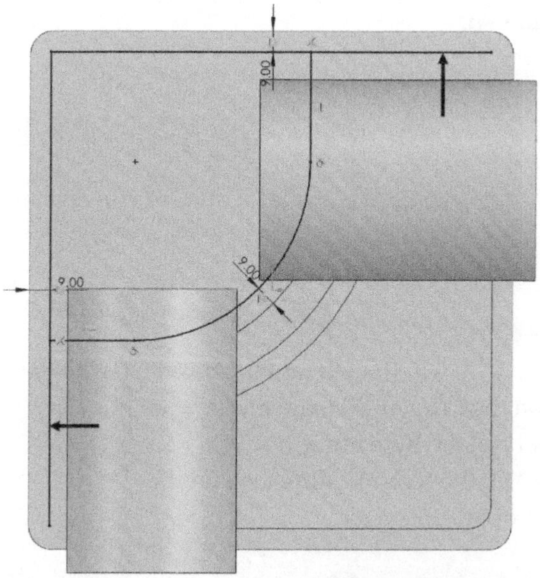

53. Click **OK** on the **Trim** PropertyManager.

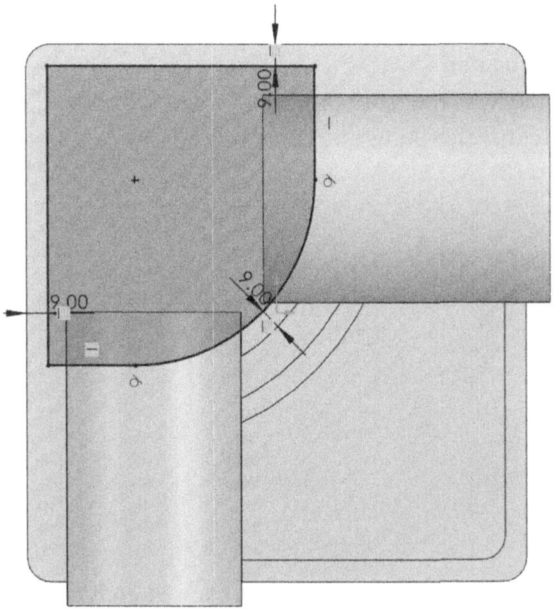

54. On the **Sketch** CommandManager, click **Exit Sketch**.
55. On the **Cut-Extrude PropertyManager**, under the **Direction 1** section, select **End Condition > Offset from Surface**.
56. Rotate the model and select the bottom face of it.

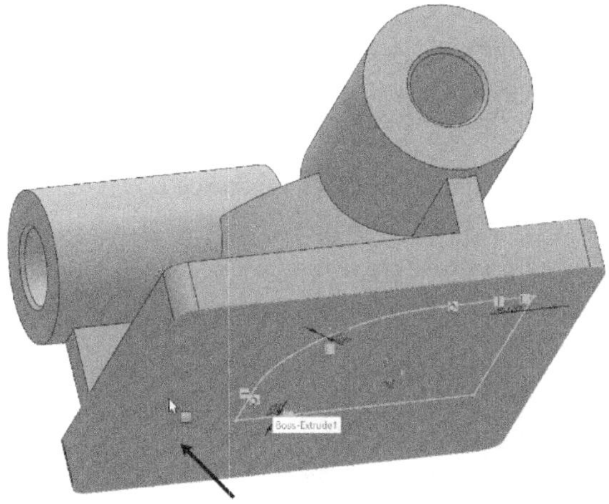

57. Type **5** in the **Offset Distance** box.
58. Click **OK** to create the cut offset from the bottom face.

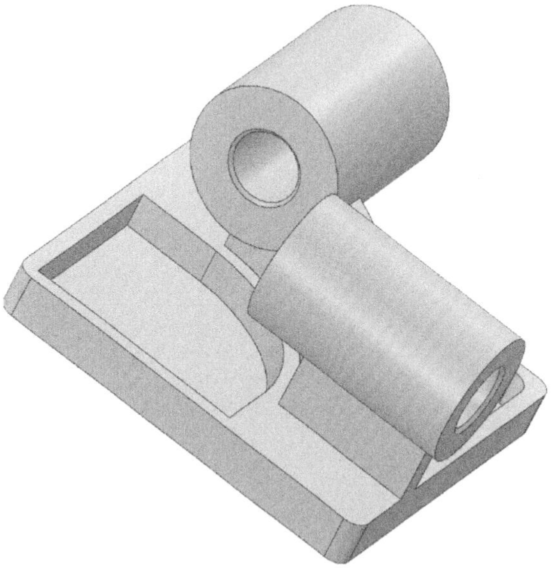

59. Click and drag the **Cut-Extrude** feature located at the bottom of the FeatureManager Design Tree.
60. Release it above the **Fillet** feature.

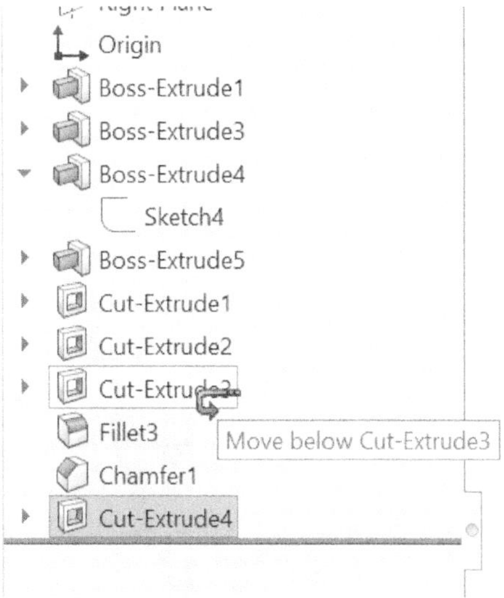

61. In the FeatureManager, right-click on the **Fillet** feature and select **Edit Feature**.

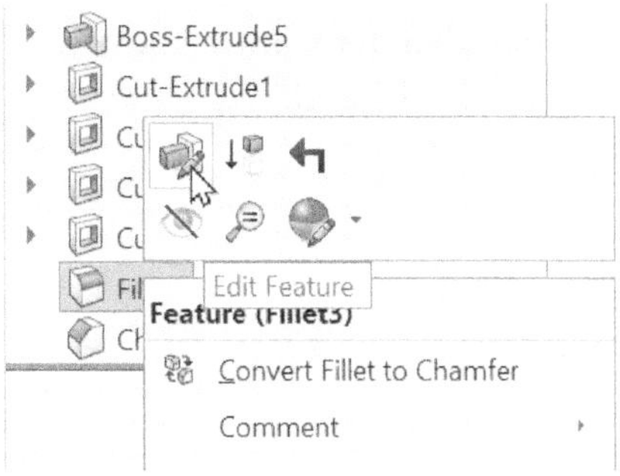

62. Select the corner vertical edges of the newly created **Cut-Extrude** feature.

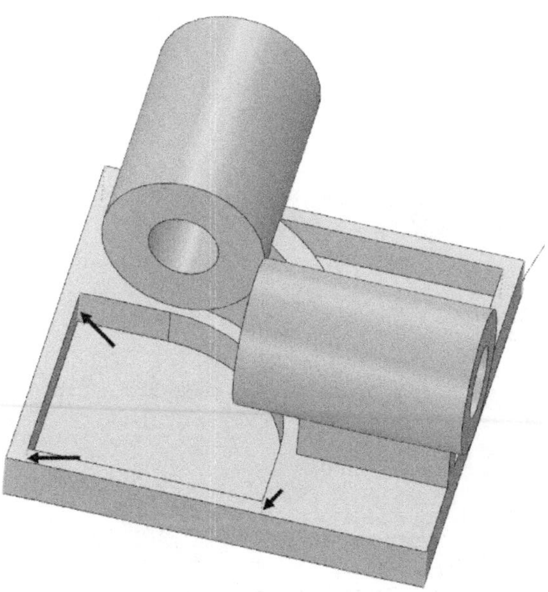

63. Click **OK**.

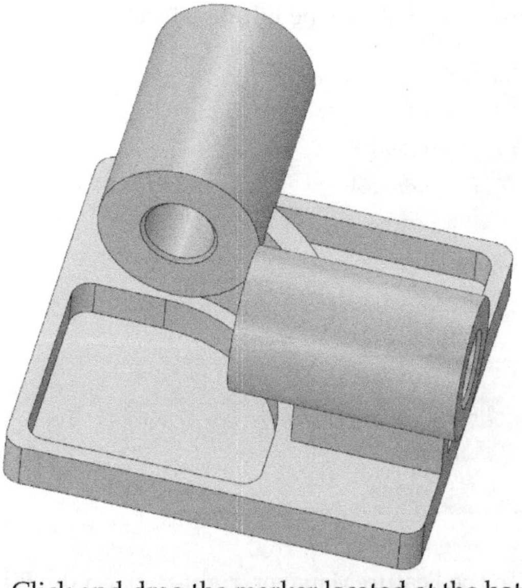

64. Click and drag the marker located at the bottom of the FeatureManager Design Tree, and then release it above the Chamfer feature; the **Chamfer** feature is suppressed.

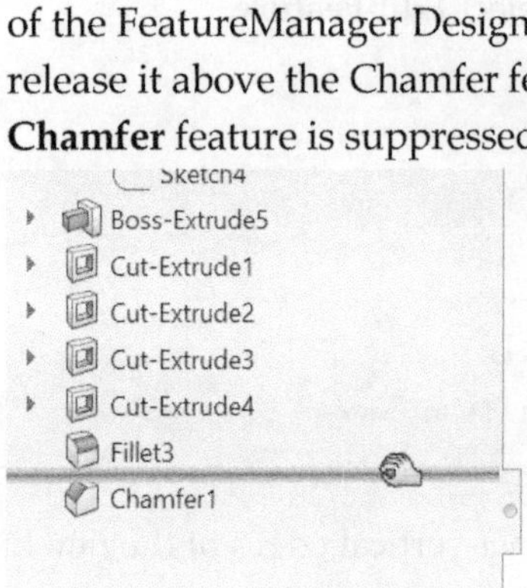

65. On the **Features** CommandManager, click **Extruded Cut** .
66. Click on the front face of the Boss-Extrude feature, as shown.

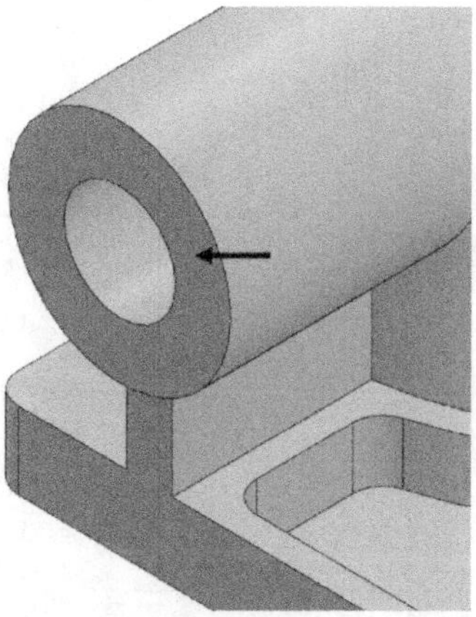

67. On the CommandManager, click **Sketch** tab > **Offset Entities**.
68. Click on the face selected to start the sketch.

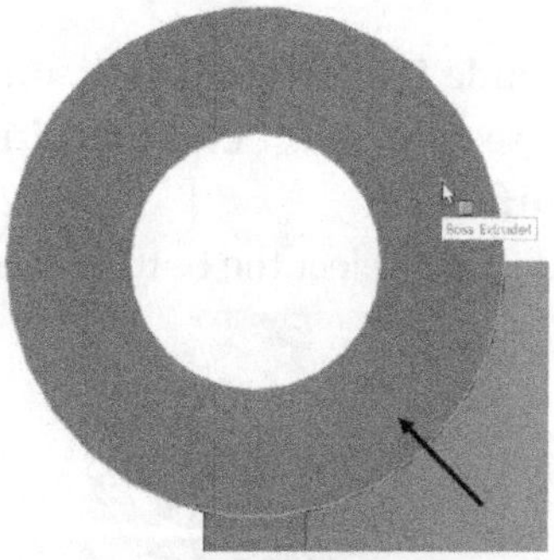

69. Type-in **10** in the **Offset Distance** box available on the PropertyManager.
70. Check the **Reverse** option.
71. Click **OK** on the PropertyManager.

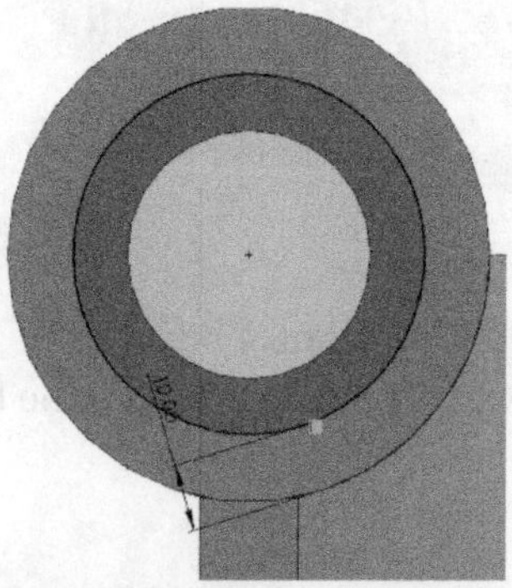

72. Click the **Line** command on the CommandManager.
73. Select the vertex of the vertical edge.

74. Move the pointer vertically upward and click on the circle.
75. Double-click to end the line chain.
76. Select the vertex of the right vertical edge.
77. Move the pointer vertically upward and click on the circle.

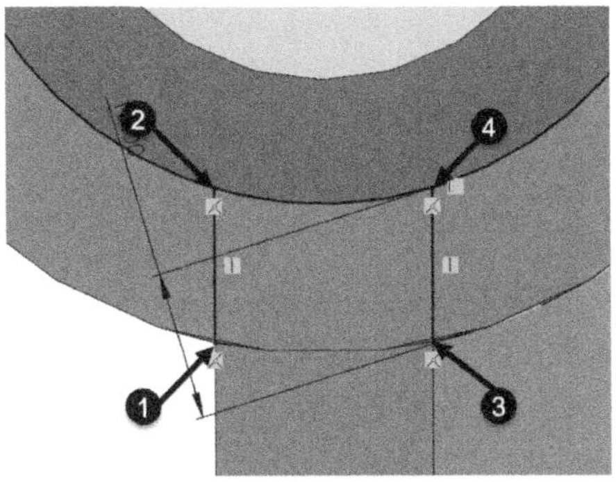

78. On the CommandManager, click the **Convert Entities** button.
79. Select the circle edge of the model, as shown.

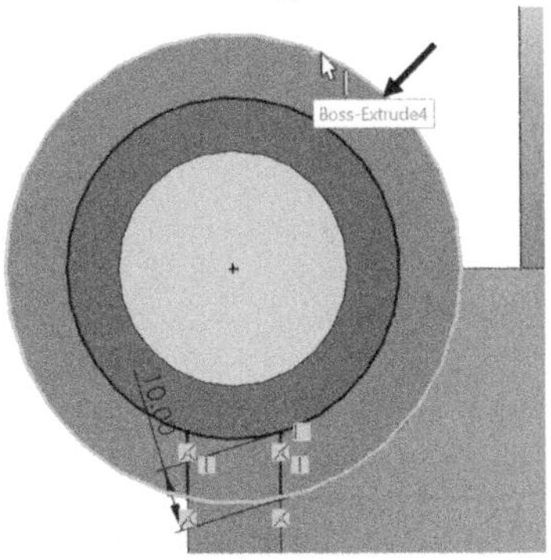

80. On the CommandManager, click **Trim Entities**.
81. Click the **Trim to Closest** button on the PropertyManager.
82. Select the portion of the circles between the two vertical lines; the portion is trimmed.

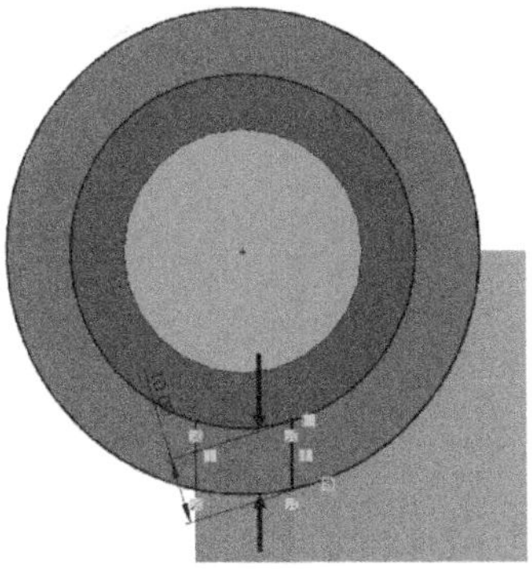

83. Click **OK** on the PropertyManager.
84. Click **Exit Sketch** on the CommandManager.
85. On the **Cut-Extrude PropertyManager**, select **From > Offset**.
86. Type **30** in the **Enter Offset Value** box located below the **From** drop-down.
87. Under the **Direction 1** section, select **End Condition > Blind**.
88. Type **30** in the **Depth** box.
89. Click **OK** to create the cut offset from the front face.

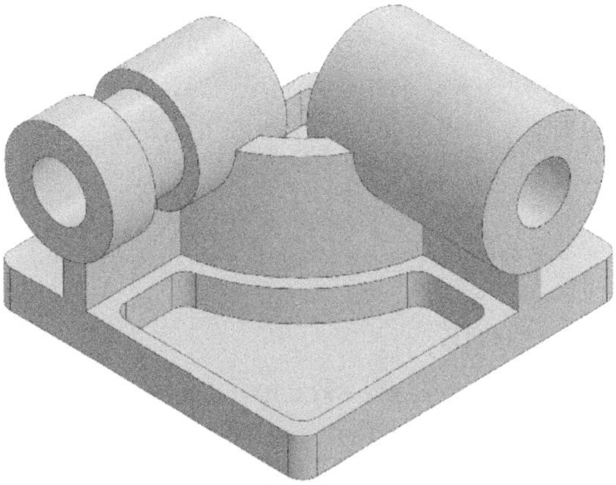

8. Click and drag the marker in the FeatureManager Design tree, and then release it below the Chamfer feature.
9. On the CommandManager, click **Evaluate > Mass Properties** ; on the **Mass Properties** dialog, the **Mass** is displayed.
10. Close the **Mass Properties** dialog.
11. Click **Save** on the Quick Access Toolbar.
12. Type **C9_Example3** in the **File name** box and click **Save**.

Updating the Model

1. In the FeatureManager Design Tree, right-click on the **Equations** node and select **Manage Equations**.
2. On the **Equations, Global Variables, and Dimensions** dialog, select the value in the **Value/ Evaluation** box next to the **A** global variable.
3. Type **229**.
4. Likewise, change the **B**, **C**, **D**, and **E** values to **217, 163,119** and **34**, respectively.

Equations, Global Variables, and Dimensions

Filter All Fields

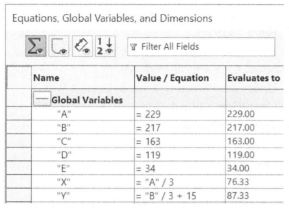

Name	Value / Equation	Evaluates to
Global Variables		
"A"	= 229	229.00
"B"	= 217	217.00
"C"	= 163	163.00
"D"	= 119	119.00
"E"	= 34	34.00
"X"	= "A" / 3	76.33
"Y"	= "B" / 3 + 15	87.33

5. Click **OK**; the model is updated.

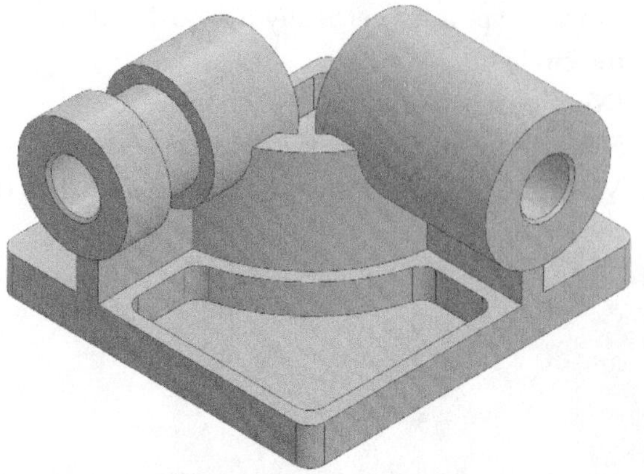

6. Click **Evaluate > Mass Properties** on the CommandManager. The **Mass** value is updated.
7. Close the **Mass Properties** dialog.
8. Save the file and close it.

Questions

1. How to modify the sketch of a feature?
2. How to modify a feature directly?
3. How to suppress a feature?

Exercises

Exercise 1

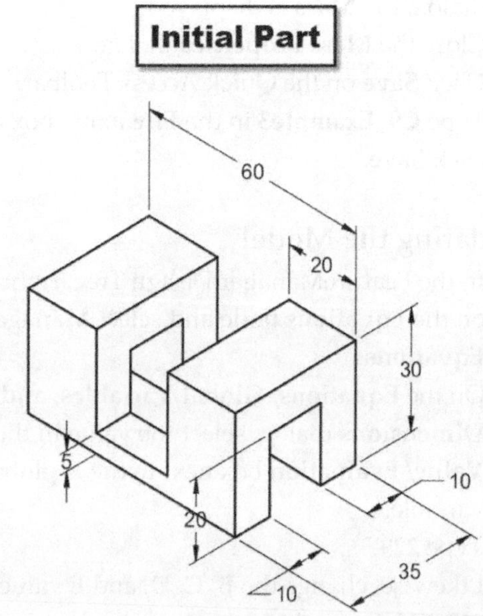

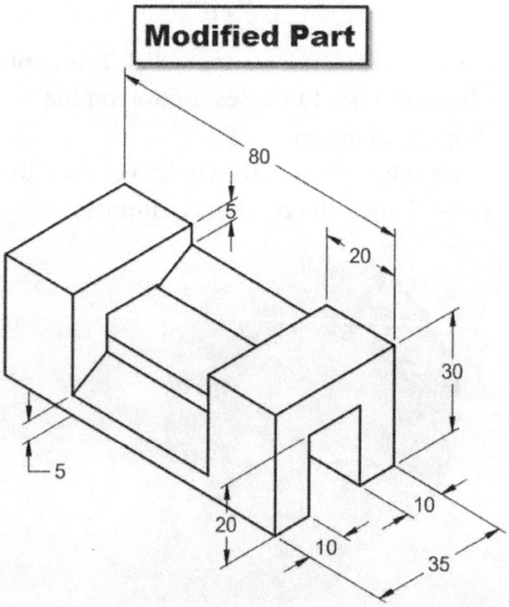

Chapter 10: Assemblies

After creating individual parts, you can bring them together into an assembly. By doing so, it is possible to identify incorrect design problems that may not have been noticeable at the part level. In this chapter, you will learn how to bring parts into the assembly environment and position them.

The topics covered in this chapter are:

- *Starting an assembly*
- *Inserting Components*
- *Adding Mates*
- *Moving and Rotating components*
- *Check Interference*
- *Copy with mates*
- *Editing Assemblies*
- *Replace Parts*
- *Pattern and Mirror Parts*
- *Create Subassemblies*
- *Dissolve assemblies*
- *Assembly Features*
- *Top-down Assembly Design*
- *Exploding assemblies*

Starting an Assembly

To begin an assembly file, click the **New** icon on the **Quick Access Toolbar** and select the **Assembly** template from the **New SOLIDWORKS Document** dialog. Next, click **OK**.

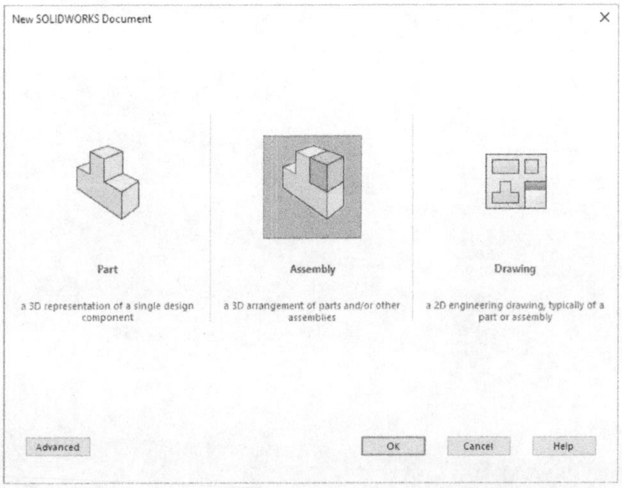

Now, you can insert parts into the assembly by using the **Begin Assembly** PropertyManager that appears immediately after opening a new assembly document. In addition to that, the **Open** dialog appears on the screen. Use this dialog to go to the location of the component to be inserted into the assembly. Next, select the component to be inserted into the assembly, and then click **Open**; the part is attached to the pointer. Click in the graphics window to position the component.

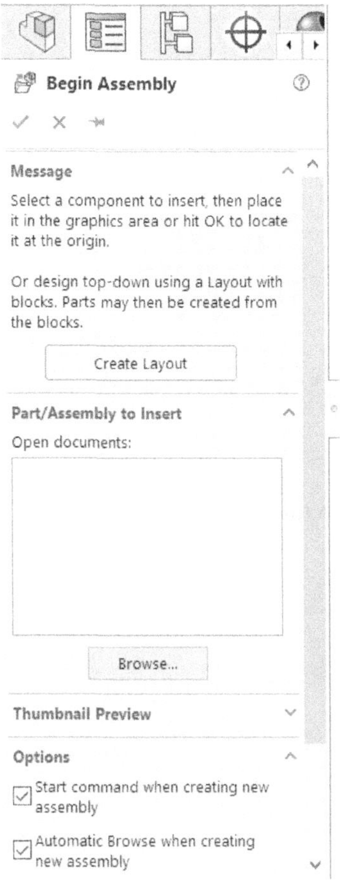

Inserting Parts

There are three different methods to insert an existing part into an assembly. The first one is to use the **Insert Components** command. Activate this command (on the CommandManager, click **Assembly > Insert Components**), and then select the component. You can also choose multiple components at a time by pressing the Ctrl key and clicking on them. Next, click **Open**, and then click in the graphics window.

The second way is to drag the component into the assembly window. To do this, open the part and assembly files in two separate windows. Next, click **Window > Tile Horizontally** or **Tile Vertically** on the Menu bar.

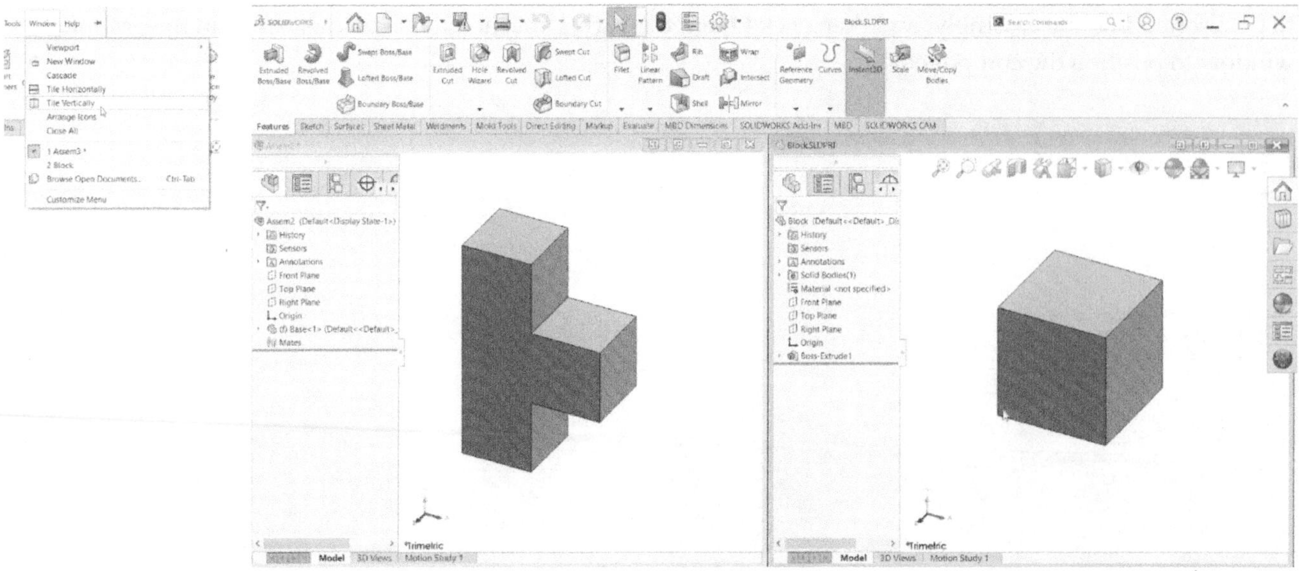

Press and hold the left mouse button on the part, and then drag it into the assembly window.

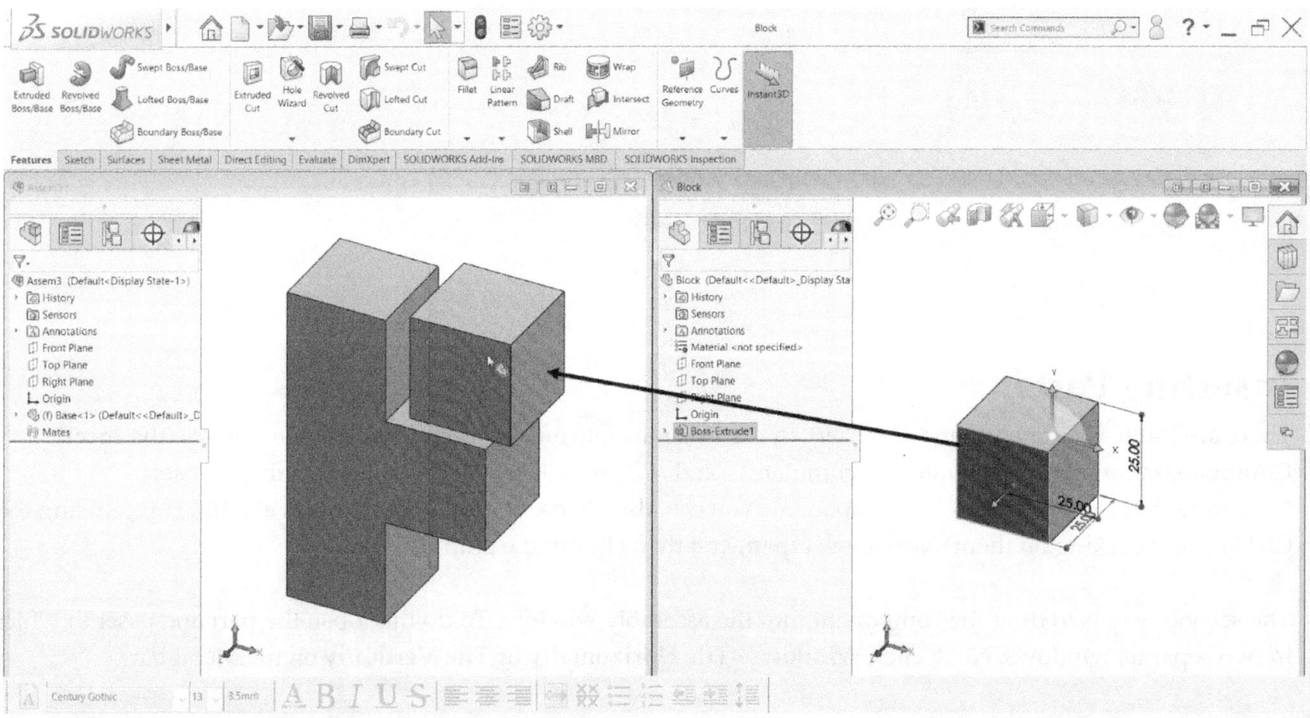

The third way is to drag the part directly from the Windows Explorer into the assembly window.

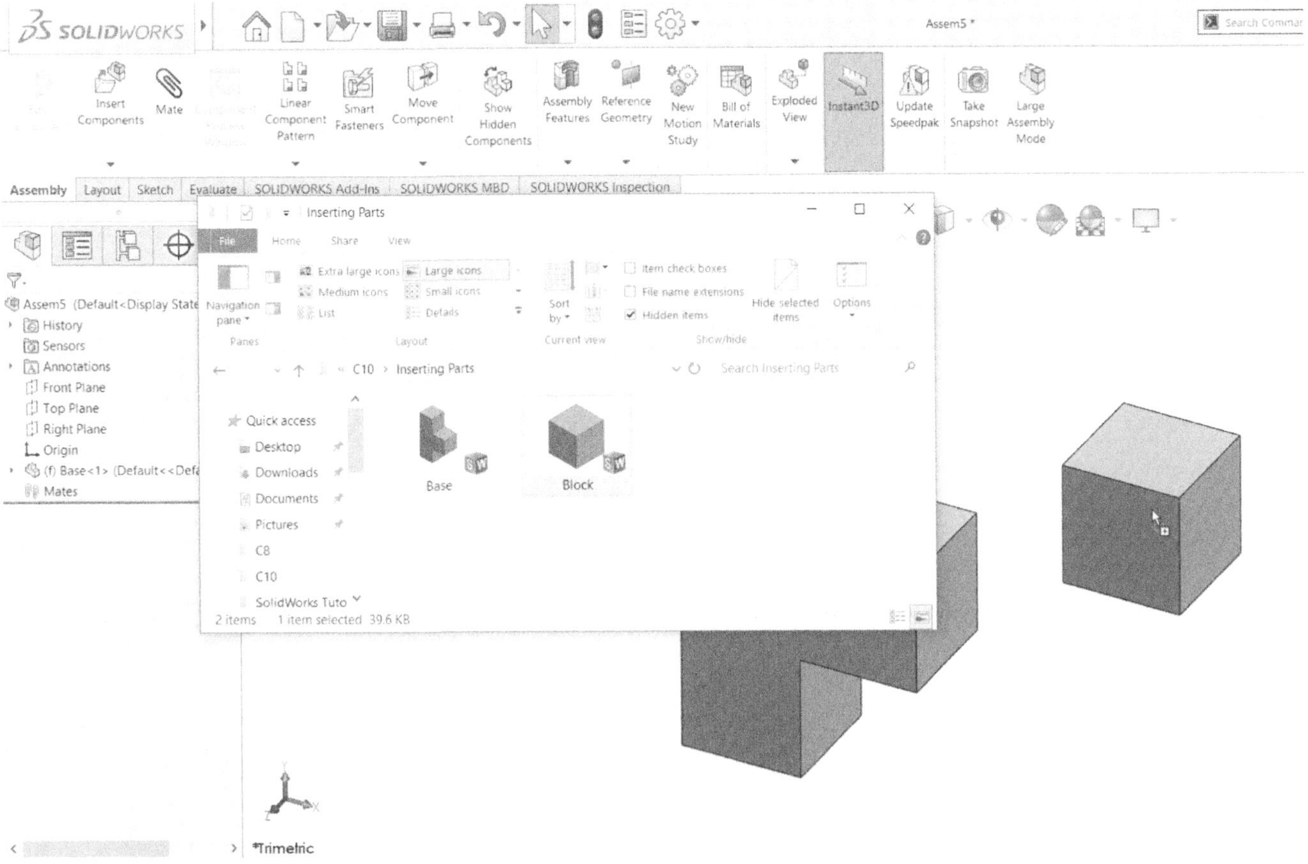

Move Components

As you insert a part into an assembly, the part will be under-constrained and free to move and rotate. You can move the part by pressing and holding the left mouse button and dragging it. However, the **Move Component** command provides a variety of options to move the component. Activate this command by clicking **Assembly > Move Component** on the CommandManager. On the **Move** PropertyManager, select the **Free Drag** option from the drop-down list available in the **Move** section. Next, press and hold the left mouse button on the under-constrained part, and then drag it.

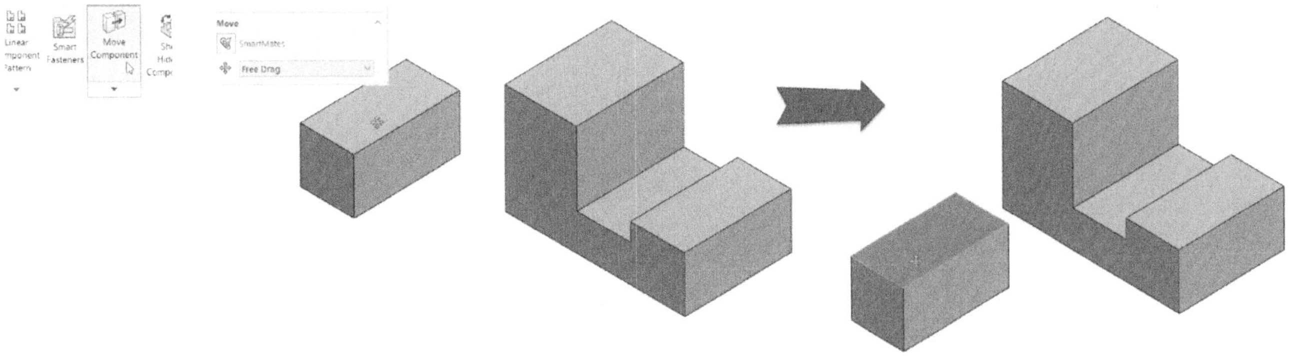

Select the **Along Assembly XYZ** option from the drop-down to move the part in the X, Y, or Z directions. For example, to move the part in the X-direction, press and hold the left mouse button on the part and drag it along the X-direction.

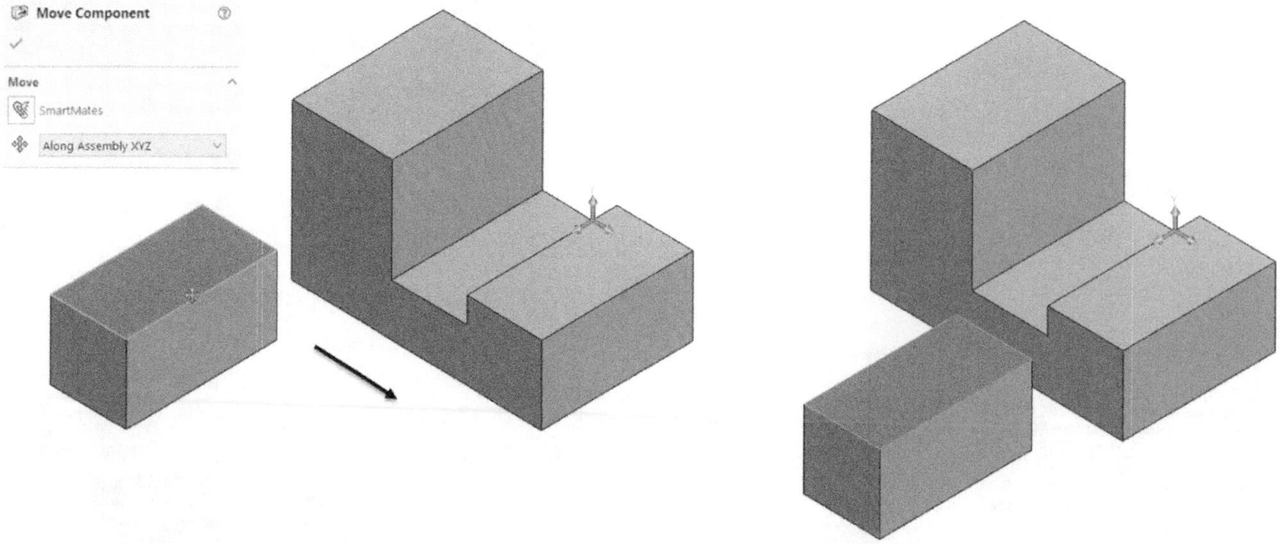

Select the **Along Entity** option from the drop-down and click in the **Selected item** selection box. Next, select an edge or face from the graphics window to define the direction of the part. Click on the part and drag it; the part is moved along the selected entity.

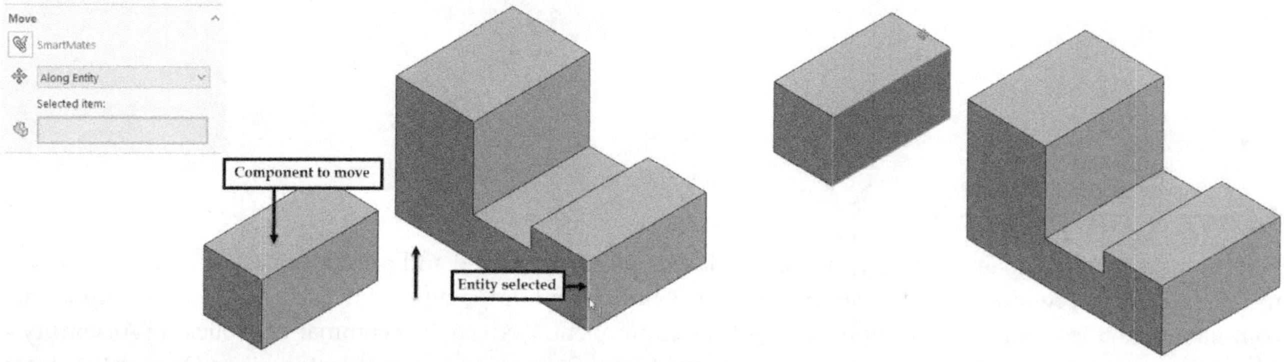

Select the **By Delta XYZ** option from the drop-down and click on the component to move. Next, type in a value in the **ΔX** box and press Enter; the specified value will move the selected component along the X-direction. Likewise, type-in values in the **ΔY** or **ΔZ** boxes and press enter to move the component along the Y or Z directions, respectively.

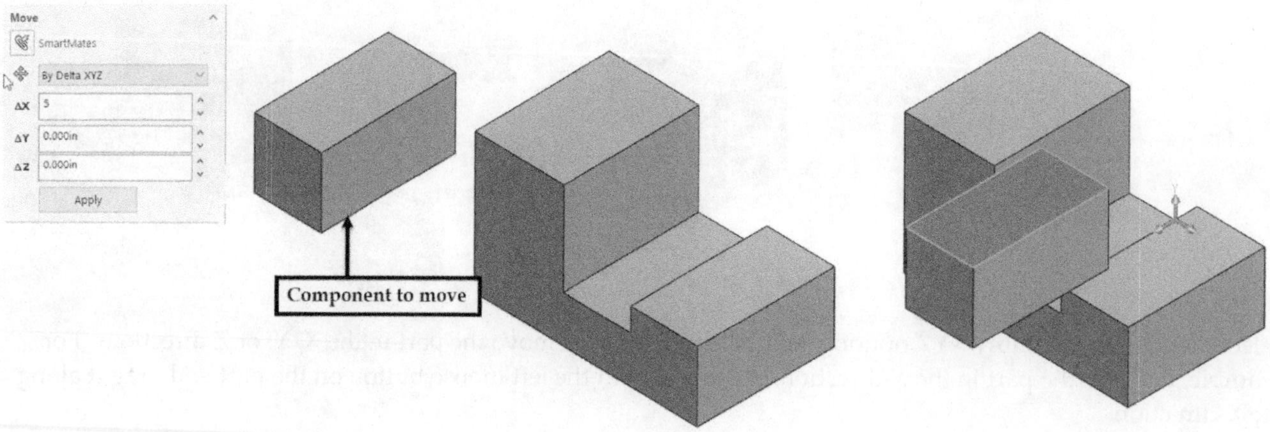

Select the **To XYZ Position** option from the drop-down and select the component to move. Next, enter values in the x, y, z boxes, and then press Enter.

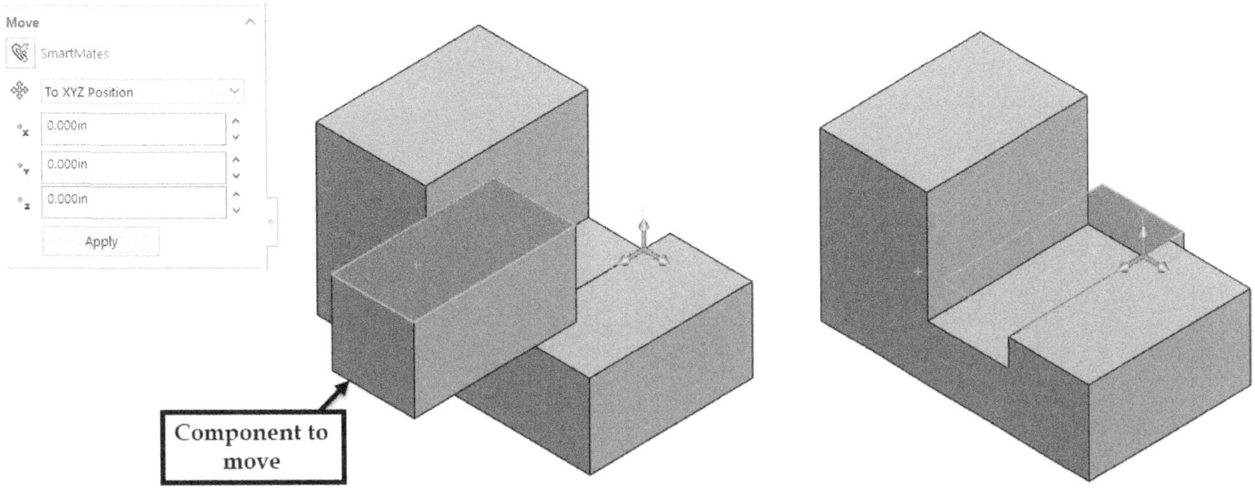

Use the **Collision Detection** option on the PropertyManager to detect collisions while moving or rotating the parts.

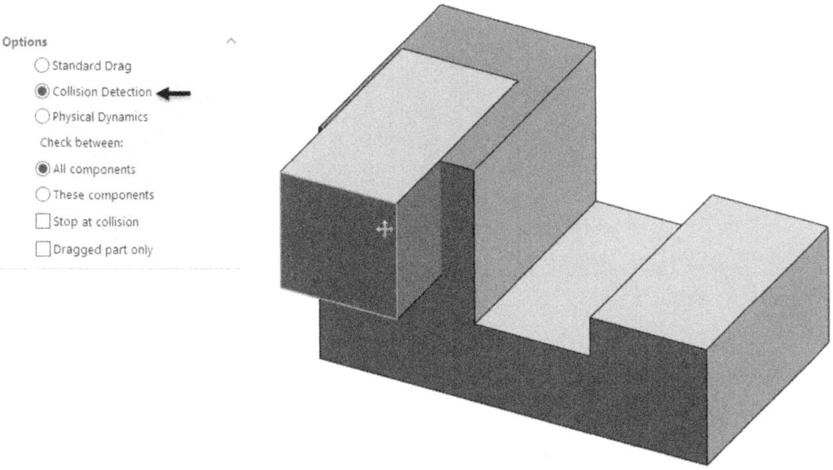

Use the **Stop at collision** option on the PropertyManager to stop the part when it collides with another part.

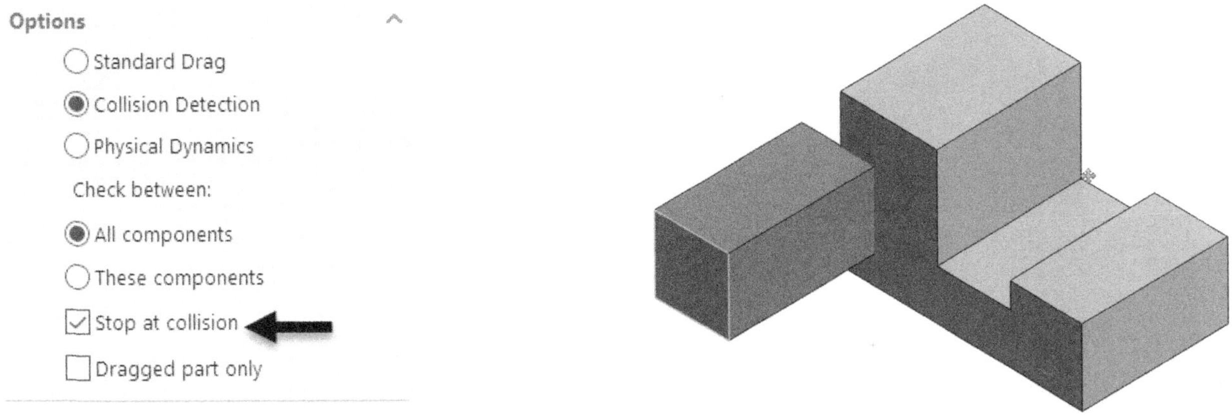

Rotate Components

You can use the **Rotate Component** command to rotate the under-constrained parts in the assembly window. Activate this command by clicking **Assembly > Move Component > Rotate Component** on the CommandManager. On the PropertyManager, select **Free Drag** from the drop-down available in the **Rotate** section. Next, select the part from the assembly window, press and hold the left mouse button, and then drag the cursor; the part is rotated.

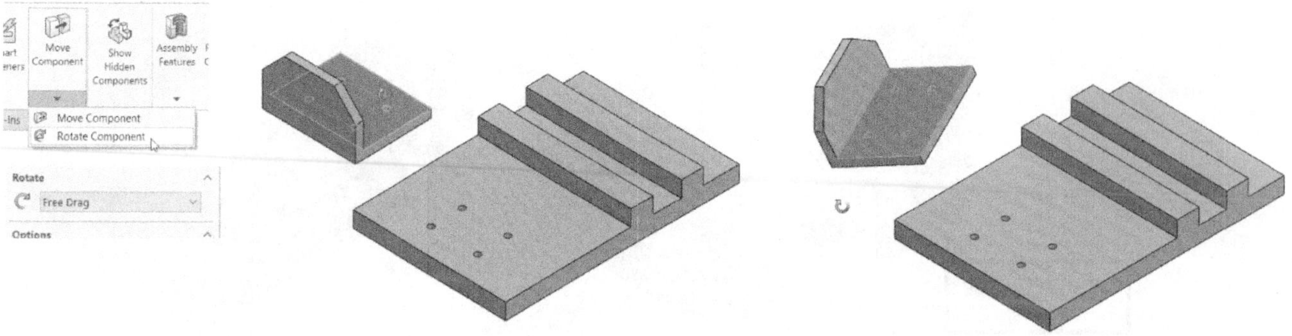

Select the **About Entity** option from the drop-down in the **Rotate** section, and then select an edge, sketched line, or axis. Next, drag the under-constrained part about the selected entity.

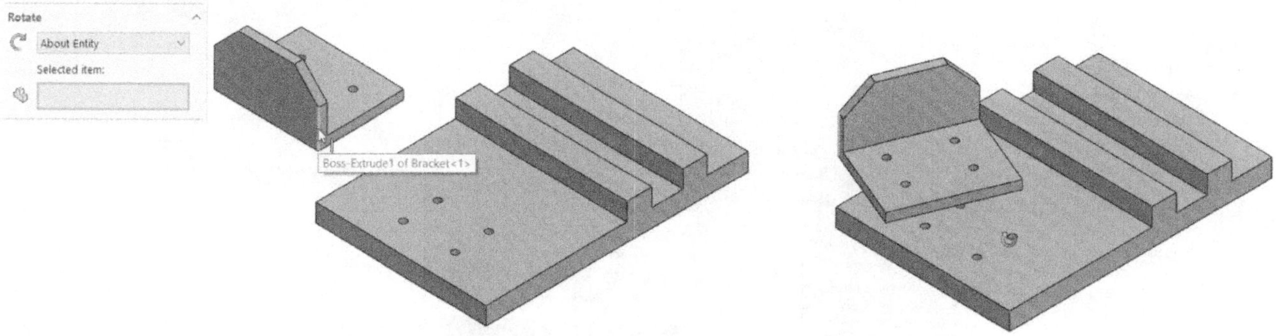

Select the **By Delta XYZ** option from the drop-down available in the **Rotate** section, and then select the component to rotate. Next, type in the rotation angle values in the **ΔX**, **ΔY**, and **ΔZ** boxes. Click **Apply** to turn the components by the specified angle values.

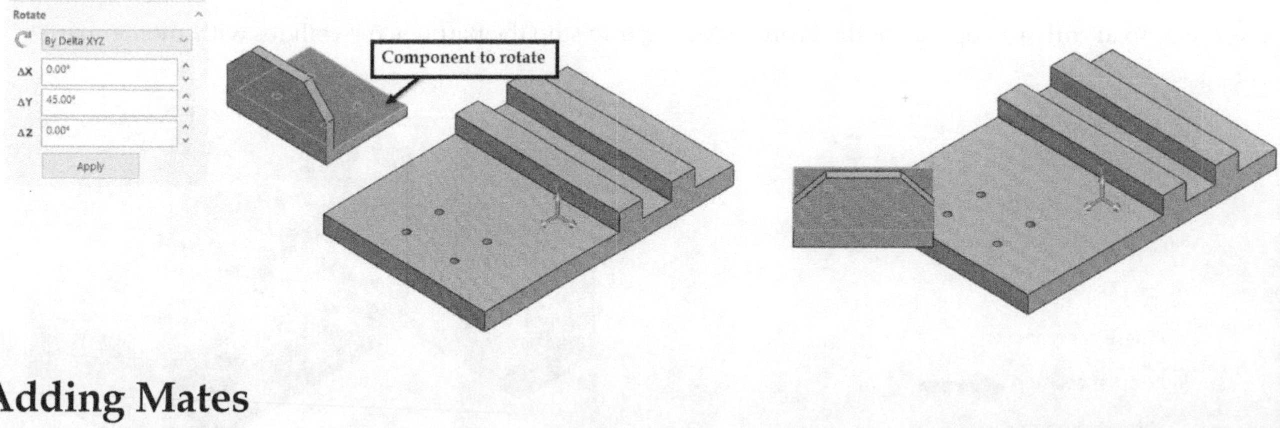

Adding Mates

After inserting parts into an assembly, you have to define mates between them. By applying mates, you can make parts to touch each other or make two round faces concentric with each other, and so on. As you add mates between components, the degrees of freedom will be removed from them. By default, there are six degrees of freedom for a part (three linear and three rotational). Eliminating degrees of freedom will make parts attached and interact with each other as in real life. Now, you will learn to add mates between parts.

The Coincident mate

The **Coincident mate** makes two faces coincident with each other. To define this mate, click **Assembly > Mate** on the CommandManager. Next, click in the **Mate Selections** selection box on the **Standard** tab of the **Mate** PropertyManager, select a face of a part and click on the face of the target part. Next, click the Coincident icon on the **Mate Type** section of the PropertyManager. The two chosen faces will mate with each other. You can change how the two selected faces are aligned with each other using the **Mate alignment** options: **Aligned** and **Anti-Aligned**. These options are explained next.

Aligned

This option makes the selected two faces flush with each other. To do this, select the faces to be aligned from two parts, as shown. Next, set the **Mate Alignment** to **Aligned** under the **Standard Mates** section. Click the **Add/Finish Mate** button on the context toolbar.

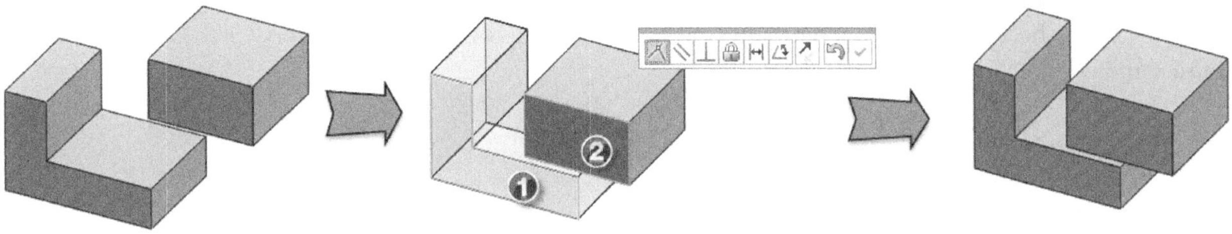

Anti-Aligned

This option makes two faces aligned opposite to each other. Select the faces of two parts. Next, set the **Mate Alignment** to **Anti-aligned** and click the **Add/Finish Mate** button on the context toolbar.

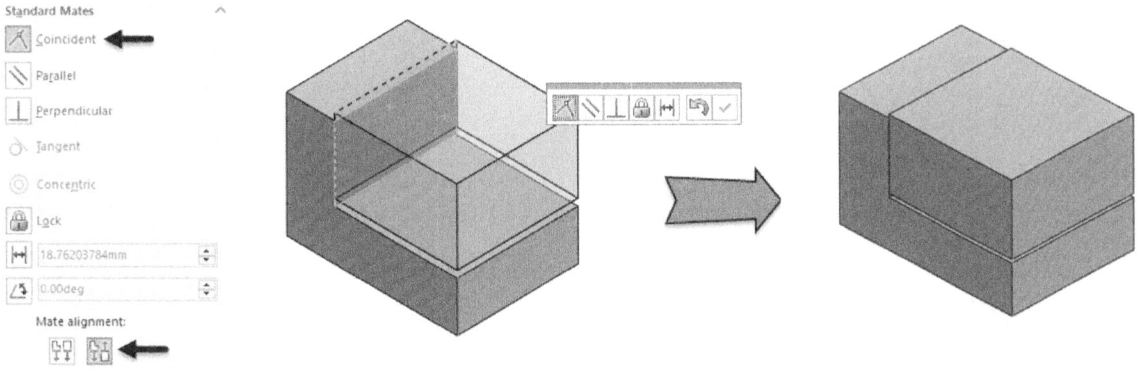

Likewise, select the upper face of the target part and the bottom face of the inserted part. Click the **Anti-aligned** button and click the **Add/Finish Mate** button on the context toolbar. Click **OK** on the **Mate** PropertyManager.

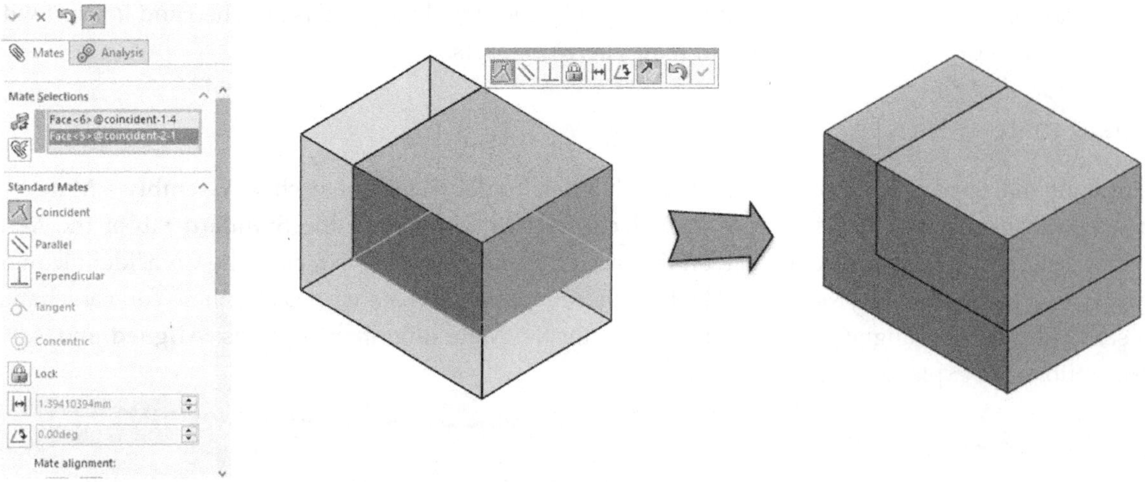

The Distance mate

The **Distance** mate positions the two selected faces or edges at the specified distance. Activate the **Mate** command and click the **Distance** icon on the PropertyManager. Next, select two faces or edges, and then type-in the **Distance** value. Check the **Flip Dimension** option if you want to reverse the direction. Click **OK** to create the **Distance** mate.

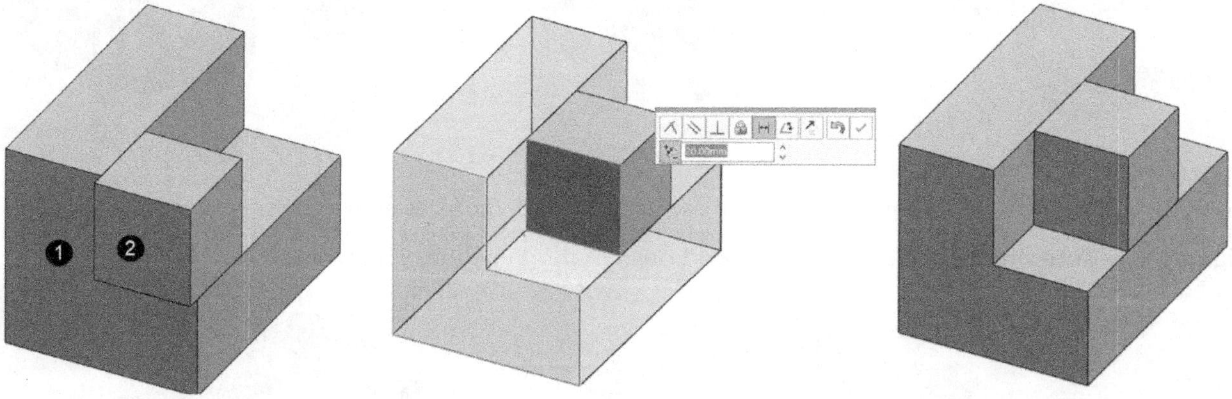

The Concentric mate

The **Concentric** mate makes two round faces concentric with each other. To apply this mate, click **Assembly > Mate** on the CommandManager. On the **Mate** PropertyManager, click the **Concentric** icon on the **Standard** tab. Select the round faces from two parts, as shown.

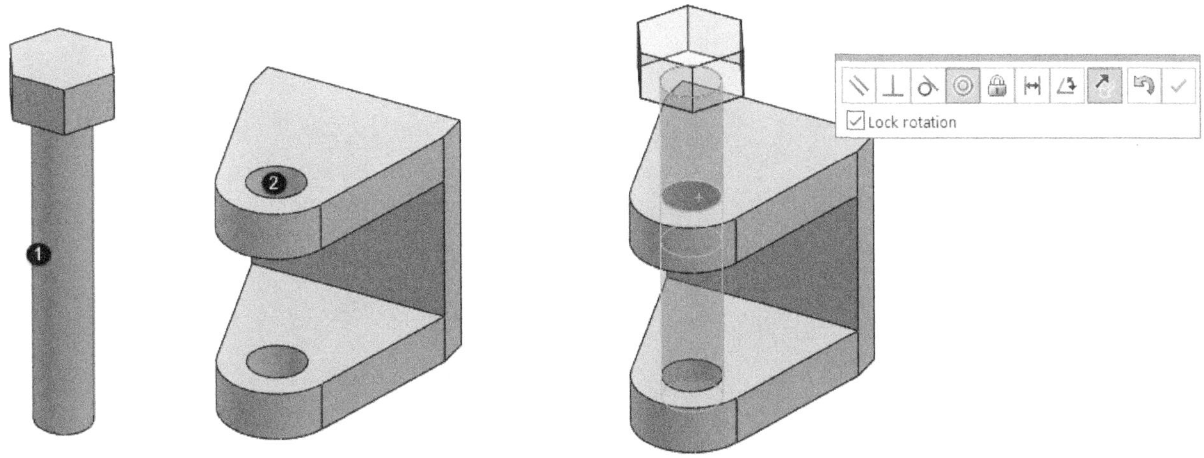

Check the **Lock rotation** option if you want to lock the rotation of the inserted part. Click the **Add/Finish Mate** button on the context toolbar. Click **OK** on the PropertyManager.

The Angle mate

The **Angle** mate is used to position the selected faces or linear edges at the specified angle. Activate the **Mate** command and click the **Angle** button under the **Mate Type** section on the **Standard** tab of the **Mate** PropertyManager. Next, specify a value in the **Angle** box. Select the straight edges or flat faces of the first and second parts, as shown. Next, click in the **Reference entity** selection box under the **Mate selections** section and select the plane on which the angle will lie. Click **Add/Finish Mate** on the context toolbar. Click **OK** on the PropertyManager.

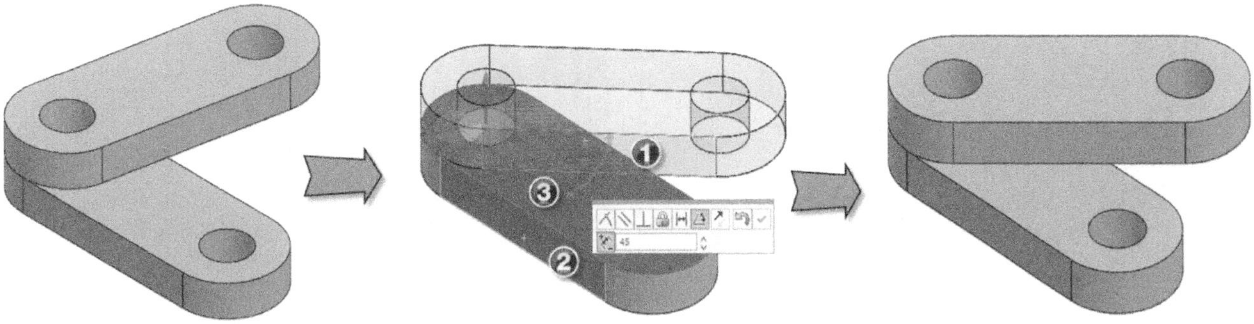

The Angle mate with limits

SOLIDWORKS allows you to apply angle mate with start and end limits. This mate will enable the model to move only between the specified angle limits. For example, the door in the assembly is free to move in 360 degrees, as shown. This movement is illogical and not possible in the real world. You need to apply the **Angle** mate to the door along with the start and end limits.

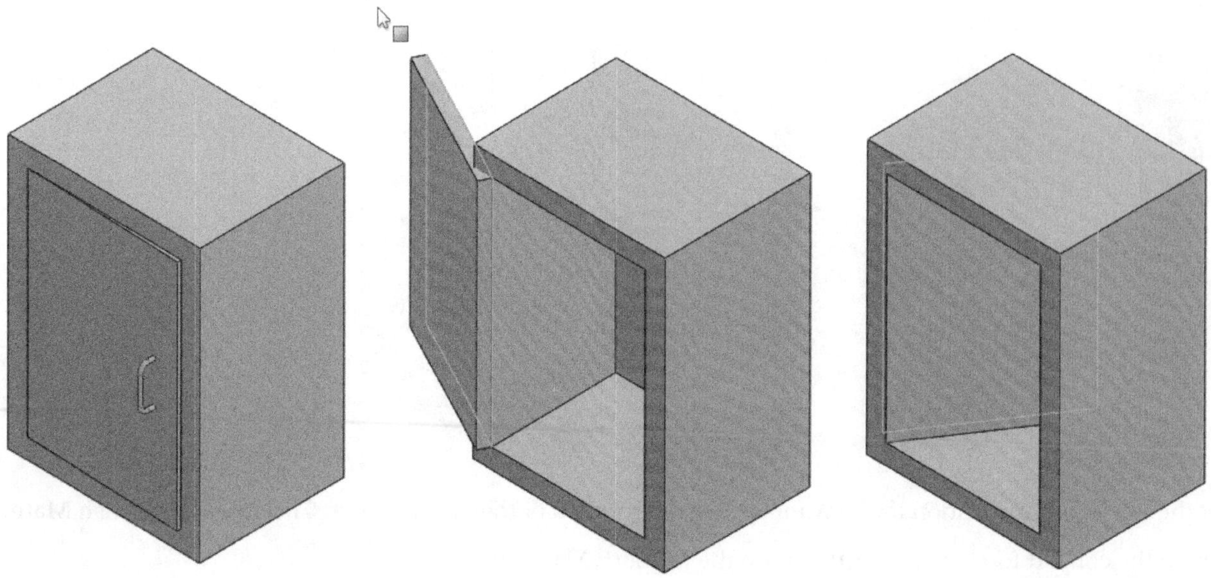

Activate the **Mate** command and click the **Angle** icon on the **Advanced** tab of the **Mate** PropertyManager. Next, select the edges of the door and base, as shown. Type **0** in the **Angle** box.

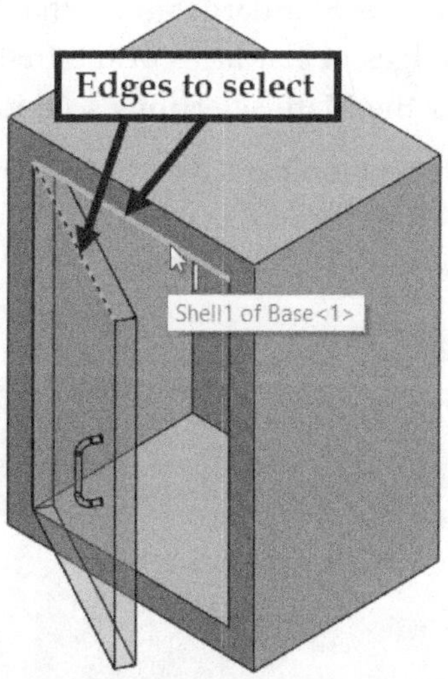

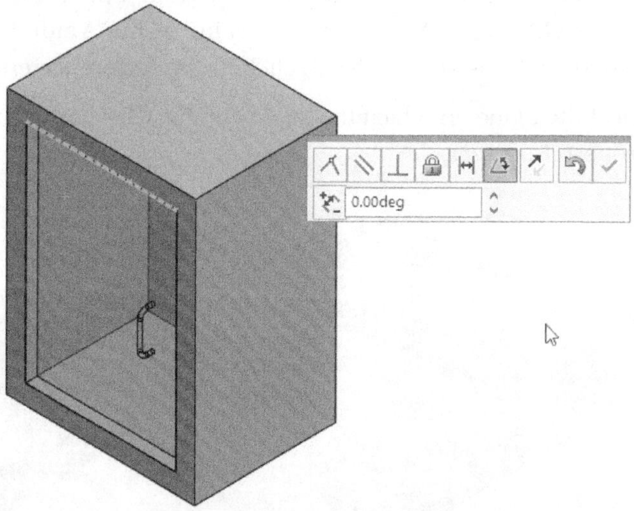

Type 90 and 0 in the **Maximum Value** and **Minimum Value** boxes, respectively. Click **OK** and drag the door. You will notice that the door moves only between the maximum and minimum limits.

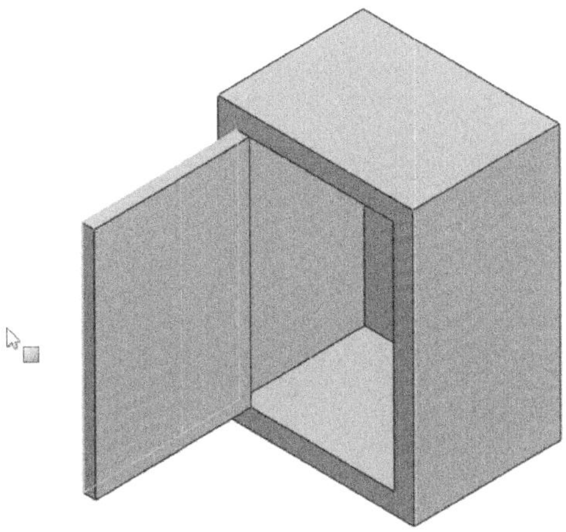

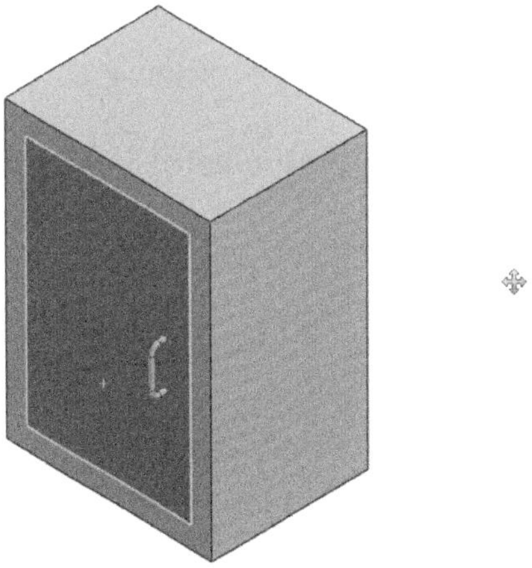

The Distance mate with limits

Similar to the **Angle** mate with limits, you can apply the Distance mate within the specified limits. For example, the shaft in the assembly shown below is free to move, as shown. You need to constrain the motion of the knob within the groove.

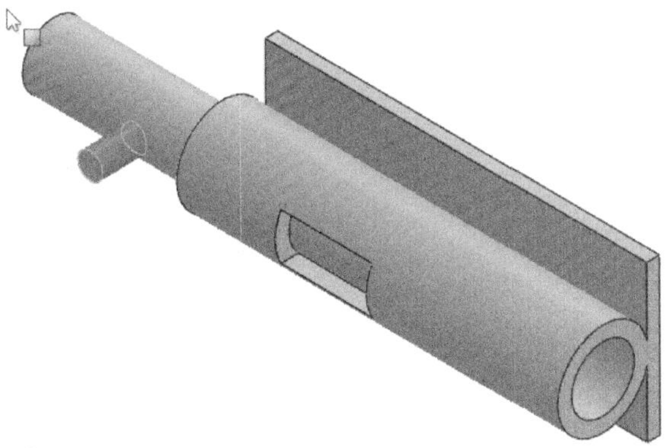

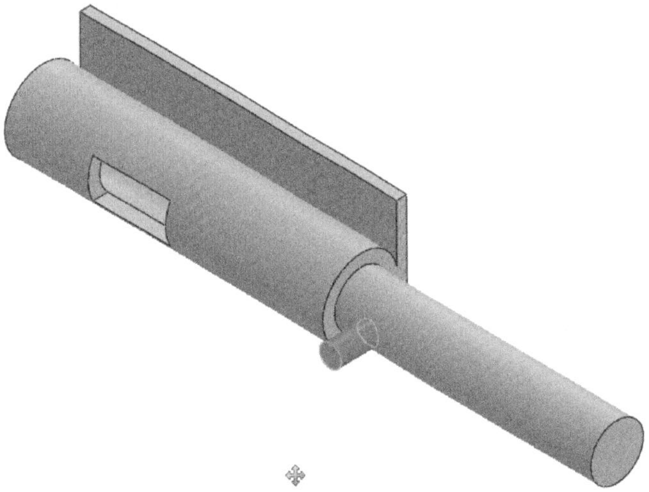

To restrict the movement of the knob within groove, first, measure the distance of the groove and the diameter of the knob.

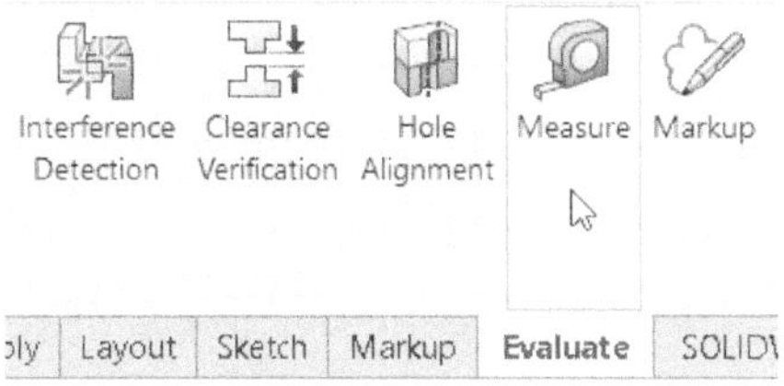

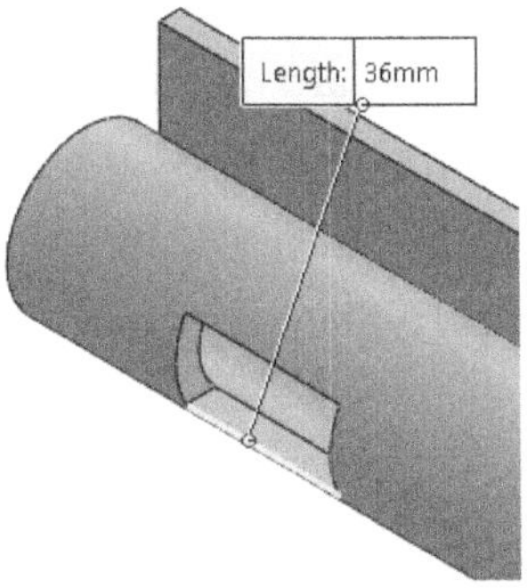

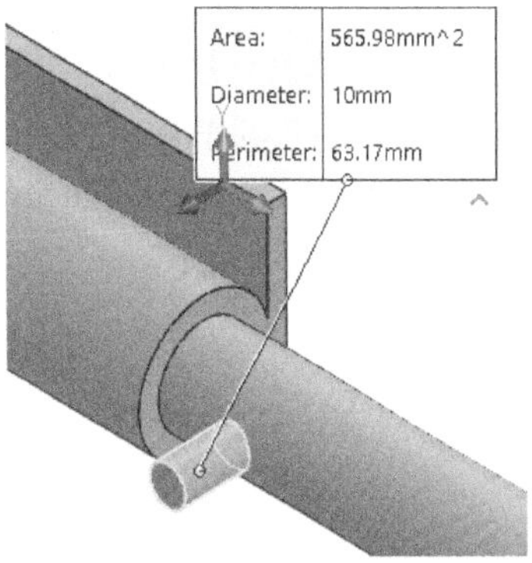

Activate the **Mate** command and click the **Distance** icon on the **Standard** tab of the **Mate** PropertyManager. Next, select the two faces of the assembly, as shown. Click the **Minimum distance** icon, and then set the **Mate alignment** to **Anti-aligned**.

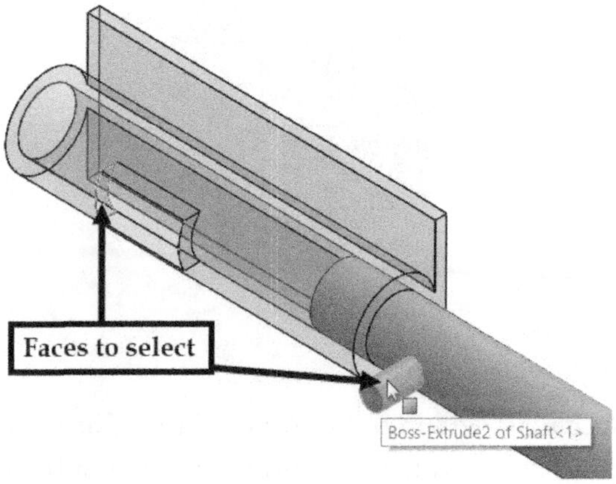

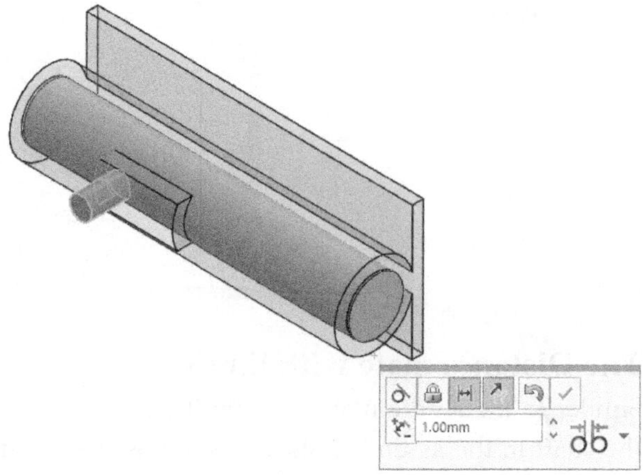

Click the **Advanced** tab and type-in 26 (Slot distance-knob diameter) and 0 in the **Maximum Value** and **Minimum Value** boxes, respectively. Click **OK** on the PropertyManager. Now, click and drag the knob, and then notice that it moves within the slot.

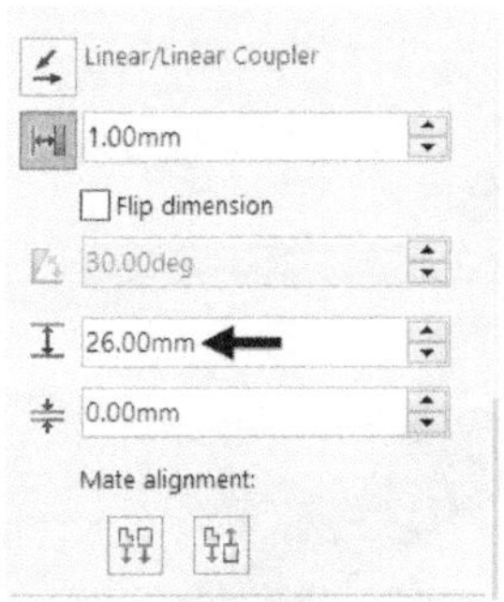

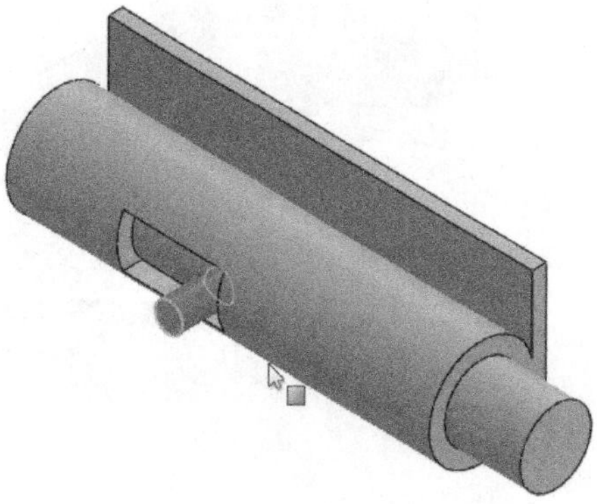

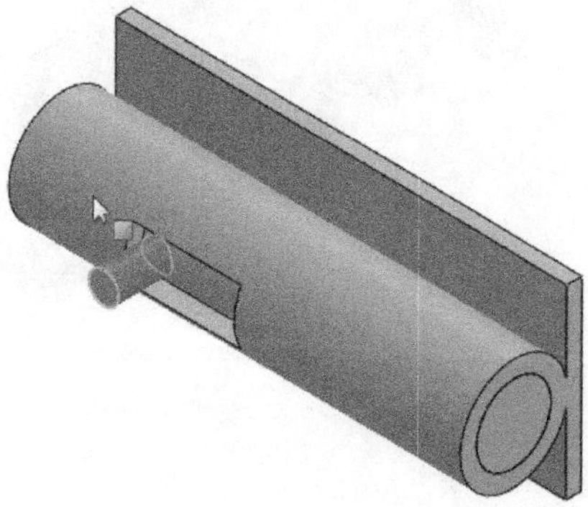

Tangent Mate

The **Tangent** mate is used when working with cylinders and spears. It causes the geometry to maintain contact at a point of tangency. To define this mate, click **Assembly > Mate** on the CommandManager. On the **Mate** PropertyManager, click the **Tangent** icon in the **Standard** tab. Next, click on the face to be made tangent. Next, click on the tangent face on the target part. Click **Add/Finish Mate** on the context toolbar.

Likewise, click on the cylindrical face and then click on the sloped face of the target part. Click **Add/Finish Mate** on the context toolbar. Click **OK** on the PropertyManager to create the tangent mate.

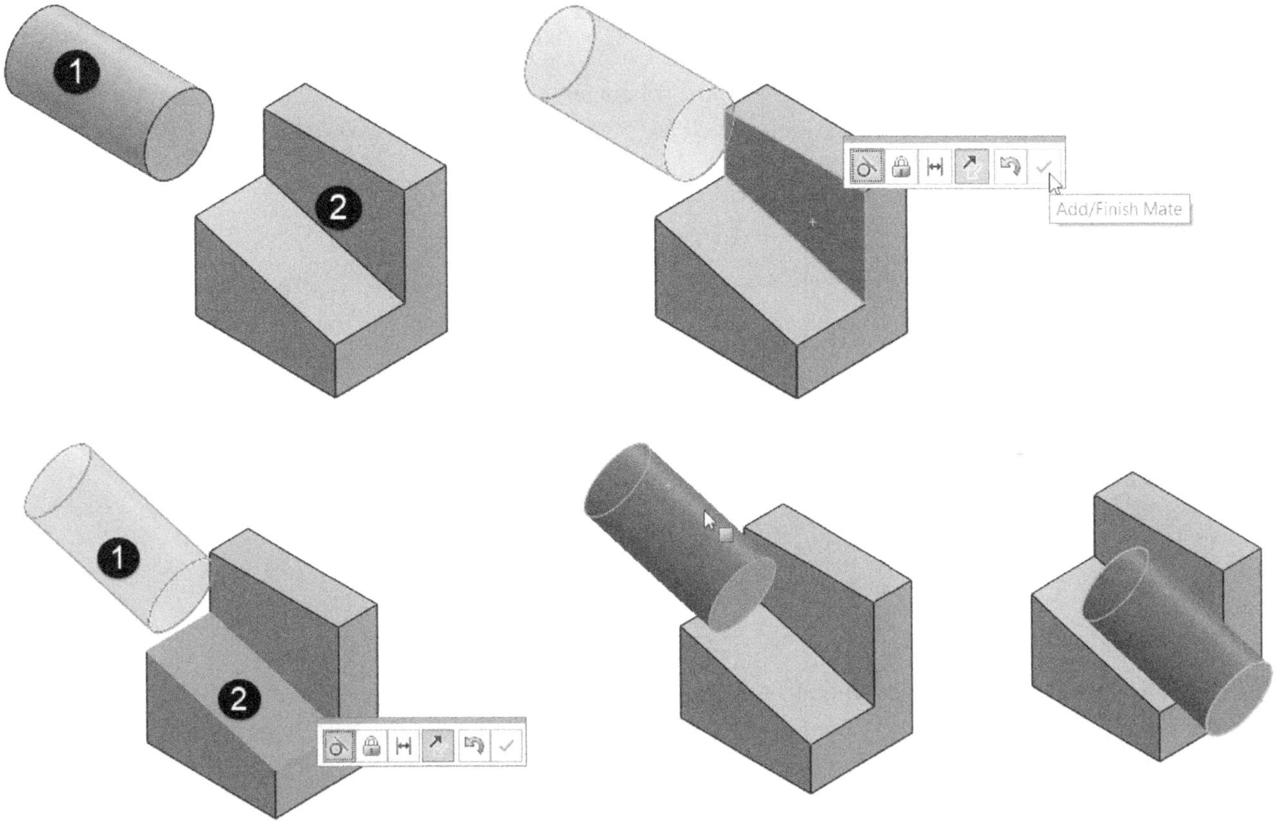

The Parallel Mate

The **Parallel** mate makes an axis or edge of one part parallel to that of another part. To define this mate, click **Assembly > Mate** on the CommandManager. On the **Mate** PropertyManager, click the **Parallel** icon in the **Standard** tab. Next, click in the **Mate Selections** selection box and click on a linear edge, flat face or cylindrical face of the first part. Click on a linear edge, flat face, or circular face of the second part, as shown. Click the **Add/Finish Mate** button on the context toolbar. Click **OK** on the PropertyManager. Note that you cannot make a flat face parallel to a cylindrical face.

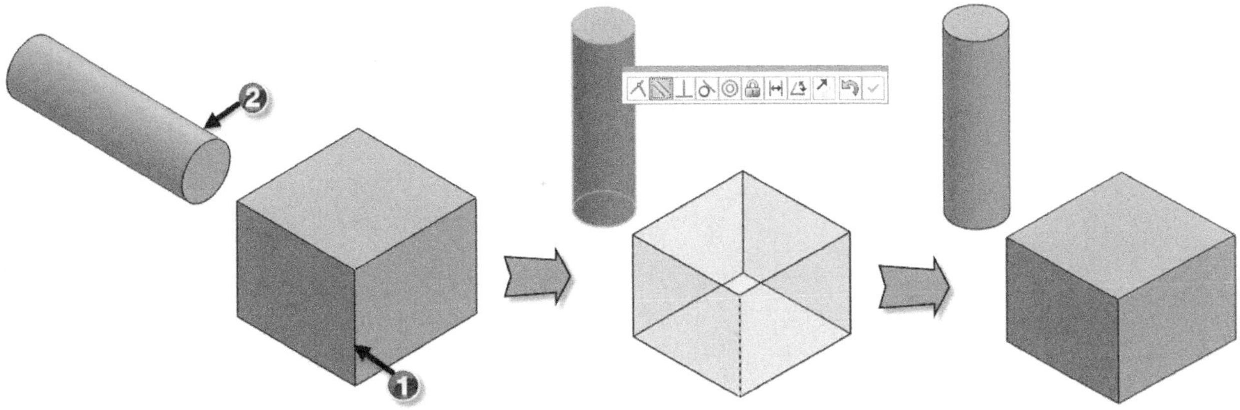

Width Mate

The **Width** mate allows you to place a part between two faces. Activate this command (click **Assembly > Mate** on the CommandManager). On the **Mate** PropertyManager, click the **Width** icon from the **Advanced** tab. Next, you need to select the faces between which the part will be positioned. These faces are called width faces.

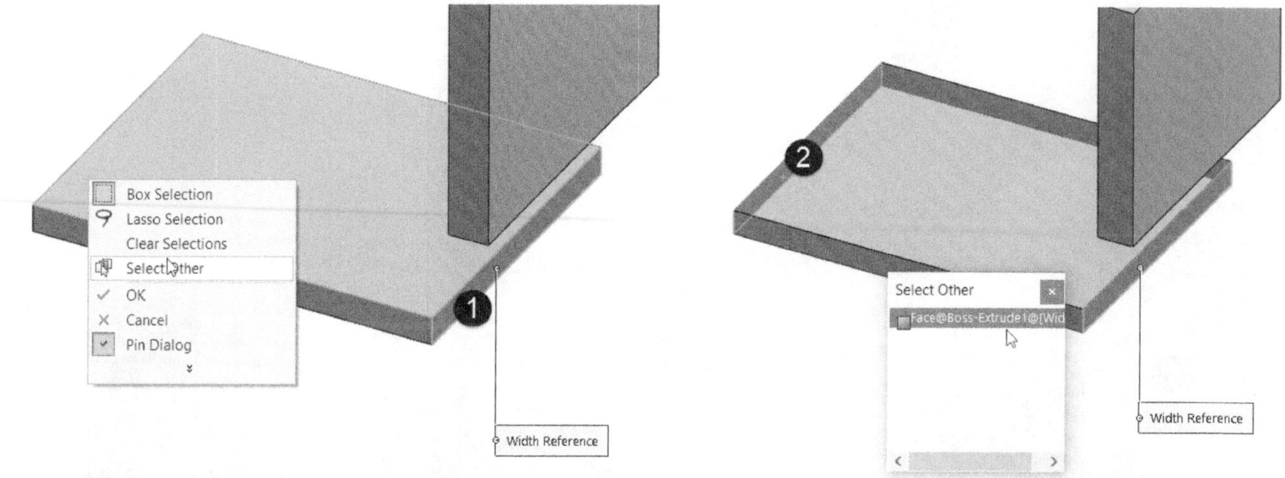

After selecting the width faces, select the faces of the part to be positioned. These faces are called tab faces. Next, select the **Centered** option from the **Constraint** drop-down available below the **Width** icon; the tab part will be positioned at the center location between the width faces.

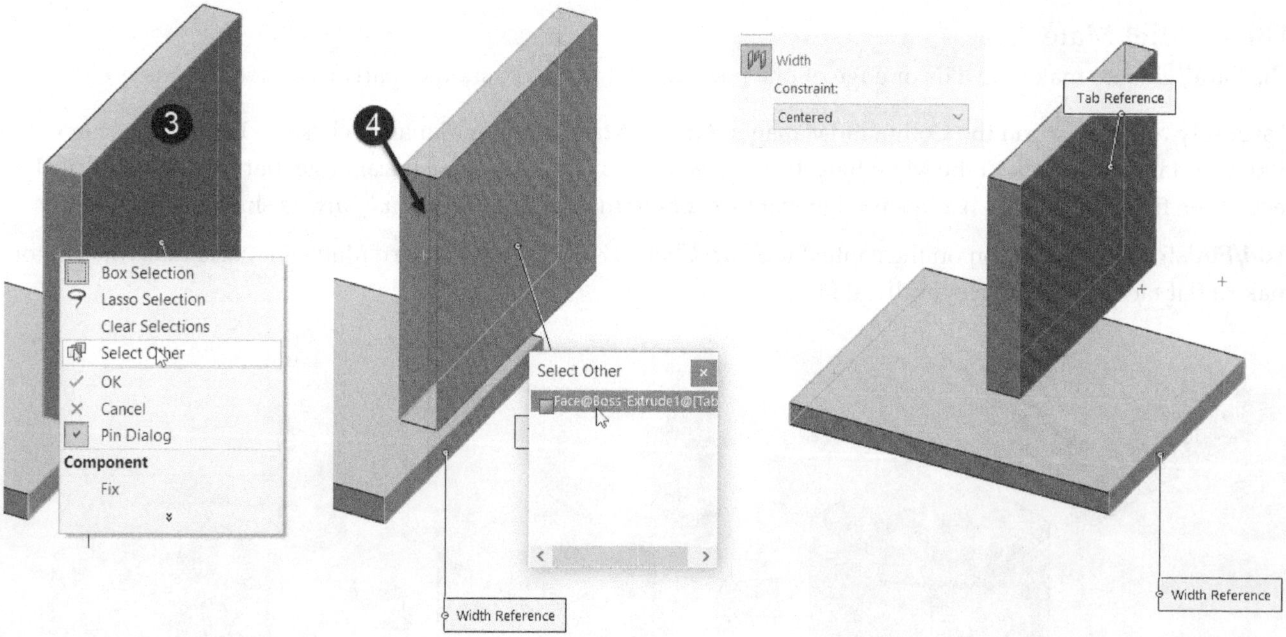

Select the **Free** option from the **Constraint** drop-down to move the part freely between the width faces.

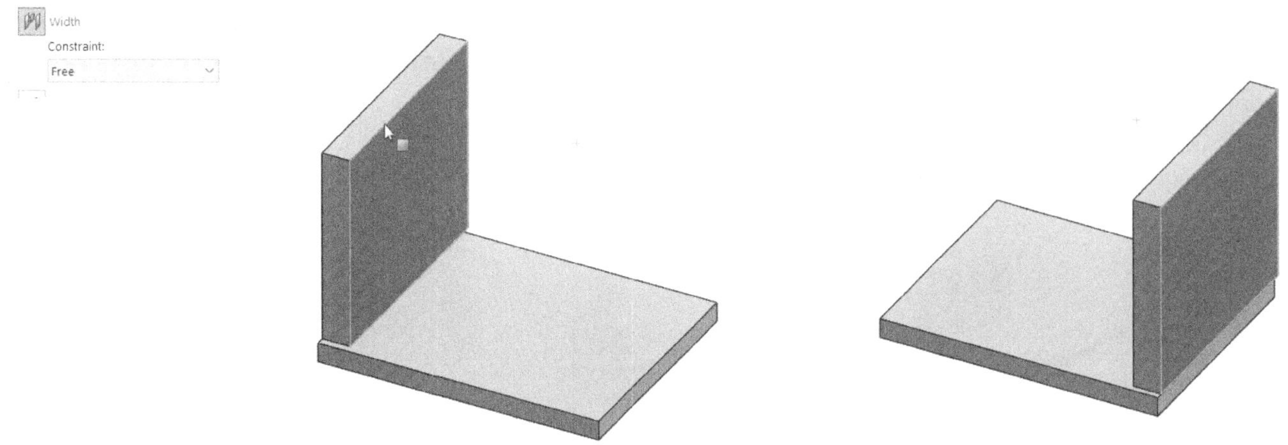

Select the **Dimension** option from the **Constraint** drop-down and enter a value in the **Distance from End** box; the part will be positioned at the specified distance from the end face. Check the **Flip dimension** option to reverse the placement direction.

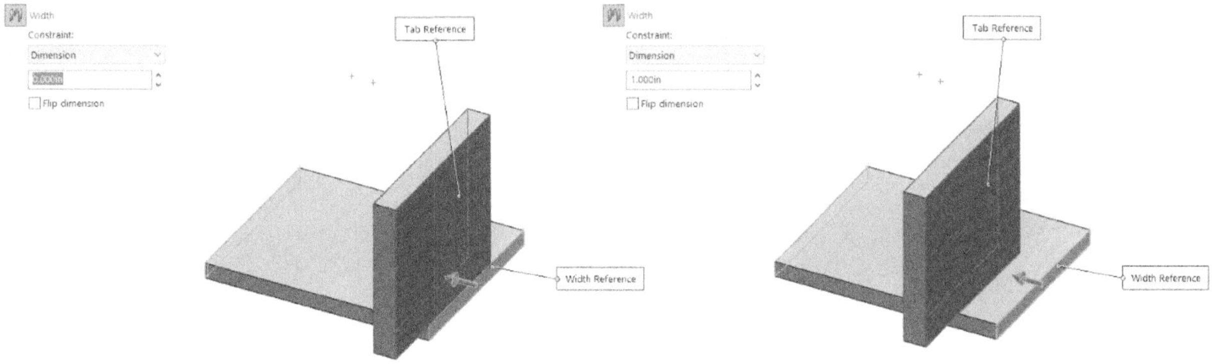

Select the **Percent** option from the **Constraint** drop-down and enter a value in the **Percentage of Distance** from the **End** box.

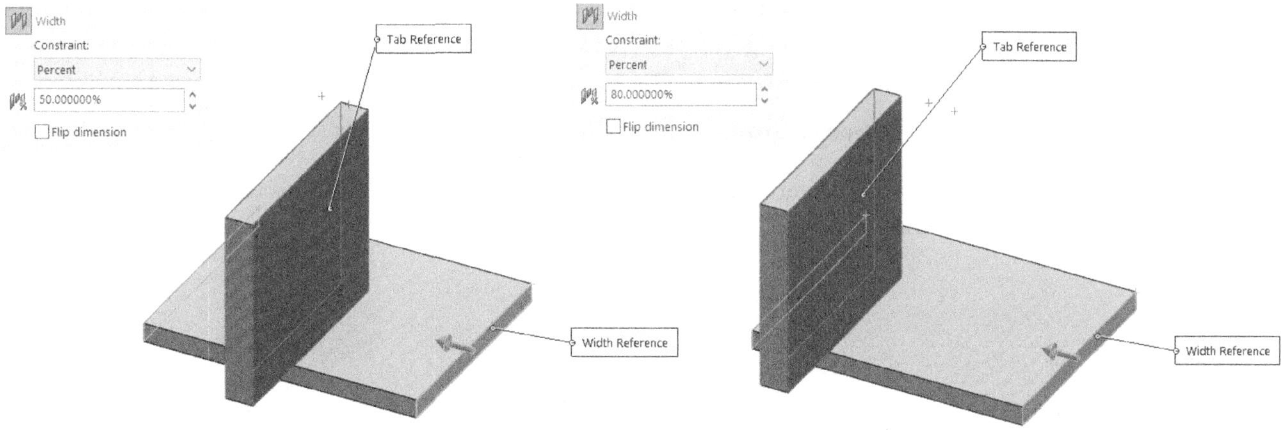

Symmetric Mate

You can create the **Symmetric** mate between two components if you want them to behave symmetrically about a plane. To do this, activate this command (click **Assembly > Mate**  on the CommandManager). On the **Mate** PropertyManager, click the **Symmetric**  icon on the **Advanced** tab and select the symmetric plane from the

FeatureManager Design Tree. Next, click in the **Entities to Mate** selection box and select the faces, edges, or vertices from the two parts between which the Symmetric mate is to be applied. Next, click **OK** on the PropertyManager to create the Symmetric mate.

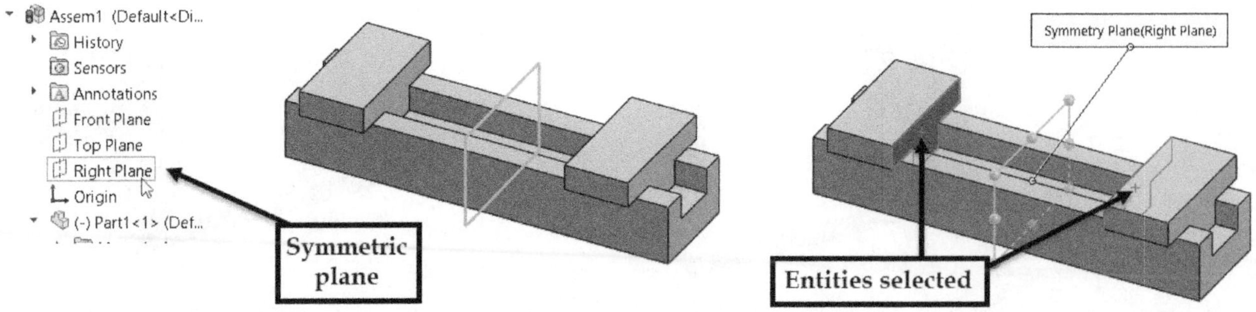

Click **Assembly > Move Component** button on the CommandManager. Now, click and drag any of the parts with the symmetric mate; the other part will also move.

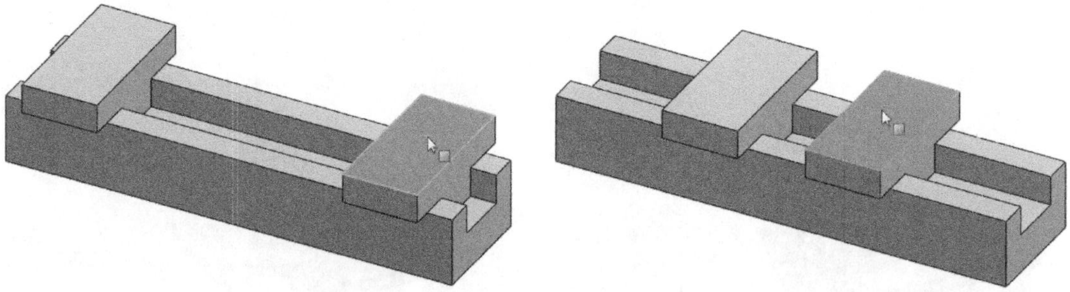

Path Mate

The **Path** mate is used to constrain a selected point along a path. To create this mate, click **Assembly** > **Mate** on the CommandManager, and then click the **Path** icon under the **Advanced** tab. Next, and click on a point or vertex to define the follower. Click on an edge or sketch to specify the path. You can also select tangentially connected edges by using the **Selection Manager**. To do this, click the **Selection Manager** button on the PropertyManager, and then click the **Select Group** icon. Click any one of the edges, and then select the **Tangent** option from the callout attached to the selected edge. Click the **OK** button on the selection manager. Leave the other default options and click **OK**.

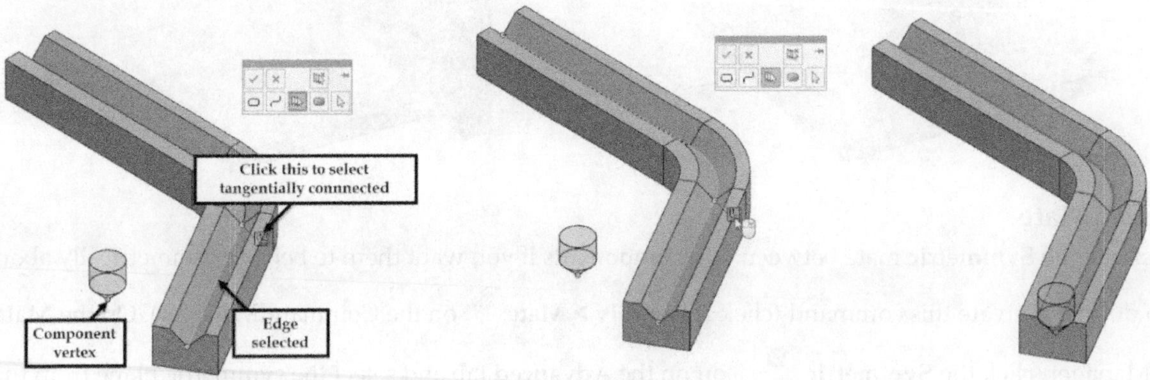

Click and drag the follower and notice that it moves along the path.

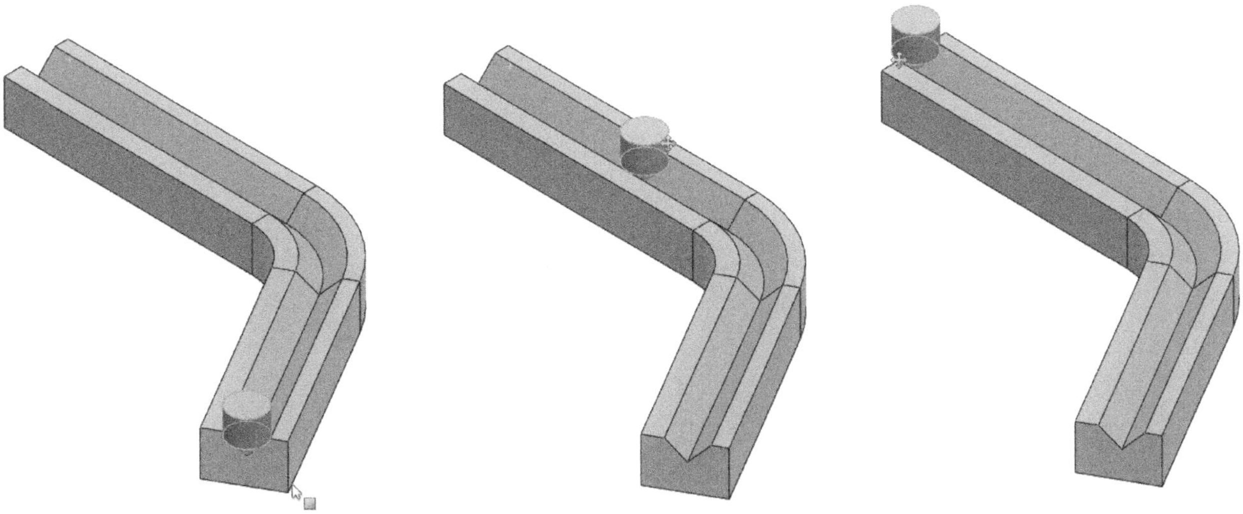

Interference Detection

In an assembly, two or more parts can overlap or occupy the same space. However, this would be physically impossible in the real world. When you add relations between components, SolidWorks develops real-world contacts and movements between them. However, sometimes interferences can occur. To check such errors, SolidWorks provides you with a command called **Interference Detection**. Activate this command (click **Evaluate > Interference Detection** on the CommandManager) and click the **Calculat**e button. The Results section shows the number of interferences detected. Click **OK** on the Interference Detection PropertyManager. If there is no interference, the **Results** section shows that there are no interferences in the assembly.

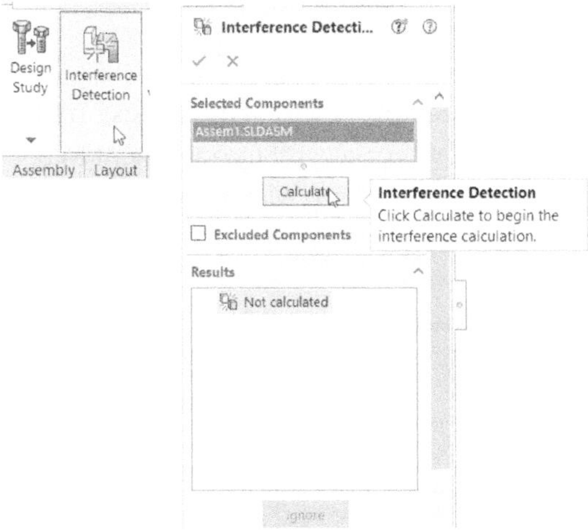

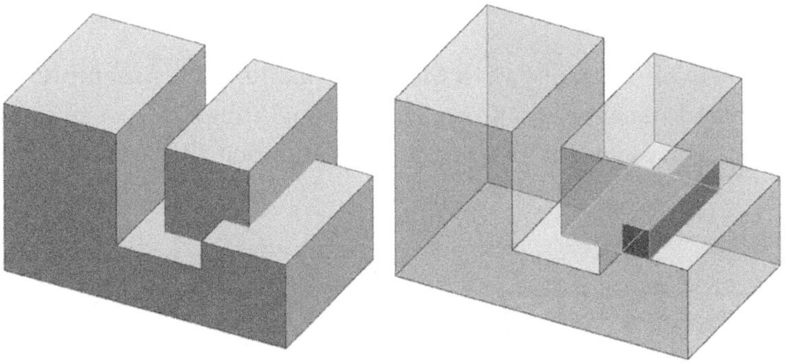

Copy with Mates

If you have an assembly where you need to assemble the same part multiple times, it would be a tedious process. In such cases, the **Copy with Mates** command will drastically reduce or eliminate the time used to assemble commonly used parts. To use this command, first, you need to create a mate or set of mates between two parts. For example, create the **Coincident** and **Concentric** mates between the screw and the hole.

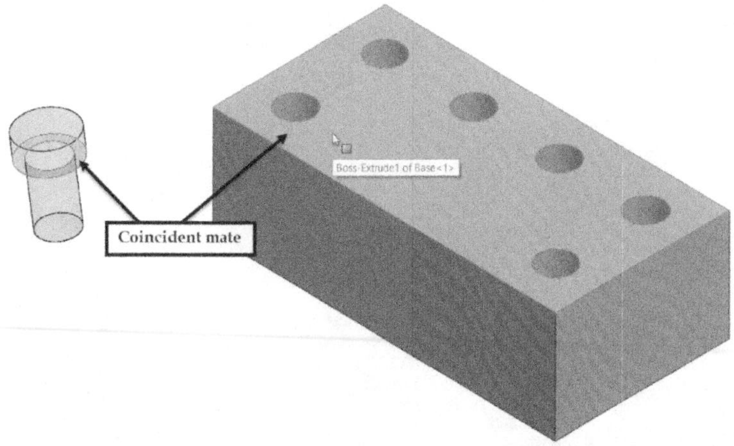

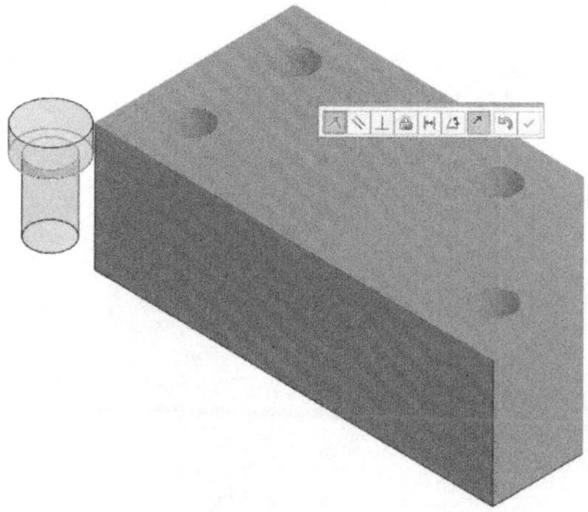

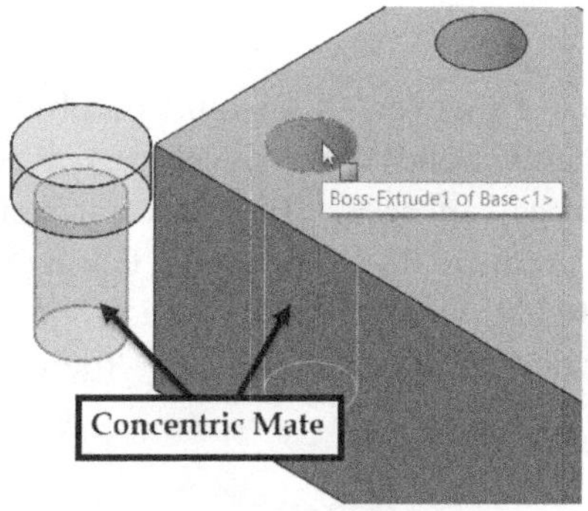

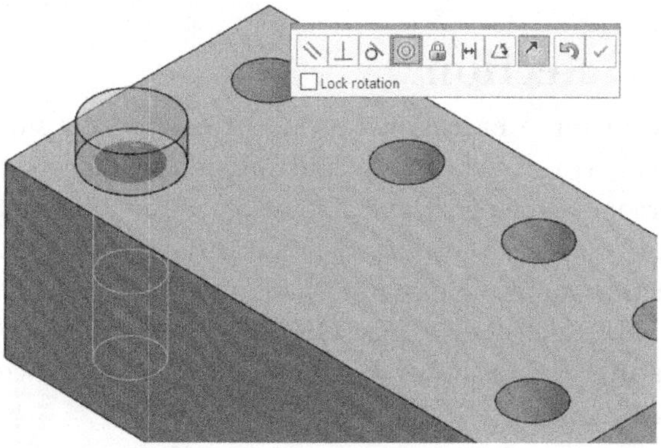

Next, activate the **Copy with Mates** command (click **Assembly > Insert Component > Copy with Mates** on the CommandManager); the **Copy with Mates** PropertyManager appears on the screen. Select the component to be copied with mates (in this case, Screw), and then click the **Next** arrow icon on the PropertyManager.

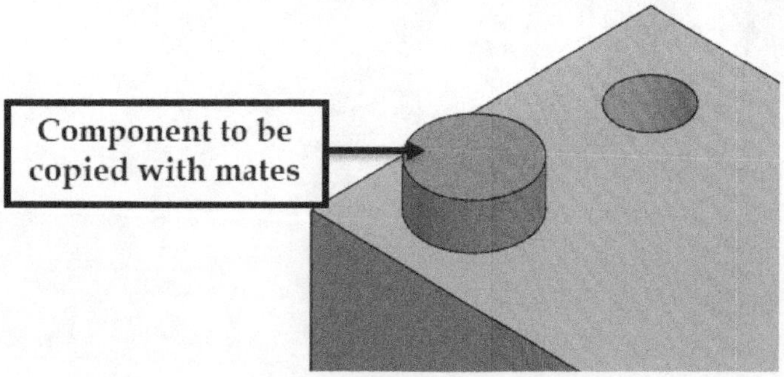

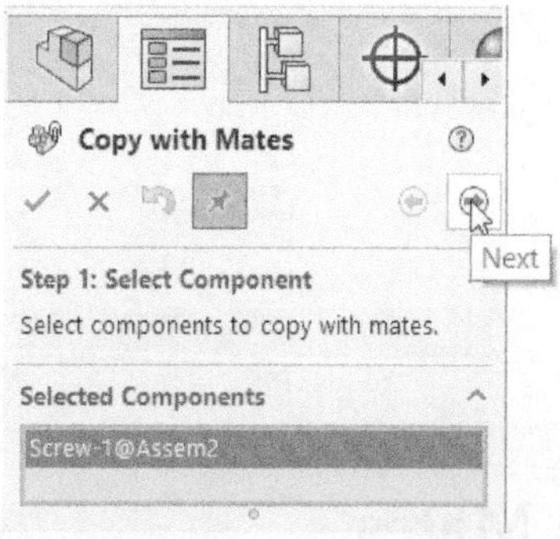

Next, you need to select the faces from the other component to create the mates. In this case, you need to choose a flat and cylindrical face from the component into which the screw will be inserted. The flat face will remain the same for all the holes. So, check the **Repeat** option in the **Coincident** section. Next, select any one of the holes from the base component, and then click **OK**; the screw will be inserted into the assembly, and you are prompted to select another hole.

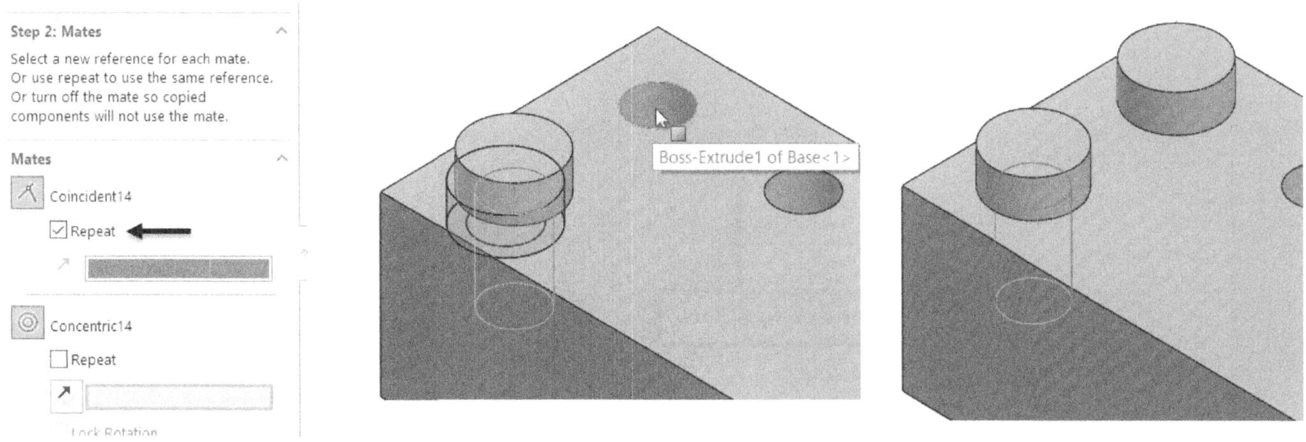

Select another hole and click **OK**. Likewise, insert the screw into the remaining holes, and then click the **Cancel** icon on the PropertyManager.

Mate References

SOLIDWORKS provides you another useful feature called the **Mate References** command to apply mates to multiple instances of a single part. This command allows you to define the mate references in the part environment itself. For example, you can easily insert a screw into the holes by creating mate references on the screw.

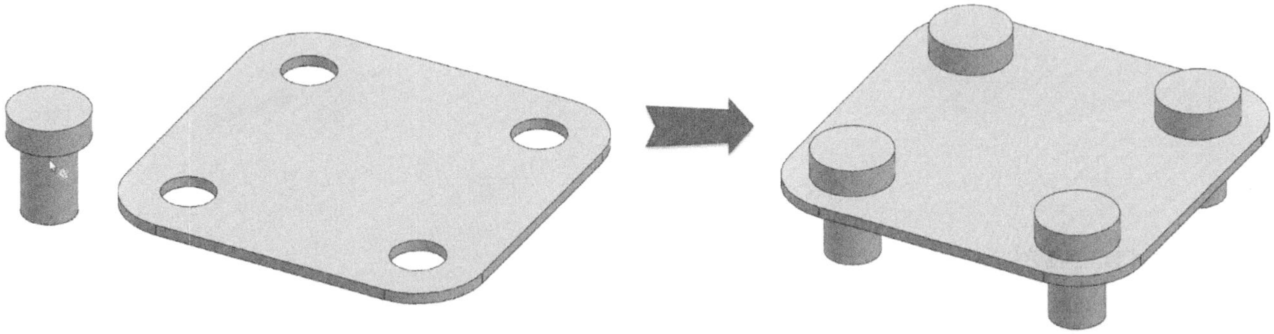

To create mate references, first, open the screw part file. Next, activate the **Mate References** command (on the CommandManager, click **Features > Reference Geometry > Mate Reference**) and select the circular edge of the model, as shown in the figure below. Select **Default** from the **Mate Reference Type** drop-down available in the **Primary Reference Entity** section. Next, select **Any** from the **Mate Reference Alignment** drop-down. Save and close the part file.

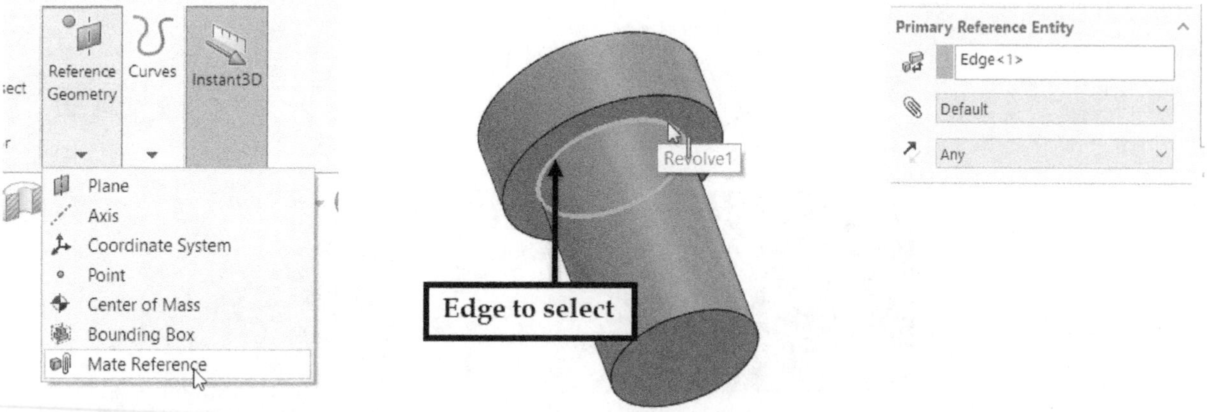

Next, start a new assembly file and insert the component with holes. Next, click **Assembly > Insert Components** on the CommandManager, double-click on the screw with mate reference; the screw is attached to the pointer. Move the pointer to the hole location and notice that the screw is inserted into it. Click the left mouse button.

Press and hold the Ctrl key and select the screw; a copy of the screw is attached to the pointer. Move the pointer to the adjacent hole, and then release it; the copy of the screw is inserted into another hole. Likewise, insert the copies of the screws into the remaining holes.

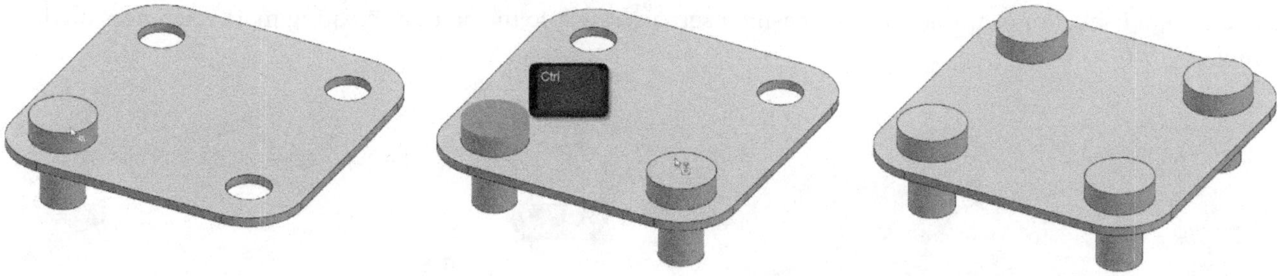

Magnetic Mates

The **Magnetic Mates** command was introduced in SOLIDWORKS 2017. This command's primary purpose is to make it easy to assemble many parts to one main component. For example, the following figure shows a box and drawer body. You need to create many mates to assemble the box into the drawer body. The **Magnetic Mates** command avoids this by helping you mate the parts by merely dragging closer to each other.

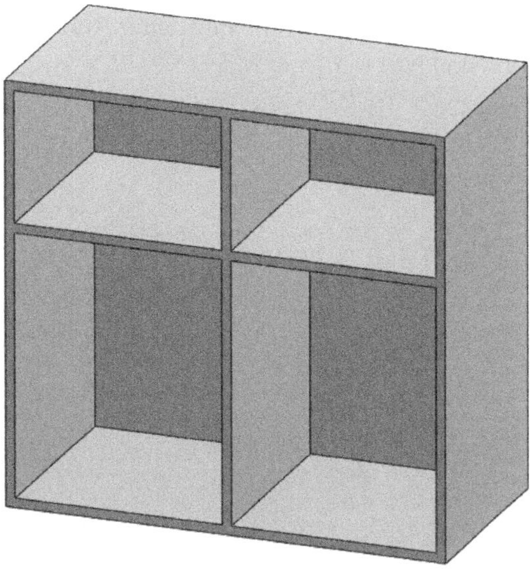

Note that you need to create the connector points before creating the Magnetic Mates. To do this, open the Box part file and start a sketch on the model's back face. Next, create a diagonal centerline, and then place a point at the midpoint. Next, click **Exit Sketch** on the CommandManager.

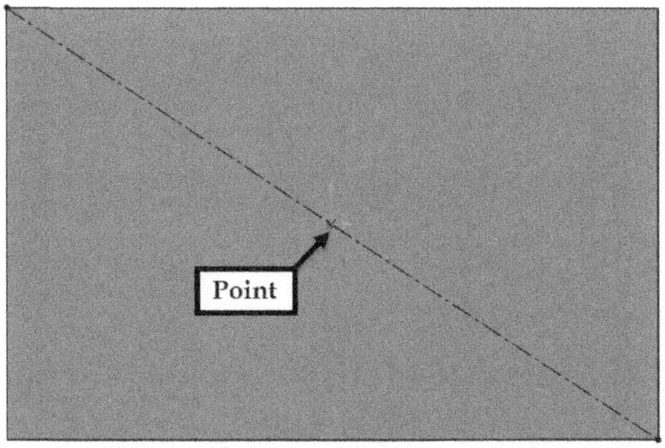

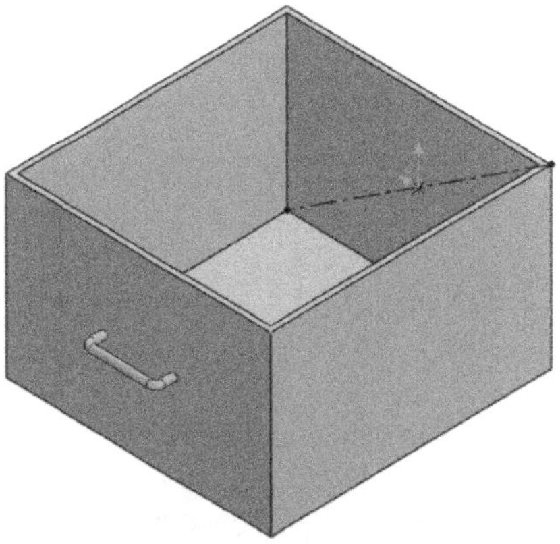

Activate the **Asset Publisher** command (on the Menu bar, click **Tools > Asset Publisher**), and then select the bottom face of the model to define the ground plane. Next, choose the sketch point to set the connector position. On the PropertyManager, click in the **Connector Direction** selection box and select the model's back face.

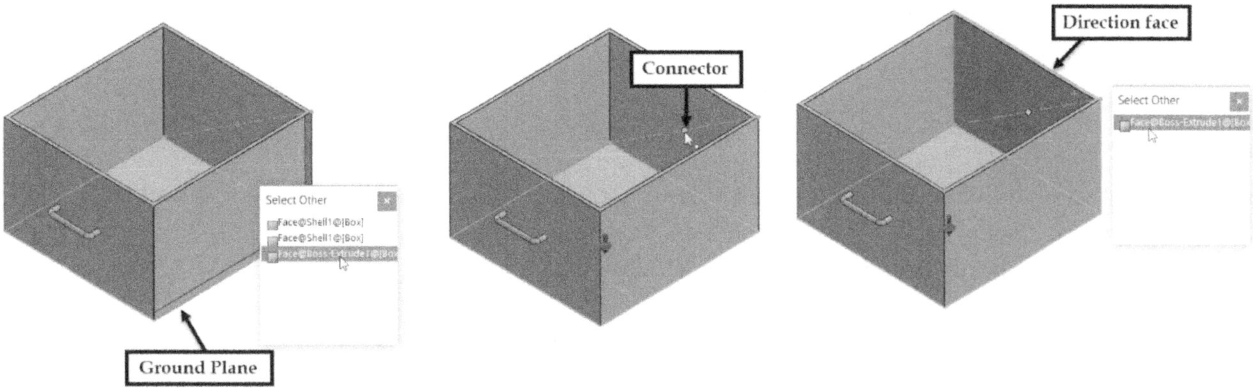

Click the **Add Connector** button on the PropertyManager; the direction arrow appears on the connector, as shown. Next, click **OK** on the PropertyManager; the **Publisher References** node is added to the FeatureManager Design Tree. Save and close the Box part file.

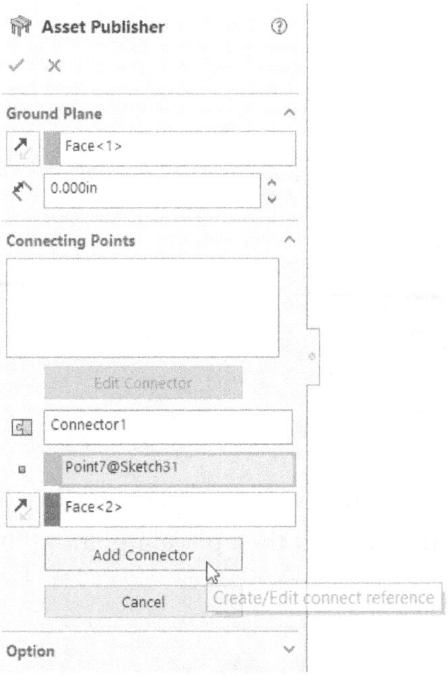

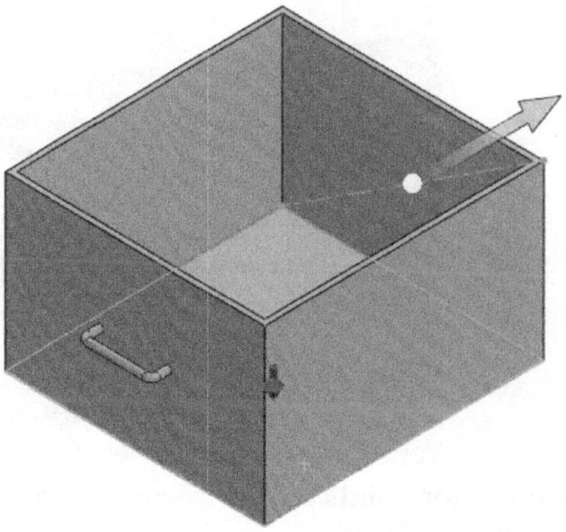

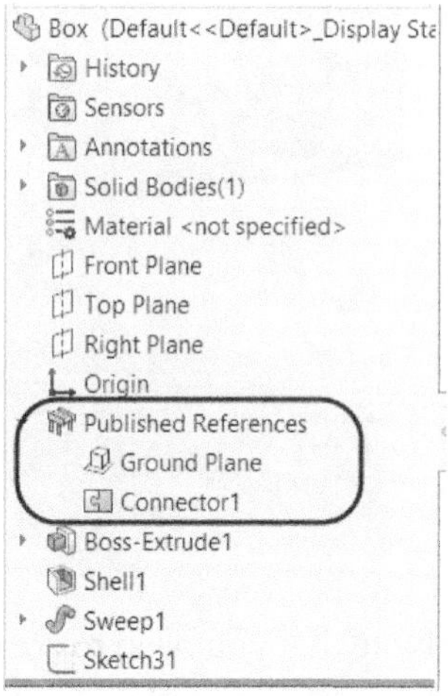

Now, open the Drawer Body part file, and then start a sketch on the Shell feature's open face, as shown. Create centerlines and points, as shown, and then click **Exit Sketch** on the CommandManager.

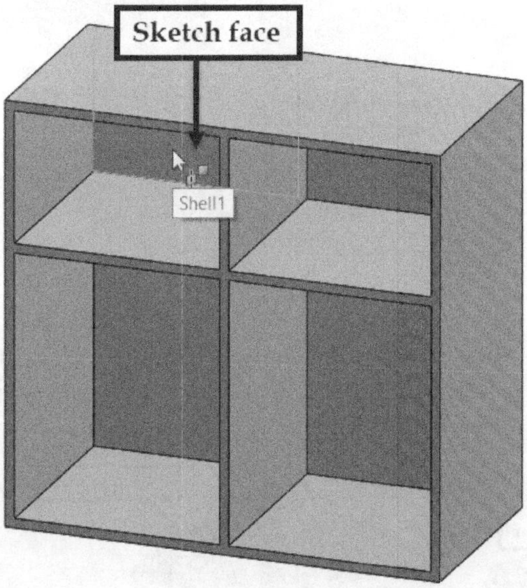

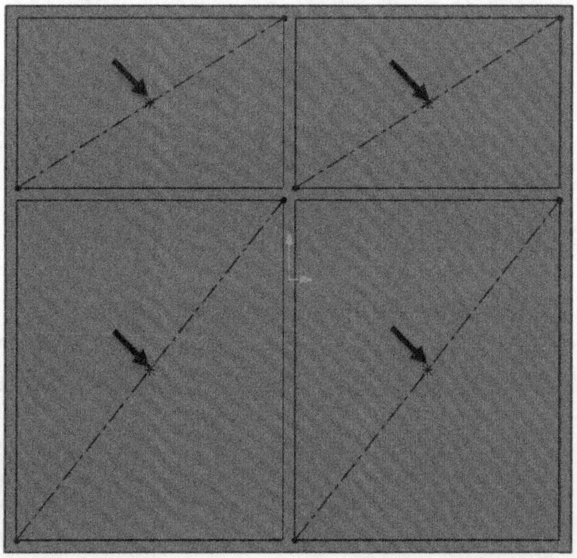

Activate the **Asset Publisher** command (on the Menu bar, click **Tools > Asset Publisher**), and then select the horizontal face to define the ground plane, as shown. Next, select the sketch point, as shown. Click on the open face of the Shell feature to specify the connector direction. Next, click the **Add Connector** button.

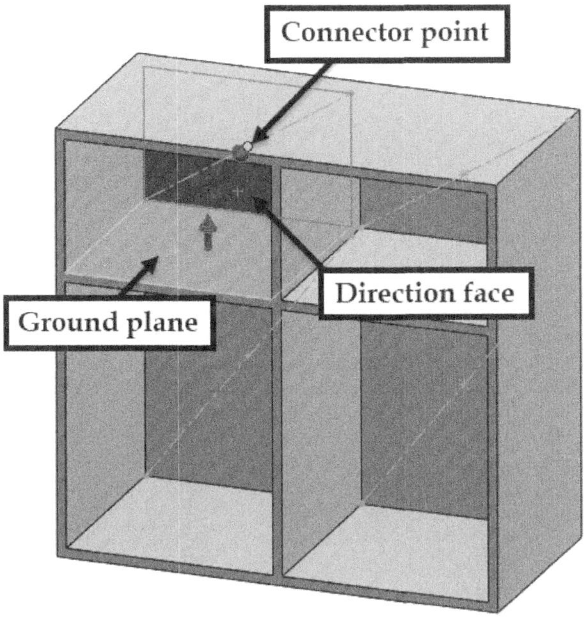

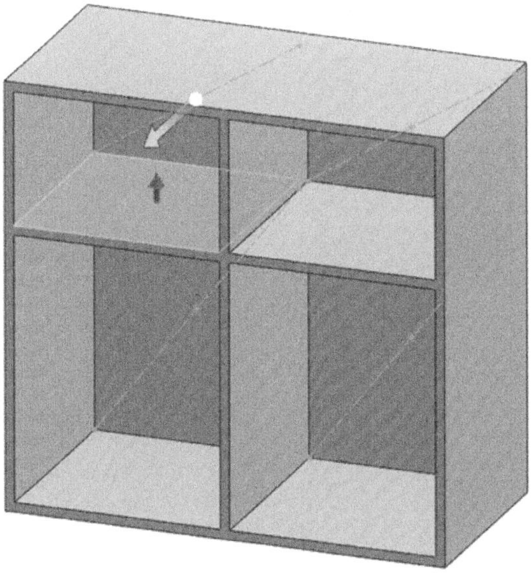

Select the next point, as shown in the figure. Next, click on the open face of the shell feature to define the connector direction. Click the **Add Connector** button to add the connector. Likewise, add two more connectors, as shown. Click **OK** on the PropertyManager, and then save and close the part file.

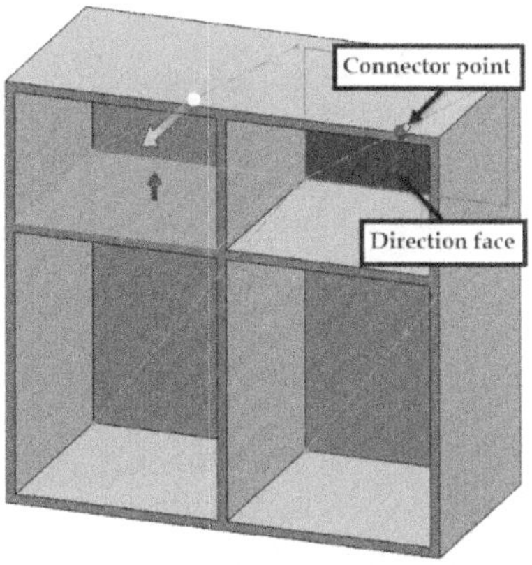

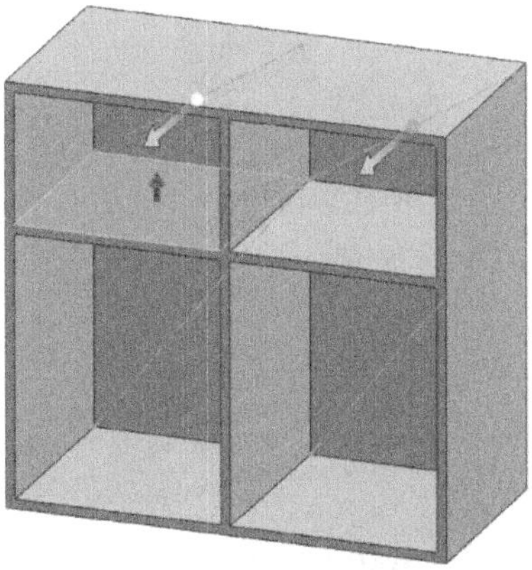

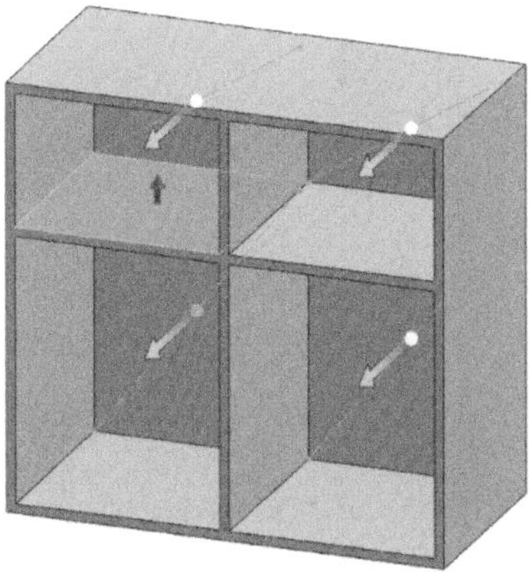

Start a new assembly file and insert the Drawer body and Box part files. Next, click and drag the box towards the top left connector point of the Drawer body; the connectors on the Box and the Drawer body are attached. Release the pointer to place the box.

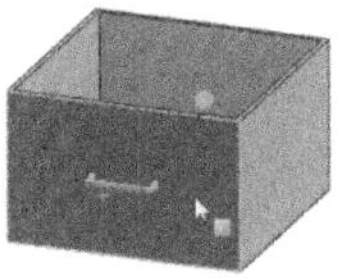

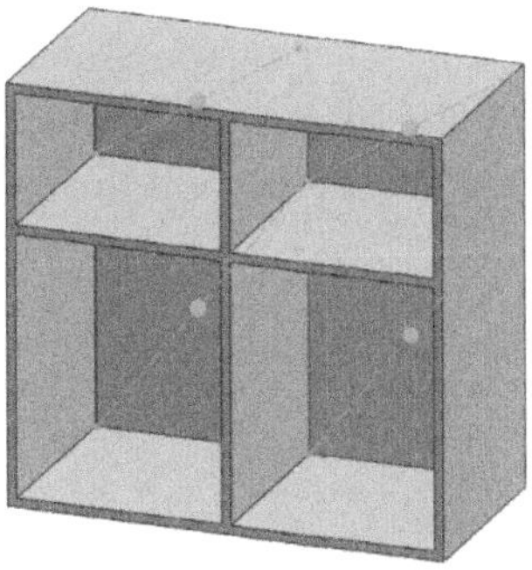

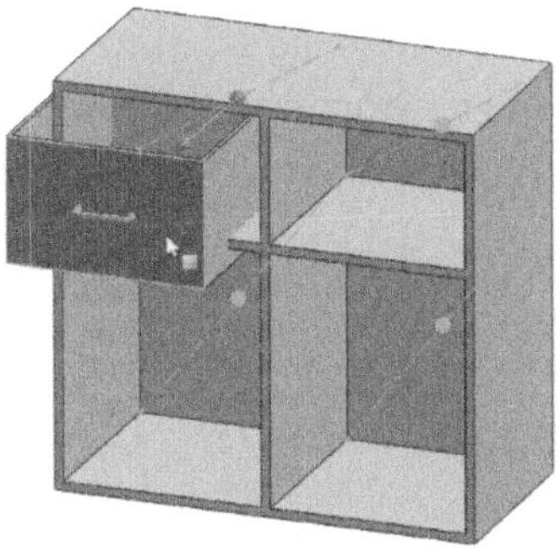

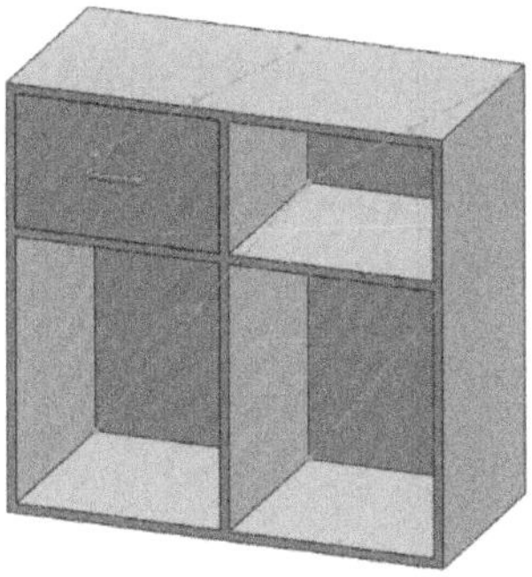

Press and hold the Ctrl key and click on the box. Next, drag the pointer to the next connector point, and then release it.

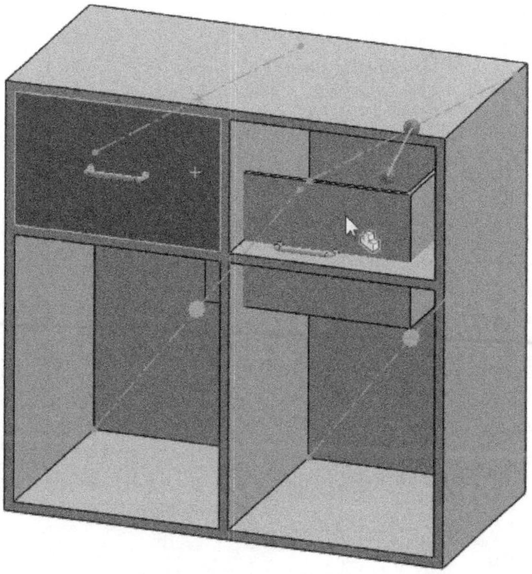

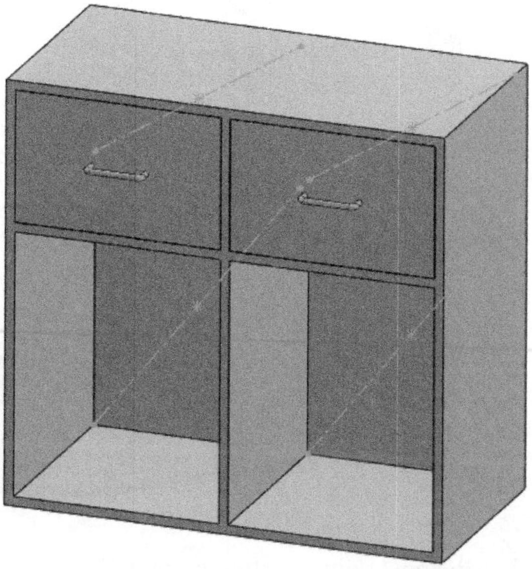

Likewise, place the box at the remaining two connectors, as shown.

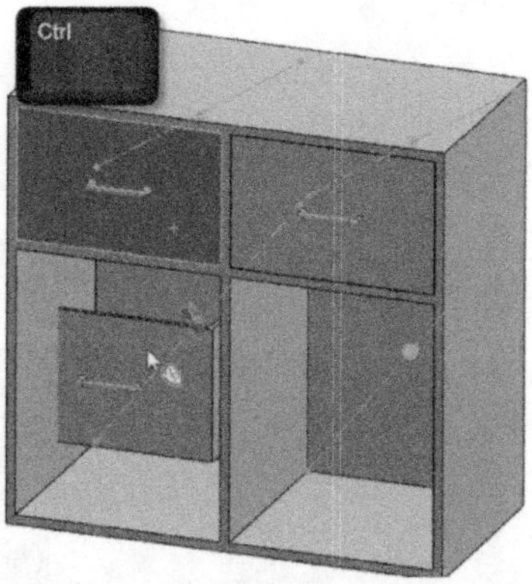

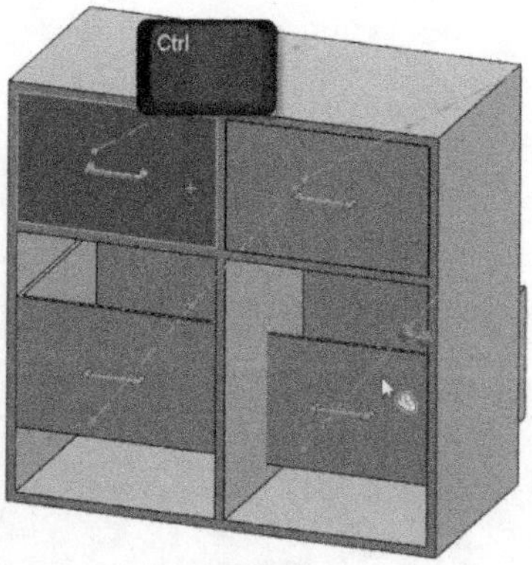

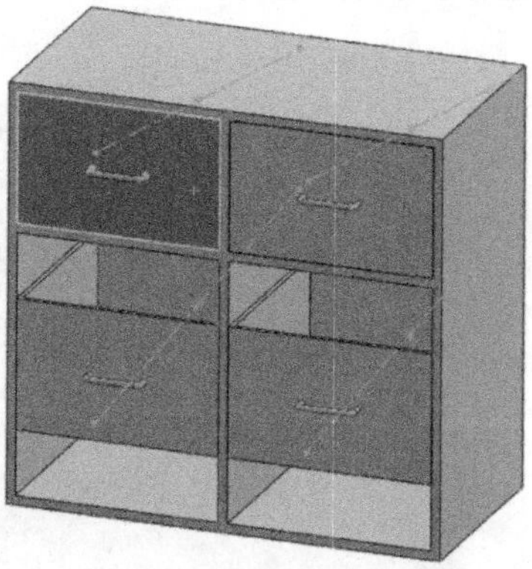

Open the Box part file and click the **ConfigurationManager** tab on the left pane. In the ConfigurationManager, select the Default configuration and press Ctrl+C on your keyboard. Next, press Ctrl+V on your keyboard to copy the configuration. Right-click on the duplicated configuration and select **Rename tree item**. Type Large and press Enter. Next, double-click on the Large configuration to activate it.

Select the model's front face to display the model dimensions – double-click on the dimension with the value of 14. Next, type 28 and click the down-arrow next to the value box on the **Modify** box. Select the **This Configuration** option, and then click **OK**. Next, save and close the Box part file, and then switch to the assembly file.

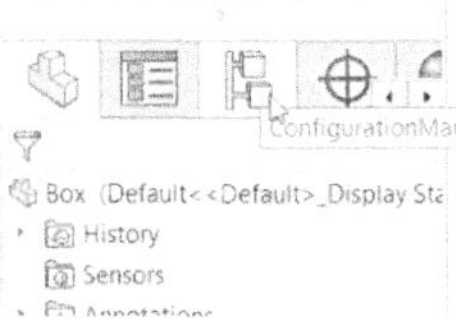

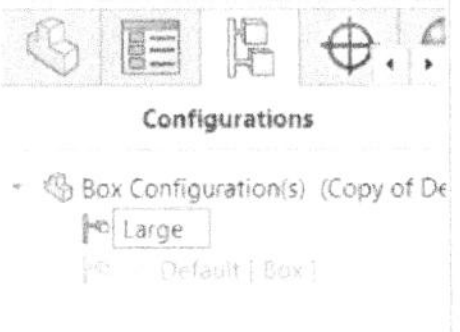

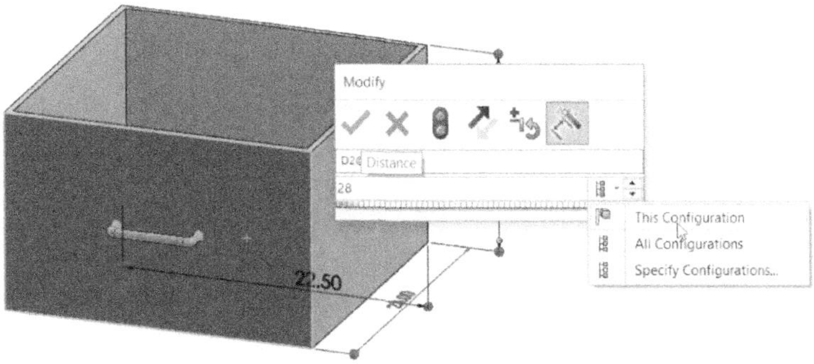

Press and hold the Ctrl key and select the front faces of the bottom two boxes. Next, select the Large option from the **Configuration** drop-down displayed on the Context toolbar. Click the green check to update the configuration.

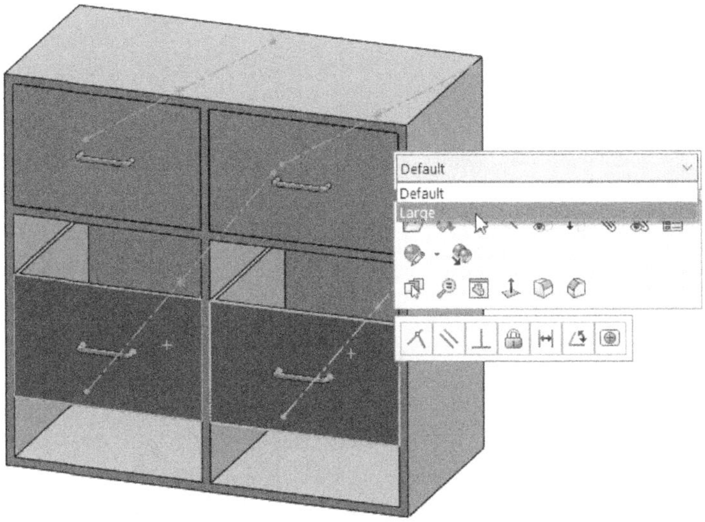

Editing and Updating Assemblies

During the design process, the correct design is not achieved on the first attempt. There is always a need to go back and make modifications. SOLIDWORKS allows you to accomplish this process very quickly. To modify a part in an assembly, click on it and select **Open Part**; the part will be opened in a separate window. Make changes to the part and save it. Next, switch to the assembly window, and then click **Yes** on the **SOLIDWORKS 2024** message box. The part will be updated in the assembly automatically.

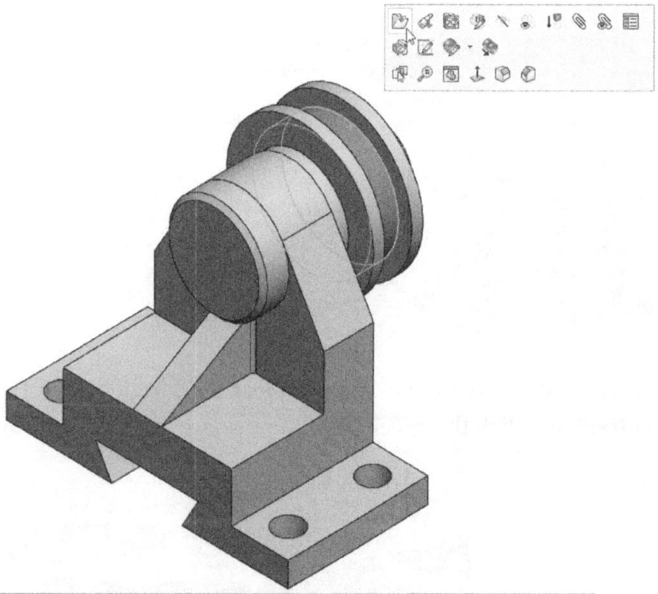

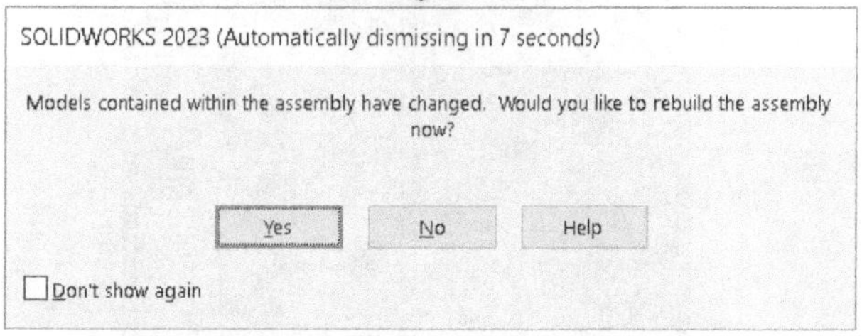
SOLIDWORKS 2023 (Automatically dismissing in 7 seconds)
Models contained within the assembly have changed. Would you like to rebuild the assembly now?
Yes
No
Help
Don't show again

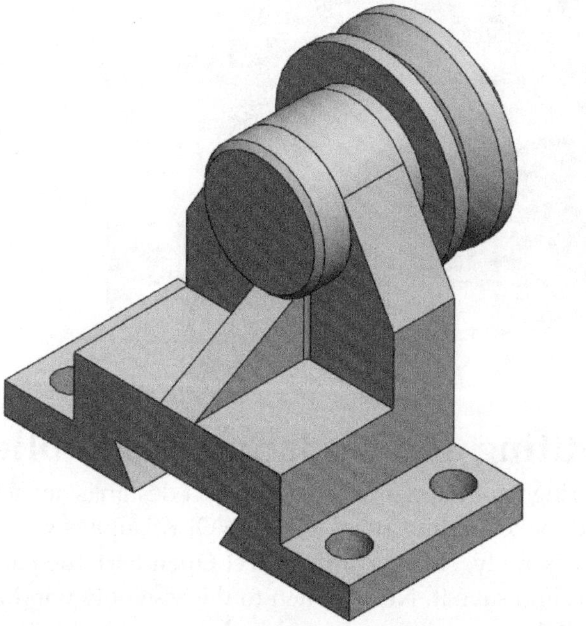

You can also edit the mates of a part in an assembly. To do this, expand the part in the **FeatureManager Design Tree**, and then expand the **Mates** folder; the mates applied to part appear. You can also view the mates applied to the part by clicking on it and selecting **View Mates** from the **Context** toolbar. Next, Identify the mate to be edited by placing the pointer on it; the entities related to the chosen mate appear. Next, click on the required mate, and then choose **Edit Feature**; the PropertyManager appears.

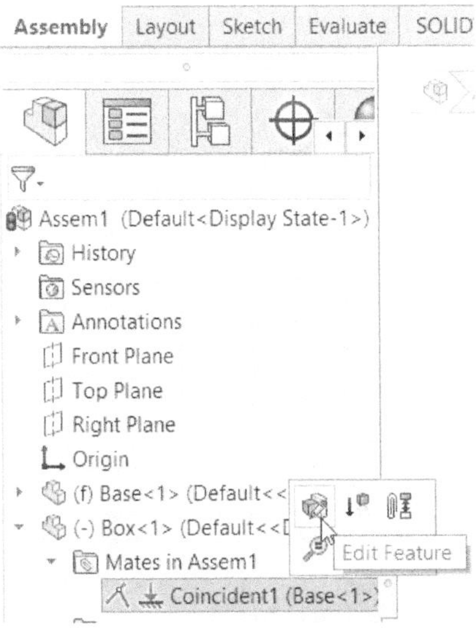

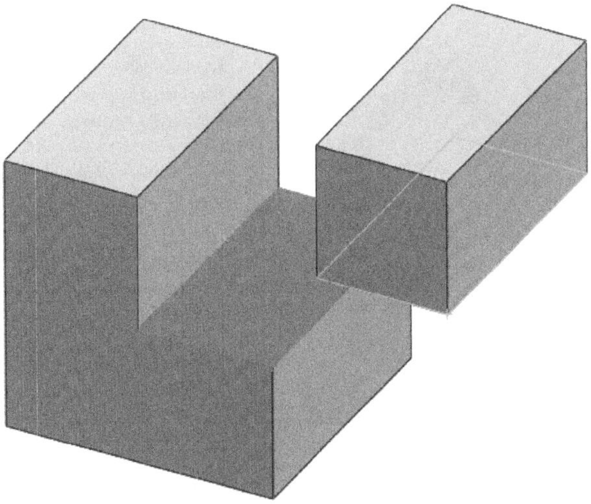

Now, you can change the mate type or the entities to which the mate is applied. To change the mated entities, right-click in the **Entities to mate** selection box available in the **Mate Selections** section, and then select **Clear Selections**. Now, pick the new set of entities to mate, and then click **Add/Finish Mate**.

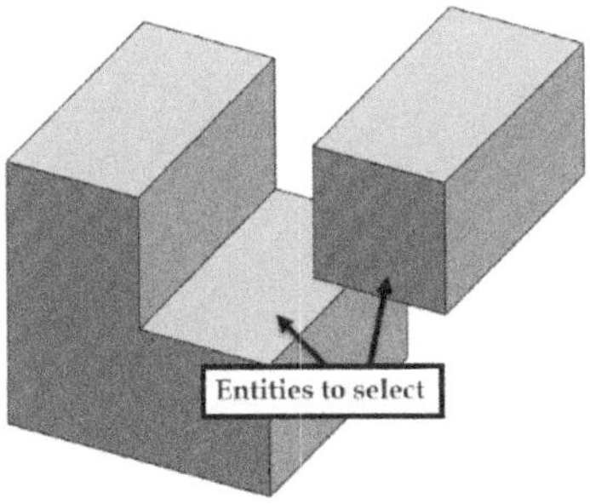

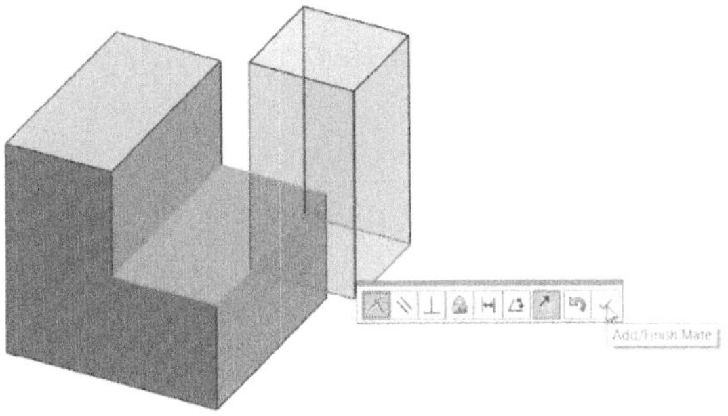

Replace Components

SOLIDWORKS allows you to replace any component in an assembly. To do this, click the **Open** icon on the Quick Access Toolbar and select the assembly file. On the **Open** dialog, click the **References** button; the **Edit Referenced file locations** dialog appears. On this dialog, double-click on the part to be replaced, and then double-click on the replacement part from the **Open** dialog. Click **OK** and **Open**; the **SOLIDWORKS** message box appears. Click the **Rebuild** button; the **What's Wrong** dialog appears. It shows the mate and other reference errors. Click **Close** on this dialog.

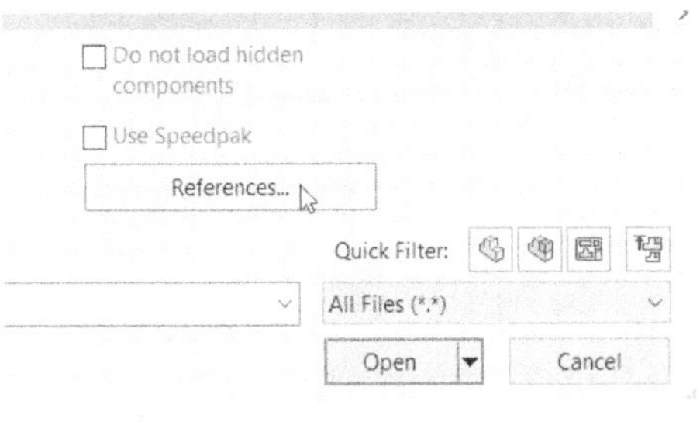

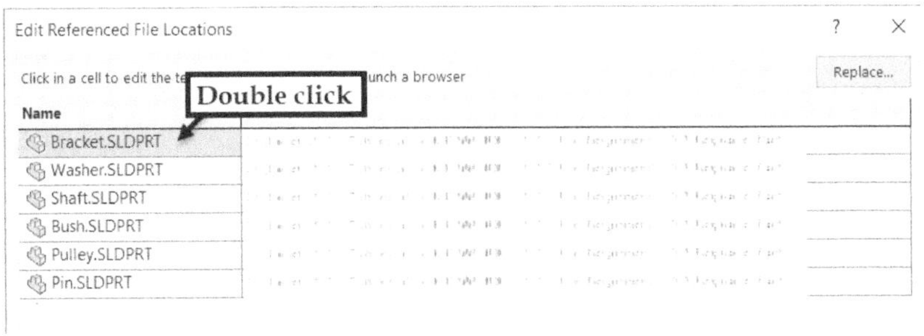

Another way to replace a component is to right-click on the component to replace. On the shortcut menu, click the arrow located at the bottom; the shortcut menu is expanded. Next, select the **Replace Component** option from the shortcut menu. On the **Replace** PropertyManager, click the **Browse** button, select the replacement part, and click **Open**. Click **OK** on the PropertyManager, and then click **Close** on the **What's Wrong** dialog.

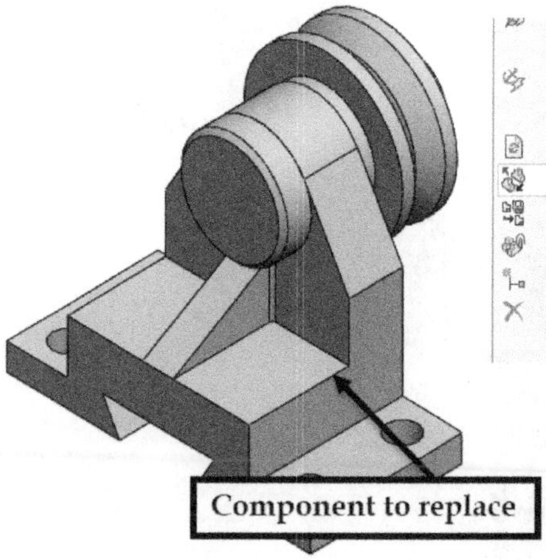

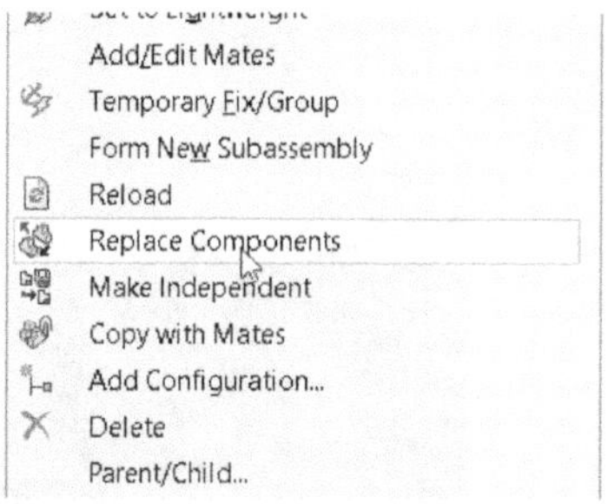

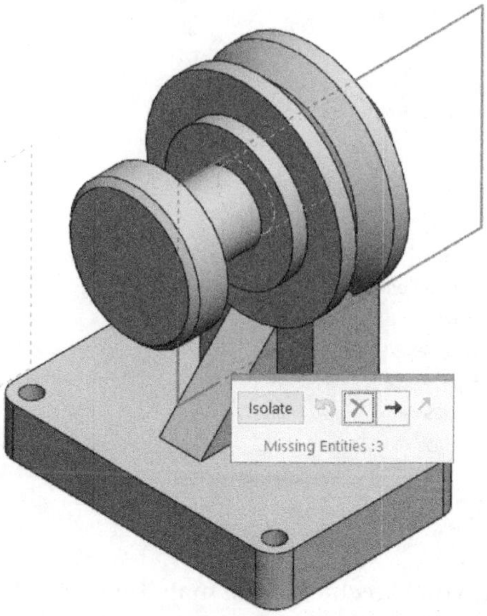

The **Mate Entities** PropertiesManager appears if there are any mate errors. On this PropertyManager, select the face from the **Mate Entities** section; the chosen face is highlighted in the replaced part window. Click in the **Replacement Mate Entity** selection box and choose the matching face from the replacement part. Likewise, solve the remaining errors and click **OK** on the **Mate Entities** PropertyManager.

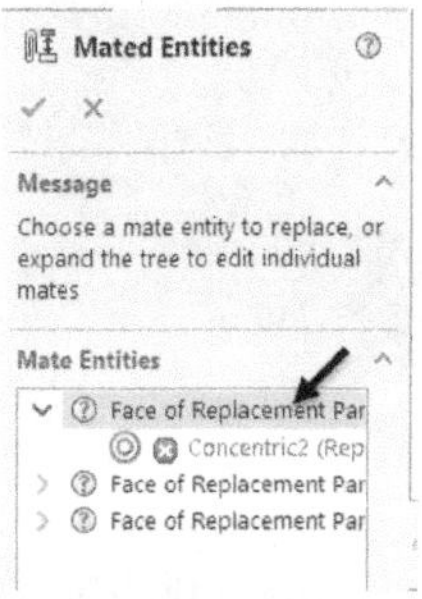

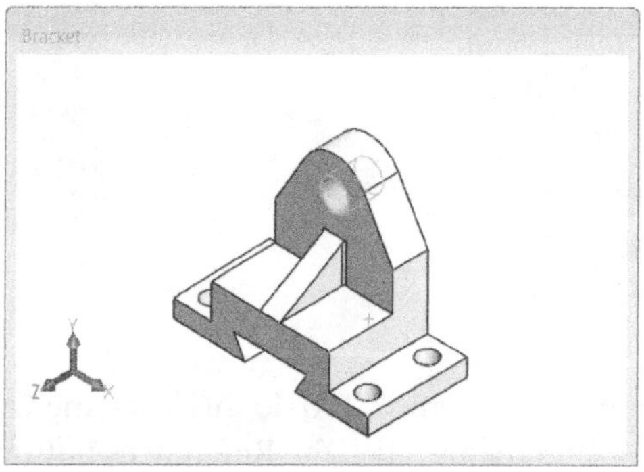

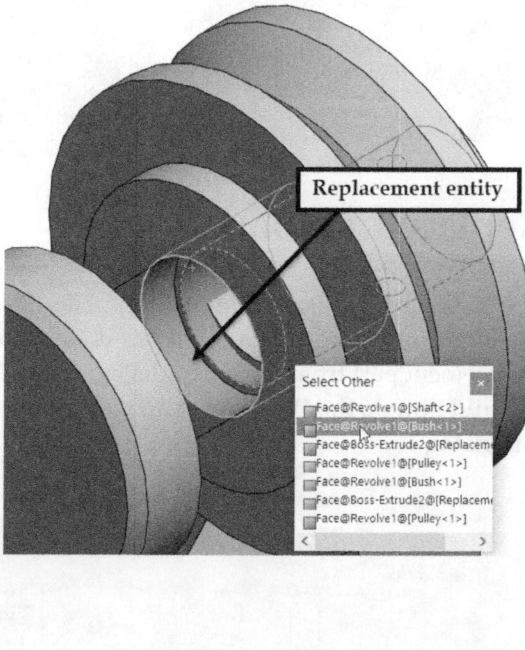

Pattern-Driven Component Pattern

The **Pattern Driven Component Pattern** command allows you to replicate individual parts in an assembly. However, instead of creating linear or circular patterns, you can select an existing pattern as a reference. For example, in the assembly shown in the figure, you can position one screw using mates and then use the **Pattern Driven Component Pattern** command to place screws in the remaining holes.

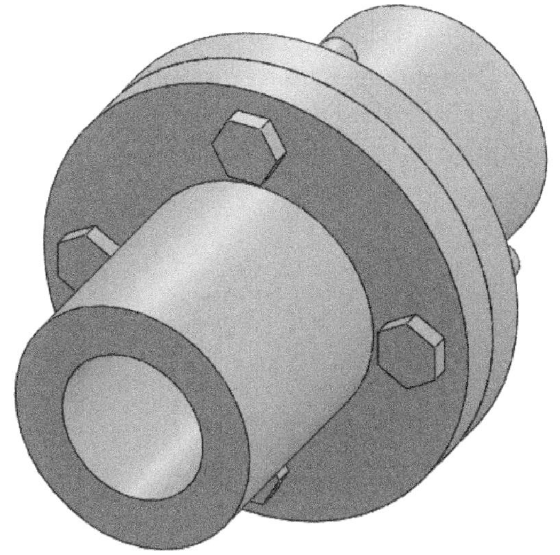

First, insert the screw in one hole using the **Mates** command. Next, activate the **Pattern Driven Component Pattern** command (click **Assembly > Linear Component Pattern drop-down > Pattern Driven Component Pattern** on the CommandManager) and click on the component to pattern. Click in the **Driving Feature or Component** selection box on the PropertyManager, and then select the pattern feature.

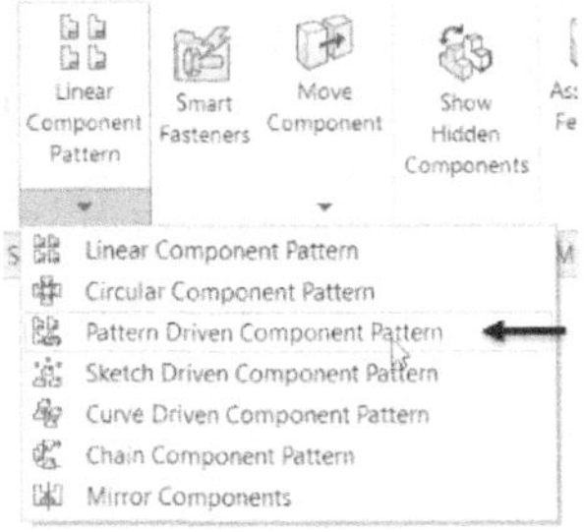

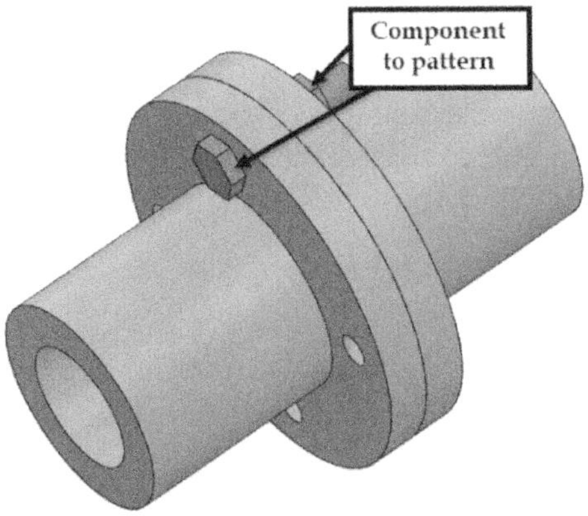

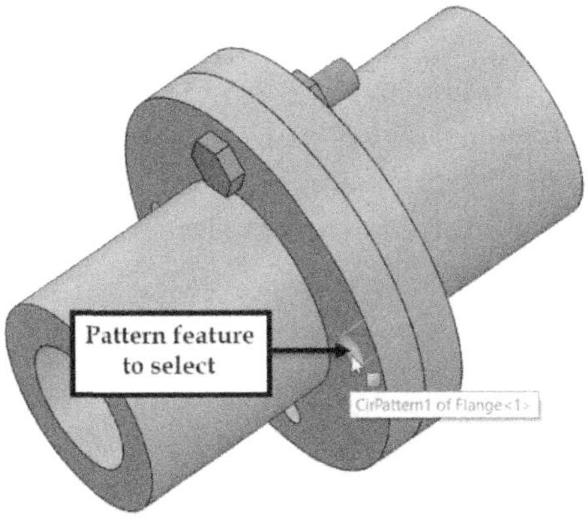

If you want to specify the seed location, click the **Select Seed Position** button in the **Driving Feature or Component** section. Next, select any one of the blue dots displayed on the pattern. Click OK on the PropertyManager.

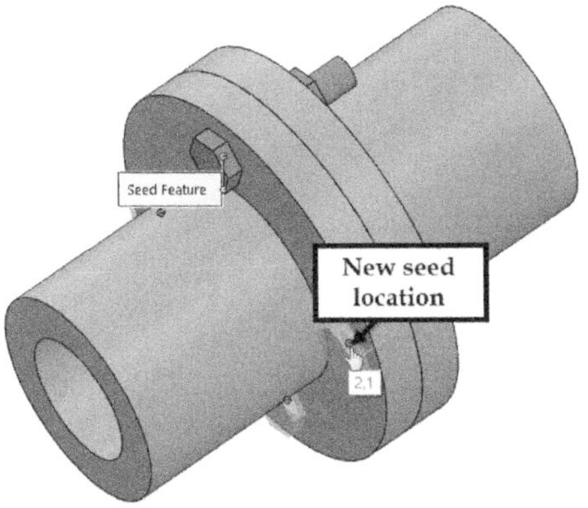

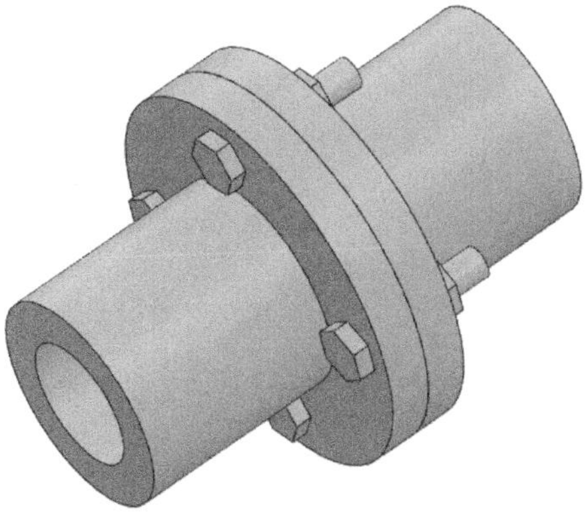

Mirror Components

When designing symmetric assemblies, the **Mirror Components** command will help you to save time and express the design intent. Activate this command (click **Assembly > Linear Component Pattern drop-down > Mirror Components** on the CommandManager) and select the mirroring plane from the FeatureManager Design Tree. Next, choose the components to be mirrored, and then click **OK** to complete the mirroring.

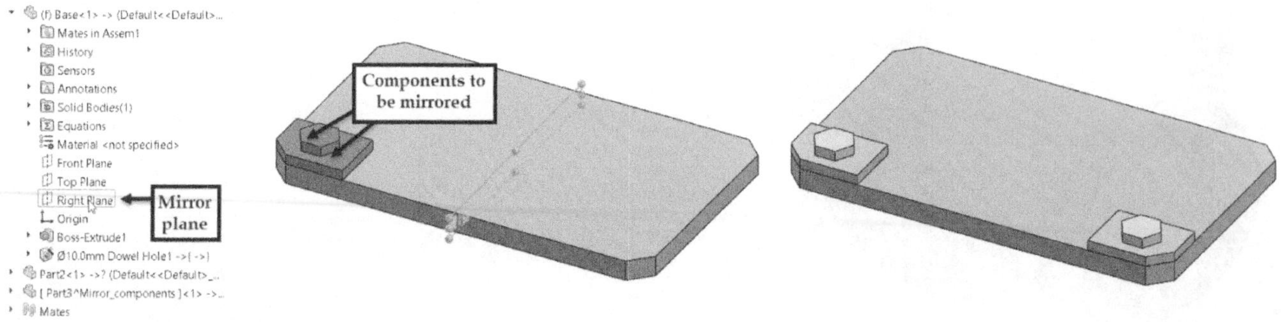

Sub-assemblies

The use of sub-assemblies has many advantages in SOLIDWORKS. Sub-assemblies make large assemblies easier to manage. They make it easy for multiple users to collaborate on a single large assembly design. They can also affect the way you document a large assembly design in 2D drawings. For these reasons, you need to create sub-assemblies in a variety of ways. The easiest way to form a sub-assembly is to insert an existing assembly into another assembly. You need to use the **Insert Components** command to insert the subassembly into the current assembly. Next, apply mates to constrain the assembly. The process of applying mates is also simplified. You are required to apply mates between only one part of a sub-assembly and a part of the main assembly. Also, you can easily hide a group of components with the help of sub-assemblies. To do this, click the right mouse button on a sub-assembly in the FeatureManager Design Tree and select **Hide**.

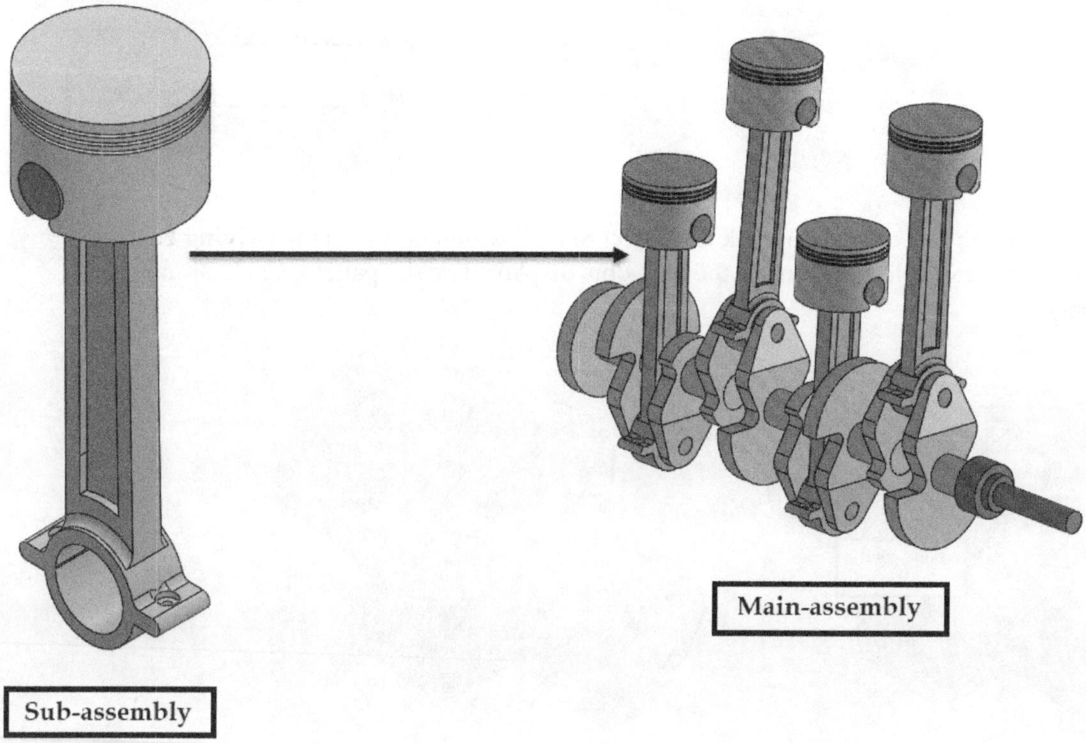

Rigid and Flexible Sub-Assemblies

By default, SOLIDWORKS treats a sub-assembly as a rigid body. When you move a single part of a sub-assembly, the entire sub-assembly will be moved. If you want to move the individual components of a sub-assembly, you must make the sub-assembly flexible. Click the right mouse button on the sub-assembly in the graphics window and select **Select Assembly**. Next, click the **Make Subassembly Flexible** icon on the Context toolbar. Now, you can move the individual parts of a sub-assembly. If you have multiple sub-assembly occurrences, each occurrence can be defined as rigid or flexible separately. SOLIDWORKS displays a different icon for each of them in the FeatureManager Design Tree to help you recognize the difference between the rigid and flexible assemblies.

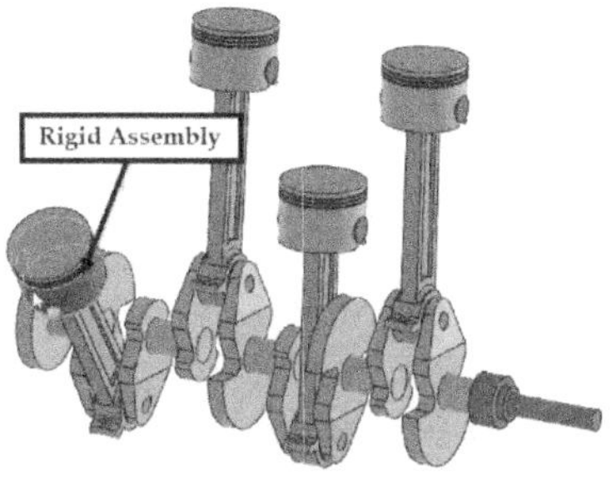

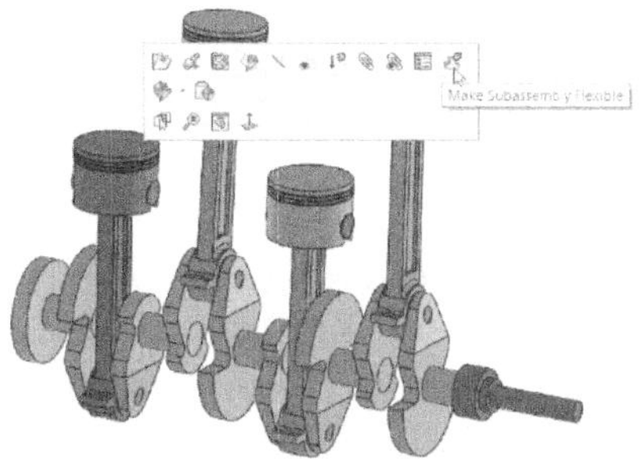

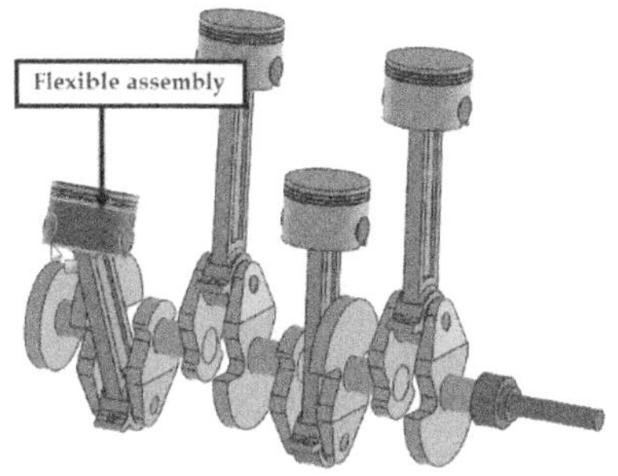

Form New Subassembly

In addition to creating sub-assemblies and inserting them into another assembly, you can also take individual parts in an assembly and make them into a sub-assembly. For example, press and hold the **Ctrl** key and select the four parts from the assembly, as shown. Next, right click and select **Form New Subassembly**.

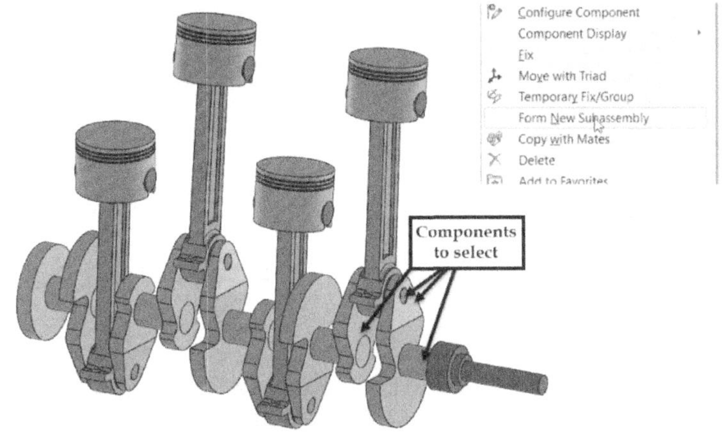

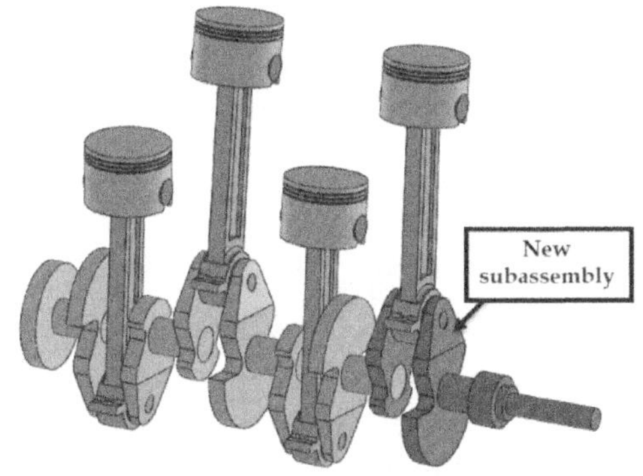

Dissolve a Subassembly

After inserting subassemblies, you may require to dissolve them into individual parts. SOLIDWORKS provides you with the **Dissolve Subassembly** option to break a subassembly into different parts. In the FeatureManager Design Tree, right click on a sub-assembly, and then select **Dissolve Subassembly**; the sub-assembly components are transferred to the main assembly.

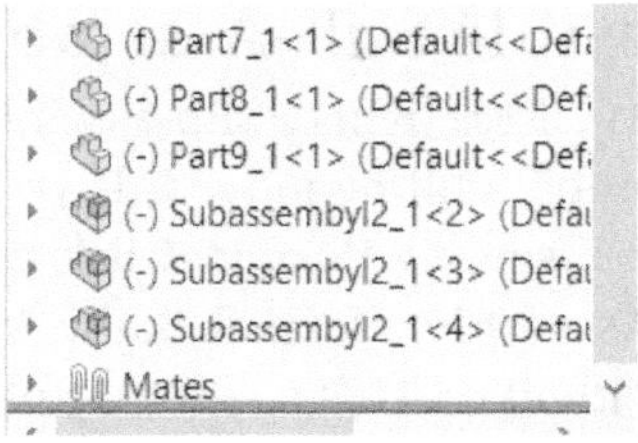

Assembly Features

Assembly features are just like regular features like extruded cuts, holes, revolved cuts, and welds. However, these features are created at the assembly level. These features are commonly created at the assembly level to represent post assembly machining. For example, to add a cut feature to the assembly shown in the figure, activate the **Extruded Cut** command (click **Assembly > Assembly Features > Extruded Cut** on the CommandManager). Next, Select the rod's top face and draw the sketch of the cut and exit the sketch. On the PropertyManager, select **Through All** from the **End Condition** drop-down. Next, choose an option (**All components** for this example) from the **Feature Scope** section, and then click **OK**.

Now, open the individual part in another window. You will notice that the cut feature does not affect the part. Check the **Propagate feature to parts** option in the **Feature Scope** section of the PropertyManager, if you want the cut feature to affect the part.

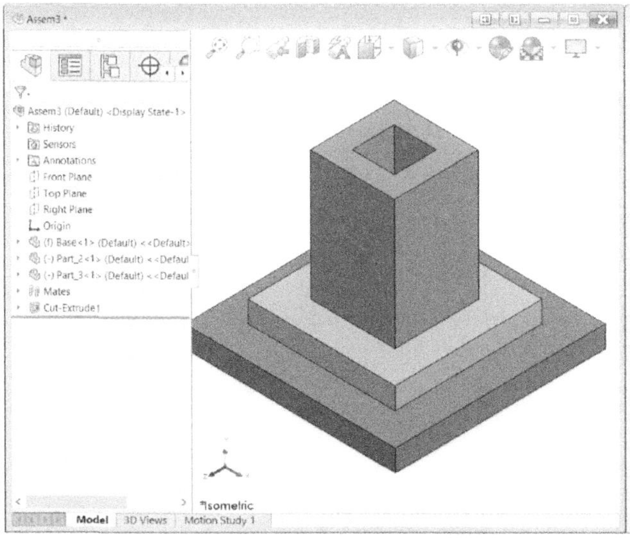

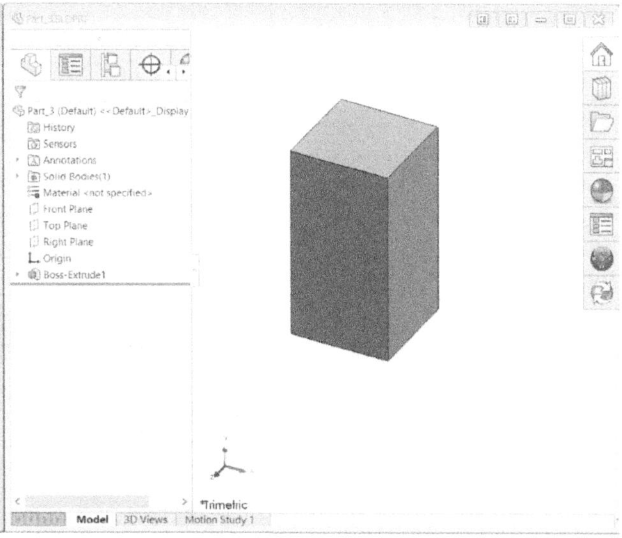

You will also notice that the **Cut-Extrude1** feature is added to the assembly file's FeatureManager Design Tree. You can edit the cut feature by clicking the right mouse button on the **Cut-Extrude1** feature and selecting **Edit Feature**. On the PropertyManager, select the **Selected components** option under the **Feature Scope** section, and then uncheck the **Auto-select** option. Next, select the parts to be excluded from the extruded cut. Click the green check on the PropertyManager; you can see that the extruded cut no longer affects the selected components.

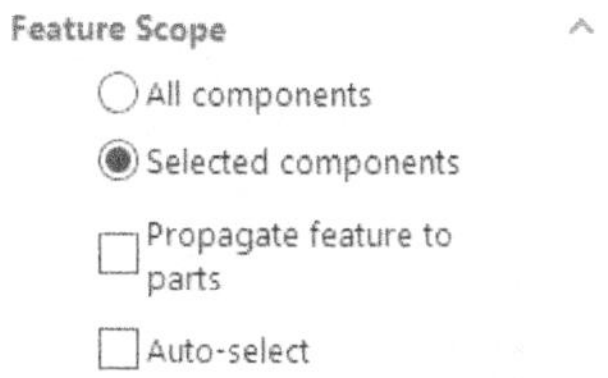

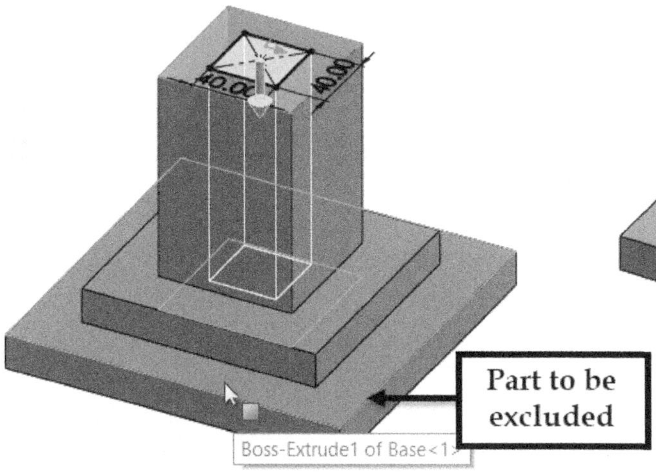

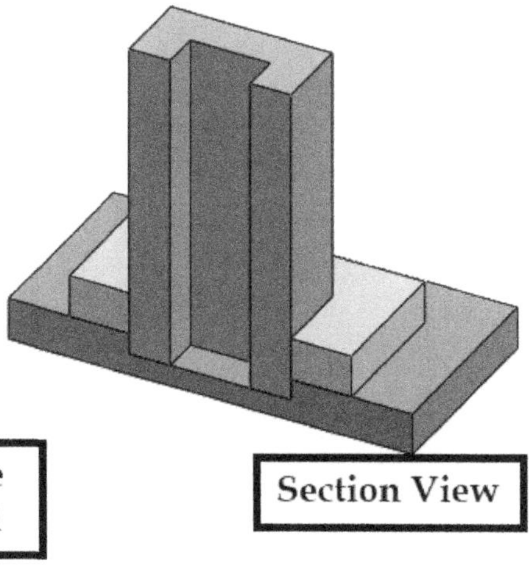

Hole Series

The **Hole Series** command is one of the important assembly feature commands. It is the advanced form of the **Hole Wizard** command, which can create holes on a series of multiple parts. Note that this command can be accessed only in the **Assembly** environment. Activate this command (on the CommandManager, click **Assembly > Assembly Features > Hole Series**) and notice the PropertyManager with four tabs. The first tab helps you to position the hole. It has two options: **Create hole new hole** and **Use existing hole**. Select the **Create new hole** option and click on the flat face of the start component.

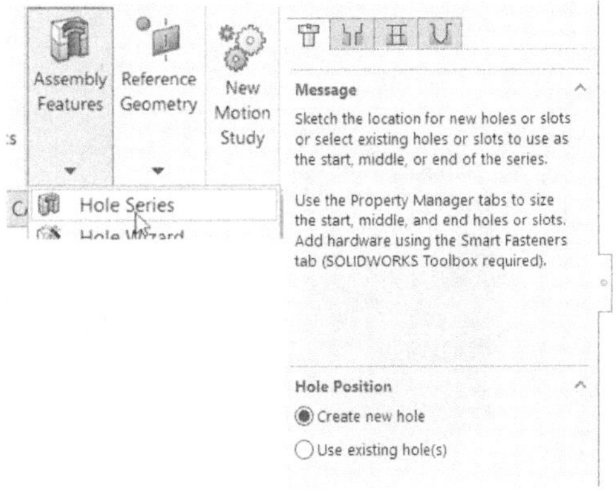

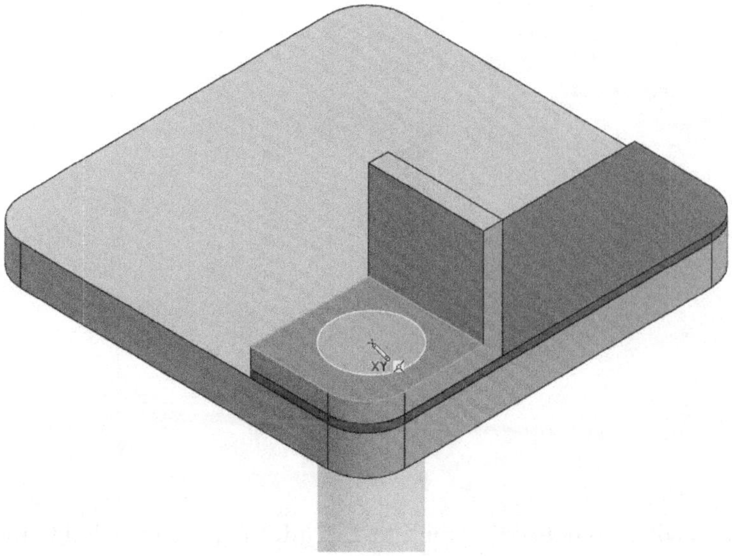

Click on the next tab and specify the hole size of the start component. Next, click the third tab of the PropertyManager and specify the hole size on the middle component. By default, the **Auto size based on start hole** option is selected. As a result, the hole size is specified automatically based on the start hole. Next, click the fourth tab and specify the hole on the end component.

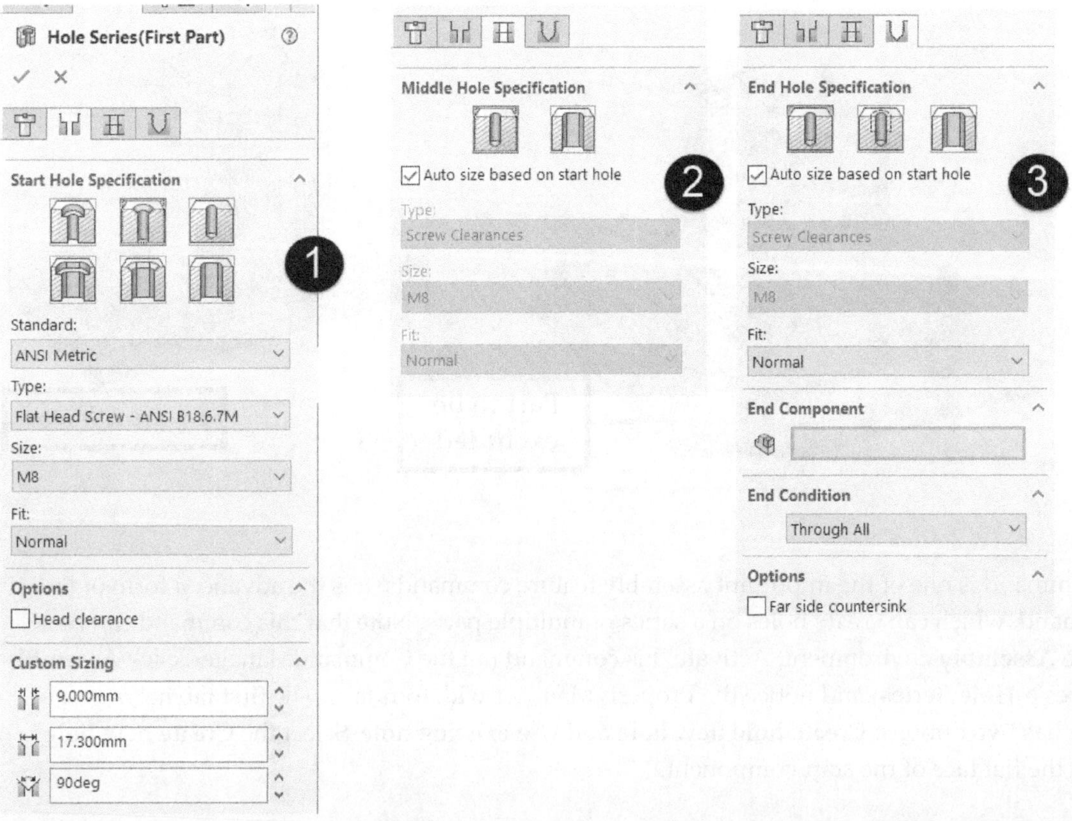

Click in the **End Component** selection box and select the rear side component. Next, click **OK** to create the hole series feature.

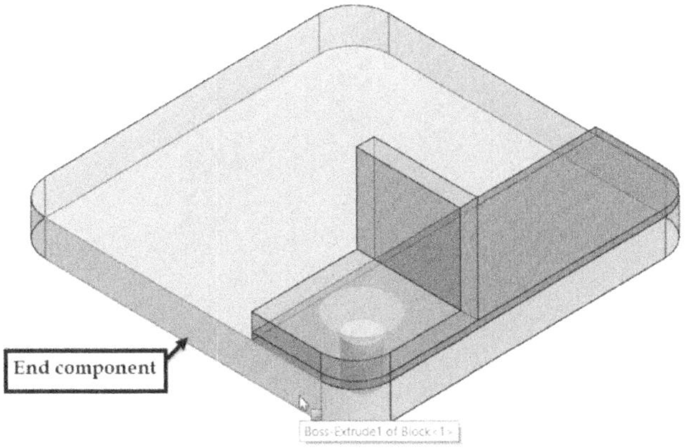

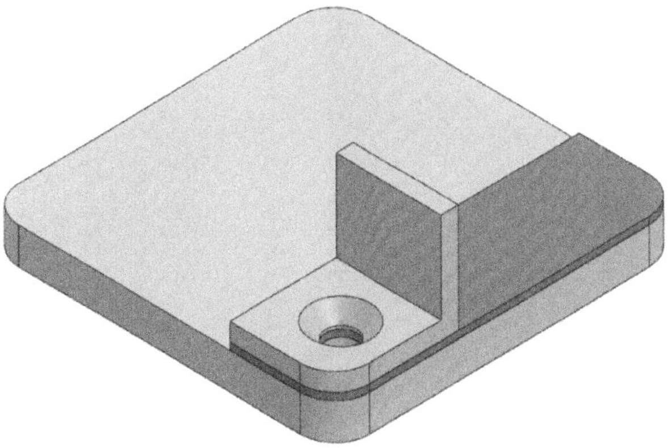

Belt/Chain

The Belt/Chain feature is one of the Assembly features available only in the assembly environment. Activate this command (click **Assembly > Assembly Features > Belt/Chain** on the CommandManager or click **Insert > Assembly Feature > Belt/Chain** on the Menu bar). On the PropertyManager, click in the **Belt Members** selection box and select the pulleys' cylindrical faces, as shown.

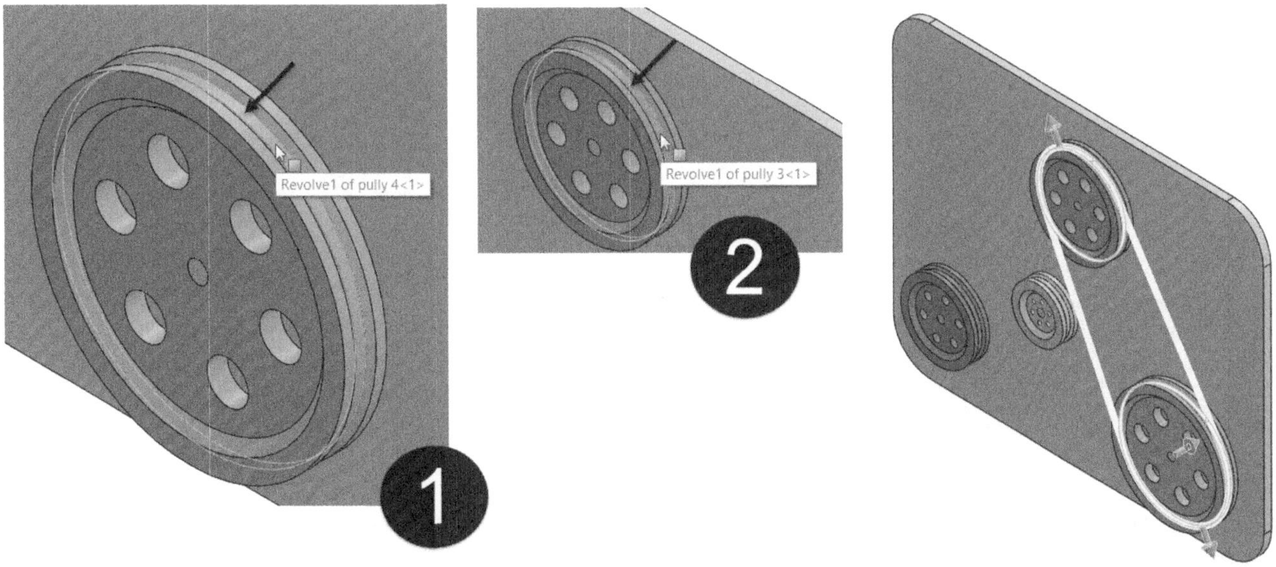

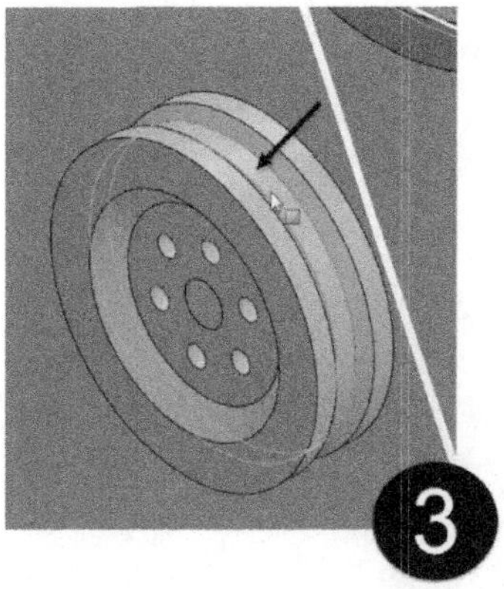

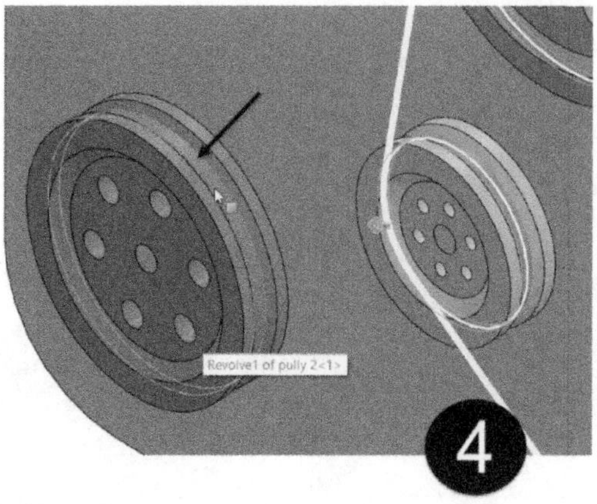

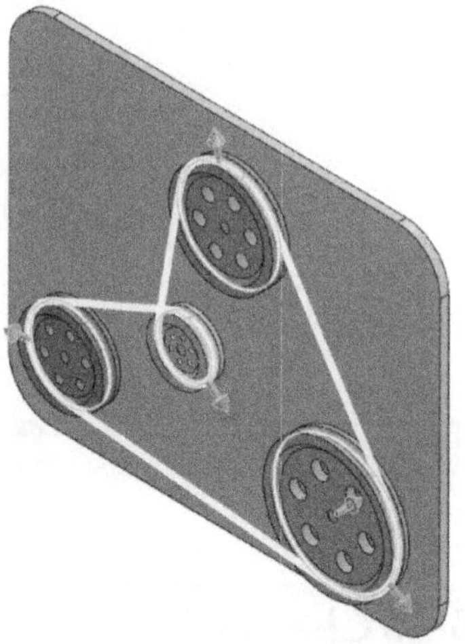

Click the arrow that is displayed on the pulley if the belt appears in the reverse direction. You can also select the pulley from the **Belt Members** selection box and then click the **Flip belt side** icon.

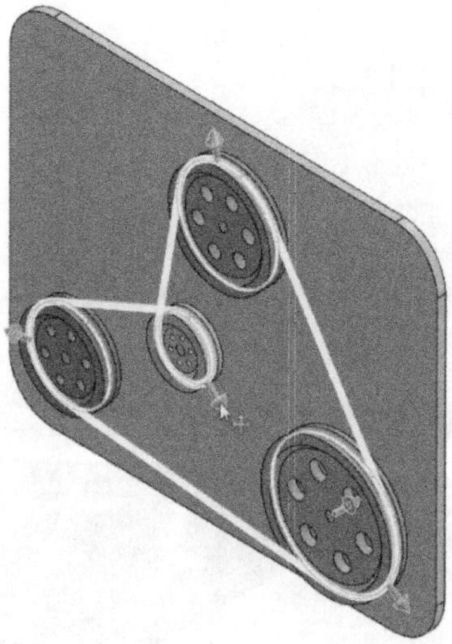

OR

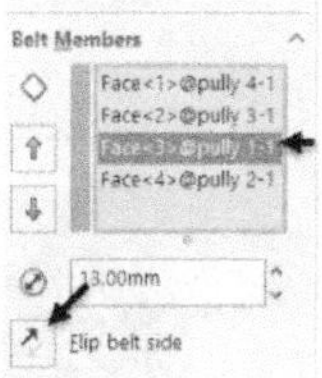

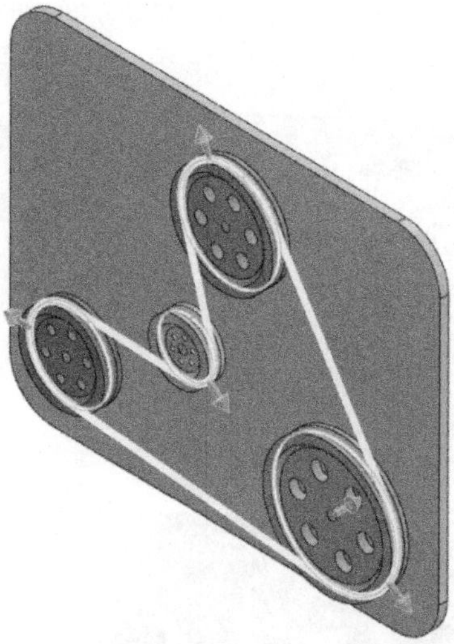

In the **Properties** section on the PropertyManager, check the **Use belt thickness** option and enter a value in the **Belt thickness** box. Make sure that the **Engage belt** option is checked. It helps you to create a link between the belt and all the pulleys touching it. As a result, all the pulleys will rotate when you rotate any pulley.

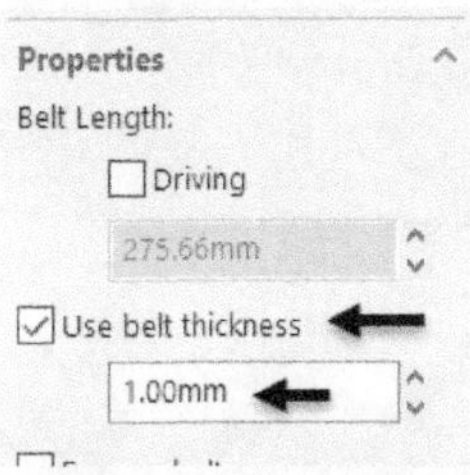

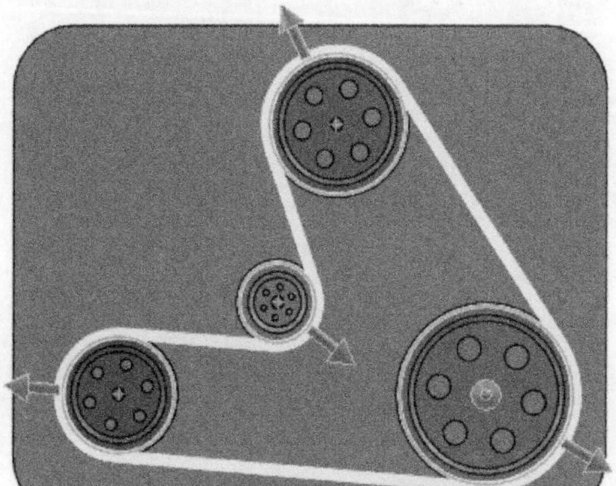

Check the **Driving** option if you want to specify a standard belt size. Note that you need to make any one of the pulleys under-constrained to make the belt driving.

Next, check the **Create belt part** option if you want to create a separate belt part. Click **OK** to create the **Belt** feature. Expand the Belt feature in the FeatureManager Design Tree, and then click on the Belt part. Next, select **Open Part** to open the belt part.

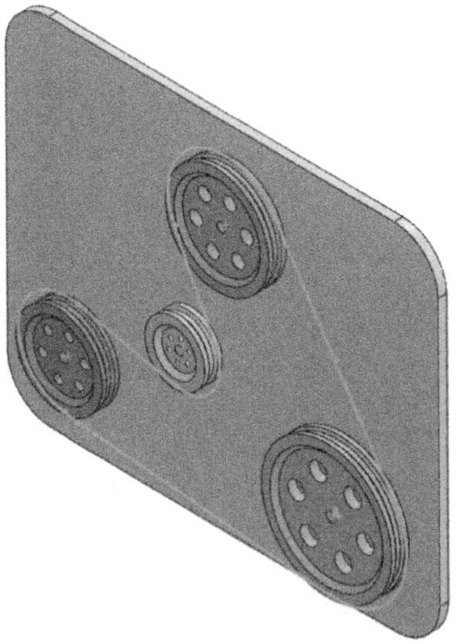

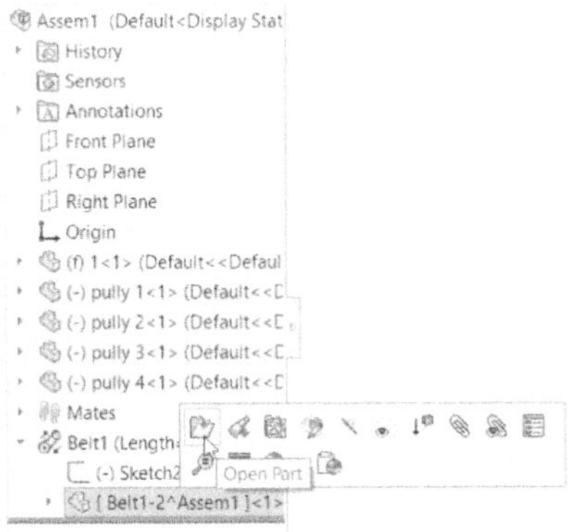

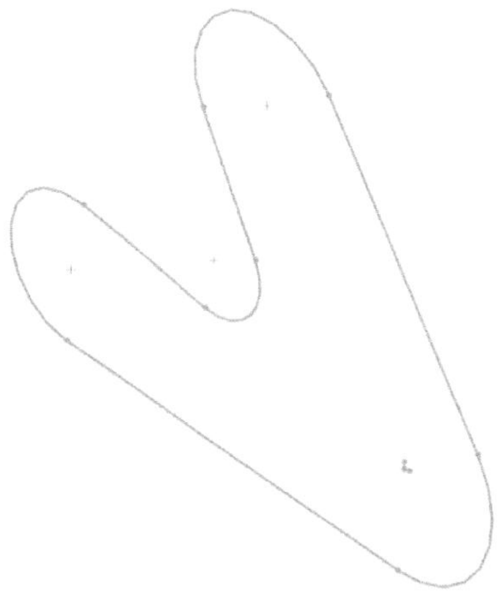

Activate the **Extruded Boss/Base** command and select the sketch. On the PropertyManager, check the Thin Feature option and select **Mid-Plane** from the drop-down. Next, type-in the Thickness and Depth values, and then click **OK**. Save the part file and switch to the assembly file.

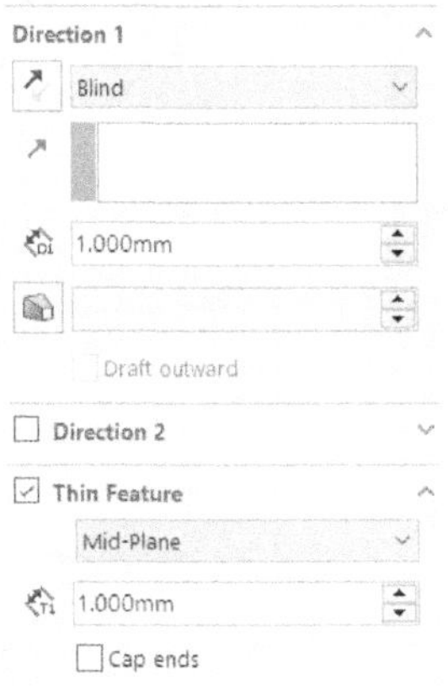

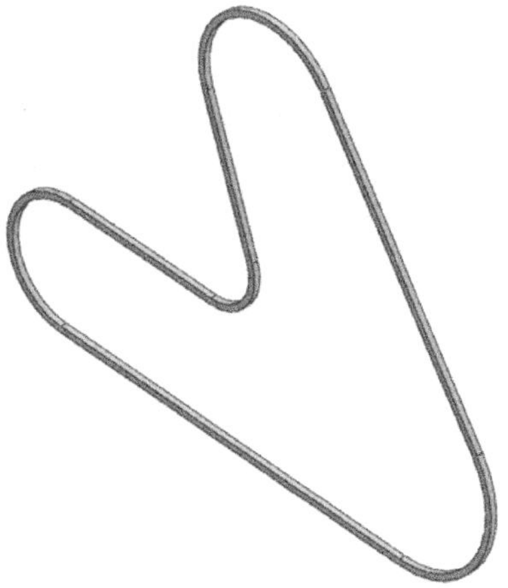

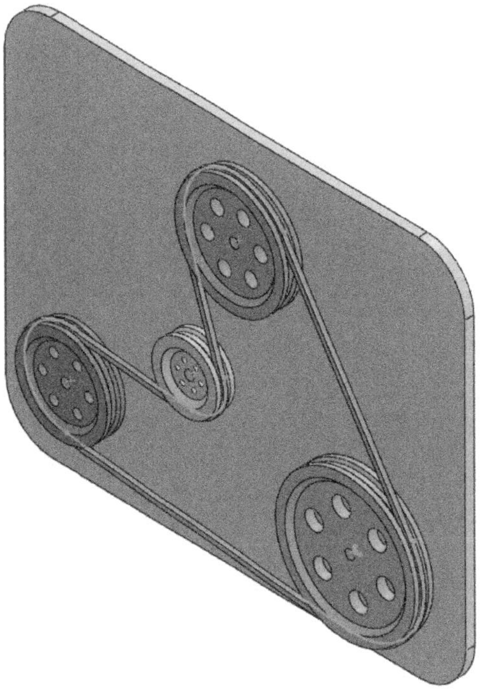

Top-Down Assembly Design

In SOLIDWORKS, there are two methods to create an assembly. The method you are probably familiar with is to create individual parts and then insert them into an assembly. This method is known as Bottom-Up Assembly Design. The second method is called Top-Down Assembly Design. In this method, you will create individual parts within the assembly environment. This method allows you to design an individual part while considering how it will interact with other parts in an assembly. There are several advantages to Top-Down Assembly Design. As you design a part within the assembly, you can be sure that it will fit properly. You can also use reference geometry from the other parts.

Create a New Part inside the assembly

Top-down assembly design can be used to add new parts to an already existing assembly. You can also use it to create entirely new assemblies. To create a part using the Top-Down Design approach, activate the **New Part** command (click **Assembly > Insert Components > New Part** on the CommandManager). Next, select the face or plane on which the part is to be positioned. Now, create the features of the part. In the example given in the figure, the **Convert Entities** command is used to convert the existing part's edges to create a sketch. The projected sketch is then extruded, making it easy to create a part using the existing part's edges.

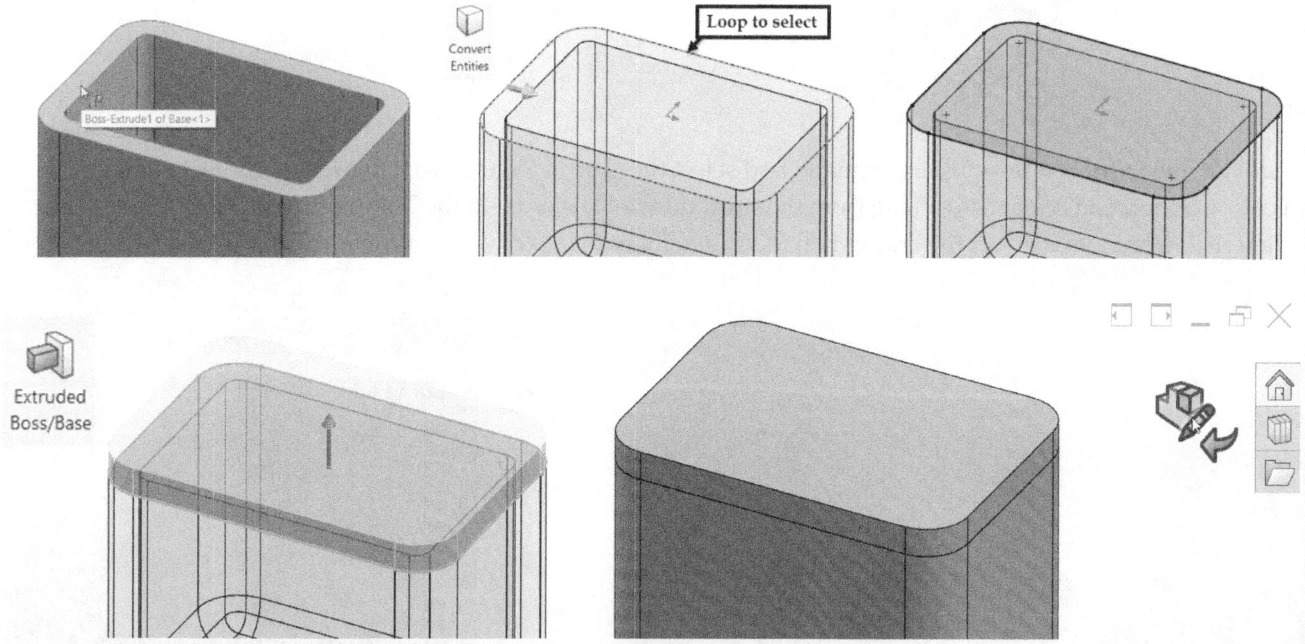

If you make changes to the first part; the second part is changed automatically.

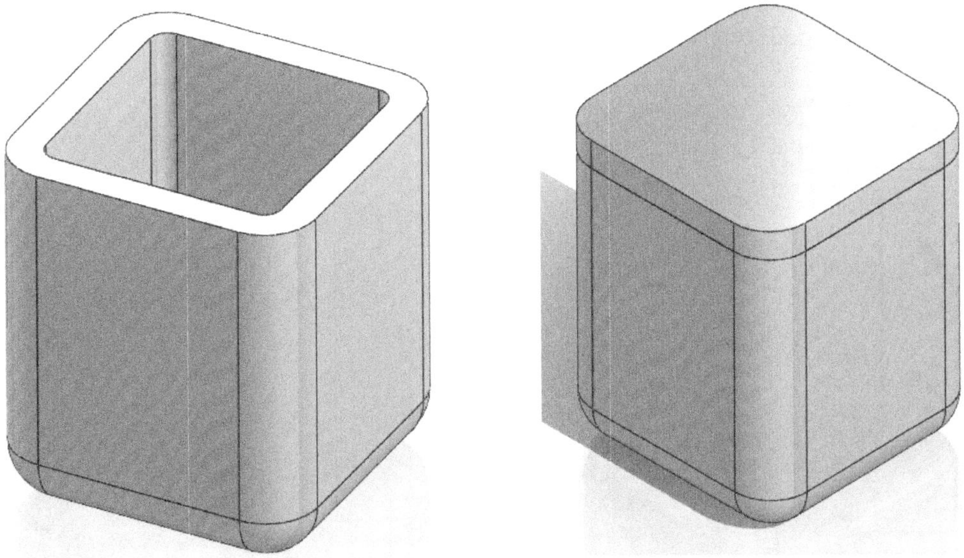

However, if you delete the first part, an error symbol appears next to the second part in the FeatureManager Design Tree.

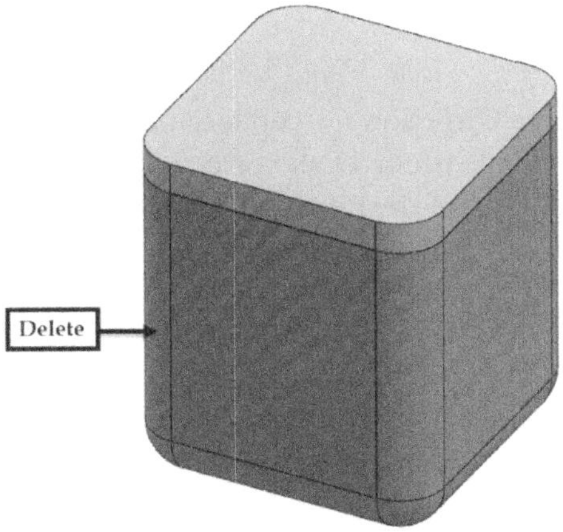

Exploding Assemblies

To document an assembly design properly, it is very common to create an exploded view. In an exploded view, the parts of an assembly are pulled apart to show how they were assembled. To create an exploded view, activate the **Exploded View** command (click **Assembly > Exploded View** on the CommandManager); the **Explode** PropertyManager appears. Click the **Regular Step (translate and rotate)** 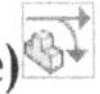 icon on the PropertyManager, select the component to be exploded; the manipulator appears on the selected part. Select any one of the arrows of the manipulator to define the explosion direction. Next, press and hold the left mouse button on the selected arrow, and then drag it. You can use the scale that appears along the manipulator to specify the exact explode distance. Next, release the mouse button; the explode step is added to the **Exploded Step** section.

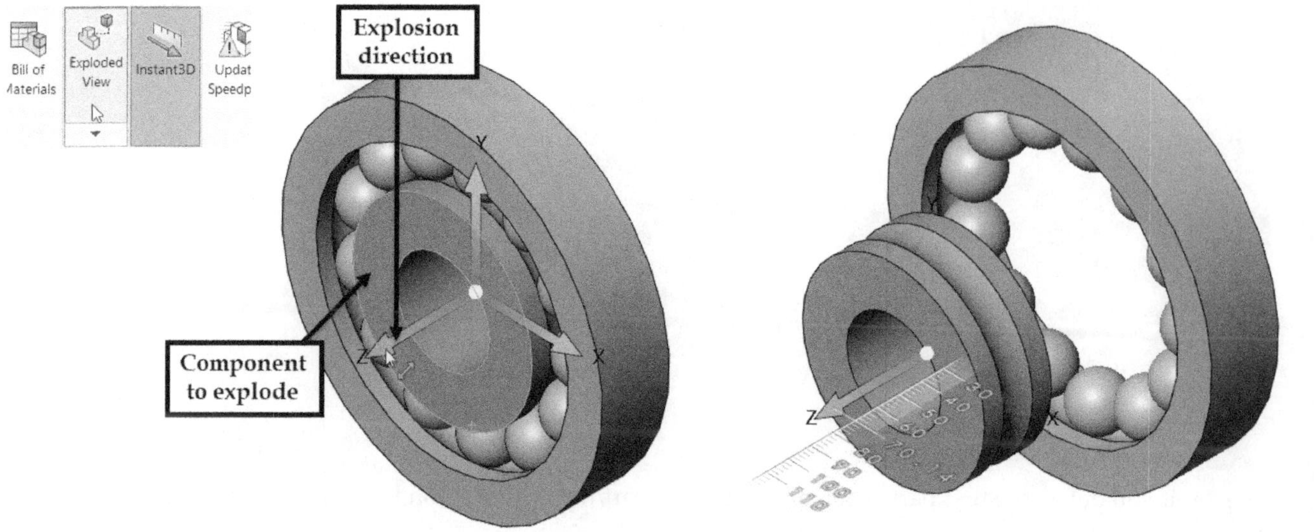

You can change the explode direction by clicking in the **Explode Direction** selection box and then selecting the required arrow from the manipulation triad. You can also click the **Reverse Direction** icon located next to the **Explode Direction** selection box to reverse the explosion's direction. Use the **Explode Distance** box to specify an explosion distance value, and then click the **Done** button located at the bottom of the **Editing Chain1** section. Likewise, explode the other component, and then click **Done**.

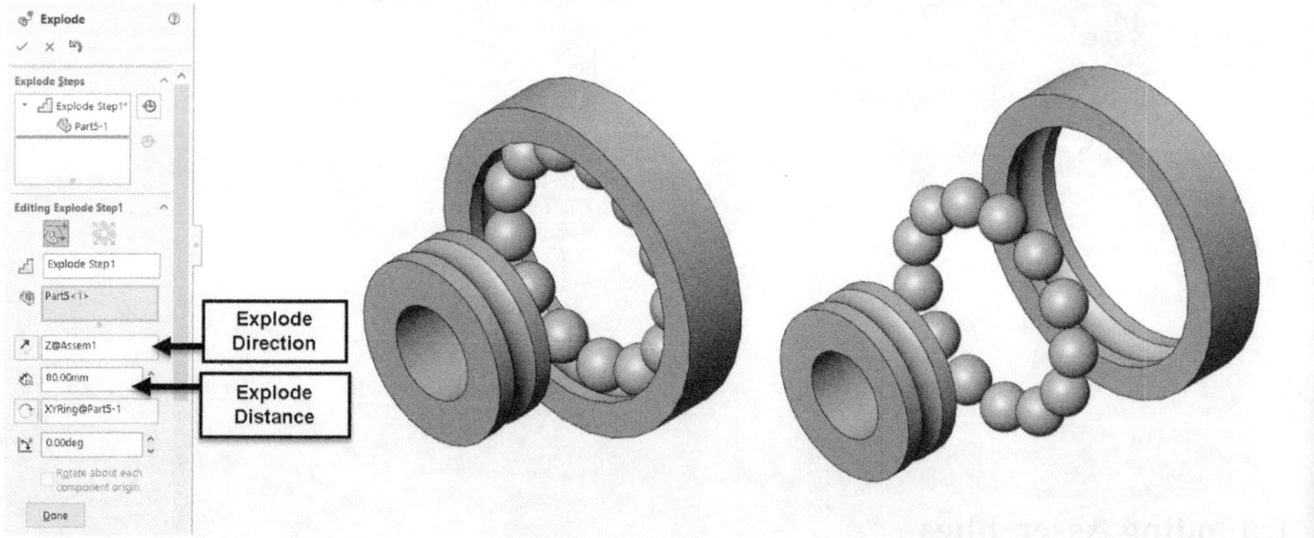

Creating Radial Explosion

Click the **Radial Step** icon on the PropertyManager to explode the components in the radial direction. Next, select the components to be exploded; an arrow appears along with a circular handle. Click and drag the arrow up to the required distance, and then release. Click **Done** and **OK** on the PropertyManager.

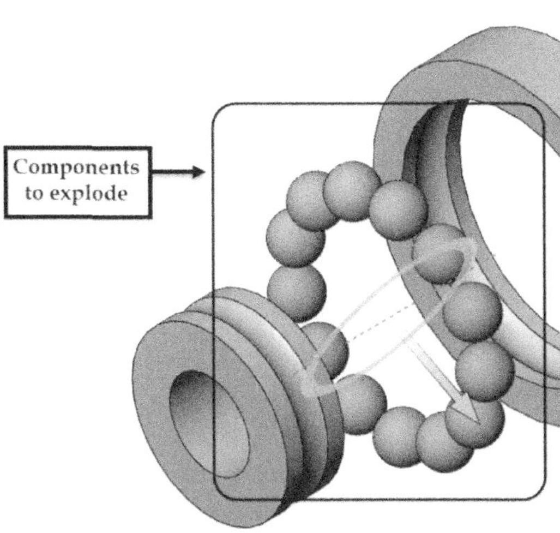

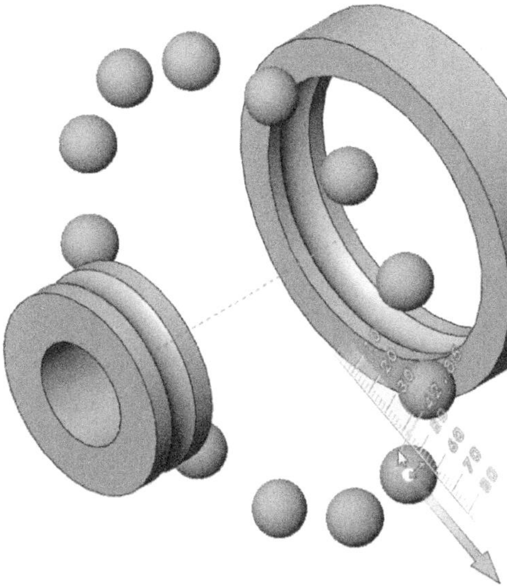

The exploded view is added to the ConfigurationManager. To view the exploded view, click the ConfigurationManager tab on the left pane, and then expand **Assembly Configuration (s) > Default [Assem 1] > Exploded View**. If you want to collapse an exploded view, right click on it, and then select **Collapse**.

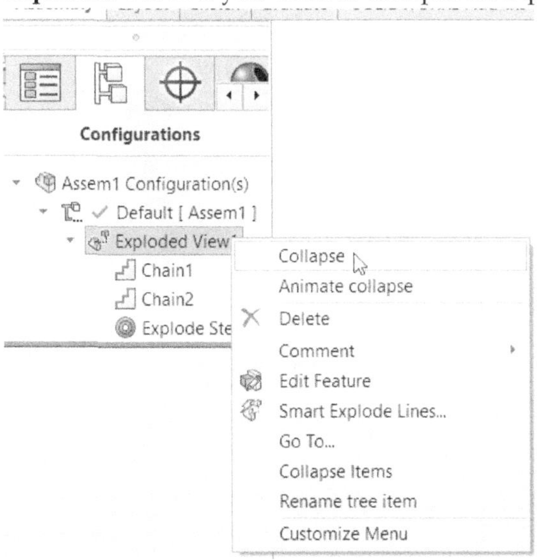

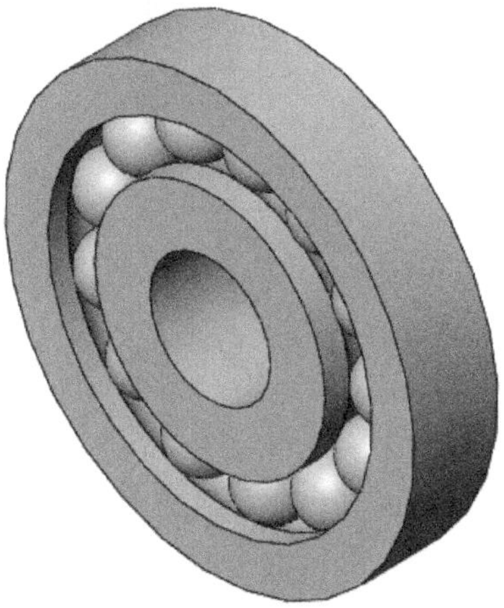

To animate the explosion, right-click on it and select **Animate explode**. Use the **Animation Controller** dialog to control the animation. You can also save the animation using the **Save Animation** icon.

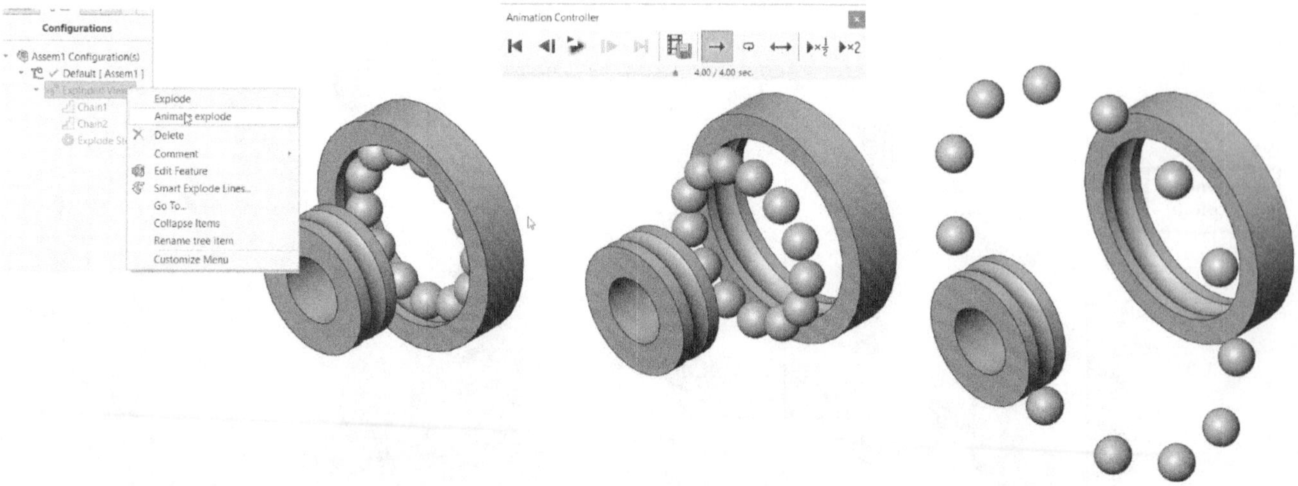

Right-click on the **Exploded View** in the ConfigurationManager, and then select **Smart Explode Lines**. On the PropertyManager, use the **Reference point** options under the **Component route line** section to specify the points from which the route lines are generated. Next, click **OK** to create the explode lines.

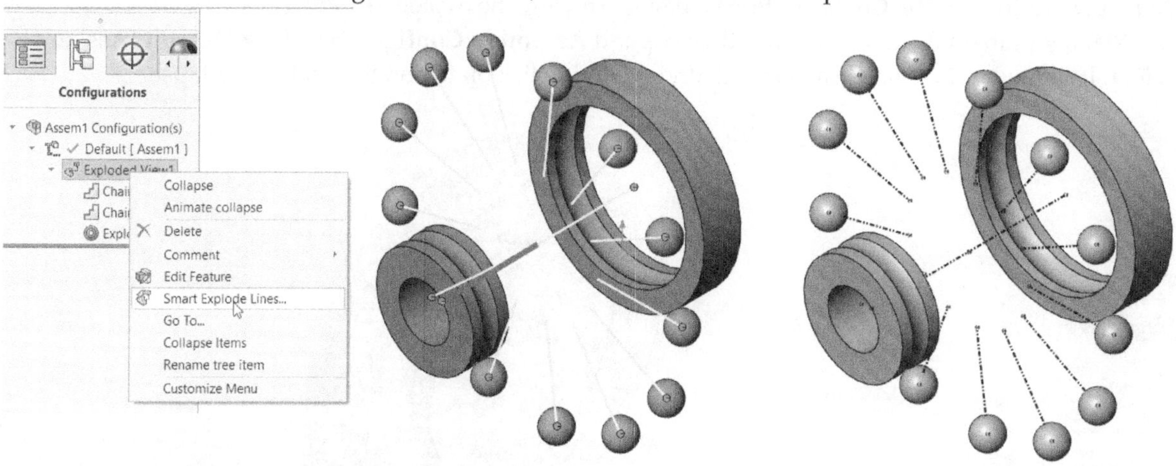

Best Practices

Planning Your Design: Before applying constraints, it is essential to meticulously plan your assembly. Understand how components will interact, considering aspects like clearance, motion, and alignment to optimize the design process.

Naming Conventions: Utilize meaningful names for parts and subassemblies to maintain an organized assembly tree structure, facilitating easier navigation and management of the project.

Component Origin: Establish a consistent origin point for components within assemblies. This practice enhances alignment precision and predictability throughout the assembly process.

Additional Recommendations

Standardization: Adhere to SOLIDWORKS' standard practices and guidelines to ensure compatibility and consistency across projects.

Mate References: Utilize mate references to streamline the assembly process by predefining component relationships, enhancing efficiency.

Configuration Management: Implement robust configuration management practices to effectively handle design variations and iterations.

General Cautions

Avoid over constraint: Over constraining can result in errors or unexpected behavior. Apply constraints judiciously to maintain design integrity and functionality.

Adaptation to Design Changes: Stay prepared to adjust assembly constraints when modifications are made to parts, ensuring seamless integration and functionality.

Documentation and Labeling: Thoroughly document assembly constraints within SOLIDWORKS projects and utilize labels for clarity and ease of troubleshooting potential issues.

Examples

Example 1 (Bottom-Up Assembly)

In this example, you will create the assembly shown next.

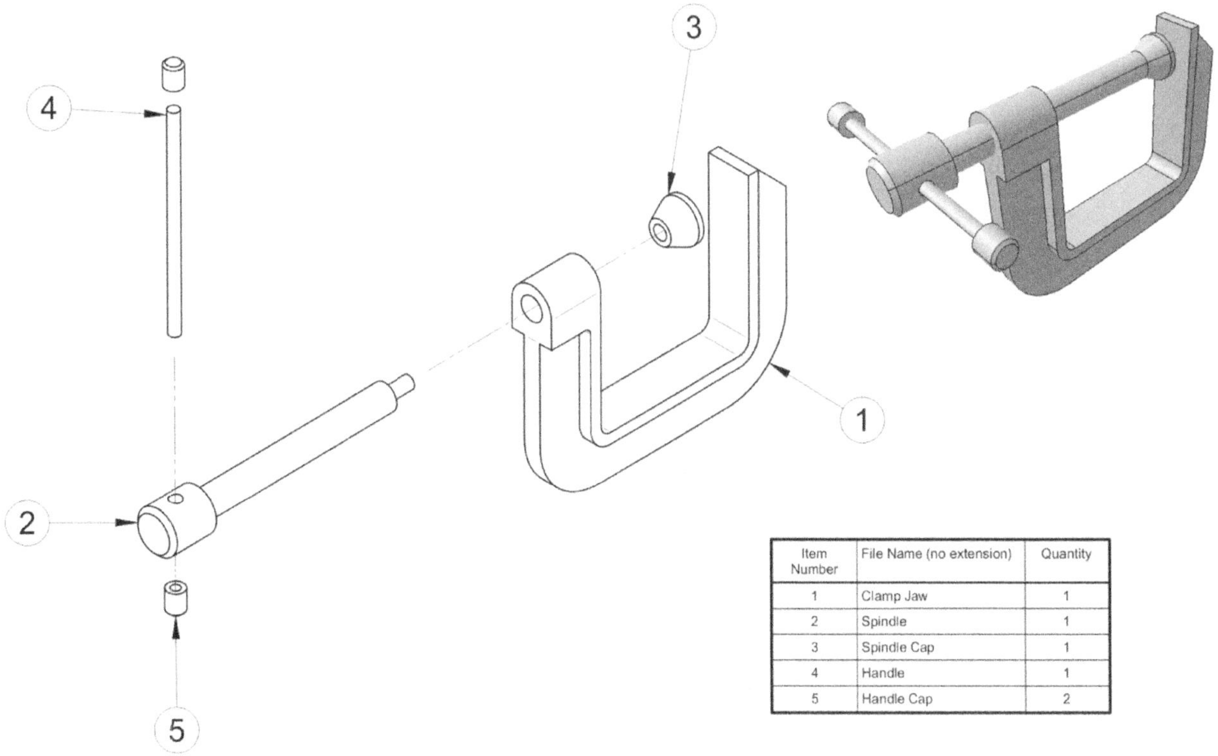

Item Number	File Name (no extension)	Quantity
1	Clamp Jaw	1
2	Spindle	1
3	Spindle Cap	1
4	Handle	1
5	Handle Cap	2

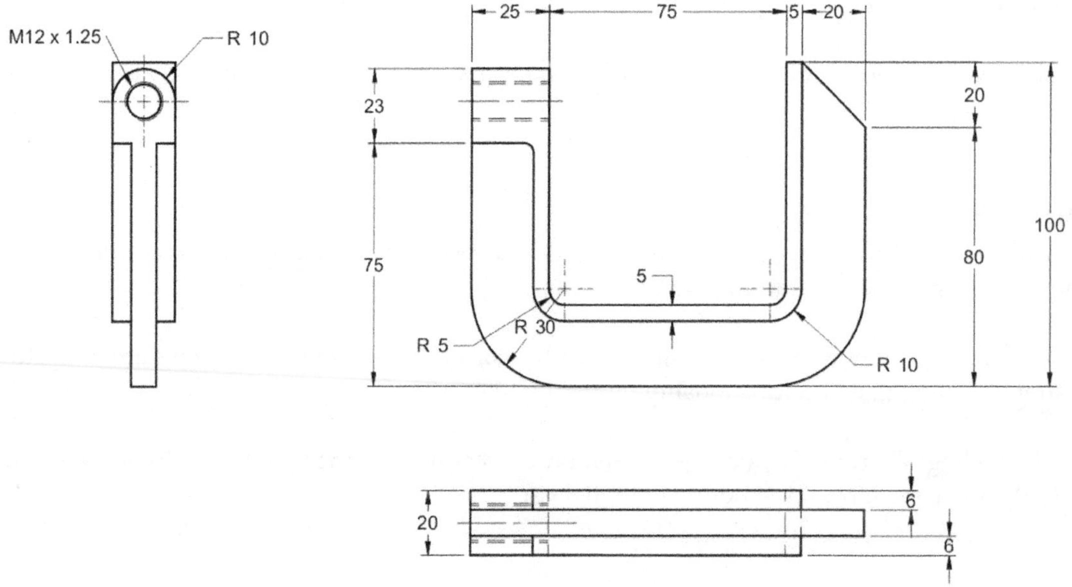

CLAMP JAW

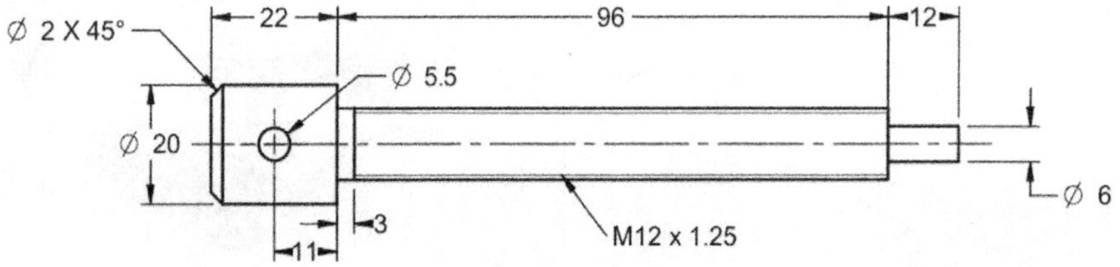

SPINDLE

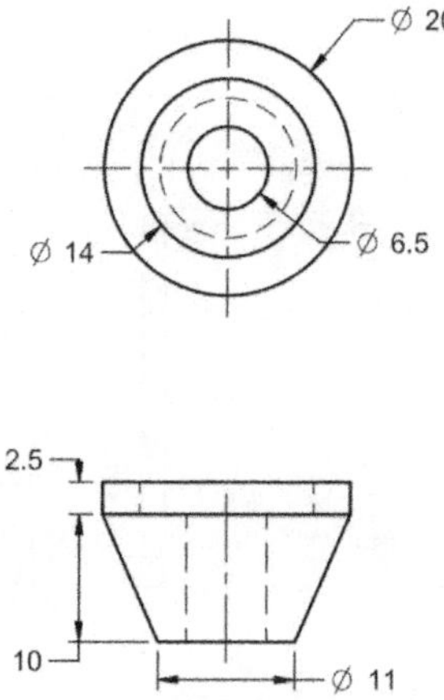

SPINDLE CAP

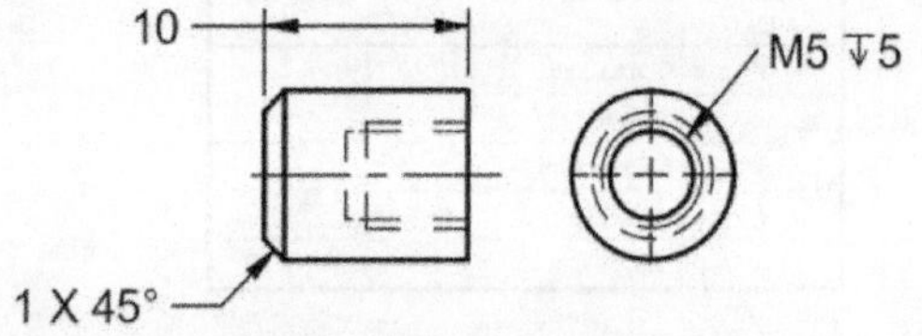

HANDLE CAP

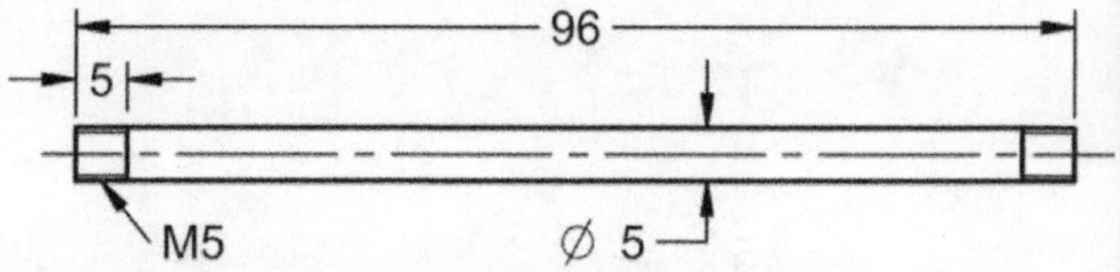

HANDLE

1. Start **SOLIDWORKS 2024**.

2. Create and save all the components of the assembly in a single folder. Name this folder as *G-Clamp*. Close all the files.
3. On the Quick Access Toolbar, click the **New** button.
4. On the **New SOLIDWORKS Document** dialog, click **Assembly**, and then click **OK; the Begin Assembly PropertyManager appears**. The **Open** dialog pops up, prompting you to select a part file to be inserted into the assembly.
5. On the **Open** dialog, go to the *G-Clamp* folder. Select *Clamp Jaw* and click **Open**.
6. Click in the graphics window; the Clamp Jaw is inserted into the assembly.

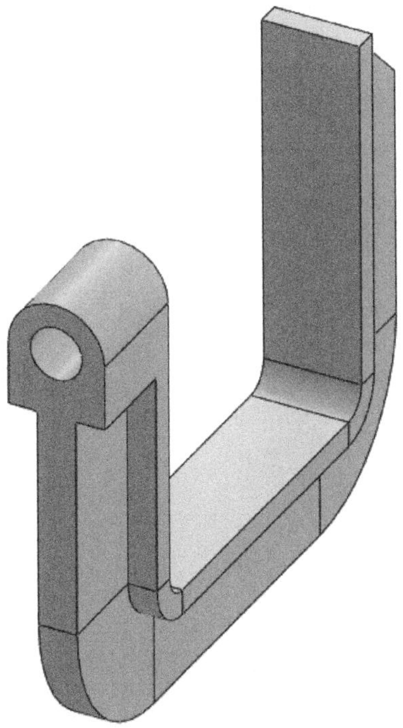

7. On the CommandManager, click the **Assembly** tab > **Insert Components**  icon (or) select **Insert > Component > Existing Part/Assembly** from the menu bar.
8. On the **Open** dialog, select *Spindle*, and then click **Open**.
9. Position the component, as shown in the figure.

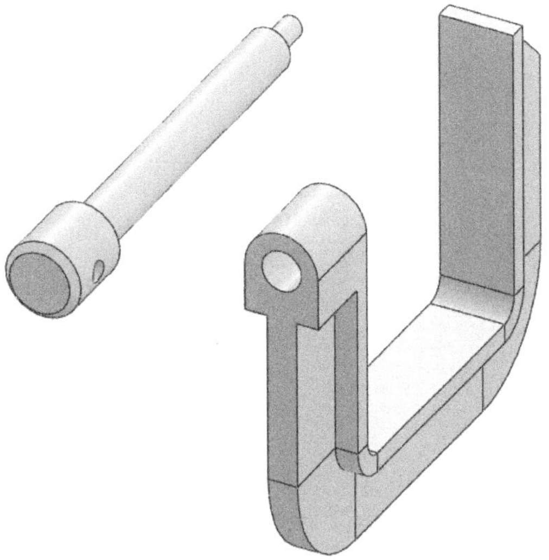

10. Likewise, insert the remaining components into the assembly.

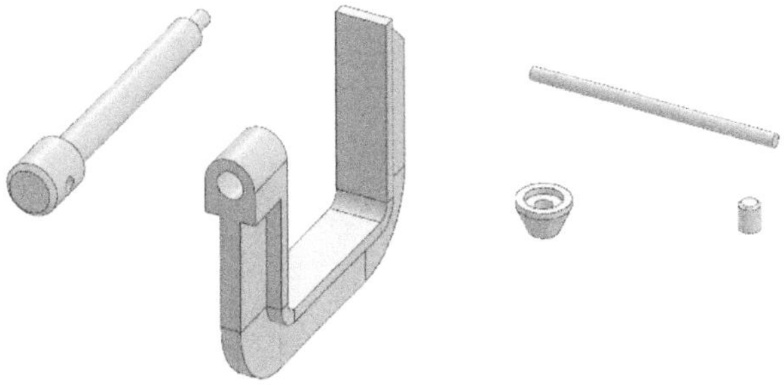

11. On the CommandManager, click the **Assembly** tab > **Mate**  icon (or) select **Insert > Mate** from the menu bar.
12. Select the cylindrical faces of the spindle and clamp jaw.

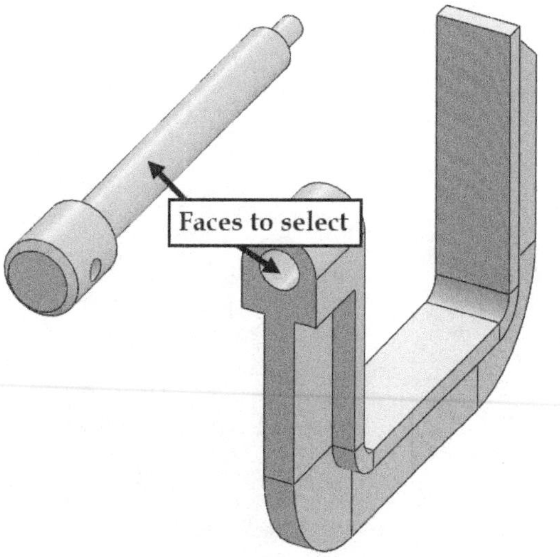

13. Click the **Concentric** icon on the pop-up dialog.
14. Click **Add/Finish Mate** to apply the mate.

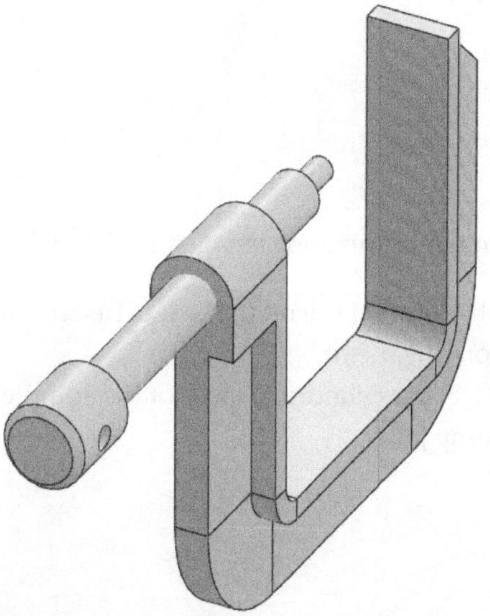

15. Click on the front face of the clamp jaw and that of the spindle.

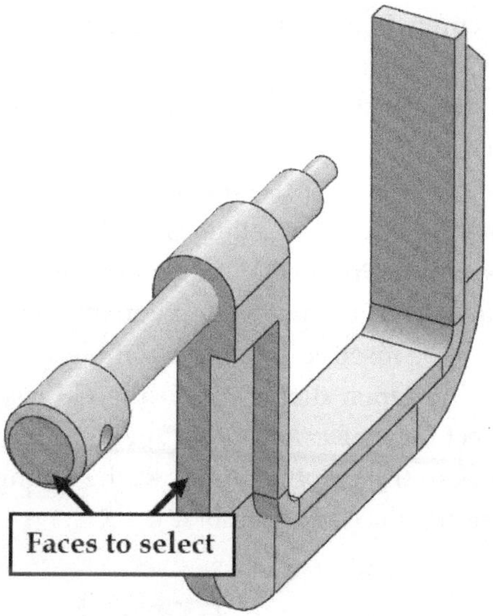

16. Click the **Distance** icon on the pop-up dialog.
17. On the flyout, type-in **40** in the **Distance** box.

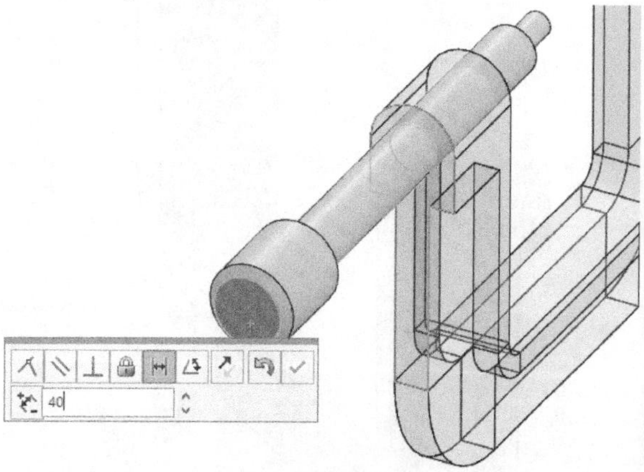

18. Click **Add/Finish Mate** .

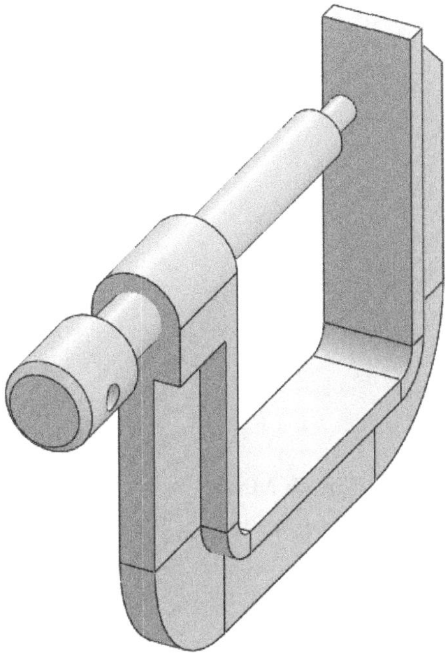

Next, you need to lock the spindle's rotation by making its horizontal plane parallel to the assembly's horizontal plane.

19. Click **Close** × on the **Mate** PropertyManager.
20. Drag a selection window across all the parts.

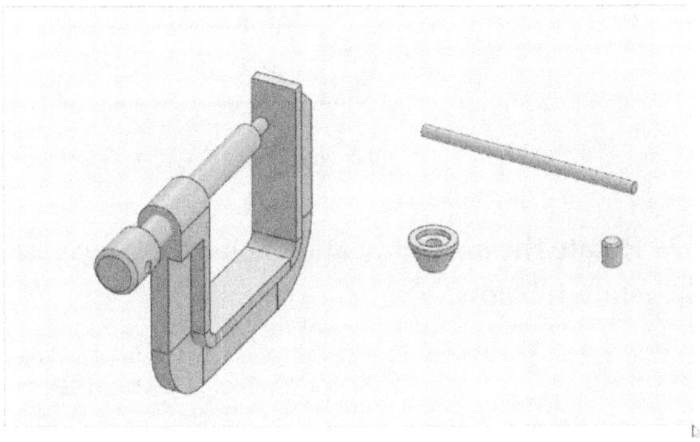

21. On the **View (Heads Up)** toolbar, click **Hide/Show Items** drop-down > **View Planes** .
22. On the **Hide/Show Items** drop-down, click the **Hide/Show Primary Planes** icon.

The primary planes of all the components are displayed.

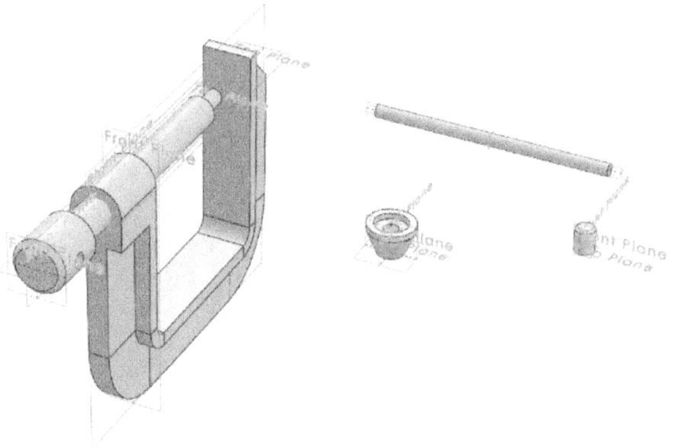

23. On the CommandManager, click **Assembly > Mate** .
24. Select the Top plane of the Spindle and the horizontal face of the Clamp Jaw.

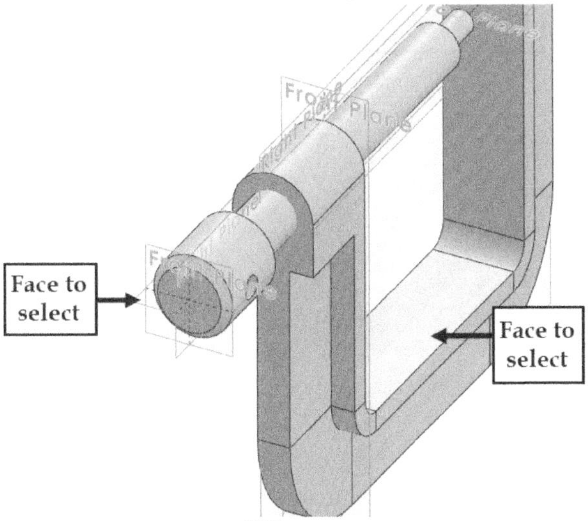

25. Click the **Parallel** icon on the pop-up dialog.
26. Click **Add/Finish Mate** .

The rotation of the Spindle is locked, and it is fully constrained.

27. Click on the cylindrical faces of the spindle and spindle cap.

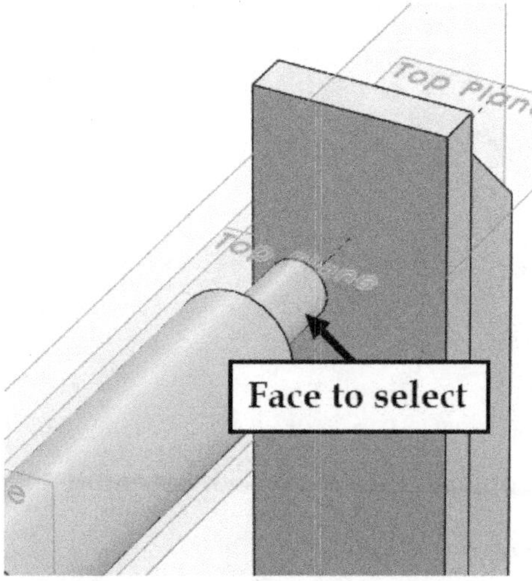

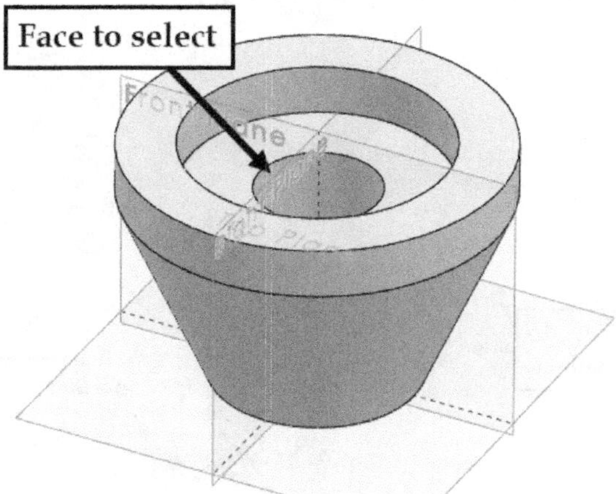

28. Click the **Concentric** icon on the pop-up dialog.
29. Click the **Flip Mate Alignment** icon on the pop-up dialog to reverse the direction, as shown (skip this step if the Spindle cap is already oriented in the given direction).

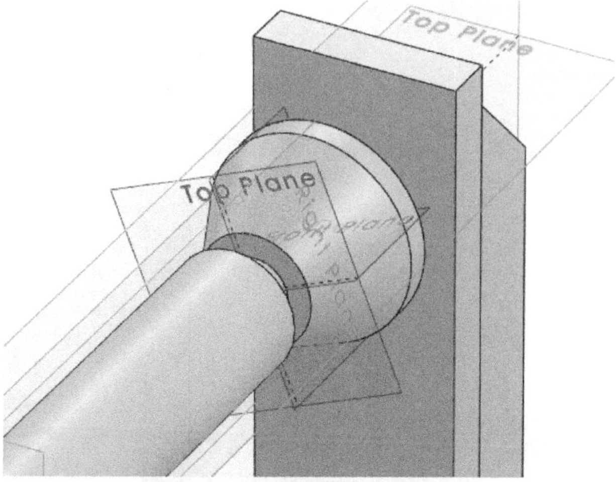

30. Click **Add/Finish Mate** .
31. Click on the bottom face of the spindle cap.

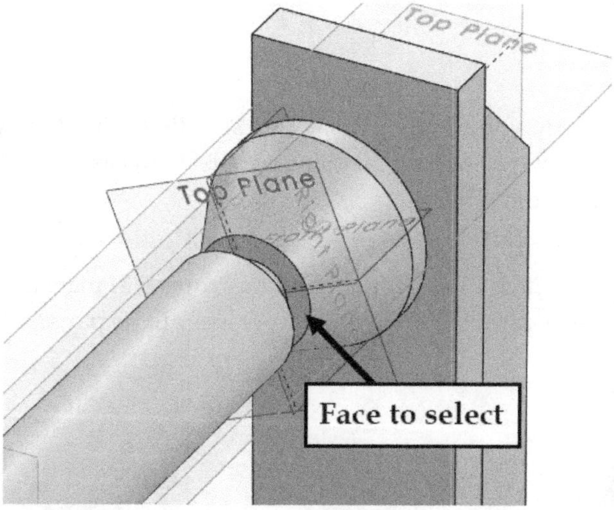

32. Rotate the assembly and click on the face, as shown below.

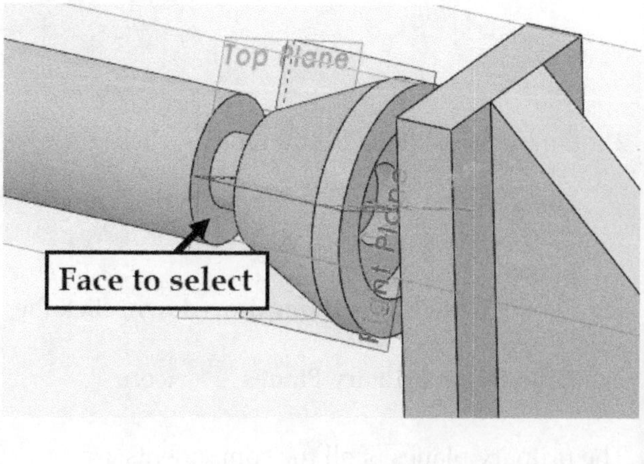

33. Click the **Coincident** icon on the Pop-up dialog.
34. Click **Add/Finish Mate** .
35. On the **View (Heads Up)** toolbar, click **View Orientation > Trimetric**.

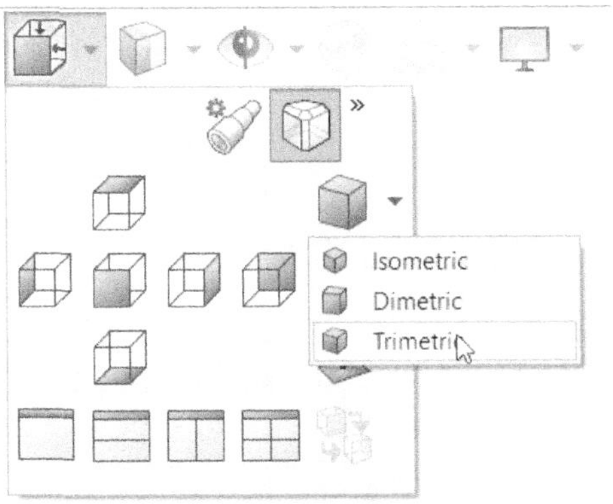

36. Select the Front plane of the spindle cap and Top plane of the spindle.

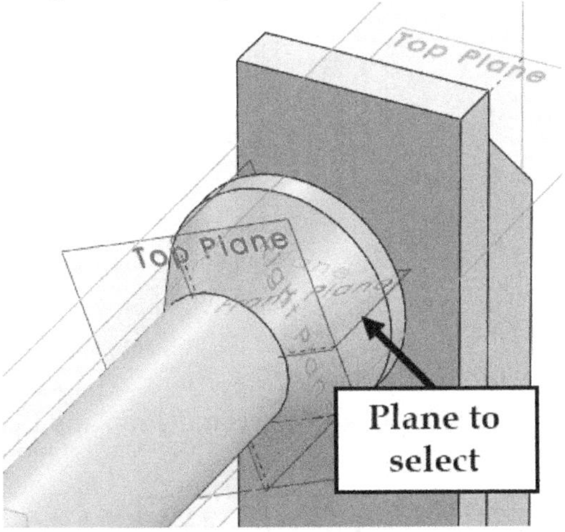

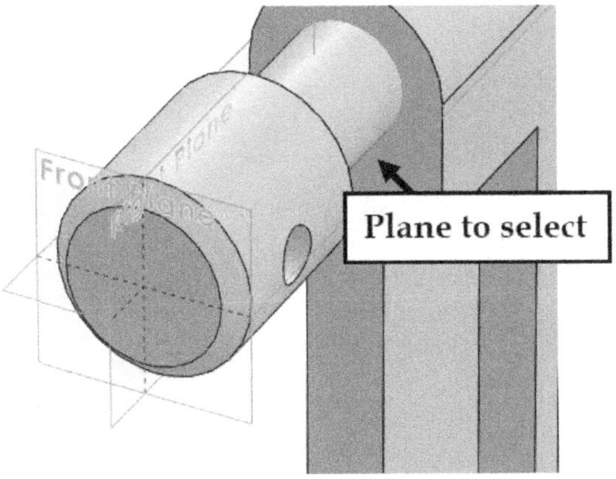

37. Click the **Parallel** icon on the pop-up dialog.

38. Click **Add/Finish Mate** .

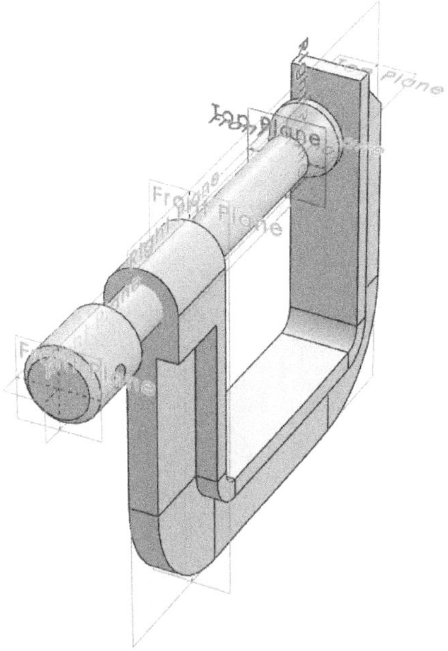

39. Select the cylindrical faces of the handle and the hole on the spindle.

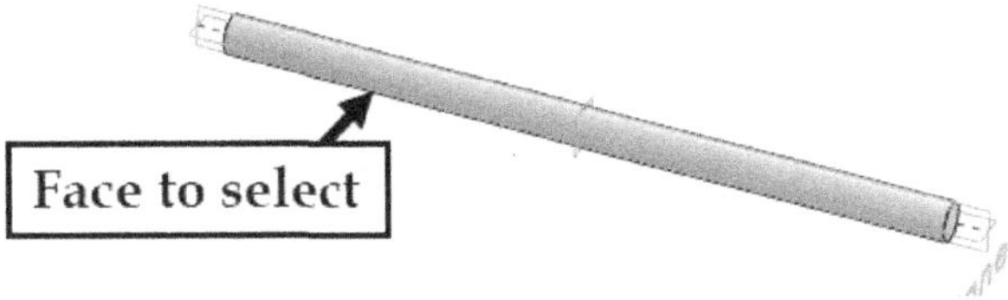

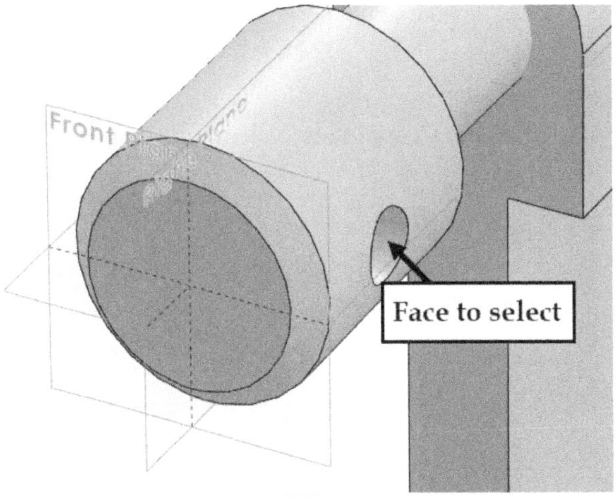

40. Click the **Concentric** icon on the pop-up dialog.
41. Click **Add/Finish Mate** .
42. Click on the Right plane of the handle and that of the spindle.

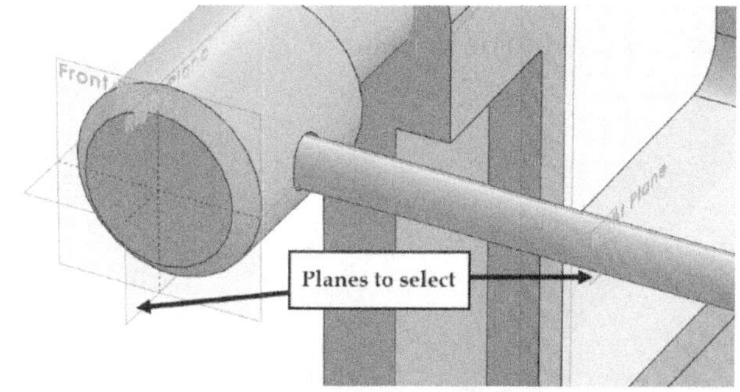

43. Click the **Coincident** icon on the Pop-up dialog.
44. Click **Add/Finish Mate** .

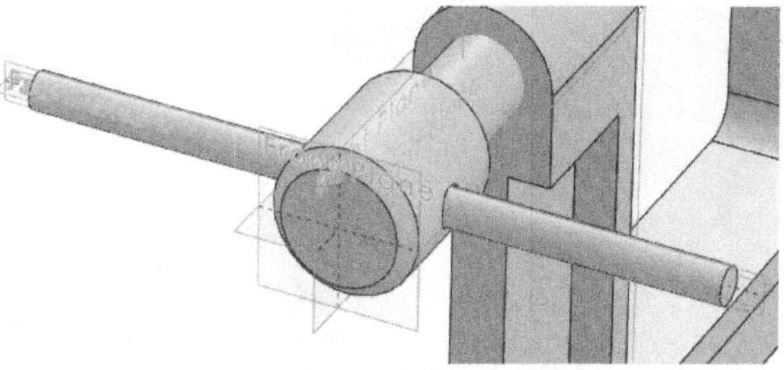

45. Click on the Top plane of the handle and that of the spindle.

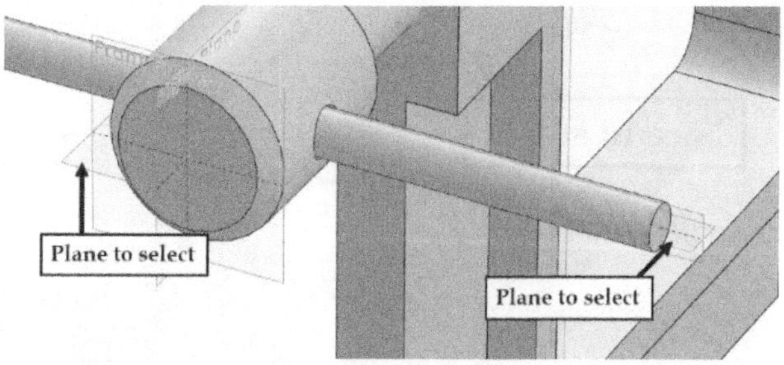

46. Click the **Coincident** icon on the Pop-up dialog.
47. Click **Add/Finish Mate** .
48. Select the cylindrical faces of the handle and handle cap.

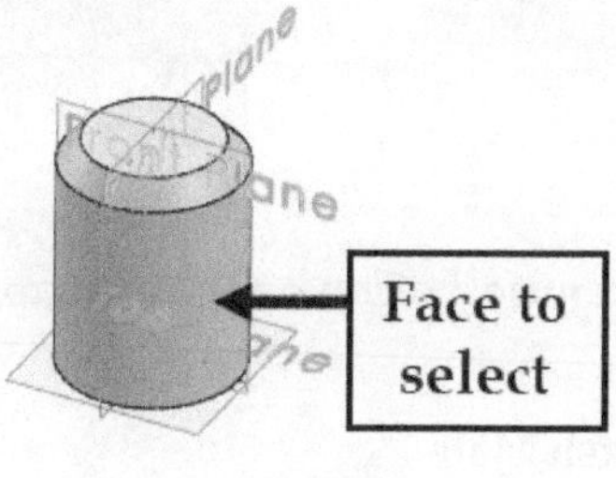

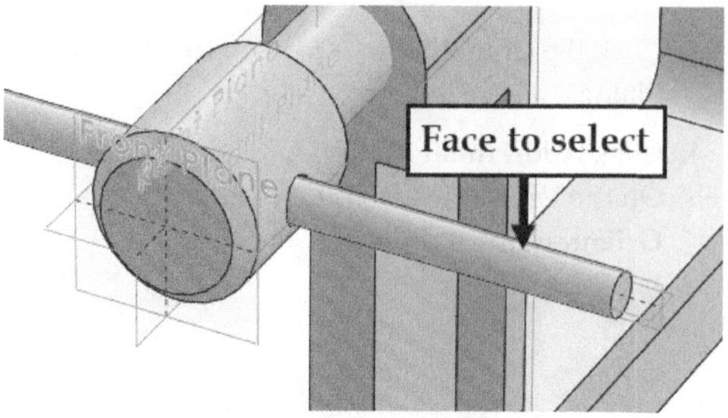

49. Click the **Concentric** icon on the pop-up dialog.
50. Click the **Flip Mate Alignment** icon on the pop-up dialog to reverse the direction, as shown.

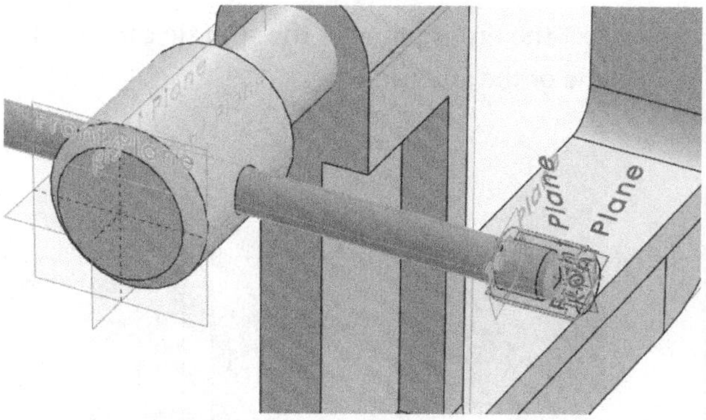

51. Click **Add/Finish Mate** .
52. Select the Right plane of the Handle cap and the Front plane of the Handle.

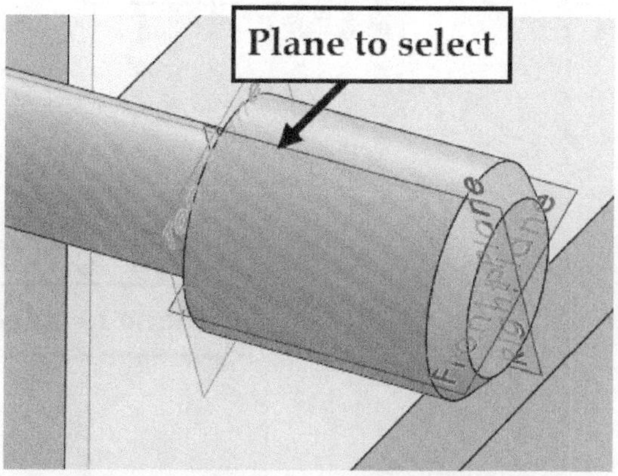

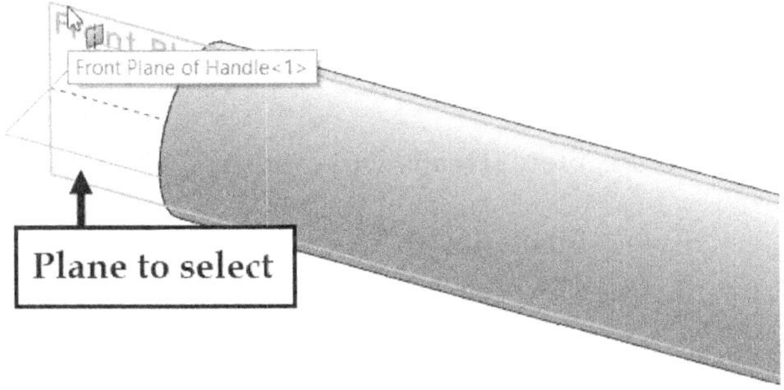

53. Click the **Coincident** icon on the Pop-up dialog.
54. Click **Add/Finish Mate**.
55. On the **View (Heads Up)** toolbar, click **Display Style > Hidden Lines Visible**.
56. Place the pointer on the hidden edge of the handle, as shown.

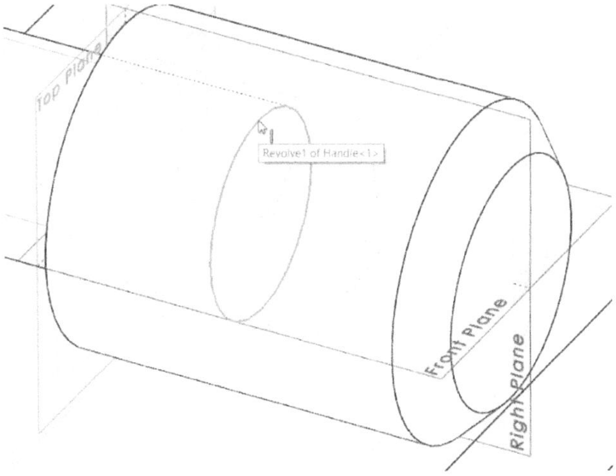

57. Right-click and select **Select Other**.

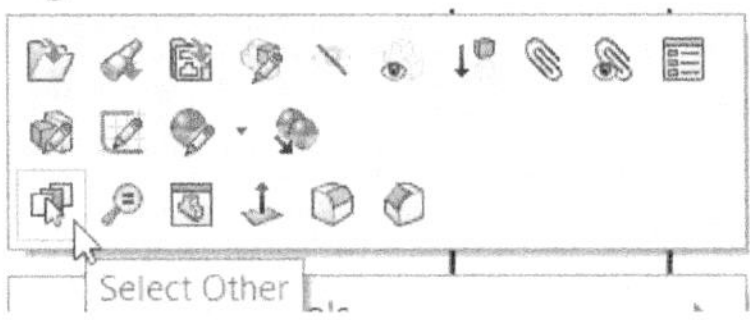

58. On the **Select Other** dialog, move the pointer over the faces displayed and then click when the handle's end face is highlighted.

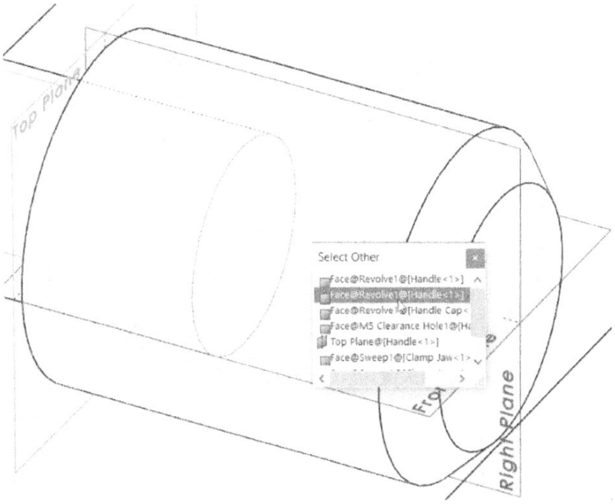

59. Select the end face of the Handle cap, as shown.

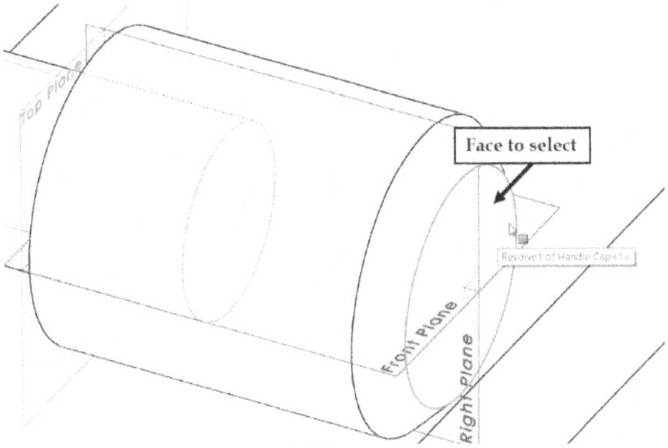

60. Click the **Distance** icon on the pop-up dialog.
61. On the flyout, type-in **5** in the **Distance** box.

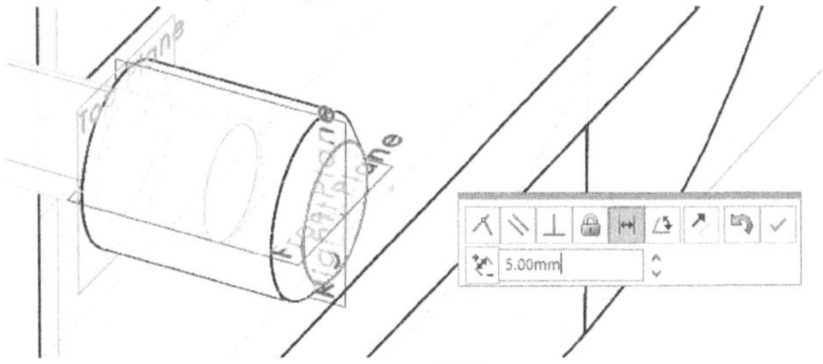

62. Click **Add/Finish Mate**.
63. Click **Close** on the **Mate** PropertyManager.

Copying Components with Mates

1. On the CommandManager, click **Assembly > Insert Components > Copy with Mates**.
2. Select the Handle Cap from the assembly.
3. Click the **Next** icon on the PropertyManager.

The selection box in the **Concentric** section is highlighted. You need to select a cylindrical face to be made concentric to the handle cap's cylindrical face.

4. Select the cylindrical face of the Handle.

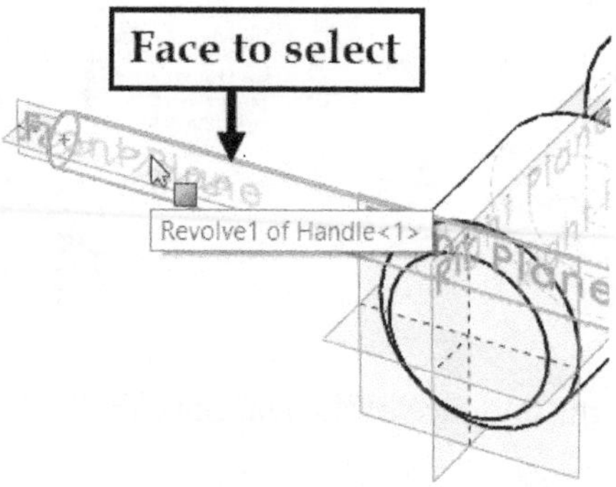

5. Select the Front plane of the handle to apply the **Coincident** mate.
6. Rotate the model and select the other end face of the handle.

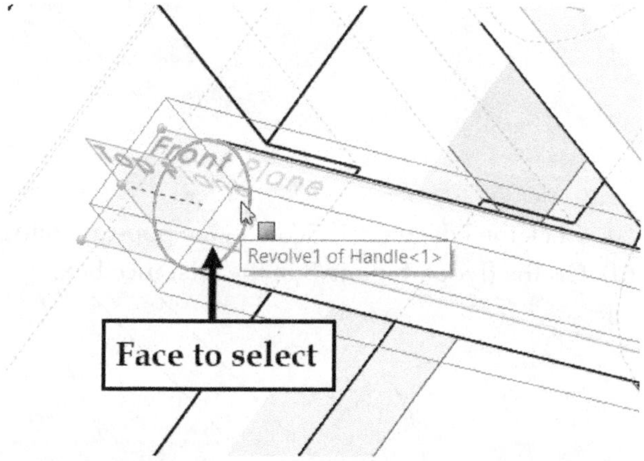

7. Click **OK** on the PropertyManager.
8. Click **Cancel** on the PropertyManager to close it.
9. On the **View (Heads Up)** toolbar, click **Display Style > Shaded with Edges**.
10. On the **View (Heads Up)** toolbar, click **Hide/Show Items > View Planes**.

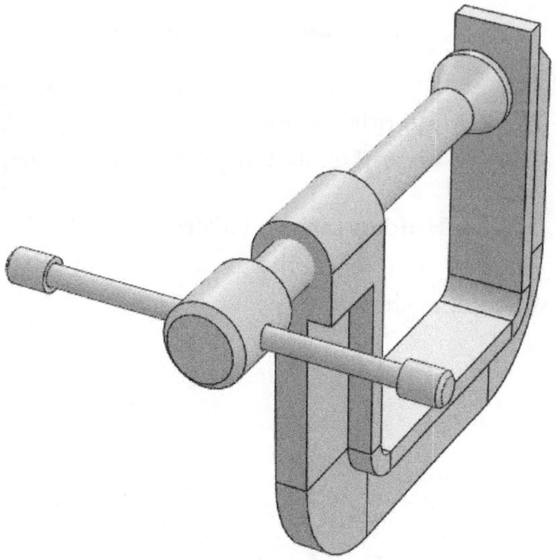

11. Save and close the assembly.

Example 2 (Top-Down Assembly)

In this example, you will create the assembly shown next.

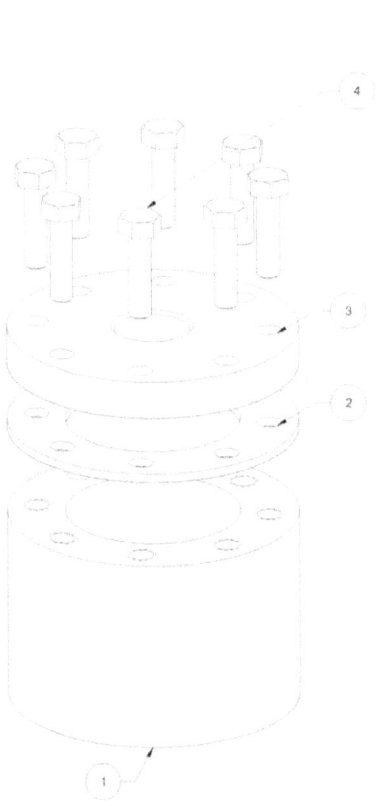

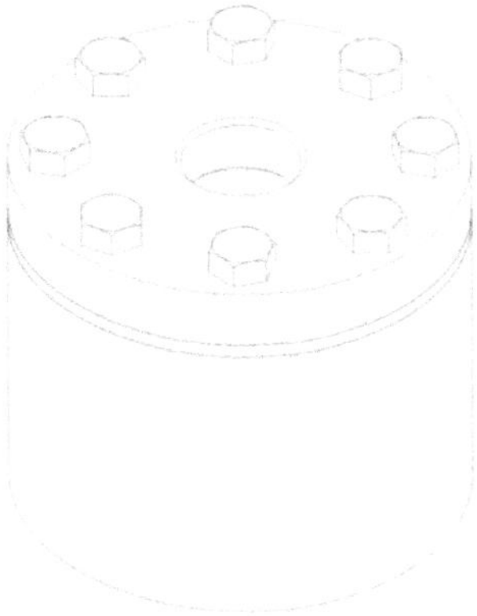

4	HEX BOLT AM,M8X1.25X30	8
3	COVER PLATE	1
2	GASKET	1
1	CYLINDER BASE	1
PC NO	PART NAME	QTY

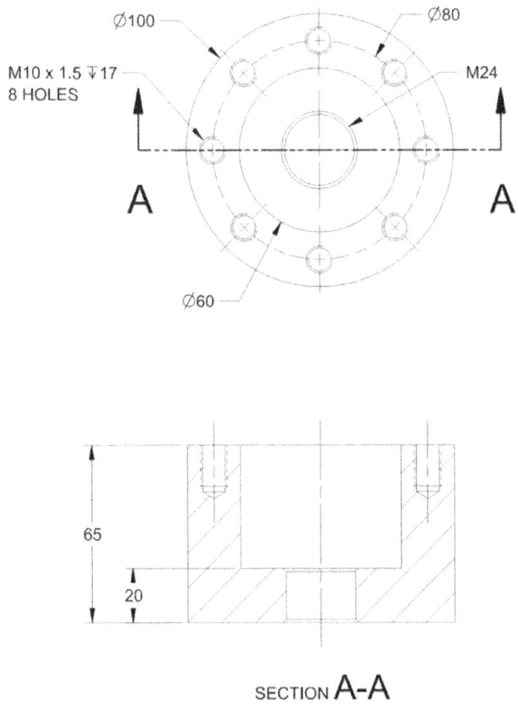

Cylinder Base

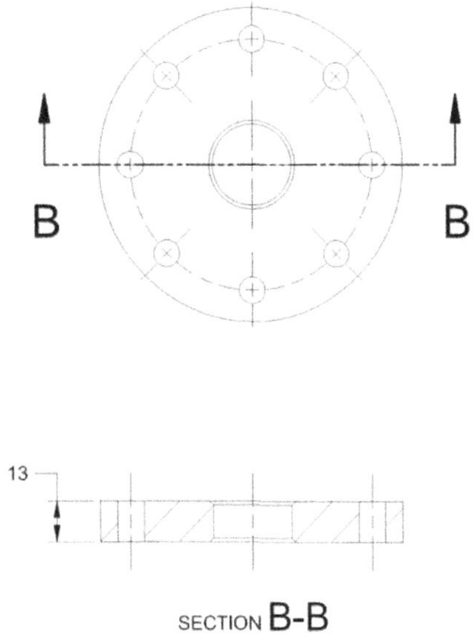

Cover Plate

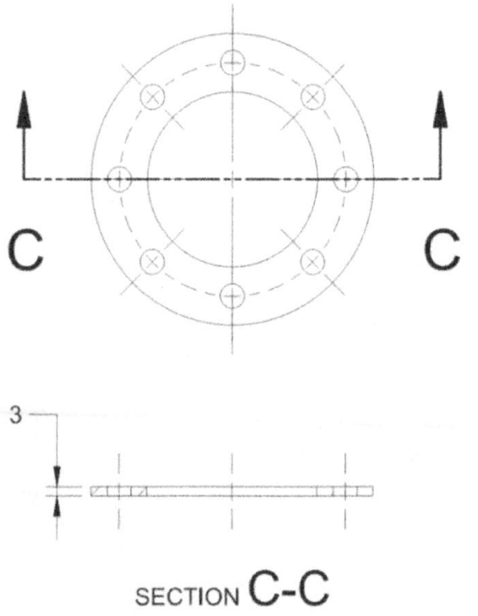

Gasket

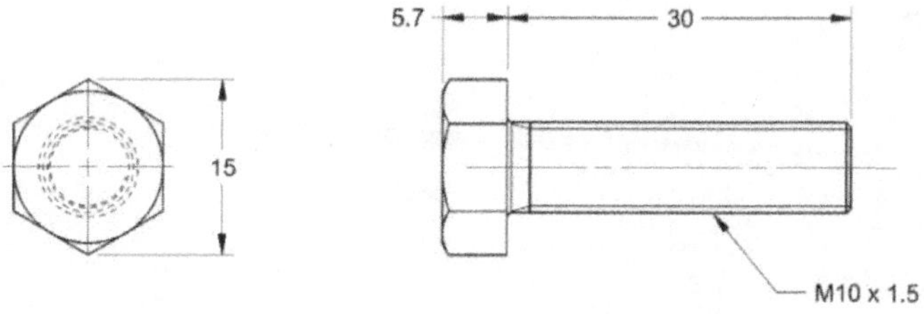

Screw

1. Start **SOLIDWORKS 2024**.
2. On the Menu bar, click the **New** icon to open the **New SOLIDWORKS Document** dialog.
3. On the **New SOLIDWORKS Document** dialog, click the **Assembly** icon, and then click **OK**.
4. Click **Cancel** on the **Open** dialog.
5. Click **Cancel** on the **Begin Assembly** PropertyManager.
6. On the CommandManager, click **Assembly > Insert Components** drop-down > **New Part** .
7. Click in the graphics window.
8. Select **Part 1** from the FeatureManager Design Tree.
9. Click **Assembly > Edit Component** on the CommandManager to activate the part mode.
10. On the CommandManager, click **Sketch > Sketch**.
11. Expand the FeatureManager Design Tree in the graphics window, and then select the Front Plane.
12. Draw a sketch, as shown below.

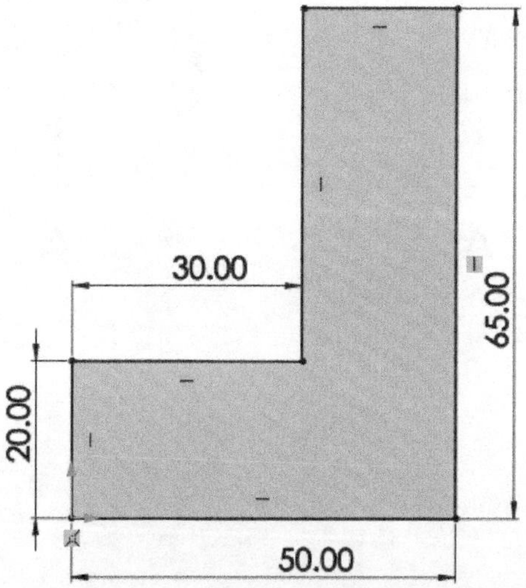

13. On the CommandManager, click **Sketch > Exit Sketch**.
14. On the CommandManager, click **Features > Revolved Boss/Base** .
15. Select the vertical line of the sketch, as shown.

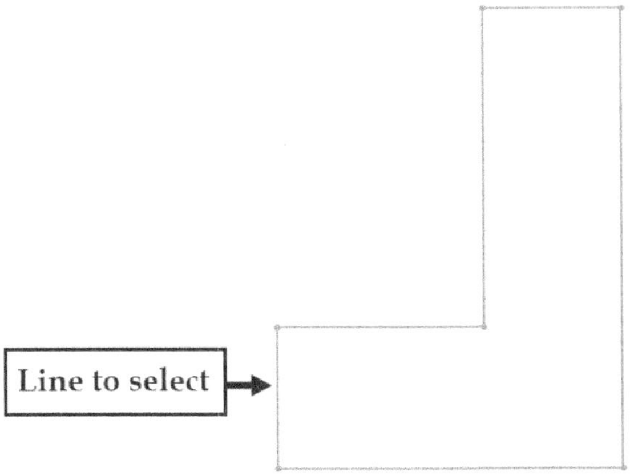

16. Leave the default options on the **Revolve** PropertyManager, and then click **OK**.

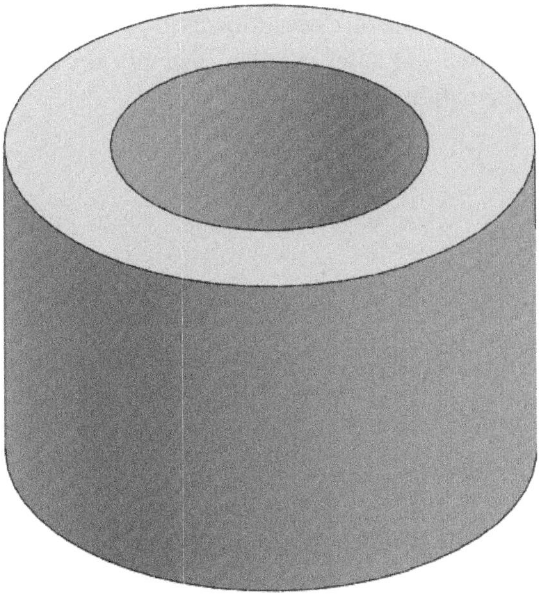

17. On the CommandManager, click **Features > Hole Wizard** .
18. On the PropertyManager, click the **Type** tab.
19. Click the **Straight Tap** icon under the **Hole Type** section.
20. Select **Standard > ISO**.
21. Select **Type > Tapped Hole**.
22. On the PropertyManager, under the **Hole Specification** section, select **Size > M10x1.25**.
23. On the PropertyManager, under the **End Condition** section, select **Blind** from the **End Condition** drop-down.
24. Type **15** in the **Blind Hole Depth** box.
25. Type **15** in the **Tap Thread Depth** box.
26. Under the **Options** section, click the **Cosmetic Thread** icon.
27. Click the **Positions** tab on the PropertyManager.
28. Click on the top face of the model.
29. Place the point on the top face at the location, as shown.

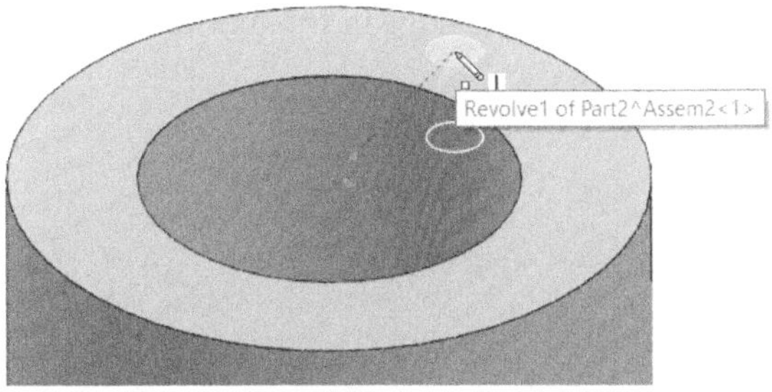

30. Fully-constrain the hole point by creating a centerline between the sketch origin and hole point. Next, add a dimension to the centerline.

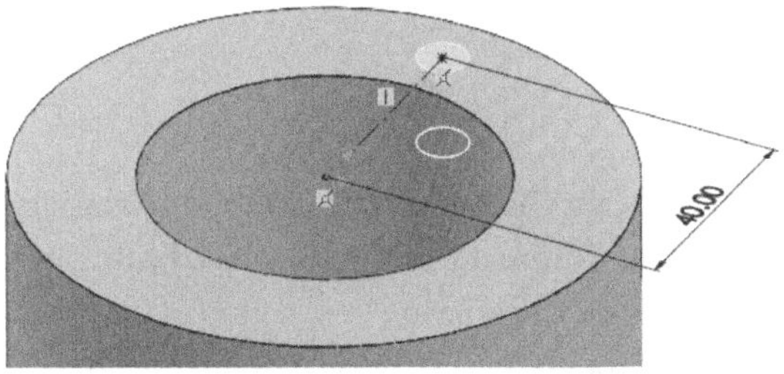

31. Click **OK** on the PropertyManager.
32. Create a circular pattern of the hole. The number of holes in the pattern is 8. The holes are equally spaced.

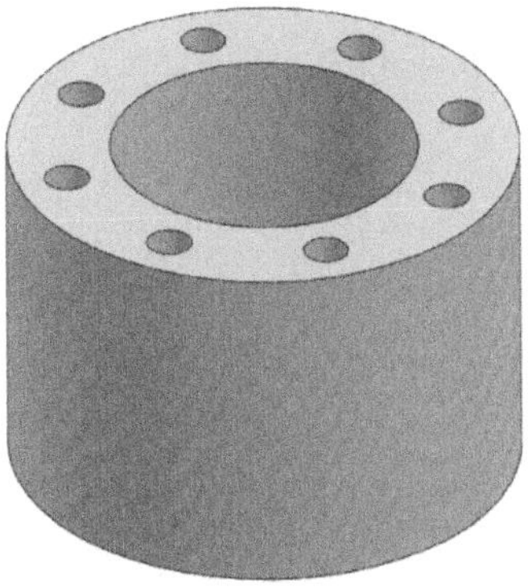

33. Click the **Edit Component** icon on the CommandManager to switch back to the **Assembly** mode.
34. On the CommandManager, click **Assembly > Insert Components** drop-down > **New Part** .
35. Click on the top face of the model.
36. On the **CommandManager**, click **Sketch > Convert Entities** .
37. Click on the top face of the model.
38. On the PropertyManager, click the **Select all inner loops** button, and then click **OK**.
39. Activate the **Convert Entities** command and select the outer edge of the model.
40. Click **OK**.

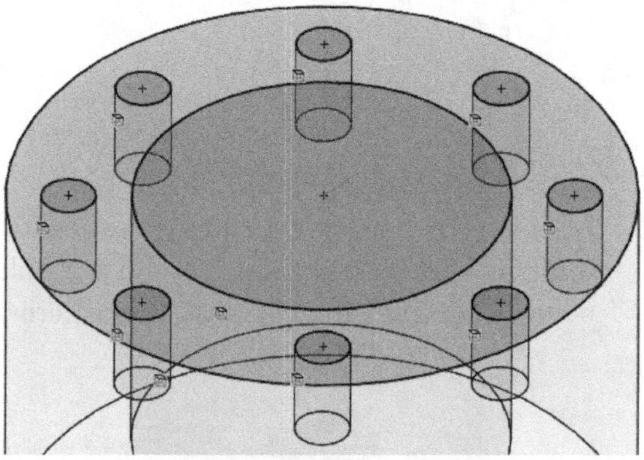

41. Click **Exit Sketch** on the CommandManager.
42. On the CommandManager, click **Features > Extruded Boss/Base**.
43. Select the newly created sketch from the graphics window.
44. On the PropertyManager, right click in the **Selected Contours** selection box located at the bottom.
45. Select **Clear Selections**.
46. Type **3** in the **Depth** box.
47. Click **OK**.

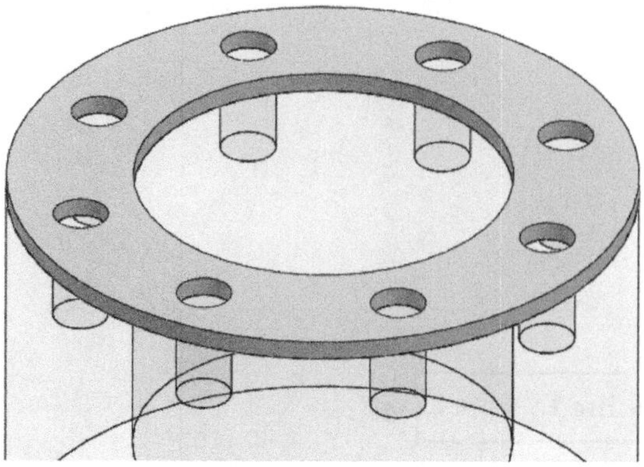

48. Click the **Edit Component** icon on the CommandManager to switch back to the **Assembly** mode.
49. On the CommandManager, click **Assembly > Insert Components** drop-down > **New Part** .
50. Click on the top face of the model.
51. Project the outer circular edge and the edges of the pattern holes.

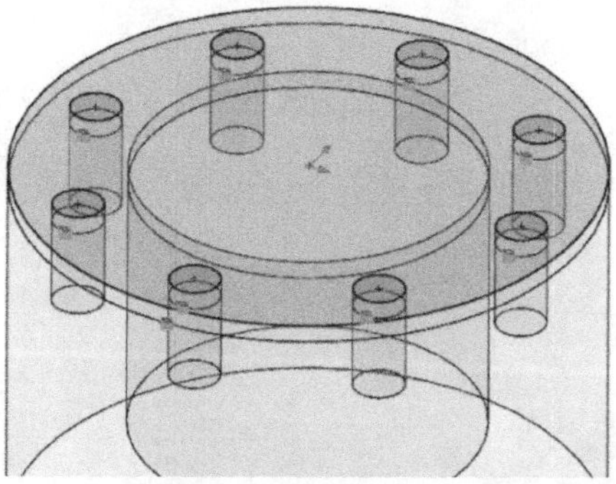

52. Exit the sketch and extrude it up to 13 mm depth.

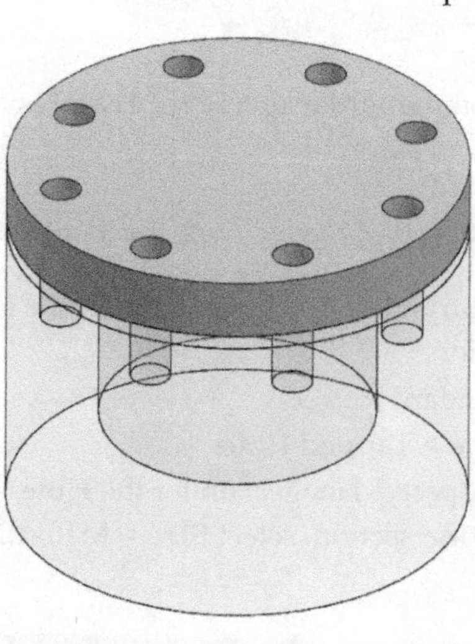

53. Click the **Edit Component** icon on the CommandManager to switch back to the **Assembly** mode.
54. On the CommandManager, click **Assembly > Insert Components** drop-down > **New Part** .
55. Click on the top face of the model.
56. Activate the **Convert Entities** command and select any one of the circular edges of the holes.
57. Click **OK** to project the selected edge.

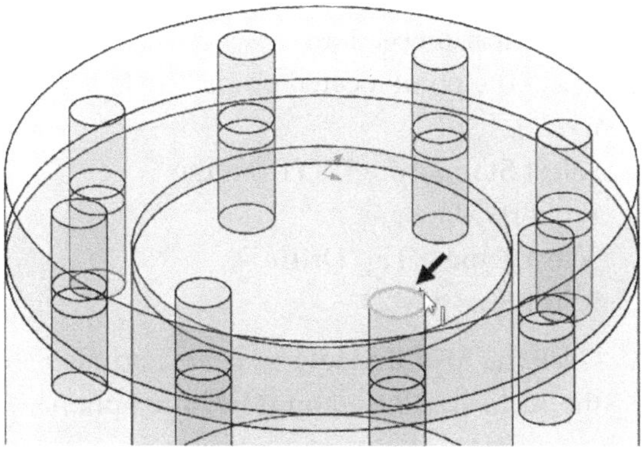

58. Exit the sketch and activate the **Extruded Boss/Base** command.
59. Extrude the sketch up to 30 mm in length in the downward direction.

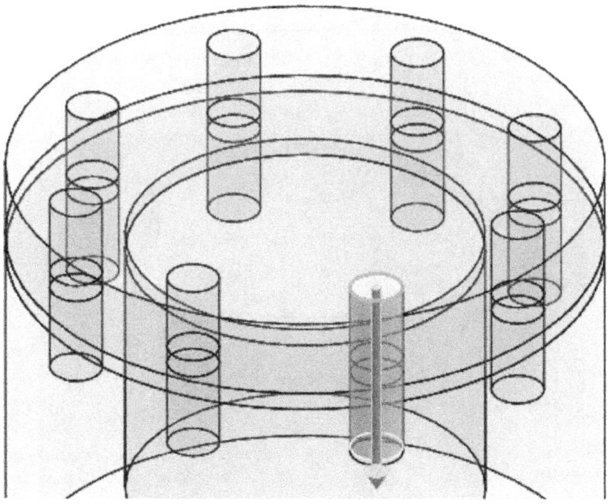

60. Start a sketch on the top face of the extruded boss/base feature.
61. On the CommandManager, click **Sketch > Polygon**, and then draw a hexagon.
62. Press and hold the Ctrl key, select the circle edge of the extrude feature and center point of the hexagon.
63. On the PropertyManager, click the **Concentric** icon.
64. Click **OK**.
65. Select any one of the edges of the hexagon.
66. On the PropertyManager, click the **Horizontal** icon, and then click **OK**.
67. Add a dimensional constraint to the hexagon.

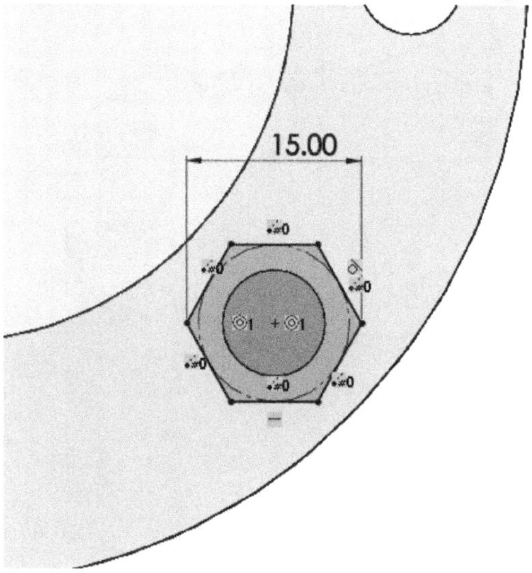

68. Exit the sketch and extrude it up to 5.7 mm in length.

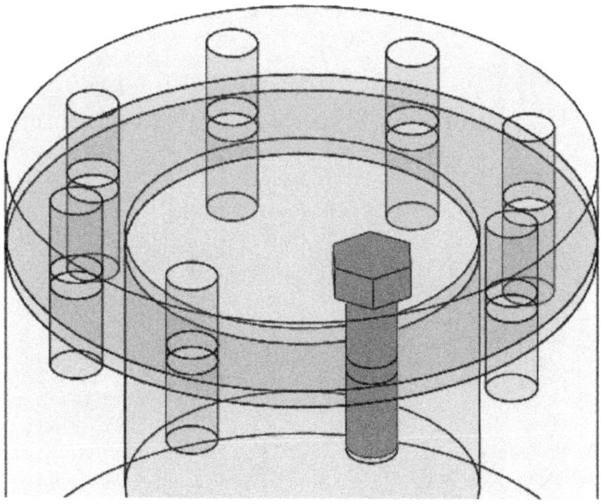

69. Click the **Edit Component** icon on the CommandManager to switch back to the **Assembly** mode.

70. On the CommandManager, click **Assembly > Pattern** drop-down > **Pattern Driven Component Pattern** .
71. Select the bolt to define the component to pattern.
72. On the PropertyManager, click in the **Driving Feature or Component** selection box.
73. In the graphics window, expand the FeatureManager Design Tree, select the **CirPattern1** of the first part.
74. Click **OK** to pattern the bolts.

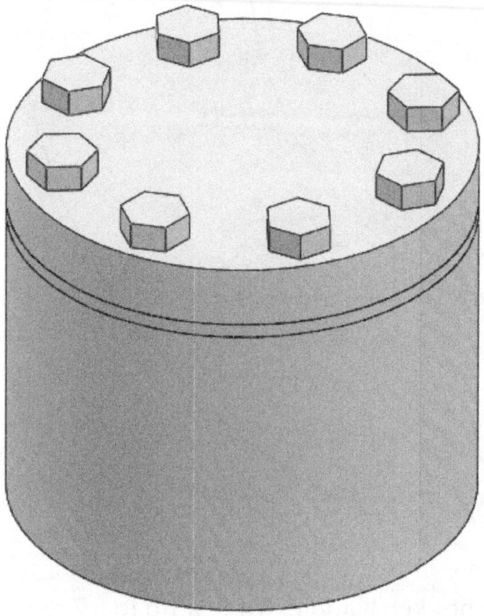

75. On the CommandManager, click **Assembly > Assembly Features drop-down > Hole Series** .
76. Click on the top face of the third component.

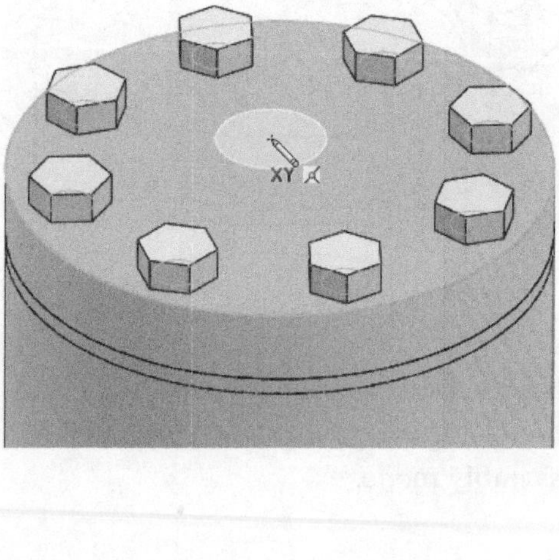

77. On the CommandManager, click **Display/Delete Relations** drop-down > **Add Relation** .
78. Select the outer circular edge of the assembly.
79. On the PropertyManager, click the **Concentric** icon.
80. Click **OK** on the **Add Relation** PropertyManager.
81. Click the **Start Hole** tab on the PropertyManager.
82. Click the **Hole** icon under the **Start Hole Specification** section.
83. Select the point located on the top face of the model.
84. Select **Standard > ISO** from the PropertyManagar.
85. Select **Type > Tap Drills**.
86. Select **Size > M24**.
87. Click the **Middle Hole** tab, and then check the **Auto size based on start hole** option.
88. Click the **End Hole** tab, and then check the **Auto size based on end hole** option.
89. Select the bottom part to define the End component.
90. Click **OK** to create the hole. You will notice a new item in the FeatureManager Design Tree.

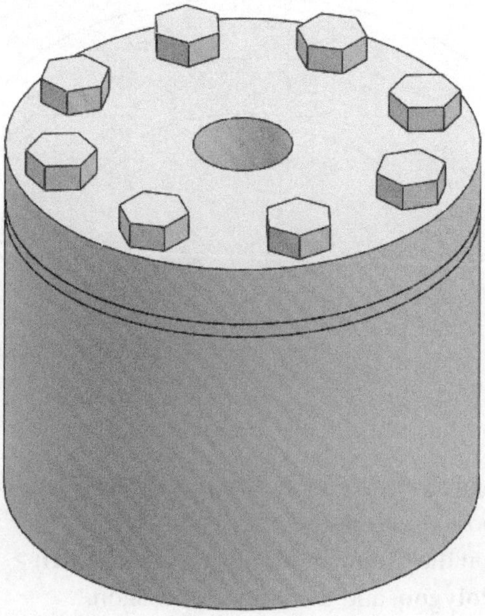

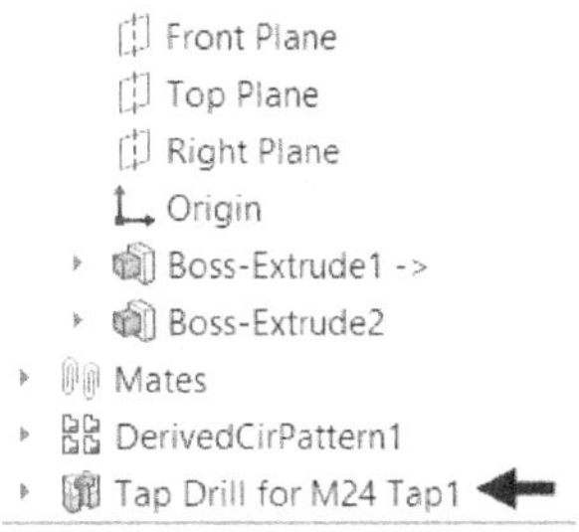

91. On the CommandManager, **Assembly > Exploded View** .
92. On the **Explode** PropertyManager, click the **Regular Step** icon.
93. Press and hold the CTRL key, and then select the bolts one-by-one.
94. Click on the Y-axis of the manipulator (axis pointing in the upward direction).

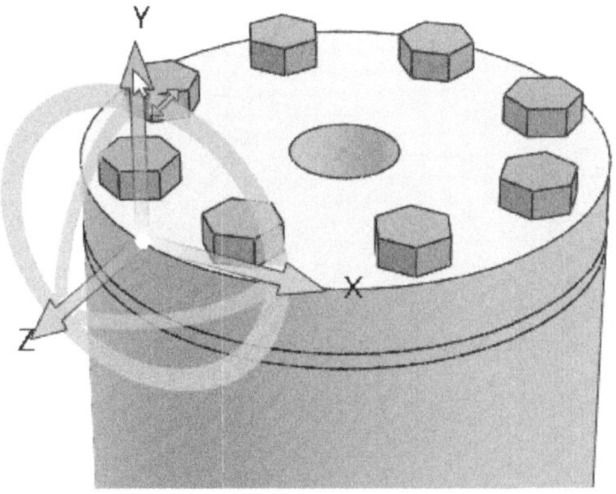

95. On the PropertyManager, type 100 in the **Explode Distance** box.
96. Click **Add Step**.

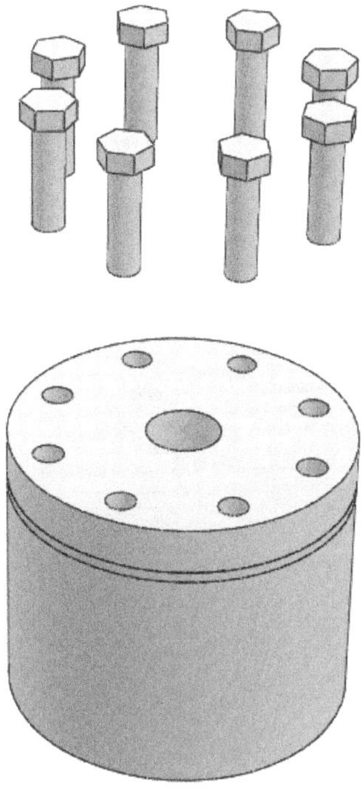

97. Select the cover plate and click on the Y-axis of the manipulator.
98. Type 50 in the **Explode Distance** box.
99. Click **Add Step**.

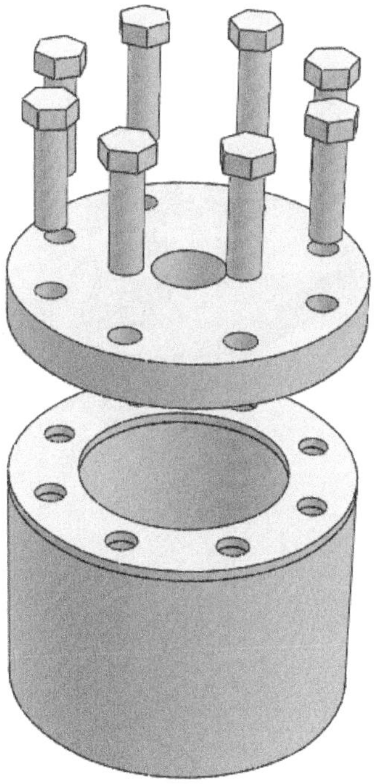

100. Likewise, explode the gasket up to 30 mm distance.
101. Click **OK** on the PropertyManager.

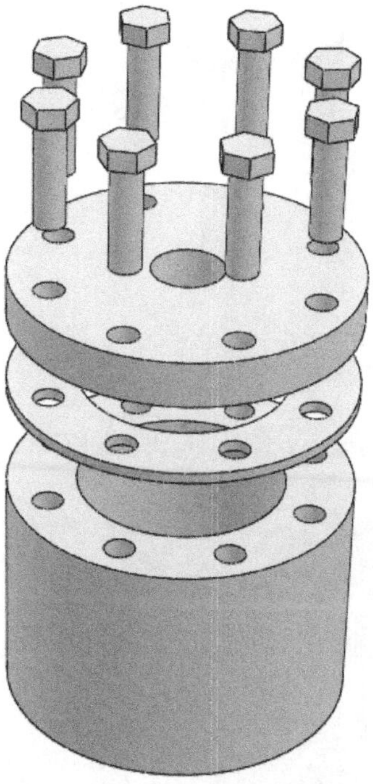

102.Save and close the assembly.

Questions

1. How do you start an assembly from an already opened part?
2. What is the use of the **Pattern Driven Component Pattern** command?
3. List the advantages of the Top-down assembly approach.
4. How do you create a sub-assembly in the Assembly environment?
5. Briefly explain how to edit components in an assembly.
6. What are the results that can be achieved using the **Mirror Components** command?
7. How do you redefine mates in SOLIDWORKS?
8. What are the uses of **Angle** mate?

Exercise 1

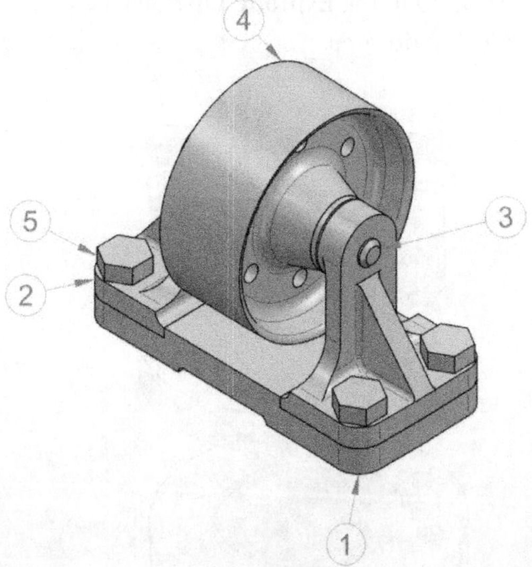

Item Number	File Name (no extension)	Quantity
1	Base	1
2	Bracket	2
3	Spindle	1
4	Roller-Bush assembly	1
5	Bolt	4

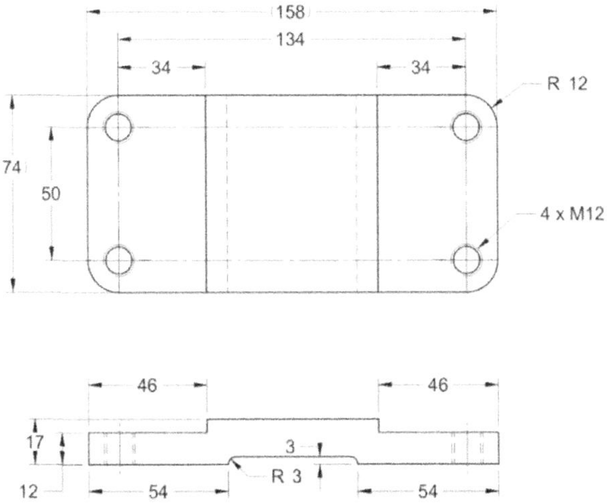

Base

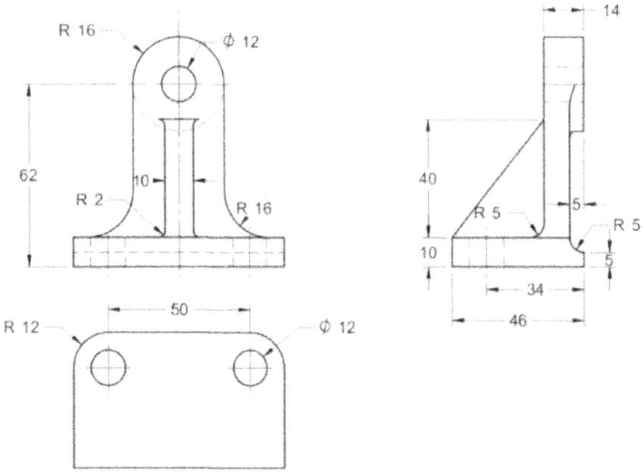

Bracket

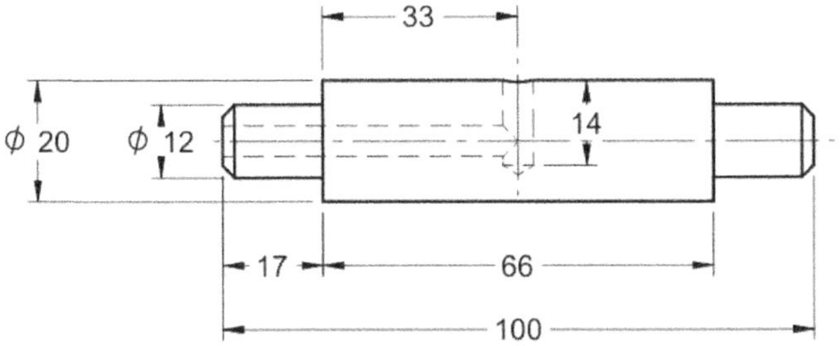

SPINDLE

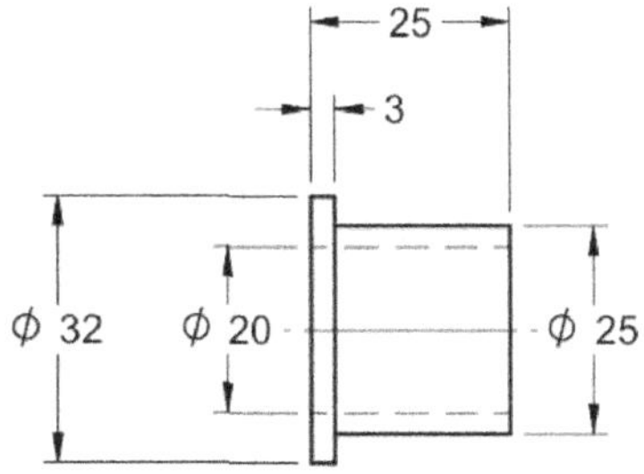

BUSH

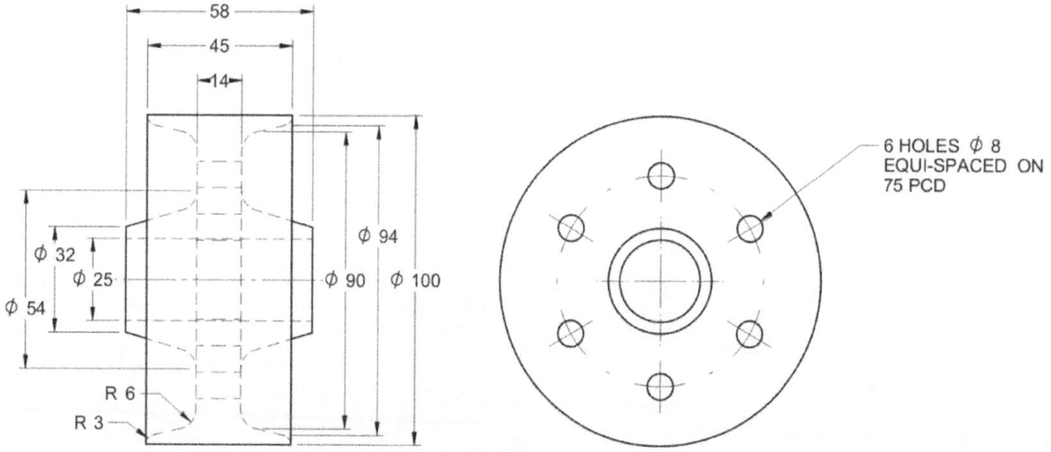
58
45
14
Φ 32
Φ 25
Φ 54
Φ 94
Φ 90
Φ 100
R 6
R 3
6 HOLES Φ 8
EQUI-SPACED ON
75 PCD

Roller

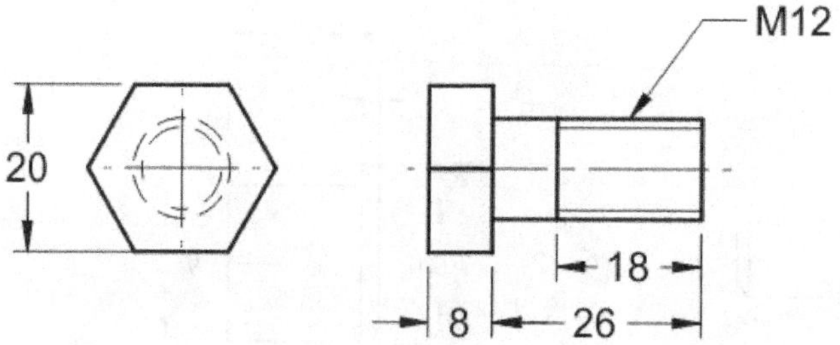
M12
20
18
8
26

Bolt

Chapter 11: Drawings

Drawings are used to document your 3D models in the traditional 2D format, including dimensions and other instructions useful for manufacturing purposes. In SOLIDWORKS, you first create 3D models and assemblies and then use them to generate drawings. There is a direct association between the 3D model and the drawing. When changes are made to the model, every view in the drawing will be updated. This relationship between the 3D model and the drawing makes the drawing process fast and accurate. Because of the mainstream adoption of 2D drawings of the mechanical industry, drawings are one of the three main file types you can create in SOLIDWORKS.

The topics covered in this chapter are:

- *Create model views*
- *Projected views*
- *Auxiliary views*
- *Section views*
- *Detail views*
- *Broken-Out views*
- *Break Lines*
- *Alternate Position View*
- *Display Options*
- *View Alignment*
- *Exploded View*
- *Bill of Materials*
- *Adding Balloons*
- *Model Items*
- *Ordinate Dimensions*
- *Chamfer Dimension*
- *Horizontal/Vertical Ordinate Dimensions*
- *Baseline Dimensions*
- *Angular Running Dimensions*
- *Center Marks*
- *Centerlines*
- *Hole Callouts*
- *Notes*

Starting a Drawing

To start a new drawing, click the **Drawing** icon on the **Welcome** window. On the **Units and Dimension Standard** dialog, select **Units > MMGS** and **Dimension standard > ISO** (the **Units and Dimension Standard** dialog appears only if you are creating a drawing for the first time). Next, click **OK**. On the **Sheet Format/Size** dialog, select the sheet size from the list box, and then click **OK.**

Starting a new drawing using the New SOLIDWORKS Document dialog

Click the **New** icon on the **Quick Access Toolbar**. On the **New SOLIDWORKS Document** dialog, click the

Drawing button, and then click **OK**; the **Sheet Format/ Size** dialog appears. If you want to start the drawing in any other standard, uncheck the **Only show standard formats** option. Next, select the required sheet size from the list box. Use the **Custom sheet size** option if you want to type-in the sheet's width and height. Click **OK** after defining the sheet size.

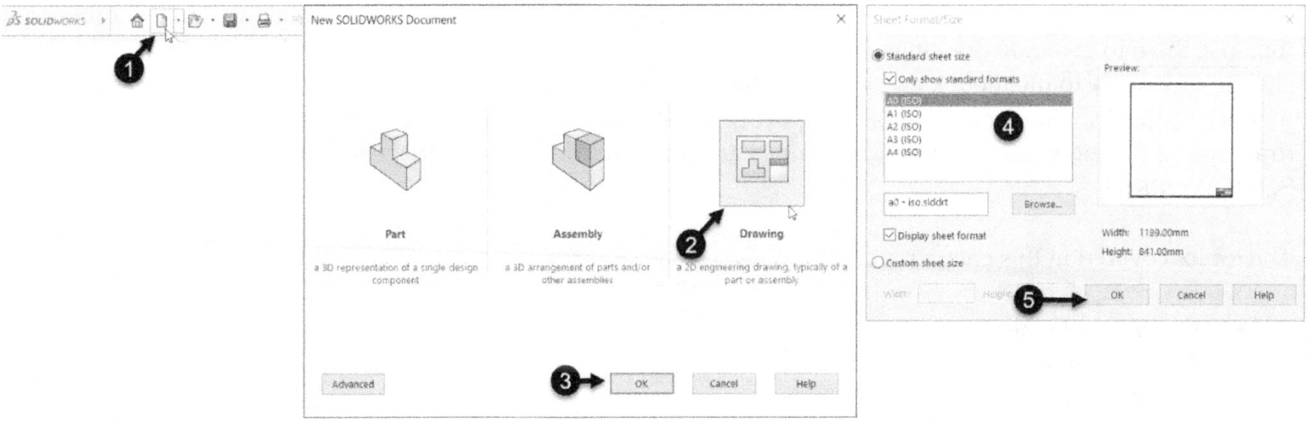

Creating a Drawing from an already opened Part or Assembly

If you already have a part or assembly opened, you can click **New > Make Drawing from Part/Assembly** on the **Quick Access Toolbar**; the **Sheet Format/Size** dialog appears. On this dialog, click the **Browse** button to access different sheet formats. Select any one of the sheet formats and click **Open**. Click **OK** on the **Sheet Format/Size** dialog to start a new drawing.

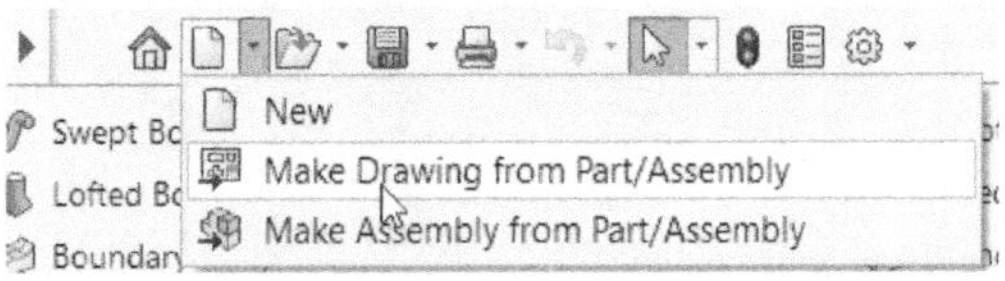

Model View

There are different standard views available in a 3D part, such as front, right, top and isometric. In SOLIDWORKS, you can create these views using the **Model View** command. This command is activated automatically if you have created a drawing with any part opened. If it is not activated, click **Drawing > Model View** on the CommandManager. The **Model View** PropertyManager appears. Click the **Browse** button, and then browse to the location of the part or assembly. Next, double-click on the part or assembly file; a model view will be attached to the pointer. Also, the **Model View** PropertyManager displays all the options.

On the **Model View** PropertyManager, under the **Orientation** section, check the **Create multiple views** option. Next, select the base view, and then click on the icons that represent the standard views that are to be created. After selecting the standard views, click **OK** on the **Model View** PropertyManager.

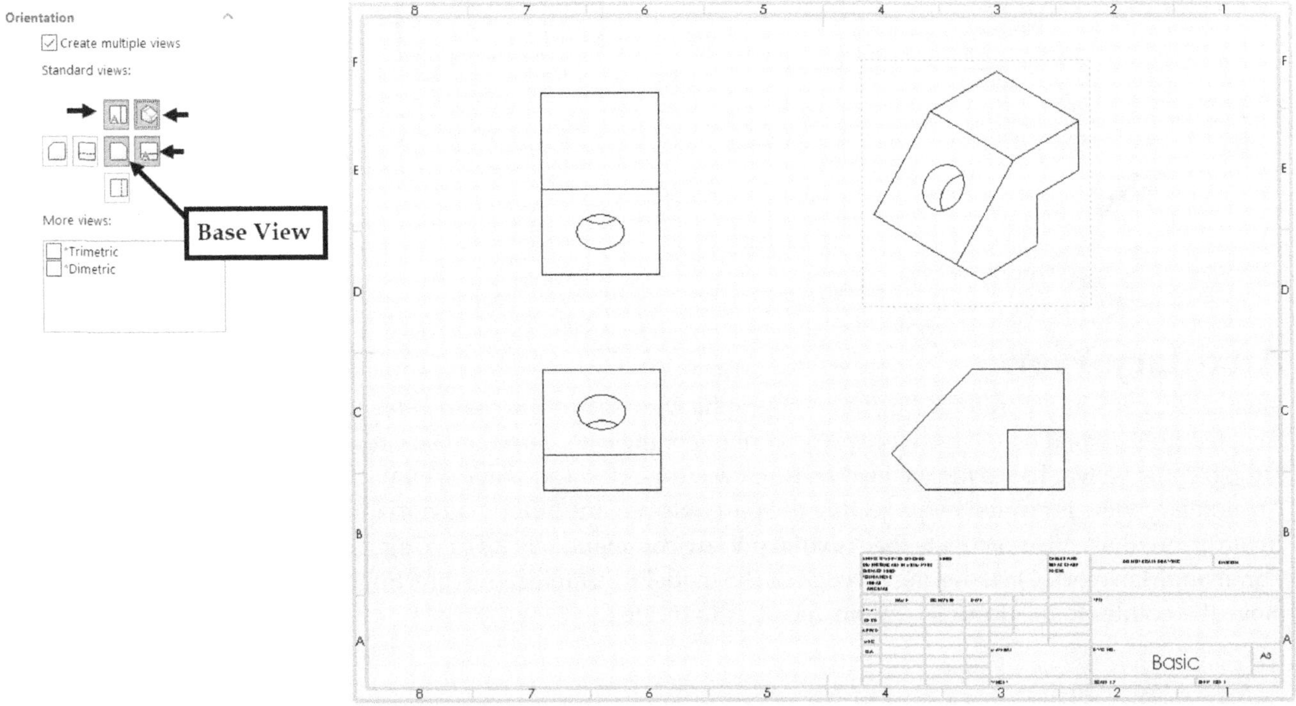

You can change the orientation of a base view even after creating all the views associated with it. To do this, select the base view and click on the required orientation on the PropertyManager. Click **Yes** on the **SOLIDWORKS** message box; the base view and all its associated views are changed. Click **Close Dialog** on the PropertyManager.

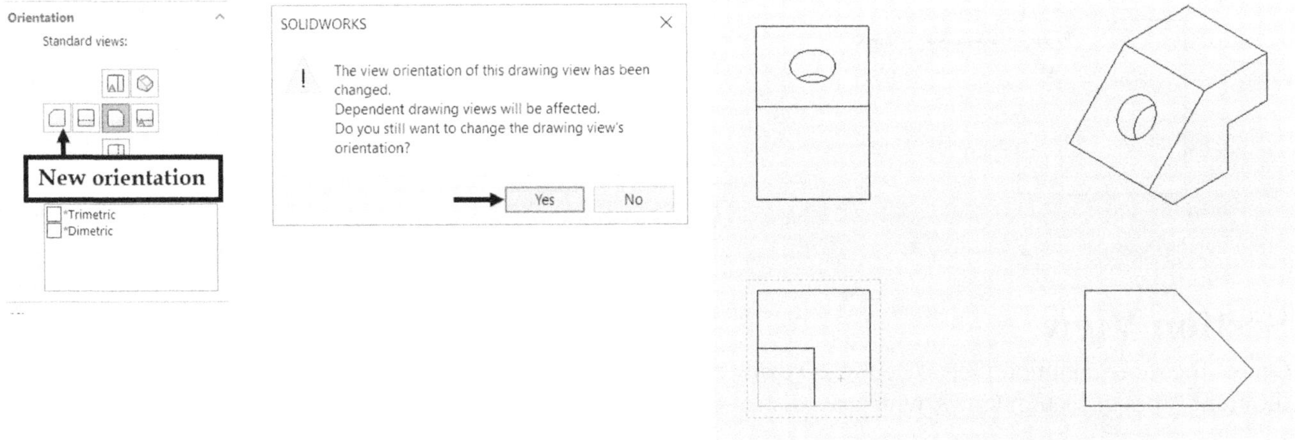

Projected View

After you have created the first view in your drawing, a projected view is one of the easiest views to create. Activate the **Projected View** command (click **Drawing > Projected View** on the CommandManager) and select a view you wish to project from. Next, move the pointer in the direction you wish to have the view be projected, and then click to specify the location; the projected view will be created. Click the right mouse button to deactivate this command.

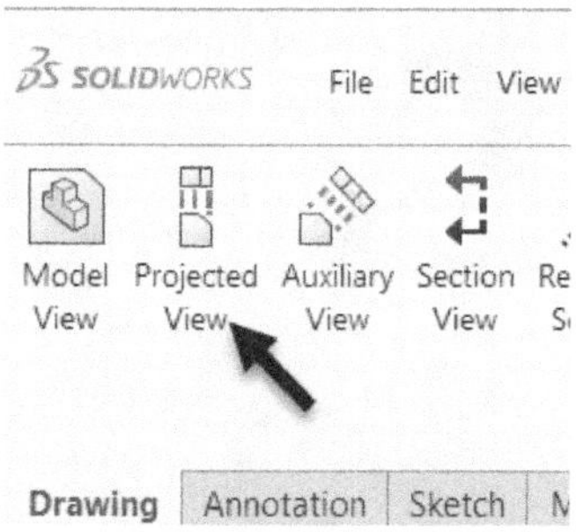

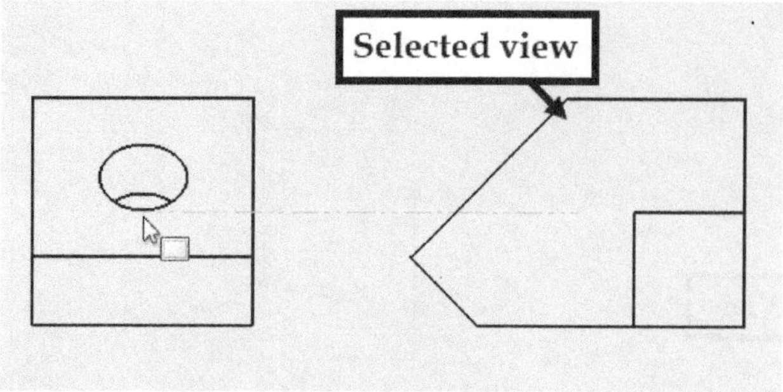

Auxiliary View

Most of the parts are represented using orthographic views (front, top and side views). However, many parts have features located on inclined faces. You cannot get the true shape and size for these features by using the orthographic views. To see the inclined features' accurate size and shape, you need to create an auxiliary view. An auxiliary view is created by projecting the part onto a plane other than horizontal, front or side planes. To create an auxiliary view, activate the **Auxiliary View** command (click **Drawing > Auxiliary View** on the CommandManager). Click the angled edge of the model to establish the direction of the auxiliary view. Next, move the pointer to the desired location and click to locate the view.

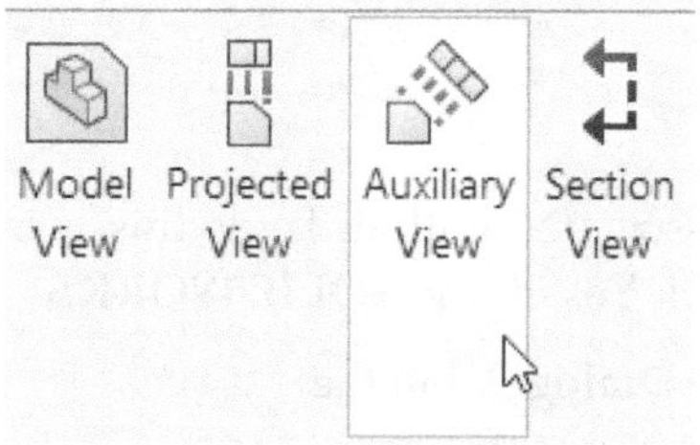

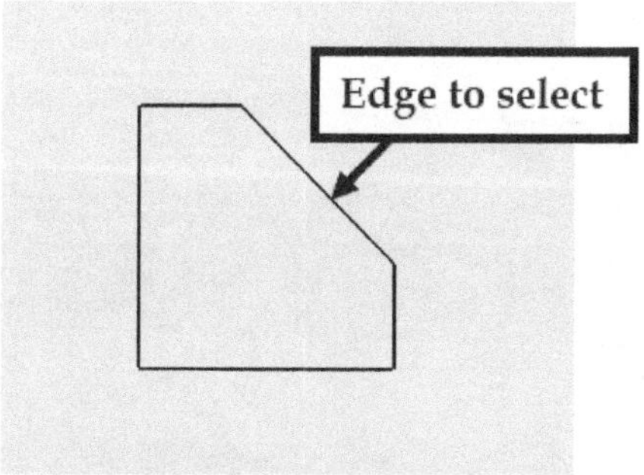

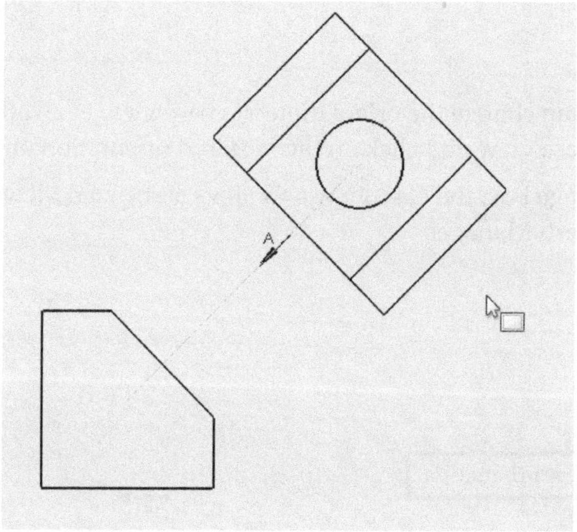

Section View

One of the more common views used in 2D drawings is the section view. SOLIDWORKS allows you to create different types of section views, which are discussed in the following sub-sections.

Create a Vertical/Horizontal Section View

Creating a vertical/horizontal section view in SOLIDWORKS is very simple. Once a view is placed on the drawing sheet, you need to activate the **Section View** command (click **Drawing > Section View** on the CommandManager) and click on the **Vertical** or **Horizontal** cutting plane icon available on the PropertyManager. Specify the position of the cutting plane on the base view. Next, click **OK** on the context toolbar. Move the pointer and click to position the section view.

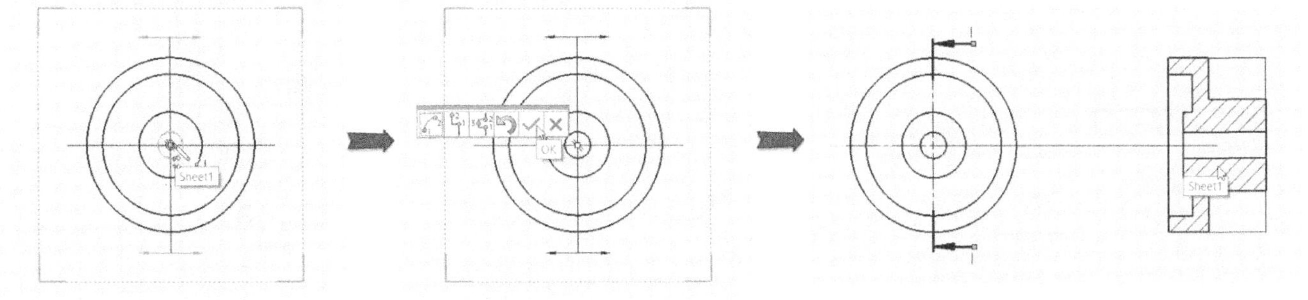

Click the **Flip Direction** button on the PropertyManager, if you want to reverse the direction in which the section view is displayed. Type-in a letter in the **Label** box, and then click **OK**.

Creating an Offset section view

The offset section view can display multiple features of a model in a single section view. Activate the **Section View** command and click the **Vertical** or **Horizontal** icon in the **Cutting Plane** section of the PropertyManager. Next, select a point on the base view to define the location of the cutting plane. Click the **Single Offset** icon on the context toolbar. Move the pointer along the cutting plane and click to define the point at which the cutting plane is offset. Select the centerpoint of the feature to be included in the section view. Click **OK** on the context toolbar, move the pointer and click to position the section view. Type-in a letter in the **Label** box, and then click **OK** on the PropertyManager.

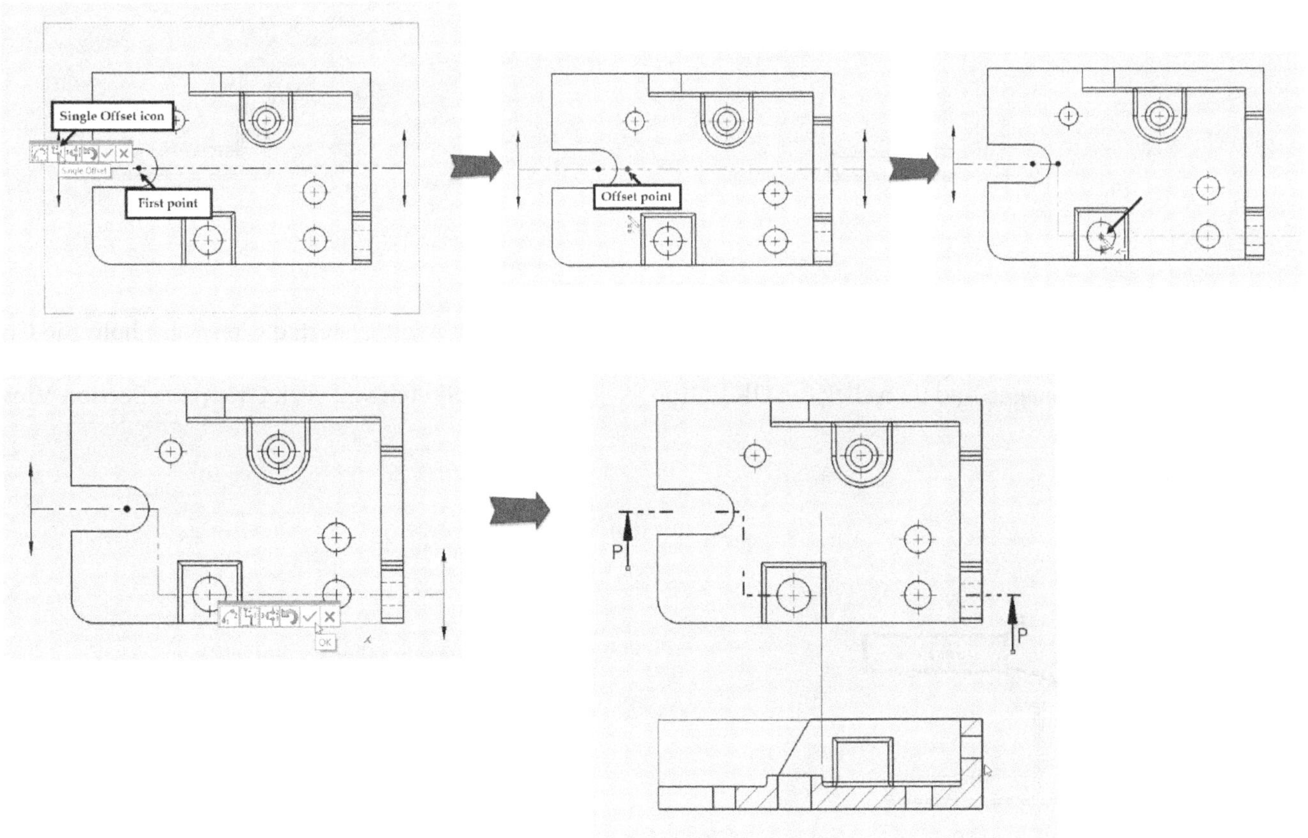

You can also use the **Notch Offset** option to create an offset section view. The selection sequence for this option is shown in the figure below.

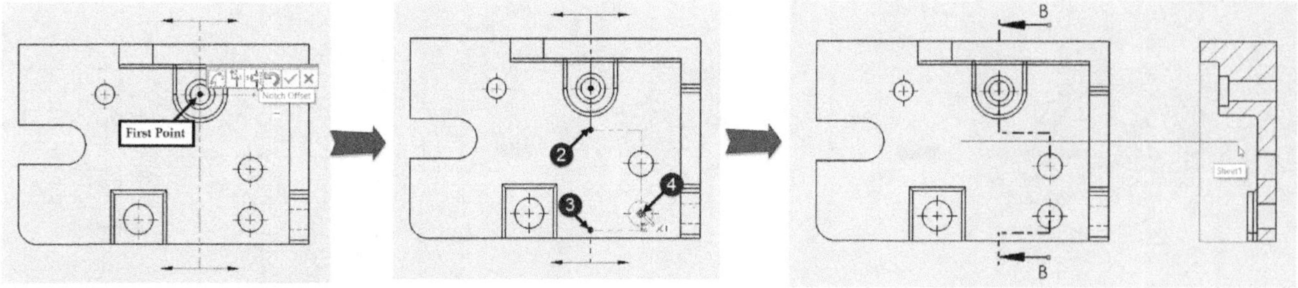

Creating an Auxiliary section view

The Auxiliary section view allows you to show the details of an inclined part clearly. To create an auxiliary section view, activate the **Section View** command (on the CommandManager, click **Drawing > Section View**). Next, click the **Auxiliary** icon under the **Cutting Line** section of the PropertyManager. Specify the location of the cutting plane. Next, move the pointer and click to specify the inclination angle of the cutting plane.

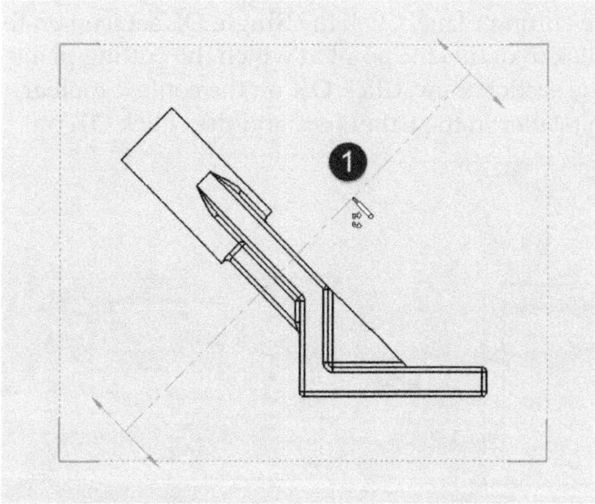

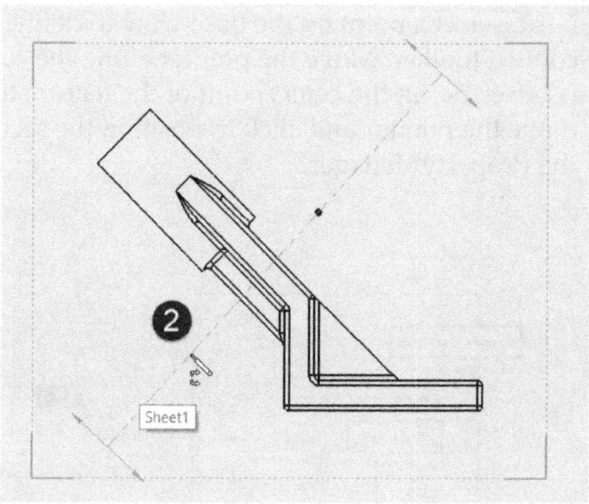

Click the **Edit sketch** button on the PropertyManager; the **Sketch** environment is started. Press and hold the Ctrl key, and then select the cutting plane. Next, click on the edge of the part, as shown. Select the **Perpendicular** icon from the PropertyManager, and then click the **OK** button. Next, click **Exit Sketch** and click **OK** on the **Section View** dialog.

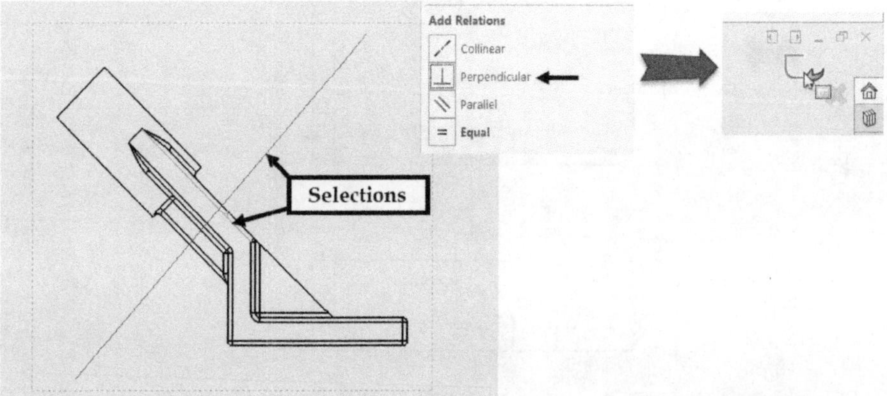

Move the pointer and click to position the auxiliary section view. Type-in a letter in the **Label** box on the PropertyManager.

Check the **Slice Section** option in the **Section View** section of the PropertyManager to display only the cutting plane's geometry. Check the **Emphasize outline** option to thicken the outline of the section view. Click **OK** on the PropertyManager.

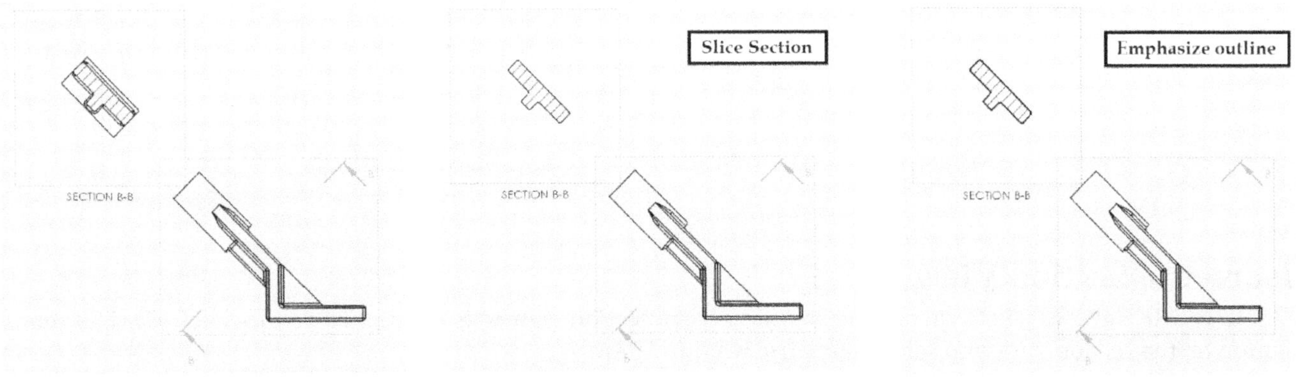

Creating an Aligned Section View

Use the **Aligned** option to create an aligned section view. Activate the **Section View** command and click the **Aligned** icon from the **Cutting Line** section. Specify the center point of the cutting plane. Specify the location of the first line of the section view. Move the pointer and specify the location of the second fold line. Click the green check. Next, move the pointer and click to position the aligned section view.

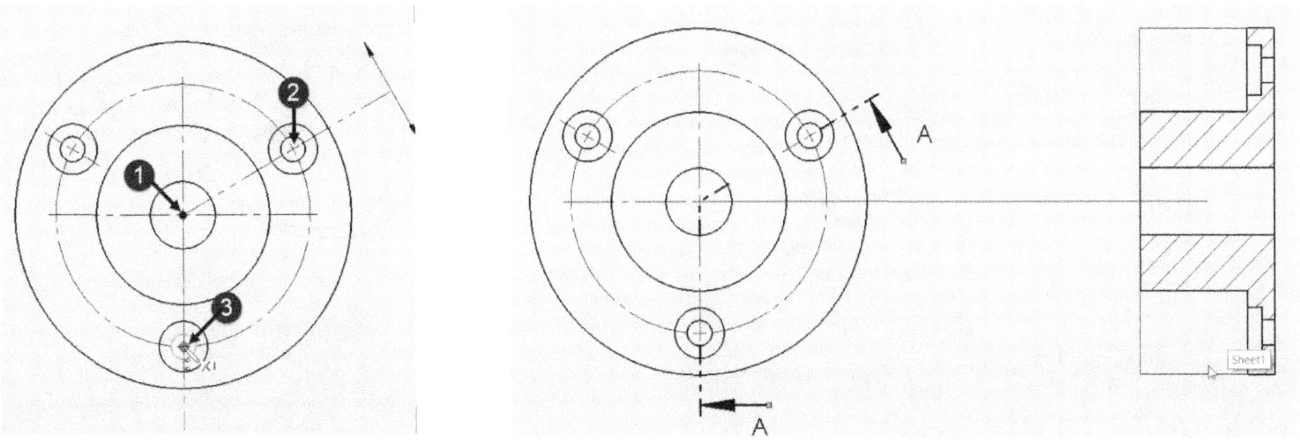

Creating a Half Section View

The Half section view is used to show half of the outside and half of the inside of an object. To create this view, activate the **Section View** command (click **Drawing > Section View** on the CommandManager) and click the **Half Section** tab on the PropertyManager. Next, click on any one of the icons available on the PropertyManager to define the **Half Section** type. The top four icons are used to create a half-section view in the horizontal direction. Whereas the bottom four icons are used to create the half-section view in the vertical direction.

Specify the location of the cutting plane by selecting a point on the base view. Next, move the pointer and click to position the half-section view.

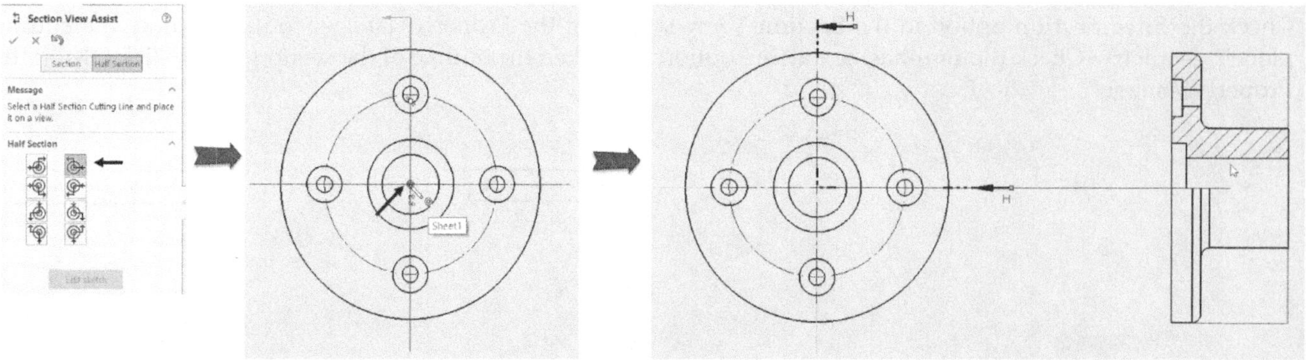

Isometric Section View

SOLIDWORKS allows you to change the orientation of a section view to Isometric. To do this, click the right mouse button on the section view and select **Isometric Section View**.

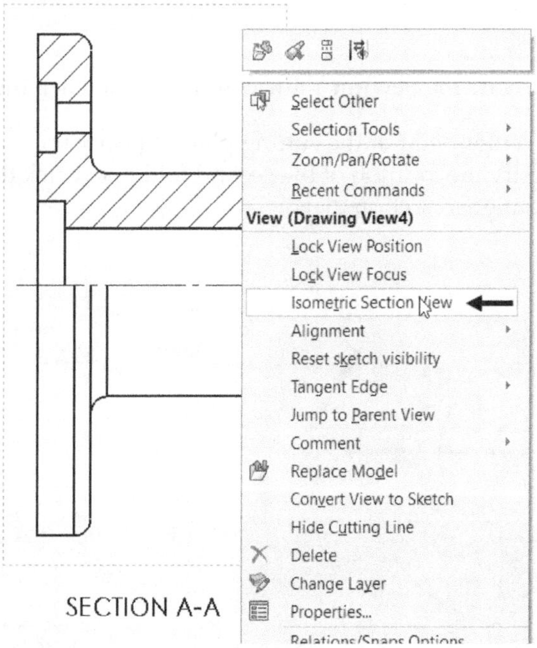

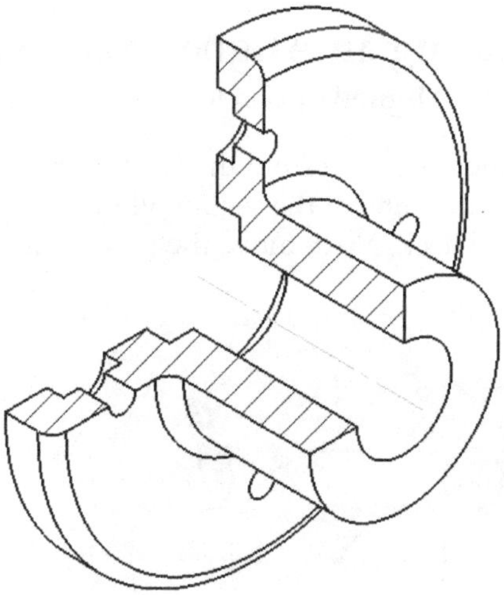

Excluding components from a section view

When creating a section view of an assembly, you can choose to exclude one or more components from the section cut. For example, to exclude the piston of a pneumatic cylinder, activate the **Section View** command, and then click the **Horizontal** icon on the PropertyManager. Specify the location of the cutting plane, as shown. Click **OK** on the Context toolbar; the **Section View** dialog appears. Click the **FeatureManager Design Tree** tab. On the FeatureManager Design Tree, expand the design in the sequence, as shown. Next, select the piston from the design, and then click **OK** on the **Section View** dialog. Move the pointer and click to locate the section view. You will notice that the piston is not cut.

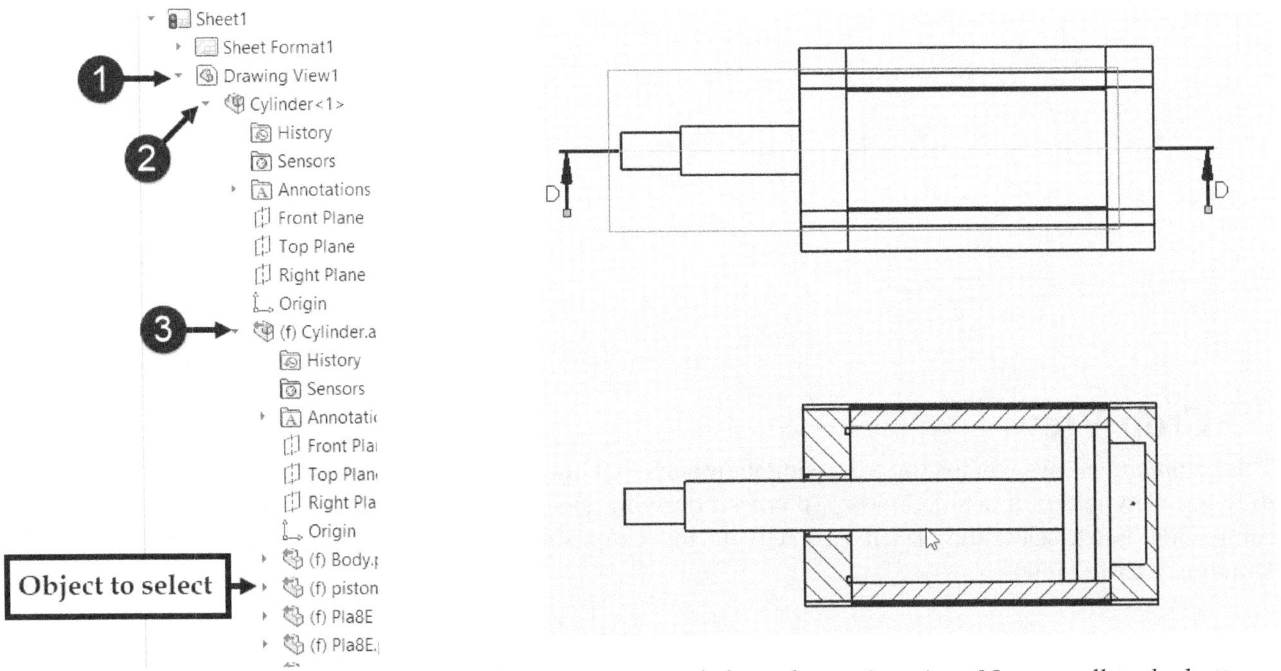

If you want to include the excluded item in the section view, click on the section view. Next, scroll to the bottom of the PropertyManager, and then click the **More Properties** button. On the **Drawing View Properties** dialog, click the **Section Scope** tab, right click on the items in the **Excluded components/rib features** list, and then select **Delete**. Next, click **OK**.

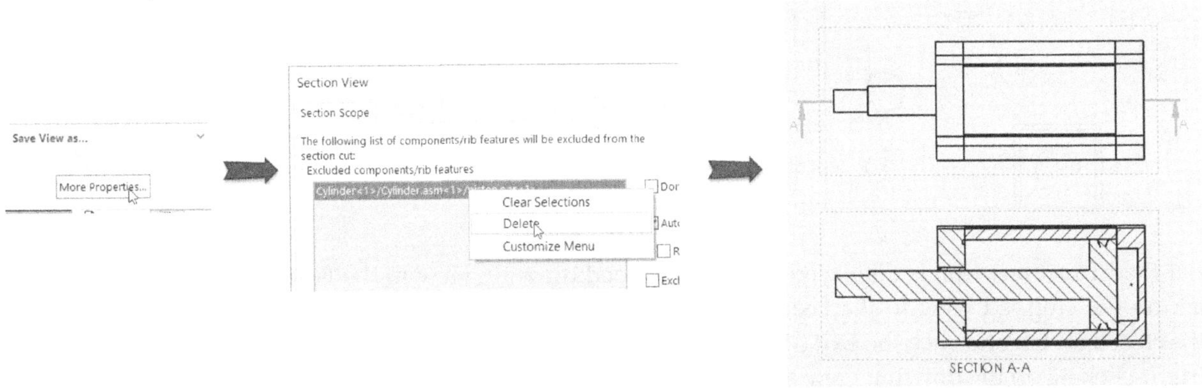

Detail View

If a drawing view contains small features that are difficult to see, a detailed view can be used to zoom in and clarify. To create a detailed view, activate the **Detail View** command (click **Drawing > Detail View** on the CommandManager); this automatically activates the circle tool. Draw a circle to identify the area that you wish to zoom into. Once the circle is drawn, set the **Scale** value on the **Scale** section of the PropertyManager. Next, move the pointer and click to locate the view; the detail view will appear with a label.

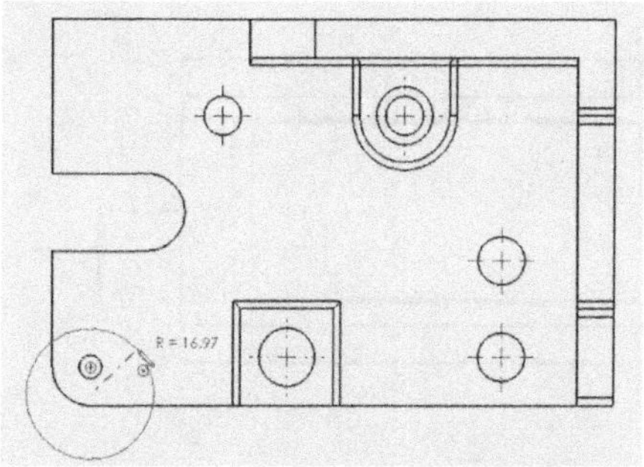

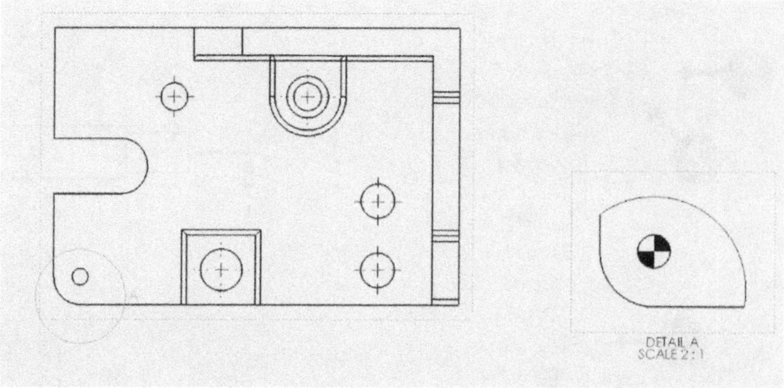

Crop View

This command allows you to crop a drawing view such that only a portion of it is visible. You can crop all types of drawing view except a detailed view. To crop a drawing view, first create a closed sketch using the sketching commands. Next, select the sketch and activate the **Crop View** command (On the CommandManager, click **Drawing > Crop view**).

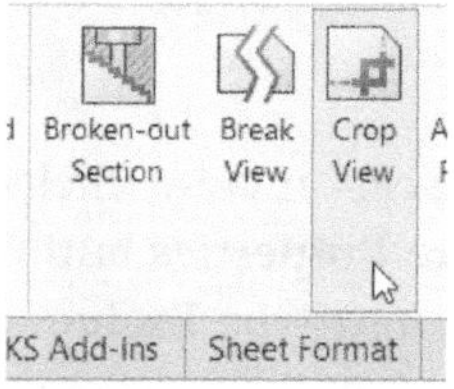

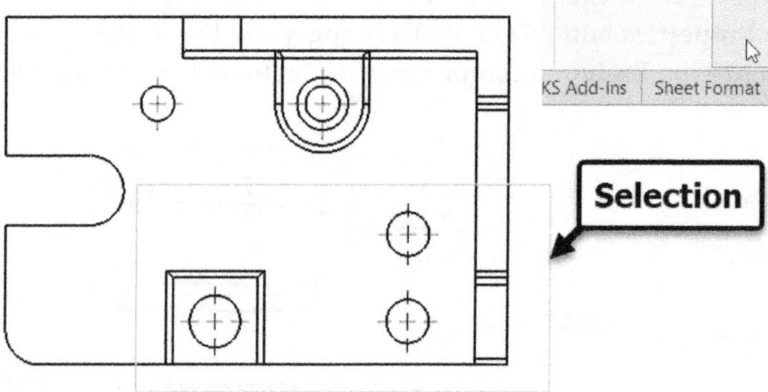

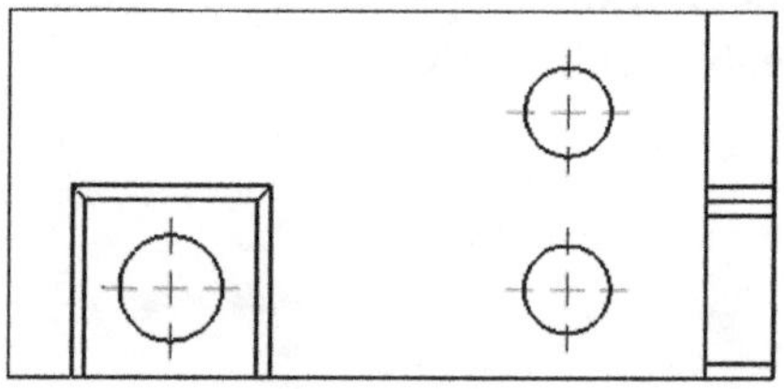

Notice that the **Crop View** icon is displayed next the cropped drawing view in the FeatureManager Design Tree. Right-click on the cropped view in the FeatureManager Design Tree and select **Crop view > Edit Crop**. Next, modify the closed sketch and click the **Exit Sketch** icon on the top-right corner. Likewise, you can remove the crop view by right-clicking on the drawing view and selecting **Crop view > Remove Crop**.

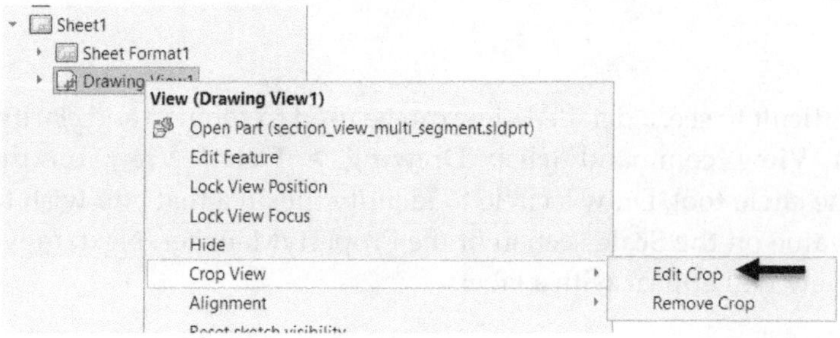

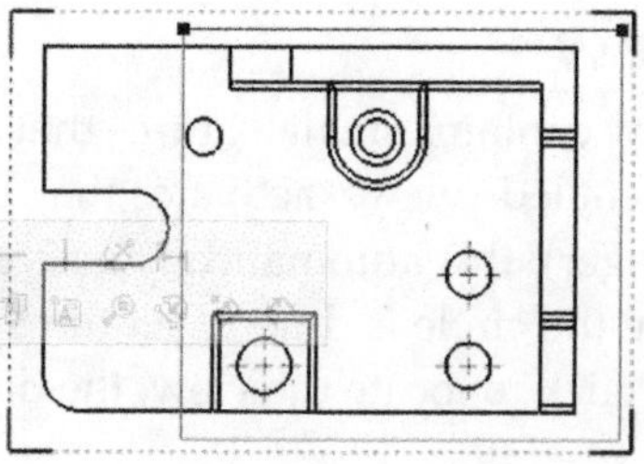

Add Break lines

Break lines are added to a drawing view, which is too large to fit on the drawing sheet. They break the view so that only important details are shown. To add break lines, select the view and activate the **Break View** command (click **Drawing > Break View** on the CommandManager); the **Break View** PropertyManager appears. On this PropertyManager, click the **Add vertical break lines** or **Add horizontal break lines** icon and define the **Cut**

direction. Type in the desired value in the **Gap Size** box and move your pointer to the area of the view where you would like to start the break. Click once to locate the beginning of the break. Move the pointer and click again to locate the end of the break. On the PropertyManager, click any one of the icons available in the **Break line style** section. Click **OK** on the PropertyManager; the view is broken.

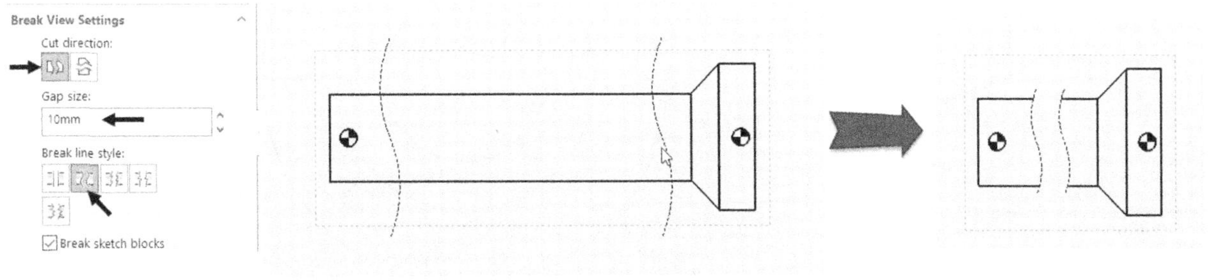

Broken Out Section View

The **Broken-Out Section** command alters an existing view to show the hidden portion of a part or assembly. This command is very useful in showing the features or parts hidden inside a part or assembly view. You need to have a closed profile to break-out a view. For example, if you want to show the inside portion of a part shown in the figure, activate the **Broken-Out** command (click **Drawing > Broken-Out Section** on the CommandManager) and specify the first point of the closed profile on the drawing view to be broken out. Next, draw a closed profile on the selected drawing view. Check the **Preview** option, and then specify the depth of the cutout in the **Depth** box available on the PropertyManager. Click **OK** to create the broken-out section view.

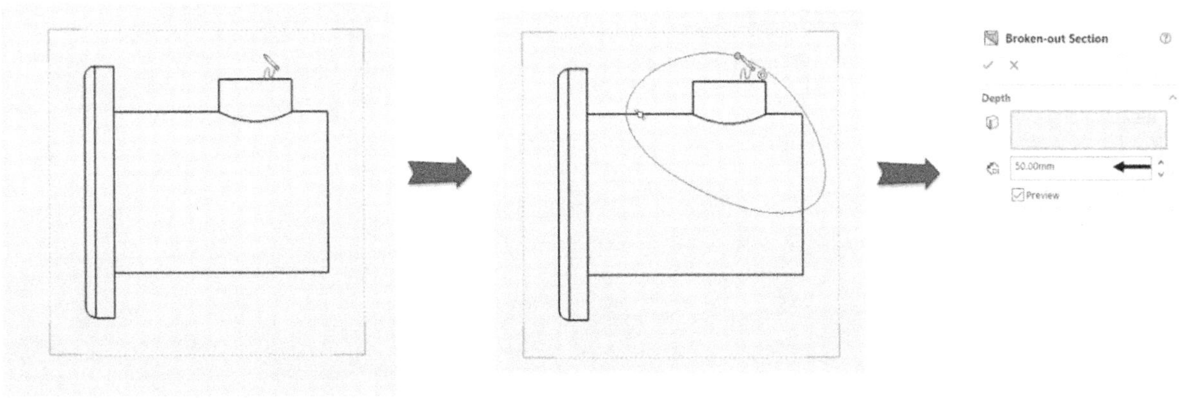

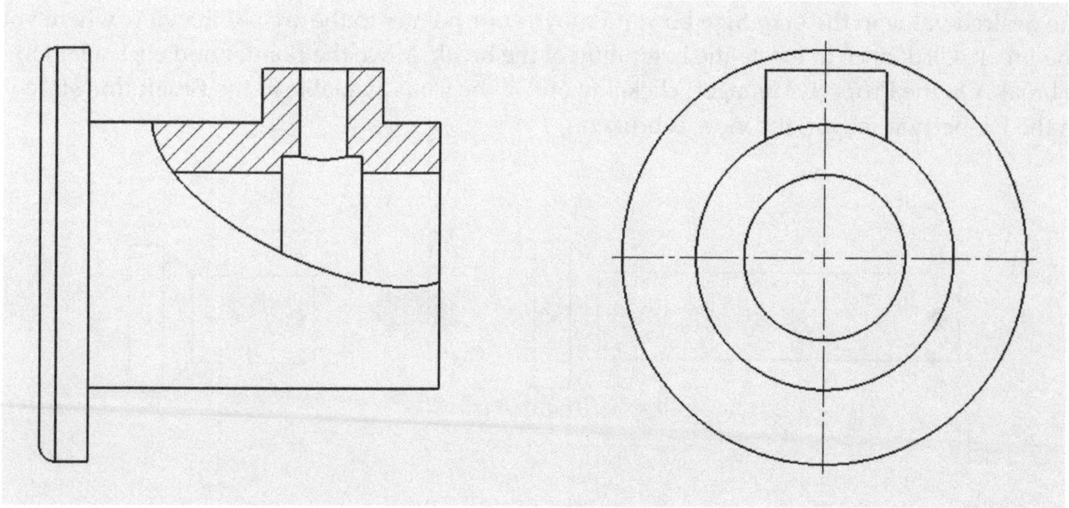

Removed Section View

The **Removed Section View** command allows you to create sliced section views. Activate this command (On the CommandManager, click **Drawing > Removed Section** (or) click **Insert > Drawing View > Removed Section** on the Menu bar) and select the side edges of the drawing view. Next, select an option from the **Cutting Line Placement** section: **Automatic** or **Manual**. For example, select the **Automatic** option and move the pointer; the cutting line is attached to the pointer and changes the orientation based on the side edges. Click to position the cutting line; the section view is attached to the pointer. Move the point and click to place the removed section view.

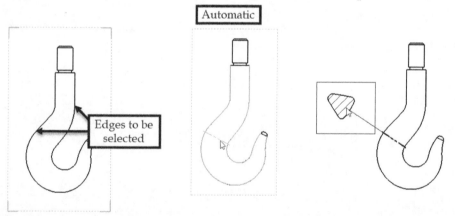

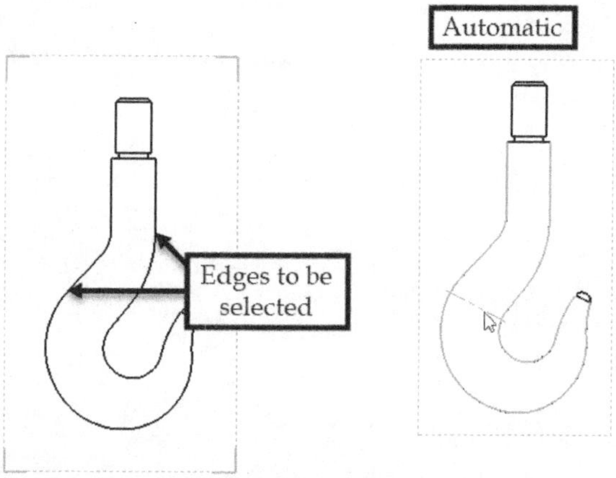

Select the cutting line and press the Delete key on your keyboard. Click **Yes** on the **Confirm Delete** dialog.

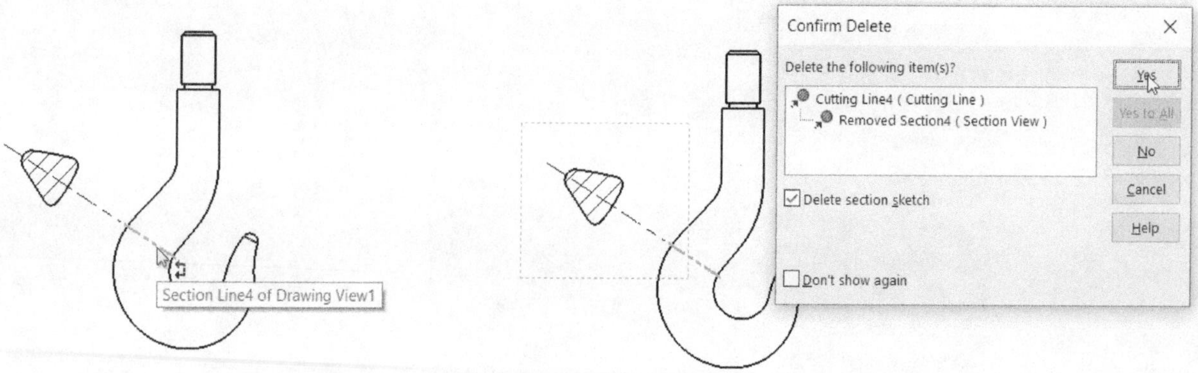

Activate the **Removed Section View** command and select the side edges of the drawing view. Next, select the **Manual** option from the **Cutting Line Placement** section. Move the pointer and select a point on the first side edge. Next, move the pointer horizontally and select a point on the second side edge; the cutting line is defined. Move the pointer horizontally and click to place the removed section view. Next, click **OK** on the **Removed Section** PropertyManager.

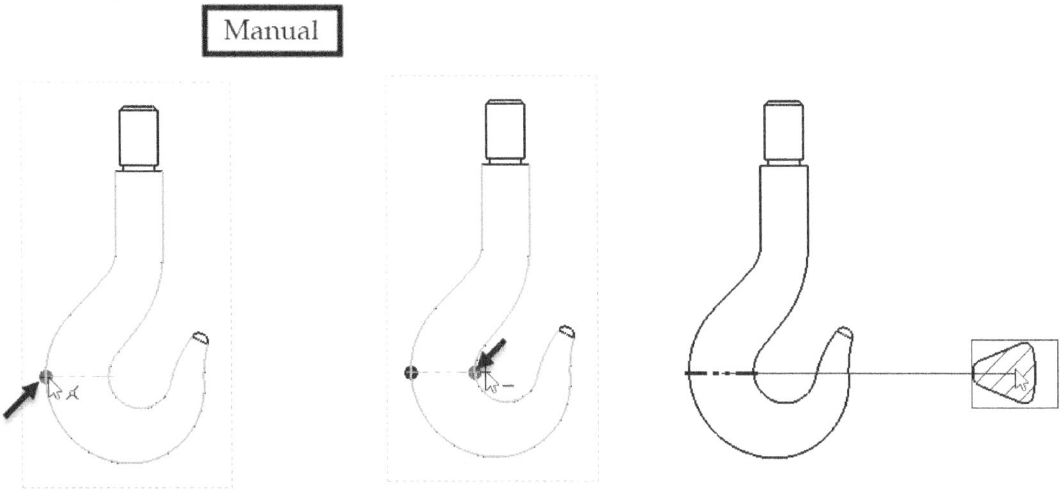

Alternate Position View

The Alternate position view is useful when you have an assembly of moving components. For example, the left plate in the assembly is located at 90 degrees to the bottom plate. You can show the plate positioned at a different angle with respect to the bottom plate. To do this, right click on the drawing view in the FeatureManager Design Tree, and then select **Open Assembly**.

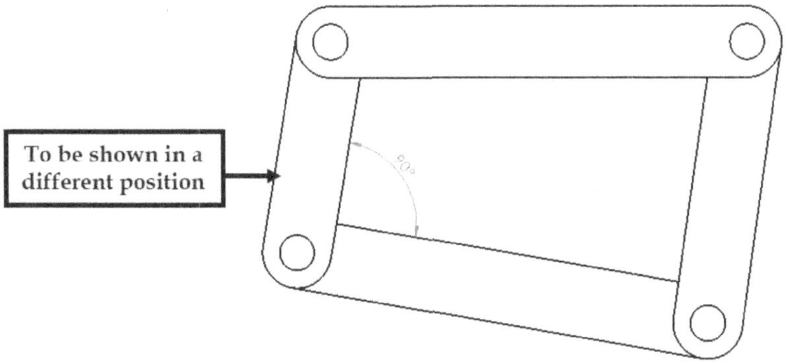

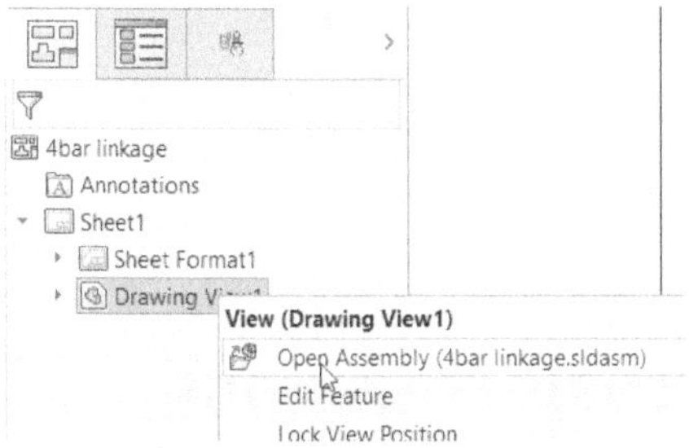

In the assembly file, expand the **Mates** node in the FeatureManager Design Tree and select the **Angle1**; the angle is displayed in the graphics window. In the graphics window, click the right mouse button on the angle dimension and select **Configure Dimension**. On the **Modify Configurations** dialog, click in the ***Creates a new configuration*** box and type 60-degree configuration. Click in the **Angle1** box next to the 60-degree configuration box and type 60. Next, click **Apply** and **OK**.

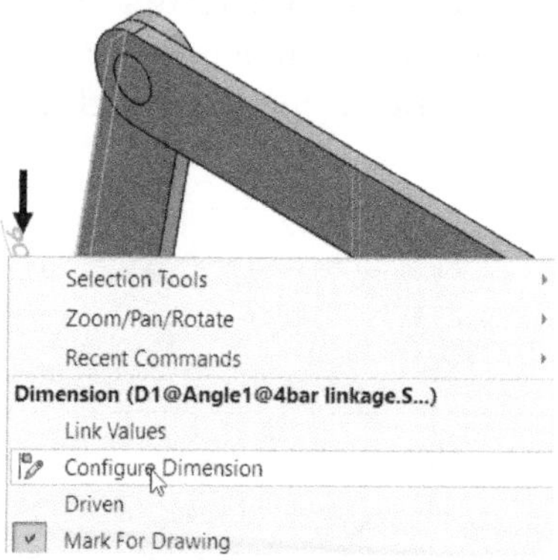

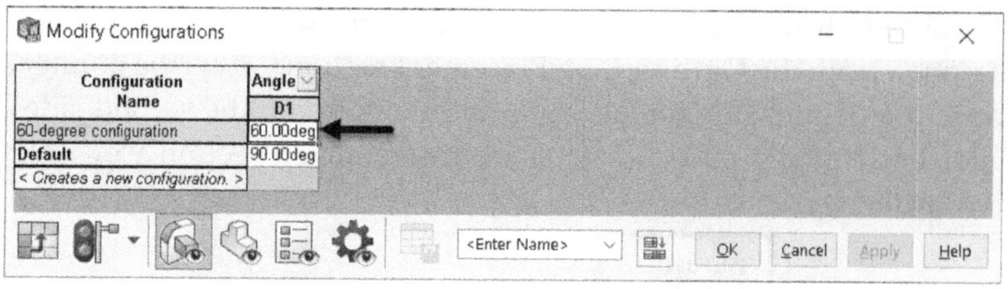

Switch to the drawing file and activate the **Alternate Position View** command (click **Drawing > Alternate Position View** on the CommandManager), and click on the drawing view. On the PropertyManager, select **Existing configuration**; the 60-degree configuration is selected in the drop-down. Click **OK** on the PropertyManager; the 60-degree configuration is displayed in the phantom lines.

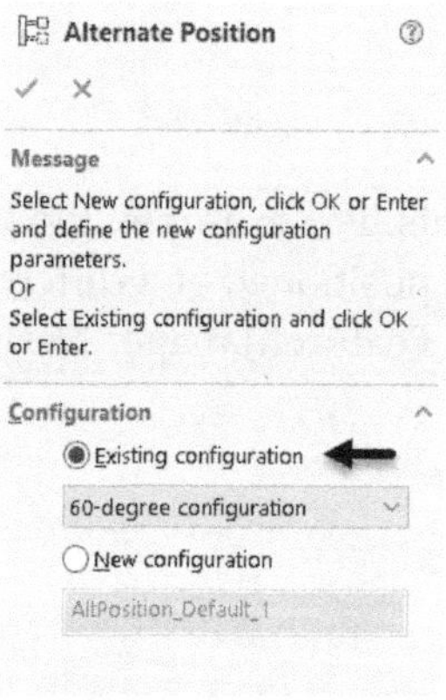

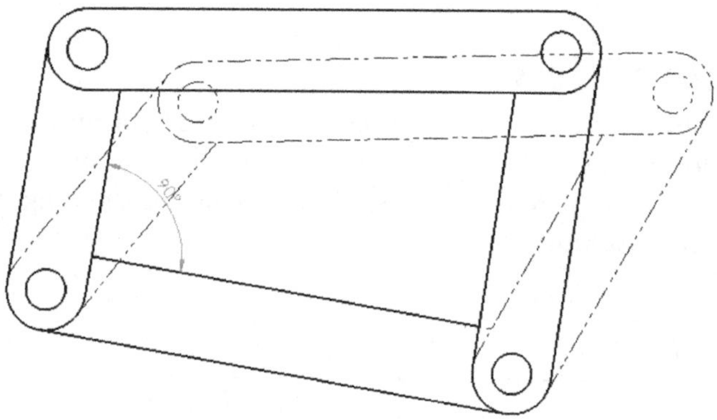

Display Options

When working with SOLIDWORKS drawings, you can control how a model view is displayed using display options. Select a view from the drawing sheet and scroll down to the **Display Style** section on the PropertyManager. Next, select the desired display type; the display type of the view will be changed.

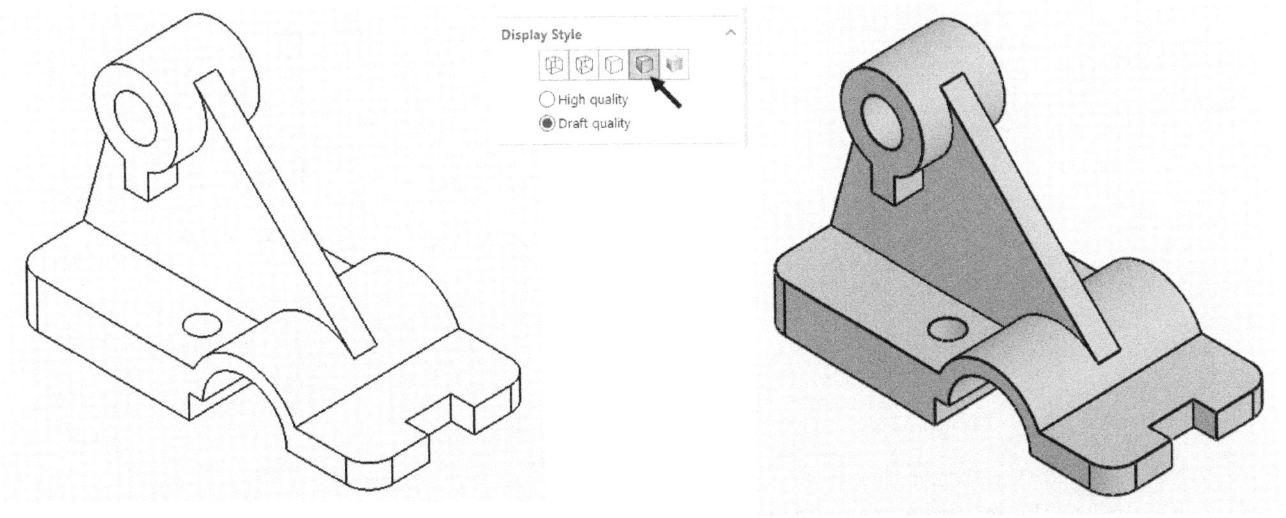

View Alignment

There are several types of views that are automatically aligned to a parent view. These include section views, auxiliary views, and projective views. If you move down a view, the parent view associated with it will also move.

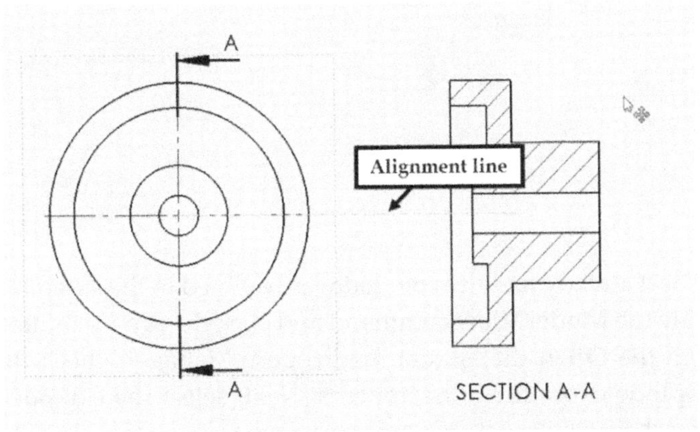

You need to break the alignment between them to move the view separately. Click the right mouse button on the dependent view and select **Alignment > Break Alignment**.

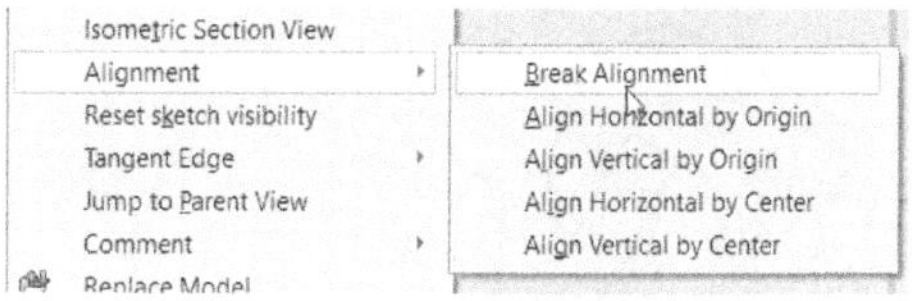

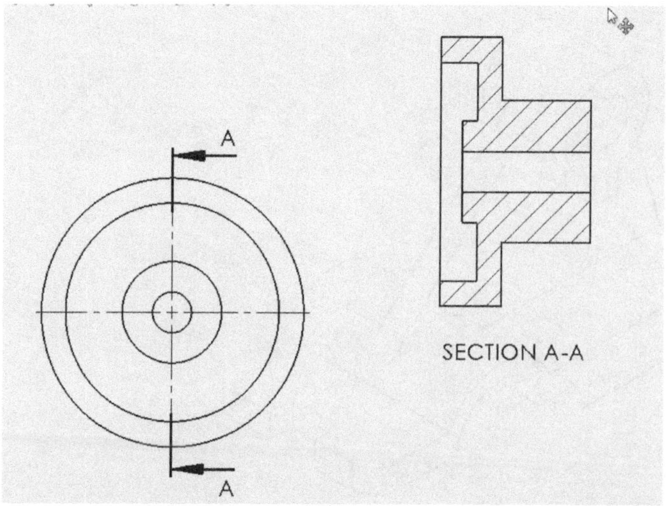

If you want to create alignment between the views, click the right mouse button on the view to be aligned and select **Alignment**; a flyout appears with four alignment options. Select the required alignment option and click on the parent view.

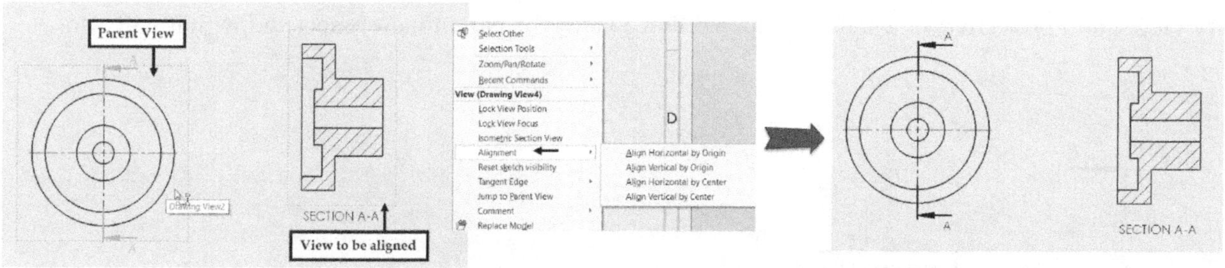

Exploded View

You can display an assembly in an exploded state as long as it already has an exploded view defined in the assembly file. If you want to add an isometric exploded view, activate the **Model View** command and click the Browse button on the PropertyManager. Next, select the assembly from the **Open** dialog. On the PropertyManager, check the **Show in exploded or model break state** option; the **Explode** drop-down list appears. Next, select the exploded view from the drop-down menu and click on the drawing sheet to place it.

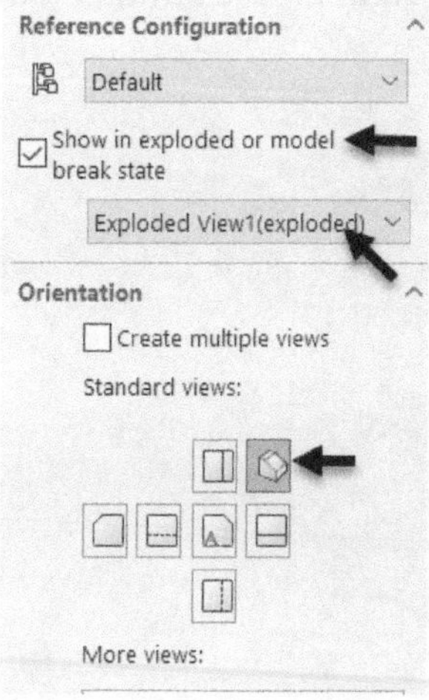

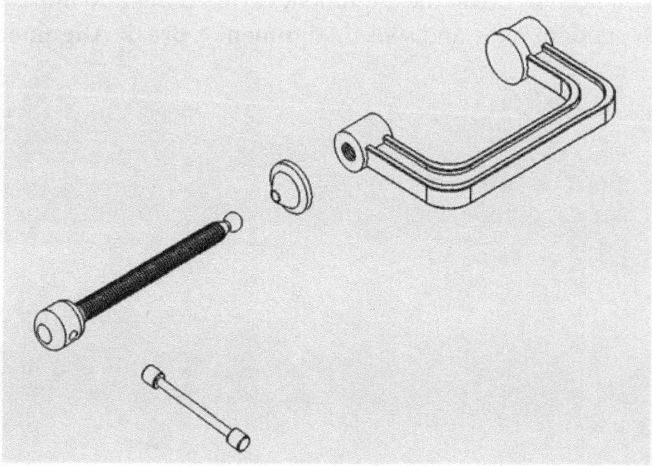

If you want to show an already existing isometric view in an exploded state, all you have to do is right-click the view and select **Show In Exploded State**.

Bill of Materials

Creating an assembly drawing is very similar to creating a part drawing. However, there are few things unique in an assembly drawing. One of them is creating a bill of materials. A bill of materials identifies the different components in an assembly. Generating a bill of materials is very easy in SOLIDWORKS. First, you need to have a view of the assembly. Next, click **Annotation > Tables > Bill of Materials** on the CommandManager, and then click on the drawing view. On the PropertyManager, select the table template from the **Table Template** section. To do this, click the **Open table template of Bill of Materials** icon and select the table template. For this case, you can use the default **bom-standard** template. Next, click the **Open** button to load the template. After selecting the template, you need to select the BOM type from the **BOM Type** section. You can select the **Top-level only**, **Parts only**, or **Indented** option. The **Top-level only** option creates a bill of materials with the sub-assemblies and parts assembled in the selected assembly file.

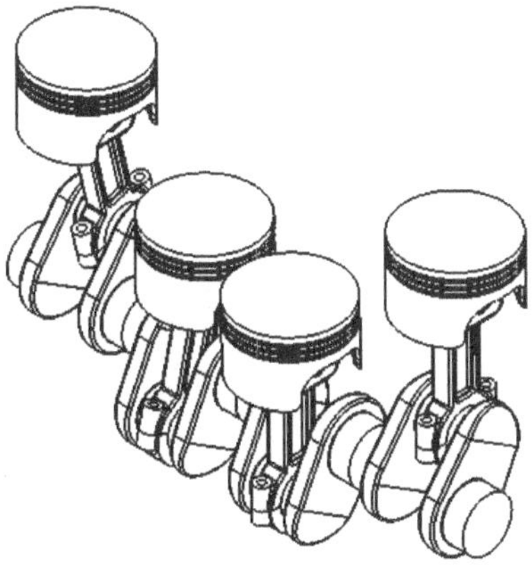

ITEM NO.	PART NUMBER	DESCRIPTION	QTY.
1	Assem4^Assem1		1
2	shaft		1
3	Assem3^Assem1		1
4	Assem2^Assem1		1
5	Assem1^Assem1		1

The **Parts only** option creates a bill of materials with all the parts available in the main assembly subassembly.

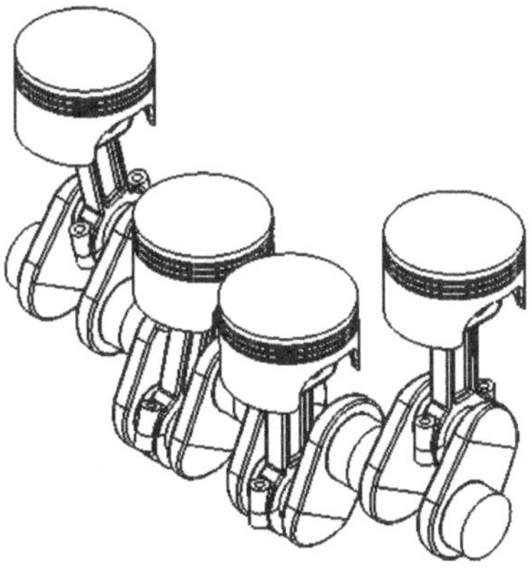

ITEM NO.	PART NUMBER	DESCRIPTION	QTY.
1	piston head		4
2	piston pin		4
3	piston rod		4
4	piston rod cap		4
5	shaft		1

The **Indented** option creates a bill of materials with subassemblies and its parts arranged in the indented form. The drop-down available below this option has three options: **No numbering**, **Detailed numbering**, and **Flat numbering**. The **No numbering** option does not assign any number to the individual parts of the subassembly. The **Detailed numbering** option creates a detailed numbering for each and every part of the subassembly. For example, if the subassembly is assigned the 1 number, the parts inside the subassembly are assigned 1.1, 1.2, and so on. The **Flat numbering** option assigns continuous numbering for the entire bill of materials.

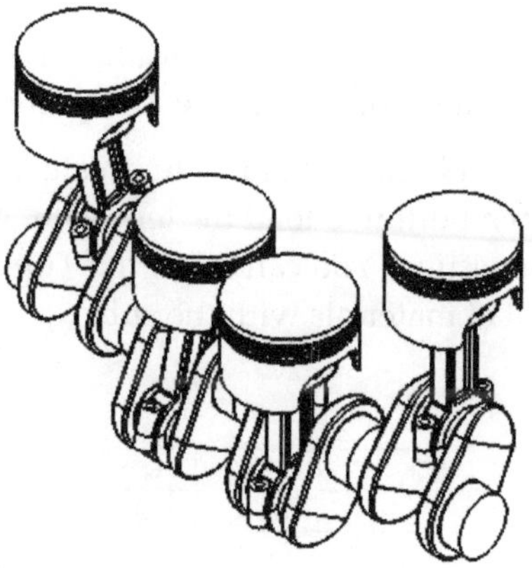

ITEM NO.	PART NUMBER	DESCRIPTION	QTY.
1	Assem4^Asseml		1
	piston head		1
	piston pin		1
	piston rod		1
	piston rod cap		1
2	shaft		1
3	Assem3^Asseml		1
	piston rod cap		1
	piston rod		1
	piston head		1
	piston pin		1
4	Assem2^Asseml		1
	piston rod cap		1
	piston rod		1
	piston head		1
	piston pin		1
5	Assem1^Asseml		1
	piston rod cap		1
	piston rod		1
	piston head		1
	piston pin		1

Click **OK** on the PropertyManager and click on the drawing sheet to position the bill of materials table. Hover the pointer on the table and notice the Move handle at the top-left corner. You can click and drag it to move the BOM table. You can also modify the properties of the table by clicking on the stationary handle. On the PropertyManager, you can check the **Attach to anchor point** option in the **Table Position** section to attach the stationary corner of the table to the anchor point.

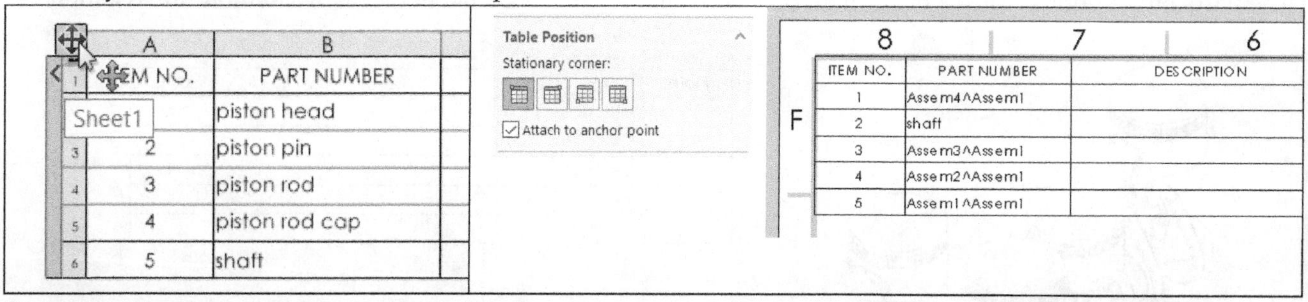

A
B
ITEM NO.
PART NUMBER
Sheet1
piston head
2
piston pin
3
piston rod
4
piston rod cap
5
shaft
Table Position
Stationary corner:
Attach to anchor point
8
7
6
F
ITEM NO.
PART NUMBER
DESCRIPTION
1
Assem4^Asseml
2
shaft
3
Assem3^Asseml
4
Assem2^Asseml
5
Asseml^Asseml

By default, the anchor point is defined at the top left corner of the drawing sheet. However, you can change the anchor point as per your requirement. To do this, right click on the drawing sheet and select the **Edit Sheet Format** option. Next, select any one of the corner points of the title block or sketch points. Next, right-click and select **Set as Anchor > Bill of Materials**. Next, click on the icon located at the top-right corner to exit the Sheet Format mode.

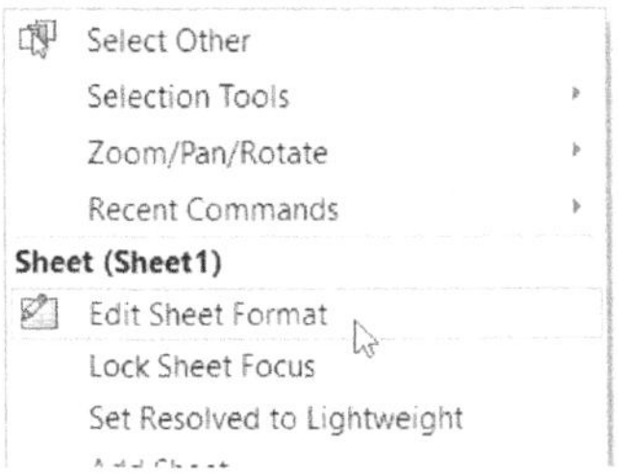

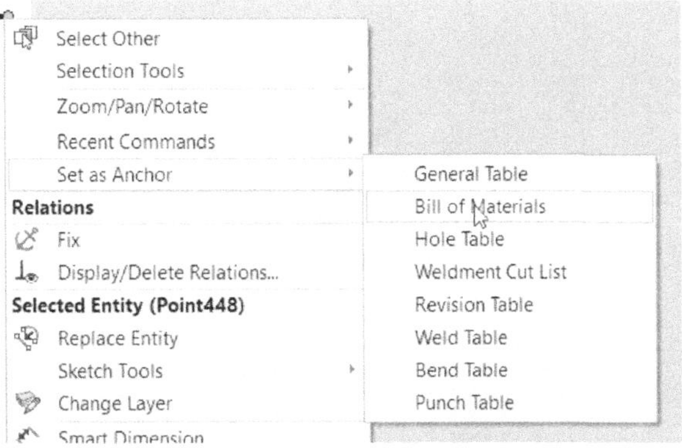

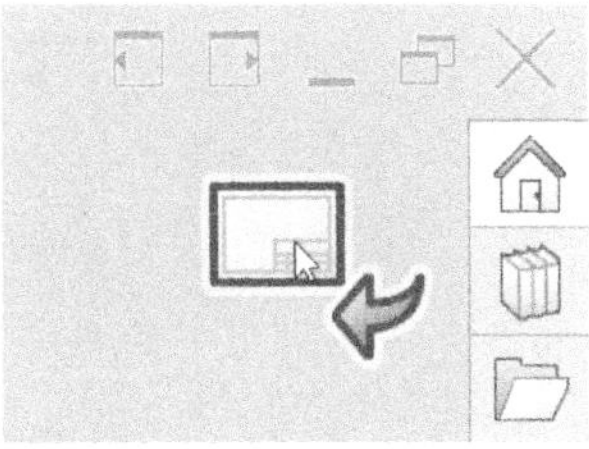

You can change the stationary corner of the bill of materials. Click on the Move handle, and then select the stationary corner from the **Table Position** section of the PropertyManager.

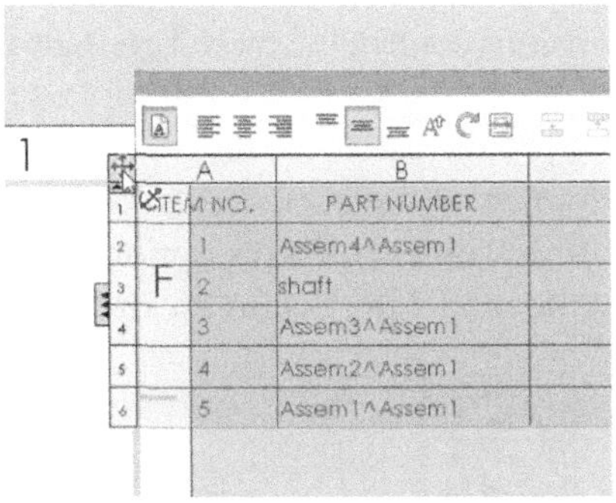

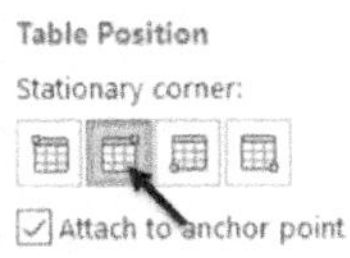

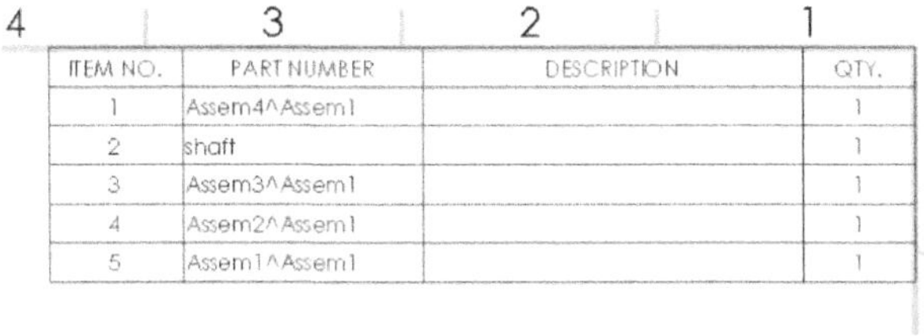

ITEM NO.	PART NUMBER	DESCRIPTION	QTY.
1	Assem4^Assem1		1
2	shaft		1
3	Assem3^Assem1		1
4	Assem2^Assem1		1
5	Assem1^Assem1		1

Adding Balloons

To add balloons, click **Annotation > Balloon** on the CommandManager and select an edge of the part from the drawing view. Next, click the **Follow assembly order** icon on the PropertyManager (note that this option is available only after inserting the Bill of Materials in the drawing). Move the pointer and click to locate the balloon. Likewise, you can add balloons to the remaining parts of the assembly. Click **OK** on the PropertyManager to complete the balloon creation.

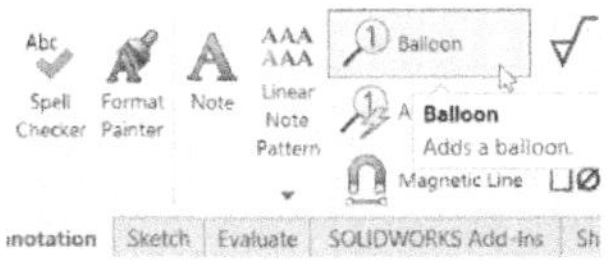

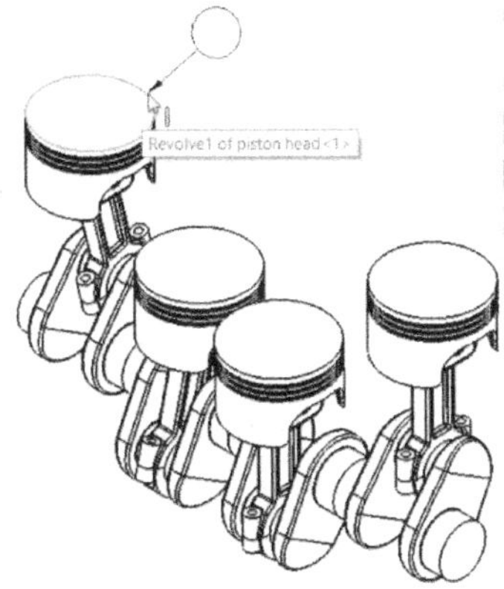

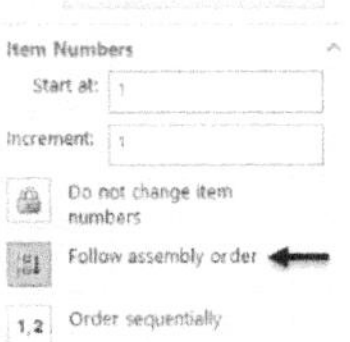

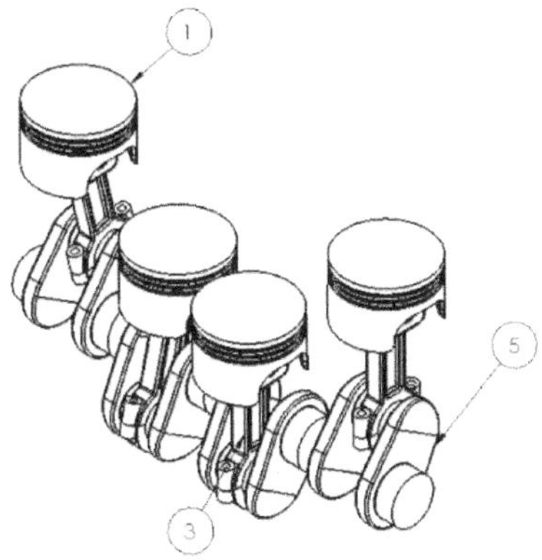

Using the Auto Balloon command

The **Auto Balloon** command adds balloons to the multiple parts of an assembly at a time. Activate this command (**Annotation > Auto Balloon**) and select the assembly view. Select any one of the pattern types from the **Balloon Layout** section: **Layout Balloons to Top, Layout Balloons to Bottom, Layout Balloons to Left, Layout Balloons to Right, Layout Balloons to Square**, and **Layout Balloons to Circular**. Next, select the balloon style and size from the **Style** and **Size** drop-downs in the **Balloon Settings** section. Click **OK** on the PropertyManager.

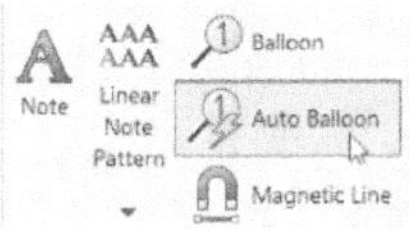

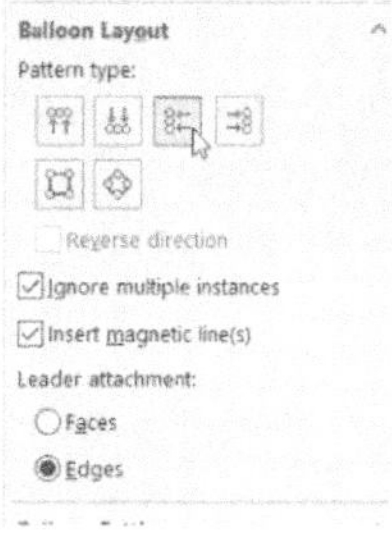

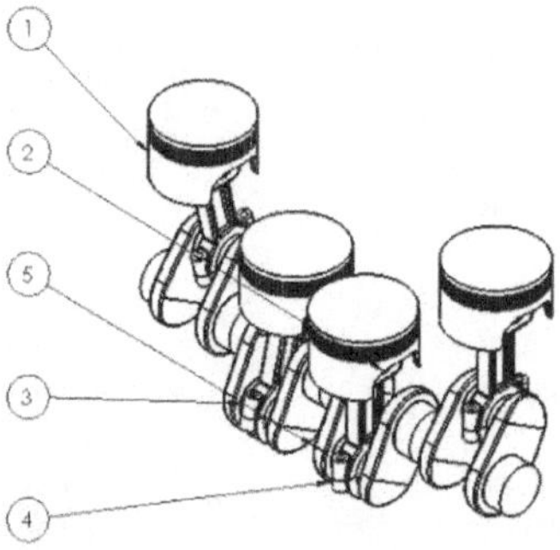

ITEM NO.	PART NUMBER	DESCRIPTION	QTY.
1	piston head		4
2	piston pin		4
3	piston rod		4
4	piston rod cap		4
5	shaft		1

Dimensions

SOLIDWORKS provides you with different ways to add dimensions to the drawing. One of the methods is to retrieve the dimensions that are already contained in the 3D part file. The dimensions retrieved from the 3D model are called driving dimensions, and they are displayed in black color. These dimensions control the model geometry. You can edit the model by modifying these dimensions in the drawing.

To retrieve the model dimensions, click **Annotations > Model Items** on the CommandManager. Next, select the **Entire Model** option from the **Source** drop-down, and then check the **Import Items into all Views** option if you want to retrieve the dimensions into all the existing views. Click **OK** to retrieve the model items into the drawing.

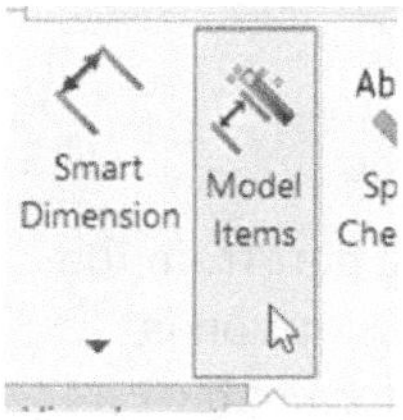

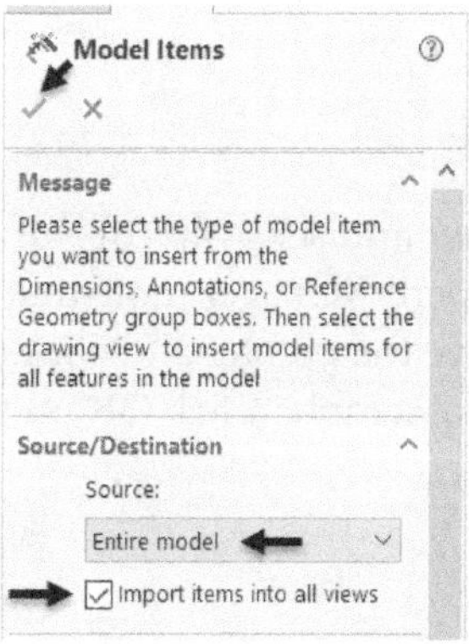

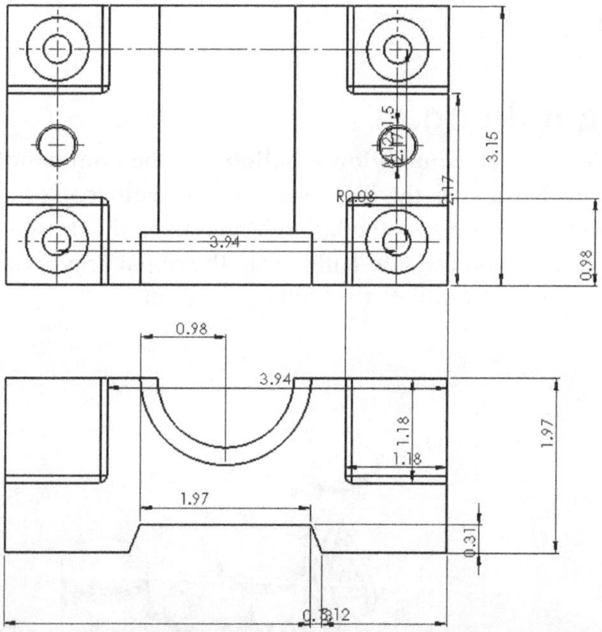

You may notice that there are some unwanted dimensions. Also, the dimensions may not be positioned properly. Press Ctrl+Z on your keyboard to undo the **Model Items** command. Again, activate the **Model Items** command and select the **Selected Feature** from the **Source** drop-down available on the PropertyManager. The **Selected Feature** option retrieves only the dimensions of the selected feature. This option is useful to retrieve the dimensions when the model geometry is complex.

Go to the **Options** section of the PropertyManager, and uncheck the **Use dimension placement in sketch** option. Select the edge of the model, as shown; the dimensions associated with the selected edge are retrieved. Also, the dimensions are arranged properly.

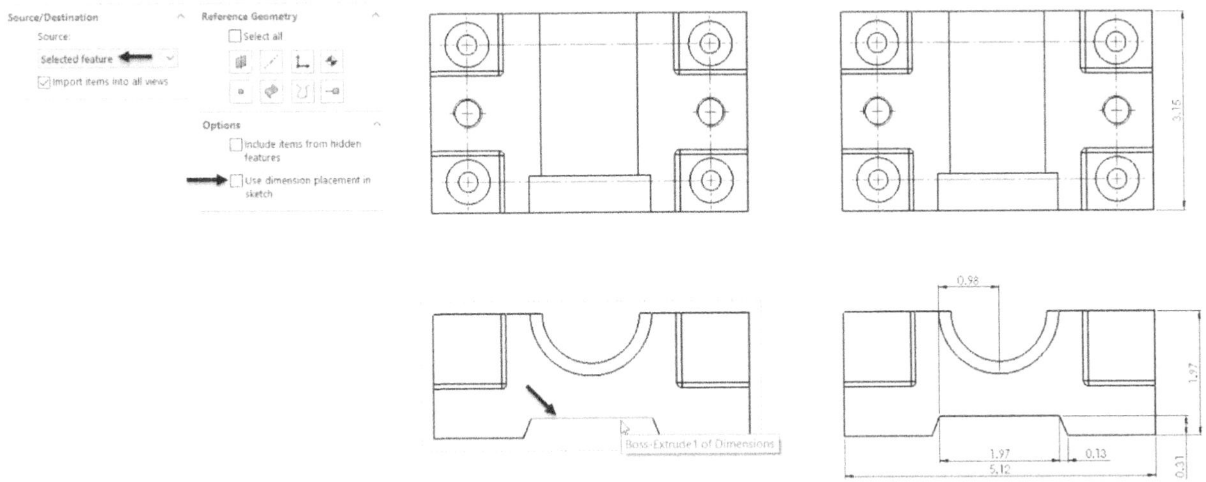

If you want to add some more dimensions necessary to manufacture a part, activate the **Smart Dimension** command and add them to the view. You can also use the **Horizontal Dimension** or **Vertical Dimension** commands to add linear dimensions. Note that the dimensions added in the drawing are driven dimensions, and they are displayed in grey.

Ordinate Dimensions

Ordinate dimensions are another type of dimension that can be added to a drawing. To create them, activate the **Ordinate Dimension** command (click **Annotation > Smart Dimension** drop-down **> Ordinate Dimension** on the CommandManager), and then click on any edge of the drawing view to define the ordinate or zero reference. A horizontal zero reference will be created if you click on a vertical edge. Likewise, a vertical zero reference will be created if you select a horizontal edge. Now, click on the points or edges of the drawing view and place the ordinate dimensions.

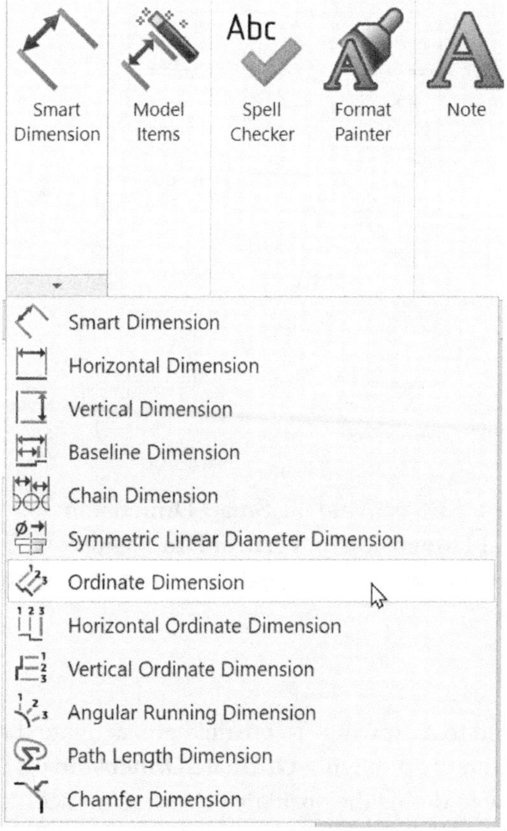

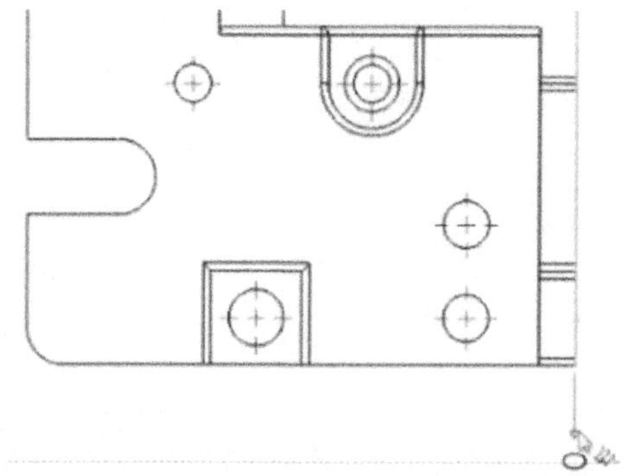

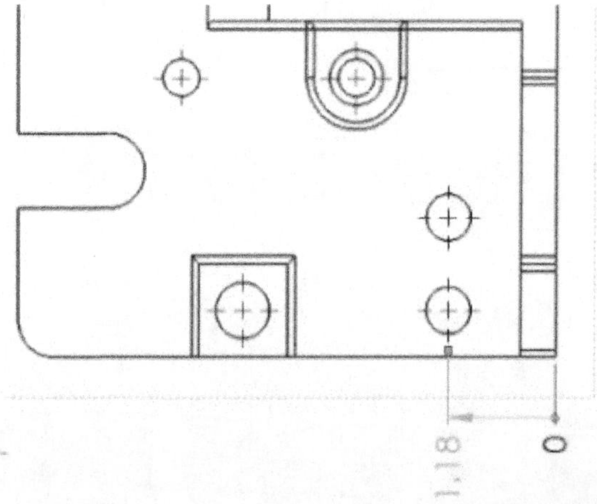

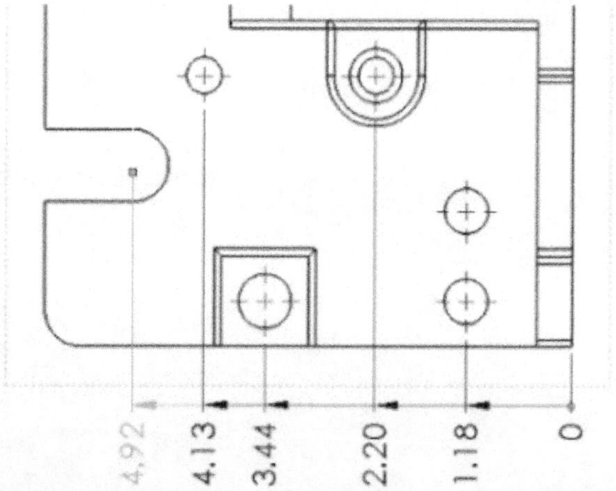

Horizontal/Vertical Ordinate Dimensions

The **Horizontal Ordinate Dimension** and **Vertical Ordinate Dimension** commands allow you to create horizontal and vertical ordinate dimensions, respectively. On the CommandManager, click **Annotation > Smart Dimension** drop-down **> Horizontal ordinate Dimension**, and then select the origin point of the ordinate dimension. Next, select the points or edges to dimension; the horizontal ordinate dimensions are created. Likewise, use the **Vertical Ordinate Dimension** command to create the ordinate dimensions in the vertical direction.

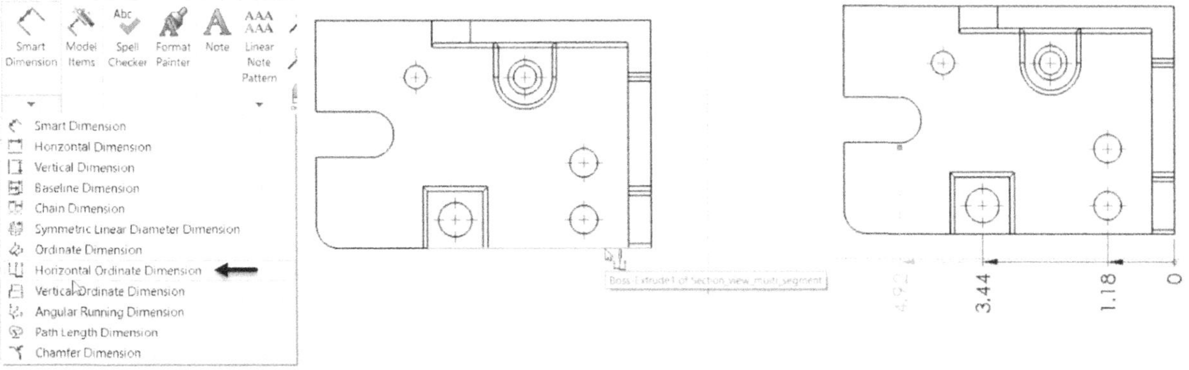

Baseline Dimension

The **Baseline Dimension** command allows you to create and arrange dimensions very quickly. Activate this command (on the CommandManager, click **Annotation > Smart Dimension** drop-down **> Baseline Dimension**), and then select a vertical or horizontal edge of the drawing view; this defines the origin of the baseline dimension. Next, select an edge or point from the drawing view. Likewise, select the remaining edges from the drawing view; the baseline dimensions are created. The baseline dimensions are created in the horizontal or vertical direction based on the edge selected as the origin. Right click and select **Select** to end the baseline dimension creation.

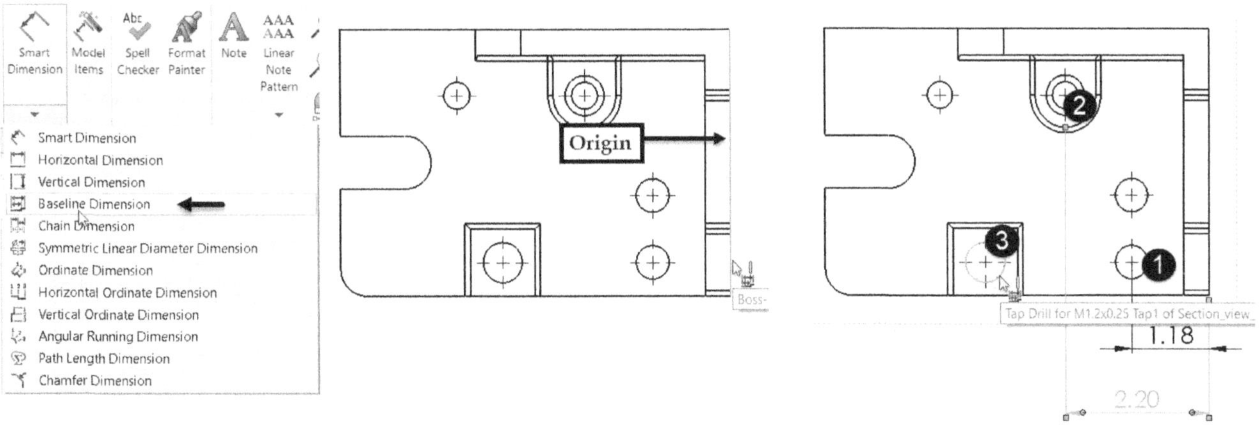

Select any one of the dimensions and right click on the dimensions; a list of options is displayed (**Hide, Hide Dimension Line, Add to BaseLine, Display Options, Break Alignment,** and **Show Alignment**). These options allow you to modify the baseline dimensions. For example, if you want to add a new dimension to the existing baseline dimension, select the **Add to BaseLine** option. Next, select an edge or point from the drawing view; a new dimension is added to the baseline dimension set. Press Esc.

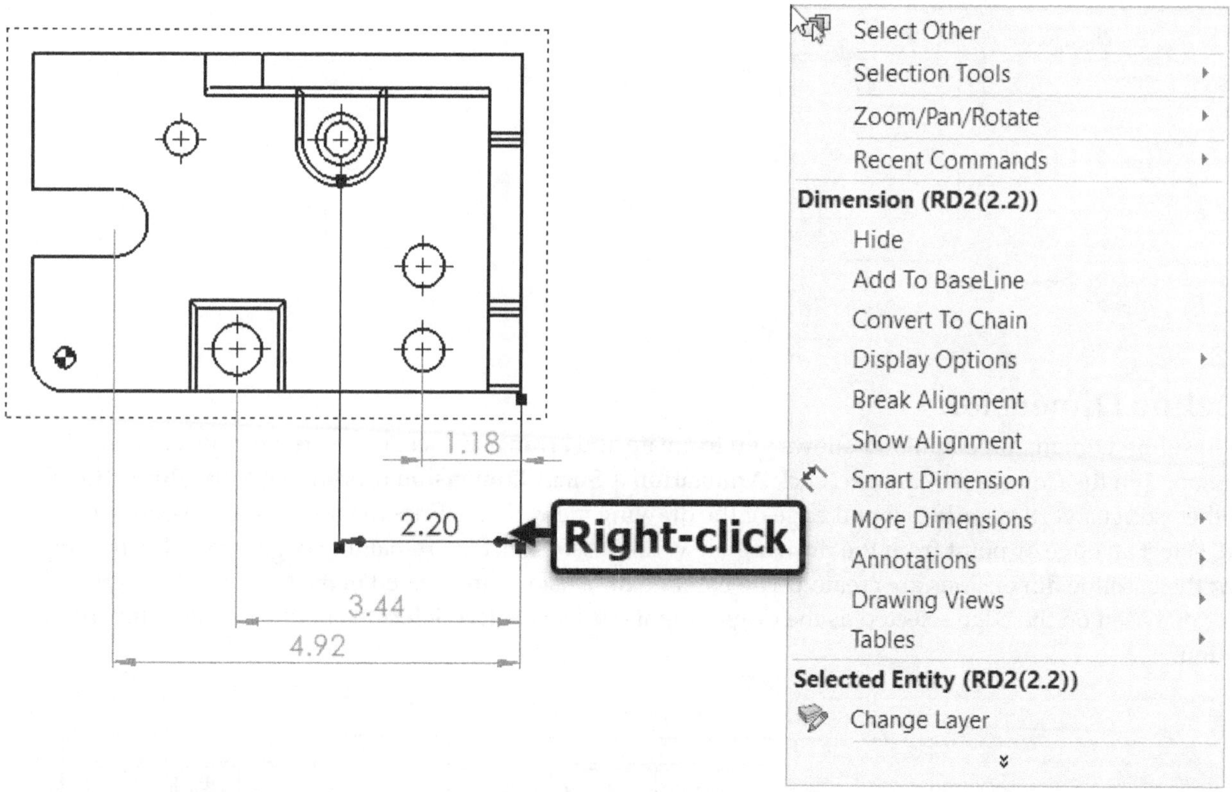

You can convert a **Baseline Dimension** into a **Chain Dimension**. To do this, right the right mouse button on any one of the chain dimensions, and then select **Convert To Chain**.

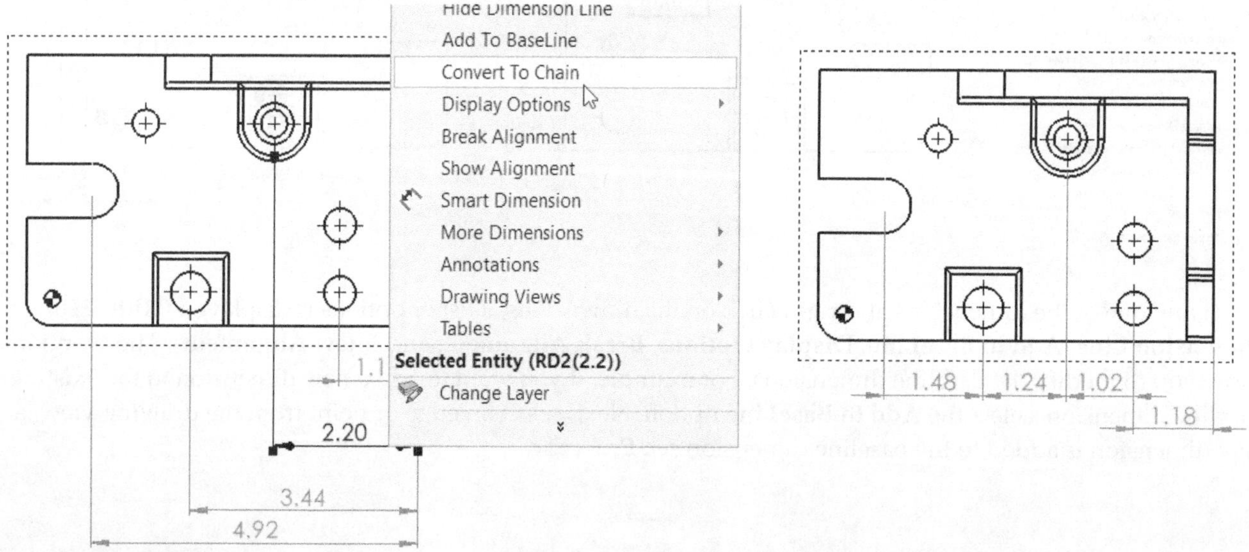

Chain Dimension

The **Chain Dimension** command allows you to create continuous dimensions easily. This command is similar to the **Baseline Dimension** command. Activate this command (on the CommandManager, click **Annotation > Smart Dimension** drop-down **> Chain Dimension**), and then select a vertical or horizontal edge of the drawing view; this defines the origin of the chain dimension. Next, select an edge or point from the drawing view. Likewise, select the remaining edges from the drawing view; the chain dimensions are created. The chain dimensions are created in the horizontal or vertical direction based on the edge selected as the origin. Right click and select **Select** to end the chain dimension creation.

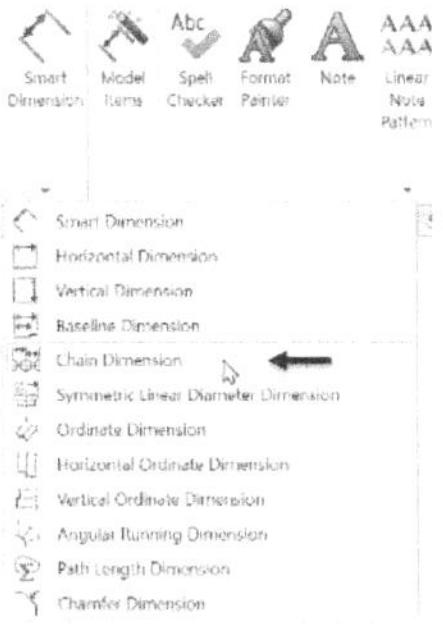

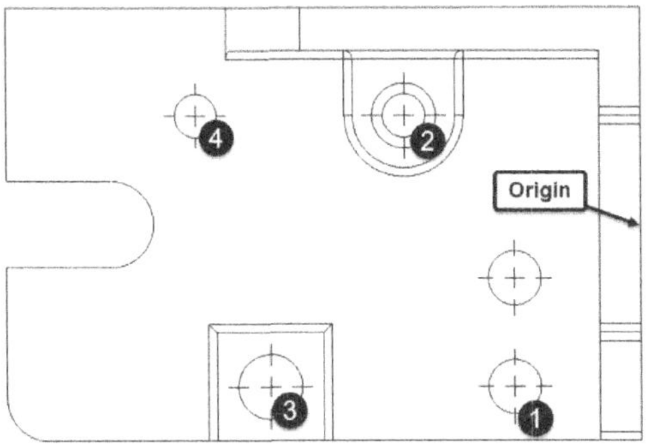

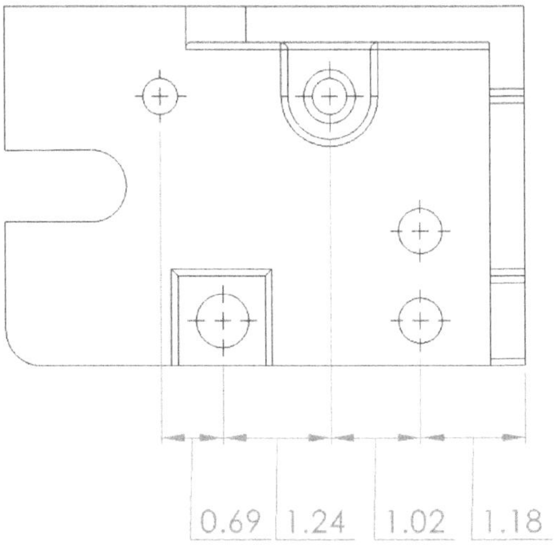

You can convert a **Chain Dimension** into a **Baseline Dimension**. To do this, right the right mouse button on any one of the chain dimensions, and then select **Convert to Base**.

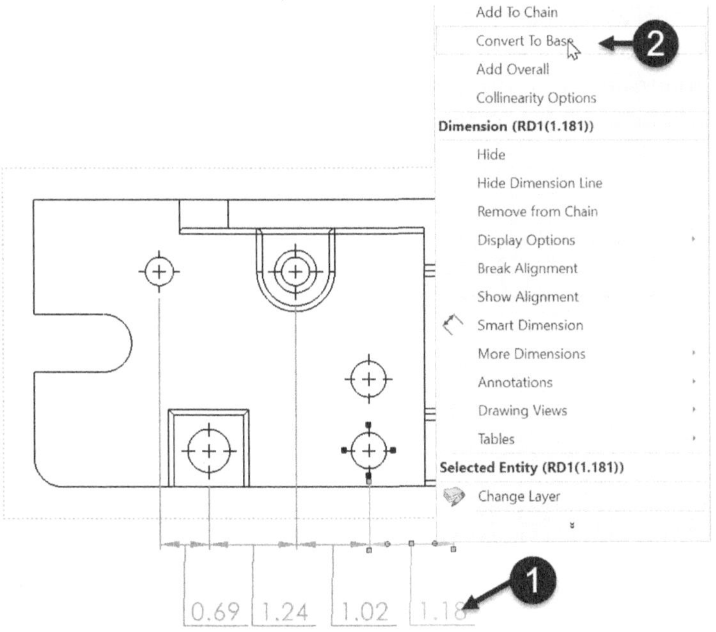

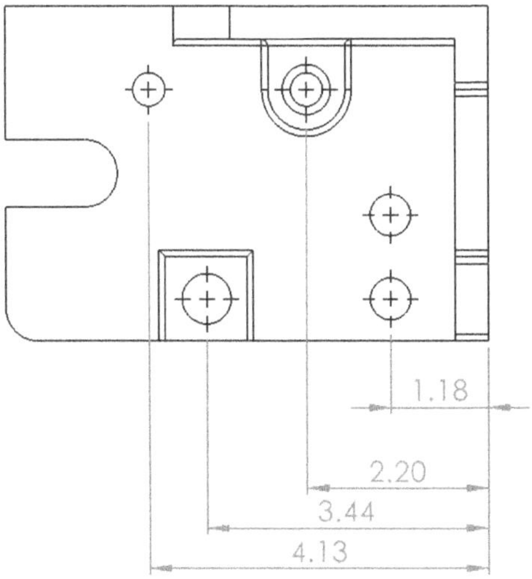

Chamfer Dimensions

The **Chamfer Dimension** command identifies the information of the beveled edges created using the **Chamfer** command in the Part modeling environment. Activate this command (click **Annotation > Smart Dimension > Chamfer**) and select the chamfered edge. Next, select the reference edge, move the pointer, and click to position the chamfer note. Right click and select **Select**.

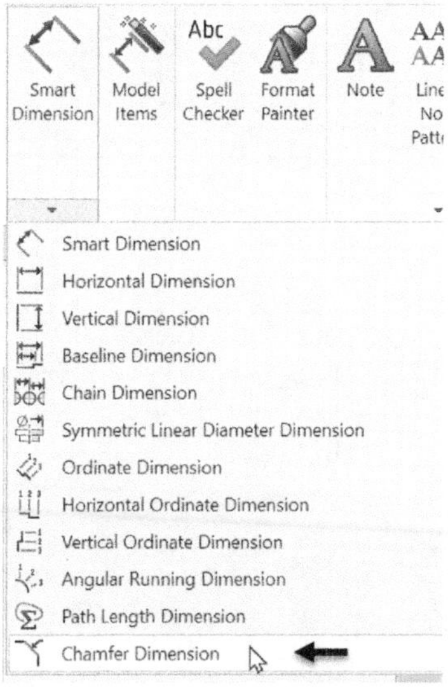

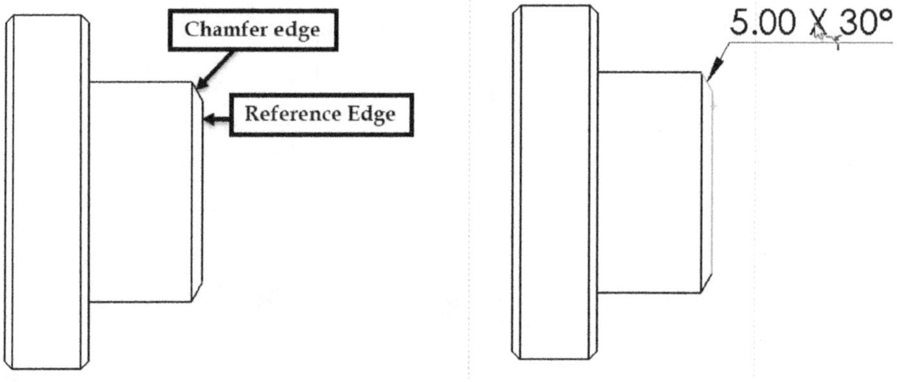

Angular Running Dimensions

The **Angular Running Dimensions** command creates a continuous angular dimension from the zero references. The angle of all the selected entities is measured from the zero references. Activate this command (on the CommandManager, click **Annotation > Smart Dimension** drop-down **> Angular Running Dimensions**), and then select a point or circular edge to define the origin point. Move the point and place the zero reference at the desired angle.

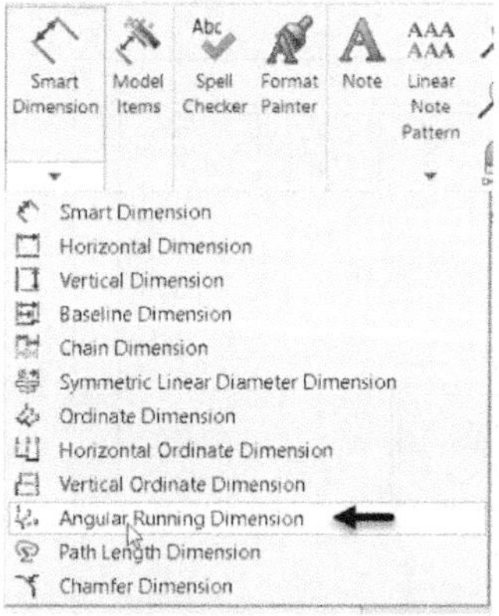

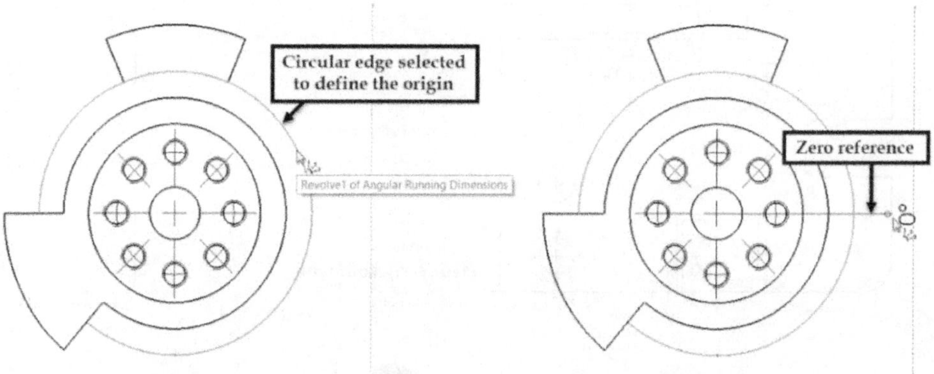

Next, select the edge located at the angle with the zero references; the angle dimension is created between the selected edge and the zero reference. Likewise, select the remaining angular edges; the angular dimensions are created. Right click and select **Select** to complete the angular running dimensions.

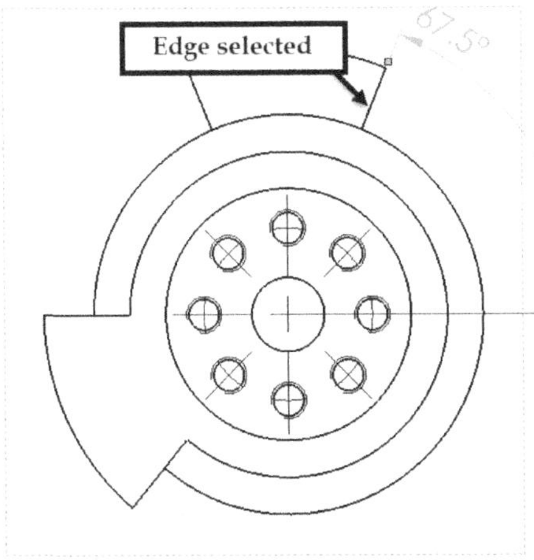

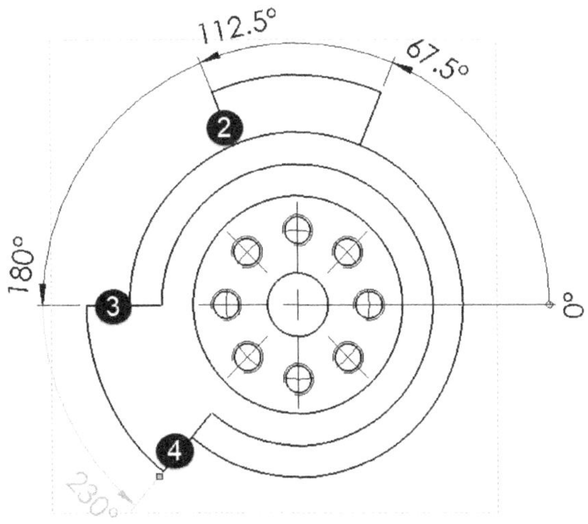

Center Marks and Centerlines

Centerlines and Centermarks are used in engineering drawings to denote hole centers and lines. By default, the centermarks are applied to the holes while the view is being created. If you want to apply the center marks manually, click the **Options** icon on the Quick Access Toolbar. Click the **Document Properties** tab on the dialog and select **Detailing** from the tree located on the left side. Next, uncheck the **Center marks – holes -part** and the **Connection lines to hole patterns with center marks** options in the **Auto insert on view creation** section, and then click **OK**.

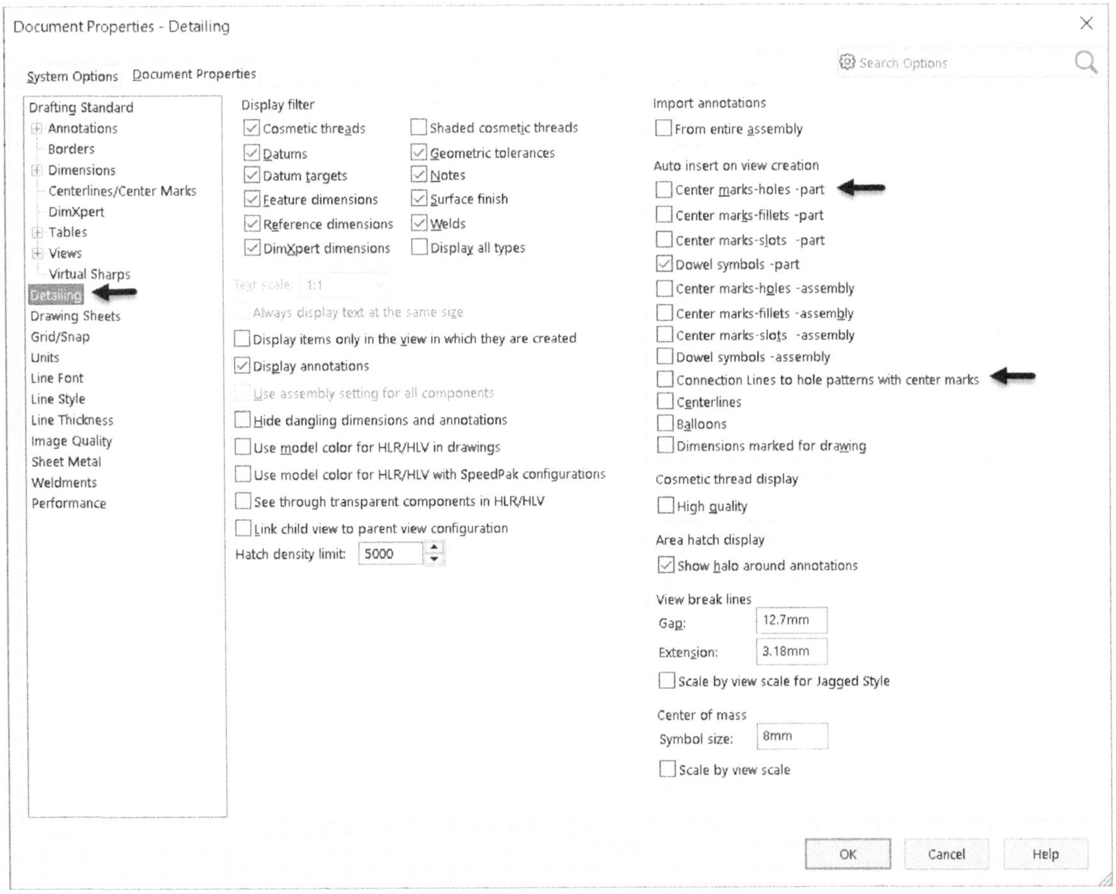

To manually add center marks to the drawing, activate the **Center Mark** command (click **Annotation > Center Mark** on the CommandManager) and click on the hole circle. The centermarks are added to the circle.

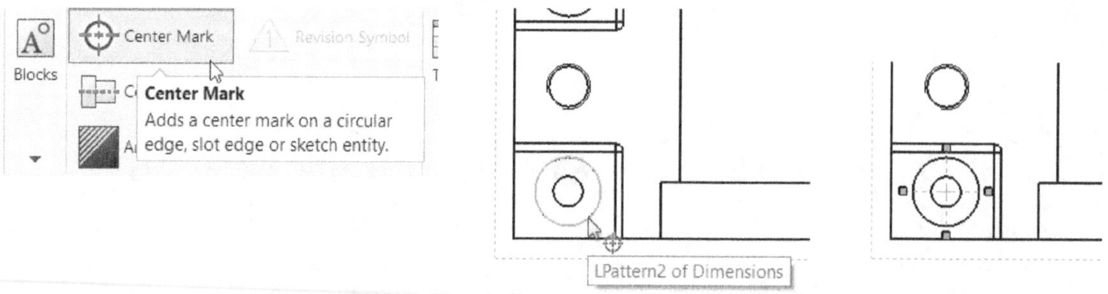

If you want to add center marks to the holes of a linear pattern, click the **Linear Center Mark** icon in the **Manual Insert Options** section of the PropertyManager. Next, select the holes of the linear pattern.

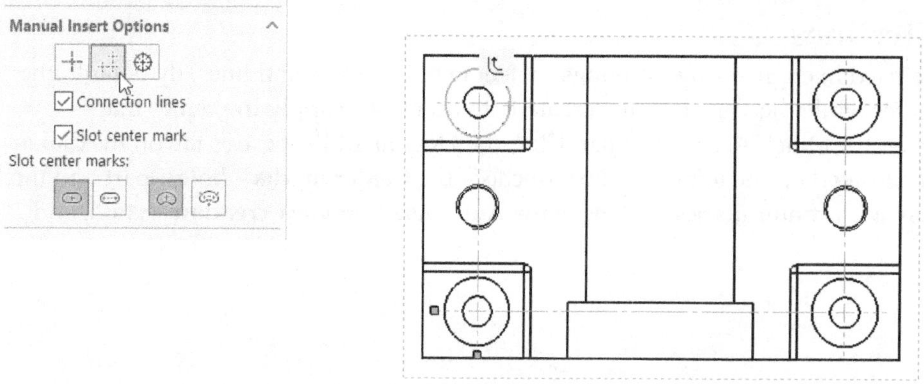

The **Circular Center Mark** option allows you to add center marks to the holes arranged circularly. Click this icon in the **Manual Insert Options** section of the **Center Mark** PropertyManager and select the circles arranged circularly; the circular center mark will be created.

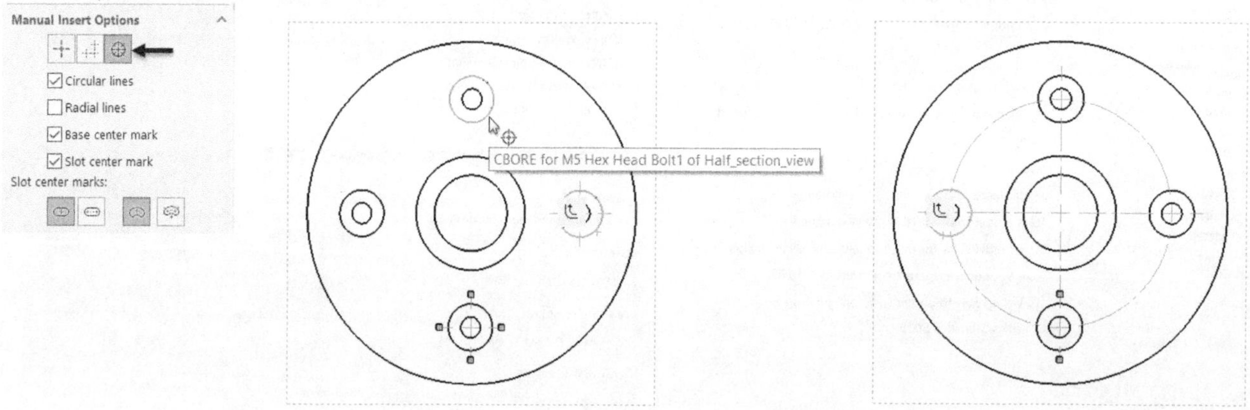

Check the **For all holes** option in the **Auto Insert** section, and then select the drawing; the center marks are added to all the holes of the drawing view.

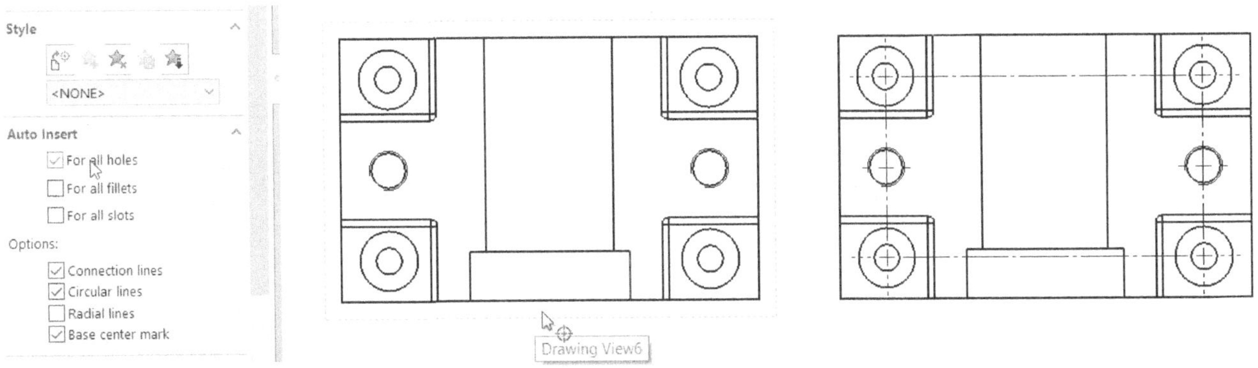

To add centerlines, activate the **Centerline** command (click **Annotation > Centerline** on the CommandManager), and then click on two parallel edges of the drawing view. A centerline will be created between the two lines.

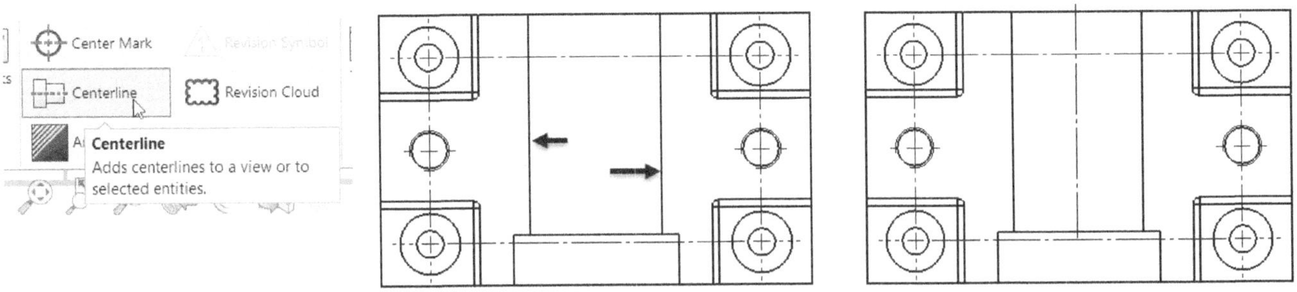

If you want to add centerlines automatically, check the **Select View** option in the **Auto Insert** section of the PropertyManager. Next, select the drawing view to add centerlines.

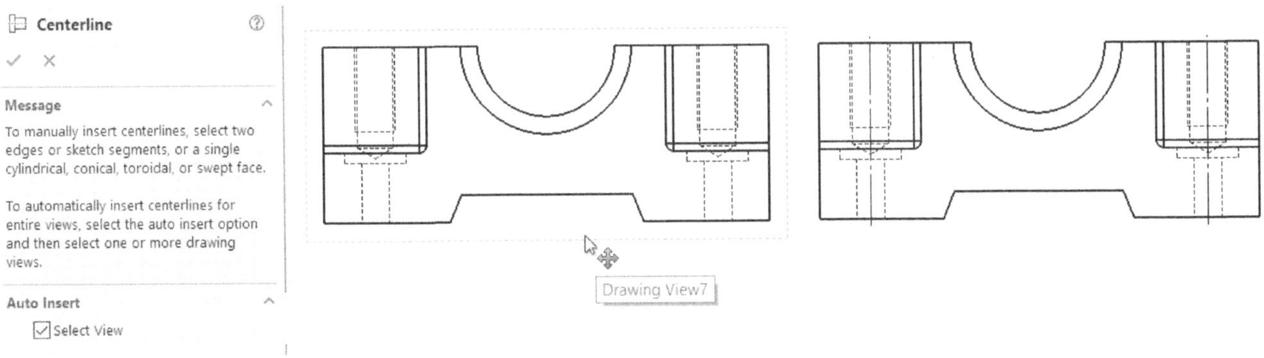

Hole Callout

The **Hole Callout** command allows you to add a note to a hole. Activate this command (click **Annotation > Hole Callout** on the CommandManager) and select a hole; the note is attached to the pointer. Notice that the **Hole Callout** command captures the information of the hole from the 3D model. Move the pointer and position the hole note.

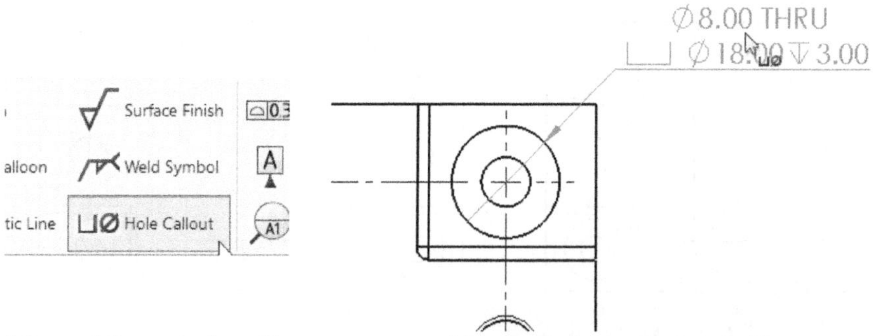

After adding a hole callout, you can add more information to it. For example, if you want to display the quantity of the holes in the hole note, place the cursor in front of the diameter symbol, and then enter the number of holes; the **SOLIDWORKS** message box appears showing that manually altering the note will break the link with the 3D model. Click **OK** on the message box. Next, click the **Leader** tab of the PropertyManager and select the required leader type.

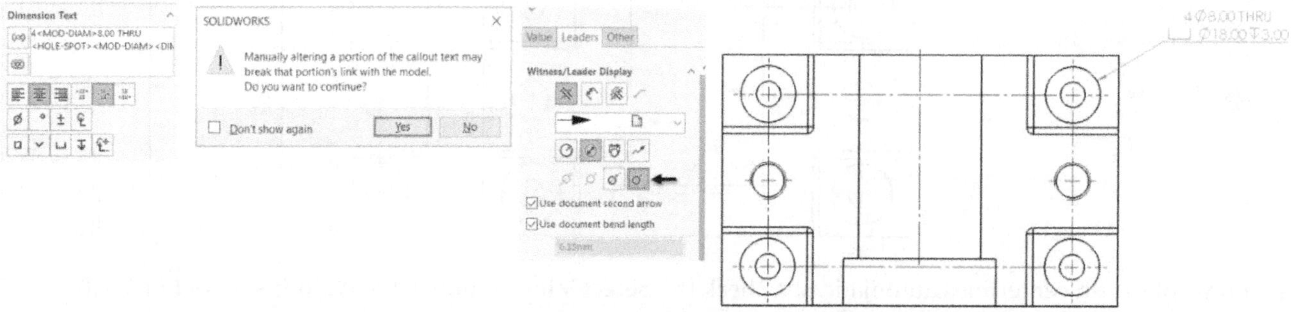

You can also add a hole callout to the holes in a section view. To do this, activate the **Hole Callout** command and select the hole's edge in the section view.

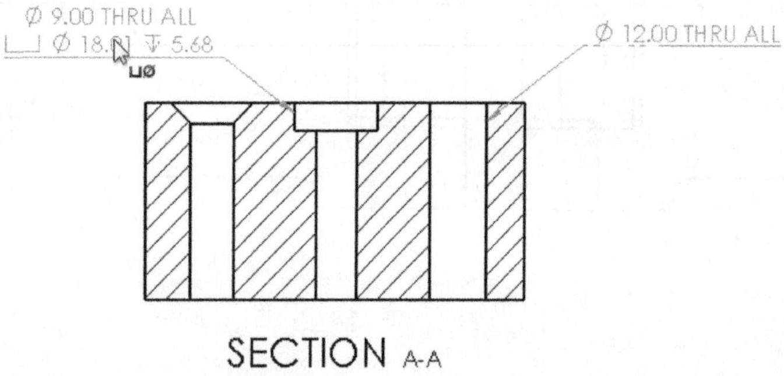

Notes

Notes are an important part of a drawing. You add notes to provide additional details, which cannot be done using dimensions and annotations. To add a note or text, activate the **Note** command (click **Annotation > Note** on the CommandManager), and then click to position the note. On the **Formatting** toolbar, select the font and font size and type the text inside the box. Next, click **OK** on the PropertyManager.

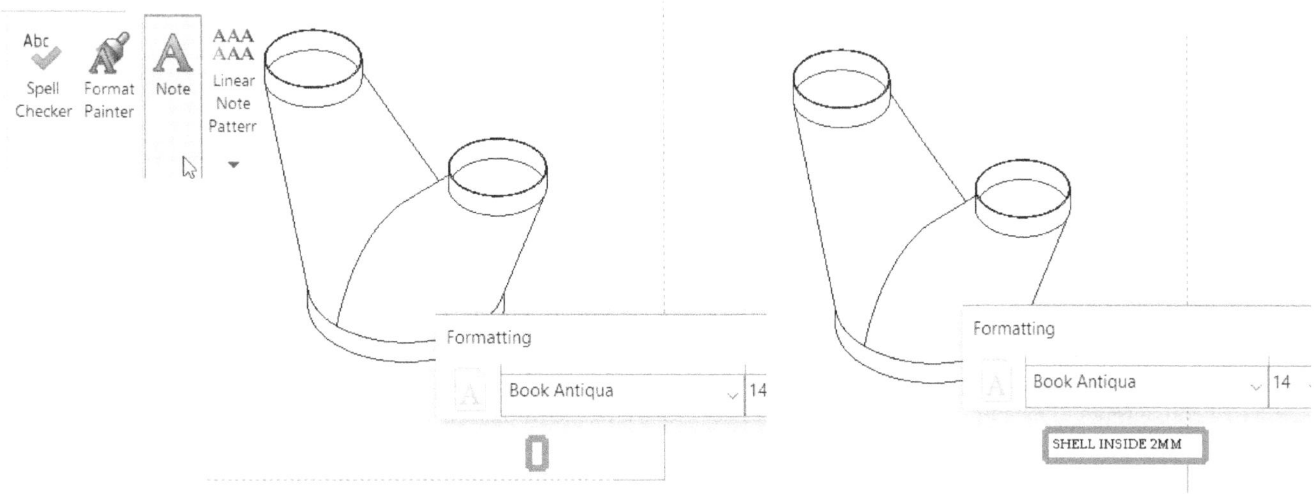

You can use the **Note** command to insert associative text (reference and property texts). To insert an associative text in the drawing, activate the **Note** command and click to define the text position. On the **Note** PropertyManager, click the **Link to Property** icon on the **Text Format** section. On the **Link to Property** dialog, select a property from the **Property name** list and click **OK**. Next, click **OK** on the **Note** PropertyManager.

Best Practices

Standards and Templates

Standards and Templates: Begin by selecting industry and international drawing standards compatible with SOLIDWORKS, such as ISO, ASME, or ANSI. Customize templates to include essential information like title blocks and company logos.

Document Properties

Document Properties: Configure document properties according to project requirements, setting units (metric or imperial), dimensioning standards, and tolerancing standards.

Views and Annotations

Orthographic Views: Utilize standard orthographic views to accurately represent part geometry, following manufacturing and inspection norms.

Projection Views: Include projected views, auxiliary views, or section views when necessary to represent complex geometry or hidden features.

Annotations: Use annotations like notes and labels to provide additional information, cautions, or special instructions clearly.

Specific Considerations for Different Types of Parts

Plastic Parts: Emphasize draft angles and wall thickness for mold design support. Specify material properties, including modulus and shrinkage factors. Call out critical features for injection molding.

Casted Parts: Identify parting lines and draft angles to aid the casting process. Include allowances for machining and surface finish requirements. Indicate potential defects and inspection points.

Forged Parts: Highlight grain flow direction for strength considerations. Specify necessary heat treatment processes. Clearly define fillet radii, blends, and flash dimensions.

Machined Parts: Specify machining processes, tool selections, and toolpath information. Detail surface finishes, tolerances, and fits for critical dimensions. Highlight datum references for CNC programming.

Precautions

Avoid over-dimensioning: Include only essential dimensions and annotations in drawings to prevent clutter and confusion.

Ensure consistency: Double-check units, tolerances, and symbols in drawings to maintain uniformity.

Minimize ambiguities: Be clear and unambiguous in language when using notes and GD&T symbols in drawings.

Proofread and review: Carefully review drawings to identify errors, typos, and inaccuracies.
Document revisions: Maintain a comprehensive record of revisions and changes made to drawings.

Examples

Example 1

In this example, you will create the 2D drawing of the part shown below.

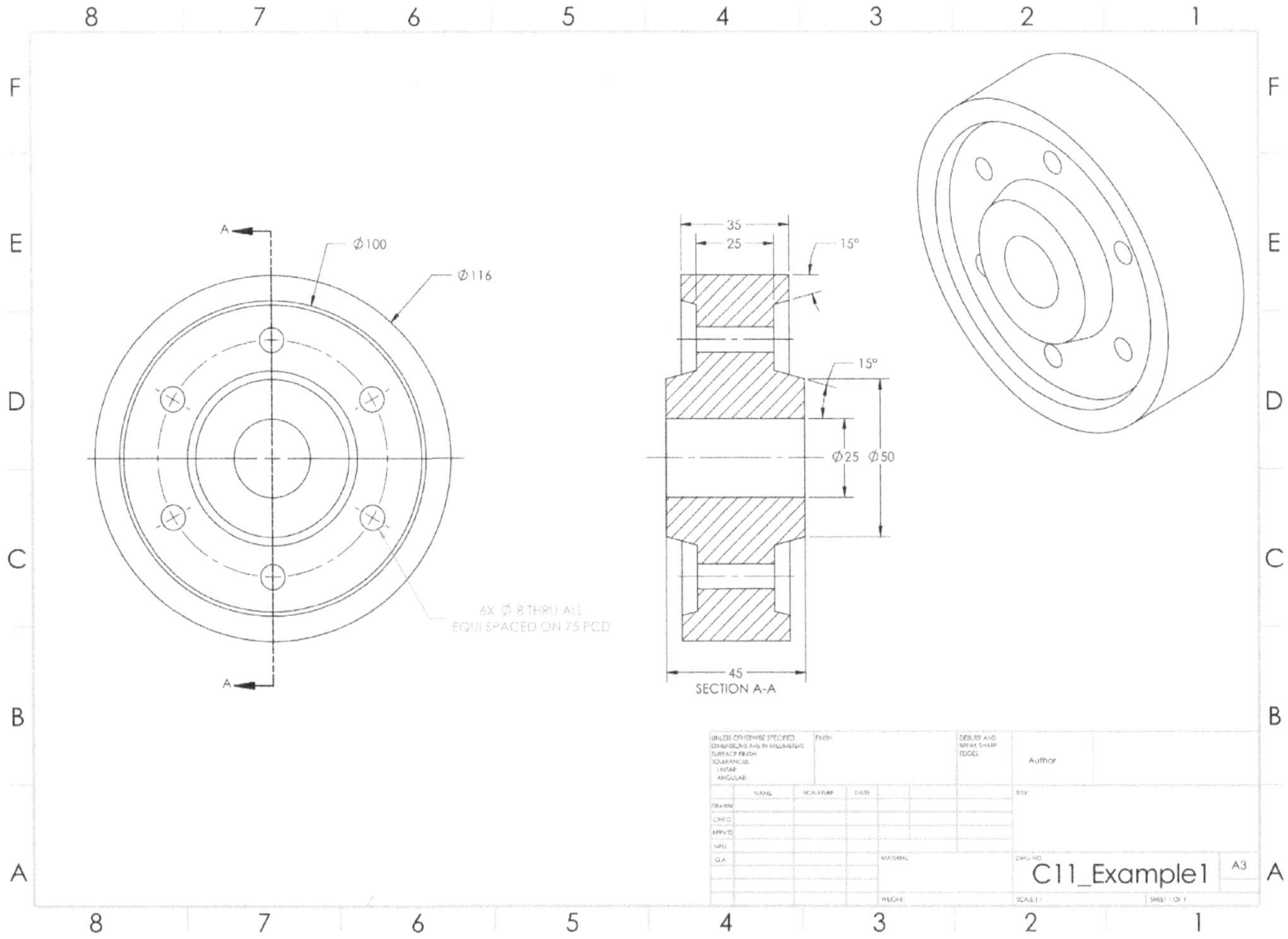

Creating a Custom Sheet Format

1. On Quick Access Toolbar, click the **New** button. Next, select **Drawing** on the **New SOLIDWORKS Document** dialog and click **OK**.
2. Select **A3 (ANSI) Landscape** from the list box in the **Sheet Format/Size** dialog. Click **OK**.
3. Close the **Model View** PropertyManager.
4. On the **FeatureManager Design Tree**, click the right mouse button on **Sheet1** and select **Properties**.
5. On the **Sheet Properties** dialog, set the **Type of projection** to **Third angle**. Click **Apply Changes**.
6. On the CommandManager, click **Sheet Format > Edit Sheet Format** .
7. Zoom to the title block, and then select the lines and notes shown in the figure. Press **Delete** on your keyboard.
8. On the CommandManager, click **Sketch > Line** and draw a line on the title block.

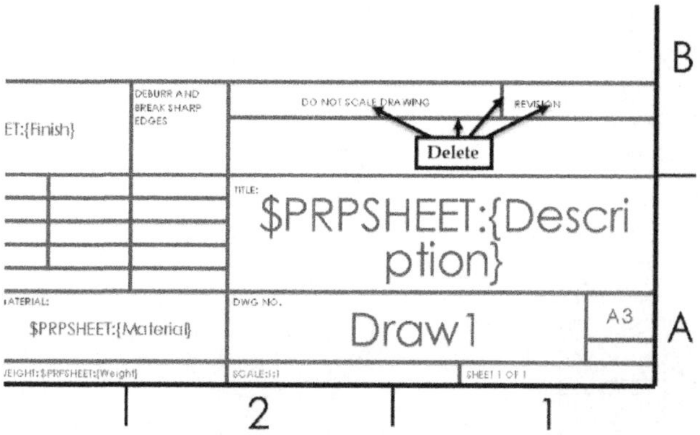

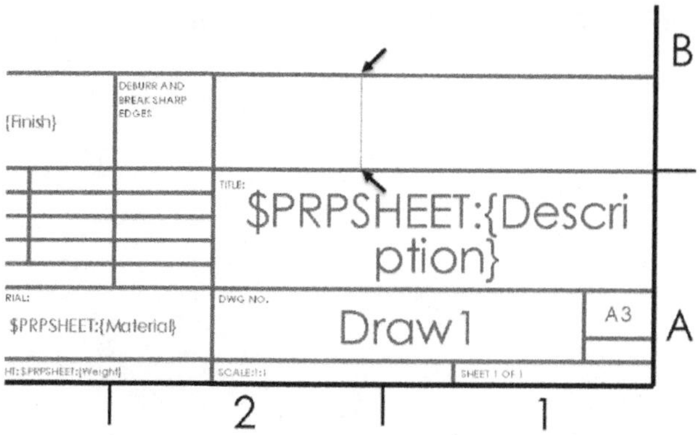

9. On the CommandManager, click **Annotation > Note**. On the PropertyManager, under the **Text Format** section, click **Link to Property**.
10. On the **Link to Property** dialog, select **Current document**. Select **SW-Author (Author)** from the **Property name** drop-down, and then click **OK**.

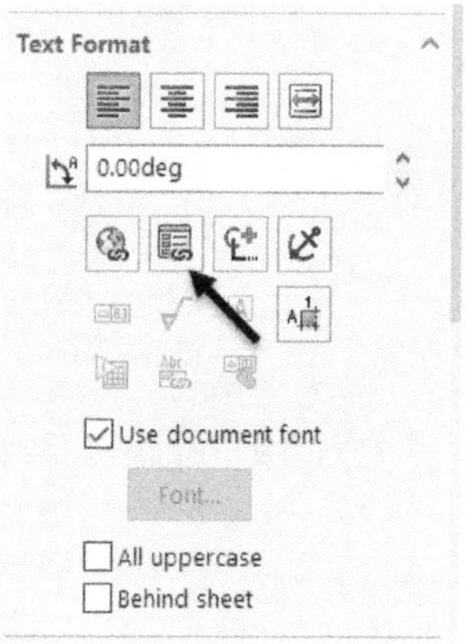

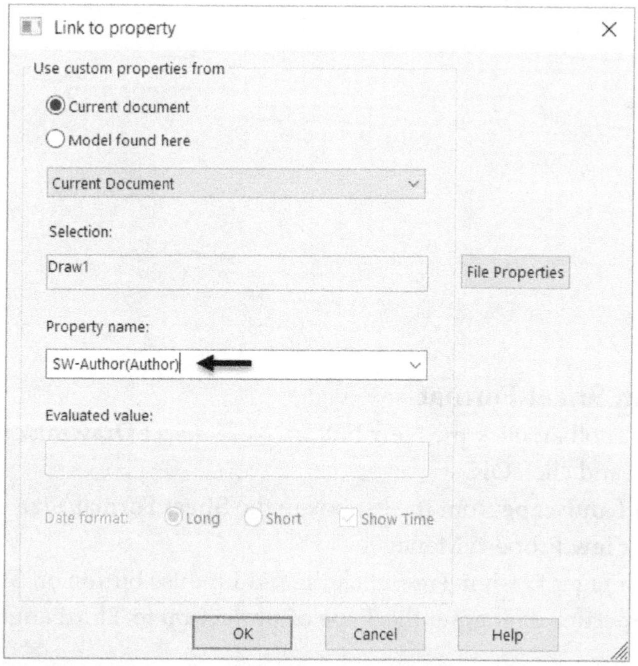

11. On the PropertyManager, under the **Leader** section, select **No Leader**. Click inside the box, as shown in the figure. On the **Formatting** toolbar, set the font size to 8 and type **Author**. Click **OK** on the PropertyManager.

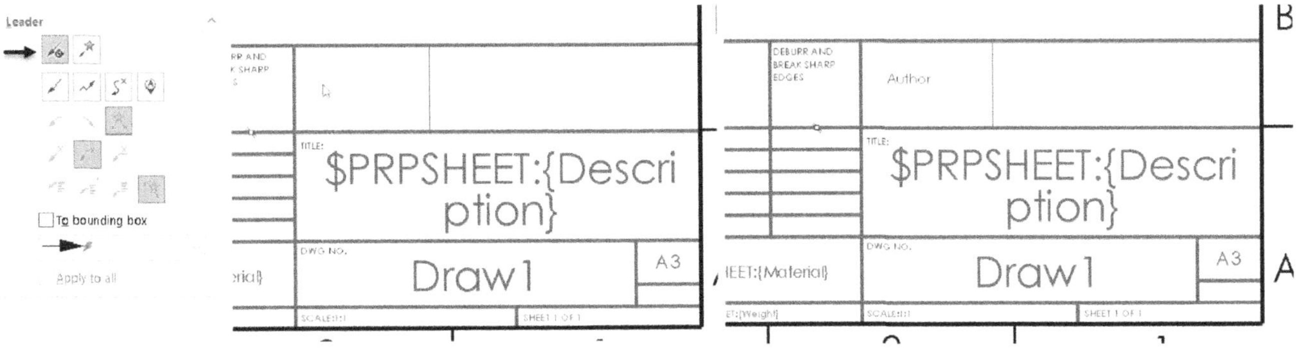

12. On the Menu bar, click **Insert > Picture**. Select your company logo and click **Open**.
13. Resize (click and drag the top right corner of the image) and position the image on the title block.
14. Click **OK** on the PropertyManager. Next, click **Exit Sheet Format**.
15. On the **Quick Access Toolbar**, click **Options**. On the dialog, under the **Document Properties** tab, click **Drafting Standard**.
16. Select **Overall drafting standard >ANSI**. Click **Units** and set the **Unit System** to **MMGS**.
17. Set the **Decimal** values to **None** in the table displayed on the **Units** page, and then click **OK**.

Type	Unit	Decimals	Fractions	More
Basic Units				
Length	millimeters	None		...
Dual Dimension Length	inches	None		...
Angle	degrees	None		
Mass/Section Properties				
Length	millimeters	.12		
Mass	grams			

18. Click **File > Save**. On the **Save As** dialog, select **Save as type > Drawing Templates**.
19. Browse to **C:\Users\Username\Documents\SOLIDWORKS For Beginners** folder. Type-in **Sample** in the **File name** box and click **Save**.
20. On the Menu bar, click **File > Close**.

Starting a New drawing

1. On the Quick Access Toolbar, click the **Options** button. On the **System Options** dialog, click **File Locations**.
2. Select **Show folders for > Document Templates** and click the **Add** button.

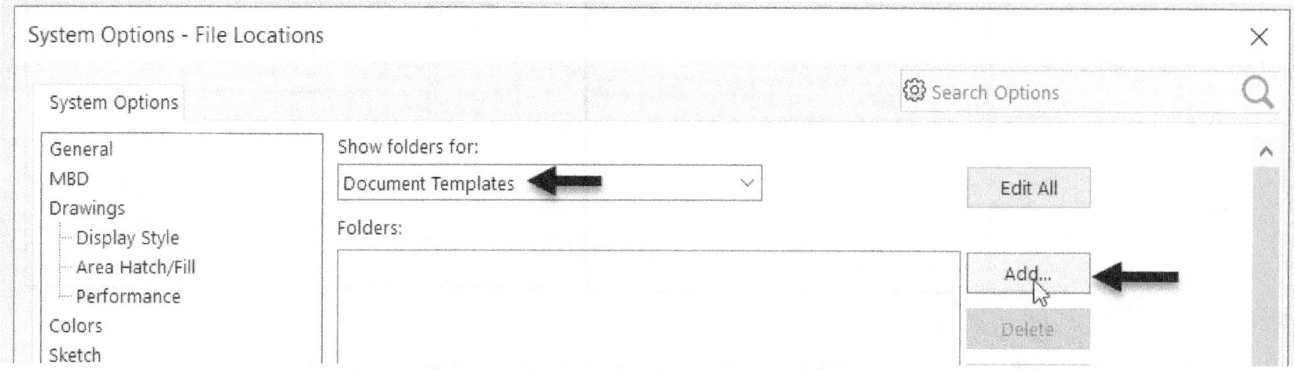

3. Browse to **C:\Users\Username\Documents\SOLIDWORKS For Beginners** folder. Click **Select Folder**.
4. Click **OK** on the **System Options** dialog, and then click **Yes**.
5. On the Quick Access Toolbar, click the **New** button. On the **New SOLIDWORKS Document** dialog, click the **Advanced** button. Under the **SOLIDWORKS For Beginners** tab, select **Sample** and click **OK**.

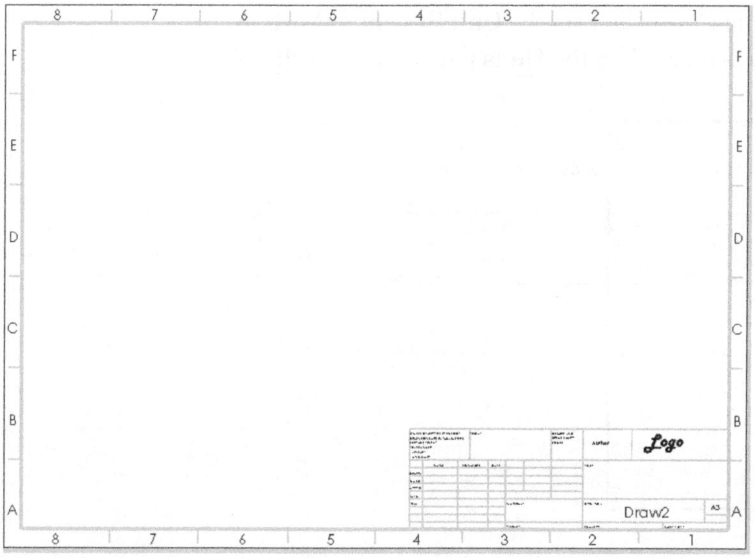

Generating Model Views

1. Download the C11_Example1 from the companion website.
2. Activate the **Model View** command (click **Drawing > Model View** on the CommandManager) if not already active.
3. Click the **Browse** button on the PropertyManager and browse to the location of the downloaded file. Next, click on the part file. Click the **Open** button.
4. On the PropertyManager, check the **Create multiple views** option in the **Orientation** section.
5. Click on the **Front** and **Isometric** view icons in the **Orientation** section. Next, uncheck all the options in the **More views** list box.
6. On the PropertyManager, select the **Use Custom Scale** option in the **Scale** section. Next, select **1:1** from the **Scale** drop-down.

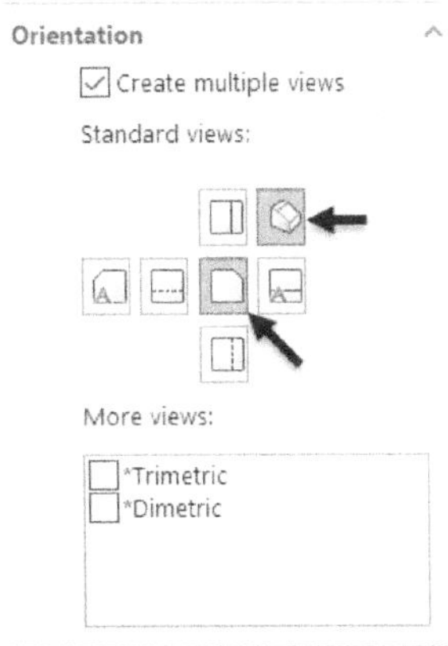

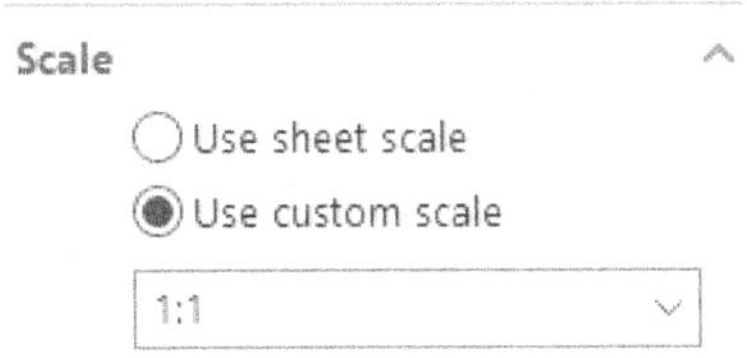

7. Click **OK** on the PropertyManager to place the drawing views. Place the pointer on the Isometric view, and then click on the dotted frame displayed around the view. Next, press and hold the left mouse button and drag the isometric view to the top right corner. Likewise, drag the front and top views slightly upward.

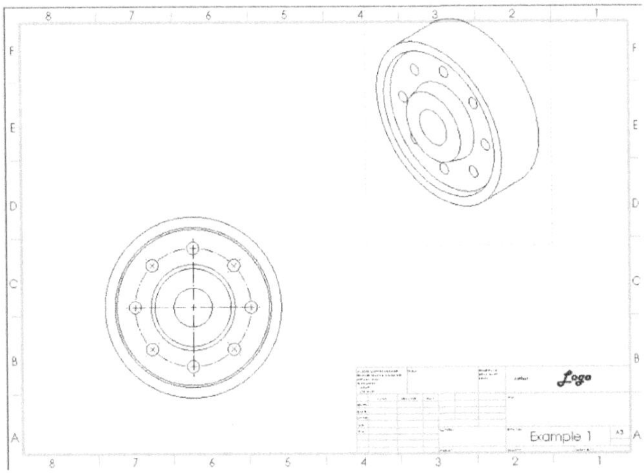

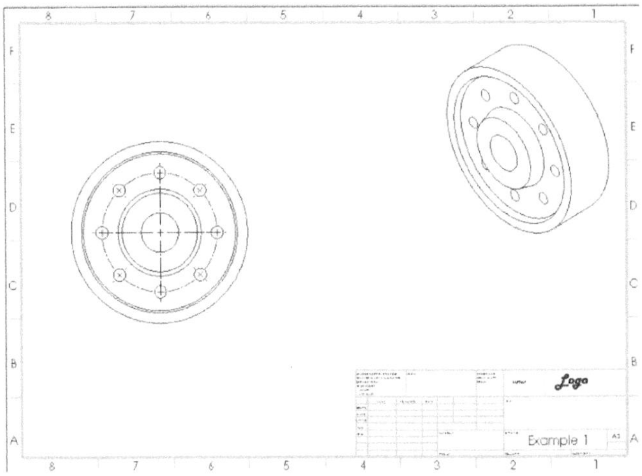

8. Activate the **Section View** command (click **Drawing > Section View** on the CommandManager) and click the **Vertical** 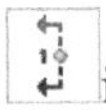 icon in the **Cutting Line** section of the PropertyManager.
9. Select the centerpoint of the front view, and then click **OK** on the context toolbar.
10. Drag the pointer toward the right and click to position the view. Click **Close Dialog** on the PropertyManager.

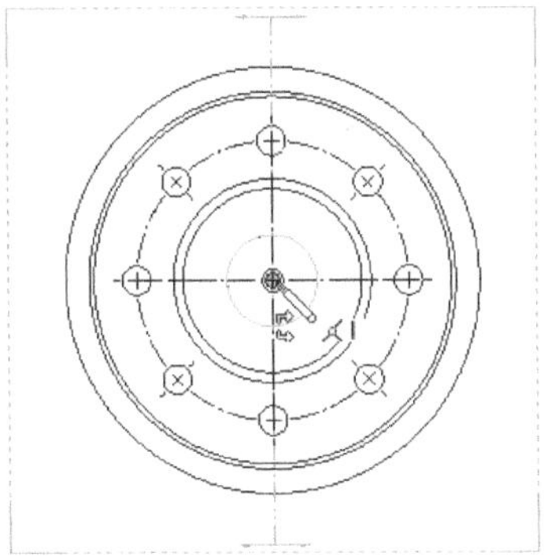

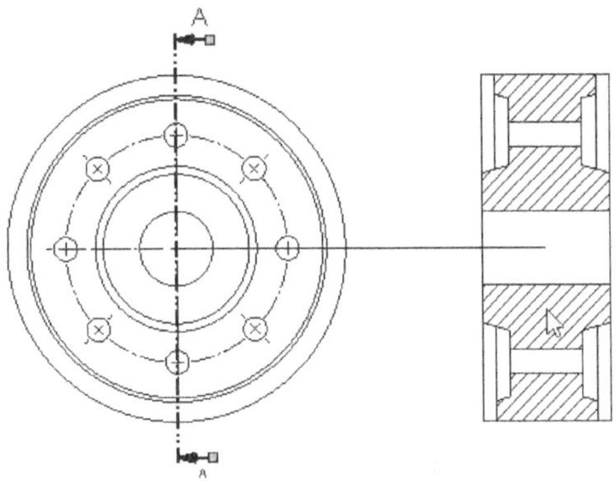

Creating Centerlines

1. Activate the **Centerline** command (click **Annotation > Centerline** on the CommandManager). On the PropertyManager, check the **Select view** option.
2. Select the section view; the centerlines are created between the hole lines. Click **OK** on the PropertyManager.

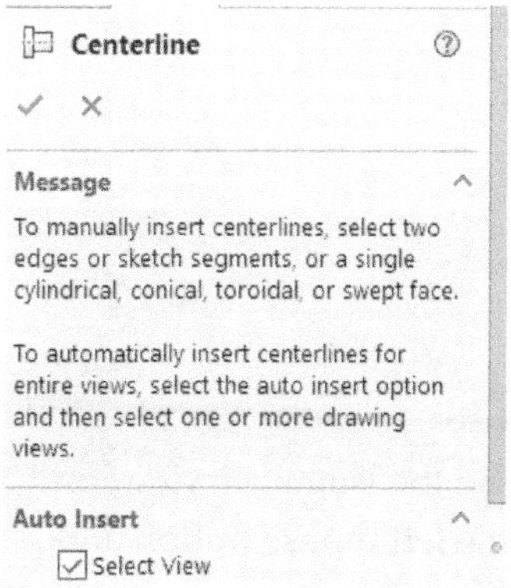

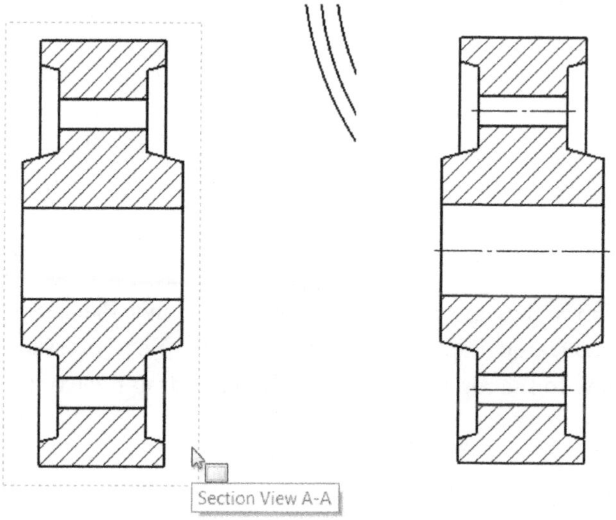

Retrieving Model Dimensions

1. Activate the **Model Items** command (click **Annotation > Model Items** on the CommandManager), and then select **Source > Entire Model** from the PropertyManager. Next, scroll to the **Options** section and uncheck the **Use dimension placement in sketch** option.
2. Check the **Import items into all views** option in the **Source/Destination** section and click **OK** on the PropertyManager.

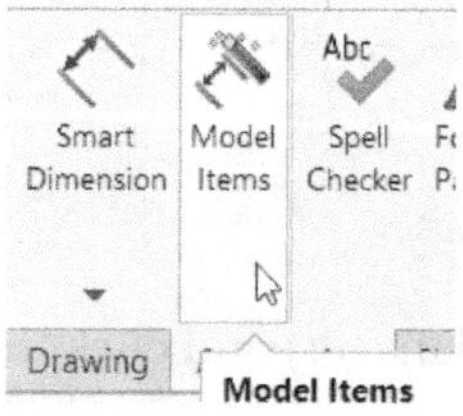

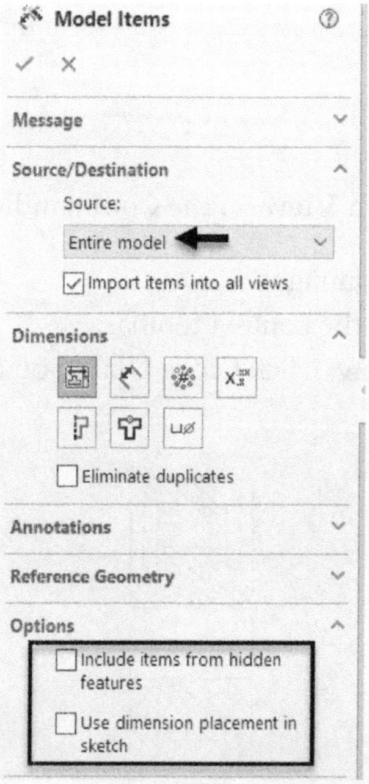

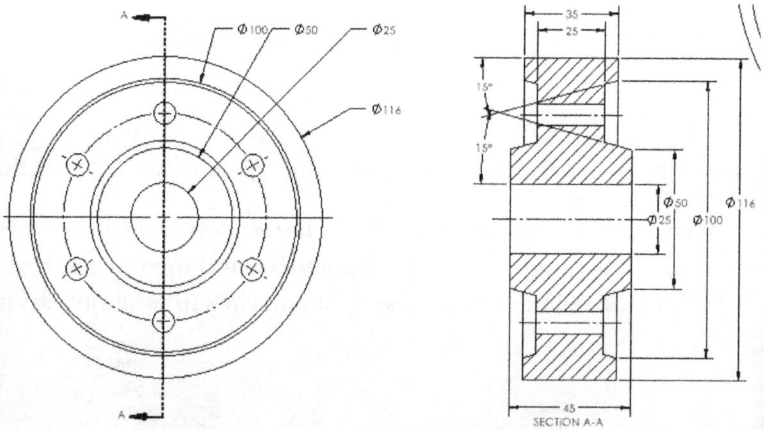

3. Delete the redundant dimensions from the two views, as shown.
4. Drag and arrange the remaining dimensions, as shown.

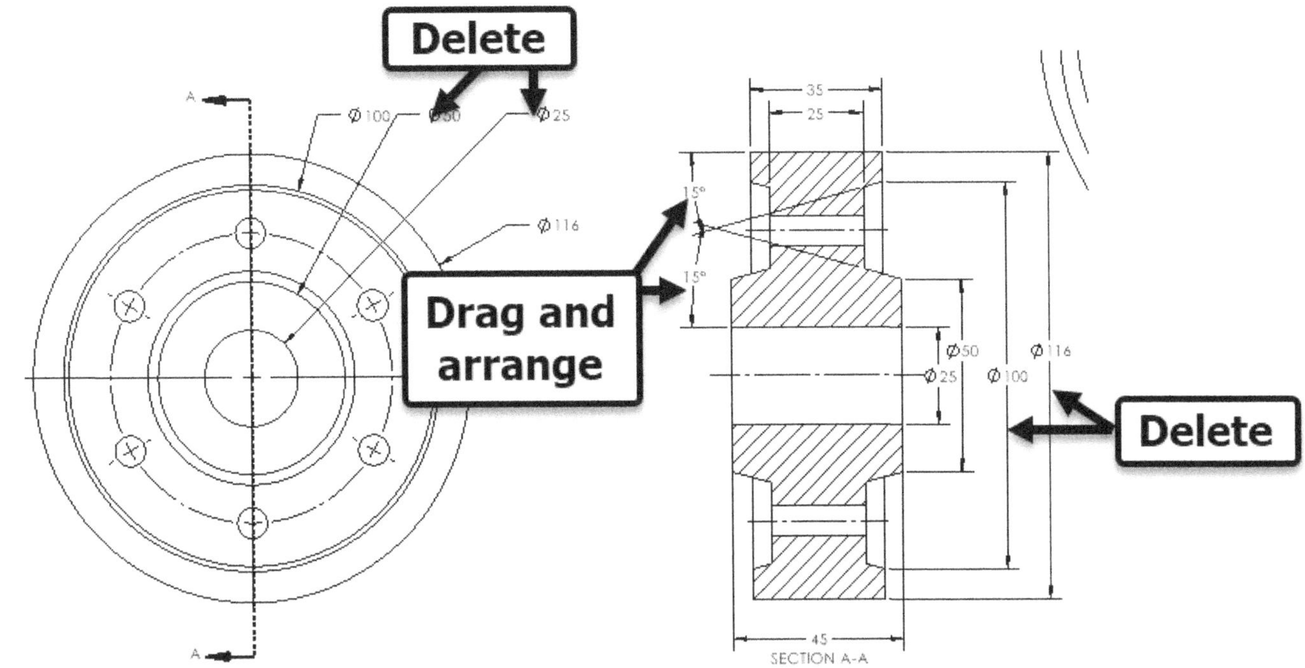

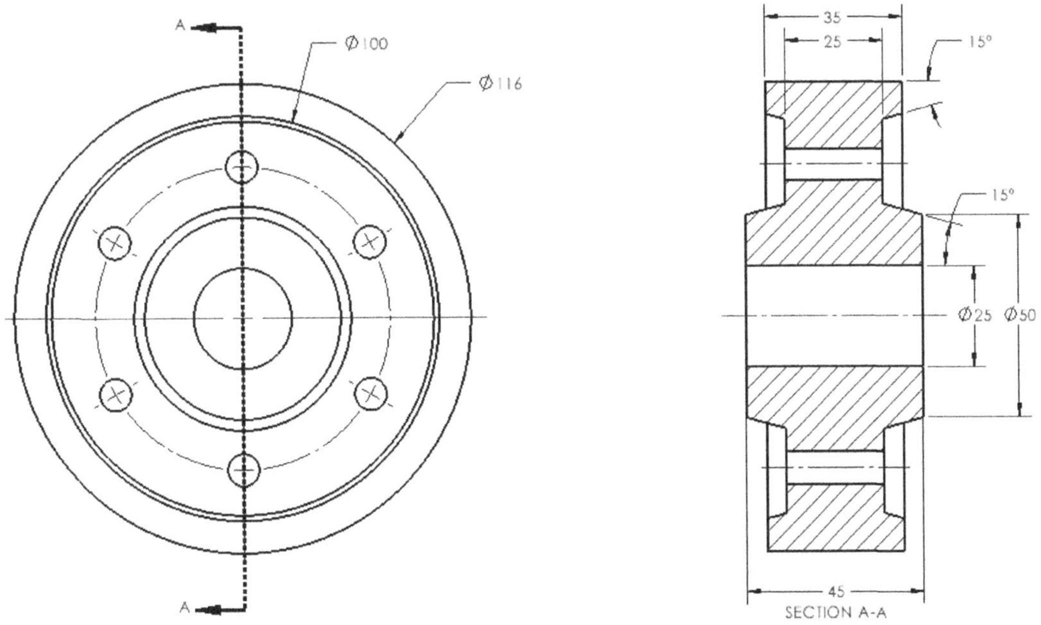

5. Activate the **Hole Callout** command (on the CommandManager, click **Annotation > Hole Callout**) and click on the small hole of the front view.
6. Move the pointer toward the right and click to position the hole callout.
7. In the **Dimension Text** dialog, click at the end of the text and press Enter. Type EQUI SPACED ON 75 PCD. Click **OK** and position the hole dimension.

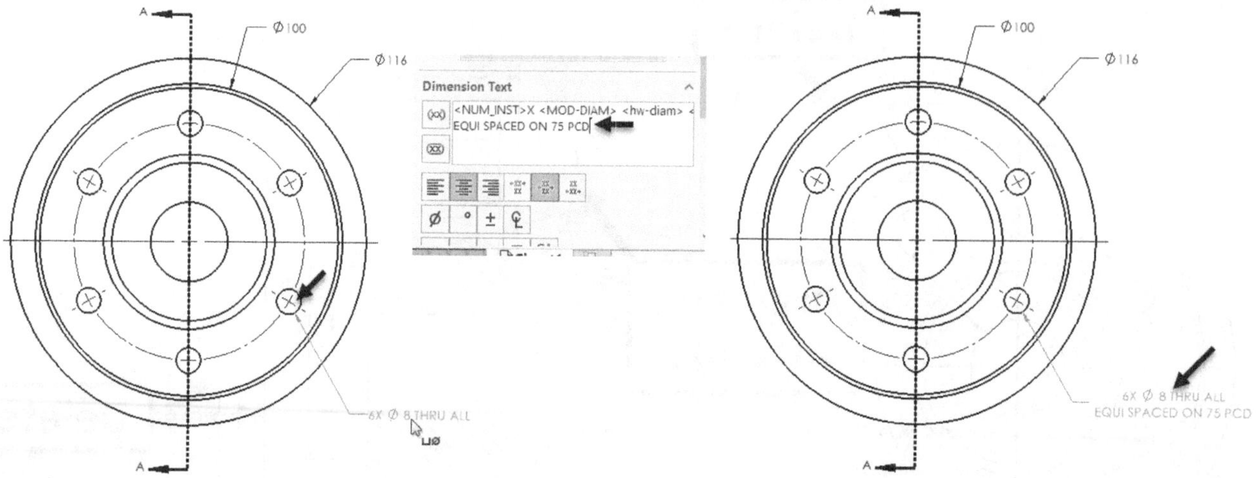

8. Save and close the drawing.

Example 2

In this example, you will create an assembly drawing shown below.

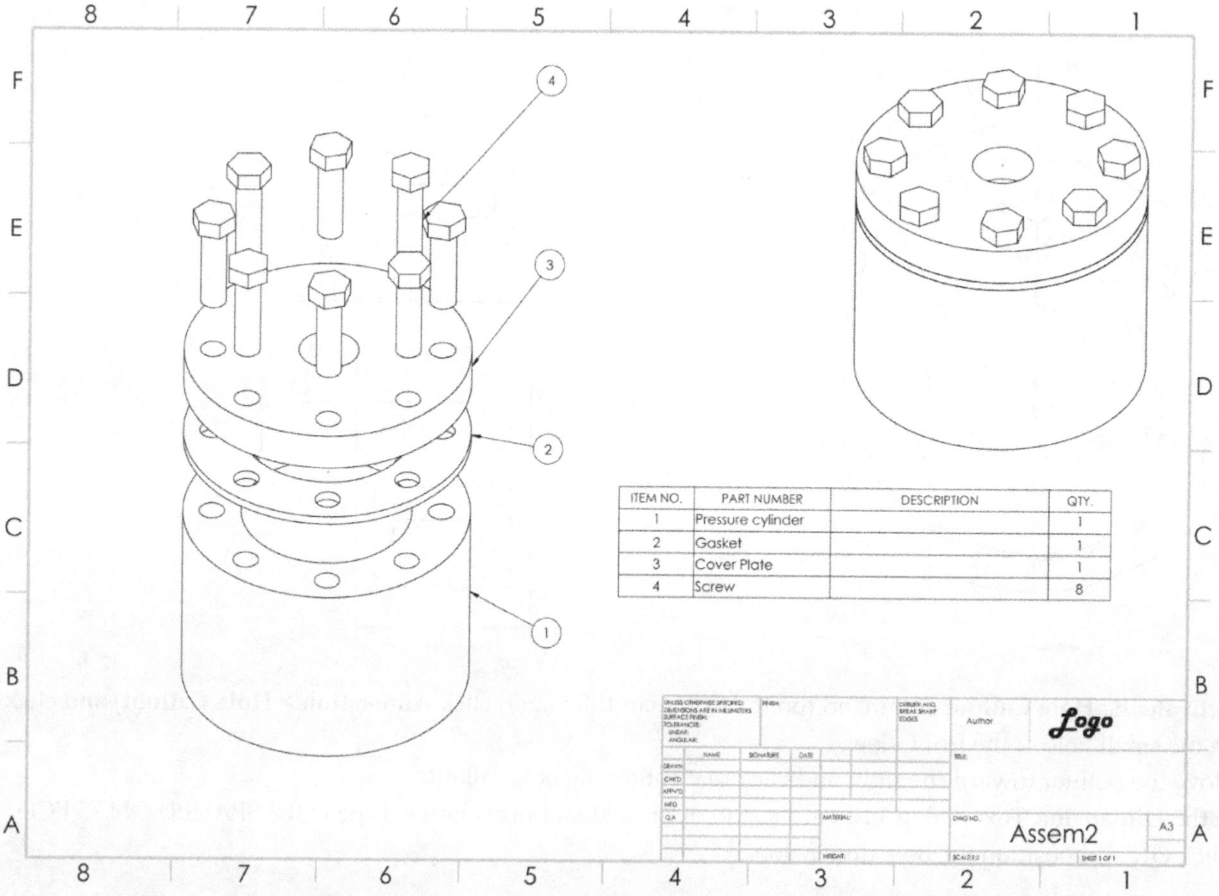

4
3
2
1
ITEM NO. | PART NUMBER | DESCRIPTION | QTY.
1 | Pressure cylinder | | 1
2 | Gasket | | 1
3 | Cover Plate | | 1
4 | Screw | | 8
Author
Logo
Assem2
A3

1. Download the Drawings chapter from the Companion website, and then unzip the files.
2. Start **SOLIDWORKS**.
3. On the Quick Access Toolbar, click the **New** button. On the **New SOLIDWORKS Document** dialog, click

the **Advanced** button. Under the **SOLIDWORKS For Beginners** tab, select **Sample** and click **OK**.

4. Activate the **Model View** command (click **Drawing > Model View** on the CommandManager). Click the **Browse** button on the **Model View** PropertyManager.
5. Browse to the location of the *Drawings* chapter files, and then select the Example 2 assembly file. Click the **Open** button.
6. Make sure that the **Show in exploded and Model break state** option is checked on the PropertyManager.
7. On the **Model View** PropertyManager, click the **Isometric** icon in the **Orientation** section.
8. In the **Scale** section, select the **Use custom scale** option and select **1:1** from the **Scale** drop-down.
9. Click on the left side to place the exploded view of the assembly. Next, click the green check on the PropertyManager.
10. Again, activate the **Model View** command and double click on the assembly file listed in the Open documents list box of the PropertyManager.
11. Uncheck the **Show in exploded and Model break state** option on the PropertyManager.
12. In the **Scale** section, select the **Use custom scale** option and select **1:1** from the Scale drop-down.
13. Click the **Isometric** icon in the **Orientation** section.
14. Click on the top right corner to place the isometric view of the assembly.

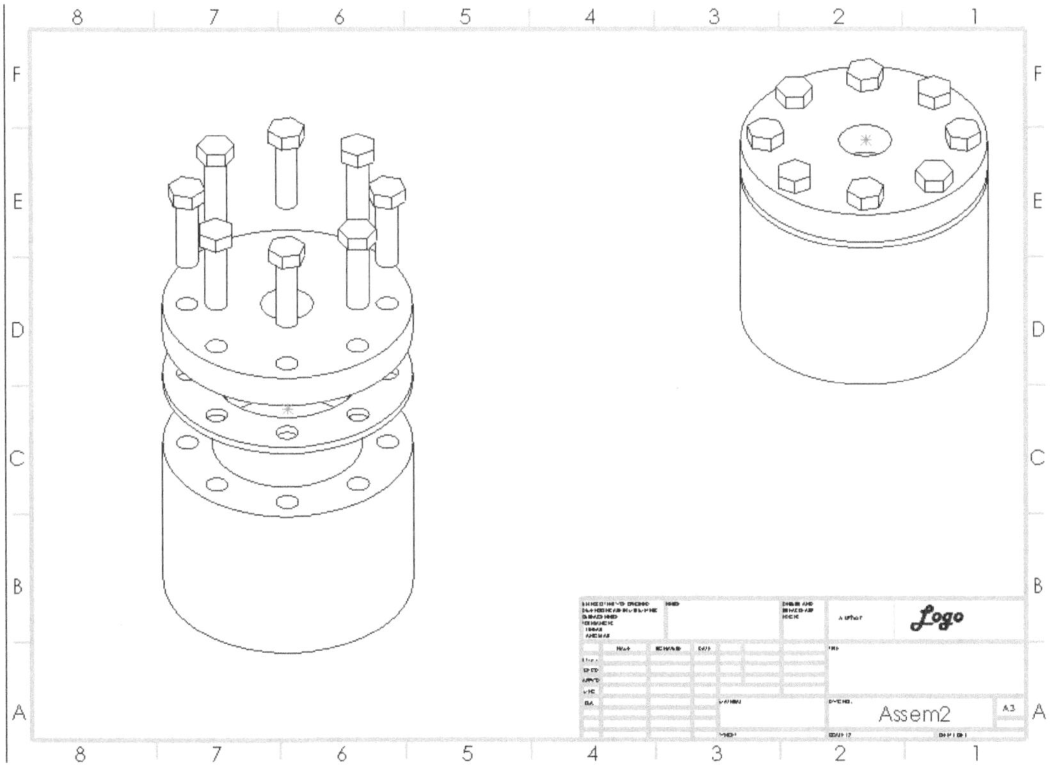

15. Activate the **Bill of Materials** command (click **Annotation > Tables > Bill of Materials** on the CommandManager) and select the exploded view.
16. Select the **Top-level only** option from the **BOM Type** section, and then click **OK** on the PropertyManager.
17. Position the table below the isometric view.

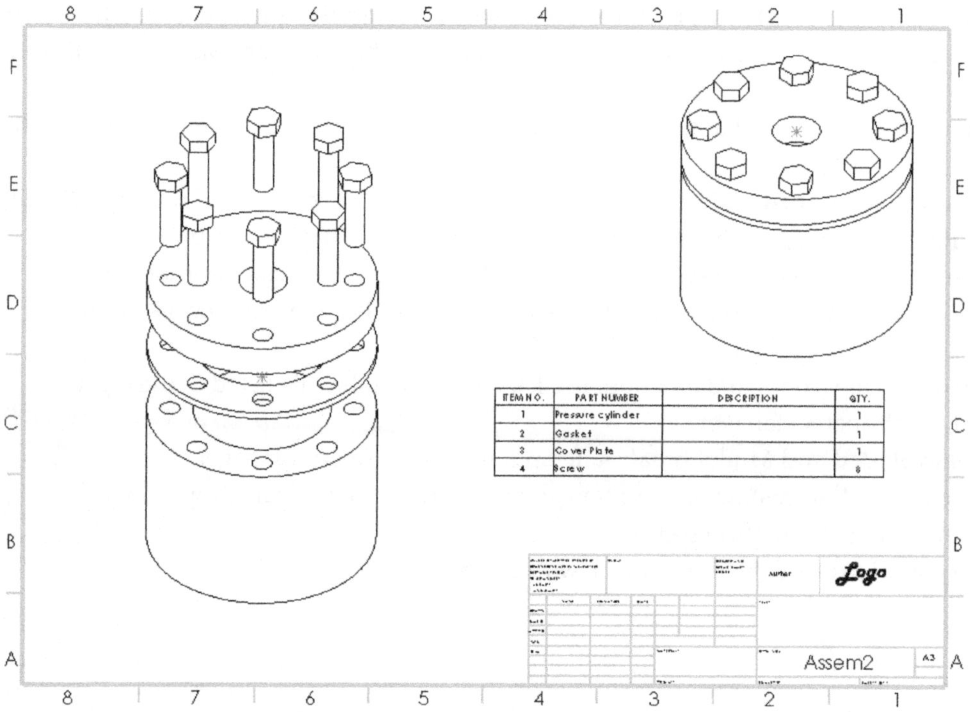
ITEM NO.
PART NUMBER
DESCRIPTION
QTY.
1 Pressure cylinder 1
2 Gasket 1
3 Cover Plate 1
4 Screw 8
Author
Logo
Assem2
A3

18. Activate the **Auto Balloon** command (on the CommandManager, click **Annotation > Auto Balloon**)
19. Click the **Layout Balloons to Right** icon on the PropertyManager, and then select the exploded view.
20. Click **OK** to create the balloons.

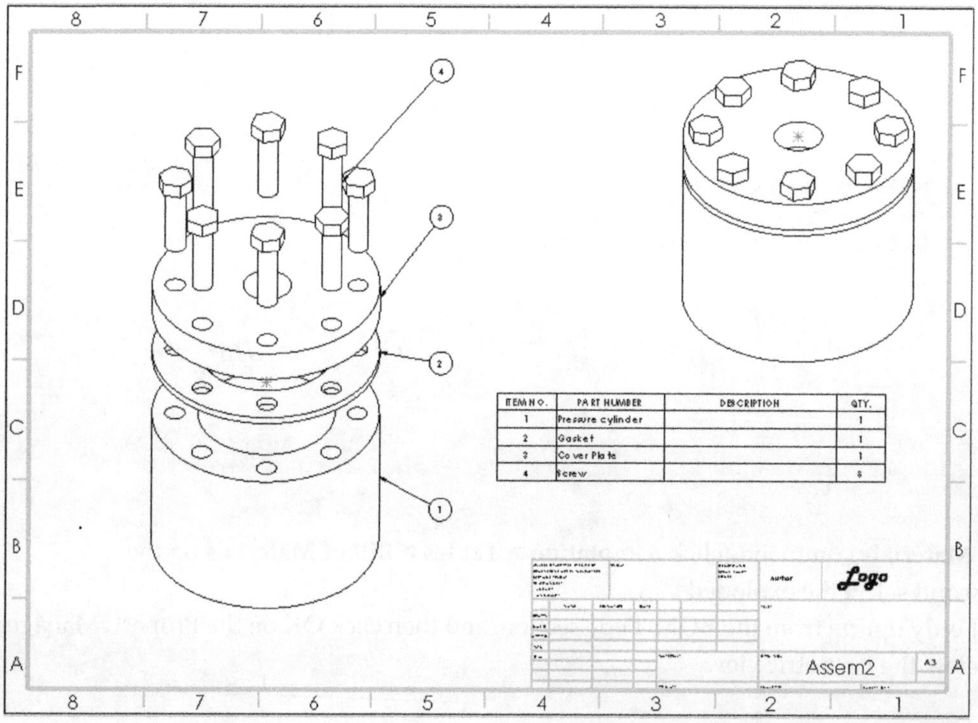

21. Save and close the drawing.

Questions

1. How to create drawing views using the **Model View** command?
2. How to change the display style of a drawing view?
3. List the commands used to create centerlines and center marks.
4. How to add symbols and texts to a dimension?
5. How do you add break lines to a drawing view?
6. How to create aligned section views?
7. How to create exploded view of an assembly?
8. How to create the alternate position view?

Exercises

Exercise 1

Create orthographic views of the part model shown below. Add dimensions and annotations to the drawing.

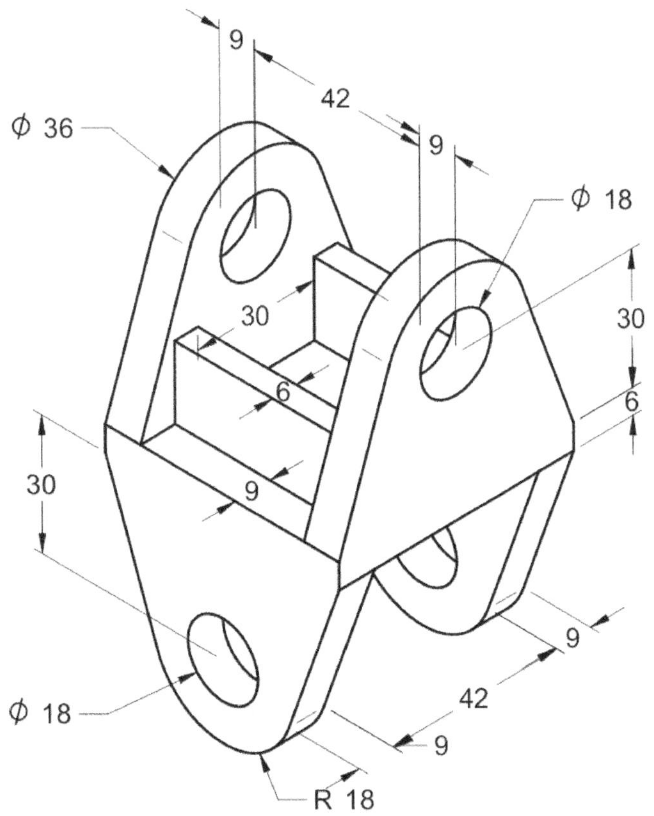

Exercise 2

Create orthographic views and an auxiliary view of the part model shown below. Add dimensions and annotations to the drawing.

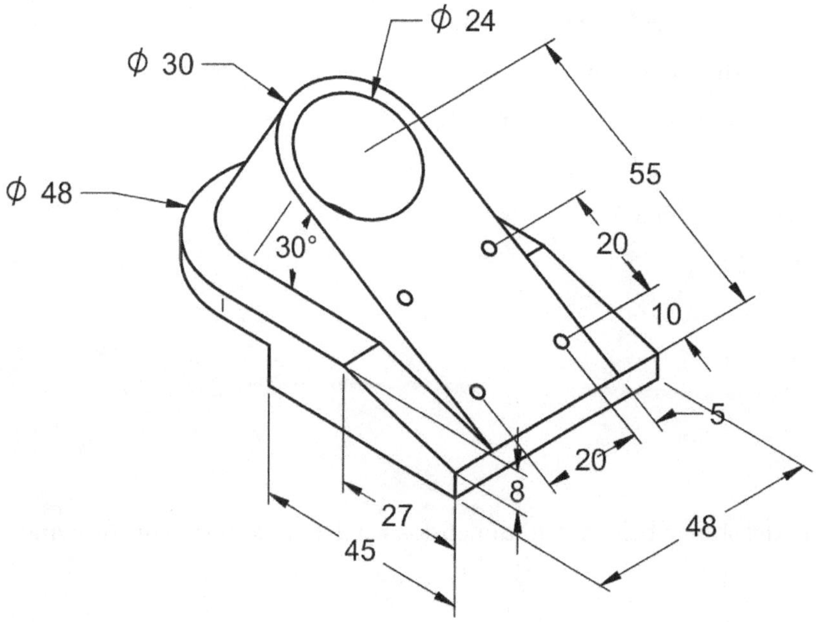

Φ 24
Φ 30
Φ 48
55
20
30°
10
5
20
8
27
45
48

Index

Made in the USA
Las Vegas, NV
19 October 2024

10071968R00240